SLOW DYNAMICS IN CONDENSED MATTER

AIP CONFERENCE PROCEEDINGS 256

SLOW DYNAMICS IN CONDENSED MATTER

PROCEEDINGS OF THE 1st TOHWA UNIVERSITY INTERNATIONAL SYMPOSIUM

FUKUOKA, JAPAN 1991

EDITORS:

KYOZI KAWASAKI
TOSHIHIRO KAWAKATSU
KYUSHU UNIVERSITY

MICHIO TOKUYAMA
TOHWA UNIVERSITY

American Institute of Physics   New York

Authorization to photocopy items for internal or personal use, beyond the free copying permitted under the 1978 U.S. Copyright Law (see statement below), is granted by the American Institute of Physics for users registered with the Copyright Clearance Center (CCC) Transactional Reporting Service, provided that the base fee of $2.00 per copy is paid directly to CCC, 27 Congress St., Salem, MA 01970. For those organizations that have been granted a photocopy license by CCC, a separate system of payment has been arranged. The fee code for users of the Transactional Reporting Service is: 0094-243X/87 $2.00.

© 1992 American Institute of Physics.

Individual readers of this volume and nonprofit libraries, acting for them, are permitted to make fair use of the material in it, such as copying an article for use in teaching or research. Permission is granted to quote from this volume in scientific work with the customary acknowledgment of the source. To reprint a figure, table, or other excerpt requires the consent of one of the original authors and notification to AIP. Republication or systematic or multiple reproduction of any material in this volume is permitted only under license from AIP. Address inquiries to Series Editor, AIP Conference Proceedings, AIP, 335 East 45th Street, New York, NY 10017-3483.

L.C. Catalog Card No. 92-53224
ISBN 0-88318-938-0
DOE CONF-911121

Printed in the United States of America.

Contents

Preface xv
Organization of Symposium xvii
Group Photograph xix
Opening Address xx
Welcoming Address xxi
Banquet Speech xxii
Tohwa Institute for Science
  and Tohwa University International Symposia xxiii
Executive Members of Tohwa University
  International Symposia xxiv
Addresses of Participants xxv

I. GLASS TRANSITION
A. Experiment

Slow Processes in Viscous Liquids: Stress and Structural Relaxation,
Chemical Reaction Freezing, Crystal Nucleation and Microemulsion Arrest,
in Relation to Liquid Fragility 3
    C. A. Angell (*invited speaker*), C. Alba, A. Arzimanoglou, R. Böhmer, J.
    Fan, Q. Lu, E. Sanchez, H. Senapati, and M. Tatsumisago
Slow Relaxation Processes in Glassy Crystals 20
    H. Suga (*invited speaker*)
Tracer Diffusion in Polymer and Organic Liquids Close to the Glass
Transition 30
    M. Lohfink and H. Sillescu (*invited speaker*)
Brillouin and Raman Scattering Spectroscopy of the Liquid–Glass Transition 40
    H. Z. Cummins (*invited speaker*), G. Li, W. M. Du, X. K. Chen, N. J. Tao,
    and A. Sakai
Dynamic Structure Factor Near the Glass Transition: Specific Features 53
    F. Mezei (*invited speaker*)
Influence of Thermodynamic Interactions on the Transport Process
of Molecular Tracer in Binary Polymer Mixtures 63
    Q. Tran-Cong, Y. Ishida, K. Meisyo, O. Yano, and T. Soen
Neutron Scattering Experiments Near the Glass Transition
of Polybutadiene 67
    R. Zorn, D. Richter, B. Frick, and B. Farago
Intrinsic Unstability and Nonergodic Behavior of a Glassy Crystal 71
    M. Descamps, J. F. Willart, and O. Delcourt
Isothermal Glass Transitions and Entropic Relaxation 75
    C. Alba-Simionesco

The Different Universality Classes, from Strong to Fragile,
of the Glass and Spin-Glass Transitions 79
    J. Souletie and D. Bertrand
Experimental Study of Non-Debye Relaxation in Glass Transition
of Glycerol 83
    S. Kojima
Production of New Alloy Phases in Bi–Mn Binary System from
Vapor Phase and Their Structure Investigation by High Resolution
Electron Microscopy 85
    K. Yoshida and T. Yamada
Glass Transition and Localization of Mobile Ions in Superionic Glasses:
Investigated by Dielectric and Thermodynamic Relaxation Techniques 87
    J. Kawamura
Structure and Dynamics of Undercooled and Glassy Aqueous Ionic
Solutions by NMR, X-Ray, and Neutron Scattering 89
    T. Yamaguchi, M. Yamagami, T. Takamuku, T. Hirano, and H. Wakita
Dynamics of Amorphous Polymers Near Glass Transition 91
    T. Kanaya, T. Kawaguchi, and K. Kaji

B. Theory

α Relaxation in Supercooled Liquids 95
    W. Götze (*invited speaker*) and L. Sjögren
β Relaxation Near Glass Transitions 105
    L. Sjögren (*invited speaker*) and W. Götze
Stochastic Dynamics in a Supercooled Fluid 115
    T. Odagaki (*invited speaker*) and Y. Hiwatari
The Glass Transition in Orientational Glasses 122
    D. Walton
Transport Properties in Ordinary and Supercooled Liquids 126
    U. Balucani, S. F. Duffy, A. Torcini, and R. Vallauri
Accelerating Glassy Relaxation in the Frenkel–Kontorova Model 130
    S. Shumway and J. P. Sethna
Mode Coupling, Universality, and the Glass Transition 134
    G. F. Mazenko
Crystallization and Nonergodic Mode Dynamics 140
    T. Munakata

C. Computer Simulation

Monte Carlo Simulations of the Polymer Glass Transition 145
    W. Paul (*invited speaker*)
Molecular-Dynamics Study of Highly Supercooled Liquids:
Dynamical Singularities Near the Liquid–Glass Transition 155
    Y. Hiwatari (*invited speaker*), H. Miyagawa, T. Muranaka, and K. Uehara
Loss of Ergodicity in Glassy Systems 165
    R. D. Mountain (*invited speaker*) and D. Thirumalai

Simulation of Supercooled Atomic and Brownian Systems:
Comparison with Mode-Coupling Theory 173
    J.-N. Roux (*invited speaker*)
On the Importance of Kinetic Inhomogeneities
in Understanding Glassy Dynamics 183
    S. Butler and P. Harrowell
Ionic Dynamics on the Potential Energy Surfaces
of Glassy Alkali Chloride Systems 185
    K. Kinugawa, K. Kadono, and H. Tanaka
A Study on the Mechanism of the Liquid-Glass Transition
in Lithium Metasilicate 187
    J. Habasaki, I. Okada, and Y. Hiwatari
Molecular-Dynamics Study on the Glass Transition of LiI
about Dynamical Singularities 189
    S. Itoh, H. Miyagawa, and Y. Hiwatari

II. POLYMER DYNAMICS
A. Experiment

Neutron Spin Echo Investigations on the Dynamics
of Dense Polymer Systems 193
    D. Richter (*invited speaker*), L. J. Fetters, J. S. Huang, B. Farago, and B. Ewen
Analysis of DNA Electrophoretic Mobility 203
    B. Chu and Z.-L. Wang
The Dynamic Properties of Silica Aerogel: A Key to Understand
the Thermal Properties of Dense Glasses? 207
    D. Posselt, J. K. Kjems, A. Bernasconi, T. Sleator, and H. R. Ott
Slow Dynamics in Polyelectrolyte Solutions 211
    M. Drifford, J. P. Dalbiez, L. Belloni, and O. Spalla
Time-Resolved Small-Angle Neutron Scattering Study
of Later Stage Spinodal Decomposition of a Polymer Blend 216
    H. Jinnai, H. Hasegawa, T. Hashimoto, and C. C. Han
Dielectric Relaxation of Isotactic Polystyrene
in Long Time Region 222
    K. Fukao, Y. Miyamoto, and H. Miyaji
Ultrasonic and Dielectric Studies on Curing Process
of Allyl-Oligomer 224
    H. Okabe, H. Kanaya, K. Hara, S. Taki, and K. Matsushige
Gelation Process of Actomyosin 226
    H. Kanaya, K. Hara, H. Okabe, K. Matsushige, S. Nishimuta, M. Muguruma, and T. Fukazawa
Frequency Dependence of the Dynamic Heat Capacity Accompanying
the Transition Phenomena 228
    Y. Saruyama
Structure Formation of Poly (Ethylene Terephthalate) During
Annealing Process 230
    M. Imai, T. Mizukami, T. Kanaya, and K. Kaji

Critical Behavior of Complex Shear Modulus
in Concentrated Polymer Solutions 232
    H. Tanaka and T. Miura

Ordering Process of Lamellar Microdomains Following Morphological
Transition from Cylindrical Microdomains in Block Copolymer Melts 234
    S. Sakurai, K. Taie, T. Momii, and S. Nomura

Crystallization of Crystalline/Amorphous Polymer Blends 236
    M. Takahashi, H. Matsuda, H. Yoshida, and Y. He

Anomalous Phase Separation and Pattern Formation in a Polymer/Water
Mixture with a Double-Well-Shaped Phase Diagram 238
    H. Tanaka

Elementary Process of Coalescence Among Droplets in
Phase-Separating Binary Polymer Mixtures: New Mechanism
of Droplet Coalescence 240
    H. Tanaka

B. Theory

Stress Relaxation in Heterogeneous Polymers 245
    T. A. Witten (*invited speaker*)

Hydrodynamic, Soft, and Relaxational Modes in Chiral Liquid
Crystal Side-Chain Polymers 253
    H. Pleiner and H. R. Brand

Viscosity and Particle Diffusion of Dilute Particle Dispersions
in Polymers 257
    W. Sung

Dynamics of Physically Crosslinked Polymer Networks 261
    F. Tanaka and S. F. Edwards

Towards the Unified Theory of Polymer Solution Dynamics 263
    Y. Shiwa

Dipole Decay Function and Dielectric Loss Curve
of Polymers in Dilute Solution 265
    H. Ogura, R. Ozao, and M. Ochiai

New Langevin Dynamics Model for Polymer Melts 267
    K. Kawasaki, T. Kawakatsu, and W. Zimmermann

C. Computer Simulation

Relaxation Behavior of Crosslinked Polymer Melts 271
    K. Kremer (*invited speaker*), G. S. Grest, and E. R. Düring

Dynamical Correlation Function of Fractal Networks: Computer Experiments 279
    T. Nakayama and K. Yakubo

The Computer Simulation of Spinodal Decomposition in Polymer Blends—
Effects of Uniaxial Compression During Spinodal Decomposition 283
    M. Takenaka, T. Hashimoto, K. Kawasaki, and T. Kawakatsu

Computer Simulations of Rheological Response in a Microphase
Separated Block Copolymer Melt 286
    T. Ohta, A. Tetsuka, Y. Enomoto, and M. Doi

Strong Anisotropy of Orientational Relaxation in Bulk Amorphous
Polymer: A Molecular-Dynamics Simulation 288
    H. Takeuchi and R. J. Roe

Amplitude Mode Around a Moving Soliton
in Polyacetylene 290
    M. Kuwabara and Y. Ono

Dynamics of DNA in Gel Electrophoresis 292
    M. Matsumoto and M. Doi

New Results on Computer Simulation of Spinodal Decomposition
in Binary Fluids 294
    T. Koga and K. Kawasaki

Acceleration of Soliton in Polyacetylene by External Electric
Field—Effect of Switching 296
    Y. Ono and M. Kuwabara

III. EMULSION
A. Experiment

Dynamic Slowing-Down in Dense Microemulsions Near
the Percolation Threshold 301
    S. H. Chen (*invited speaker*), F. Mallamace, J. Rouch, and P. Tartaglia

Mean-Field Behavior at Phase Separation in Three-
Component Microemulsion System 318
    H. Seto, S. Komura, D. Schwahn, and K. Mortensen

Dynamic Universality in Microemulsion System 320
    Y. Harada and M. Tabuchi

Experimental and Theoretical Studies of Critical
Phenomena in Supramolecular Fluid Systems 322
    C. Cametti, P. Codastefano, J. Rouch,
    and P. Tartaglia

B. Theory

Aggregation and Coalescence of Water Droplets in a Dielectric
Liquid Phase in an Electric Field 327
    M. Yamaguchi, T. Ise, and T. Katayama

C. Computer Simulation

Phase Separation in Binary Mixtures with Surfactants 331
    T. Kawakatsu and K. Kawasaki

IV. COLLOIDAL SUSPENSION
A. Experiment

Dynamic Light Scattering Studies of the Glass Transition
in Colloidal Suspensions 335
    W. van Megen (*invited speaker*), S. M. Underwood, and P. N. Pusey

Rheological Properties of Silica Suspensions in Aqueous
Cellulose Derivatives Solutions 350
    Y. Ryo and M. Kawaguchi
Observation of a Finite Wavelength Instability
in Crystallization 352
    K. Schatzel and B. J. Ackerson
Superlong Range Attractive and Repulsive Interactions Between
Colloid Particles 354
    S. Yoshino

B. Theory

Transport Properties of Concentrated Colloidal Suspensions 359
    E. G. D. Cohen (*invited speaker*) and I. M. de Schepper
Brownian Motion of Suspensions 370
    B. U. Felderhof (*invited speaker*)
A Model for Repulsive Hard Spheres with Surface Adhesion 378
    D. Bedeaux and G. J. M. Koper
Aggregation Kinetics in Electro-Rheological Fluids 382
    H. See and M. Doi
Time-Dependent Self-Diffusion Coefficient of Interacting
Brownian Particles 384
    B. Cichocki and B. U. Felderhof
Precipitation in a Particle in a Colloidal Suspension 386
    K. Kishi, A. Yoshida, and T. Yoshida

C. Computer Simulation

Microstructured Fluids: Structure, Diffusion, and Rheology
of Colloidal Dispersions 391
    T. Phung and J. F. Brady (*invited speaker*)
Dynamics of Deformable Incompressible Colloidal Particles
in Dense Systems. Computer Simulation 401
    T. Pakula and H. Nilgens

V. SPIN GLASS
A. Experiment

Nonequilibrium Dynamics in Spin Glasses 407
    L. E. C. Lundgren (*invited speaker*)
Hierarchical Aspects of the Slow Dynamics in Spin Glasses 417
    E. Vincent, J. Hammann, M. Ocio, and F. Lefloch
Time-Dependent Magnetic Phenomena in Dilute Antiferromagnets
$Fe_{1-x}Mg_xCl_2$ Exhibiting Random Field Using Model and Spin
Glass Behaviors 421
    K. Iio, A. Kitazawa, and K. Nagata
Small Angle Neutron Scattering Studies on $Fe_{1-x}Al_x$
Reentrant Spin Glass 423
    J. Suzuki, Y. Endoh, M. Arai, and M. Furusaka

B. Computer Simulation

Relaxation Phenomena of the Fuzzy-Spin Model 427
    S. Miyashita and T. Kawasaki

VI. OTHER RELATED TOPICS
A. Experiment

Kinetics of Ordering in the Percolation Magnet 433
    H. Ikeda (*invited speaker*)

Size and Spatial Distributions of Precipitates in the Ostwald
Ripening in Some Alloys 441
    T. Eguchi, K. Arita, and Y. Tomokiyo

Nonlinear Slow Fluctuation and Relaxation in the Ordering
of Some Complex Magnetic Systems 445
    M. Matsuura and M. Hagiwara

Crossover Between Nonequilibrium Relaxation to Equilibrium
Relaxation in CDW Ground State 449
    K. Biljaković, J. C. Lasjaunias, and P. Monceau

Relaxation Processes in 2-Butoxyethanol Aqueous Solutions
Near the Critical Region 453
    G. D'Arrigo and A. Paparelli

Enhanced Effect of Salts on Polymer Transport in Structured Flow 457
    H. Maeda, T. Mashita, and S. Sasaki

Experimental Studies in a Phase-Separating Mixture Under
Shear Flow 459
    K. Hamano, S. Yamashita, K. Kubota, N. Kuwahara, and J. V. Sengers

Pressure Dependence of the Ferroelectric Soft Mode of KDP 461
    T. Yagi, A. Sakai, and M. Arima

Dielectric Dispersion Associated with the DC-Electric-Field
Enforced Ferroelectric Phase Transition in Antiferrorlectrics 463
    N. Yasuda

Light Scattering and Turbidimetric Study of Gelling Tungstic Acid 465
    K. Hara, H. Kanaya, H. Okabe, and K. Matsushige

X-Ray Studies on Structures and Phase Transition in Evaporated
Films of Liquid Crystals 467
    Y. Yoshida, T. Horiuchi, and K. Matsushige

An Explosive Hydrodynamic Flow Induced by Phase
Diffusion Wave in the BZ Solution Layer 469
    H. Miike and H. Yamamoto

Dynamical Property of Ferroelectric Domain Wall Near
the Curie Point 471
    K. Hamano, H. Sakata, and J. Zhang

Observation of Slow Dynamic Process During
the Main Transition in DSPC by Time-Dependent
Heat Capacity Measurement 473
    K. Ema, H. Yao, and Y. Kawase

Thermal Diffusivity and Surface Tension
of Liquid Nitrogen Near its Critical Point 475
    T. Shigenari, M. Mogi, K. Abe, M. Suzuki, N. Itagaki, and A. Sato
Growth Pattern of the Surface of Fungus *Aspergillus* Colony 477
    S. Matsuura and S. Miyazima
An *In Situ* Study of the Dynamics of Oxygen Precipitation in Si 479
    A. Magerl, K. D. Liss, J. R. Schneider, and W. Zulehner
Diffusion Limited Aggregation in a Flow Field 481
    K. Oota, K. Okumura, K. Maruyama, and S. Miyazima

B. Theory

Time Scale Invariance in Transport and Relaxation 485
    H. Scher (*invited speaker*)
Electron Transport Process Induced by Protonation in Alpha-
Helical Protein 495
    S. Ichinose and T. Minato
Two-Time Fourier Convolution Theorem and its Applications 497
    H. Akama
On Rotating Spiral Waves in Reaction-Diffusion Systems 499
    S. Koga
Spinodal Decomposition in Tetragonal TiO_2–SnO_2 System 501
    S. Nambu, A. Sato, and D. A. Sagala
Concentration Profile of Polymers Near a Spherical Surface 503
    T. Taniguchi, T. Kawakatsu, and K. Kawasaki
Long Time Tails in Diffusion-Controlled Recombinations 505
    T. Ohtsuki
Relaxation of Crystal Shape Profiles Near the Facet Edge 507
    T. Yamamoto, N. Akutsu, and Y. Akutsu
Void Fraction Dynamics in Fluidization 509
    S. Sasa and H. Hayakawa
Calculation of Rotational Tunneling States of the Methyl
Group in the Higher Order of Potential 511
    Y. Ozaki
Fluctuations of Hydrogen Bond Network in Liquid Water 513
    M. Sasai
Hole Conduction in the One-Dimensional Molecular Conductor
Nickel-Rich Cobalt Phthalocyanine Iodide 515
    A. Mishima
Variation Principle for Stochastic Processes—Generalization
of Path Probability Method 517
    M. Kaburagi, K. Wada, A. Suzuki, R. Kikuchi, and H. Sato
Calculation of Phonons in the Amorphous Structure 519
    H. Sato, A. Ishida, and M. Itoh
Long Tail Behaviors of Random Medium 521
    H. Hara and J. Koyama
Crossover Phenomenon in Slow Particle Growth
on a Substrate 523
    M. Tokuyama and Y. Enomoto

Modulated Patterns and Pinning Effect in Phase-Separating Alloys 525
    A. Onuki and H. Nishimori

C. Computer Simulation

Self-Avoiding Walks on a Percolation Cluster
in Four Dimensions 529
    S. B. Lee and Y. J. Song

Role of Locally Distributed Properties on the Relaxation Dynamics
in Disordered Systems 533
    J. L. Bocquet and Y. Limoge

Dynamics of Pattern Formation of Antiphase Ordered Domain
in Alloys 537
    K. Shiiyama, K. Horai, and T. Eguchi

Logarithmically Slow Coarsening in Nonrandomly Frustrated
Models 541
    J. D. Shore, J. P. Sethna, M. Holzer, and V. Elser

A DLA Model of Interactive Particles 545
    M. Nakagawa and K. Kobayashi

Flow-Induced First Order Transition of the Aggregation
in a Diffusion Field 547
    Y. Saito, M. Uwaha, and S. Seki

Metastability and Transfer-Matrix Finite-Range Scaling 549
    P. A. Rikvold, B. M. Gorman, and M. A. Novotny

Rheological Properties of Foams: Computer Simulation of Vertex
Model for Two-Dimensional Random Cellular Structures 551
    T. Okuzono, K. Kawasaki, and T. Nagai

Ostwald Ripening in Open Systems 553
    A. Nakahara, T. Kawakatsu, and K. Kawasaki

Relaxation Dynamics of Phasons in Quasicrystals 555
    Y. Ishii

A Numerical Modeling of Convective Motions
in Granular Materials 557
    Y. Taguchi

Growth Mechanism of Homogeneous Diffusion-Limited Aggregation ... 559
    S. Ohta

Structural Characterization of Molten Calcium Chloride by
Molecular-Dynamics Simulation 561
    N. Umesaki

Slow Relaxation Processes in the Ferromagnetic State 563
    H. Nakanishi

Forced Two-Dimensional Patterns in Anisotropic Convective
Systems 565
    A. Ogawa, W. Zimmermann, K. Kawasaki, and T. Kawakatsu

Monte Carlo Study of Vesicles 567
    S. Komura and A. Baumgärtner

Domain Growth in Quenched Random Impurities 569
    H. Hayakawa and T. Iwai

Clustering Motion in Conservative Coupled Map Systems 571
    T. Konishi and K. Kaneko
Glassy Entrainment in a Large Population of Limit-Cycle
Oscillators with Random and Frustrated Interactions 573
    H. Daido
Structural Relaxation Process in a Two-Dimensional XY
Clock Model 575
    S. Komura, H. Kobayashi, S. Ueno, and T. Takeda
Dynamics of Random Cellular Structures in Three Dimensions 577
    S. Ohta, T. Nagai, and K. Kawasaki
Sintering of Crystalline Solids: New Modelization Techniques 579
    A. Pavlovitch and G. Martin

VII. AFTERWORDS

Gold Seal and Kyushu Dynasty: Unsolved Mysteries
of Ancient Japan 583
    M. Kamimura

Author Index 595

PREFACE

Nonequilibrium phenomena occurring in nature are generally combinations of processes characterized by widely different time scales. The slowest processes among them are the subject of macroscopic physics like hydrodynamics. On the other hand, microscopic details enter the fastest processes which involve atomic and molecular physics. The traditional nonequilibrium statistical mechanics bridges these two types of processes having extreme time scales, thus providing microscopic foundations of macroscopic physical laws. In many cases macroscopic physical laws take classical well-established simple forms like the Navier–Stokes equation for viscous fluids, Fourier's law for heat conduction, Fick's law for diffusion, etc.

However, there are cases where this simple picture, which is a consequence of wide separation of the two time scales (microscopic and macroscopic), breaks down. A notable example is the critical dynamics where a part of microscopic time scales undergoes slowing down (the so-called critical slowing down) and sometimes overlap with macroscopic time scales. This results in breakdown of simple macroscopic laws, which are replaced by more complicated macroscopic laws characterized by memory effects and spatial nonlocality. It is here that the mode coupling theory (MCT) first appeared. This theory, when combined with the renormalization group idea of critical phenomena, completely elucidated critical dynamics. On the other hand, interests in critical phenomena, though fundamental, are rather limited once they are understood. They occur only in very small regions in phase diagrams and the frontier of research is directed towards more minute details and higher precision.

Fortunately nature is still full of the cases where breakdown of the simple picture mentioned above occurs and our understanding is still quite inadequate. The archetypical among them is the glass transition. Here one observes enormous slowing down of a part of the "microscopic" degrees of freedom and it is still debated whether or not this has to do with a real phase transition in thermodynamic sense. Other examples are provided by dynamics of complex fluids like concentrated polymer solutions, microemulsions, and colloidal suspensions. Here, from the outset one has to deal with extended objects of mesoscopic scales like long chains, liquid droplets, etc. and inevitably their dynamics are slow and rich. It appeared as though that no single theory can describe well any of these complicated dynamical phenomena.

Therefore, it came as a great surprise to find that basically the same formalism of MCT was applied to the glass transition problem and predicated the existence of a new transition, the so-called ergodic to nonergodic transition, and furthermore the prediction appeared to be supported by experiments for a variety of systems exhibiting glass transitions. It must be noted however that although the formalism used is similar, the new MCT has a markedly distinct character from the old one. The new MCT is concerned with long time behavior of short wavelength fluctuations and no singularity is assumed for static quantities entering the theory. The transition point is identified with the bifurcation point of the self-consistent MC equation. The MCT by its construction does not contain *ad hoc* phenomenological parameters and all the parameters in the theory can, in principle, be traced to a microscopic starting model system although in practice this can be implemented only for the simplest systems like hard sphere and Lennard-Jones systems. This is indeed quite remarkable and is worthwhile to know more about it.

This was the circumstance that motivated the initial choice of the topics when the symposium chairman was asked by Tohwa University to plan the first Tohwa University International Symposium. The scope of the symposium was subsequently broadened and modified by the inputs from members of the organizing committee and the international advisory committee whose names are listed elsewhere in this volume. Thus, besides glass transitions we now include polymers, emulsions, colloidal suspensions, and other related subjects which come under the broad title of slow dynamics in condensed matter. Also, in view of the relatively modest role played by theory in this general area we have emphasized experiments and computer simulations on the equal footing as theory.

Thus the first Tohwa University International Symposium on Slow Dynamics in Condensed Matter took place at Tohwa University, Fukuoka, Japan Nov. 4–8, 1991. The present volume contains both invited and contributed papers presented at the Symposium. The content is divided into six parts: Part I (glass transition), Part II (polymer dynamics), Part III (emulsion), Part IV (colloidal suspension), Part V (spin glass), and Part VI (other related topics). Each part is further subdivided into Sec. A (experiment), Sec. B (theory), and Sec. C (computer simulation).

As an afterward we include the record of the after-dinner talk by Professor M. Kamimura on the ancient history of the conference site, namely, the city of Fukuoka and its vicinity.

Although the symposium covered many different fields as explained above, the topics which drew most attention was the MCT. There now appears to be a general (though by no means unanimous) consensus that the MCT is a kind of mean field theory and as such works reasonably well for the ergodic to nonergodic transition occurring above the glass transition, but may not be adequate to describe the glass transition itself. Still it is certainly amazing to see how much new wine can flow out of an old bag. But the old bag needs occasional repairs and may eventually have to be replaced by a new one. This reminds us of the recent history of critical phenomena. The old mean field theory for equilibrium critical phenomena was replaced by the renomalization group theory and the MCT for critical dynamics was superseded by the dynamical renormalization group theory. If this analogy holds, now that we have a reasonable mean field theory we can expect exciting new developments in our understanding of glass transitions and also of slow dynamics in condensed matter (in general) in the near future, although the situation here appears to be much tougher than the case of critical phenomena which, in retrospect, look relatively simple.

On behalf of all those who gathered for this Symposium, both participants and organizers, we would like to express our sincere gratitude to Dr. Toshinami Fukuda, President of Tohwa University, for initiating the Tohwa University Symposia, carrying much of the financial burden and providing the conference rooms and facilities. The Symposium was also aided in one form or another by the supporting agencies listed elsewhere in this volume, which are gratefully acknowledged. Organization of the Symposium was expedited by the research meetings sponsored by the Yukawa Institute for Theoretical Physics, Kyoto.

In planning the Symposium the organizers received encouragement and pertinent advice from members of the international advisory committee whose names are listed elsewhere and a few other individuals (Professor W. Götze and Professor H. Frauenfelder), which undoubtedly played crucial roles in the success of the Symposium.

Last but not least, we wish to thank Ms. Yukimi Itoh, Ms. Kiyoshi Moriyama, and all the others who took such good care of the million chores needed to be done for the smooth operation of the Symposium.

Kyozi Kawasaki
Michio Tokuyama
Toshihiro Kawakatsu

ORGANIZATION OF SYMPOSIUM

CHAIRMAN:
Kyozi Kawasaki
Department of Physics
Faculty of Science
Kyushu University 33
Fukuoka 812, Japan

SECRETARY GENERAL:
Michio Tokuyama
Tohwa Institute for Science
Tohwa University
Fukuoka 815, Japan

ORGANIZING COMMITTEE:
M. Doi	(Nagoya University)
T. Hashimoto	(Kyoto University)
Y. Hiwatari	(Kanazawa University)
A. Ikushima	(HOYA Corporation)
T. Kawakatsu	(Kyushu University)
T. Odagaki	(Kyoto Institute of Technology)
A. Onuki	(Kyoto University)
H. Takayama	(Tsukuba University)
F. Yonezawa	(Keio University)

LOCAL COMMITTEE:
H. Akama	(Tohwa University)
T. Nagai	(Kyushu Kyoritsu University)
H. Ninomiya	(Fukuoka University)
S. Ohta	(Kyushu University)
Y. Shiwa	(Kyushu Institute of Technology)
T. Yamaguchi	(Fukuoka University)

INTERNATIONAL ADVISORY COMMITTEE:
C. A. Angell	(Arizona State University)
K. Binder	(Universität Mainz)
S. H. Chen	(M.I.T.)
S. F. Edwards	(Cambridge University)
J.-P. Hansen	(Ecole Normale Superieure de Lyon)
L. Lundgren	(Uppsala University)
A. Sjölander	(Chalmers University of Technology, Göteborg)

SUPPORTING AGENCIES:
The Physical Society of Japan
The Chemical Society of Japan
The Japan Society of Applied Physics
Fukuoka City
Fukuoka Prefecture
The HOYA Corporation
The Asahi Shimbun
The Mainichi Newspaper
The Nishinippon Newspaper (Western Japan Daily)
Japan Broadcasting Corporation
Fukuoka Broadcasting Corporation
Kyushu Asahi Broadcasting Company
RKB-Mainichi Broadcast Corporation
Television Nishinippon Corporation
Nippon Telegraph and Telephone Corporation
Nishi-Nippon Rail Road Co., Ltd.
Gakujutsu Tosyo Publishing Corporation
Coca-Cola Company

The 1st Tohwa University International Symposium
November 4-8, 1991, Fukuoka, Japan

OPENING ADDRESS

On behalf of Tohwa University, I welcome you to this Symposium.

It is well-known that the world has changed extensively in the areas of science, culture, and economy, and advanced countries have become richer and wealthier, but on the other hand, there are food shortages in many countries. There is a mass exodus of the poor from their countries. The state of things in the world is very precarious. In a state like this, I think, it is very important for us in advanced countries to endeavor to secure a stabilized living condition in the world, and to bestow the benefits from our advanced science on the people of the Earth.

Fortunately Japan has rapidly progressed since the war, and now we can lend support to other countries in order to provide better living conditions, and also, with our scientific know-how we can provide assistance for poorer countries.

Considering this world situation, Tohwa Unversity decided to sponsor the International Scienific Symposium once a year, in order to find ways to increase scientific advancement in all nations, and this year we decided to hold an International Symposium on Slow Dynamics in Condensed Matter. We hope to contribute to the peace and happiness of human beings through scientific progress.

I am very delighted to see that so many scientists from all over the world have come to participate in this important cause for world development.

I am sure with so many talented scientists here that this Symposium will be successful.

Thank you.

Toshinami Fukuda
President, Tohwa University

WELCOMING ADDRESS

Ladies and Gentlemen,

As we commence the First Tohwa University International Symposium, it is a great pleasure and honor to have this opportunity to welcome our distinguished guests to Fukuoka on behalf of the citizens and City Council.

The autumn season is now upon us, beautifully decorating Fukuoka's abundant natural areas in an array of colors.

Fukuoka has truly been blessed by Mother Nature in its natural resources and ideal location, therefore enabling Fukuoka to develop as the leading city in Kyushu and western Japan.

Another reason for Fukuoka's success is due in part to the youth and vitality of the students found studying in one of the eleven four-year universities or nine junior colleges in Fukuoka.

Fukuoda is a prominent academic city in Japan, partially because we, the citizens of Fukuoka, value the role unviersities play in educating our youth, developing scientific research and our city.

One of Fukuoka's objectives is the promotion of scientific and international exchanges, therefore, it seems only natural that we are welcoming you here to discuss the fundamentals of scientific studies.

I sincerely hope that this symposium serves in cultivating ideas and fruitful results in each respective field. Also, may you have a very pleasant and memorable experience during your stay.

Atsuko Kato
Deputy Mayor
of Fukuoka City

Banquet Speech

We heartily welcome all of you from overseas countries and various parts of Japan to the Fukuoka prefecture.

There are many attractive points in the Fukuoka prefecture, including historic spots, cultural assets, scenic beauty, and urban culture.

The universities, research institutes, and the number of undergraduate and postgraduate students in Kyushu account for 10% in Japan. Most of the universities and research institutes in Kyushu gather in the northern part of Kyushu, centering in the Fukuoka prefecture. The Fukuoka prefecture is the region full of these young brains and vitality.

In addition, the number of passengers embarked/disembarked at Fukuoka International Airport in the year 1990 was some 1 500 000 in total, which exceeds the population of Fukuoka city. The number of students sent to Fukuoka for study from various countries of the world is increasing year by year. Thus, internationalization in Fukuoka has progressed remarkably.

Through the promotion of policy based on these internationalizing potentials, we are aiming at constructing an international prefecture open to the world, starting from Asian countries.

Under these circumstances, it is very significant to hold such an international and authorized symposium here in Fukuoka leading other prefectures.

I hope that this symposium is fruitful and promotes mutual friendship.

Wishing you further success.

With best regards.

Hachiji Okuda
Governor of Fukuoka Prefecture
7 November 1991

TOHWA INSTITUTE FOR SCIENCE
AND
TOHWA UNIVERSITY INTERNATIONAL SYMPOSIA

The Tohwa Institute for Science was established in April 1990 in Fukuoka with the aid of Dr. Toshinami Fukuda. Dr. Fukuda is President of Tohwa University. The principal purpose of the Tohwa Institute for Science is to devote itself to the exchange of scientific and technological knowledge in well-established, interdisciplinary fields and thus to contribute to the progress of science for happiness of human beings. The main activity of the Tohwa Institute for Science is to assist in the exchange of visiting scientists and also to encourage international symposia in most stimulating and timely fields.

In this context the Tohwa Institute for Science sponsors one Tohwa University International Symposium every year in Fukuoka, Japan.

**EXECUTIVE MEMBERS
OF
TOHWA UNIVERSITY INTERNATIONAL SYMPOSIA**

OFFICERS:

Toshinami Fukuda
Yoshiaki Nagano
Akira Katase
Michio Tokuyama
Junichiro Tsutsumi

Addresses of Participants

PARTICIPANTS

Bruce J. Ackerson	Department of Physics Oklahoma State University Stillwater, OK 74078
Shin Akahoshi	Department of Applied Physics, Faculty of Science Fukuoka University Fukuoka 814, Japan
Hachiro Akama	Physics Department Tohwa University Fukuoka 815, Japan
Christiane Marie Alba-Simionesco	Laboratoire de Chimie-Physique 11 rue Pierre et Marie Curie, 75231 Paris Cedex 05, France
Shlomo Alexander	Department of Chemical Physics Weizmann Institute of Science Rehovot 76100, Israel
Charles Austen Angell	Department of Chemistry Arizona State University Tempe, AZ 85287
Toshio Aoyagi	Department of Physics, Faculty of Science Kyoto University Kyoto 606, Japan
Toshihico Arimitsu	Institute of Physics University of Tsukuba Ibaraki 305, Japan
Umberto Balucani	Istituto di Elettronica Quantistica CNR 50127 Firenze, Italy
Dick Bedeaux	Gorlaeus Laboratories University of Leiden P.O. Box 9502, 2300 RA Leiden, The Netherlands
Katica Biljakovic	Institute of Physics of the University P.O. Box 304, 41001 Zagreb, Yugoslavia
Jean-Louis Bocquet	CEREM/DTM/SRMP-C.E. Saclay 91191 Gif sur Yvette Cedex, France
John F. Brady	Department of Chemical Engineering California Institute of Technology Pasadena, CA 91125
Sow Hsin Chen	Department of Nuclear Engineering M.I.T. Cambridge, MA 02139

Benjamin Chu	Department of Chemistry State University of New York at Stony Brook Stony Brook, NY 11794-3400
Bogdan Cichocki	Institute of Theoretical Physics Warsaw University 00-681 Warsaw, Poland
Paolo Codastefano	Dipartimento di Fisica Università di Roma "La Sapienza" I-00185 Roma, Italy
E. G. D. Cohen	Rockefeller University New York, NY 10021
Herman Z. Cummins	Department of Physics City College of CUNY New York, NY 10031
Hiroaki Daido	Department of Physics, Faculty of Engineering Kyushu Institute of Technology Kitakyushu 804, Japan
Giovanni D'Arrigo	Dipartimento di Energetica Università di Roma "La Sapienza" Via A. Scarpa, 14-00161 Roma, Italy
Marc Descamps	Laboratoire de Dynamique et Structure des Matériaux Moléculaires Bâtiment P5-231 Université Lille I 59655 Villeneuve d'Ascq Cédex, France
Masao Doi	Department of Applied Physics, School of Engineering Nagoya University Nagoya 464-01, Japan
Maurice Drifford	Service de Chimie Moléculaire CEA-CE Saclay 91191 Gif sur Yvette Cédex, France
Tetsuo Eguchi	Department of Applied Physics Fukuoka University Fukuoka 814-01, Japan
Kenji Ema	Department of Physics Tokyo Institute of Technology Tokyo 152, Japan
Yasuo Endoh	Department of Physics Tohoku University Sendai 980, Japan
Yoshihisa Enomoto	Department of Physics, Faculty of Science Nagoya University Nagoya 464-01, Japan
B. Ubbo Felderhof	Institut für Theoretische Physik A RWTH Aachen 5100 Aachen, Germany

Kazuhiro Fuchizaki	Department of Physics, Faculty of Science Kyushu University Fukuoka 812, Japan
Satoru Fujime	Graduate School of Integrated Science Yokohama City University Yokohama 236, Japan
Hirokazu Fujisaka	Department of Physics Kyushu University Fukuoka 812, Japan
Koji Fukao	Department of Physics, College of Liberal Arts and Science Kyoto University Kyoto 606, Japan
Hiroshi Furukawa	Faculty of Education Yamaguchi University Yamaguchi 753, Japan
Wolfgang Götze	Physik-Department Technische Universität München D-8046 Garching, Germany
Gary S. Grest	Exxon Research and Engineering Corporation Annandale, NJ
Junko Habasaki	Department of Electronic Chemistry Tokyo Institute of Technology at Nagatsuta Yokohama 227, Japan
Makoto Hagiwara	Department of Electronics and Information Science Kyoto Institute of Technology Kyoto 606, Japan
Katsumi Hamano	Department of Physics Tokyo Institute of Technology Tokyo 152, Japan
Kenzi Hamano	Department of Biological and Chemical Engineering Faculty of Technology Gunma University Kiryu 376, Japan
Nozomu Hamaya	Department of Physics, Faculty of Science Ochanomizu University Tokyo 112, Japan
Hiroaki Hara	Department of Engineering Science Tohoku University Sendai 980, Japan
Kazuhiro Hara	Department of Applied Science, Faculty of Engineering Kyushu University Fukuoka 812, Japan
Yoshifumi Harada	Department of Applied Physics, Faculty of Engineering Fukui University Fukui 910, Japan

James L. Harden	Department of Applied Physics, School of Engineering Nagoya University Nagoya 464-01, Japan
Peter Harrowell	Theoretical Chemistry Section University of Sydney Sydney N.S.W. 2006, Australia
Takeji Hasimoto	Department of Polymer Chemistry, Faculty of Engineering Kyoto University Kyoto 606, Japan
Hisao Hayakawa	Department of Physics Tohoku University Sendai 980, Japan
Yasuaki Hiwatari	Department of Physics Kanazawa University Kanazawa 920, Japan
Mizuhiko Ichikawa	Department of Physics, Faculty of Science Hokkaido University Sapporo 060, Japan
Shinichi Ichinose	Nara University Nara 631, Japan
Katsunori Iio	Department of Physics, Faculty of Science Tokyo Institute of Technology Tokyo 152, Japan
Hironobu Ikeda	BSF, National Laboratory for High Energy Physics Oho 1-1, Tsukuba 305, Japan
Akira Ikushima	Materials Research Laboratory Hoya Corporation 3-3-1 Musashino, Akishima, Tokyo 196, Japan
Masayuki Imai	Research Center Toyobo Co., Ltd. Otsu, Shiga 520-02, Japan
Yoshiro Inoue	Department of Chemical Engineering Osaka University Toyonaka 560, Japan
Akira Ishida	Department of Physics, Faculty of Science Shimane University Matsue 690, Japan
Yasushi Ishii	Department of Material Science Himeji Institute of Technology Akou-gun, Hyogo 678-12, Japan
Manabu Ishimaru	Department of Materials Science and Technology Kyushu University Kasuga-shi, Fukuoka 816, Japan
Masaki Itoh	Department of Physics, Faculty of Science Shimane University Matsue 690, Japan

Sumiko Itoh	Department of Electronics and Information Science The Nishi-Tokyo University Yamanashi 409-01, Japan
Hiroshi Jinnai	Department of Polymer Chemistry Kyoto University Kyoto 606, Japan
Wolfram Just	Department of Physics Kyushu University Fukuoka 812, Japan
Makoto Kaburagi	College of Liberal Arts Kobe University Kobe 657, Japan
Keisuke Kaji	Institute for Chemical Research Kyoto University Uji, Kyoto 611, Japan
Haruichi Kanaya	Department of Applied Science, Faculty of Engineering Kyushu University Fukuoka 812, Japan
Toshiji Kanaya	Institute for Chemical Research Kyoto University Uji, Kyoto 611, Japan
Masami Kawaguchi	Department of Chemistry for Materials Faculty of Engineering Mie University Tsu, Mie 514, Japan
Toshihiro Kawakatsu	Department of Physics, Faculty of Science Kyushu University Fukuoka 812, Japan
Junichi Kawamura	Department of Chemistry, Faculty of Science Hokkaido University Sapporo 060, Japan
Kyozi Kawasaki	Department of Physics Kyushu University Fukuoka 812, Japan
Tatuo Kawasaki	Physics Department, College of Liberal Arts and Sciences Kyoto University Kyoto 606, Japan
Hatsuo Kimura	Department of Applied Physics Nagoya University Nagoya 464-01, Japan
Wataru Kinase	Department of Physics, School of Science and Engineering Waseda University Tokyo 169, Japan
Kenichi Kinugawa	Government Industrial Research Institute Osaka Osaka 563, Japan

T. R. Kirkpatrick	University of Maryland College Park, MD 20742
Kiyoshi Kishi	Faculty of Science Science University of Tokyo Toko 162, Japan
Kunihiro Kitamura	Research Center Taisho Pharmaceutical Co., Ltd. Saitama 330, Japan
Hideki Kobayashi	Faculty of Integrated Arts and Sciences Hiroshima University Hiroshima 730, Japan
Shinji Koga	Osaka Kyoiku University Osaka 543, Japan
Tsuyoshi Koga	Department of Physics, Faculty of Science Kyushu University Fukuoka 812, Japan
Seiji Kojima	Institute of Applied Physics University of Tsukuba Ibaraki 305, Japan
Shigehiro Komura	Faculty of Integrated Arts and Sciences Hiroshima University Hiroshima 730, Japan
Shigeyuki Komura	Institute of Physics, College of Arts and Sciences University of Tokyo Tokyo 153, Japan
Tetsuro Konishi	Department of Physics, School of Science Nagoya University Nagoya 464-01, Japan
Toshiyuki Koyama	Department of Materials Science and Engineering Metals Section Nagoya Institute of Technology Nagoya 466, Japan
Kurt Kremer	Institut für Festkörperforschung-KFA 5170 Jülich, Germany
Masaki Kurokiba	Department of Applied Physics Fukuoka University Fukuoka 814-01, Japan
Makoto Kuwabara	Department of Physics Toho University Chiba 274, Japan
Sang Bub Lee	Department of Physics Kyungpook National University Taegu 702-701, Korea

Leif Elis Christian Lundgren	Institute of Technology Uppsala University Box 534, S-75121 Uppsala, Sweden
Hiroshi Maeda	Department of Chemistry, Faculty of Science Kyushu University Fukuoka 812, Japan
Andreas Magerl	Institut Laue Langevin 38042 Grenoble Cedex, France
Takeo Matsubara	Department of Applied Physics Okayama University of Science Okayama 700, Japan
Mitsuhiro Matsumoto	Department of Applied Physics, School of Engineering Nagoya University Nagoya 464-01, Japan
Motohiro Matsuura	Department of Electronics and Information Science Kyoto Institute of Technology Kyoto 606, Japan
Shu Matsuura	School of High Technology for Human Welfare Tokai University Sizuoka 410-03, Japan
Gene F. Mazenko	James Franck Institute and Department of Physics University of Chicago Chicago, IL 60637
William van Megen	Department of Applied Physics Royal Melbourne Institute of Technology Melbourne, Victoria 3001, Australia
Ferenc Mezei	Hahn-Meitner-Institut Pf. 390128, D-1000 Berlin 39, Germany
Hidetoshi Miike	Department of Electrical Engineering Yamaguchi University Ube 755, Japan
Akiomi Mishima	Kanazawa Institute of Technology Ishikawa 921, Japan
Hiroo Miyagawa	Research Center Taisho Pharmaceutical Co., Ltd. Saitama 330, Japan
Kenji Miyakawa	Faculty of Science Fukuoka University Fukuoka 814, Japan
Yoshihito Miyako	Department of Physics, Faculty of Science Osaka University Toyonaka 560, Japan
Seiji Miyashita	Physics Department, College of Liberal Arts and Sciences Kyoto University Kyoto 606, Japan

Sasuke Miyazima	Department of Engineering Physics Chubu University Kasugai Aichi 487, Japan
Tsuyoshi Mizuguchi	Department of Physics, Faculty of Science Kyoto University Kyoto, Japan
Raymond D. Mountain	Thermophysics Division National Institute of Standards and Technology Gaithersburg, MD 20899
Toyonori Munakata	Department of Applied Mathematics and Physics Kyoto University Kyoto 606, Japan
Tatsuzo Nagai	Physics Department Kyushu Kyoritsu University Kitakyushu 807, Japan
Masahiro Nakagawa	Department of Electrical Engineering Faculty of Engineering Nagaoka University of Technology Nagaoka, Niigata 940-21, Japan
Akio Nakahara	Department of Physics Kyushu University Fukuoka 812, Japan
Hiizu Nakanishi	Department of Physics, Faculty of Science and Technology Keio University Yokohama 223, Japan
Tsuneyoshi Nakayama	Department of Applied Physics Hokkaido University Sapporo 060, Japan
Shinji Nambu	Central Research Laboratory Kyocera Corporation Kokubu, Kagoshima 899-43, Japan
Satoru Nasuno	Department of Physics Kyushu Institute of Technology Kitakyushu 804, Japan
Hiroshi Ninomiya	Department of Applied Physics, Faculty of Science Fukuoka University Fukuoka 815, Japan
Moyuru Ochiai	Department of Electronics North Shore College of SONY Institute Atugi 243, Japan
Takashi Odagaki	Department of Liberal Arts and Sciences Kyoto Institute of Technology Kyoto 606, Japan

Atsushi Ogawa	Department of Physics Kyushu University Fukuoka 812, Japan
Ryuichiro Oguma	Department of Applied Physics, Faculty of Science Fukuoka University Fukuoka 814, Japan
Masaharu Oguni	Faculty of Science Tokyo Institute of Technology Tokyo, Japan
Hiroshi Ogura	Department of Electronics North Shore College of SONY Institute Atugi 243, Japan
Shigetoshi Ohta	Physics Department Kyushu Kyoritsu University Kitakyushu 807, Japan
Shonosuke Ohta	Department of Physics, College of General Education Kyushu University Fukuoka 810, Japan
Takao Ohta	Department of Physics Ochanomizu University Tokyo 112, Japan
Toshiya Ohtsuki	Department of Applied Physics Fukui University Fukui 910, Japan
Hirotaka Okabe	Department of Applied Science, Faculty of Engineering Kyushu University Fukuoka 812, Japan
Tohru Okuzono	Department of Physics, Faculty of Science Kyushu University Fukuoka 812, Japan
Yoshiyuki Ono	Department of Physics Toho University Chiba 274, Japan
Akira Onuki	Research Institute for Fundamental Physics Kyoto University Kyoto 606, Japan
Yoshiaki Ozaki	Department of Chemistry Nagoya Institute of Technology Nagoya 466, Japan
Tadeusz Pakula	Max-Planck-Institut für Polymerforschung Postfach 3148, 6500 Mainz, Germany
Giervila Marta Pasenkiewicz	Department of Molecular Science, Research Center Taisho Pharmaceutical Co., Ltd. Saitama 330, Japan

Wolfgang Paul	Institute for Physics Johannes Gutenberg University D-6500 Mainz, Germany
Andre Pavlovitch	Service de Recherches de Metalluroie Physique C. E. Saclay 91191 Gif sur Yvette Cedex, France
Harald Pleiner	FB7 Physik Universität Essen D4300 Essen 1, Germany
Dorthe Posselt	Institut für Physik Johannes-Gutenberg Universität Staudingerweg 7, D-6500 Mainz, Germany
Peter N. Pusey	Physics Department Edinburgh University Edinburgh, United Kingdom
Dicter O. Richter	Institut für Festkörperforschung KFA Jülich, Germany
Per Arne Rikvold	Department of Physics Florida State University Tallahassee, FL 32306
Jacques Rouch	CPMOH-University Bordeaux 1 351 Cours de la Libération 33405 Talence, France
Jean-Noël Roux	Laboratoire Central des Ponts et Chaussées Service de Chimic, SLR, point 10 58, boulevard Lefébvre, 75732 Paris Cedex 15, France
Yukio Saito	Department of Physics Keio University Yokohama 223, Japan
Hidetsugu Sakaguchi	Department of Physics, College of General Education Kyushu University Fukuoka 810, Japan
Hideaki Sakata	Department of Physics Tokyo Intitute of Technology Tokyo 152, Japan
Taiichi Sakaya	Takatsuki Research Laboratory Sumitomo Chemical Co., Ltd. Takatsuki 569, Japan
Shinichi Sakurai	Department of Polymer Science and Engineering Kyoto Institute of Technology Kyoto 606, Japan
Yasuo Saruyama	Faculty of Textile Science Kyoto Institute of Technology Kyoto 606, Japan
Shinichi Sasa	Department of Physics Kyoto University Kyoto 606, Japan

Masaki Sasai	Department of Chemistry, College of General Education Nagoya University Nagoya 464-01, Japan
Shigeo Sasaki	Department of Chemistry, Faculty of Science Kyushu University Fukuoka 812, Japan
Hirokazu Sato	Department of Physics Aichi University of Education Kariya 448, Japan
Klaus Schätzel	Institut für Angewandte Physik der Universität D-2300 Kiel, Germany
Harvey Scher	BP Research 4440 Warrensville Center Road Cleveland, OH 44128
Howard Tomio See	Department of Applied Physics, Faculty of Engineering Nagoya University Nagoya 464-01, Japan
James P. Sethna	Laboratory of Applied Physics University of Denmark Lyugby 2800, Denmark
Hideki Seto	Faculty of Integrated Arts and Sciences Hiroshima University Hiroshima 730, Japan
Kaoru Shibata	Institute for Materials Research Tohoku University Sendai 980, Japan
Takeshi Shigenari	Department of Applied Physics and Chemistry The University of Electro-Communications Chofu-shi, Tokyo 182, Japan
Kenichi Shiiyama	Department of Applied Physics Fukuoka University Fukuoka 814-01, Japan
Yasuhiro Shiwa	The Physics Labs., Kyushu Institute of Technology Fukuoka 820, Japan
Michael F. Shlesinger	Office of Naval Research Physics Division Arlington, VA 22217
Joel David Shore	Physics Department Cornell University Ithaca, NY 14853
Shelly L. Shumway	Materials Science Division Argonne National Laboratory Argonne, IL 30439
Hans Sillescu	Institut für Physikalische Chemie Universität Mainz Jakob-Welder-Weg 15, D-6500 Mainz, Germany

Lennert Sjögren	Institute of Theoretical Physics Chalmers University of Technology S-41296 Göteborg, Sweden
Jean Souletie	Centre de Recherches sur les Très Basses Températures CNRS, BP166X, 38042 Grenoble-Cédex, France
Hiroshi Suga	Department of Chemistry and Microcalorimetry Research Center, Faculty of Science Osaka University Osaka 560, Japan
Wokyung Sung	Department of Physics Pohang Institute of Science and Technology Pohang, Korea
Jun-ichi Suzuki	Department of Physics Tohoku University Sendai 980, Japan
Masaru Suzuki	Division of Natural Science The University of Electro-Communications Chofu-shi, Tokyo 182, Japan
Yoshihiro Taguchi	Department of Physics Tokyo Institute of Technology Tokyo 152, Japan
Masato Takahashi	Faculty of Fiber Science Shinshu University Nagano 386, Japan
Yashuhiro Takahashi	Faculty of Science Osaka University Toyonaka 560, Japan
Hajime Takayama	Institute of Physics University of Tsukuba Ibaraki 305, Japan
Takayoshi Takeda	Faculty of Integrated Arts and Sciences Hiroshima University Hiroshima 730, Japan
Mikihito Takenaka	Department of Polymer Chemistry, Faculty of Engineering Kyoto University Kyoto 606, Japan
Hisao Takeuchi	Research Center Mitsubishi Kasei Corporation Midori-ku, Yokohama 227, Japan
Fumihiko Tanaka	Department of Physics, Faculty of General Education Tokyo University of Agriculture and Technology Toyko 183, Japan
Hajime Tanaka	Department of Applied Physics and Applied Mechanics Institute of Industrial Science University of Tokyo Tokyo 106, Japan

Toshijiro Tanaka	Department of Physics, Faculty of Science Kyushu University Fukuoka 812, Japan
Takashi Taniguchi	Department of Physics, Faculty of Science Kyushu University Fukuoka 812, Japan
Piero Tartaglia	Dipartimento di Fisica Università di Roma "La Sapienza" Piazzale Aldo Moro 2, I-00185 Roma, Italy
Michio Tokuyama	Tohwa Institute for Science Tohwa University Fukuoka 815, Japan
Qui Tran-Cong	Department of Polymer Science and Engineering Kyoto Institute of Technology Kyoto 606, Japan
Satoru Ueno	Faculty of Integrated Arts and Sciences Hiroshima University Hiroshima 730, Japan
Norimasa Umesaki	Material Physics Department Governmental Industrial Research Institute Osaka 563, Japan
Sylvia M. Underwood	ICI Australia Research Ascot Vale, Victoria 3032, Australia
Makio Uwaha	Institut Laue-Langevin 156X, 38042 Grenoble Cedex, France
Eric Vincent	S.P.S.R.M.-C.E.N. Saclay Orme des Merisiers 91191 Gif sur Yvette Cedex, France
Dereic Walton	Physics Department McMaster University Hamilton, Ontario L8S 4M1, Canada
Stanley Windwer	Department of Chemistry Adelphi University Garden City, NY 11530
Thomas A. Witten	James Franck Institute University of Chicago Chicago, IL 60637
Toshirou Yagi	Research Institute of Applied Electricity Hokkaido University Sapporo 060, Japan
Kousuke Yakubo	Department of Applied Physics Hokkaido University Sapporo 060, Japan
Tomoji Yamada	Division of Electronic Physics Kyushu Institute of Technology Kitakyushu 804, Japan

Manabu Yamaguchi	Department of Chemical Engineering Faculty of Engineering Science Osaka University 560 Osaka, Japan
Toshio Yamaguchi	Department of Chemistry Fukuoka University Fukuoka 814-01, Japan
Takao Yamamoto	Department of Physics, Faculty of General Studies Gunma University Gunma 376, Japan
Tsunetaka Yamamoto	FSY Project Asahi Chemical Industry Co., Ltd. Nobeoka-shi, Miyazaki 882, Japan
Naohiko Yasuda	Electrical Engineering Department Gifu University Gifu 501-11, Japan
Fumiko Yonezawa	Faculty of Science and Technology Keio University Yokohama 223, Japan
Kentaroh Yoshida	Faculty of Engineering Kobe University Kobe 657, Japan
Yuji Yoshida	Department of Applied Science, Faculty of Engineering Kyushu University Fukuoka 812, Japan
Shigeo Yoshino	Department of Physics, School of Science Nagoya University Nagoya 464-01, Japan
Walter Zimmermann	IFF Theorie III Forschungszentrum Jülich D-5170 Jülich, Germany
Reiner Zorn	Forschungszentrum Jülich IFF Postfach 1913, D-5170 Jülich, Germany

I. GLASS TRANSITION

A. EXPERIMENT

SLOW PROCESSES IN VISCOUS LIQUIDS: STRESS AND STRUCTURAL RELAXATION, CHEMICAL REACTION FREEZING, CRYSTAL NUCLEATION AND MICROEMULSION ARREST, IN RELATION TO LIQUID FRAGILITY.

C. A. Angell, C. Alba,* A. Arzimanoglou,# R. Böhmer,‡
J. Fan, Q. Lu, E. Sanchez, H. Senapati, and M. Tatsumisago.+
Department of Chemistry, Arizona State University, Tempe, AZ 85287

ABSTRACT

We review a variety of measurements on model systems in the medium viscosity range which seem consistent with both thermodynamical (entropy vanishing) and dynamical (mode coupling) origins of glassy behavior and then examine behavior near and below T_g to seek relations between liquid fragility and the non-exponential and non-linear aspects of liquid relaxation processes. We include the model ionic system $Ca(NO_3)_2$–KNO_3 and analogs, van der Waals systems, and the covalently-bonded system Ge-As-Se in which the relation of liquid properties to the vector percolation concepts of Phillips and Thorpe can be conveniently studied.

With some basic phenomenology in the liquid state itself thereby established, we turn attention to longer length-scale processes occurring in viscous liquid media. Among these will be the kinetics of nucleation of crystals, the freezing of microemulsion droplet sizes during continuous cooling of temperature sensitive microemulsions, and the freezing of chemical reactions during continuous cooling or continuous evaporation of solvent. The latter freezings can occur at temperatures which are far above the solvent glass transition temperature depending on solvent fragility, which may be a consideration in the strategies adopted by nature in preservation of plant and insect integrity in cold and arid climates.

Finally we consider the slowing down which occurs in liquids with density maxima like water and SiO_2 which appear to have, as their low temperature metastable limits, spinodal instabilities (with associated divergences in physical properties) in place of the usual ideal glass transitions. So far little studied for lack of tractable slow systems, these offer a new and challenging arena for relaxation studies.

THE MODERATELY-VISCOUS DOMAIN AND A SCIENTIFIC IRONY.

For much of the time that the phenomenology of the glass transition has held scientific, as opposed to merely technological, attention a leading question has concerned the validity and meaning of the three-parameter Vogel-Fulcher equation[1,2] for the glassforming liquid viscosity,

$$\eta = \eta_o \exp B/(T-T_o) \qquad (1)$$

---

*Laboratoire de Chimie Physique, 11 Rue Pierre et Marie Curie, 75231 Paris Cedex 05, FRANCE
#Now at Famar, S. A., Athens, GREECE
‡Institut für Experimentalphysik, University of Darmstadt, Darmstadt, GERMANY
+Department of Applied Chemistry, University of Osaka Prefecture, Sakai, Osaka 591, JAPAN

(η_0, B and T_0 constants, $T_0 > 0K$). Within two years of the publication of the empirical equation, Tammann studying molecular liquids recognized the close relation between T_0 of this equation and the glass transition temperature T_g and suggested that T_0 represented the temperature at which some limiting structure would be achieved if the glass transition did not intervene.[3,4]

The author's involvement in this fascinating field was provoked by (i) the simple and appealing derivation of Eq. (1) using free volume concepts by Cohen and Turnbull,[5] and (ii) his own experimental finding[6,7,8,9,10] that not only the temperature dependence but also the composition and pressure dependences of electrical conductivity of a system containing very simple particles K^+, Ca^{++} (isoelectronic with argon) and the triangular anion NO_3^-, could be simply rationalized by an Eq. (1) analysis. Furthermore, based on this analysis, predictions of then-unmeasured properties could be made. We summarize these findings in the following in order to present the comparison with mode coupling analysis of this system's behavior.[11,12,13,14] Indeed the irony we point out in this introduction is that it is data for the same systems and from the same temperature ranges in these systems which were used originally to argue the credibility of Eq. (1) which are now being used, very effectively, to establish the validity of an entirely different body of theory based on mode-coupling arguments. It is interesting to compare the two cases directly because they attribute fundamental significance to quite different singular temperatures: in one case, the singular temperature (T_0 of Eq. (1)) lies below the experimental glass transition temperature T_g and in the other case (T_c of the mode coupling theory power law, see below) it lies above it.

In Fig. 1(a) electrical (equivalent) conductivity data for $(Ca(NO_3)_2$-$KNO_3)$ (CKN) solutions are shown as a function of composition for several temperatures,[21] and the strong composition dependence is noted. Fig. 1(b) shows how all these data collapse to a single straight line when plotted on the basis of Eq. (2) which is derived from Eq. (1) on the assumption that T_0 of Eq. (1) is the only composition dependent variable and that it varies linearly with composition (at 2.1K/mole% $Ca(NO_3)_2$),

$$\sigma = \sigma_0 \exp(B/[t-px]) \qquad (2)$$

Here x is the mol.% of $Ca(NO_3)_2$ in the solution, p is dT_0/dx and t is the difference between T_0 and the temperature of the isotherm at each composition. The variation of T_0 with composition is compared with the corresponding variation of T_g, determined in different lower T experiments, in Fig. 2, which gives the phase diagram for this system. Similar correlations may be obtained for viscosity data which yield the same T_0 values within 2K [15] at each composition.

Fig. 3 provides a stringent test of the form of Eq. (1). It shows how data for the instantaneous slope of the Arrhenius plot of specific conductivity (i.e. the "activation energy" $E_\sigma + 1/2RT^{(15)*}$) for a wide range of compositions in the system $Ca(NO_3)_2 + KNO_3$ (CKN) yield a straight line passing through the origin when plotted vs. $(1-T_0/T)^{-2}$ as predicted by a slightly modified version of Eq. (1) (see footnote), with T_0 the only variable, linear in composition, see Fig. 3 inset.

Even more pleasing was the behavior under pressure, reproduced in Fig. 4. Here all data could be collapsed to single straight line by means of Eq. 1 again using only linear variations of the T_0 parameter (see inset). In this case, the pressure

---

*RT is added to E_σ to give the quantity expected for a diffusion process, since the Cohen-Turnbull theory was formulated for the diffusivity.

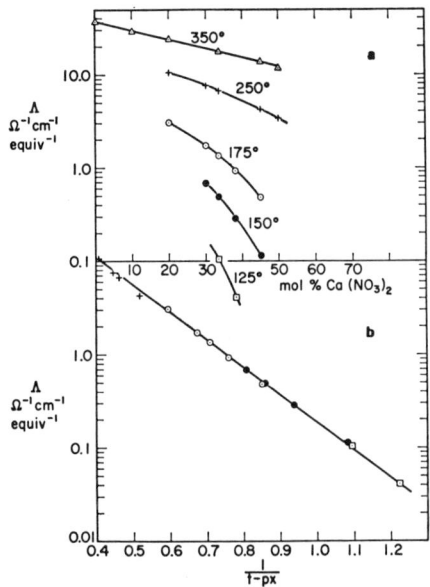

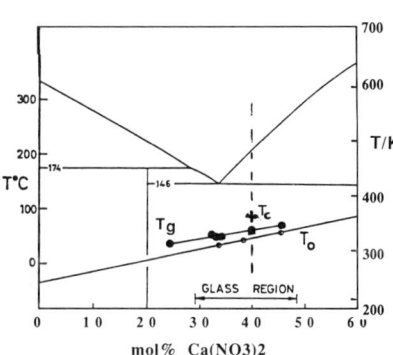

Fig. 1. Equivalent conductance in the system $Ca(NO_3)_2 + KNO_3$. (a) Dependence on mole% $Ca(NO_3)_2$ at different temperatures. (b) Correlation of composition and temperature effects by the Eq. (2) factor $[1/(t - px)]$.

Fig. 2. Phase diagram for $Ca(NO_3)_2$ KNO_3 system, showing relation of T_g to T_o of Eq. (1) and T_C of mode coupling theory.

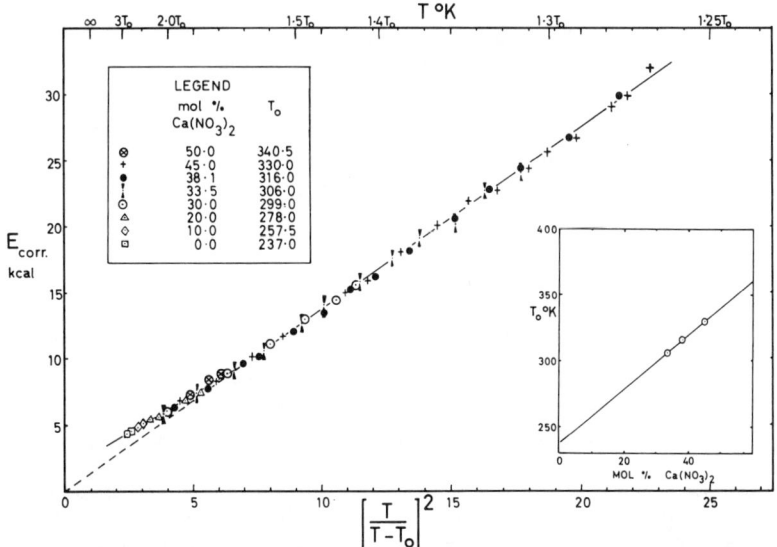

Fig. 3. Plot of conductivity activation energy (corrected to diffusivity equivalent) for various compositions in CKN system, versus $(T/[T-T_o])^2$. Note reduction to single straight line plot passing through origin for T_o linear in composition, see inset.

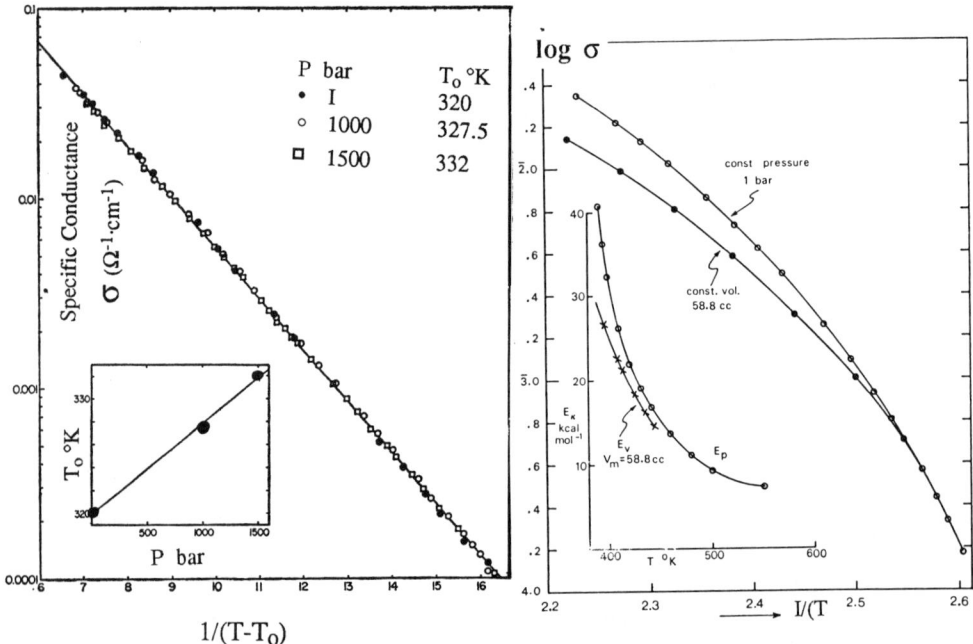

Fig.4. Conductivity data for 2:3 CKN at three pressures, showing reduction to single line by Eq. (1) with T_o linear in pressure, see inset.

Fig. 5. Conductivity, and activation energy (inset), of 2:3 CKN at constant volume compared with same at constant pressure.

dependence observed for T_o, 8 ±2 K/kbar,[8a] (later refined to 7 ±1 K/kbar[8b]) could be used together with expansivity data through T_g, to predict the change in heat capacity at T_g, ΔC_p, which has since been measured many times.[16,17,18] The relation used for this prediction is one of the Ehrenfest equations for second order thermodynamic transitions shown by Davies and Jones,[19] and Goldstein[20] to apply to the glass transition if the excess entropy of liquid over crystal remains constant at T_g. The relation is

$$(\partial T_g/\partial P)_{Sc} = V_g T_g \Delta\alpha/\Delta C_p \qquad (3)$$

if $(\partial T_g/\partial P)_{Sc}$ is assumed to be equal to dT_o/dP then the value of ΔC_p predicted is 0.10 ±.02 cal/gK which compares with 0.105 ±0.002 cal/gK subsequently measured.[17,18] Subsequent experiments measured dT_g/dP directly and obtained 6.4 K/kbar.[21]*

An important feature of this study which has never been properly reported is the

---

*It is worth noting that the small differences between measured and fitted dT_o/dp and also dT_o/dx (Fig. 2) values would be removed if the one-variable-parameter fits had been made using Eq. (5) below rather than Eq. (1) i.e. keeping the "strength" parameter D rather than B of Eq. (1) constant. It is significant in this respect that a later independent and computer-aided analysis of the Fig. 1 data yielded a T_o value of 203 K at the composition 33.3% $Ca(NO_3)_2$ (C. A. Angell and C. T. Moynihan, in *Molten Salts; Characterization and Analysis*, Ed. G. Mamantov, Marcel Dekker, New York, June 1979, p. 315). For the significance, see next footnote.

behavior encountered when temperature is changed at constant volume. When compressibility measurements by Pollard et al.[22] are used to assess the electrical conductivity at constant volume, a result quite incompatible with the free volume-based derivation of Eq. (1) is obtained. Instead of the conductivity varying only as $T^{-1/2}$, a strong temperature dependence is found. In fact, the activation energy assessed at a constant volume of 58.8 cc/mole is found to be only a little less than the constant pressure activation energy assessed at the same temperature. The two are compared directly in Fig. 5. This finding implies that the second Ehrenfest-like relation,

$$(\partial T_g/\partial P)_V = \Delta\alpha/\Delta\kappa_T \qquad (4)$$

which can be derived on the assumption that volume is constant at T_g,[19,20] is quite inapplicable at the glass transition, as was subsequently shown to be the case.[23]

The fact that the findings of an Eq. (1) analysis of the pressure dependence of conductivity could be rationalized by equations based on the assumption that the most important extensive variable is the excess entropy of liquid over crystal, was consistent with the idea[24] that the temperature T_0 might be identified with Kauzmann's vanishing excess entropy temperature[25] now symbolized as T_K. Unfortunately, this could not be tested by direct calorimetric investigation for CKN because of the lack of a congruently melting compound in the glassforming range. However, the related nitrate hydrate system $Ca(NO_3)_2 \cdot 4H_2O$ melts congruently at the convenient temperature 42°C. Viscosity and conductivity studies[26,27] in the same viscosity range as for CKN proved to give values of T_0 by Eq. (1) which were almost identical, $T_0 = 204\pm2$, and 201 ± 2 respectively, and which indeed accorded very well with the value $200\pm4K$ determined by direct calorimetric measurement and short (20K) extrapolation.[10] The latter result has recently been confirmed by an independent calorimetric study.[28] A comparable level of agreement was obtained for the analog system $Cd(NO_3)_2 \cdot 4H_2O$.[10] The heat capacity vs. temperature data even showed approximately the hyperbolic temperature dependence needed to derive Eq. (1) from the entropy based Adam-Gibbs theory of relaxation.[29]*

Thus for the glassforming nitrate systems, which are all very "fragile" liquids, studies of liquid dynamics in the moderate viscosity range have yielded results which conform rather convincingly to the pattern expected from a Gibbsian configurational entropy-based cooperative-rearrangement transport viewpoint.[24,29] Macedo and coworkers[30] and others[31] subsequently demonstrated that this attractive set of consistencies is seriously diminished when the transport measurements are pursued into the more viscous domain. However, their findings apply only to behavior near T_g (and below T_c of the mode coupling theory) and therefore do nothing to relieve the irony of the situation we now face as a result of recent sophisticated studies of dynamics in CaKN.[12,13,32,34] This is that the behavior of these *same* liquids in the *same* moderate

---

*We quote here a new result obtained in the days just before this meeting. The analog system $Cd(NO_3)_2$-KNO_3, contains a compound $Cd(NO_3)_2 \cdot 2KNO_3$ in the glassforming range (Thilo, Wieker and Wieker, Silic. Tech., **15**, 109 (1964)) hence is suitable for thermodynamic analysis. We have now studied the thermal characteristics of liquid, glass and crystal (J. Fan and C. A. Angell (to be published)) and find the Kauzmann temperature T_K to lie at 283K, 29.6K below the T_g value of 312.6K, a result comparable to that quoted for $Ca(NO_3)_2 \cdot 4H_2O$. To obtain the T_K value appropriate to the CKN system, we scale by the ratio of T_g values at this composition 1.033, to obtain T_K (CKN) = 201K, just 2K below the best fit Eq. (1) T_0 value (see previous footnote and Fig. 3).

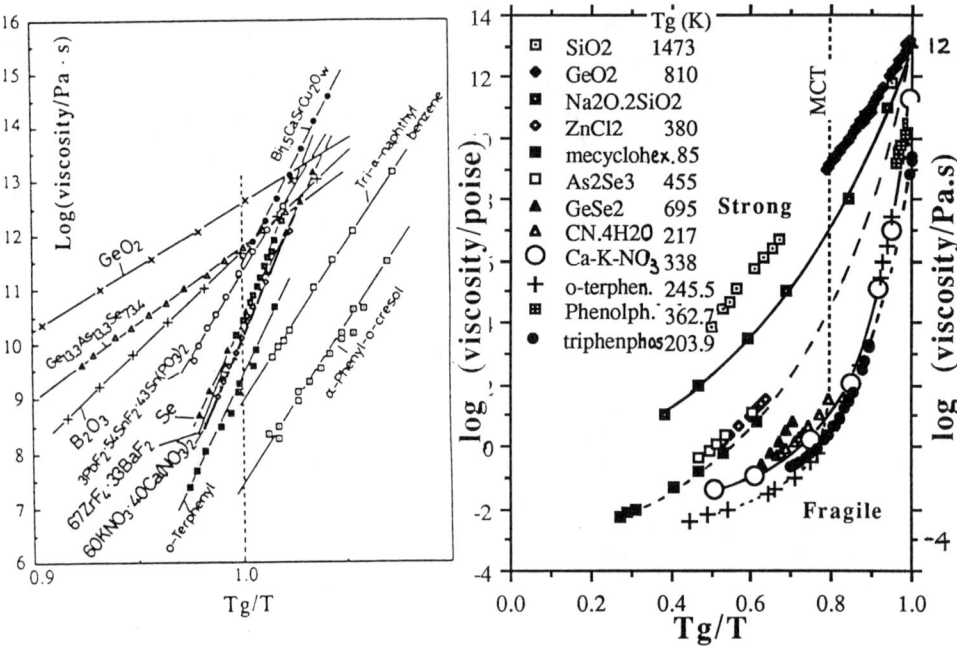

Fig. 6. (a) Arrhenius variation of viscosity near Tg for liquids of differing type and fragility. Tg is the temperature of the calorimetric glass transition temperature at which the enthalpy relaxation time has a constant value of about 200s(23).

(b) Tg-scaled Arrhenius plot for viscosity of various liquids including recently studied highly fragile case, tri-phenyl phosphite and convenient (T_g < ambient) molecular system, phenolphthalein.

viscosity range is, in detail,(12-14) that expected from the recent mode-coupling theory of transport processes in moderately viscous liquids.(11)

As others will show later,(11,33,34) mode-coupling arguments insist that dynamic processes observed in the viscosity range under discussion are all topologically constrained to follow the structure factor towards a singularity at a temperature far above T_g (and even farther above the Kauzmann temperature) and that this constraint is <u>only</u> lifted at a lower temperature where activated processes become a more efficient method of relaxing applied stress. The findings described in Figs. 1-5 and their consistency with entropy vanishing studies must, in the mode-coupling theory viewpoint, be no more than coincidence.

Support for the latter contention can indeed be obtained by turning attention to fragile organic liquids in which the attractive forces between constituent particles are of much shorter range. In these cases, fits of Eq. (1) to the viscosity in the moderate viscosity range yield T_0 values which are not only much greater than the Kauzmann temperatures but are also greater than the T_g values, which is unphysical and clearly demonstrates a failure of Eq. (1) at the lower temperatures. *These same liquids, however, yield behavior in good accord with the predictions of mode-coupling theory* insofar as the relevance of the dynamical singularity temperature of mode-coupling theory T_c ($T_c > T_g > T_K$) is equally well indicated by measurement of the slow mode at $T > T_c$ and of the fast mode at $T < T_c$.(35,36,37) The consistency with predicted power law viscosity behavior above T_c is shown in Fig. 13 (postponed until later for comparison with behavior of supercooled water).

On the other hand, if the fragile molecular liquids are studied with respect to slow (α) relaxations at temperatures far below the mode-coupling transition (i.e. in the domain where activated processes are expected to dominate the slow process), then both enthalpy and dielectric relaxation times again conform to Eq. (1), with T_o close to T_K,[37,38,39] even though *viscosity* data had earlier indicated a return to Arrhenius behavior.[31] It now seems clear that in such liquids the shear relaxation modes can split off from the slower enthalpy relaxation modes. A detailed study of this subject by two of us[40] has so far revealed no clear pattern for this decoupling though it seems not to occur in the cases of intermediate or strong liquids. Data obtained near T_g illustrating this important aspect of viscous liquid behavior are shown in Fig. 6 (a).

The positions of certain of these latter liquids in the overall strong/fragile liquid behavior pattern[41] are indicated in the extended range plot in Fig. 6 (b), in which the calorimetric T_g (at which the enthalpy relaxation time is ~200 sec[22]) is used as the scaling temperature. The T_g values used in the scaling are listed beside the figure (errors in the value for o-terphenyl listed, and used, in Fig. 2 of Ref. 41b have been corrected in Fig. 6 (b)). As shown elsewhere,[41b] this pattern can be reproduced, except for the decoupling near T_g for very fragile liquids, by variation of a single parameter, D, in a modification of Eq. (1)

$$\eta = \eta_o \exp\left(\frac{DT_o}{[T-T_o]}\right) \qquad (5)$$

The behavior of fragile liquids predicted by the simplest form of mode-coupling theory, viz., that which yields power law divergences, is shown by the heavy dashed line in Fig. 6 (b). It is this temperature which is indicated consistently by both <u>slow</u> mode measurements above T_c and by <u>fast</u> mode measurements below T_c as the most significant temperature for liquid state dynamics considerations.[11,14,33,35] While the analysis of the behavior of a number of liquids, particularly of the scaling properties of CKN found in recent and very sophisticated experiments,[12-14] give rather convincing support of mode-coupling theory, it is rather evident from Fig. 6b that there is a great range of slow relaxation data which falls outside the reach of the mode-coupling approach. Since this domain is phenomenologically rich as well as technologically important we make it the focus of the next section.

PHENOMENOLOGY OF VERY VISCOUS LIQUIDS NEAR AND BELOW THE GLASS TRANSITION TEMPERATURE

In the high viscosity regime approaching T_g, it is generally agreed that relaxation proceeds by processes which are "activated" in some way. The barriers being crossed in these processes are believed by many to be barriers between separate potential energy minima of various depths on the (3N + 1) dimensional hypersurface characteristic of each multi-particle system. The manner in which the energy spacing and multiplicity of these minima relate to the thermodynamic properties of the liquid has been discussed elsewhere.[41b]

We have already seen, through Fig. 6, that near the glass transition temperature a great variety of shear relaxation times exist, and it would appear that the behavior could be very complex. What is not at all clear from data presented so far is whether

10 Slow Processes in Viscous Liquids

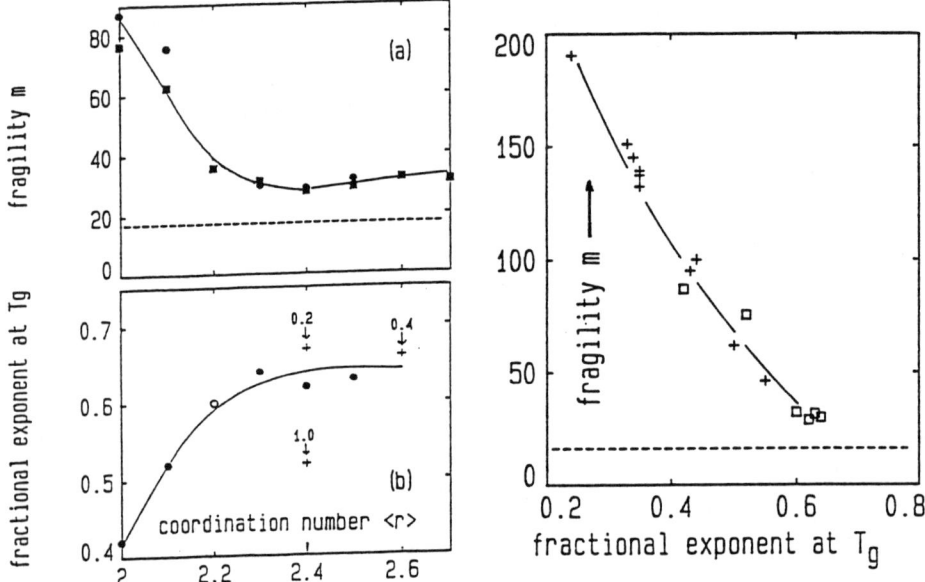

Fig. 7. Variation of fragility m and non-exponentiality of mechanical relaxation with average cordination number <r> in Ge-As-Se glasses.

Fig. 8 Correlation of fragility m with non-exponentiality of relaxation β, showing consistency of behavior of covalent and chain polymer systems (data from Refs. 49, 50).

these differences are to be regarded as a result of different discrete responses to different stresses or simply of selections, by different stresses, of different portions of a broad but continuous spectrum of relaxation times.

To obtain insight into this matter, it is obviously necessary to obtain frequency or time dependence information on the different responses. This can be undertaken in a variety of ways[12-14,42] but, for economy, we will concentrate on one way, which is applied directly in the time domain, and will deal with only one system which is chosen for the simplicity of the particles constituting the relaxing liquid. In contrast with the complex, multi-atom (even multi-phenyl) ring molecules on which the most sophisticated measurements have been made, our system of choice contains only atoms of essentially identical mass, which interact with one another either through directionally specified single covalent bonds, or van der Waals interactions — either of which can be thermally disrupted. The system we consider is the chalcogenide glass system Ge-As-Se, now studied by many groups.[43,44,45,46] The behavior of different alloy compositions in this system may be very different depending on the number of the bonds in a given N atom sample. Indeed at the average coordination number predicted by Phillips[47] and Thorpe[48] to give the optimum bond distribution, the behavior observed is that of liquids near the "strong" extreme of the overall pattern of viscous liquid behavior,[44] Fig. 6. As this optimized network is broken up by addition of Se, i.e. as the average coordination number decreases towards 2.0, the fragility increases dramatically and, at the limit of pure Se, behavior approaching that of CKN is found (the effect of breaking up the Se chains by iodine additions has yet to be investigated). A simple way to characterize the fragility from measurements made near T_g is to use the slope of the Arrhenius representation of the data about T_g on the T_g-reduced plot to define a dimensionless "fragility" m by

$$m = E_a / RT_g \tag{6}$$

in close analogy to the suggestion of Tatsumisago.[40] The same quantity has been called the "steepness index" by Plazek et al[49] in an analysis of chain polymer relaxation to be referred to below. m is simply related to the parameter D of Eq. 5 by

$$m = 590(D^{-1} + 0.027) \tag{7}$$

and to the T_g/T_o ratio by

$$T_g/T_o = (1 - 16/m)^{-1} \tag{8}$$

at least for cases in which the viscosity varies 16 orders of magnitude between the limits $T_g/T = 0$ and 1.0.

It was earlier shown that not only the fragility m, but the jumps in the specific heat and expansivity at T_g, varied systematically with connectivity $<r>$.[44] It has recently been found[50] that not only m but also the other characteristics of relaxation near T_g, viz., the departure from exponential relaxation, and the thermodynamic state dependence of relaxation, vary systematically with $<r>$ though the latter variation has not yet been fully characterized. The variations of m and β with $<r>$ are shown in Fig. 7.

If m and β are both correlated with the connectivity, they must be correlated with each other, and such a correlation diagram could offer a route to defining an "effective connectivity" for systems where none is obvious. In Fig. 8 we show m vs. β for the small variations observed in Ref. 50 and combine them with the wider variations seen by Plazek and Ngai[49] for chain polymers also using mechanical relaxation measurements. According to this plot, the fragility and spectral width for relaxation near T_g are indeed closely related. Accordingly it seems reasonable to interpret the decoupling of viscosity from enthalpy relaxation in terms of the shear stress being relaxed by short-time elements of the total structural relaxation spectrum.* All this implies that the <u>fragility</u> of a given liquid would best be determined from the fundamental enthalpy relaxation temperature dependence since, when viscosity is very decoupled (as in some cases in Fig. 6) m determined from viscosity will appear small.

It is interesting to follow this line of thought through to the question of what happens to viscous relaxation at temperatures well below T_g (where the main part of the structure can be considered to have become frozen). This question has been little investigated partly because of the pervading view that viscous flow in a system of constant structure doesn't make sense. However, creep viscosity exhibiting a different temperature dependence from normal viscosity has been measured for both metallic glasses[51] and oxide glasses.[52] It is naturally of interest to examine this question for the most fragile systems known (excepting, of course, entangled polymers which have a different origin of viscosity).

---

*The viscosity, in turn, has a temperature dependence which is greater than that of the diffusivity according to recent studies by Sillescu and colleagues (Ehlich, D. and Sillescu, H., Macromol., **23**, 1600 (1990); Sillescu, H. et al., in press), as anticipated by Stillinger's "tear and repair" viscosity mechanism (Stillinger, F. H., Phys. Rev. B, **41**, 2409 (1990); J. Chem. Phys. **89**, 6461 (1988)).

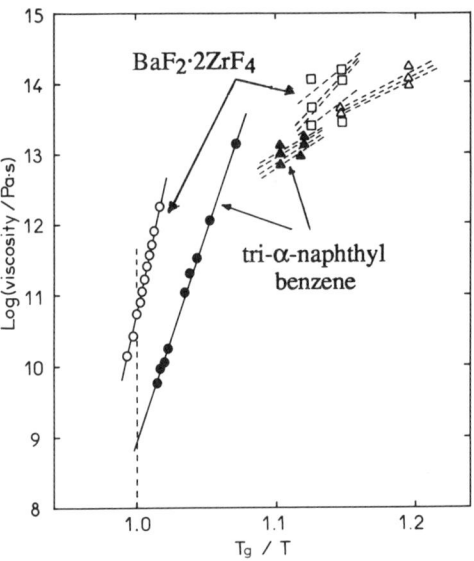

Fig. 9. Variation of viscosity with temperature in the **equilibrated** supercooled liquid state, and below T_g in the **"constant structure"** (but slowly relaxing) state obtained by cyclic T,η,beam-bending measurements for two fragile liquids.

Recently two of us[53] performed such measurements on two of the most fragile liquids which can be studied above ambient, viz., the ionic fluoride glass $BaF_2 \cdot 2ZrF_4$ and the molecular glass, tri-α-naphthyl benzene. These measurements, which are being reported in more detail elsewhere,[53] were performed using a standard beam bending method, and steady velocities characteristic of the usual equilibrium viscosity measurements were indeed obtained after short transient times. However, when viscosity was determined at two different temperatures cyclically, it was found that the viscosity was slowly changing during temperature equilibration such that the whole viscosity temperature curve was moving "up" in a manner which depended on thermal history. If averages of successive pairs of points taken at one temperature were combined with one point at a different temperature and intermediate time, then a "constant structure" viscosity could be defined (see dashed lines in Fig. 9). Although subject to considerable uncertainty, these results continue the trend observed with liquids of lesser fragility[51,52] viz. to give viscosity vs. temperature plots with temperature dependence increasingly smaller than for the equilibrium viscosity. The observations, which are summarized in Fig. 9, are certainly consistent with the idea that the shear stress on a glass can be relaxed by short-time elements of the total structure even though the longer range structure is completely frozen ($\tau_{enthalpy} > 10^{24}$ sec).

FREEZING OF DYNAMIC PROCESSES AT $T > T_g$

It can be argued that, irrespective of the precise mechanism by which the glass constituent particles move to permit viscous flow, the extent of particle movement during the relaxation process must be very tiny—the average particle moves only fractions of a molecular diameter in one relaxation time[41b]. We now go on to discuss briefly some cases of interest in which the relaxation time becomes so long before T_g is

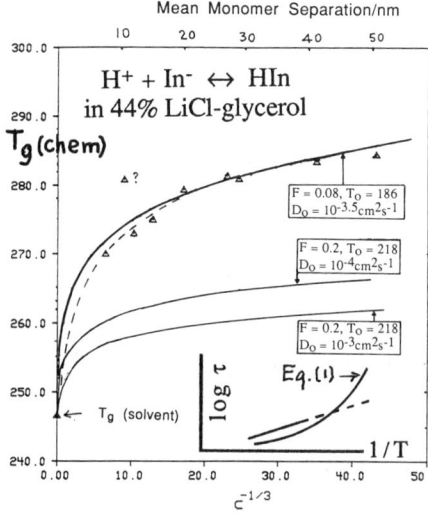

Fig 10. Variation of chemical reaction glass transition temperature with reactant separation. Solid lines show diffusion model predictions for solvents of two different fragilities. Inset shows how Eq. (1) behavior of solvent leads to diffusion control.

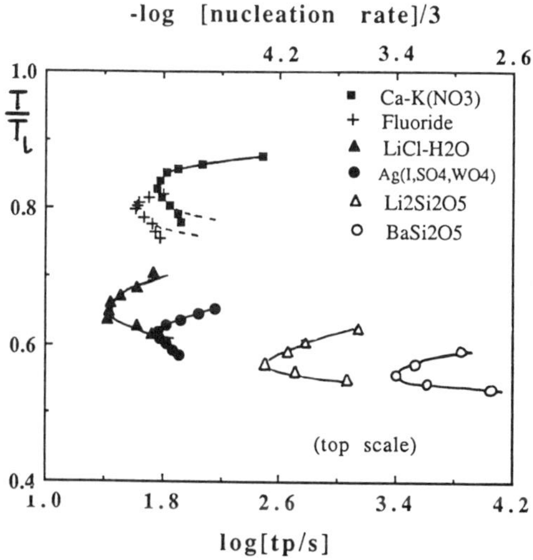

Fig. 11. Nucleation rate as function of T/T_l, for liquids of different fragility (from ref. 59).

reached that the process becomes "frozen," i.e. the system becomes non-ergodic with respect to that process while the liquid is, by other criteria, in its equilibrium state.

The best known of such phenomena is of course the liquid→rubber transition encountered in liquids with chain molecules long enough to entangle (hence to provide a viscosity mechanism indepedent of the degrees of freedom which determine the glass transition). We consider three other cases briefly, all of which "freeze" because they demand diffusion to occur over non-local distances in order to establish equilibrium.

1. Freezing of equilibrium in diffusion-controlled chemical reactions.

Chemical processes, which might normally exhibit an Arrhenius temperature dependence due to the presence of an energy barrier, can become diffusion-controlled at lower temperatures due to the non-Arrhenius nature of the structural relaxation time. This is illustrated in Fig. 10, inset. When a chemical process with a temperature-dependent equilibrium constant comes under diffusion control, a sort of glass transition due to the freezing of the equilibrium on cooling (or its restoration on heating) can be observed.[54] The temperature at which this "chemical glass transition" occurs in a viscous solvent relative to the normal glass transition is of interest for a variety of reasons.[55] An important one is that it is relevant to the suspension of life processes in cooling or desiccating cellular systems. The temperature at which the chemical glass transition occurs depends on reactant concentrations (because this determines the diffusion distance if they are homogeneously distributed) and on the solvent fragility. An example of the effect of concentration on the chemical glass transition in a protonation equilibrium[55] is shown as a function of mean reactant separation $\langle\lambda\rangle$ in Fig.10. The glassforming solvent in this case is concentrated solution of LiCl in glycerol, the fragility of which is known from independent viscosity studies. The solid line passing through the points is a plot of the equation

$$<\lambda> = \left[1000 D_o \exp\left\{ -\left(F\left[\frac{T_{g,\text{ spec}}}{T_o} - 1 \right] \right)^{-1} \right\} \right]^{\frac{1}{2}} \quad (9)$$

where F is a fragility parameter [F = (D of Eq.5)$^{-1}$] and Do is a diffusivity preexponent. Eq. (7) may be derived[55] by treating the chemical process as a relaxation mode, as in Zener's anelasticity theory,[56] of relaxation time τ_i

$$\tau_i = \lambda_i^2/D \quad (10)$$

where D is the self diffusivity which follows Eq (5) via the the Stokes-Einstein relation to solvent viscosity. The only free parameter in Eq. (7) is the pre-exponent of the Vogel-Fulcher equation for diffusivity, D_o, which needs the value $10^{-3.5}$ cm^{2}sec^{-1} to fit the data. The important result from Eq. (9), illustrated in Fig. 10, is that in strong liquids, chemical processes even in quite concentrated solutions can become frozen on long time scales at temperatures which are far above the solvent glass transition temperature. The extent to which nature uses this strategy to preserve life under unfavorable conditions (desiccation etc.) remains to be established.

2. (a) Nucleation of crystals from viscous liquids.

It is well known that many liquids can be vitrified if cooled more rapidly than the "critical cooling rate," due to the fact that there is a temperature between the liquidus and glass transition temperatures where the probability of crystals forming (in response to the thermodynamic drive to stability) is a maximum. Various techniques to determine the nucleation rate or alternatively, the form of the nucleation rate vs. T curve, have been described.[57,58] Recently the results for several cases were presented at temperatures reduced by the temperature at which the liquid first becomes metastable, viz. the liquidus temperature T_l, and it was noted that strong liquids withstand supercooling over much wider reduced temperature ranges than do fragile liquids.[59] The relationships are illustrated in Fig. 11.

(b) Nucleation of clusters in liquids.

Very recently, evidence has been accumulating that diaphanous clusters of unknown nature form reproducibly in a variety of molecular liquids.[60,61] These may be destroyed by ultrafiltration or ultrasonification, but reform after a certain period of time has elapsed when the temperature is maintained within appropriate limits. This interesting phenomenon is susceptible to study by the same 1-step and 2-step time-temperature sequences used to obtain the TTT curves seen in Fig. 10, using light-scattering rather than calorimetry as the transformation detection scheme. The finding of separable nucleation and growth kinetics would do much to clarify the phenomenology of this interesting new long-range ordering feature of supercooled liquids.

3. Freezing of microemulsion structures.

Microemulsion systems which are stable against freezing of the constituent phases have recently been described, both for O/W[62] and W/O[63] cases. This development opens the door to studies of the mechanisms by which their droplet are modified by changing external conditions. In a recent study[64] it was found that during slow cooling, the peak in the neutron scattering intensity stopped moving suddenly at temperature $T_{g,ME}$ far above the matrix T_g. This is presumed to be a result of slow diffusion over the several nm separating the individual droplets. Studies of $T_{g,ME}$ as a function of droplet separation are yet to be performed.

SLOW DYNAMICS NEAR SPINODAL BOUNDARIES IN SUPERCOOLED LIQUIDS.

In liquid systems with density maxima, computer simulation studies,[65] equation of state predictions,[66,67] lattice theories,[68] and physical measurements[69] all suggest important qualitative differences from normal behavior. Normally the metastable liquid state is bounded in the superheated and (at lower temperatures) stretched (negative pressure) regimes by a line of mechanical instabilities which is known as the spinodal. The spinodal calculated from equations of state such as the van der Waals equation moves to always-increasing negative pressures as temperature decreases. Where density maxima exist, however, it appears that the spinodal boundary at negative pressures must reverse its temperature dependence on crossing the extrapolated density maximum and come back to positive pressures as an absolute mechanical stability limit on the supercooled liquid range. What the metastable liquid first transforms to on passing this boundary is unclear at this time, but for our

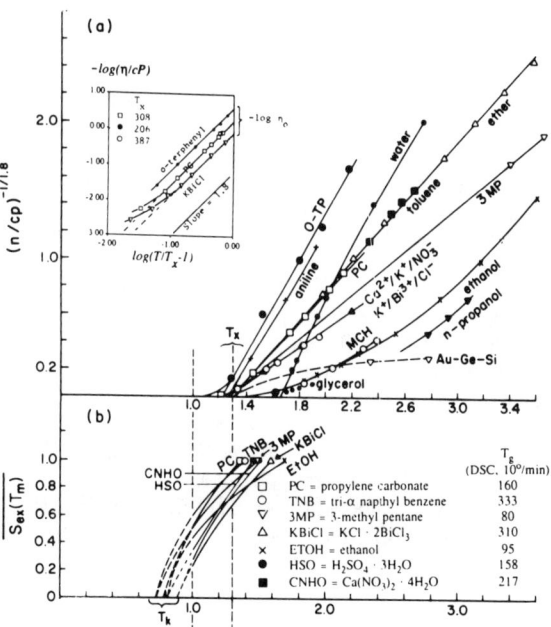

Fig. 12. (a) Plots of the mode coupling theory viscosity expectation log (viscosity η)$^{-1/1.8}$ vs. T/T_g where T_g is the laboratory calorimetric glass transition temperature, showing predicted linear behavior for low and moderate viscosity data at $T>T_c$ in simple liquid cases. Note water behaves non-linearly in the opposite sense to other hydrogen-bonded liquids on this plot. Inset shows usual log-log plot for selected cases.
(b) Variation of excess entropy of various liquids showing fraction of excess entropy of fusion remaining when cross-over from mode-coupling domain to activated process (energy landscape) control occurs during supercooling (adapted from Fig. 2 of Ref. 78).

present purposes, this is not important. Rather it is the slow dynamics which develop as the system approaches the spinodal which we wish to examine and compare with that observed near the glass transition.

At the moment, the best studied case is pure water for which power law divergences in thermodynamic as well as mass transport properties have been observed as the spinodal is approached.[70-74] The viscosity divergency in this case is sharper $[(T-T_c)^{1.6}]$ than any predicted by mode coupling $[(T-T_c)^{1.8}$ at the limit of exponential decay[75]] as is evident from the comparison with MCT fits of viscosity data for a variety of liquids in Fig. 12. The viscosity power law for water is demonstrated[73] in Fig. 13, indicating divergence at 227K for normal pressure measurements. Unfortunately, water is very prone to crystallization in the interesting region. However, since the diverging component is independent of the background component, there seems no reason why cases could not be found where the process can be studied in more leisurely fashion. Liquid silica, for instance, has a density maximum,[76] and computer simulation studies indicate large and growing fluctuations during equilibration at low temperatures.[77] Unfortunately, SiO_2 is at the opposite extreme from water, and we need more tractable intermediate cases. We believe a number may be found in nature and that this phenomenology will prove very attractive in future work.

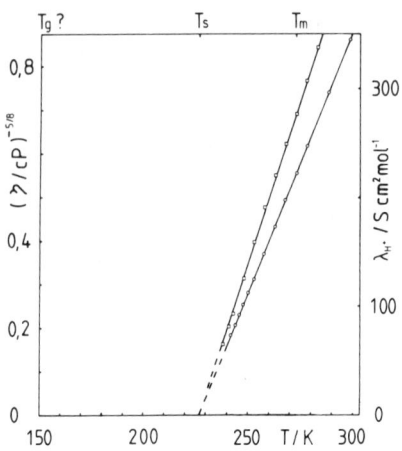

Fig. 13. Demonstration of power law $(T-T_s)^{1.6}$ for viscosity of water, and comparison with behavior of proton conductivity. T_s is the temperature of the spinodal, -45°C, at ambient pressure.

CONCLUDING REMARKS

We have surveyed at different degrees of superficiality, a variety of processes in liquids which exhibit slow dynamics to the point of ergodicity-breaking with decreasing temperature. All are tied, directly or indirectly, to the diffusive slow-down responsible for the glass transition observable in all classes of liquids. The most disconcerting aspect of our review is that we are unable to understand how data on the same model system, CKN, studied in the same temperature range, can be simultaneously consistent with two very different dynamic theories (mode coupling and entropy theories), characterized by widely different singular temperatures. The resolution of this dilemma stands as a challenge to future workers in this area.

ACKNOWLEDGEMENT

This work was supported by the U.S. N.S.F. under Solid State Chem. Grant Nos. DMR-9108028 and 8744945, and DOE Grant No. DE-FG02-89ER45398.

REFERENCES

1. Vogel, H., Physik. Z., **22**, 645 (1921)
2. Fulcher, G. S., J. Am. Ceram. Soc., **8**, 339 (1925).
3. Tamann, G. and Hesse, W., Z. anorg. allgem. Chem., **156**, 245 (1926).
4. Tammann, G., *Der Glaszustand*, Leopold Voss, Leipzig, 1933.
5. Cohen, M. H. and Turnbull, D., J. Chem. Phys. **31**, 1164 (1959).
6. Angell, C. A., J. Phys. Chem., **68**, 218 (1964).
7. Angell, C. A., J. Chem. Phys. **46**, 4673 (1967).
8. (a) Angell, C. A., Pollard, L. J. and Strauss, W., J. Chem. Phys., **50**, 2694 (1969). (b) Pollard, L. J., Ph.D. thesis, Univ. of Melbourne, 1970.
9. Moynihan, C. T., Smalley, C. R., Angell, C. A. and Sare, E. J., J. Phys. Chem., **73**, 2287 (1969).
10. Angell, C. A. and Tucker, J. C., J. Phys Chem., **78**, 278 (1974).
11. Götze, W., in *Liquids, Freezing, and the Glass Transition*, Eds. Hansen, J.P. and Levesque, D., (Les Houches 1989) and references cited therein (see also Götze, W., these proceedings).
12. (a) Mezei, F., Knaak, W. and Farago, B., Phys. Rev. Lett. **58**, 571 (1987). (b) Mezei, F. in *Liquids, Freezing and the Glass Transition*, Eds. J. P. Hansen, D. Levesque and J. Zinn-Justin, North Holland, 1991, p. 631.
13. Tao, N. J. and Cummins, H., Phys. Rev. Lett., **66**, 1334 (1991); Phys. Rev. B (in press).
14. Cummins, H. (these proceedings).
15. (a) E. Rhodes, Smith, W. E. and Ubbelohde, A. R., Trans. Far. Soc., **63**, 1943 (1967). (b) Tweer, H., Laberge, N. and Macedo, P. B., J. Am. Ceram. Soc., **54**, 121 (1971).
16. J. de Neufville, 1969, unpublished work described in Ref. 8.
17. Rao, K. J., Angell, C. A. and Helphrey, D. B., Phys. Chem. Glasses, **14**, 26 (1973).
18. DeBolt, M.A., Easteal, A.J., Macedo, P.B. and Moynihan, C.T., J. Am. Ceram. Soc. **59**, 16 (1976).
19. Davies, R.O. and Jones, G.O., J. Adv. Phys. **2**, 370 (1953).
20. Goldstein, M., J. Chem. Phys., **39**, 3369 (1963); J. Phys. Chem. **77**, 667 (1973).
21. Williams, E. and Angell, C. A., J. Phys. Chem., **81**, 232 (1977).
22. Pollard, L. J., Crowe, M. L., Strauss, W., J. Chem. Eng. Data, **16**, 134 (1971).
23. Moynihan, C.T. et al., Ann. N.Y. Acad. Sci. **279**, 15 (1976).
24. Gibbs, J.H. In *Modern Aspects of the Vitreous State*, ed. J.D. McKenzie (Butterworths, London, 1960) ch. 7.
25. Kauzmann, W., Chem. Rev. **46**, 219 (1948).
26. Moynihan, C. T., J. Chem. Ed., **44**, 531 (1967).
27. Angell, C. A. and Bressel, R. D., J. Phys. Chem., **76**, 3244 (1972).
28. Hepler, L. et al., J. Chem. Thermodynamics (in press).
29. Adam, G. and Gibbs, J. H., J. Chem. Phys. **43**, 139 (1965).
30. Weiler, R., Blaser, S. and Macedo, P.B., J. Phys. Chem. **73**, 4147 (1969).
31. Laughlin, W.T. and Uhlmann, D.R., J. Phys. Chem. **76**, 2317 (1972).
32. Cheng, L. -T. and Nelson, K. A., Phys. Rev. B, **39**, 9437 (1989).
33. Mezei, F. (these proceedings).
34. Fujara, F. and Petry, W., Europhys. Lett., **4**, 921 (1987).
35. Sillescu, H., J. Non-Cryst. Sol., **131-133**, 378 (1991) (discussion section).
36. There is a great need of molecularly simple van der Waals liquids which are glassforming to help clarify some of these issues. We give elsewhere (Senapati, H., Berger, Choi and Angell, C. A. (to be published)) basic information on systems containing only simple and readily available molecular constituents, e.g. CS_2 and CBr_4 or at least large mole fractions of them. Application of the $T_b/T_m = \geq 2$ rule suggests that the simplest glassforming systems may be

OF$_2$ and/or O$_3$-O$_2$ mixtures which, unfortunately, have their TgS well below liquid nitrogen temperature.

37 Birge, N. and Nagel, S., Phys. Rev. Lett. **54**, 2674 (1985).
38 Birge, N.O., Phys. Rev. **B34**, 1641 (1986).
39 Dixon, P., Phys. Rev., **B42**, 8179 (1990).
40 (a) Tatsumisago, M. and Angell, C. A., Proc. 1st Japan Conf. Glass. (b) Tatsumisago, M and Angell, C. A. (in preparation).
41 Angell, C. A. (a) in *Relaxations in Complex Systems*, ed. K. Ngai and G. B. Wright, National Technical Information Service, U. S. Department of Commerce, Springfield, VA 22161 (1985), pg 3. (b) J. Non-Cryst. Sol., **131-133**, 13 (1991).
42 See for instance Wong, J. and Angell, C. A., *Glass: Structure by Spectroscopy*, Marcel Dekker, New York, New York (1976) chapter 11.
43 Webber, P. J. and Savage, J. A., J. Non-Cryst. Sol., **20**, 27 (1976).
44 Tatsumisago, M., Halfpap, B. L., Green, J. L., Lindsay, S. M. and Angell, C. A., Phys. Rev. Lett., **64**, 1549 (1990). Note, however, that the extremum at $<r>$ = 2.4 is not found in some chalcogenide systems lacking Ge. [Lucas, J., Ma, H. L., Zhang, X. H., Senapati, H., Böhmer, R. and Angell, C. A., J. Phys. Chem. Sol. (in press Nov. 1991)] hence may contain a bonding dimensionality factor as well as the connectivity factor.
45 Yun, S. S., Hui, Li, Cappelletti, R. L., Enzweiler, R. N. and Boolchand, P., Phys. Rev. B, **39**, 8702 (1989).
46 Tanaka, K., Phys. Rev., **B39**, 1270 (1989).
47 Phillips, J.C., J. Non-Cryst. Sol. **34**, 153 (1979).
48 Thorpe, M.F., J. Non-Cryst. Sol. **57**, 355 (1983).
49 Plazek, D. K. and Ngai, K.-L., Macromol., **24**, 1222 (1991).
50 Böhmer, R. and Angell, C. A. (a) (submitted to Phys. Rev. B). (b) (in press) Proc. 6th European Conf. on Internal Friction and Ultrasonic Attenuchou, Cracow, Poland (1991). These results are consistent with the pattern suggested some time ago by Ngai (Solid State Ionics, **5**, 27 (1981) for polymer melts and ionic liquids.
51 Greer, A. L. and Spaepen, F., Ann. N. Y. Acad. Sci., **371**, 218 (1981).
52 Mazurin, O.V., Startsev, Yu. K. and Stoljar, S. V., J. Non-Cryst. Sol., **52**, 105 (482).
53 Tatsumisago, M. and Angell, C. A. (to be published).
54 Barkatt, A. and Angell, C. A., J. Chem. Phys., **70**, 901 (1979).
55 Arzimanoglou, A. and Angell, C. A., J. Chem. Phys. (submitted).
56 Zener, C., *Elasticity and anelasticity of Metals*, Univ. of Chicago, Chicago, 1948, p. 76 et. seq.
57 (a) Uhlmann, D. R., in *Advances in Ceramics*, Vol. 4, *Nucleation and Crystallization in Glasses*, Edited by J. H. Simmons, D. R. Uhlmann and G. H. Beall, American Ceramic Society, Columbus, OH 1982. (b) Nielsen, G. F. and Weinberg, M. C., J. Non-Cryst. Sol., **34**, 137 (1979).
58 MacFarlane, D. R., Kadiyala, K. and Angell, C. A., J. Chem. Phys., **79**, 3921 (1983).
59 Senapati, H. and Angell, C. A., J. Amer. Ceram. Soc., **74**, 2659 (1991).
60 Fischer, E. W., Meier, G., Rabenau, T., Patkowoki, A., Steffan, W., Thonnes, W., J. Non-Cryst. Sol., **131-133**, 134 (1991).
61 Meier, G. (private communication).
62 Dubochet, J., Adrian, M., Teixeira, J., Kadiyala, R. K., Alba, C. M., MacFarlane, D. R. and Angell, C. A., J. Phys. Chem., **88**, 6727 (1984).
63 Green, J. L., J. Phys. Chem., **94**, 5647 (1990).
64 Alba-Simionesca, C., Teixeira, J. and Angell, C. A., J. Chem. Phys., **91**, 395 (1989).
65 Kataoka, Y., Hamada, H., Nose, S., Yamamoto, T., J. Chem. Phys., **77**, 5699 (1982).
66 Speedy, R. J., J. Phys. Chem., **86**, 982 (1982).
67 Green, J. L., Durben, D. J., Wolf, G. H. and Angell, C. A., Science, **249**, 649 (1990).

68. Debenedetti, P. G., J. Phys. Chem., **95**, 4540 (1991). Debenedetti, P. G. and D'Antonio, M. C., Am. Inst. Chem. Eng. J., **34**, 447 (1988).
69. Zheng, Q., Durben, D. J., Wolf, G. H. and Angell, C. A., Science (in press) Nov. 30 issue (1991).
70. Speedy, R. J. and Angell, C. A., J. Chem. Phys., **65**, 851 (1976).
71. Kanno, H. and Angell, C. A., J. Chem. Phys., **73**, 1940 (1980).
72. Oguni, M. and Angell, C. A., J. Chem. Phys., **73**, 1948 (1980).
73. Cornish, B. D. and Speedy, R. J., J. Phys. Chem., **88**, 1888 (1984).
74. Angell, C. A., Ann. Rev. Phys. Chem., **34**, 593-630 (1983).
75. Götze, W. (private communication).
76. Brückner, R., J. Non-Cryst. Sol., **5**, 281 (1971).
77. Angell, C. A., Cheeseman, P. A. and Phifer, C. C., Mat. Res. Soc. Symp. Proc., **63**, 85 (1986).
78. Angell, C. A., J. Phys. Chem. Sol., **49**, 863 (1988).

SLOW RELAXATION PROCESSES IN GLASSY CRYSTALS*

H. Suga
Department of Chemistry and Microcalorimetry Research Center
Faculty of Science, Osaka University, Toyonaka, Osaka 560 Japan

ABSTRACT

This article reviews slow dynamics of relaxation processes of glassy crystals in which the reorientational motion of the constituent molecules is frozen out, keeping the translational periodicity with respect to the centers of molecular mass. The glassy crystals around their glass transitions exhibit relaxational behavior which resembles closely those of glass-forming liquids. Sub-glass transitions occasionally appear below and above the primary Tg. In this way, the glassy crystals show universal features characteristic of the glassy liquids in all respects. Glassy crystals are thus promizing model substance for deeper understanding of the glasses in the sense that the translational invariance of the molecular arrangements renders their study more tractable.

INTRODUCTION

Several kinds of mesophase exist between anisotropic crystals and isotropic liquids. One of them is plastic crystals[1] in which the molecules are disordered with respect to their orientation but ordered with respect to the centers of mass. They are characterized thermodynamically by large entropies of transition into ordered phases on cooling, in contrast to small entropies of fusion into isotropic liquids on heating. Structural feature of the plastic crystals is their high crystallographic symmetry, usually fcc or bcc structure. If a plastically crystalline phase is cooled rapidly enough to by-pass the transformation into an ordered phase, the supercooled phase shows behavior similar to glass transition of liquids and has a definite residual entropy. These phenomena are due to freezing out of the large-amplitude reorientational motion of the molecules in the crystal and correspond to freezing process of configurational degree of freedom in liquids. The frozen-in states of the orientationally disordered crystals were designated as glassy crystals[2].

The glassy crystal undergoes a progressive departure from the metastable equilibrium supercooled crystal as the temperature is lowered below Tg. Consequently, the excess Gibbs energy biases the reorientational motion of molecules in the lattice. This bias drives the density and energy fluctuation averages toward their equilibrium values, thus recovering the equilibrium structure. Among many structure sensitive properties of the crystal, enthalpy relaxation can be followed precisely over a long period of time by an adiabatic calorimeter because of its high thermal stability and temperature resolution. The residual heat leakage is of the order of 50 erg s^{-1} and the temperature resolution is less than 0.1 mK in most of the temperature range[3]. The calorimeter is operated in a discontinuous heating mode so that the measurement is composed of energyzing and equilibration periods. Any spontaneous temperature drift of a specimen beyond that due to residual heat leakage can be correlated to the enthalpy relaxation of the system. The observation of the drift rate combined with heat capacity data gives the rate of enthalpy relaxation as a function of time. Thus the adiabatic calorimeter works as a time-domain spectrometer while being engaged in the heat capacity determination.

DESULTS AND DISCUSSION

The first paramount example of the glassy crystals was given by cyclohexanol[4]. The liquid cyclohexanol crystallizes into an fcc system at 300 K. The glass transition is observed at around 150 K with a drastic heat capacity jump of 40 $JK^{-1}mol^{-1}$. The supercooled fcc phase above Tg tends to recrystallize into a low temperature phase (LTP) which returns to the fcc phase at 265 K on heating. The LTP has proved to obey the third law of thermodynamics by comparison of the calorimetric and spectroscopic entropies of gas in the ideal state. Enthalpy recovery from a non-equilibrium to the equilibrium supercooled fcc crystal is clearly observed around Tg. A loop calculation along the different thermodynamic paths shows that the glassy crystalline phase of fcc has the residual entropy of 4.7 $JK^{-1}mol^{-1}$. The extrapolated excess entropy over that of the stable crystal shows existence of the Kauzmann temperature, as in the case of conventional glass. Dielectric measurement reveals the existence of primary and secondary relaxations in the fcc phase. The temperature at which the maximum dielectric dispersion takes place at 10^{-3} Hz corresponds roughly to Tg value. In all respects, the supercooled fcc crystal exhibits universal features of glassy liquids. Among many examples of the glassy crystal, ethyl alcohol provides an interesting system in the sense that both of the glassy liquid and glassy crystal can be realized by controlling the cooling rate of the liquid[5]. The results are summarized in Fig. 1. Both of the glass transitions take place at essentially the same temperature region, indicating that the reorientational motion of molecules governs primarily the relaxational process. The residual entropy of the glassy crystal is a little less than that of glassy liquid, reflecting the existence of positional order in the glassy crystal. For the realization of the glassy liquid it is necessary to cool a liquid specimen with a rate more than -50 $Kmin^{-1}$. This situation makes it difficult to judge the actual state of ethanol used frequently as a matrix

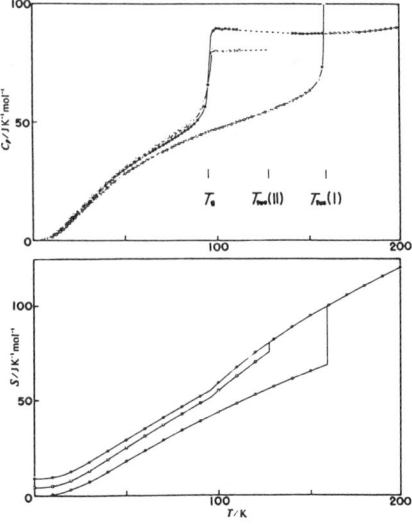

Fig. 1 Heat capacity and entropy of ethyl alcohol.

described in many literatures.

Figure 2 shows the heat capacity of isocyanocyclohexane ($C_6H_{11}NC$ ICCH) which has been recently measured[6]. ICCH is a liquid at room temperature and forms a plastic phase on cooling. The plastic phase easily supercools and exhibits three anomalous regions of the heat capacity at 55 K, 134 K and 156 K, respectively. The drastic heat capacity change at 134 K is associated with primary relaxation and a tiny anomaly at 55 K is due to secondary relaxation. The third anomaly occurring at 156 K above the main Tg may be of the intramolecular origin. This kind of sub-glass transition events are one of current topics in amorphous polymers. The same problem was encountered in the case of 1,1-difluoro-1,1,2,2-tetrachloroethane ($CFCl_2$-$CFCl_2$,DFTCE) as revealed by an adiabatic calorimetry[7]. For this substance, the plastic phase (bcc) exhibits three regions of anomalous heat capacity accompanied by spontaneous temperature drifts. The main Tg occurs at 90 K and the β-relaxation at 60 K. A small heat capacity jump around 130 K is believed to be due to freezing process of conversion between the trans and gauche conformers, which has been known to exist from the previous IR and NMR experiments[8]. These results reveal that the trans conformer is stable energetically by 0.7~0.8 kJmol^{-1}. The contribution from this internal rotational degree of freedom to the heat capacity can be evaluated by a Schottky function with 1:2 degeneracy scheme. By fitting this model function to the observed heat capacity jump at 130 K, the energy difference can be calculated to be (790±60) kJmol^{-1}, which is in reasonable agreement with the spectroscopic data.

The exothermic temperature drift of DFTCE observed at 125 K can be represented well by the following equation.

$$T(t) = A + Bt - C\exp(-t/\tau), \qquad (1)$$

where, T(t) is the calorimetric temperature at time t, (A-C) the initial temperature, B the constant drift rate due to residual heat leakage, C the amplitude of the relaxation, and τ the relaxation time. This means that the enthalpy relaxation follows the Debye-type relaxation with single characteristic time τ = 10.4 ks. The process of the trans-gauche conversion can be treated as a unimolecular reaction between the two species separated by an average barrier height. With

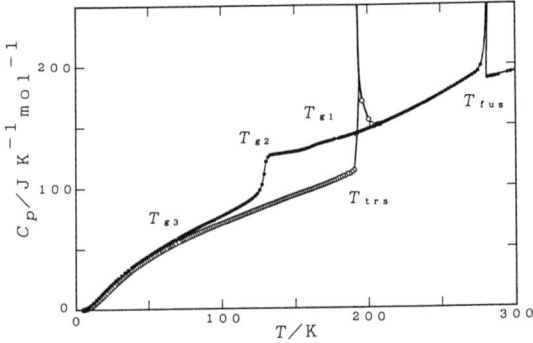

Fig. 2   Heat capacity of isocyanocyclohexane. ● ; Liquid, plastically crystalline and glassy crystalline states. ○ ; Stable LTP

the known value of the torsional frequency for the internal-rotational mode, the average hindering potential is determined to be 39.8 kJmol^{-1} which is in good agreement with that derived from the spectroscopic data.

The same method was applied to ICCH crystal in which the plastic phase is known to be composed of two kinds of molecules, axial and equatorial chair conformers in about the same proportion, as revealed by an IR measurement. Again, the sub-glass transition is related reasonably to the freezing of the two conformers in the ICCH lattice. The rate of enthalpy relaxation is reproduced by an exponential function, and the activation energy of 51 kJmol^{-1} is obtained by an Arrhenius plot of the relaxation time data.

Simultaneous measurements of enthalpy and polarization relaxation of ICCH crystal were carried out around the main Tg of 134 K by using a polaro-calorimeter newly designed for a comparison of relaxation processes of two independent quantities. The apparatus is essentially an adiabatic calorimeter with built-in electrodes in the sample cell. A dc voltage 50 V is applied to the sample initially. The cell is cooled to a measuring temperature keeping the voltage constant. At the measuring temperature, the electric field is turned off to short-circuit the electrodes by a sensitive electrometer. Total change of the polarization is given by the integration of the depolarization current with respect to time.

Figure 3 draws the enthalpy and polarization relaxation data measured at three representative temperatures. The data are reproduced fairly well by an empirical function refered to as KWW function[9].

$$X(t) = X(0) \exp[-(t/\tau)^{\beta}], \qquad 0 < \beta < 1 \qquad (2)$$

where X is the enthalpy or polarization at time t, τ an effective relaxation time, and β a non-exponential parameter. It is clear from the figure that relaxation of both the quantities takes place faster at higher temperature and this is obviously reasonable for a thermally activated process. It is interesting to note that the electric polarization relaxes faster than the enthalpy and the non-exponential parameter of the polarization is smaller than that of the enthalpy. The polarization relaxation concerns with only the recovery of the dipolar molecules toward the equilibrium value, while the enthalpy relaxation involves all the processes that bring about enthalpy change of the whole crystal.

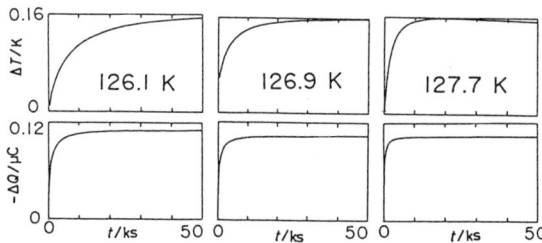

Fig. 3 Enthalpy and polarization relaxation in isocyanocyclohexane measured at three temperatures.

The value β ranges between 0.4~0.8. This is typical values[10] observed in many glass-forming liquids. It is not unreasonable to seek the origin of the small β value of the present system in the variety of various environments in which each molecule is located. Since the conversion between the axial and equatorial conformers is frozen out, the glassy crystalline phase is actually composed of two chemical species with different molecular shapes and intermolecular forces. The situation results in a wide distribution of the relaxation times.

There is another category of glassy crystals, which can be realized without by-passing the transformation into an ordered LTP. It has the same situation of orientational disorder existing in the crystal at high temperatures. Owing to a prolonged relaxation time of the reorientational motion on cooling, the relevant degree of freedom is frozen-in either (A) before the crystal reaches the completely ordered state after exhibiting an ordering transition or (B) before the crystal reaches the hypothetical transition temperature. In any way, the glass transition takes place in the stable crystalline phase, in contrast to the metastable phase in the previous category. Occurrence of the two cases is drawn schematically in Fig. 4 in a form of the orientational order parameter ξ and the relaxation time τ plotted as a function of temperature. In the case (B), the relaxation time becomes of the order of 1 ks at the temperature at which a short-range orientational order develops. The experimental time scale of a single heat capacity determination is approximately 1 ks. In the case (A), the relaxation time crosses the experimental time scale at the temperature where the long-range order is developed but not completely. In this case, the enthalpy relaxation can be observed likely but the residual entropy is too small to be confirmed definitely. In principle, however, glass transition and residual entropy should exist because of the non-ergodic nature[11] of the frozen system. It is this non-ergodic nature which distinguishes the equilibrium crystals and the glassy crystals or glassy liquids.

The classical example of the case (B) is provided by hexagonal ice. The residual entropy observed in ordinary ice has been interpreted in terms of reorientational disorder of water molecules among six orientations under the constraints of the Bernal-Fowler' ice conditions. All the possible orientations have nearly the same energy,

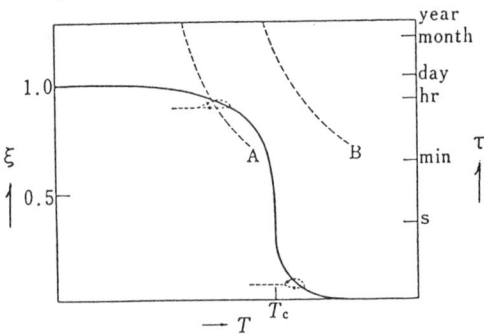

Fig. 4 Ordering transition, orientational order parameter ξ, and relaxation time τ.

but one of them must be more stable than the others. Therefore, the absence of any heat capacity anomaly that can be regarded as due to a phase transition has puzzled many scientists in spite of their effort.

It has been conjectured that the ice conditions constrain severely the reorientational motion of the water molecules and to give it a highly cooperative nature so that the motion will freeze out before the crystal reaches a hypothetical transition temperature[12]. The dielectric dispersion of ice due to orientation polarization extrapolated to the probed frequency of 10^{-3} Hz suggests that the freezing will occur around 100 K. In fact, a glass transition[13] was observed around 100 K with a small heat capacity jump depending on the prehistory of specimen. Spontaneous temperature rise observed below Tg is described approximately by Eq. (1). A typical result obtained for the longest annealing experiment is reproduced in Fig. 5. The relaxation time is 37 ks at 99.4 K, 72 ks at 94.4 K and 0.52 Ms at 89.4 K,

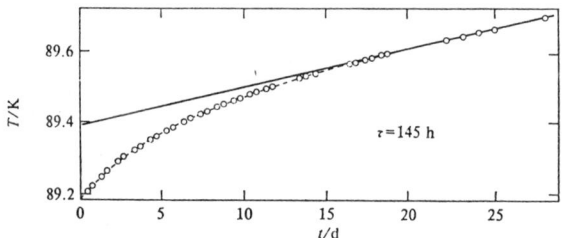

Fig. 5 Spontaneous temperature changes of ice observed below Tg.

respectively. In this way, the adiabatic calorimeter can have a role of a time-domain spectrometer in the range between mHz and µHz. The range can be extended by higher qualification of calorimeter and patience of experimenter.

In order to reveal the equilibrium heat capacity of ice, it is necessary to release the freezing process of water reorientational motion. This may be done by introducing a small amount of impurity which might hopefully relax the severe conditions and enhance the orientational mobility to recover the equilibrium properties within the experimental time. Among many dopants examined, alkali hydroxides[14] were found to shorten drastically the relaxation time. Thus, an ice specimen doped with 10^{-4} mole fraction of KOH exhibited a first order phase transition at 72 K. This is shown typically in Fig. 6. The temperature is independent of the nature and amount of cation but the associated entropy change depends on the concentration of the dopant. A substantial fraction of the residual entropy is removed by the phase transition. The new ordered ice is designated as ice XI. The equilibrium phase diagram[15] determined by a high-pressure adiabatic calorimeter is also depicted in Fig. 6. The transition temperature is found to increase by 2.4 K by the application of 159 MPa, so that dP/dT = (0.015 ± 0.001) KMPa^{-1}. Application of the Clausius-Clapeyron equation gives the volume change of transition to be 0.051 cm^{3}mol^{-1}, indicating rather small volume change compared to the entropy change at the transition.

Similar kind of disorder exists in clathrate hydrates[16], which is

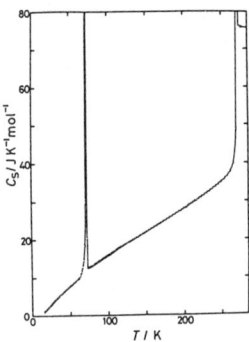 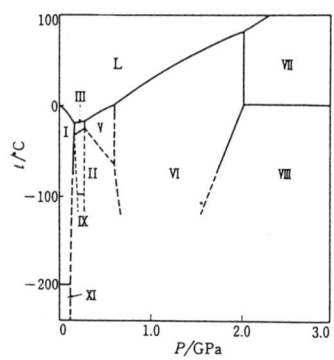

Fig. 6   Equilibrium heat capacity and phase diagram of ice.

one of the most common kind of clathrate compounds. The host lattice forms various types of Archimedes' polyhedra made of water molecules through hydrogen bondings and encages various kinds of guest molecules of suitable size and shape. The hydrogen-bond networks in the hydrate crystals are similar to that of hexagonal ice. Molecules of spherical shape enclathrated in these cages undergo rattling motion in almost spherical environments. Dipolar guest molecules undergo reorientational motion in addition to the rattling. Thus there are two kinds of orientational disorder in the clathrate hydrates[17]. For example, tetrahydrofuran (THF) hydrate crystal is known to show two kinds of dielectric dispersion. The high-temperature dispersion occurring around 120 K at 1 kHz is similar to that of ice, and can be related to water reorientational motion. The low-temperature one occurring around 25 K at 1 kHz is ascribed to the guest THF motion. The two dielectric dispersions mean that both the reorientational motions freeze out at some lower temperatures before each kind of disorder disappears by their hypothetical ordering transitions.

Figure 7 shows the heat capacities of pure and KOH-doped THF hydrate crystals based on the water mole. For the pure specimen, there appears a clear glass transition around 85 K associated with freezing out of the water motion. The heat capacity data agree reasonably to those reported by MacLean et al[18] (• mark), but they did not mention the existence of the anomalous behavior. Analysis[19] of the enthalpy relaxation data gives an activation energy of (19.5 ± 1) kJmol^{-1}, which is almost the same to that of ice (22 ± 2) kJmol^{-1}.

Doping of KOH in the THF hydrate lattice results in appearance of a first order transition at 62 K. This is the first successful observation of the ordering transition of host lattice in the clathrate hydrates. The entropy of transition is 2.56 JK^{-1} per mole of water. A dielectric measurement of pure and KOH-doped THF hydrate crystals shows that the guest molecules order concomitantly at the phase transition[20]. The audio-frequency dielectric constant does not indicate any dispersion phenomenon in the LTP and the value is close to ε_∞ of the crystal. It may be that the ordering of water dipoles produce a strong electric field at the guest site which gives a preferred orientation to the dipolar molecule. In this way, the electrostatic interaction among all the dipolar species seems to play an important role in the ordering transition. The dielectric measurement shows a signi-

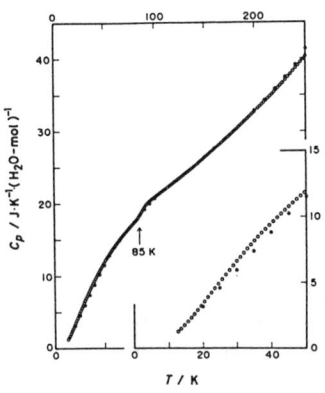 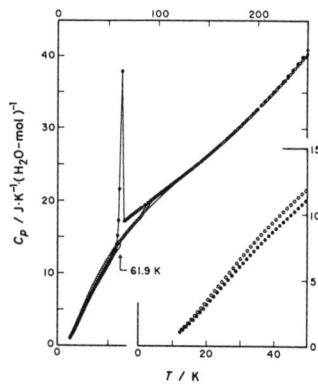

Fig. 7   Heat capacities of pure and KOH-doped THF clathrate hydrates.

ficant effect of the relaxation time by doping of KOH even in the mole fraction of 10^{-3}. The dopant shortens the relaxation time by a factor of 10^{10} at 70 K and also reduces the activation energy.

Thiophene is another example of glassy crystals belonging to the case (A). Thiophene, a heterocyclic five-membered molecule, is known to crystallize into an orthorhombic system which is the same as that of benzene having 6-fold symmetry. This means that significant orientational disorder exists in the thiophene crystal. In fact, thiophene has seven crystalline modifications, as shown in the following table.

$$V \xrightleftharpoons{112.35 K} IV \xrightleftharpoons{138.5 K} III \xrightleftharpoons{170.70 K} II \xrightleftharpoons{175.03 K} I \xrightleftharpoons{235.02 K} l$$

$$II_2 \xrightleftharpoons{90.76 K} II_1 \xrightleftharpoons{139.2 K}$$

The transition II-to-III is easily by-passed during continuous cooling which leads to metastable phase sequence. Our calorimetric measurement shows that both of the lowest temperature phases have glass transition and retain their residual entropies[21]. The glass transition takes place at 42 K for the stable V and at 37 K for the metastable II_2 phase, respectively. Since thiophene and benzene are mutually

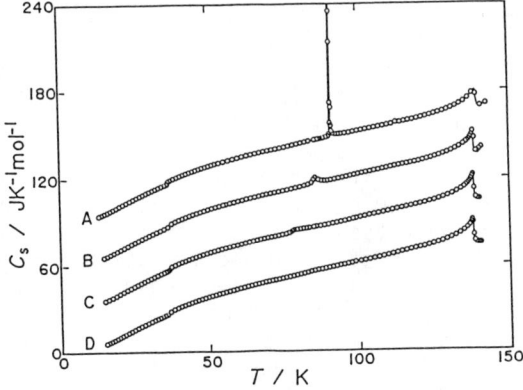 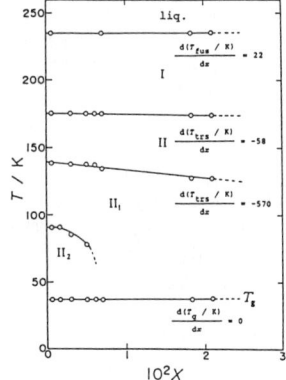

Fig. 8   Heat capacities and phase diagram of $(1-x)C_4H_4S + xC_6H_6$ solid solution in their metastable phase sequence in the region $x < 0.03$.

soluble not only in liquid but also in solid, it may be interesting to examine how the transitions in thiophene are affected by benzene doping[22]. Formation of the solid solution over the whole concentration range will be based on the same crystal symmetry and similar unit cell volumes of the component crystals.

A drastic change in the behavior was found only for the II_1-to-II_2 transition. As shown in Fig. 8, the sharp transition changes dramatically as the mole fraction x of benzene is increased. Thus the sharp transition disappears completely at $x=6.1\times10^{-3}$. In this figure, the curves A, B, C, and D correspond to x=0.0016, 0.0030, 0.0050 and 0.0061, respectively. Each curve is shifted upwards by 30 $JK^{-1}mol^{-1}$ step by step for the sake of clarity. Based on the observation, the phase diagram in the thiophene-rich region of this binary system is drawn also in Fig. 8.

The phases II and II_1 are known to be incommensurate[23]. McMillan has applied the Landau theory to charge-density waves and predicted[24] the sequence of phases, normal-incommensurate-commensurate with decreasing temperature, separated by first order transitions. He discussed a mechanism through which impurities pin the incommensurate modulation wave and stabilize the incommensurate phase. From the phenomenological similarity to the present binary system, it is considered that the benzene as an impurity hinders the incommensurate II_1 phase to transform into a commensurate phase upon cooling, leading to freezing of the incommensurably modulated structure. The lock-in transition is known to disappear by a slight doping of particular impurity in some dielectrics.

The glass transition phenomenon is also affected by the dopant. The rate of enthalpy relaxation below Tg is described by Eq (2). The non-exponential parameter β is 1.0 for x=0.0006, 0.93 for 0.0030, 0.90 for 0.0050, 0.87 for 0.0061. The results do not contradict to the increased variety of the environments in which the thiophene molecules are placed. For x=0.021, the observed temperature drift rate behaves as if a pair of glass transitions takes place consecutively with temperature. This means that there are at least two kinds of thiophene molecules in the solid solution which have slightly different kinetic parameters for the relaxation process. In any way, the present system provides a new example of multi-glass-transition which is different in nature from that usually observed in many glass-forming liquids.

It is worth considering an impurity effect on the phase transition under conditions of low concentration. In the case of ice, trace amounts of alkali hydroxides doped into the lattice induced a first order phase transition in an otherwise frozen-in system. The dopant creates a kind of orientational defect which infringes the ice conditions and breaks the chain-like motion for the rearrangement of the water molecules in the lattice. This resuls in the dramatic shortening effect of the relaxation time for the motion. The situation is in good contrast to thiophene in which one of the phase transitions II_2-to-II_1 disappears by a slight doping of benzene. The dopant enhances the stability of the phase II_1 and freezes the incommensurate structure down to the lowest temperatures. The phase II_1 is forced to persist without restoring the translational periodicity, probably through pinning effect of benzene. We can not control the time, but we can control the rate of various degrees of freedom by a particular impurity. We may modify the properties and dynamics of a pure compound by a slight doping of particular impurity in converting it from an equilibrium to a non-equilibrium state or vice versa. Thus the "doping chemistry" is undoubtedly a new pathway to enrich our world of materials science.

ACKNOWLEDGEMENTS

The author would like to express his sincere thanks to many collaborators and colleague who have actually engaged in the experiments reviewed here. These studies were supported financially by the Japanese Ministry of Education through the Grant-in-Aid for Fundamental Scientific Research, The Japan Society for Promotion of Science, The Takeda Science Foundation, The Nissan Science Foundation, to whom the author's thanks are due.

REFERENCES

* Contribution No 46 from the Microcalorimetry Research Center.
1 N.G.Parsonage and L.A.K.Staveley, Disorder in Crystals (Clarendon Press, Oxford, 1978).
2 H.Suga and S.Seki, J.Non-cryst.Solids, 16, 171 (1974).
3 H.Suga and T.Matsuo, Pure & Appl.Chem., 61, 1123 (1989).
4 K.Adachi, H.Suga and S.Seki, Bull.Chem.Soc.Jpn., 1073 (1968).
5 O.Haida, H.Suga and S.Seki, J.Chem.Thermodyn., 9, 1133 (1977).
6 I.Kishimoto, J.-J.Pinvidic, T.Matsuo and H.Suga, Proc.Jpn.Acad., 67B, 66 (1991).
7 K.Kishimoto, H.Suga and S.Seki, Bull.Chem.Soc.Jpn., 51, 1691 (1978)
8 R.A.Newmark and E.R.Graves, J.Phys.Chem., 72, 4299 (1968).
9 G.Williams and D.C.Watts, Trans Faraday Soc., 66, 80 (1970).
10 S.Brawer, Relaxation in Viscous Liquids and Glasses (Am.Ceram.Soc., Columbus, 1985).
11 J.Jäckel, Rep.Prog.Phys., 49, 171 (1986).
12 W.Kauzmann, Rev.Mod.Phys., 14, 12 (1942).
13 O.Haida, T.Matsuo, H.Suga and S.Seki, J.Chem.Thermodyn., 6, 815 (1974).
14 Y.Tajima, T.Matsuo and H.Suga, Nature, 299, 810 (1982).
15 O.Yamamuro, T.Matsuo and H.Suga, J.Chem.Phys., 86, 5137 (1987).
16 D.W.Davidson, Water - A Comprehensive Treatise (Plenum Press., N.Y. 1973) Vol.2, 115.
17 O.Yamamuro and H.Suga, J.Therm.Anal., 35, 2025 (1989).
18 M.J.MacLean and M.A.White, J.Phys.Chem., 89, 1380 (1985).
19 N.Kuratomi, O.Yamamuro, T.Matsuo and H.Suga, J.Chem.Thermodyn., 23, 485 (1991).
20 O.Yamamuro, T.Matsuo and H.Suga, J.Incl.Phenom., 8, 33 (1990).
21 P.Figuiere, H.Szwarc, M.Oguni and H.Suga, J.Chem.Thermodyn., 23, 485 (1985).
22 N.Okamoto, M.Oguni and H.Suga, J.Phys.Chem.Solids, 50, 1285 (1989).
23 D.André, A.Dworkin, P.Figuiere, A.H.Fuchs and H.Szwarc, J.Phys. Chem.Solids, 46, 505 (1985); J.Physique, 47, 61 (1986).
24 W.L.McMillan, Phys.Rev., B12, 1187 (1975); B14, 1496 (1976).

TRACER DIFFUSION IN POLYMER AND ORGANIC LIQUIDS CLOSE TO THE GLASS TRANSITION

M. Lohfink, H. Sillescu
Institut für Physikalische Chemie der Johannes Gutenberg-Universität Mainz, Jakob-Welder-Weg 15, D-6500 Mainz, Germany

ABSTRACT

Translational diffusion coefficients D of photochromic dye molecules have been measured by forced Rayleigh scattering in glass forming liquids and polymers at temperatures ranging to below the glass transition T_g. In mixtures where the polymer and diluent component have the same T_g, D is investigated over the whole composition range from pure polymer to pure liquid matrices. In the supercooled liquids, the size of the tracers is similar to that of the surrounding liquid molecules. Thus, we mimic self diffusion which is found to become much faster than predicted by the Stokes-Einstein relation as the temperature is reduced below a characteristic temperature $T_c > T_g$. The crossover temperature T_c agrees with the corresponding temperature determined from neutron scattering experiments and by a power law plot of the shear viscosity as suggested by mode coupling theory. Our finding that translational diffusion behaves different from rotational diffusion which follows the Debye relation down to T_g is discussed in terms of cooperative molecular motion.

INTRODUCTION

The diffusion of small molecules in polymer glass formers has been studied for a long time by permeation and sorption techniques [1]. The measured diffusion coefficients D are typically in a range of 10^{-5} to 10^{-10} cm^{2}s^{-1} and have an Arrhenius-type temperature dependence with apparent activation energies E_A of 5 - 20 kJ mol^{-1}. For larger diffusant molecules, an increase is observed at the glass transition temperature T_g and the T dependence of diffusion can be described by a Vogel-Fulcher-Tamman (VFT) equation

$$D = D_0 \exp[-\xi B_0/(T - T_\infty)] \qquad (1)$$

for $T \geq T_g$ [1,2] where the parameters B_0 and T_∞ are obtained from the corresponding VFT equation for the shear viscosity η of the polymer matrix and $\xi \leq 1$ accounts for the decoupling of diffusion and viscosity. ξ depends upon diffusant and matrix properties as has been discussed in some detail in ref. [2]. At temperatures $T \sim T_g$ an increase of the diffusant's size by a factor ~ 10 from rare gas atoms to large organic dye molecules causes a decrease of D by about 9 decades from 10^{-5} to $10^{-14} cm^2 s^{-1}$ in typical polymers as polystyrene or polymethylmethacrylate [2]. Thus, it is apparent that the Stokes-Einstein (SE) relation $D = kT / 6 \pi \eta r_s$ where r_s is some "hydrodynamic" diffusant radius, does not apply in polymer glass formers. On the other hand, the SE relation is well established in typical van der Waals liquids though r_s is found smaller than the geometric molecular radius [3]. We have performed diffusion studies in order to find out whether a decoupling of diffusion and viscosity similar to that in polymers can be observed in supercooled liquids on approaching the glass transition. In supercooled orthoterphenyl (OTP) where we have measured tracer diffusion coefficients D of two photochromic dyes (TTI and ACR)[4, 5] having about the same size as OTP molecules as well as OTP self diffusion coefficients (using a NMR magnetic gradient method)[5] all D values at $T > T_g$ could be described by a modified SE relation

$$D = \frac{kT}{6\pi\eta f_\xi r_s} \; ; \; f_\xi = (\frac{\eta_c}{\eta})^{1-\xi} \; ; \; \begin{matrix} \xi = 1 \text{ for } T \geq T_c \\ \xi < 1 \text{ for } T_g < T < T_c \end{matrix} \qquad (2)$$

η_c is the shear viscosity at the temperature T_c separating the low temperature from the high temperature regime in supercooled liquids (cf. fig. 5). The power law $D \propto \eta^{-\xi}$ implicit in eq. (1) has been found previously in various liquids [2, 3, 6]. If $f_\xi r_s$ is interpreted as an "effective hydrodynamic radius" [6] we obtain an extremely small value in OTP where $r_s = 0.21$ nm (determined at $T > T_c$) and $f_\xi = 6 \cdot 10^{-3}$ at T_g. In polymer glasses, the effective radii are further reduced by some decades. [2] This appears rather puzzling in view of many indications of cooperative motion at the glass transition [7-9] which should be associated with effective radii larger than r_s. We shall come back to this problem in the discussion where we offer an interpretation of translational and rotational diffusion which implies cooperative motion. In the following section, we also present results on tracer diffusion in mixtures of a polymer and a liquid component where T_g remains constants over the whole composition range. Thus, we avoid the high diluent mobility below T_g which is typical for many polymer plasticizer systems studied previously by diffusion techniques [1, 10].

RESULTS

In Figs. 1 and 2, we show some results of an investigation of tracer diffusion in mixtures of poly(methylphenylsiloxane) (PMPS) and 1,1 - bis (p-methoxyphenyl)cyclohexane (BMC). Within experimental accuracy of the DSC method $T_g = 243$ K is the same in all mixtures. The diffusion coefficients D of the photochromic tracer TTI[2] were measured by forced Rayleigh scattering [2]. Since a detailed description of our results and their interpretation in terms of free volume theory will be

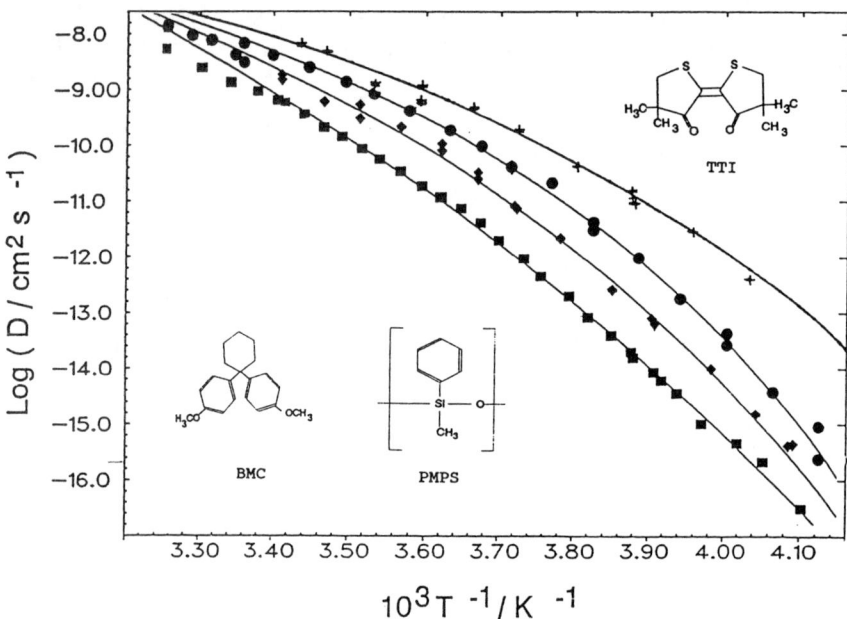

Fig. 1. Diffusion coefficients D of TTI in mixtures of 0, 3,2, 52 and 100 wt. % (from above) BMC in PMPS

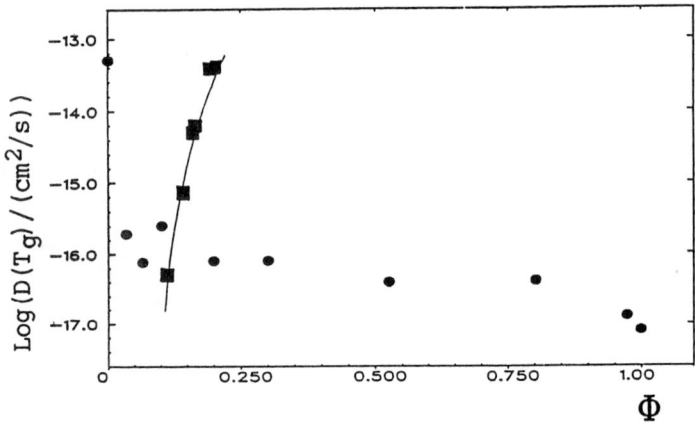

Fig. 2 Diffusion coefficients D of ACR in BMC/PMPS (dots) and in toluene / polystyrene (squares)[10] at $T = T_g$. Φ denotes the solvent volume fraction.

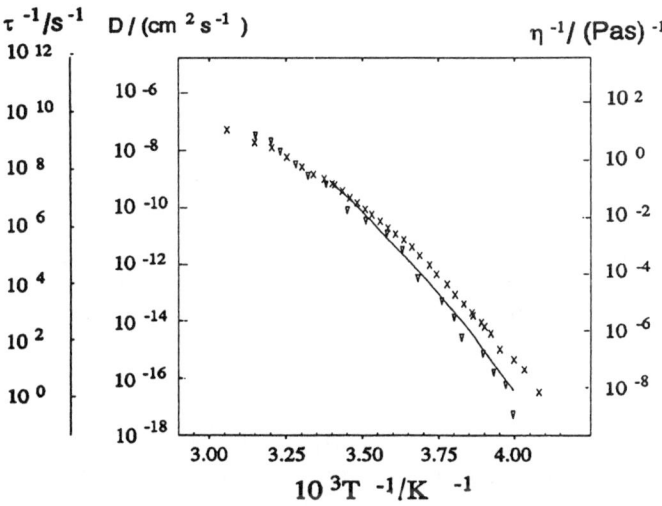

Fig. 3. Diffusion coefficients D(x) of TTI in BMC, inverse dielectric relaxation times τ^{-1} ($\triangledown$), and shear viscosities η^{-1} (full line) of BMC).

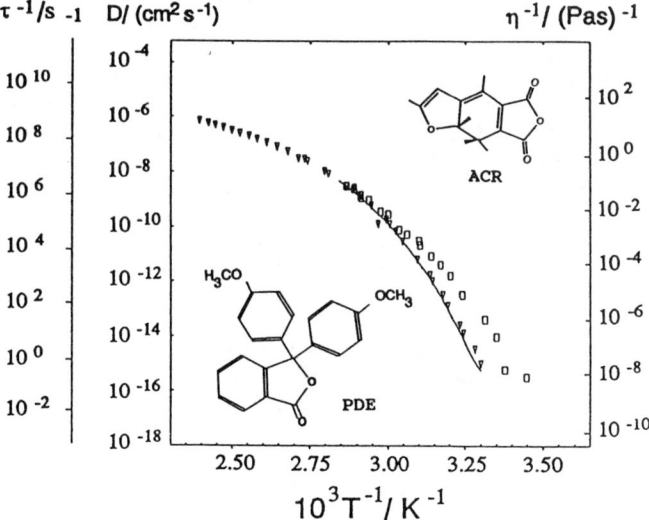

Fig. 4. Diffusion coefficents D ($\square$) of ACR in PDE, inverse dielectric relaxation times τ^{-1} ($\triangledown$), and shear viscosities η^{-1} (full line) of PDE.

published elsewhere [4, 11] we concentrate here on some aspects concerning the difference between diffusion in polymers and low molecular weight liquids. The full lines in Fig. 1 are a fit of eq. (1); the D values obtained from the fit curves of all mixtures investigated at T_g are shown in Fig. 2. In Fig. 3, the D values of the same TTI tracer [2] (see formula in Fig. 1) in pure BMC are compared with the shear viscosity and the dielectric relaxation time of BMC (called BCDE in ref. 12a). The same is shown in Fig. 4, however, for the supercooled liquid phenolphthaleindimethylether (PDE, T_g = 294 K) and the photochromic dye Aberchrome 540 (ACR) [10]. The dielectric relaxation times have been determined from averages over the α-process fitted by a Havriliak-Negami distribution [12].

DISCUSSION

It is apparent from Fig. 1 that the tracer diffusion coefficients are rather similar in the polymer and low molecular weight liquids at temperatures far above T_g. At T_g, a few percent of diluent suffice to reduce D by almost 3 decades to values typical for other supercooled liquids (Fig. 3, 4). This appears plausible in view of the "dual sorption" behavior found in many polymers by sorption experiments indicating that a small fraction of the sorbed molecules is filling up special sites within the polymer [1]. The different behavior found for diffusion of ACR in the polystyrene / toluene system is probably caused by the high toluene mobility in the mixed glass below T_g of the mixture because of the much lower T_g of pure toluene. Since we have found a value of log D (T_g) = - 14.1 for TTI diffusion at T_g of pure polystyrene [2] and the same D values for TTI and ACR in OTP (see Fig. 5) we expect a minimum of D(T_g) at a toluene concentration where the slowing down due to "filling up holes" and the speeding up by mobile diluent molecules just balance. Finally, we note the decrease of the Vogel temperature T_∞ and the increase of ξ defined in eq. (1) on changing over from the pure polymer to the pure liquid in Fig. 1 [4, 11].

In Figs. 3 and 4 the temperature dependence of D, τ^{-1}, and η^{-1} can be superimposed in the high temperature range where the SE relation, eq. (2) for ξ = 1, and the corresponding Debye relation

$$D_R = \frac{kT}{8\pi\eta\, r_D^3} \; ; \; r_D \approx r_S \tag{3}$$

for the rotational diffusion coefficient D_R = $(2\tau)^{-1}$ are applicable. At lower temperature, the Debye relation is still valid for rotational diffusion whereas translational diffusion decouples and a fit to eq. (1) yields ξ = 0.86 and 0.69 for TTI in BMC and ACR in PDE, respectively. The same is shown in Fig. 5 for diffusion in OTP [5]. Here D_R was obtained from depolarized dynamic light scattering experiments, $D_R = (6 \tau_{VH})^{-1}$, where $\tau_{\alpha VH}$ is the mean decay time determined from stretched exponential fits [13, 14]. It should be noted that due to different time and frequency windows of experimental techniques the D_R averages to be compared with eq. (3) may differ somewhat in their absolute number and their T dependence if extracted from different experiments. However, the approximate agreement with eq. (3) is outside experimental error, in particular, for the results shown in Figs. 3 and 4 where the frequency window includes the full α-peak. The simultaneous validity of eq. (2) with $\xi < 1$ and eq. (3) is a highly nontrivial result which should contain important information upon microscopic mobility at the glass transition. An interpretation in terms of cooperative molecular motion as given below must be considered speculative though in the tradition of previous treatments [7-9]. However, we emphasize that cooperative motion is <u>assumed</u> rather than proven in these treatments.

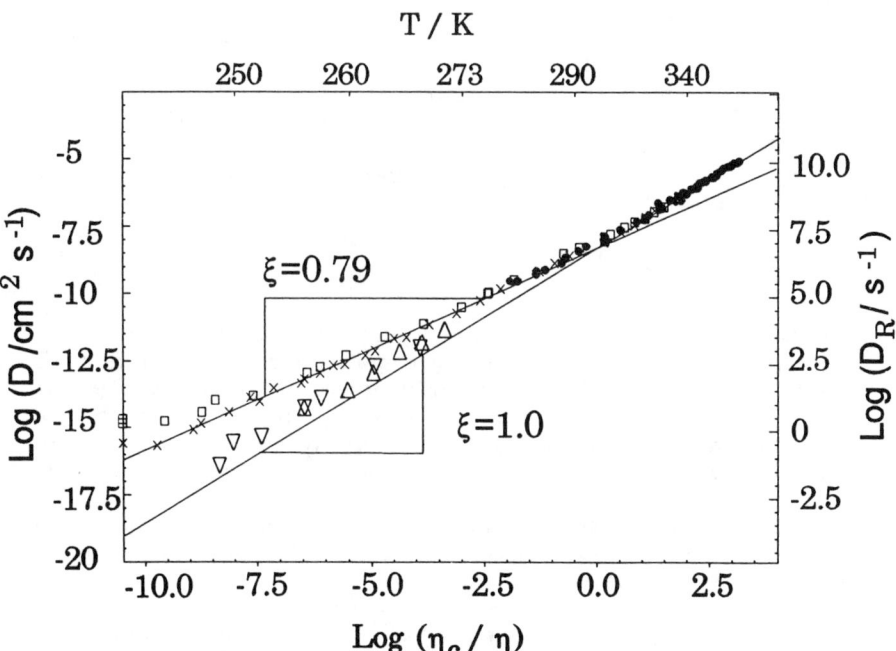

Fig. 5. Translational (D) and rotational (D_R) diffusion coefficients in OTP. Tracer diffusion: □ ACR, x TTI, ● Self diffusion, see ref. 5. Rotational diffusion △ ref. 13. ▽ ref. 14.

If we insist that translational and rotational diffusion occurs through cooperative motion we can <u>define</u> an aggregate radius r_a and a local viscosity η_a by

$$D = \frac{kT}{6\pi\eta_a r_a} \quad ; \quad D_R = \frac{kT}{8\pi\eta_a r_a^3} \tag{4}$$

By combination with eqns. (2-3) we obtain ($r_D \approx r_S$):

$$r_a = \left(\frac{3D}{4D_R}\right)^{1/2} = \left(\frac{r_D^3}{f_\xi r_S}\right)^{1/2} \approx f_\xi^{-1/2} r_S \tag{5}$$

$$\eta_a = \frac{kT D_R^{1/2}}{\pi(3D)^{3/2}} = \left(\frac{f_\xi r_S}{r_D}\right)^{3/2}\eta \approx f_\xi^{3/2}\eta \tag{6}$$

By defining a liquid of aggregates through eqns. (4) we run into the problem that usual, say, hard sphere liquids have a self diffusion coefficient of the order $kT/6\pi\eta r_a$ where η is the macroscopic shear viscosity [3]. Thus, the aggregates diffuse with $D_a = D\eta_a/\eta$, slower than the molecules since $f_\xi < 1$, see eq. (6). In order to save the aggregate picture it is <u>necessary</u> to make a further assumption for explaining why the molecules can diffuse faster than the aggregates. We assume that the <u>life time τ_a of a molecule in an aggregate</u> is of order of the rotational correlation time:

$$\tau_a \approx D_R^{-1} \tag{7}$$

Now, a molecule can enter an aggregate, rotate with it by an angle π, and leave it after being displaced by a distance of $2 r_a$ which yields the <u>molecular</u> translational diffusion coefficient $D \sim r_a^2/\tau_a \sim r_a^2 D_R$ in harmony with eqns. (4). In this way, translational and rotational diffusion were also related in a treatment of solvent molecule motion in solvation spheres [15].

Some remarks seem appropriate in order to bring the aggregate picture drawn above in perspective with other approaches to glass formation:

- An aggregate radius $r_a = 12.6 \, r_S = 2.8$ nm is obtained for OTP at T_g from eq. (5) with $r_S \approx r_D \approx 0.22$ nm [5]. This is rather close to the value of 2.2 nm estimated by Donth in cyclohexanol at T_g within a somewhat different picture of cooperativity. However, r_a is reduced to only $2.1 \, r_S$ at 273 K in OTP which is about half-way between T_g and T_c. Here, aggregation is apparently much less pronounced, and a picture of rotating gears is perhaps more appropriate for relating translational and rotational motion.

- The remarkable result that the correlation times τ_{VH} and τ_{VV} obtained from depolarized and polarized light scattering are found equal in supercooled liquids [13, 14] has been interpreted by assuming a strong coupling of molecular reorientation and the structural relaxation related with the α-process [13]. This fits well with the picture of cooperative translation rotation coupling given above which is also consistent with the "tear and repair mechanism" discussed by Stillinger in a model of relaxation in glass forming liquids [16]. It should further be noted that the assumption of rotating aggregates can explain the result from dielectric relaxation [17], time resolved optical anisotropy [18], and NMR [19] experiments that the rotational correlation times of the investigated tracer molecules become size independent as T_g is approached.

- The ξ parmeter introduced in eqns. (1-2) has been interpreted in free volume theory as the ratio of the critical molecular volume of the solute jumping unit over that of the solvent [20, 21]. This assumption, which was already shown not to apply to flexible solutes [2, 21], is also in contrast to our result that ξ < 1 in OTP since it requires ξ = 1 for self diffusion in pure liquids. The fact that ξ < 1 for self diffusion (see Fig. 5) proves the inconsistency of free volume theory as applied to the glass transition.

- The crossing of two straight lines at the temperature T_c shown in Fig. 5 clearly defines a characteristic temperature that has an important influence upon glassy dynamics at all temperatures $T < T_c$ which is apparent from the dependence of f_ξ upon $\eta_c = \eta(T_c)$ in eq. (2). In OTP, T_c has also been determined by neutron scattering [22] and by a power law fit of η at $T > T_c$ suggested by the mode coupling theory of the glass transition [23]. In this theory, α-relaxation at $T < T_c$ is accounted for by "medium assisted activated hopping processes" originating from a "backflow term" that couples density and current correlation [23, 24]. Although this backflow implies cooperative motion [24] we should note that the aggregate picture discussed above for coupling of molecular reorientation and structural relaxation requires nonspherical molecules whereas present mode coupling theories are formulated for spherical molecules which rotate much faster than predicted by eq. (3) even in normal liquids [3].

- Finally, let us discuss whether the decoupling of translational diffusion can be related with β-relaxation which has been studied in OTP by dielectric [25] and NMR [26] relaxation and has been associated with fast rotational diffusion within an angular range of $\lesssim 15°$ [26]. Since the β-process [25] has a weak T dependence and a very broad distribution of correlation times ($\sim 10^{-9} - 10^{-6}$s) it may, indeed, have an influence upon the ξ parameter. In this case, we should expect a strong molecular shape dependence of ξ, and molecular reorientation should become faster than predicted by eq. (3) if the angular range of fast rotational diffusion is larger for more spherical shapes. Probably, the fast β-process defined in mode coupling theory [23] and detected on a ps time scale in neutron scattering [22] has little influence upon translational diffusion at $T < T_c$ where transport is governed by the "backflow term" [24].

In conclusion, we find that all experimental results on translational and rotational diffusion in supercooled liquids can be explained within a picture of cooperative molecular motion guided by application of Stokes-Einstein and Debye relations. However, though many experiments provide strong evidence for cooperative motion [13, 14, 17-19, 27, 28] a direct proof is still lacking. Thus we should be prepared that future theoretical developments based on mode coupling theory will provide an alternative picture.

ACKNOWLEDGEMENTS

Support by the Deutsche Forschungsgemeinschaft (SFB 262) is gratefully acknowledged. We thank Dr. F. Kremer and Dr. T. Pakula, Max-Planck-Institut für Polymerforschung, Mainz, for suppling yet unpublished dielectric relaxation times and shear viscosities of PDE.

REFERENCES

1. J. Crank, G.S. Park; Diffusion in Polymers, Academic Press, London 1968
2. D. Ehlich, H. Sillescu, Macromolecules 23, 1600 (1990)
3. H.J.V. Tyrrell, K.R. Harris, Diffusion in Liquids, Butterworths, London 1984
4. M. Lohfink, Diplomarbeit, Universität Mainz 1988, Dissertation Universität Mainz 1991
5. M. Lohfink, F. Fujara, H. Sillescu, G. Fleischer, submitted for publication
6. R. Zwanzig, A.K. Harrison, J.Chem. Phys. 83, 5861 (1985)
7. G. Adam, J.H. Gibbs, J.Chem. Phys. 28, 373 (1965)
8. E.J. Donth, J. Non-Cryst. Solids 53, 325 (1982)
   E.J. Donth, Glasübergang, Akademie Verlag, Berlin 1981
9. G.F. Fredrickson, Ann. Revs. Phys. Chem. 39, 149 (1988)
10. T.S. Frick, W.J. Huang, M. Tirrell, T.P. Lodge, J. Polym. Sci. (Phys. Ed.) 28, 2629 (1990)
11. M. Lohfink, H. Sillescu, manuscript in preparation
12. a) G. Meier, B. Gerharz, D. Boese, E.W. Fischer, J. Chem. Phys. 94, 3050 (1991)
    b) A. Schönhals, F. Kremer, E. Schlosser, Phys. Rev. Lett. 67, 999 (1991)
13) G. Fytas, C.H. Wang, D. Lilge, T.H. Dorfmüller, J. Chem. Phys. 75, 4274 (1981)
14. J.V. Hagenah, Dissertation, Universität Mainz 1988
15. L. Endom, H.G. Hertz, B. Thül, M.D. Zeidler, Ber. Bunsenges. Phys. Chem. 71, 1008 (1967)

16. F.H. Stillinger, J. Chem. Phys. 89, 6461 (1988)
17. G. Williams, M.F. Shears, J. Chem. Soc. Faraday II 69, 608 (1973)
18 P.D. Hyde, T.E. Evert, M.D. Ediger, J. Chem. Phys. 93, 2274 (1990)
19. E. Rössler, K. Börner, M. Pöschl, J. Tauchert, M. Taupitz, Ber. Bunsenges. Physik. Chem., in press
20. J.S. Vrentas, J.L. Duda, J. Polym. Sci, (Polym. Phys. Ed.) 15, 403 (1977)
21. J.S. Vrentas, J.L. Duda, A.C. Hou, J. Appl. Polym. Sci., 31, 739 (1986)
22. W. Petry, E. Bartsch, F. Fujara, M. Kiebel, H. Sillescu, B. Farago, Z. Phys. B-Condensed Matter 83, 175 (1991)
23. L. Sjögren, W. Götze, Springer Proceedings in Physics 37, 18 (1989)
24. L. Sjögren, Z. Phys. B. Condensed Matter 79, 5 (1990)
25. G.P. Johari, M. Goldstein, J. Chem. Phys. 53, 2372 (1970)
26. W. Schnauss, Dissertation, Universität Mainz 1991.
    W. Schnauss, F. Fujara, H. Sillescu. manuscript in preparation
27. W. Schnauss, F. Fujara, K. Hartmann, H. Sillescu, Chem. Phys. Lett. 166, 381 (1990)
28. K. Schmidt-Rohr, H.W. Spiess, Phys. Rev. Lett. 66, 3020 (1991)

BRILLOUIN AND RAMAN SCATTERING SPECTROSCOPY OF THE LIQUID-GLASS TRANSITION

H.Z. Cummins, G. Li, W.M. Du and X.K. Chen
Department of Physics, City College of the City University of New York
New York, NY 10031

N.J. Tao
Department of Physics, Arizona State University, Tempe, AZ 85287

A. Sakai
Dept. of Electrical Engineering, Muroran Institute of Technology,
Muroran, JAPAN

ABSTRACT

The liquid-glass transition in calcium potasssium nitrate (CKN), salol, and aqueous LiCl solutions was investigated by laser light scattering spectroscopy. Combining data obtained with a Sandercock tandem Fabry Perot interferometer and a Spex tandem Raman spectrometer provided spectra spanning more than four decades in frequency. The data was analyzed to test both qualitative and quantitative predictions of the mode coupling theory of the glass transition, and produced generally excellent agreement. For CKN, we found that the MCT crossover temperature T_c is $\sim 378K$, 45^0 above T_g.

INTRODUCTION

When a liquid is cooled below its melting temperature T_m rapidly enough to avoid crystallization, its viscosity and relaxation time increase until, at a calorimetric glass transition temperature T_g (often defined as the temperature at which the viscosity reaches 10^{13} poise) the supercooled liquid solidifies into an amorphous solid. This liquid-glass vitrification transition has been studied extensively for many years, but remains incompletely understood.

Recently, renewed interest in the glass transition has been stimulated by the development of new theoretical approaches embodying mode-coupling and percolation concepts. In particular, the self-consistent mode coupling theory (MCT) of the glass transition,[1,2,3] developed in a series of papers starting in 1984, primarily by W. Götze, L. Sjögren and their coworkers, leads to quantitative predictions that are directly accessible to experimental investigation, and several experiments have been undertaken recently to test some of these predictions. In the MCT, nonlinear interaction between density fluctuation modes causes the slowing down and structural arrest, leading to an ergodic-nonergodic transition at a crossover temperature T_c somewhat above T_g.

Recent experimental studies of the glass transition have exploited several powerful new techniques such as neutron spin echo spectroscopy[4] and impulsive stimulated light scattering.[5] Our light scattering experiments on $CaKNO_3$, $LiCl-H_2O$

and salol, to be discussed below, employed conventional Brillouin and Raman scattering spectroscopy. The major new feature of our work is the use of a Sandercock tandem Fabry-Perot interferometer in conjunction with a Raman spectrometer which has permitted us to analyze light scattering spectra covering a frequency range of more than four decades.

BRILLOUIN SCATTERING

Numerous Brillouin scattering studies of the glass transition in various materials have been reported,[6] generally with results similar to our CKN [$0.4\text{Ca}(\text{NO}_3)_2$-$0.6\text{KNO}_3$; $T_g=333$K, $T_m=438$K] (VV polarized) Brillouin data shown in Fig. (1). At temperatures $T \gg T_m$, the spectra exhibit two symmetrically placed Brillouin components with Lorentzian lineshapes, and an unresolved Rayleigh line consisting of quasielastic scattering from entropy and concentration fluctuations as well as parasitic elastic scattering. As the sample is cooled (from top to bottom in Fig. (1)) the Brillouin shift increases monotonically, while the Brillouin linewidth increases to a maximum at $T \sim 458$K, and then decreases, becoming very narrow at low temperatures. As the Brillouin components broaden, an additional central component appears and narrows continuously as T decreases, finally disappearing into the unresolved Rayleigh line. Such spectra have been analyzed in the past with various forms of phenomenological viscoelastic theory which we will review briefly.

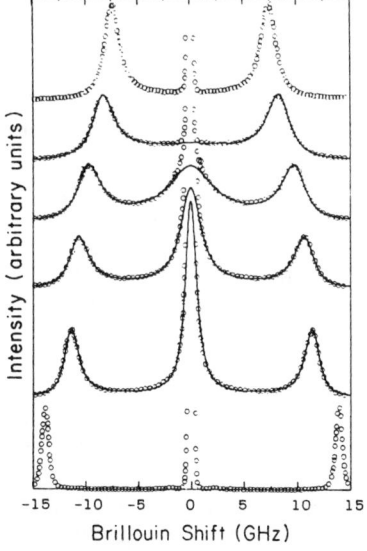

Figure 1. VV polarized Brillouin spectra of CKN at (top to bottom) 598, 528, 478, 448, 428 and 348K. The circles are the experimental data. The solid lines are fits to Eqs. (3) and (7) with $M_\Gamma(q,t) = V^{(2)}(q,q_0)\phi_{q_0}^2(t)$, where $\phi_{q_0}^2(t)$ was taken from the neutron spin-echo results of Mezei et al. (Ref. 4).

In a simple structureless fluid, the linearized hydrodynamic equations predict five modes for each (small) wavevector q: two transverse viscous shear modes (which do not couple to light), a narrow quasielastic thermal diffusion mode, and two propagating longitudinal acoustic modes.[7] If we ignore the contributions of temperature fluctuations (which are not resolvable in Brillouin scattering), the simplified equation of motion for ρ_q, the qth Fourier component of the fluctuation in the density ρ, is

$$\ddot{\rho}_q + \omega_0^2 \rho_q + \gamma_0 \dot{\rho}_q = 0 \qquad (1)$$

where $\omega_0^2 = G_L q^2/\rho$ (G_L is the adiabatic compressional modulus) and $\gamma_0 = \eta_L q^2/\rho$ ($\eta_L = \frac{4}{3}\eta_S + \eta_B$ is the longitudinal viscosity).

As a fluid is cooled below T_m its thermodynamic properties become frequency dependent, signalling the increasingly long relaxation time of the evolving structural order in the supercooled melt. Structural relaxation processes can be incorporated into the hydrodynamic equations in three equivalent ways: (1) the compressional modulus G_L can be generalized to $G_L(\omega)$; (2) the longitudinal viscosity η_L can be generalized to $\eta_L(\omega)$; (3) another relaxing degree of freedom $\psi(r,t)$ can be introduced to represent structural order, and can be linearly coupled to ρ. These three approaches all lead to results that are formally equivalent, and can be most easily represented by replacing the damping constant γ_0 in (1) by a generalized damping or friction function $\Gamma(\omega)$:

$$\Gamma(\omega) = \Gamma'(\omega) + i\Gamma''(\omega) = \gamma_0 + \int_0^\infty e^{i\omega t} M_\Gamma(t) dt \qquad (2)$$

where $M_\Gamma(t)$ is the memory function for the damping. The light scattering spectrum $I_q(\omega)$, which is (aside from a proportionality constant) equal to the dynamic structure factor $S(q,\omega)$, is given by the fluctuation-dissipation theorem as:

$$I_q(\omega) = \frac{I_0}{\omega} \text{Im}[\omega_0^2 - \omega^2 - i\omega\Gamma(\omega)]^{-1} \qquad (3)$$

In the simplest phenomenological single relaxation time approximation (Maxwell viscoelasticity) the memory function $M_\Gamma(t) = \Gamma_R e^{-t/\tau}$ and Eq. (3) reduces to

$$I_q(\omega) = \frac{I_0}{\omega} \text{Im}[\omega_0^2 - \omega^2 - i\omega(\gamma_0 + \frac{\Gamma_R \tau}{1 - i\omega\tau})]^{-1} \qquad (4)$$

Eq. (4), which was widely used in the 1970's to analyze central peaks observed in the neutron or light scattering spectra of crystals near structural phase transitions,[8] has also been used to analyze the Brillouin spectra of supercooled liquids such as $ZnCL_2$[9] and $LiCl-H_2O$.[6] Eqs. (2) and (3) can be easily generalized to more complicated memory functions such as the stretched exponential (Kohlrausch or KWW) function $M_\Gamma(t) = \Gamma_R e^{-(t/\tau)^\beta}$ or the closely related Cole-Davidson function $\Gamma(\omega) = \gamma_0 + \Gamma_R(\frac{1}{1-i\omega\tau})^{\beta_{CD}}$ which was used by Borjesson et al. in their Brillouin scattering study of the liquid-glass transition in propylene carbonate.[10]

The viscoelastic generalization of $\Gamma(\omega)$ can be incorporated into the equation of motion (1) for $\rho_q(t)$ as

$$\ddot{\rho}_q(t) + \omega_0^2 \rho_q(t) + \gamma_0 \dot{\rho}_q(t) + \int_0^t M_\Gamma(q, t-t')\dot{\rho}_q(t')dt' = 0 \qquad (5)$$

The corresponding equation of motion for the time correlation function $F(q,t) = <\rho_q^*(0)\rho_q(t)>$, or its normalized form $\phi_q(t) = F(q,t)/S(q)$, is:

$$\ddot{\phi}_q(t) + \omega_0^2 \phi_q(t) + \gamma_0 \dot{\phi}_q(t) + \int_0^t M_\Gamma(q, t')\dot{\phi}_q(t-t')dt' = 0 \qquad (6)$$

In the mode coupling theory (MCT), the memory function $M_\Gamma(q,t)$ describes nonlinear interactions between density fluctuation modes expressed as

$$M_\Gamma(q,t) = \Sigma_{q_1} V^{(1)}(q,q_1)\phi_{q_1}(t) + \Sigma_{q_1 q_2} V^{(2)}(q,q_1,q_2)\phi_{q_1}(t)\phi_{q_2}(t) + \cdots \qquad (7)$$

As a first qualitative test of MCT concepts, we analyzed the CKN Brillouin data of Fig. (1) with a simple version of MCT which includes only the term $V^{(2)}(q,q_0)\phi_{q_0}^2(t)$ in Eq. (7) where q_0 is the wavevector corresponding to the first peak of the static structure factor $S(q)$.[11] This approximation corresponds to the schematic diagram shown in Fig. (2a) in which the long-wavelength acoustic mode ρ_q probed by light scattering decays into two q_0 modes. The indicated interaction between q_0 modes is absorbed into $\phi_{q_0}(t)$ whose Fourier transform is the physical $S(q_0,\omega)$. For $\phi_{q_0}(t)$ we used the neutron spin echo results of Mezei et al.[4] The resulting fits with $V^{(2)}$ as a fitting parameter, shown by the solid lines in Fig. (1), were excellent. Actually, since the neutron data are well described by a stretched exponential $\phi_{q_0}(t) = ae^{-(t/\tau)^\beta}$ with $\beta=0.58$, the mode coupling result is *identical* to the viscoelastic result of Eqs. (2) and (3) with $M_\Gamma(q,t) = V^{(2)}(q,q_0)a^2 e^{-2(t/\tau)^\beta}$. The essential difference is that τ and β, rather than being treated as free parameters of the phenomenological viscoelastic memory function, are physical quantities in MCT which can be determined experimentally by neutron scattering. Furthermore, the values of $V^{(2)}(q,q_0)$ found from our fits agreed within a factor of four with the values computed for a hard-sphere fluid using the Percus-Yevick equation.

Our Results show that the Brillouin spectra of CKN can be explained by MCT in the approximation that long wavelength acoustic modes decay preferentially into pairs of q_0 density fluctuation modes. Although this result supports the correctness of the physical mechanism underlying MCT, it does not allow the form of the memory function to be determined independently, primarily because the Brillouin spectra span a very limited frequency range. We have tried fitting both real and simulated Brillouin data using Eqs. (2) and (3) with a stretched exponential for $M_\Gamma(t)$ and found that it is extremely difficult to determine β accurately. A far more sensitive approach to the analysis of Brillouin data will be

44 Brillouin and Raman Scattering Spectroscopy

discussed below.

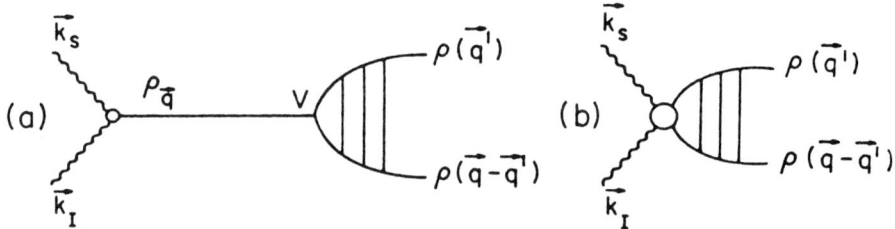

Figure 2. Interaction of light with q_0 modes (at or near the peak of the static structure factor $S(q)$). (a) Light scatters from an acoustic mode (q) which decays into two (interacting) modes with $q' \sim q^0$. (b) Light scatters directly from two q_0 modes. Figures adapted from R. Klein, "Phonon Transport Theory and the Central Mode", in: **Anharmonic Lattices, Structural Transitions and Melting** edited by T. Riste (Noordhoff-Leiden, 1974) p. 161.

RAMAN SCATTERING

The Raman spectra of glasses exhibit characteristic low-frequency structure not observed in the corresponding crystals. The glass spectra usually exhibit a broad "Boson peak" somewhere between $\sim$20 and 80 cm^{-1} which extends all the way to the origin.[12] In the past, these spectra have generally been analyzed with a disordered-crystal model in which the static frozen-in disorder characteristic of glasses presumably breaks the usual crystal momentum selection rules, permitting acoustic modes with any wavevector q to contribute to the Raman spectrum. This model, as formulated by Martin and Brenig,[13] leads to a predicted low frequency spectrum $I(\omega) \propto \omega^2$ and a "Boson peak" position determined by the correlation length of the static disorder. The additional intensity usually observed at low frequencies, designated as the "light scattering excess", therefore requires an additional mechanism such as the two-state defect scattering first discussed by Winterling.[14]

Liquids also produce pronounced broadband low frequency Raman scattering, and it has been observed that the liquid spectrum evolves continuously into the glass spectrum as a material is cooled through the glass transition, suggesting that the low frequency Raman spectra of glasses and liquids may have a common origin. In Fig. (3) we show a series of (VV polarized) Raman spectra of a 15 mole% aqueous LiCl solution (T_g=140K) demonstrating this evolution.[15] The "Boson peak" near 50 cm^{-1} is clearly resolved at low temperatures, but already appears as a weak shoulder 50^0 above T_g. We have found that the predictions of the Martin-Brenig disordered crystal-model produce very poor agreement with the low-frequency low-temperature spectra of this and several other materials, particularly for the depolarizatiion ratio which, for LiCl-H$_2$O, is $\sim$0.8 and is independent of both frequency and temperature.

The low frequency Raman spectra of liquids have been successfully modelled by a collision-induced scattering theory intially developed for gases, and extended to fluids by Stephen.[16] Subsequently, much more work has been done on this problem, much of it by Madden and his coworkers who have made extensive use of molecular dynamics calculations.[17]

Stephen's theory is based on the dipole-induced-dipole (DID) mechanism, which is one of four interactiions included in Madden's analysis. We will follow Stephen's simpler approach here which always leads to a depolarization ratio of 3/4.

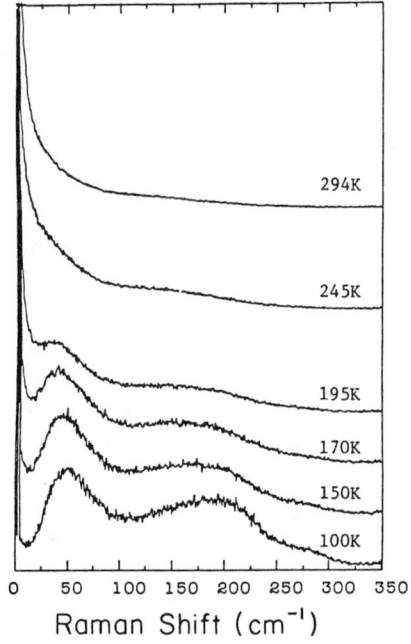

Figure 3. VV polarized Raman spectra of 15 mole % aqueous LiCl solution at temperatures between 294K and 100K ($T_g \sim 140K$).

In Stephen's analysis, the optical spectrum is given by
$I(K,\omega) = I_0^{(1)}(\hat{E}_S \cdot \hat{E}_0)^2 S(K,\omega) + I^{(2)}(K,\omega)$, where the $I_0^{(1)}$ term produces the usual first-order Brillouin spectrum, and

$$I^{(2)}(K,\omega) = I_0^{(2)}[1 + \frac{1}{3}(\hat{E}_S \cdot \hat{E}_0)^2] \times \int\int d^3q d\omega' S(q,\omega') S(K-q, \omega-\omega') \quad (8)$$

$\hat{E}_0$ and $\hat{E}_S$ are unit vectors in the direction of the incident and scattered electric fields and K is the magnitude of the scattering vector $K = k_0 - k_S$.

Eq. (8) gives the second-order optical spectrum as a convolution of $S(q,\omega')$ with $S(K-q, \omega-\omega')$. This second-order scattering process is indicated schematically in Fig. (2b). If we assume that the integral in Eq. (8) is dominated by modes with

$q \approx q_0$, then

$$I^{(2)}(K,\omega) = BS(q_0,\omega) \otimes S(q_0,\omega) \qquad (9)$$

As a second qualitative application of MCT, we have evaluated Eq. (9) using the simplified one-mode (or F_2) model first discussed by Leutheusser[2] and Bengtzelius *et al.*[1] in which Eq. (7) (for $q=q_0$) is reduced to $M_\Gamma(q_0,t) = V^{(2)}(q_0)\phi_{q_0}^2(t)$ so that Eq. (6) becomes:

$$\ddot{\phi}_{q_0}(t) + \omega_0^2\phi_{q_0}(t) + \gamma_0\dot{\phi}_{q_0}(t) + \int_0^t V^{(2)}(q_0)\phi_{q_0}^2(t')\dot{\phi}_{q_0}(t-t')dt' = 0 \qquad (10)$$

Eq. (10) is identical to Leutheusser's Eq. (1) if we set $V^{(2)}(q_0) = 4\Lambda\omega_0^2$. The ergodic to non-ergodic transition then occurs with increasing Λ at $\Lambda_c = 1$, where $\phi_{q_0}(t \to \infty)$, the non-ergodic fraction, increases discontinuously from 0 to 1/2.

We solved Eq. (10) numerically with $\gamma_0 = \omega_0 = 160 cm^{-1}$ for different values of Λ, obtaining results similar to Fig. (1) of Ref. (2). The convolution $S(q_0,\omega) \otimes S(q_0,\omega)$ in Eq. (9) was obtained from the computed $\phi_{q_0}(t)$ by squaring $\phi_{q_0}(t)$ and then performing a numerical FFT. The results are shown in Fig. (4) for values of Λ between 0.8 and 2.0. Comparison of the theoretical spectra of Fig. (4) with the experimental Raman data of Fig. (3) below 100 cm^{-1} indicates good qualitative agreement, including the low-frequency "excess Raman intensity". Although only qualitative, the evolution of the low frequency Raman spectrum predicted by combining Stephen's DID theory with MCT is intriguing, and suggests that the hydrodynamic theory of light scattering, when combined with a more realistic version of MCT, may well provide a unified explanation of the Raman spectra in both the liquid and glass phases.

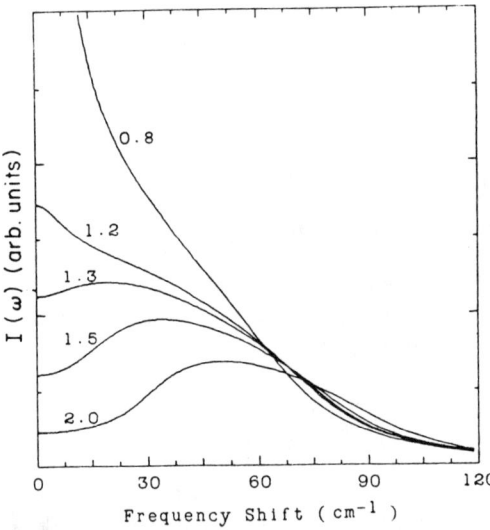

Figure 4. Theoretical Raman spectra predicted by Eq. (9). The convolution $S(q_0,\omega) \otimes S(q_0,\omega)$ was obtained by Fourier transformation of $\phi^2(q_0,t)$ found from Eq. (10) with values of Λ between 0.8 and 2.0. ($\Lambda_c = 1$)

COMBINED BRILLOUIN-RAMAN SPECTROSCOPY

In Fig. (5) we show polarized (a) and depolarized (b) 90° Brillouin spectra of CKN at T = 418K = T_g+85K.[18] In these spectra the central peak due to coupling of the LA mode to α-relaxation (a) has nearly disappeared under the Rayleigh line. Beyond the Brillouin lines in (b), additional weak depolarized scattering can be seen. When the free sprectral range of the interferometer was increased as shown in (c) and (d), the shape of the spectrum did not appear to change.

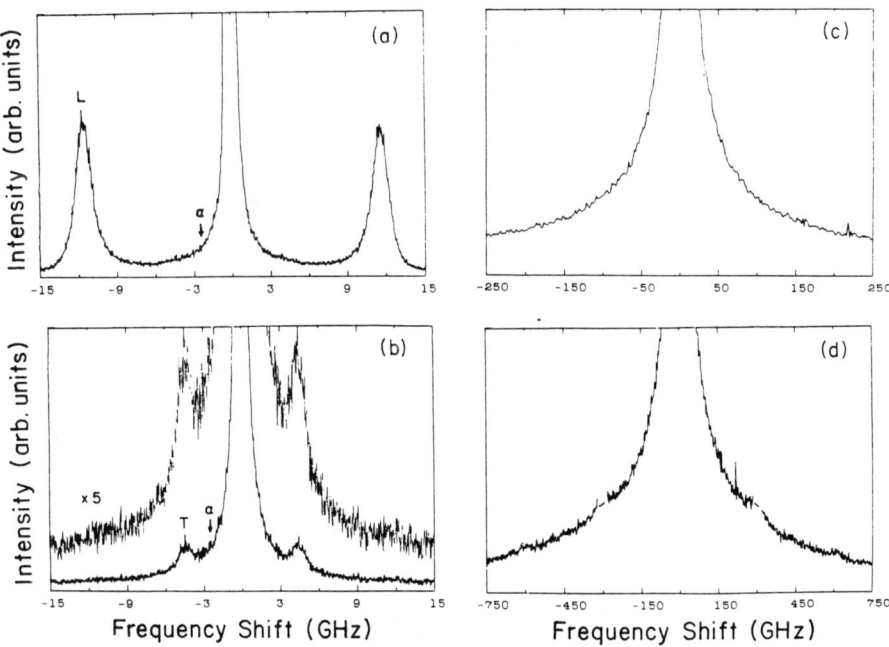

Figure 5. (a) Polarized and (b)-(d) depolarized right-angle Brillouin spectra of liquid $[Ca(NO_3)_2]_{0.4}-(KNO_3)_{0.6}$ at 418K. The longitudinal and transverse acoustic phonons and the central peak due to α relaxation are labeled in (a) and (b) by L, T, and α, respectively. In (c) and (d) the frequency range has been expanded relative to (b) by 17 times (c) and 50 times (d) to show the self-similar form of the β-relaxation spectra.

This scale invariance or spectral self-similarity implies a power-law spectrum $I(\omega) \propto \omega^{-(1-a)}$, in agreement with a central prediction of the MCT. To extend these measurements, we combined Raman spectra and interferometeric spectra recorded with four different free spectral ranges to produce the composite data shown on a log-log plot in Fig. (6). Note that this data covers a frequency range of more than four decades. The spectra were obtained in VH scattering geometry at $\theta = 173^0$ to avoid the contribution of the transverse acoustic mode at T< 400K, and therefore result from second-order scattering processes (which may include some contribution from NO_3 rotational dynamics).[11] The data shown in Fig. (6) was converted to an effective susceptibility $\chi''(\omega)$ by dividing $I(\omega)$ by the Bose factor $[n(\omega)+1]$. The results for $383 \leq T \leq 468K$ are shown in Fig. (7).

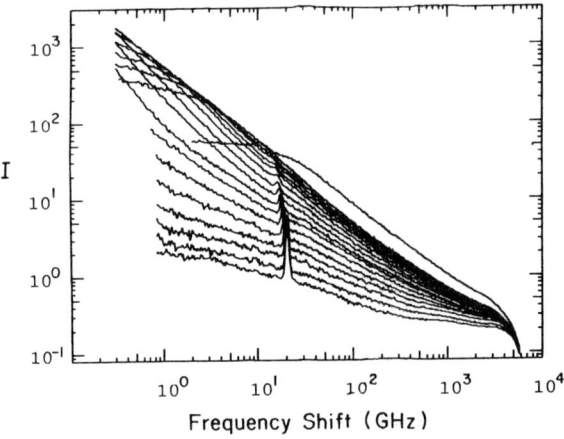

Figure 6. $I(\omega)$ vs ω for depolarized scattering from CKN. The data were obtained by combining spectra obtained with the tandem Fabry-Perot interferometer and Raman specta. Temperatures range from 578K (top) to 296K (bottom).

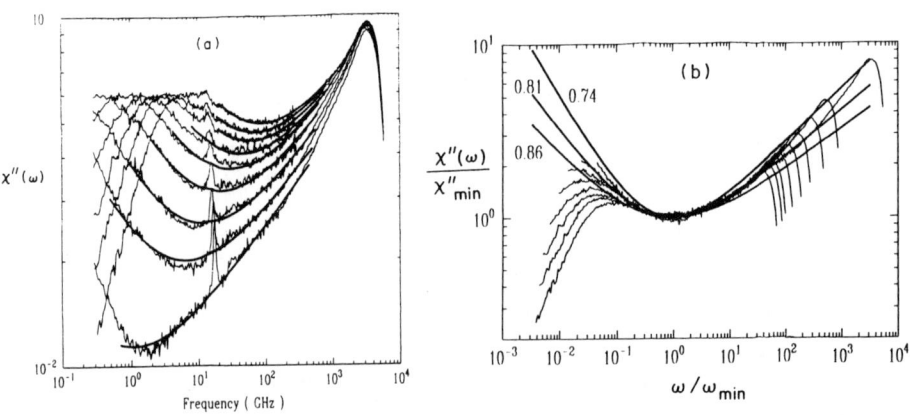

Figure 7. (a) $\chi''(\omega)$ vs ω for CKN obtained from the $I(\omega)$ data of Fig. 6 by division by the Bose factor $[n(\omega)+1]$. The solid lines are fits to Eq. (11) with $\lambda=0.815$, $a=0.27$ and $b=0.45$. (b) The data of (a) scaled by plotting $\chi''(\omega)/\chi''_{min}$ vs ω/ω_{min}. The solid lines are theoretical predictions for (top to bottom) $\lambda=0.74$, 0.81, and 0.86

The $\chi''(\omega)$ data of Fig. (7a) exhibit the characteristic low-frequency α-relaxation peak with a maximum at $\omega_\alpha = 1/<\tau_\alpha>$ which decreases rapidly with decreasing temperature, and a high-frequency peak at the microscopic frequency ω_0 (in Eq. (6)) which is weakly temperature dependent.

In the frequency range between these two peaks ($\omega_\alpha << \omega << \omega_0$) MCT predicts that the spectrum $I(\omega) = S(q_0,\omega)/S(q)$ should have two power law regimes: $I(\omega) \propto \omega^{-(1+b)}$ at low frequencies, and $I(\omega) \propto \omega^{-(1-a)}$ at high frequencies, separated by a crossover frequency $\omega_\epsilon = \omega_0|\epsilon|^{1/2a}$ where the separation parameter $\epsilon = (T_c - T)/T_c$.[19] The two critical exponents a and b are fixed by a single exponent parameter λ through

$$\Gamma^2(1-a)/\Gamma(1-2a) = \Gamma^2(1+b)/\Gamma(1+2b) = \lambda \quad (11)$$

where $\frac{1}{2} \leq \lambda < 1$, $0 < a < \frac{1}{2}$, $0 < b < 1$, and $\Gamma(x)$ is the gamma function.

In these two power-law regions, $\chi''(\omega) = \omega I(\omega)$ has the form ω^{-b} and ω^a which can be combined to form the approximate interpolation formula[20]

$$\frac{\chi''(\omega)}{\chi''_{min}} = [b(\frac{\omega}{\omega_{min}})^a + a(\frac{\omega}{\omega_{min}})^b]/(a+b) \quad (12)$$

Fits of Eq. (12) to the $\chi''(\omega)$ data of Fig. (7a), shown by the solid lines in the figure, gave $\lambda = 0.81$, a=0.273, b=0.459. The same data are replotted in scaled form in (7b), with $\lambda = 0.74$, 0.81, and 0.86 shown to indicate the sensitivity of the fit to λ.

MCT also predicts that χ''_{min}, the minimum of $\chi''(\omega)$, is proportional to $|\epsilon|^{1/2}$. We therefore plotted $(\chi''_{min})^2$ vs T which gave an excellent linear fit with T_c=378K, 45°K above T_g. Since the crossover frequency ω_ϵ is predicted to be proportional to ω_{min}, the frequency at which $\chi''(\omega) = \chi''_{min}$, MCT predicts that $\omega_{min} \propto (T-T_c)^{1/2a}$. In Fig. (8) we show ω_{min}^{2a} vs T, with a=0.273, again indicating a good linear fit. The frequency of the maximum of the α-relaxation peak, ω_α, is predicted by MCT to be $\omega_\alpha \propto (T-T_c)^\gamma$ with $\gamma = \frac{1}{2}(\frac{1}{a} + \frac{1}{b})$. In Fig. (8) we also show $\omega_\alpha^{1/\gamma}$ vs T, again indicating an excellent linear fit, with T_c=379K.

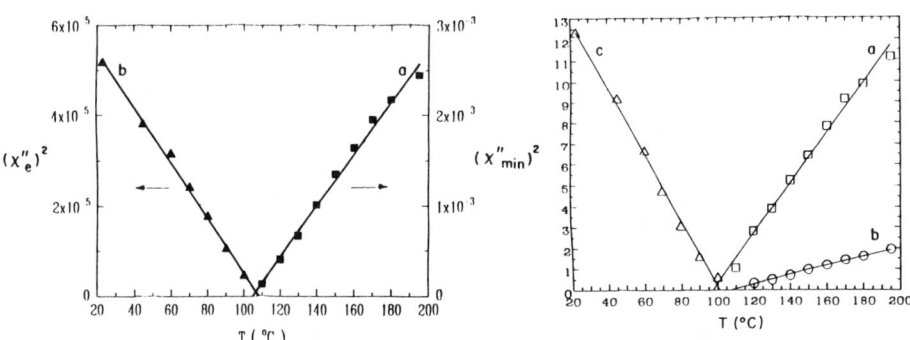

Figure 8. Left: (a) χ''_{min} vs T, (b) $(\chi''_\epsilon)^2$ vs T. Right: (a) ω_{min}^{2a}, (b) $\omega_\alpha^{1/\gamma}$, and (c) ω_ϵ^{2a} (T<T_c) vs T.

For T<T_c, our $\chi''(\omega)$ data (not shown in Fig. (7)) increase monotonically with ω since the α-relaxation occurs at frequencies below the lower limit of our resolution. Since MCT predicts that for T<T_c the $\chi''(\omega)$ data in the β-relaxation region should fall on a master curve if ω is scaled to ω/ω_ϵ and $\chi''(\omega)$ to $\chi''(\omega)/\chi''(\omega_\epsilon)$, we performed a computer search for optimum scaling, which gave $\omega_\epsilon(T)$ (apart from an unknown proportionality constant). In Fig. (8) we also show this scaling frequency for T<T_c plotted as ω_ϵ^{2a} vs T. Again, with a=0.273, linear T dependence is found, with T_c=377K. The agreement with MCT predictions is remarkable, strongly suggesting that T_c is a physically meaningful quantity.[21]

For the α-relaxation region, a reasonable approximation to the MCT prediction is:

$$\chi''(\omega) = (A\omega/T)\text{Re}\{\text{F.T.}[e^{(-t/\tau_\alpha)^\beta}]\} \qquad (13)$$

where β is a temperature-independent stretching constant. In Fig. (9a), we show a single depolarized CKN Brillouin spectrum at T=468K. In (9b) we show the susceptibility data $\chi''(\omega) = \omega I(\omega)$ for 8 spectra at temperatures between 468K and 393K, scaled as $\chi''(\omega)/\chi''_\alpha$ vs ω/ω_α. All the data fall on a single master curve, as predicted. The solid line through the data was obtained from Eq. (13) with β=0.55. Fits of the individual $\chi''(\omega)$ vs ω data at each temperature showed no systematic T-dependence of β.

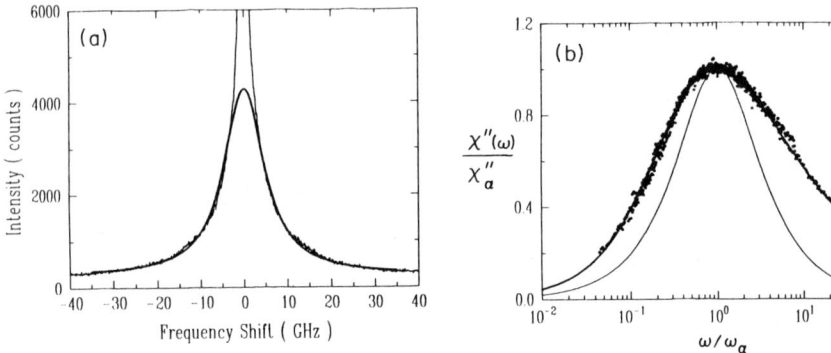

Figure 9. (a) Depolarized $\theta = 173°$ Brillouin spectra of CKN at 468K. The solid line is a Lorentzian fit with half-width of 5.4 GHz plus a background. (The $\chi''(\omega)$ vs ω plot of this spectrum has a maximum at $\nu_{\max} = 7.1$ GHz.) (b) $\chi''(\omega)/\chi''_\alpha$ plots of data like that in (a) at 8 temperatures from 468K to 393K. The solid line through the data represents stretched-exponential relaxation with β=0.55. The inner solid line represents Debye relaxation (β=1).

Note that in (a), a Lorentzian with half-width 5.4 GHz gives a good fit to the data, while in (b), the equivalent β=1 curve is clearly incompatible with the data. This difference illustrates the fact that $\chi''(\omega)$ provides a far more sensitive test of α-relaxation stretching than the direct $I(\omega)$ data, primarily because of the increase in dynamic range combined with the elimination of the influence of parasitic scattering near $\omega=0$.

To proceed further with the analysis of our data we should note that (1) the spectra shown in Fig. (6) result from a second order scattering process. In the spirit of our discussion of Raman scattering, the spectrum is given approximately by $I(\omega) \propto S(q_0,\omega) \otimes S(q_0,\omega)$ as in Eq. (9). (2) The simple F_2 model used so far is not adequate since it leads to the prediction that in the α-relaxation regime $\phi(q_0,t)$ is single exponential, i.e. $\beta=1$, which disagrees with the data.

We have therefore begun to implement the next simplest schematic model, Götze's F_{12} model, in which[3]

$$M(q_0, t) = \omega_0^2[\lambda_1 \phi(q_0 t) + \lambda_2 \phi^2(q_0, t)] \tag{14}$$

λ_1 and λ_2 are given in terms of the separation parameter ϵ and exponent parameter λ by Eqs. (3.3a) and (3.3b) of Ref. (22). In our calculations, we will evaluate $\phi(q_0,t)$ numerically from Eq. (6) using $M(q_0,t)$ of Eq. (14) with $\lambda = 0.811$ for which a = 0.273, b = 0.458. Squaring the resulting $\phi(q_0,t)$, Fourier transforming, and dividing by $[n(\omega)+1]$ may then produce an effective susceptibility similar to the plots shown in Fig. (7), although the F_{12} model may be too limited to properly describe CKN. This analysis is currently in progress and will be presented in a forthcoming publication. One piece of the analysis which is already complete concerns the α-relaxation region. Because of the convolution, the effective susceptibility in the low-frequency region should be given by Eq. (13), but with $e^{-(t/\tau_\alpha)^\beta}$ replaced by $e^{-2(t/\tau_\alpha)^\beta}$, or equivalently by $e^{-(t/\tau'_\alpha)^\beta}$ where $\tau'_\alpha = \tau_\alpha/2^{1/\beta}$.

Fitting our CKN data for $\chi''(\omega)$ in the α-relaxation regime to Eq. (13) therefore yields τ'_α, while the neutron spin echo data of Mezei et al.[4] give τ_α. Using the τ_α values reported by Mezei et al. along with their value of for $\beta = 0.58$, we have computed $\tau_\alpha/2^{1/\beta} = \tau_\alpha/3.3$. The results agree within experimental error with the τ'_α values found from our data.

ACKNOWLEDGEMENTS

We wish to thank W. Götze for many helpful suggestions concerning the data analysis, C.A. Angell for helpful discussions about glasses, and F. Mezei for generously providing unpublished CKN spin-echo data. This research was supported by the U.S. National Science Foundation under grant no. DMR-9014344.

References

[1] U. Bengtzelius, W. Gotze and A. Sjolander, J. Phys. C **17**, 5915 (1984).

[2] E. Leutheusser, Phys. Rev. A **29**, 2765 (1984).

[3] for a recent comprehensive review of MCT, see W. Götze, in **Liquids, Freezing and the Glass Transition**, (Les Houches, 1989), J.P. Hansen, D. Levesque and J. Zinn-Justin, editors (North Holland, 1991), p. 287.

[4] F. Mezei, W. Knaak and B. Farago, Phys. Rev. Lett. **58**, 571 (1987).

[5] L.T. Cheng, Y.X. Yan and K.A. Nelson, J. Chem. Phys. **91**, 6052 (1989).

[6] for a review, see N.J. Tao, G. Li and H.Z. Cummins, Phys. Rev. B **43**, 5815 (1991).

[7] R.D. Mountain, Rev. Mod. Phys. **38**, 205 (1966).

[8] J.D. Axe, S.M. Shapiro, G. Shirane and T. Riste, in **Anharmonic Lattices, Structural Transitions and Melting**, edited by T. Riste (Noordhoff, Leiden, 1974), p. 23. (Several excellent discussions of the central peak problem can be found in this volume.)

[9] M. Soltwisch, J. Sukmanowski and D. Quitmann, J. Chem. Phys. **86**, 3207 (1987).

[10] L. Börjesson, M. Elmroth and L.M. Torell, Chem. Phys. **149**, 209 (1990).

[11] N.J. Tao, G. Li and H.Z. Cummins (submitted to Phys. Rev. B, June 1991).

[12] V.K. Malinovsky and A.P. Sokolov, Solid State Commun. **57**, 757 (1986).

[13] A.J. Martin and W. Brenig, Phys. Status Solidi B **64**, 163 (1974).

[14] G. Winterling, Phys. Rev. B **12**, 2432 (1975). The existence of the low frequency excess Raman scattering has, however, been questioned by some authors. [K.B. Lyons, P.A. Fleury, R.H. Stolen and M.A. Bösch, Phys. Rev. B **26**, 7123 (1982).]

[15] N.J. Tao, G. Li, X. Chen, and H.Z. Cummins, Phys. Rev. A (in press).

[16] M.J. Stephen, Phys. Rev. **187**, 279 (1969).

[17] P.A. Madden and K. O'Sullivan, J. Phys. Condens. Matt. **2**, SA257 (1990).

[18] N.J. Tao, G. Li and H.Z. Cummins, Phys. Rev. Lett. **66**, 1334 (1991).

[19] Spectra exhibiting such behavior have been found in inelastic neutron scattering studies by W. Knaak, F. Mezei, and B. Farago, Europhys. Lett. **7**, 529 (1988), and by W. Doster, S. Cusack, and W. Petry, Phys. Rev. Lett. **65**, 1080 (1990).

[20] W. Gotze and L. Sjogren, J. Phys. C **21**, 3407 (1988).

[21] Evidence for a crossover near T_c has also been found in a study of diffusion by E. Rössler, Phys. Rev. Lett. **65**, 1595 (1990).

[22] W. Gotze and L. Sjogren, Z. Phys. B **65**, 415 (1987).

DYNAMIC STRUCTURE FACTOR NEAR THE GLASS TRANSITION: SPECIFIC FEATURES

F. Mezei
BENSC, Hahn-Meitner-Institut, Pf. 390128, D-1000 Berlin 39

ABSTRACT

The dynamic structure factor in the supercooled liquid state just above the glass transition temperature reveals a number of characteristic universal features in the so-called "fragile" class of glassy systems. A theory independent, unbiased analysis shows that, compared to the behaviour expected in the ordinary liquid state, one can observe: (a) a two-step structural relaxation process in time, the slower of which scaling with the bulk viscosity, (b) a relatively small, oscillatory wavenumber dependence of the relaxation dynamics, and (c) approximate master functions of unconventional shapes for both relaxation steps. These qualitative features are in agreement with predictions of recent mode coupling theories and, in addition, the quantitative behaviour is also compatible with the theoretically predicted scaling properties.

INTRODUCTION

Recent years have wittnessed a considerable, conjugated experimental and theoretical effort to understand the basic physics of the glass transition. The studies concentrated on the peculiar dynamics of atomic motions in glassformers around and primarily above the conventional glass transition temperature T_g in the supercooled liquid state. The observed dynamics represents archetypical examples of slow atomic motions as compared to the characteristic atomic collision times of 10^{-12}-10^{-13} sec. Their exploration by neutron scattering, which gives the most direct information in space and time, has been made possible by the evolution of high resolution spectroscopic methods, such as Neutron Spin Echo (NSE), which makes the 10^{-11}-10^{-8} sec time domain accessible.

The available bulk of neutron scattering data[1-6] on a large variety of samples of the so-called "fragile" glassformer cathegory[7] reveal a remarkable compatibility with predictions of recent mode coupling (MC) theories[8-10]. "Compatibility" means that by an adequate choice of parameters it is possible to fit experimental results within error to the predicted scaling relations. Unfortunately, the number of fit parameters is considerable, and therefore reduces the significance of these fits in view of the inherently limited statistical accuracy of neutron scattering data. A particular problem is the lack of a clear signature of the predicted critical temperature $T_0>T_g$. This enhances the ambiguity of the comparison of theory and experiment, although the absence of any real divergence at the presumed critical temperature is well explained by the presence of the activated, jump-diffusion type of motions[11].

It is the aim of the present analysis to point out the unusual features in the experimental results, which cannot be understood by analogy to the ordinary liquids or amorphous solid states[12]. It will be shown that these unusual qualitative features specific of the vicinity of the glass transition are well explained by the MC predictions, and in fact, no other theoretical interpretation is available as of today.

MAIN UNUSUAL EXPERIMENTAL FEATURES: "β" RELAXATION

In what follows, we shall analyse neutron scattering results[1,2,6] of K-Ca-NO$_3$(KCN), which appears to be a representative example of the variety of samples innvestigated by now.

One most remarkable experimental finding is the anomalous increase of the apparent density of states at energies below 5 meV starting at $T_g \simeq 60°C$ with increasing temperature, while the rest of the spectra remain unchanged (Fig. 1), as revealed by Time-of-Flight results. The proportional increase of the scattering intensity at various energies (lower two curves in Fig. 2) and also independent higher resolution NSE results show that this is not due to the broadening of the ω = 0 quasielastic central line, but to the increasing amplitude of a component with invariant line shape. The so-called β relaxation process predicted by MC theories corresponds to this observation with a lineshape $\omega^\alpha/\omega_\epsilon^\alpha$. Experimentally α is compatible with values around -0.7 in KCN, with a large error of 0.7, however. The sharp increase of the 50 μeV spectral density in Fig. 2 above some 90°C can be

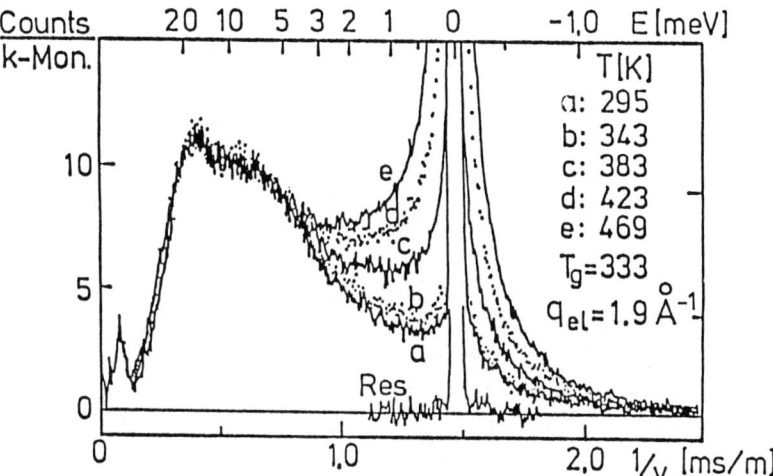

Fig. 1. Time-of-flight neutron scattering spectra in KCN at various temperatures[2]. The spectra are normalized by the Bose temperature factor in order to remove the part of the temperature dependence characteristic for constant density of states. The neutron energy change is indicated on the top scale.

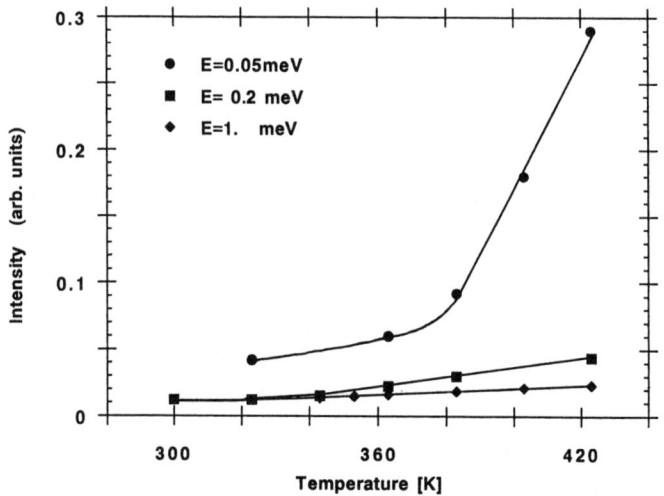

Fig. 2. Temperature dependence of the spectral intensities in Fig. 1. at two energies (lower curves) and at a smaller energy as observed by higher resolution backscattering spectroscopy. The upturn of the two lower curves around 333 K is due to the onset of the "β" relaxation, and that of the uppermost curve at about 365 K is due to the "α" relaxation. The lines are guides to the eye.

attributed to the beginning of the broadening of the $\omega = 0$ line (which is expected to have a complex lineshape) and it is also confirmed by the lack of similar upturn at higher energies, which in fact become affected by the line-broadening at higher temperatures only.

Another important feature of this anomalous scattering contribution, which we shall refer to as "β" relaxation in what follows, is that its relative amplitude $h(q)$ in the structure factor $S(q) \simeq \int S(q,\omega)d\omega$ is rather independent of the wavenumber in a broad range (Fig. 3) as determined by the NSE method (which actually measures this energy integral within the mainly relevant energy range of -1 meV to 4 meV). More precisely[13], $h(q)$ has a slight minimum at the maximum of the structure factor (also shown in Fig. 3) as predicted by MC. This relative q independence of $h(q)$ is of primary significance. Assumptions, that this process be due to some extra low energy excitations would lead to $h(q) \propto q^2$ or $q^2/S(q)$, respectively, depending on whether extended or localized modes are assumed. Due to the presence of several kinds of atoms in KCN the structure factor is rather substantial down to small q's of 0.1-0.2 Å^{-1}, i.e. only an order of magnitude smaller than the peak in the atomic density structure factor $S(q)$ at $q_0 \simeq 1.8$ Å^{-1} corresponding to the nearest atomic neighbours. A plateau ("prepeak") around 0.8 Å^{-1} corresponds to the physically really characteristic length of the system, the cation-cation (anion-anion) distance, i.e. to the peak of the charge density correlation function. This fortunate situation allows us to observe $h(q)$ at rather small q's, where the MC prediction is expected to be valid best. Wavenumbers of the order of q_0 and larger should be considered with caution, because at these values scattering contributions from local atomic vibrations become gradually dominant, and these cannot be expected to follow the universal predictions of MC theories.

Thus, the "β" relaxation process as observed here at $T \lesssim 90°$ C has the characteristic of a first, faster step of the structural relaxation, i.e. of the central $\omega = 0$ quasielastic line. This observation is only explained by MC theories up to now.

THE "α" RELAXATION PROCESS

The other remarkable unusual feature is the lineshape behaviour of the bulk (second step) of the structural relaxation at

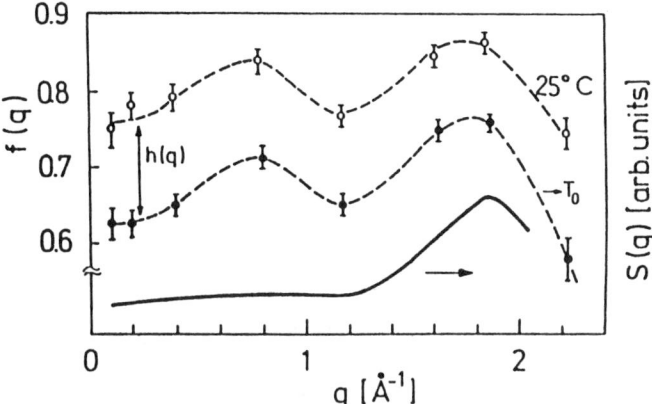

Fig. 3. Wavenumber dependence of the relative amplitudes of the "α" and "β" relaxation steps in KCN f(q) and h(q), respectively. h(q) is obtained as the difference between f(q) at room temperature (where no "β" relaxation component can be observed and the "α" relaxation is practically infinitely slow) and at the estimated critical temperature T_o, above which the two processes overlap significantly[13].

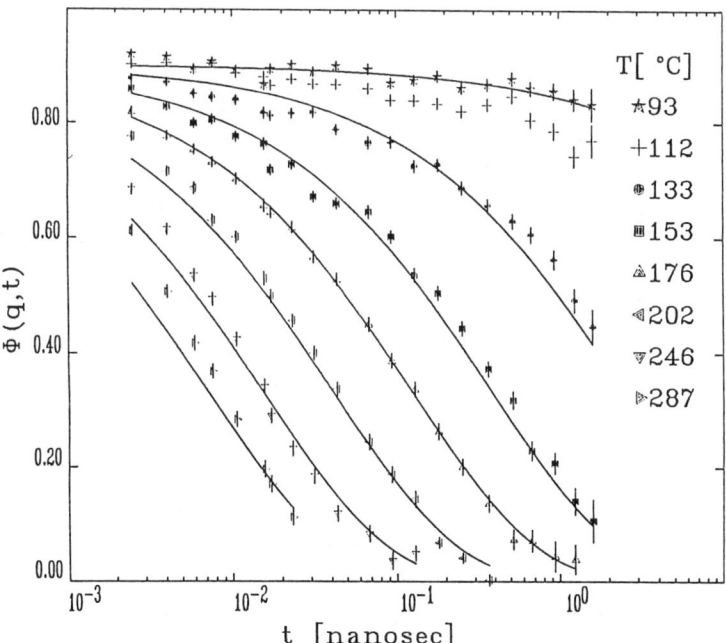

Fig. 4. Extended range NSE results on the temperature dependence of the time relaxation function (intermediate scattering function) at the peak of the structure factor q_o in KCN. The lines were obtained by shifting the time scale of a fit to the 176°C data proportionally with the change of bulk viscosity[6]. Note that the uppermost line corresponds to 112°C.

higher temperatures, the "α" process, as observed by NSE spectroscopy directly in the time domain at $q \sim q_0$ (Fig. 4).

The lines in the figure represent a common lineshape function of the Kohlrausch type

$$\Phi(q,t) = f\, e^{-(t/\tau)^\beta}$$

where τ was taken proportional to the measured bulk viscosity $\tau = B\eta$ and the parameters f, B, and β have been chosen to best fit the T=176° C curve (f=0.9 and β=0.58). This master function, which is very different from the usual exponentiql decay (Lorentzian spectral lineshape: β=1, quite well describes most of the data, with clear deviations at both the highest and lowest temperatures (the upper line is calculated for 112°, not for 93° C!). No more permissive fitting would completely cure these "problems", however common fits to all temperatures give mathematically very small errors for f and β which are often heavily misinterpreted. A critical look at the errors on the lineshape parameters f and β, which were obtained by individually fitting the data at different temperatures, independently of other temperatures, is rather sobering: 3σ errors (99% confidence level) run to some 0.05 and 0.12, respectively, at best. By taking the best common value of $f \simeq 0.90$, it is clear that the other lineshape parameter β become significantly temperature dependent. One reason for this is the admixture of the "β" relaxation process which is expected to follow a steeper initial slope (von Schweindler regime) than the Kohlrausch curve at the smallest times. This is well signified by the faster drop of the 112° C data points in Fig. 4. than the calculated line. The non-zero slope of the 93° C data is entirely due to the "β" relaxation process, assuming that τ continues to scale with η and that the exponent β remains larger than 0.3, which is not quite obvious in Fig. 5. Adding further parameters to the fit in order to include the "β" contribution would blow up the errors, and the results of an analogous attempt[6] to correct for this "α" and "β" mixing is shown in the bottom part of Fig. 5. A constant lineshape with $\beta \simeq 0.61$ now becomes possible over a temperature range of 100 °C, however this hardly looks like a prove. Note, that the temperature dependence of β observed in macroscopic and light scattering measurements is much more rapid[14, 15], and this discrepancy between the two kinds of results remains, no matter what kind of fit is used.

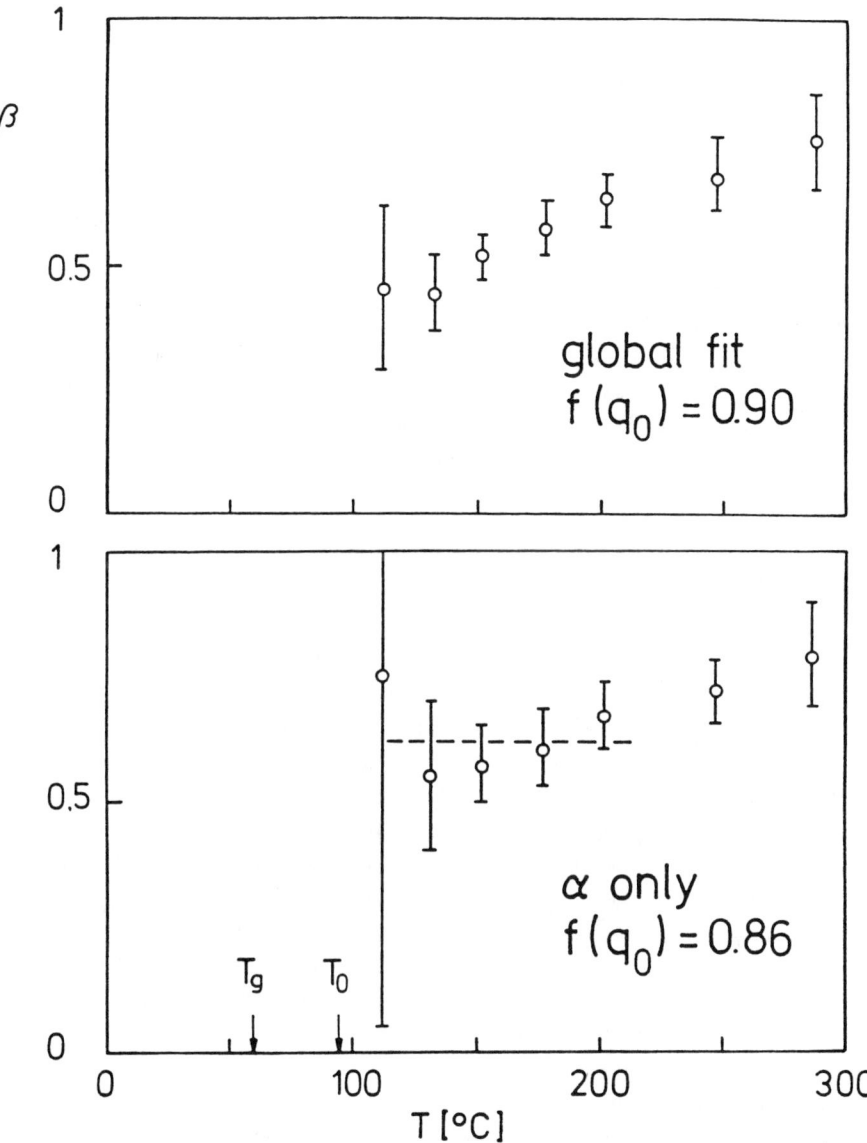

Fig. 5. Temperature dependence of the Kohlrausch lineshape exponent β obtained by fitting the data in Fig. 4. at each temperature independently, but assuming a temperature independent amplitude f. Top: all data points considered; bottom: data points strongly influenced by the "β" relaxation step disregarded[6]. The error bars indicate 3σ deviations, i.e. 99 % confidence level.

MC theories predict an assymptotically constant lineshape of the "α" relaxation process upon approaching a critical temperature T_0, which is estimated to be at about 95 °C in this case. It is seen, that the results are not incompatible with this assumption, however they do not lend much support either, except for the fact that the lineshape is strongly non-exponential. This very unusual prediction for a simple liquid without a network structure has also been suggested by other considerations[14]. Another theoretical expectation, the relative q-independence of the effective relaxation times is well born out, with a pronounced de Gennes type narrowing effect, however, around the wavenumber of the charge density correlation peak (~ 0.8 Å^{-1}), cf. Fig. 6, showing that this wavenumber indeed corresponds to the real characteristic dynamic length scale of the system. Usually this narrowing occurs at the peak of the structure factor instead.

CONCLUDING REMARKS

In quantitative respects the theory predicts a number of scaling relations between exponents describing the "β" relaxation lineshape, the temperature dependence of ω_ϵ and of η or τ. These relations can be forced to be fulfilled[16] by an appropriate choice of T_0. This, however, is again a compatibility check only, and provides little extra evidence for the theory. The problem is, that around T_0 thermally activated jump processes dominate the structural relaxation and no clear and reliable signature of T_0 remains. Actually, one should not try to fit data around T_0 by MC predictions, because just around this temperature theory breaks down. Including the jump diffussion explicitely, on the other hand, would dramatically increase the number of free parameters and thus the ambiguity of interpretation.

In sum we can conclude that MC theory - by now alone - correctly predicts the strikingly new phenomenon of a <u>two step structural relaxation</u> dymamics in supercooled liquids just above the glass transition, and correctly describes the type of lineshape for both of these two processes at temperatures where they do not overlap in time. This is a strong qualitative evidence in favour of the theory. More quantitative comparisons between theory and experiment would only be meaningful, if total lineshape ("α" plus "β") calculations were available with a reduced number of free parameters. Since such type of results go beyond the

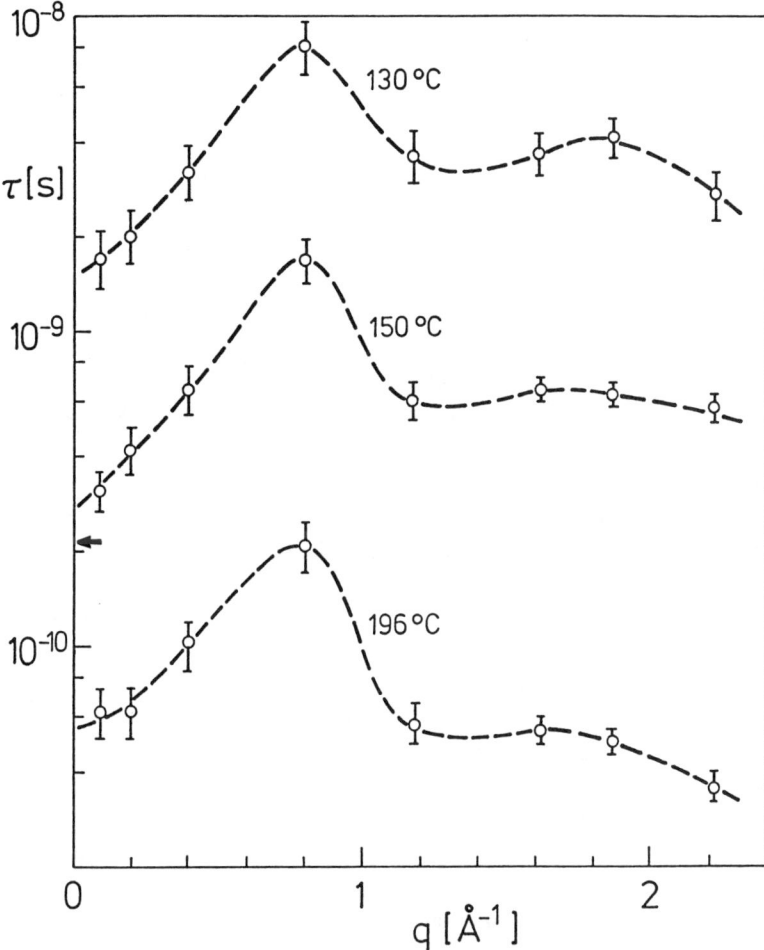

Fig. 6. Wavenumber dependence of the effective relaxation time τ of the "β" relaxation step in KCN[1]. The arrow indicates the value obtained at 150°C and $q=0$ by Brillouin light scattering[15]. The lines are mere guides to the eye.

universal features of the theory and become model dependent, realistic molecular dynamic calculations might be necessary.

REFERENCES

1. F. Mezei, W. Knaak and B. Farago, Phys. Rev. Lett. **58**, 363 (1987), and Phys. Scripta **19**, 571 (1987).
2. W. Knaak, F. Mezei and B. Farago, Europhys. Lett. **7**, 529 (1988).
3. D. Richter, B. Frick and B. Farago, Phys. Rev. Lett. **61**, 2465 (1988) and **64**, 2921 (1990).
4. F. Fujara and W. Petry, Europhys. Lett. **4**, 921 (1987); W. Petry, M. Kiebel and H. Sillescu, in Dynamics of Disordered Materials, D. Richter et. al. eds. (Springer Verlag, Heidelberg, 1989), p. 58.
5. W. Doster, S. Cusack and W. Petry, ibid. p. 120.
6. F. Mezei, J. Non-Cryst. Solids, **131-133**, 317 (1991).
7. C.A. Angell, J. Phy. Chem. Solids, **49**, 863 (1988).
8. E. Leutheusser, Phys. Rev. **A29**, 2765 (1984); U. Bengtzelius, W. Götze and A. Sjölander, J. Phys. C. **17**, 5915 (1984); S.P. Das, G.F. Mazenko, S. Ramaswamy and F.F. Toner, Phys. Rev. Lett. **54**, 118 (1985).
9. W. Götze, in Liquids, Freezing and the Glass Transition, J.P. Hansen, D. Levesque and J. Zinn-Justin, eds. (North Holland, Amsterdam, 1991), p. 287.
10. W. Götze and L. Sjögren, in this volume; L. Sjögren and W. Götze, in this volume.
11. L. Sjögren, in Basic Features of the Glassy State, J. Colmenero and A. Alegria, eds. (World Scientific, Singapore, 1990), p. 297
12. F. Mezei, in Liquids, Freezing and the Glass Transition, J.P. Hansen, D. Levesque and J. Zinn-Justin, eds. (North Holland, Amsterdam, 1991), p. 629.
13. F. Mezei, in Dynamics of Disordered Materials, D. Richter et. al. eds. (Springer Verlag, Heidelberg, 1989), p. 164.
14. K.L. Ngai, A.K. Rajagopal and C.Y. Huang, J. Appl. Phys. **55**, 1714 (1984).
15. M. Grimsditch and L.M. Torell, in Dynamics of Disordered Materials, D. Richter et. al. eds. (Springer Verlag, Heidelberg, 1989), p. 196.
16. F. Mezei, in: Basic Features of the Glassy State, J. Colmenero and A. Alegria, eds. (World Scientific, Singapore, 1990), p. 152

INFLUENCE OF THERMODYNAMIC INTERACTIONS ON THE TRANSPORT PROCESS OF MOLECULAR TRACER IN BINARY POLYMER MIXTURES

Q. Tran-Cong, Y. Ishida, K. Meisyo, O. Yano and T. Soen
Department of Polymer Science and Engineering
Kyoto Institute of Technology, Matsugasaki, Sakyo-ku,
Kyoto 606, Japan

ABSTRACT

Diffusion-controlled reaction kinetics of a molecular tracer was investigated in the miscible region of binary polymer mixtures. The reaction kinetics does not follow the Kohlrausch-Williams-Watts(KWW) mechanism and is controlled by polymer segmental free volumes in the context of classical free-volume theory. Upon approaching the phase boundary, it was found that the reaction rates were affected by the concentration fluctuations with the wavelengths comparable to the size of the tracer in accord with the results obtained by small-angle neutron-scattering experiments.

INTRODUCTION

Information on the transport processes of small molecules in the bulk state of polymers is not only important for the understanding of the correlation between the polymer local structures and the corresponding relaxation process such as physical aging but can also provide some insights into improving the long-term stability of polymers used for practical purposes such as nonlinear optical materials. So far the transport process of small molecules in bulk homopolymer matrix has been extensively investigated and experimental data have been analyzed, for most cases, in the context of free volume theory[5]. In this work, the diffusion-controlled reaction of a molecular tracer in the miscible region of binary polymer mixtures is examined together with the effects of thermodynamic interactions in the blends on this reaction kinetics.

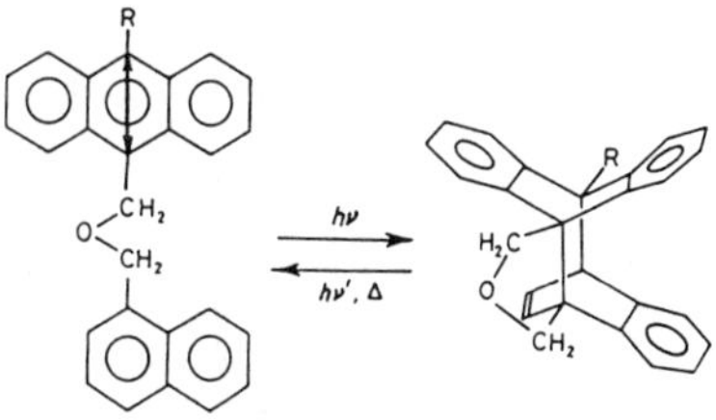

Fig. 1. Intramolecular photo-dimerization of HNMA. R=-CH$_2$OH.

I. Experimental Section:

A) Samples:

The characteristics of polymers used in this work are summarized in Table 1. The molecular tracer 9-(hydroxymethyl)-10-[(Naphthylmethoxy)methyl]anthracene (HNMA, Figure 1) was synthesized according to the procedure described previously[1]. In liquids and bulk polymer matrices, the intramolecular photodimerization of HNMA is known as a diffusion-controlled reaction[2]. HNMA is dispersed in these blends at the concentration ca. 10^{-4} mole/l. Samples are obtained by casting benzene solutions of these polymer

Table 1. Characteristics of polymers used in this work.

	Mw	Mw/Mn
PS	1.4x10^5	1.3
PVME	1.2x10^5	2.5
PEO	9.8x10^4	1.1
PMMA	1.4x10^5	1.7

PS:Polystyrene; PVME: Poly(vinyl methyl ether); PEO: Poly(ethylene oxide); PMMA: Poly(methyl methacrylate).

mixtures and subsequently dried under vacuum over several nights prior to the experiments.

B. Experimental Procedure:
Irradiation was performed with a 250W mercury lamp. The change in absorbance at 393.5nm of the anthracene moieties of HNMA was followed after different irradiation times by using a UV photometer. Decay curves of absorbance were analyzed by fitting the experimental data to appropriate model functions using the nonlinear least square regression program SALS.

C. Experimental Results and Discussion:
1) Phase behavior of PS/PVME and PEO/PMMA mixtures:
The cloud point curves of PS/PVME and PEO/PMMA mixtures were measured by light scattering with the heating rate 0.2°C/min. The melting points of PEO/PMMA blends were measured by differential scanning calorimetry (DSC) with a heating rate of 5°C/min and were confirmed by depolarized scattering. The phase behavior of these two mixtures is shown in Figures 2 and 3. Above the melting points, PEO/PMMA mixtures are found to be miscible in the temperature range of this experiment (up to 200°C).

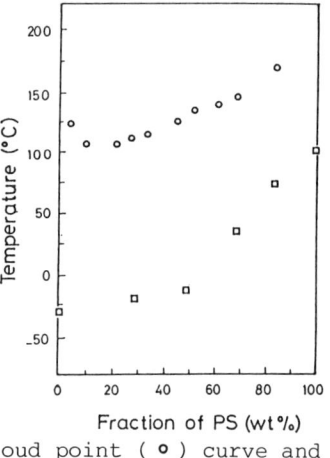

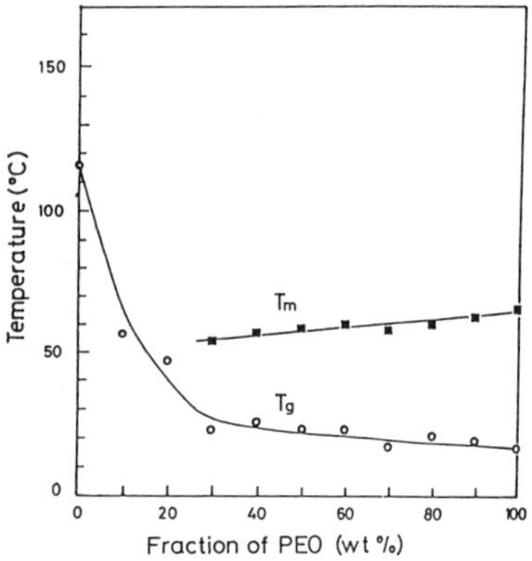

Fig.2. Cloud point (○) curve and glass transition temperature (□) of PS/PVME mixtures.

2) Reaction kinetics and temperature dependence of the intramolecular photodimerization of HNMA:
The time dependence of absorbance of HNMA on irradiation times does not follow the first-order kinetics. As an example, the decay obtained for PEO/PMMA(10/90) at 100°C is shown in Figure 4. An attempt to fit these decay curves to the KWW (stretched exponential) was failed. However, these decay processes can be well represented by the sum of two exponential functions of time:

$$OD(t) = OD_0 \{ A \cdot \exp(-k_1 t) + B \cdot \exp(-k_2 t) \}$$

where OD_0, A and B are the initial absorbance, the fractions of the fast decay process (k_1) and the slow decay process (k_2) respectively.

Taking into account that the slow rate constant k_2 is about 3 orders of magnitude smaller than k_1 and does not significantly

Fig. 3. Melting point (□) and glass transition temperature (○) of PEO/PMMA mixtures.

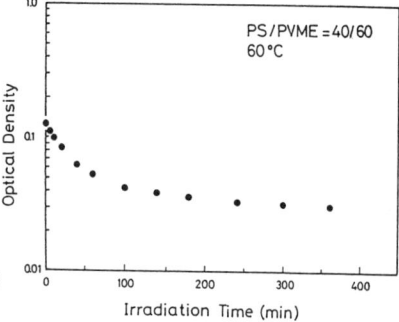

Fig.4. Dependence of optical density of HNMA on irradiation time.

depend on temperature, only the reaction rate k_1 is used for further analysis.

The temperature dependence of k_1 does not follow the Arrhenius behavior. k_1 dramatically drops upon approaching the glass transition temperature of the blends. In order to elucidate the correlations between polymer segmental relaxation and the conformational transitions of HNMA, the empirical Williams-Landel-Ferry (WLF) relationship[3] was used to interprete the temperature dependence of k_1. For this purpose, the dimensionless reaction rate $a_T = k_1(T_0)/k_1(T)$ is plotted against $1/(T - T_0)$ where T_0 is a reference temperature. The linear relationship observed from this plot implies that the reaction kinetics of HNMA in these blends is controlled by polymer segmental free volumes in the context of the WLF relationship. It is worth noting that the similar behavior has been observed for the temperature dependence of the translational diffusion of anthracenophane, a homologue of HNMA in bulk polymer matrices obtained by the forced Rayleigh scattering[4]. The fractional free volumes and thermal expansion coefficients estimated from the WLF plot (Fig. 5) are somewhat different from those obtained by rheological experiments[3]. There are several reasons responsible for this inconsistency such as free-volume distribution[5], the difference in the frequency or time domain probed by different experimental techniques. This is not clearly understood at this moment.

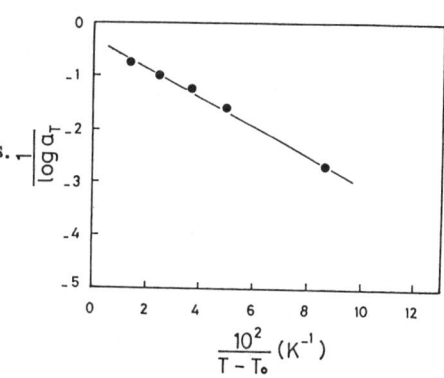

Fig.5. WLF plot for k_1 of HNMA in PS/PVME(40/60) with T_0= 40°C.

3) <u>Influence of thermodynamic interactions</u>:

At constant temperature, the reaction rate k_1 decreases with increasing the composition of the polymer component with higher T_g (PS for PS/PVME and PMMA for PEO/PMMA blends). To compensate the effects of the variation of T_g with the

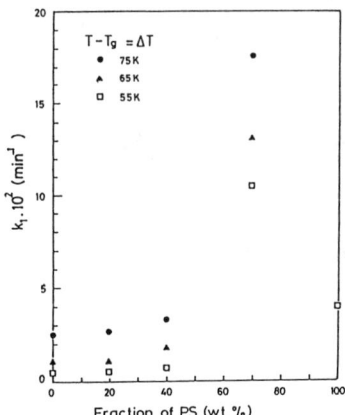

Fig.6-a. Composition dependence of k_1 in PS/PVME mixtures under the condition $T - T_g$ = constant.

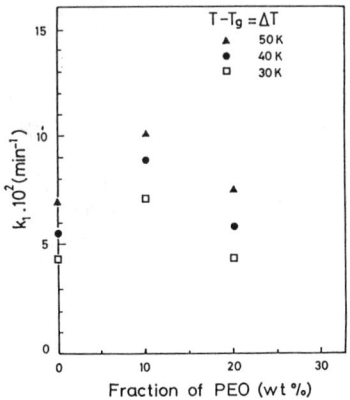

Fig.6-b. Composition dependence of k_1 in PEO/PMMA mixtures under the condition $T - T_g$ = constant.

blend composition on the reaction kinetics, k_1 is plotted versus the composition in Figures 6-a and 6-b for PS/PVME and PEO/PMMA blends under the condition (T-Tg)= constant. For the case of PS/PVME blends, k_1 is almost unchanged up to 40wt% of PS. Above 40wt%, k_1 markably increases with increasing PS composition and eventually drops to the value close to that of HNMA in pure PVME. By comparing these results with the correlation lengths ξ of PSD/PVME obtained by small-angle neutron scattering (SANS)[6], it was found that at the conditions under which the onset of SANS critical scattering is observed, k_1 exhibits the sudden change and ξ obtained from SANS experiments is comparable to the size of HNMA. On the other hand, for PEO/PMMA mixtures, k_1 does not show significant dependence on PEO composition up to 20wt% (amorphous region). Taking into account that the experiment temperatures are close to the phase boundary of PS/PVME and far away from that of PEO/PMMA mixtures[7], the composition dependence of k_1 in these two mixtures indicates that the intramolecular photodimerization of HNMA reflects the concentration fluctuations with the wavelengths comparable to the size of HNMA (ca. 20Å - 30Å). Upon approaching the phase boundary of PS/PVME blends, the wavelengths as well as the amplitudes of the concentration fluctuations in the mixtures become larger. As the concentration fluctuations with the wavelengths larger than the size of HNMA exist, there may be some HNMA in the PVME-rich regions and some in the PS-rich regions. Photodimerization in the former case will contribute to the fast decay process (k_1) whereas the reaction in the later case will affect the slower decay process. That explains why the rate constant k_1 remarkably increases as the PS composition exceeds 40wt%.

Acknowledgment: The financial support from the Ministry of Education, Science and Culture, Japan for Encouragement of Young Scientists (to Q. T-C, Grant-In-Aid No. 01750824) is gratefully appreciated.

REFERENCES AND NOTES

1) Q. Tran-Cong, T. Kumazawa, O. Yano and T. Soen Macromolecules 23, 3002 (1990).
2) Q. Tran-Cong, H. Yoshizawa, K. Ashikaga and M. Yamamoto Am. Chem. Soc. Polym. Prepr. Div. Polym. Chem. 28(2), 70 (1987); Polymer 30, 534 (1989).
3) J. D. Ferry "Viscoelastic Properties of Polymers", 3rd. Ed., John Wiley, New York (1980), Chapter 11.
4) H. Kim, D. A. Waldow, C. C. Han, Q. Tran-Cong and M. Yamamoto Polym. Commun. 32, 108 (1991).
5) For example, see K. Horie and I. Mita Adv. Polym. Sci. 88, 77 (1989).
6) C. C. Han, B. J. Bauer, J. C. Clark, M. Muroga, M. Okada, M. Matsushita, Q. Tran-Cong, T. Chang and I. C. Sanchez Polymer 29, 2002 (1988).
7) Recently, there are some works reporting that PEO/PMMA blends undergo phase separation around 250°C (For example, A. C. Fernades, J. W. Barlow and D. R. Paul J. Appl. Polym. Sci. 32, 5481 (1986)).

NEUTRON SCATTERING EXPERIMENTS NEAR THE GLASS-TRANSITION OF POLYBUTADIENE

R. Zorn, D. Richter
Forschungszentrum Jülich, IFF, Postfach 1913, D-5170 Jülich, Germany

B. Frick, B. Farago
Institut Laue-Langevin, BP 156X, F-38042 Grenoble Cedex, France

ABSTRACT

Recently, the mode-coupling theory (MCTh) of Götze et al.[1] has gained much importance in the interpretation of experimental data from glassy systems. Here, we present inelastic and quasi-elastic neutron scattering experiments viewed under the aspect of this theory. The experiments have been carried out on polybutadiene with random microstructure. A main result emerging from different experiments is the existence of a transition temperature T_c in the range 213...220 K well *above* the conventionally defined glass-temperature $T_g = 181$ K. Around this temperature power-law singularities of the types predicted by MCTh appear in the non-ergodicity parameter of the glass, the minimum of the susceptibility, and the viscosity.

MATERIAL AND METHODS

The polymer investigated in this work is fully deuterated *cis-trans*-vinyl (47 : 46 : 7) polybutadiene $(-CD_2-CD=CD-CD_2-)_n$. It contains randomly distributed C-bond orientations (*cis* and *trans*). For this reason crystallisation is prevented, and the system is known as a good glass former. The molecular weight is $M_w = 93\,000$ with a polydispersity $M_w/M_n = 1.03$. The glass transition temperature was determined to 181 K with a width of 2 K by calorimetric measurements. Diffraction experiments have shown that the static structure factor shows a first peak corresponding to interchain correlations at 1.48 Å^{-1} followed by a minimum at 2.1 Å^{-1} (at 160 K)[2]. The use of fully deuterated molecules results in predominant coherent scattering originating from two-particle correlations which are the primary subject of MCTh.

Two types of scattering experiments have been performed to explore the dynamics in a temperature range 2...280 K: 1. energy-resolving methods like time-of-flight technique (TOF) which give the scattering function $S(Q,\omega)$ and 2. neutron-spin-echo spectroscopy (NSE) producing the normalized intermediate scattering function $S(Q,t)/S(Q,0)$. The latter is the Fourier transform and reflects the decay of correlations in the time domain.
Representative for the first type we will present data only from the TOF spectrometer IN6 at the ILL, Grenoble. The NSE measurements have all been carried out on the IN11 at the same institution. The two instruments do not only differ in the type of data obtained but also in their dynamical range. While the time range of 9 ps...2.8 ns covered by NSE corresponds to extremely low energy transfers 70 μeV...0.2 μeV the energy range of IN6 extends from the lower limit given by the extension of the resolution function at about 100 μeV to some meV.

RESULTS

Fig. 1 shows NSE spectra obtained at 1.48 Å^{-1} i.e. near the maximum of the static structure factor[3]. The data are well discribed by a Kohlrausch law $f \cdot \exp(-(t/\tau)^\beta)$. Since $S(Q,t)/S(Q,0)$ does not extrapolate to 1 for $t \to 0$ but to a value $f = 0.917 \pm 0.008$ distinctly smaller a second relaxation process must take place at times smaller than those observed by NSE but within the bandpass of the instrument which extends to about 10^{-13} s (≈ 4 meV).

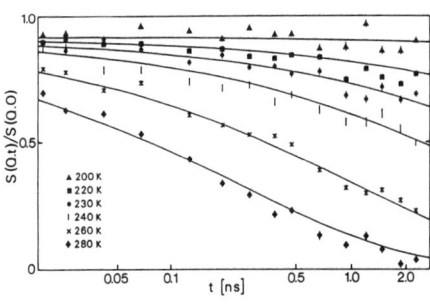
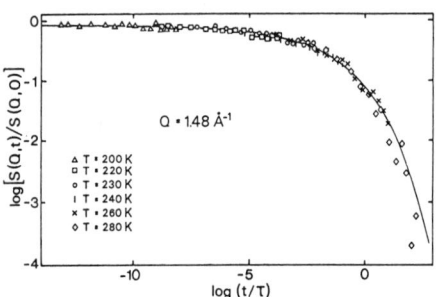

Fig. 1. Neutron-spin-echo spectra at $Q = 1.48$ Å^{-1} for various temperatures. The solid lines are the result of a combined fit by a Kohlrausch law.

Fig. 2. NSE data for $Q = 1.48$ Å^{-1} with the time rescaled according to a macroscopic viscosity measurement.

Spectra of different temperatures can be rescaled to a single master-curve using the macroscopic viscosity data. Here, the Vogel-Fulcher expression for the monomeric friction coefficient (which is proportional to the viscosity) $\zeta = \zeta_0 \exp(1/\alpha(T-T_0))$ with $\zeta_0 = 1.26 \times 10^{-11}$ dyne s/cm, $T_0 = 128$ K, and $\alpha = 7.12 \times 10^{-4}$ K^{-1} given by Berry et al[4] has been applied. Fig. 2 presents the result of rescaling the time to t/τ with $\tau = \tau_0 \cdot \zeta(T)/T$ where τ_0 is a prefactor common to all temperatures*. For more than 7 orders of magnitude in the rescaled time the NSE data follow a Kohlrausch law.

In order to investigate the wave vector dependence of the α relaxation parameters NSE spectra at different Q values for the same temperature have been taken (Fig. 3)[3]. When compared to the static structure factor $S(Q)$ they show the qualitative behaviour predicted by MCTh: $f(Q)$ has its maximum at the peak of the structure factor while $1/\tau(Q)$ goes through a minimum there. The latter fact can also be understood in terms of the De Gennes narrowing: Correlations which are preferred by the system, build up the maximum in $S(Q)$ and consequently decay more slowly[8].

According to MCTh, in the idealized case the α relaxation should arrest at a certain critical temperature T_c when cooling down. However, in reality this relaxation will persist below T_c due to additional ergodicity-restoring processes like thermally activated hopping which are not included in the simpler version of the theory. But in any case a prediction of the non-ergodicity parameter remains valid:

$$f(Q,T) = f^0(Q) + \begin{cases} h(Q) \cdot \sqrt{\epsilon} + O(\epsilon) & \text{for } T < T_c \\ O(\epsilon) & \text{for } T > T_c \end{cases} \quad (1)$$

where $\epsilon = (T_c - T)/T_c$. Here, f either takes the rôle of the pre-exponential factor in the Kohlrausch law or the constant value of $S(Q,t)/S(Q,0)$ if the relaxation is completely arrested (at least in the time-range of the experiment). Thus, in contrast to the scaling plots above, the NSE data here either were fitted with *free* f or the level of the plateau was taken as f when there was no observable relaxation. The compilation of the results of different runs show the stated behaviour with $T_c = 216 \pm 1$ K (Fig. 4)[6]. The same figure shows the viscosity data[4] fitted with the power law dependence resulting from MCTh[5]: $\zeta(T) = A(T - T_c)^\gamma$, yielding $\gamma = -3.2$ and $T_c = 214 \pm 3$ K, to demonstrate the coincidence of T_c.

* It has to be noted that only data from temperatures $T \geq 200$ K have been included here. Recent measurements[7] show that a decoupling of microscopic and macroscopic time-scales occurs at lower temperatures. This means that a time-temperature superposition of the NSE is still possible at lower temperature but the characteristic time is no longer proportional to $\zeta(T)/T$.

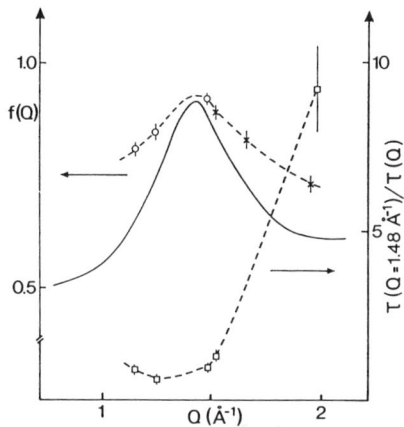
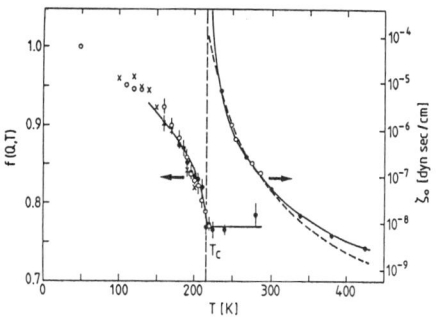

Fig. 3. Q dependence of the α relaxation strength $f(Q)$ and the ratio of the relaxation rates $\tau(Q = 1.48\,\text{Å}^{-1})/\tau(Q)$. The solid line represents the static structure factor $S(Q)$ at 230 K.

Fig. 4. Left-hand side: temperature dependence of the non-ergodicity parameter $f(Q,T)$ at $Q = 1.88\,\text{Å}^{-1}$, solid line: result of a fit with equation (1). Right-hand side: monomeric friction coefficient of PB, solid line: fit by power law, broken line: fit by Vogel-Fulcher law.

The TOF data have mainly be used to calculate the dynamic susceptibility $\chi''(Q,\omega) = \frac{\omega}{k_B T} S(Q,\omega)$. This calculation involves several steps: The raw data were converted to constant-Q scans, the elastic line has been subtracted, and finally phonon contributions which are supposed not to be included in the MCTh have been subtracted. For this purpose, data taken at 120 K were scaled up to the actual measuring temperatures by Bose and Debye-Waller factors. Fig. 5 shows the results in double logarithmic presentation.

MCTh states that a minimum in $\chi''(\omega)$ should occur at the frequency $\omega_{\min}$ where the α relaxation crosses over to the faster β relaxation. The minimum position should obey the following power laws asymptotically when T approaches T_c:

$$\omega_{\min} \sim |\epsilon|^{1/2a} \qquad (2)$$
$$\chi''_{\min} \sim |\epsilon|^{1/2} \qquad (3)$$

where ϵ is defined as above and a is a material dependent parameter in the range $0 < a \lesssim 0.39$.

To investigate the scaling prediction (3) the squares of the minima of the susceptibility have been plotted vs. temperature in Fig. 6. Those should be proportional to $|\epsilon|$ and thus form straight lines intersecting the T axis at T_c. One can see that linearity is fulfilled only in the range 220...250–260 K with $T_c \approx 213$ K in agreement with the values derived above. The deviation for large T can be explained by a breakdown of the validity of the asymptotic laws. The fact that $\chi''_{\min}$ does not tend to zero when T goes below T_c is obviously related to the non-ideality of the transition mentioned above.

We note that the low frequency part of the spectrum may be affected by uncertainties in the subtraction of the elastic line. So the shift in $\omega_{\min}$ cannot be detected with any accuracy. However, since the minimum is consicerably broad it's height will not be affected by imperfections at low frequency.

SUMMARY

By different approaches we have found evidence for the existence of a critical temperature T_c about 35 degrees above the calorimetric glass-transition temperature T_g. From below T_c a

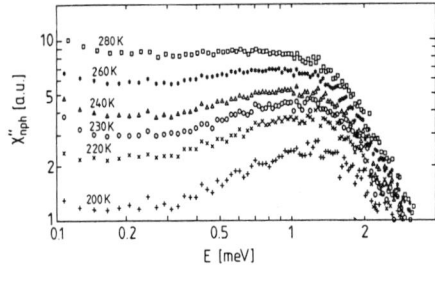
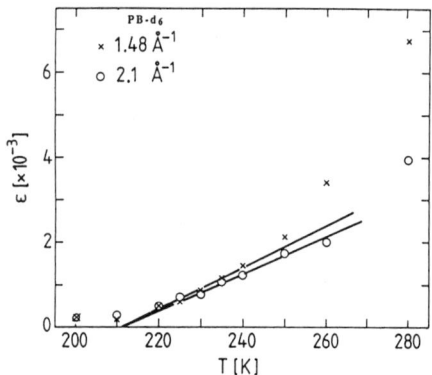

Fig. 5. Log-log-plot of the dynamic susceptibility at $Q = 1.48$ Å^{-1} for different temperatures after subtraction of elastic line and phononic contribution.

Fig. 6. Temperature dependence of the square of the minimum in the dynamic susceptibility χ''_{min} for $Q = 1.48$ Å^{-1} and 2.1 Å^{-1}.

square-root singularity was found in the temperature dependence of the non-ergodicity parameter f. From above T_c the decrease of the dynamic susceptibility and a power law fit of viscosity data yield comparable results for T_c. The values are also compatible with $T_c = 220 \pm 5$ K derived earlier from the mean-square displacement of the protons in non-deuterated PB[9]. The difference between T_g and T_c as well as the the fact that the susceptibility minimum does not tend to 0 at T_c show the existence of ergodicity-restoring processes not accounted for in MCTh unless "δ-terms" are included[10].

Concerning the spectral shape we found a pronounced Kohlrausch behaviour in the time domain for the α relaxation by NSE spectroscopy. The Kohlrausch function is also a good approximation to the result of the MCTh for the α relaxation[11]. The results in the frequency domain show a minimum of the dynamic susceptibility but because of resolution effects on IN6 the scaling properties could only be tested partially.

Finally, the Q-dependence of the α relaxation parameters could be proved to show the correlation with $S(Q)$ predicted by MCTh.

REFERENCES

[1] W. Götze, "Aspects of Structural Glass Transition", Les Houches Summer School 1989

[2] B. Frick, D. Richter, Cl. Ritter, Europhys. Lett. **9**, 557 (1989)

[3] D. Richter, B. Frick, B. Farago, Phys. Rev. Lett. **61**, 2465 (1988)

[4] G. C. Berry, T. G. Fox, Adv. Polym. Sci. **5**, 261 (1968)

[5] L. Sjögren, Z. Phys. B – Condensed Matter **79**, 5 (1990)

[6] B. Frick, B. Farago, D. Richter, Phys. Rev. Lett. **64**, 2921 (1990)

[7] D. Richter, R. Zorn, B. Farago, B. Frick, L. J. Fetters, to be published in Phys. Rev. Lett. (1991)

[8] P. G. de Gennes, Physica (Utrecht) **25**, 825 (1959)

[9] B. Frick, D. Richter, W. Petry, U. Buchenau, Z. Phys. B – Condensed Matter **70**, 73 (1988)

[10] W. Götze, L. Sjögren, Z. Phys. B – Condensed Matter **65**, 415 (1987)

[11] W. Götze, L. Sjögren, J. Phys. C: Solid State Phys. **20**, 879 (1987)

INTRINSIC UNSTABILITY AND NON ERGODIC BEHAVIOUR OF A GLASSY CRYSTAL

M.DESCAMPS, J.F. WILLART, O.DELCOURT

Laboratoire de Dynamique et Structure
des Matériaux Moléculaires (UA CNRS 801)
Université de Lille I, Bât. P.5 -
59655 VILLENEUVE D'ASCQ Cedex - FRANCE

INTRODUCTION

There are two different kinds of molecular systems which, upon cooling, lead to the freezing of the orientational degrees of freedom only

- On the one hand side those crystals for which the dilution of some non rotating "impurity" is a prerequisite to form a glass[1] (prototype is the mixed KCN-KBr). They can be connected with spin glasses while the analogy to these systems begins only to be clarified[2].

- On the other hand the glassy crystals[3] which behave exactly as conventional glasses. In particular they display a clear thermodynamic signature of the ergodicity breaking at Tg. The frustration in these systems is not induced by a dilution and the equilibrium first order transition (at T_t) to the stable non rotator phase is bypassed on cooling (eventually quenching).

In the context of the actively studied conventional glasses the trends remain controversally divided into purely dynamic approachs (mode-mode coupling[4]) and analysis based on the possible existence of a thermodynamic singularity[5]. Its position with regard to Tg could make the distinction between "fragile" and "strong" glasses[6]. The manifestation of a singularity associated with some kind of low temperature phase transition is very difficult to detect with glass-forming liquids since its static manifestations would only appear through the measurement of an S(Q) function which is spacially averaged. Owing to the underlying lattice, glassy crystals, offer in principle a good opportunity to follow spatial correlations. Obviously, conclusive results can be expected from diffraction experiments if the whole domain of non equilibrium (T<T_t) can be investigated in single crystals in good conditions of space and time resolution. These requirements seem to be incompatible with the mechanism of ordering which is basically achieved through molecular flips. The kinetics of transformation are not controled by volume diffusion and described by non conserved order parameters. These special classes of kinetics are intrinsically rapid[7]. It is the case in particular in the unstable region of the phase diagram where one expects a spontaneous amplification of the order parameter fluctuations. The rate of transformation in a system not limited by diffusion is then typically a microscopic frequency and so far there has been no chance to observe the so called "spinodal ordering".

THE APPROACH OF A SINGULARITY BELOW T_g

We have found very suitable experimental conditions in the study of Cyanoadamantane (CNa) and mixed compounds of CNa with Chloroadamantane $(Cla)_x (CNa)_{1-x}$. Furthermore their structure is simple enough to be modelized [8]. At room temperature (R.T.) $(Cla)_x (CNa)_{1-x}$ crystallize in isomorphous disordered plastic f.c.c phases [9]. for all values of x, the orientations of the molecular dipoles are preferentially and randomly along the [100] directions. For x <.5, the steric constraints are so strong that the dynamics of dipolar tumbling are exceptionally slow ($\tau \cong 10^{-7}$ s at R.T.). This facilitates the undercooling and, ultimately a glassy crystal state is reached. Depending on the concentration a jump in the specific heat is found at Tg in the range 150K-175K.

The mixed single crystals with the concentrations x=0 and x =.25 were more intensively studied by time resolved X-ray diffraction[10,11]. Qualitatively similar results have been found for both concentrations around Tg while studies were more easy and complete with x =.25. We review some recent, mostly unpublished, results.

On cooling, the ordering transition is by passed at $T_t \cong 237K$. An antiferroelectric short range order (S.R.O.) of the dipoles has been detected at R.T. and could be followed right down to Tg, both in the stable and metastable disordered state. This S.R.O. manifests itself through lorentzian superstructure peaks situated at the X positions of the Brillouin zone. Their intensity $I_{s.r.o.}(X)$ and width show a clear diverging behaviour which is arrested at Tg. The associated susceptibility $\chi(X) = I_{s.r.o.}(X)/T$ has been found to fit with $\chi \sim (T-T_C)^\gamma$; $\gamma = 1$ and $T_C \cong 160K$; so that $T_g-T_C \cong 15K$. The fit was excellent in a temperature range exceeding 120K in which 70K fall in the domain of metastability. Obviously the S.R.O. peaks could be followed reversibly as a function of temperature only during the life time of metastability (L.T.M.)

The nature of the virtual singularity at T_C was made clear by an analysis of the isothermal kinetics of ordering themselves. At moderate undercooling ($\Delta T = T_t - T$) the transformations have been found to obey the classical nucleation laws[12] : A minimum in the L.T.M. has been found for $\Delta T \cong 40K$ which is explained by the competition between the driving force of transformation and the slowing down of the molecular motions. On approaching Tg the L.T.M. was found to decrease a new to finally cancel at a temperature close to T_C. This marks the drop of the nucleation barrier and the entrance of the crystal in a situation of unstability. At lower temperatures the ordering was extremely slow but immediat. It followed the exponential Cahn-Hilliard amplification in the early stages and continued in an extremely weak time dependant evolution so that a saturation could never be reached.

The experimental results can be coherently interpreted by the fact that T_C is the pseudo-spinodal temperature in the context of mean field theories[7]. It is the first time that the absolute limit of metastability is so clearly seen experimentally and accurately located in a non diffusive system.

On approaching the metastability limit at T_C one would expect a downwards curvature of the $\chi^{-1}(T)$ plot signalizing an extra onset of easy nucleation. In contrary an upwards curvature has been observed. The temperature position of this curvature was found to depend on the cooling rate. It is the manifestation of the entrance of the disordered state into a non-ergodic situation.

This information could be most probably obtained by virtue of the unique conjunction of strong steric hindrance and dipolar interactions. This resulted in well defined metastable states under large undercooling and a deep mean field regime even quite near T_C.

CONCLUDING REMARKS

Our results show that there is an essential singularity associated with the first order transition. Its approach can be followed in all the domain of ergodicity and its virtual position is located about 15K below T_g. We believe that the approach of T_C is fundamental in the glass behaviour. Several scenarios can be imagined :

1) There would be a critical slowing down near the effective 2nd order transition occuring at T_C.

2) In the context of the "mode coupling" theory the glass transition could be triggered by the approach of T_c.

3) In the spirit of the Michel's theory[13] the transition to a glass could be associated to the coupling between molecular orientations and random fields generated by some heterophase grains unavoidably nucleated in the close proximity of T_c.

A comparison between T_c, T_k (Kauzmann temperature) and T_0 (Vogel Fülcher temperature) in this system would be highly desirable for any possible unification of these temperatures. In any case, the proximity of T_c and T_g seems to be consistent with the fact that the C_p jump is sizable.

REFERENCES

1 - U.T. HÖCHLI, K. KNORR, A. LOIDL, Adv. Phys. 39, 405 (1990)
2 - J. HESSINGER, K. KNORR, Phys. Rev.; Lett 65, 2674 (1990)
3 - H. SUGA, S. SEKI, J. of non Cryst. Sol. 16, 171 (1974)
4 - W. GÖTZE, "Aspects of structural glass transition" in Liquids Freezing and the Glass transition", J.P. HANSEN, D. LEVESQUE, J. ZINN JUSTIN Ed. (North Holland, 1990)
5 - J. SOULETIE, J. Phys. (France) 51, 883 (1990)
P. SETHNA, J.D. SHOVE, M. HUANG Phys. Rev. B (1991).
6 - C.A. ANGELL, J. Phys. Chem. Solids, 49, 863 (1988)
7 - K. BINDER Report Prog. Phys. , 50, 783 (1987)
J.D. GUNTON, M. SAN. MIGUEL, P.S. SACHNI, in Phase Transitions and Critical Phenomena ed by C. Domb. and J.L. LEBOWITZ (Vol 8, Academic London 1983).
8 - J.F. WILLART, J. NAUDTS, M.DESCAMPS Phas. Trans. 31, 261 (1991)
9 - M. FOULON, J.P. AMOUREUX, J.L. SAUVAJOL, J. LEFEBVRE, M. DESCAMPS, J. Phys. C. L 265 (1983).
10 - M. DESCAMPS, C. CAUCHETEUX, J. Phys. C : Sodil State Phys., 20, 5073 (1987)
11 - M. DESCAMPS - J.F. WILLART to be published.
12 - M. EL ADIB - M. DESCAMPS - N.B. CHANH, Phase transitions 14,85 (1989).
13 - K. H. MICHEL, Phys. Rev. B.35, 1414 (1987).

ISOTHERMAL GLASS TRANSITIONS AND ENTROPIC RELAXATION

C. Alba-Simionesco
Laboratoire de Chimie-Physique,11rue Pierre et Marie Curie,75231 PARIS cedex 05
FRANCE

ABSTRACT

We report the first calorimetric characterization of isothermal glass transitions on a so called "fragile" liquid and introduce the notion of the entropic relaxation consecutive to annealing effects in the compressed glass at P> Pg. The piezothermal method involves measurements of the heat of compression of a sample enclosed in a hydrostatic pressure cell under isothermal and reversible conditions. Approximate relaxation times and relaxation functions near the glass transition are reported.

INTRODUCTION

For a great variety of liquids, the glass transition phenomenon has been extensively studied since many years from the experimental and theoretical point of view. The tendancies to interpret its rather rich phenomenology are shared between the dynamic aspects and the relaxation processes involved and the thermodynamic underpinning of the phenomenon. Most of these investigations use the conventional technique of liquid cooling at constant pressure, mainly the atmospheric pressure. But working with the temperature as an intensive variable has a twofold action in the kinetic energy and the volume, while the adequacy of pure density effects should be tested by measurements changing the volume, i.e. the pressure at different constant temperatures. The method of the heat of compression performed here gives the direct measurement of the expansivity α or of the product αV, where V is the molar volume, which has the same hierarchic importance as the heat capacity:

$$dS = C_p \, d\ln T - \alpha V \, dP \qquad (1)$$

This opens the field of the study of isothermal glass transitions and entropic relaxation. Focussing our attention on the so-called "fragile" liquids, we have choosen the molecule of m-fluoroaniline (mFA)[1,3] , which is a good glassformer without crystallization risks and enables the observation of a liquid in its supercooled and overcompressed state.

METHOD AND APPARATUS

The piezothermal method introduced by TerMinassian and co[2] consists in measuring the heat evolved by a sample upon application of a known excess pressure at constant temperature ; it is based on the thermodynamic equations of Maxwell:

$$(\partial S / \partial P)_T = -(\partial V / \partial T)_P = -\alpha V \qquad (2)$$

and
$$\delta q = T \, dS = -T \, (\partial S / \partial P)_T \, dP = -\alpha V T \, dP \qquad (3)$$

where δq is the quantity of heat liberated by one mole of the substance. According to the experimental arrangement, the total quantity of heat ΔQ may refer to a constant effective volume Ve of the vessel containing the sample:

$$\Delta Q = n \, \delta q = Ve/V \, (-\alpha V) T \, \Delta P = -\alpha \, Ve \, T \, \Delta P \qquad (4)$$

where n=Ve/V is the number of mole involved. It may also refer to a constant quantity of matter of masse m, then n=m/M

$$\Delta Q = -n \alpha V T \Delta P \quad \Rightarrow \quad \alpha V = \Delta Q / (T \, \Delta P \, n \, V) \qquad (5)$$

Equation (4) is the original form of the piezothermal method[4], associating the techniques of calorimetry and high pressure. Equation (5) is the basis of the results presented here.Some additional contributions have to be introduced. Usually experimentalists in the field of viscous liquids apply all pressure changes in the liquid state and then cool the sample, a process known to minimize stress gradients. Here we have choosen a distinct thermodynamic path : the sample is enclosed in a pressure cell in Teflon compressed and depressed hydrostically by a known liquid, the n-pentane at a constant temperature T of the calorimeter; this technique allows a reproductibility and a reversibility for compressing and depressing data better than 1%. These additional quantities due to the teflon, the vessel and the n-pentane are carefully calibrated as a function of P andT and may be substracted.

The experiments although simple in concept are difficult to do and require calorimetric measurements of unusually high precision. The calorimeter employed is based on a pneumatique principle and has been developped in our laboratory; its working temperature is between 100K and 330K and the high pressure part of the device may reach pressure as high as 4 to 6 kbar; full details may be found in the litterature[5].The liquid is first enclosed in the teflon cell, then in the autoclave and the temperature is lowered down to the working temperature. The system is allowed to rest for a few hours to reach thermal equilibrium, then by discrete steps ΔP of 80 to 120 bar the pressure is increased up to the signature of the G.T.. At each step, the quantity of heat is measured between an initial and final equilibrium state ,which is accessible after a waiting time of length Δt of 300 secondes well below and above Pg, and up to a maximum of 1000 sec in the transformation range. The average compressing rate would be then $R = \Delta P / \Delta t$. We have applied the same waiting times for all the isotherms presented in figure n°1

RESULTS AND MODEL

We report data of 3 different isotherms at 171.5K, 182.4K, 191.8K for a sample of mFA of masse m= 0,3663gr; at Patm, the calorimetric Tg is about 173K[1]. By increasing the pressure, (αV) is a monotonously decreasing function untill a jump which greatest slope locates Pgcomp; by decreasing pressure steps, the same data are recovered untill a pressure Pgdet above Pgcomp, at which a relaxational peak appears.The most prominent aspect is the hysteresis observed between both thermodynamic paths over a pressure range of 600 bar through Pg.This is the signature of a relaxing process accompagning the glass transition, induced by the arrest of some diffusive motion and structural relaxation on the experimental time scale. All the data are collected in the table n°1

Table 1 Data analysis of the experimental curves of figure 1

T_K	Pg_{compr} bar	Pg_{det} bar	V_g cm-3.M-1	$\Delta\alpha V$	$\Delta\alpha 10^4 K^{-1}$	$\Delta\beta_T\ 10^6 bar^{-1}$
171,5	305	400	85,8	1,636	5,785 ;	4,8
182,4	1370	1580	82,3	0,922	3,389 ;	2,8
191,8	2750	2850	78,0	0,783	3,046 ;	2,5

This behavior suggests a certain similarity with a second order glass transition, in the sense of Ehrenfest; still keeping in mind the non-equilibrium nature of the phenomenon, from its relations $(dTg/dP) = \Delta\beta_T / \Delta\alpha = TVg \Delta\alpha/\Delta Cp$ and the slope

$(dTg / dp)_\tau = 8,3 \, °/\text{kbar}$, we have deduced the variation of the isothermal compressibility and a Cp jump at atmospheric pressure of 103J.K-1M.-1, greater than the measured value of $70 \text{J.K}^{-1}\text{M.}^{-1}$[1] This slope is much smaller than that determined by the more common scan method of 22°/kbar for "fragile" liquids under isobaric conditions. It reveals that other degrees of freedom are frozen along these 2 paths. In our case, it suggests that the free volume rather than the entropy or the enthalpy determines the transition[6].

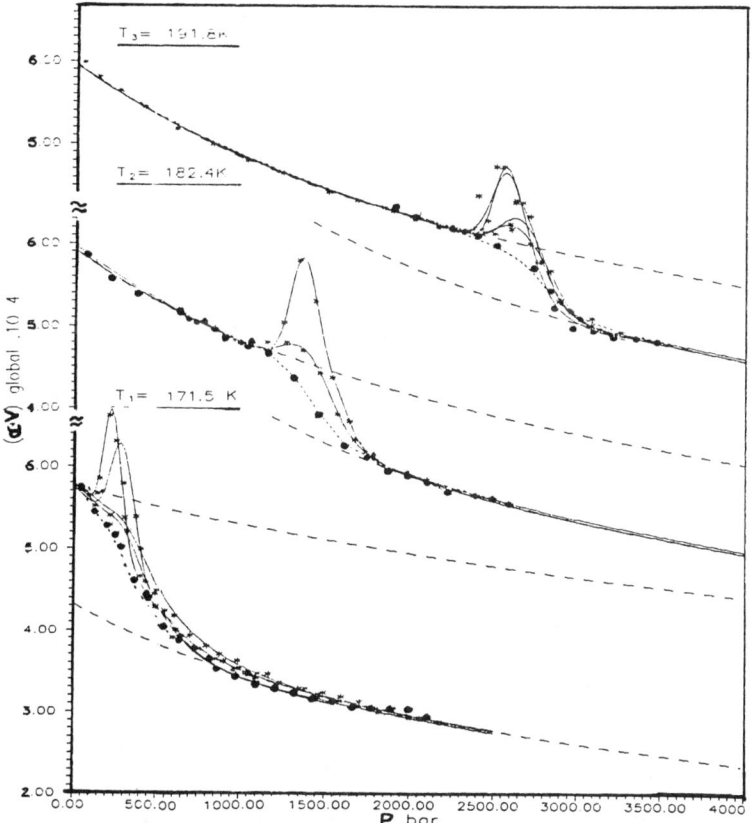

Fig. 1. Total (αV) data as a function of pressure for compressing (●) and decreasing (*) pressure steps. Full curves are ad hoc analytical fits for different annealing times and annealing pressure; doted curves are compressing fits and dashed lines are extrapolations of the liquid and the glassy states

Another aspect illustrated on figure 1 is the entropic relaxation phenomena: the recovery of the entropic lost during annealing at pressure above Pg is manifested by an increase of the area under the peak and a decreasing value of the Pg det., when the annealing time increases or the annealing pressure decreases. The relaxation function $\Phi_S(t)$ according to Brawer(1984)[7] is introduced:

$$\Delta S(t) = -(\alpha V_\infty + (\alpha V_0 - \alpha V_\infty)(1 - \Phi_S(t)))\Delta P \qquad (6)$$

This function should be analysed with the following model.

We have adapted to our compressing data the empirical and phenomenological model developped first by Moynihan [8], then by Hodge and co.[9]. It is based on Boltzmann superposition principle of non exponential responses to the pressure steps; the non-exponentiality is introduced via the Kolraush-Williams-Watt law for convenience: $\Phi(t) = \exp(-(t/\tau_0)^\beta)$ with $0<\beta<1$ (7)
The departure from equilibrium, i.e., the time dependence of the relaxation time, is taken into account via the parameter x and the fictive pressure P_f:

$$\tau_i = \tau_0 \exp(\Delta V^{\ne}/RT (x P_i + (1-x) P_{f,i-1})) \quad \text{for the } i^{th} \text{ step} \quad (8)$$

$$P_{f,i} = P_0 + \Sigma_{m=1}^{i} \Delta P_m (1 - \exp(-(\Sigma_{j=m}^{i}(\Delta t_j/\tau_j)^{\beta(P,T)}))) \quad (9)$$

Full details of this model and the fitting parameters will be described elsewhere [10]. The resulting data are however shown in figure 2 and compared to the experimental data within a 2% error, in the following form :

$$dP_f/dP = (\alpha V_{mes} - \alpha V_{liq})/(\alpha V_{liq} - \alpha V_{glass}) = (\alpha V)^N(P) \quad (10)$$

The main achievements are:

- pressure and temperature dependences of the exponent β
- the Pg representation as a two step transformation using 2 sets of data, above and below Pg.

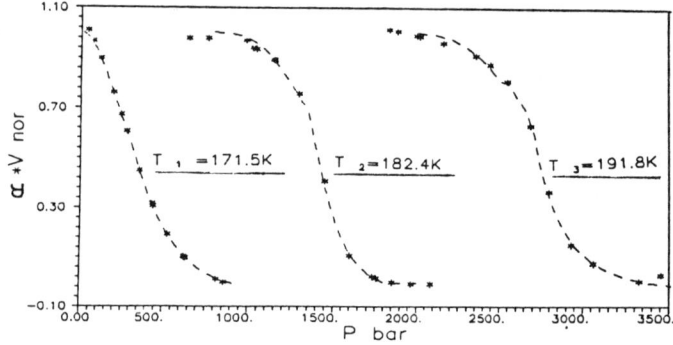

Fig.2. Fit with a Boltzmann superposition principle with $\beta=f(P)$, according to 2 relaxational regimes for the 3 isotherms(- - -) and normalized experimental points(*)

ACKNOLEDGEMENTS

We would like to thank Dr TerMinassian and Prof.C.A.Angell for very helpful and stimulating discussions, and F.Milliou and P.Chargelegue for their technical assistances

REFERENCES

1. C.Alba-Simionesco, C.A.Angell, submitted for publication
2. L.TerMinassian and co.,J.Chem.Thermod. 6,1139,(1974); 51,1123,(1979)
3. C.Alba-Simionesco H-Bonded Liquids JCDore JTeixeira ed.1991,Kluwer A. P.
4. C.Alba,L.TerMinassian,and co., J.Chem.Phys. 82,384,(1985)
5. L.TerMinassian,F.Milliou,J.Phys. E:Sci. Instrum. 16,450,(1983)
6. M. Goldstein J.Chem.Phys. 39,3369,(1963);J.Phys.Chem.77,667,(1973)
7. S.Brawer Relaxation in viscous liquids and glasses,(N.Y. Am.Ceram.Soc.)1983
8. C.Moynihan and co,J.Am.Ceram.Soc.59,16,(1976)
9. I.Hodge,Berens Macromolecules,16,371,(1983);15,762,(1982)
10. C.Alba-Simionesco, to be published

THE DIFFERENT UNIVERSALITY CLASSES, FROM STRONG TO FRAGILE, OF THE GLASS AND SPIN-GLASS TRANSITIONS

J. Souletie
CRTBT-CNRS, BP 166, 38042 Grenoble-Cédex 9,
laboratoire associé à l'Université Joseph Fourier de Grenoble
D. Bertrand
Laboratoire de Physique des Solides,
INSA, Avenue de Rangueil, 31077 Toulouse-Cédex

On figure 1 we have represented in (a) the logarithm of the viscosity of different glass-forming liquids and in (b) the relaxation time τ of several spin glasses vs. T_g/T where T_g is such that $\eta(T_g) = 10^{13}$ Poise and $\tau(T_g) = 10^4$ s. The plot of figure 1a is also a plot of the shear relaxation time vs T_g/T if we read it with the scale of figure 1b because the viscosity η and τ_s are proportional to each other $\eta = G_\infty \tau_s$ with $G_\infty \sim 10^9$ and the two log scales have been shifted by 9 decades accordingly. The figure shows that the classification "from strong to fragile"[1] which is currently used with glass-forming liquids can also be applied to spin glasses. Thus the "2 dimensional" Ising spin glass $FeMgCl_2$ follows Arrhenius behaviour and can be labelled strong like SiO_2 (or GeO_2, etc ...). On another hand the anisotropic Ising system $Rb_2Cu_{0.78}Co_{0.22}F_4$ has an intermediate behaviour which is reminiscent of that observed in amorphous Se as well as in glycerol or in PdCuFe.[2] The other more classical 3 dimensional Heisenberg spin glasses would compare to the fragile systems propylene carbonate, toluene or salol.

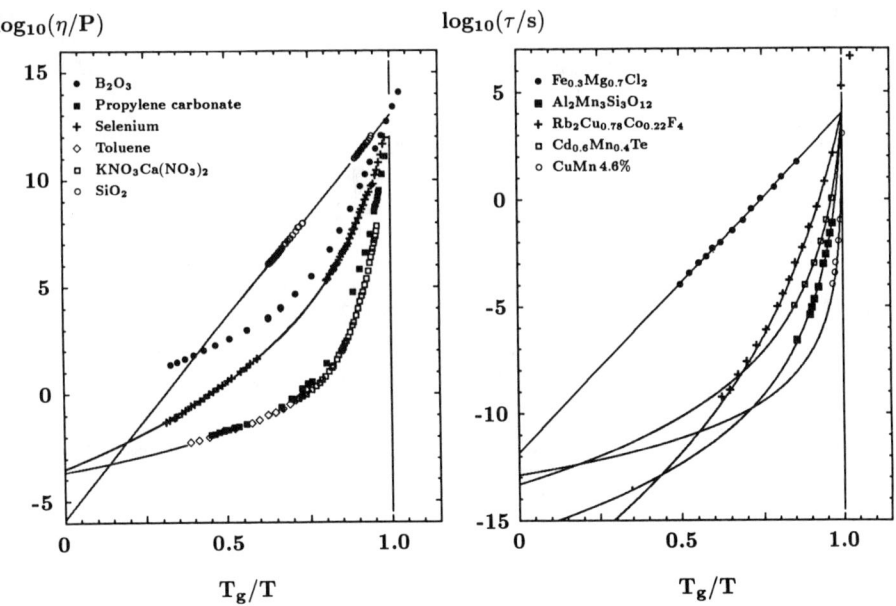

Fig. 1 Fragile and strong behaviours in glass-forming liquids (a) and in spin glasses (b) as shown by the T_g scaled Arrhenius plots of the viscosity η and of the relaxation time τ. Here T_g is the temperature at which η reaches 10^{13} Poise and $\tau = 10^4$ s. (The data which we have used for the figures come from references 3-14).

Classically the spin-glasses are discussed in a different framework: that of phase transitions and the divergence of the relaxation times is interpreted with the slowing down equation $\tau/\tau_0 = (\xi/\xi_0)^z = [(1-(T_c/T)]^{-zv}$. We propose to differentiate this equation to obtain $P_\tau(T) = -(\partial \ln T/\partial \ln \tau) = (T-T_c)/\theta = (T-T_c)/zvT_c$ which contains only two parameters T_c and zv (or T_c and θ). They can readily be obtained from a $P_\tau(T)$ vs T plot which should provide a straight line whose intersections with the axes are respectively T_c and $-1/zv$. The figure 2 shows the $P_\tau(T)$ plot for the data of figure 1. We see from figure 2b that the fragile spin glasses can be interpreted as belonging to a common universality class for which $zv \sim 11\pm2$. The intermediate system $Rb_2Cu_{0.78}Co_{0.22}F_4$ would be the (so far unique) prototype of an intermediate class with $zv \sim 25$ and $Fe_{0.3}Mg_{0.7}Cl_2$ is interpreted as a system at its lower critical dimensionality for which $T_c = 0$ and $zv \to \infty$.[2] We observe conversely that the glass forming liquids can be described within the same framework. SiO_2 would be representative of systems at their lower critical dimension. There would be an "intermediate" class with a $zv \sim 15$ which would include Se, glycerol, etc ... and the fragile systems in their high temperature regime can be described with a still smaller exponent $zv \sim 6.2$. We notice however that most of these fragile systems cross over towards stronger regimes when the temperature is decreased and before they reach T_c and while we are still at temperatures larger than T_g where the system is still at equilibrium.

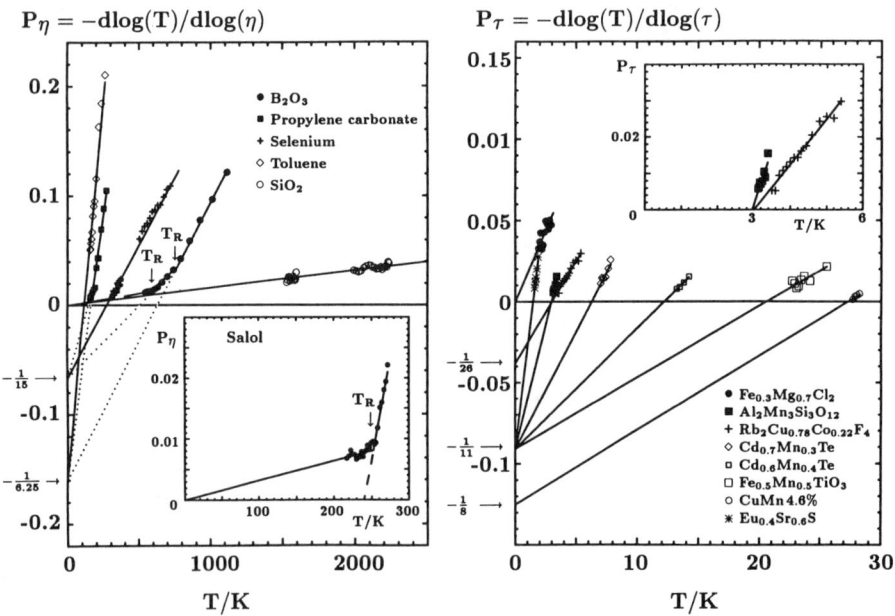

Fig. 2 In a representation of $P_\eta(T)$ vs. T (see text) a set of straight lines which aim the same point of the $T = 0$ axis means power laws with the same exponent. The fragile to strong behaviour thus hides a finite number of different universality classes. Cross-overs from fragile to stronger regimes are observed in glass-forming systems as shown for salol in the insert.

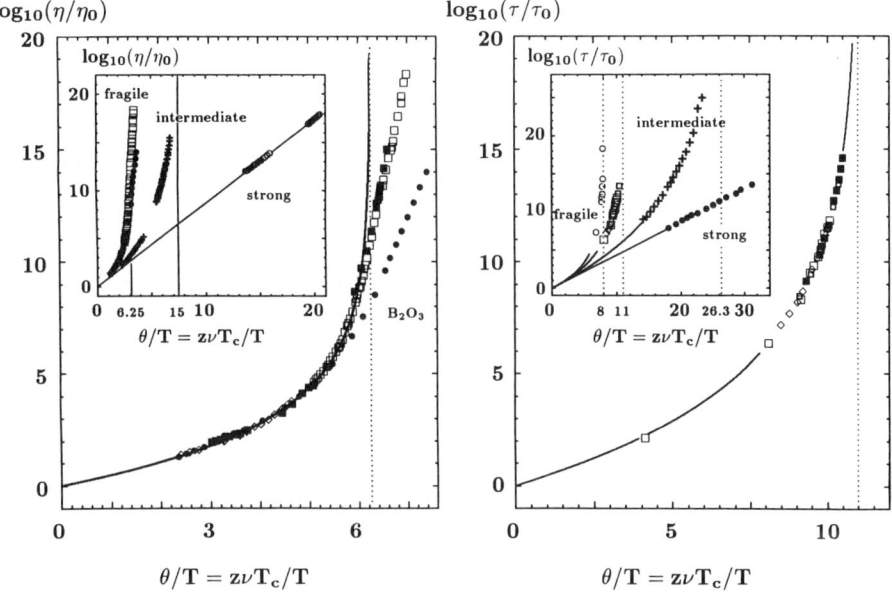

Fig. 3 The viscosity or the relaxation time diverge at $\theta/T = z\nu$ in these θ scaled Arrhenius plots where $\theta = z\nu T_c$. The cross-over towards a stronger behaviour occurs at a higher temperature in B_2O_3 than in other reputedly fragile systems. No cross-over is noticed with spin glasses in the ranges explored.

This is seen in more detail on the figure 3. In the insert we have represented $\ln(\tau/\tau_0)$ vs. θ/T with τ_0 and $\theta = z\nu T_c$ which best fit the data for the average exponent values which have been determined using the plots of figure 2. In this representation τ diverges at $\theta/T = z\nu$. On the main figure we observe that the data of very different systems superimpose in the fragile regime but B_2O_3 crosses over to a stronger regime at T_R much before (for a higher T_R/T_c ratio) propylene carbonate or salol and accordingly appears stronger than these other systems. The figure shows (see insert) that it is possible to describe the intermediate and strong cases as the result of a cross-over which has occurred at a still higher T_R/T_c ratio. General thermodynamic considerations impose that the crossover occurs without discontinuity of the slope of the $\eta(T)$ dependence and that $P_\tau(T)$ therefore should be continuous. Such crossovers, in spin glasses, would signal a change of the space or spin dimensionality : For example, we would expect a similar effect in a thin layer when the correlation length which increases first in 3d, becomes of the order of the thickness and the system is then constrained to develop in 2 d.

1. C.A. Angell, Nuclear Physics B (Proc. Suppl.) 5A, 69 (1988) and references therein.
2. J. Souletie, J. Physique (France) 51, 883 (1990) ; J. Souletie and D. Bertrand, J. Physique (France) 1 (novembre 1991) ; and references therein.
3. P.B. Macedo and A. Napolitano, J. Phys. Chem. 49, 1887 (1968).

4. A. Bondeau and J. Huck, J. Physique (France) 46, 1717 (1986).
5. R. Weiler, S. Blaser and P.B. Macedo, J. Phys. Chem. 73, 4147 (1969); R. Weiler, R. Bose and P.B. Macedo, J. Chem. Phys. 53, 1258 (1970).
6. R. Bruckner, J. Non-Cryst. Sol. 6, 177 (1971); E.H. Fontana and W.H. Plummer, Phys. Chem. Glasses 7, 139 (1966).
7. K. Uberreiter and H.J. Orthmann, Kolloid Z. 123, 84 (1951); J.C. Perron, J. Rabbit and J.F. Riallaud, Phil. Mag. 46, 321 (1982).
8. W.T. Laughlin and D.R. Uhlmann, J. Phys. Chem. 76, 2317 (1972).
9. P. Beauvillain, J.P. Renard, M. Matecki and J.J. Préjean, Europhys. Lett. 2, 128 (1987); J.J. Préjean, in Change and Matter, ed. J. Souletie, J. Vannimenus and R. Stora, North-Holland Amsterdam 1986, p. 557.
10. D. Bertrand, J.P. Redoulès, J. Ferré, J. Pommier and J. Souletie, Europhys. Lett. 5, 271 (1988).
11. C. Dekker, A.F.M. Arts, H.W. de Wijn, A.J. van Duyneveldt and J.A. Mydosh, Phys. Rev. Lett. 61, 1780 (1988).
12. S. Geschwind, A.T. Ogielski, G. Devlin, J. Hegarty and P. Bridenbaugh, J. Appl. Phys. 63, 3291 (1988); Y. Zhou, C. Rigaux, A. Mycielski, M. Menant and N. Bontemps, Phys. Rev. B 40, 8111 (1989).
13. H. Aruga, T. Tokoro and A. Ito, J. Phys. Soc. Japan 57, 261 (1988).
14. N. Bontemps, J. Rachenbach, R.V. Chamberlin and R. Orbach, Phys. Rev. B 30, 6514 (1984).

EXPERIMENTAL STUDY OF NON-DEBYE RELAXATION IN GLASS TRANSITION OF GLYCEROL

Seiji Kojima
Institute of Applied Physics, University of Tsukuba, Tsukuba-city, Ibaraki 305, Japan

INTRODUCTION

The relaxation process of a fragile liquid has been poorly understood up to this time. Alcohol is one of typical materials of a fragile liquid. It undergoes a glass transition easily when the number of the carbon atoms of a molecule is more than two. This paper reports the measurements of thermal and dielectric dispersions, and the Raman scattering of the O-H stretching vibrational modes of glycerol above the glass transition point.

THERMAL AND DIELECTRIC DISPERSIONS

The thermal and the dielectric dispersions have been studied separately by the different investigators in any material. Whereas in the present work the both two types of dispersions were measured by using the same specimen. The thermal dispersion[1] was measured by the gas-microphone detection of the photoacoustic method.[2] The dielectric dispersion was measured by the conventional LCR meters.

At first the temperature dependences of a relaxation frequency were determined from the maximum temperatures of the imaginary parts of the heat capacity and the dielectric constant. It is found that the temperature dependence of the thermal relaxation agrees with that of the dielectric one within the experimental accuracy.

Secondly the distribution of the relaxation time were studied by the Cole-Cole plot as shown in Fig.1. These plots cannot be understood by the Debye relaxation, and the data points were fitted by the Davidson-Cole formula. The obtained parameter of the formula of the thermal relaxation was found to be smaller than that of the dielectric one. It strongly suggests that the distribution of the thermal relaxation time is broader than that of the dielectric one.

In general the dielectric dispersion is caused by a relaxation of a polarization. Whereas the thermal dispersion includes not only the degree of freedom of a polarization but also that of other various excitations. Therefore the distribution of a relaxation time is broader than that of the dielectric dispersion. This means that the parameter of the Davidson-Cole formula of the thermal dispersion is always less than that of the dielectric one. In this case the difference can be attributed to the degree of freedom of nonpolar atomic motions, which are weakly coupled to the motion of hydrogen bonded hydroxyl groups.

RAMAN SCATTERING OF O-H STRETCHING MODE

Raman scattering spectra were measured to investigate the temperature dependence of the hydrogen bond of glycerol when the

© 1992 American Institute of Physics

temperature approaches to a glass transition temperature from the room temperature. In alcohols, the O-H stretching vibration mode of a free molecule shows a sharp peak at about 3620cm^{-1}. While the hydrogen bonds exists, the mode frequency becomes lower remarkably and the peak width becomes broader. In Fig.2, the arrows show the broad peaks of the O-H stretching mode of the hydrogen bonded molecules. The peak shows remarkable softening when the temperature decreases.

The broad peak of the O-H mode has been analyzed by the introduction of the three modes. One is a stretching vibrational mode in dimeric units, and others are that in oligomers with the aggregation number greater than two.[3] From the three modes analysis it is found that the number of hydrogen bonds increases as the temperature approaches to a glass transition point from the higher temperature. This fact means that the length of the chain of hydrogen bonded molecules become longer as the temperature decreases. On the other hand, the C-H vibrational modes scarcely depends on temperature. Therefore the dielectric dispersion can be attributed essentially to the motions of the hydrogen bonded hydroxyl groups. The Raman scattering study of nonpolar motions is now in progress to clarify the origin of the difference between thermal and dielectric dispersions.

REFERENCES

1. N.O.Birge, Phys. Rev, B34, 1631 (1986)
2. S.Kojima, Proc. of IEEE Ultrasonic Symp. 3, 1321 (1991)
3. A.D'aprano, D.I.Donato, P.Migliardo, F.Aliotta and C.Vasi, Mole. Phys., 58, 213 (1986)

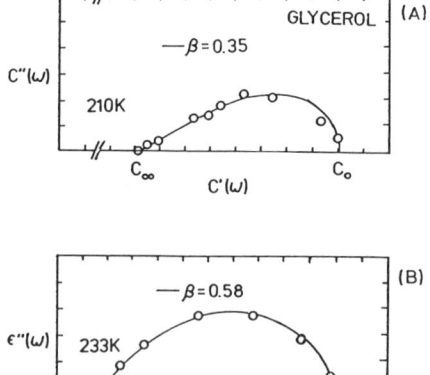

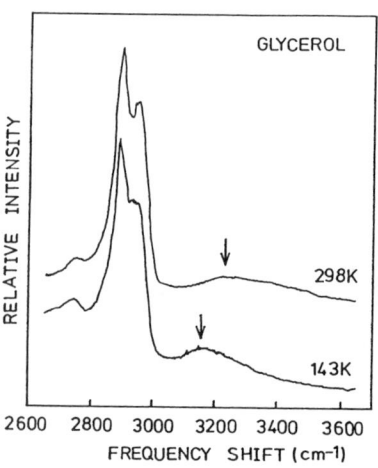

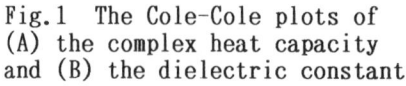

Fig.1 The Cole-Cole plots of (A) the complex heat capacity and (B) the dielectric constant.

Fig.2 Raman scattering spectra of glycerol. The arrows show the O-H stretching modes.

PRODUCTION OF NEW ALLOY PHASES IN Bi-Mn BINARY SYSTEM FROM VAPOUR PHASE AND THEIR STRUCTURE INVESTIGATION BY HIGH RESOLUTION ELECTRON MICROSCOPY

Kentaroh Yoshida and Takashi Yamada
Faculty of Engineering, Kobe University
Rokkodai, Nada, Kobe 657, Japan

INTRODUCTION

It has been well known that Bi and Mn are hardly miscible with each other and only a ferromagnetic phase, MnBi, forms by a peritectic reaction at 445° C between the solid Mn and the melt.[1] The present authours are trying to mix these two atom species by vacuum deposition. Two kinds of the deposition were employed, i.e. simultaneous and successive ones, followed by a suitable heat treatment. The products were investigated by high resolution electron microscopy and more than three kinds of new long period superlattices were found.

EXPERIMENTALS

In the simultaneous vacuum deposition, Bi and Mn were deposited at the same time onto the same carbon substrates at room temperature. Degree of vacuum during the deposition was on the order of 10^{-7} Torr and the films of 600 to 900 Å in thickness were prepared by this co-deposition for only 2 to 3 seconds.

In the successive deposition, a Bi layer of 300 Å in thickness was deposited first on the carbon substrates and then a Mn layer of 200 Å on top surface of the Bi layer. Degree of vacuum during these deposition was $\sim 1 \times 10^{-5}$ Torr and it was important to deposit the Mn layer as soon as possible after the Bi deposition so that the two layers are in good contact without any contaminations. Thus prepared double layer thin films were then heated at 265° C, just below the melting point of Bi, 271° C, in the same vacuum for 50 to 200 hr.

The specimen films were observed by an electron microscope, JEM-200CX, whose resolving power is 2.6 Å.

RESULTS OF THE INVESTIGATION

By the simultaneous deposition, the as-deposited films showing a few diffuse halos in electron diffraction patterns are obtained when their composition is in a range 85 to 65 at% Mn. When warmed up at 180° C, crystallization of the type 'nucleation and grain growth' takes place in the film and such large single crystals as long as 2 micron in diameter are easily obtained. Their net diffraction patterns were successfully constructed into three dimensions, transformed into real space and give a lattice of

$$\text{hexagonal} \; ; \; a = 19.97 \text{ Å}, \; c = 4.49 \text{ Å} , \qquad (1)$$

which has never been reported before. These crystals are metastable since they transform very quickly into QHTP of the system when heated at 220° C. From extinction rule of their reciprocal lattice points, intensity distribution over them and from contrast features in their high resolution electron micrographs, atomic arrangement in the lattice of (1) was constructed. The result shows that this metastable alloy phase has a space group of R$\bar{3}$m and 36 Mn and 12 Bi atoms are contained in the unit cell of (1), giving chemical formula of Mn$_3$Bi.[2] Reasonable agreements were obtained between its high resolution micrographs taken with several small amounts of defocus of the microscope and computer simulated images of the above structure which were inputted with the corresponding defocus values.[3]

By the low temperature heating of the double layer thin films, solid-state alloying will proceed at the interface of the two layers and more than four other kinds of lattices than that of MnBi have been observed. The most prominent one of them, tentatively called long period tetragonal lattice, has lattice constants of[4]

$$\text{tetragonal} \; ; \quad a = 17.26 \text{ Å}, \quad c = 10.21 \text{ Å} . \quad (2)$$

A high resolution micrograph of the double layer film containing this lattice region is shown in Fig. 1. White contrast features forming a square lattice can be seen in the right handed half of Fig. 1, which corresponds to the lattice (2) projected along [001]. White doughnut-shaped contrasts are, on the other hand, randomly arranged in the left half. The diffraction patterns from the tetragonal lattice regions such as in Fig. 1 show some similarity to those of Sigma-phase in transition-transition metal binary alloys, but with a few distinct differences. Arrangement of atoms in this long period tetragonal lattice is now intensively being analyzed, possibly as some derivative of the Sigma-phase structure. From regions of the doughnut-shaped contrasts as in the left half of Fig. 1, diffraction patterns were obtained which are concluded as of 12-fold symmetric quasicrystal.[5] No regular networks of the diffraction spots can be defined in those patterns. Atomic configuration of this quasicrystalline structure will have a close structural relationship to that of the lattice (2).

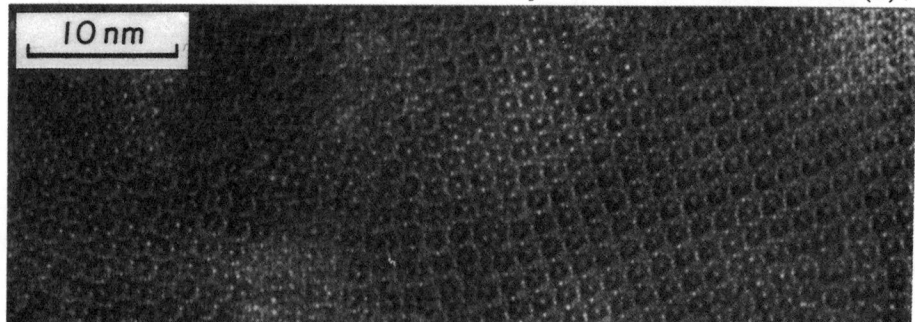

Fig. 1, HREM of the long period tetragonal lattice along [001].

REFERENCES
1. M.Hansen & K.Anderko, Constitution of Binary Alloys (McGrwaw,1958)
2. K.Yoshida, T.Yamada & Y.Furukawa, Acta metall.34,969(1986). p.318.
3. K.Yoshida, T.Yamada & Y.Furukawa, J.Electron Microsc.36,139(1987).
4. K.Yoshida, T.Yamada & Y.Taniguchi, Acta Cryst. B45, 40 (1989).
5. K.Yoshida & Y.Taniguchi, Phil. Mag. Letters, 63, 127 (1991).

GLASS TRANSITION AND LOCALIZATION OF MOBILE IONS IN SUPERIONIC GLASSES: INVESTIGATED BY DIELECTRIC AND THERMODYNAMIC RELAXATION TECHNIQUES.

Junichi Kawamura
Department of Chemistry, Faculty of Science, Hokkaido University, Sapporo, 060, Japan.

ABSTRACT

In the cooling process of multicomponent Coulomb liquids, the localization of constituent ions is accomplished through various processes. In the superionic system, only a part of the constituent ions freezes at T_g and residual "mobile" ions still remains in the liquid state. The localization of the mobile ions occurs at temperature region far below Tg. The localization of less-mobile ion is detectable by thermal relaxation technique such as DSC, but no dispersion is seen in dielectric response. On the other han, the localization of mobile ions can be detected only by dielectric response as a power-low frequency dependence of conductivity $\sigma[\omega] \sim (i\omega\tau_c)^n$ (0<n<1), and increase in the relaxation time τ_c.

Self-consistent semi quantitative analysis is given, based on the excess-free-volume model, to calculate the temperature and frequency dependence of conductivity and permittivity, as well as specific heat change through T_g. Also, a brief discussion is given on the mode-coupling approach by Sjögren.

Superionic glass such as $AgI-Ag_2MoO_4$ system is a unique system in its glass transition nature, which shows anomalously high ionic conductivity in the glassy state even at room temperature [1,2]. Through usual glass transition process from the molten salt to the glassy state, only a part of constituent ions (ex. MoO_4^{2-} and I^-) freeze at T_g to form glass framework. Residual ions such as Ag^+ diffuse very fast through the glass framework like in the liquid state.

Thus, the total localization process of constituent ions should be discussed in two different processes: normal glass transition and the localization of mobile ions. Such anomalous glass transition process has been discussed by some authors[-3], by coupling-decoupling concept [1], excess-free-volume model [2], coupled-Langevin equation theory [2], cluster-bypass model [3] and mode-coupling approach [4].

Temperature dependencies of d.c. conductivity of $AgI-Ag_2MoO_4$ system is shown in figure 1 with other glass forming system; hatching shows the temperature region where frequency dependence is observed in low-frequency conductivity (<10 MHz). In this region, the ac conductivity is well expressed by the following equation:

$$\sigma[\omega] = \sigma[0][1+(i\omega\tau_c)^n]$$

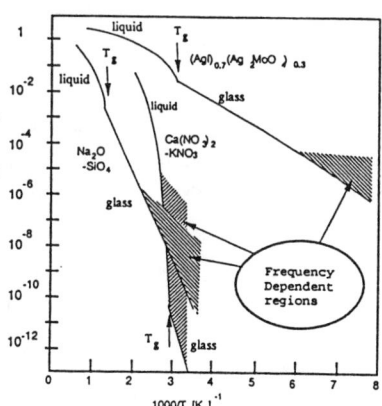

Fig.1 Temperature dependencies of superionic glass $AgI-Ag_2MoO_4$ system, and other glass forming systems. Hatching shows the temperature region where frequency dependence is seen in ac conductivity.

where σ[0] is the d.c. conductivity, τ_c is the average relation time and n (0<n<1) is the power-low index or the distribution parameter. The value of n is typically 0.75~0.85 depending on the glass composition, and is a slight decreasing function of temperature.

Although no frequency dependence was seen near Tg in SIG, it is found at microwave region as shown in figure 2. It is interestingly very common that the real and imaginary parts of them crossed over near T_g, as shown in figure 3. Although no significant change is seen at low temperature where mobile ions localize, a pronounced peak is seen as conventional glass transition at Tg, whose hight depends on the thermal history of the sample.

Experimental values in figure 2-3 are fairly well fitted by the excess-free-volume theory by the author and M.Shimoji [2], although somewhat systematic deviation are seen near the glass transition region since the single relaxation time approximation is used for the structural relaxation

Sjögren [4] proposed a theory of impurity diffusion in liquid and glass based on the mode-coupling theory, introducing two coupling parameters. The power law dependence of conductivity on both temperature and frequency expected from the theory is fairly in good agreement with the experiments: see fig. 5, although the exponent 1.5 is somewhat lower than the theoretical value 1.765 [11]. Moreover, the deviation from power-law near T_g is naturally understood from the theory without artificial introduction of other hopping mechanism.

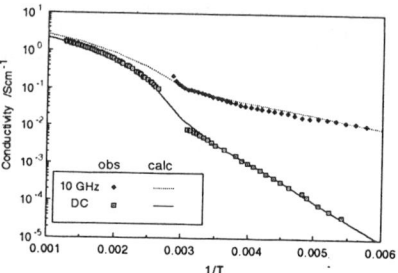

Fig. 2 DC and microwave conductivity of AgI-Ag_2MoO_4 glass and its supercooled liquid state. Lines are calculated values by EFV theory.

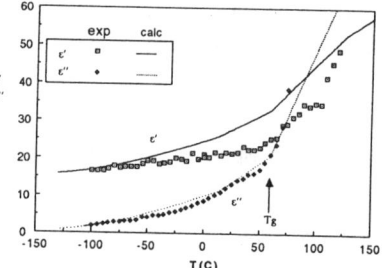

Fig. 3 Real and Imaginary parts of permittivity of AgI-Ag_2MoO_4 glass through glass transition region. Lines are calculated values.

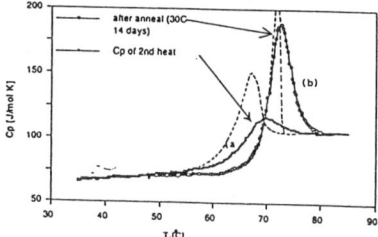

Fig. 4 Specific heat by DSC of AgI-Ag_2MoO_4 glass; Lines are calculated values.

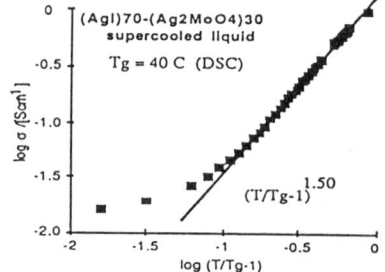

FIG. 5 Ionic conductivity of AgI-Ag_2MoO_4 system in the supercooled liquid state, plotted as the power-law expected in mode-coupling theory.

REFERENCES

1. C. A. Angell., Solid State Ionics., 9&10 3 (1983), 18&19 72 (1986).
2. J. Kawamura and M. Shimoji,Mater.Chem.Phys.,23 99 (1989), J.Non-Cryst.Solids.,88 281,295 (1986).
3. M. D. Ingram., Phil.Mag.B., 60 729 (1989), Mater.Chem.Phys., 23 51 (1989)/ M. D. Ingram. M. A. Mackenzie, W. Müller and M. Torge., Solid State Ionics., 40/41 671 (1990).
4. L. Sjogren., Phys.Rev.A., 33 1254 (1986).

STRUCTURE AND DYNAMICS OF UNDERCOOLED AND GLASSY AQUEOUS IONIC SOLUTIONS BY NMR, X-RAY AND NEUTRON SCATTERING

T. Yamaguchi,* M. Yamagami, T. Takamuku, T. Hirano, and H. Wakita
Department of Chemistry, Fukuoka University, Johan-ku, Fukuoka 814-01

ABSTRACT

X-ray, neutron, and Raman scattering experiments have been performed on highly concentrated aqueous solutions of LiCl and ZnX_2 (X=Br and I) in the undercooled and glassy states as well as at room temperature. The T_1 relaxation time and the self-diffusion coefficient of the water protons in the aqueous the $ZnBr_2$ solutions have been measured at various temperatures toward the glass transition. For the concentrated LiCl solution, both X-ray and neutron data have revealed stronger hydration of chloride ions (an increase in the hydration number and a decrease in the first neighbor Cl-D distance) in the undercooled and glassy solutions than at ambient temperature. The X-ray data have clearly shown the enhancement of the hydrogen bonded network of water in the undercooled and the glassy states. X-ray and Raman data of the ZnX_2 solutions have demonstrated the formation of both $[ZnX_4]^{2-}$ and $[Zn(OH_2)_6]^{2+}$ species in the undercooled and glassy solutions, in contrast to the presence of $[ZnX_2]$ and $[ZnX_3]^-$ as main species at room temperature. The ^{1}H-NMR data have revealed that the motion of water molecules in the $ZnBr_2$ solutions slows down with increasing solute concentration and lowering temperature. The translational motion of the water molecules is more retarded than the rotational motion with the decrease in temperature.

INTRODUCTION

Undercooled and glassy solutions have recently attracted their attention in investigation of the crystallization process of electrolyte solutions and in cryobiology for conservation of biological materials. In spite of an extensive study of a number of glassy solutions by Angel and his co-workers,[1] there are relatively few investigations on equilibria between chemical species and their structure of undercooled and glassy solutions. In the present study we report the microscopic structure and dynamic properties of undercooled and glassy LiCl and ZnX_2 solutions revealed by X-ray, neutron and Raman scattering and FT-NMR techniques.[2]

EXPERIMENTAL

Two isotopically different 8.6 mol dm^{-3} LiCl solutions in D_2O were prepared using ^{35}Cl and ^{37}Cl. An aqueous LiCl solution in normal water was also prepared. Concentrated solutions of $ZnX_2 \cdot RH_2O$ (R=5 and 10) were prepared. Neutron scattering measurements were made with the HIT instrument at the National Laboratory for High Energy Physics (KENS). X-ray experiments were performed on an Rigaku $\theta - \theta$ diffractometer using the MoKα radiation. Raman spectra were recorded on a JEOL JRS-400T spectrometer using the 5145 Å line. The T_1 and the self-diffusion coefficient of the water protons were measured, respectively, by the inversion recovery method and by the pulsed-gradient spin-echo method on a JEOL JNM-FX200 spectrometer.

RESULTS AND DISCUSSION

The neutron first-order difference functions at various temperatures are shown in Fig. 1. The first peak is ascribed to the Cl-D(1) interactions. In the second peak are superimposed both Cl-O and Cl-D(2) ones within the hydrated chloride ions. The analysis of the Cl-D(1) peak has revealed that the hydration number of Cl$^-$ changes from ~6 at room temperature to ~7 in the undercooled (173 K) and the glassy (94 K) states, accompanied with an decrease in the Cl-D(1) distance from 2.29 to 2.26 Å in the glassy state. The X-ray total radial distribution functions are shown in Fig. 2. The pronounced first peak at 3.2 Å is due to the Cl-O interactions within the hydrated Cl$^-$ ions; the Cl-O distance is independent of the temperature. An interesting feature for the undercooled (243 and 193 K) and the glassy (138 K) solutions is the growing second and third peaks around 4.4 and 7.0 Å, respectively, with decreasing temperature, which demonstrate a gradual enhancement of the three-dimensional H-bonded network in the solution.

Fig. 3 shows Raman spectra for the ZnBr$_2$ solution with R=10. Peaks at 172, 185, and 207 cm^{-1} are assigned to the ν_1(Zn-Br) band for [ZnBr$_4$]$^{2-}$ [ZnBr$_3$]$^-$, and [ZnBr$_2$] species, respectively. With lowering temperature, the peak at 172 cm^{-1} increased, but those at 185 and 207 cm^{-1} decreased. It has also been observed that the ν_1(Zn-OH$_2$) band at 390 cm^{-1} increased with the decrease in temperature. The results have shown that both [ZnBr$_4$]$^{2-}$ and [Zn(OH$_2$)$_6$]$^{2+}$ species are stable, while [ZnBr$_2$] and [ZnBr$_3$]$^-$ species are not in the undercooled and the glassy solution. X-ray data have also confirmed this finding. This is probably caused by that water molecules dispelled from the first coordination shell of the di- and the tribromo complexes are stabilized in the H-bonded network enhanced at the undercooled temperatures.

The T_1 and the self-diffusion coefficient of the water protons are given in Table I for the aqueous ZnBr$_2$.10H$_2$O solution. The intermolecular and intramolecular components were calculated according to the literature.[3] Apparently the motion of water molecules slows down with decreasing temperature. The smaller value of $(1/T_1)_{inter}$ and the larger correlation time τ_{inter}, compared with the intramolecular ones, show that the translational motion of the water molecules is retarded more rapidly than the rotational motion with the decrease in temperature.

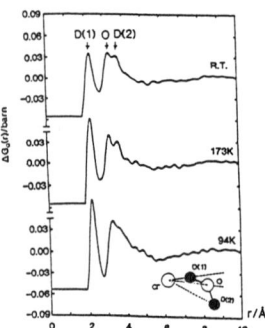

Fig. 1. Neutron $\Delta G_{Cl}(r)$ for the LiCl-D$_2$O solution.

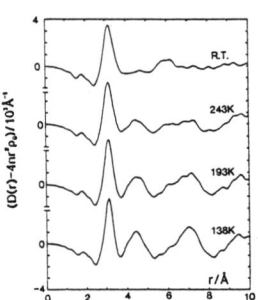

Fig. 2. X-ray $D(r) - 4\pi r^2 \rho_0$ for the LiCl-H$_2$O solution.

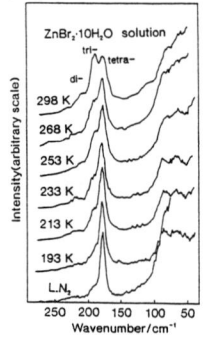

Fig. 3. Raman spectra for the ZnBr$_2$.10H$_2$O solution.

Table I Smoothed T_1 and D values and the evaluated components of the protons for the aqueous ZnBr$_2$ solution of $R=10$ at various temperatures

T (K)	T_1 (s)	$10^5 D$ (cm^{2}s^{-1})	$(1/T_1)_{inter}$ (s^{-1})	$(1/T_1)_{intra}$ (s^{-1})	τ_{inter} (ps)	τ_{intra} (ps)
293	1.70	0.894	0.286	0.301	42.6	4.32
278	0.845	0.478	0.506	0.677	79.7	9.74
273	0.702	0.403	0.587	0.837	94.6	12.0
268	0.564	0.326	0.702	1.07	117	15.4
253	0.287	0.154	1.28	2.20	248	31.7
243	0.186	0.080	1.98	3.40	477	48.8
233	0.127	0.033	2.98	4.90	1150	114

1. E. J. Sare and C. A. Angel, J. Solution Chem. **2**, 53 (1973).
2. T. Yamaguchi, Pure Appl. Chem. **62**, 2251 (1990).
3. E. W. Lang and H.-D. Lüdemann, in NMR Basic Principles and Progress (Springer, Berlin, 1990), Vol. 24, p. 129.

DYNAMICS OF AMORPHOUS POLYMERS NEAR GLASS TRANSITION

T. Kanaya, T. Kawaguchi and K. Kaji

Institute for Chemical Research, Kyoto University, Uji, Kyoto-fu 611, Japan

Dynamics of amorphous polymers near glass transition temperature Tg has been studied by quasielastic neutron scattering in the ω-range of 0.02 to 10 meV. It was found that two modes of motion appear near Tg in the energy range of ca. 1 meV and 10-20 μeV. The faster and slower modes arise from 30-50K below Tg and 10-20 K above Tg, respectively.

Glass transition is a universal phenomenon in all amorphous materials and has been investigated by many techniques. Recently, microscopic theory, so-called "mode coupling theory", has been presented to explain the phenomenon as a dynamical process. Stimulated by the theory, some experimental works have been performed by quasi- and inelastic neutron scattering. Amorphous polymers are one of typical amorphous materials, and there exists amorphous region even in crystalline polymers. In this report, we have studied dynamics of some amorphous polymers near glass transition by quasi- and inelastic neutron scattering.

Neutron scattering measurements have been carried out with the quasi- and inelastic scattering spectrometers LAM-40 and LAM-80ET [1] installed at the National Laboratory for High Energy Physics (KEK), Tsukuba, Japan. The energy resolutions are ca. 0.2 and 0.02 meV, respectively. The samples used for the scattering experiments are cis-1,4-polybutadiene (PB) and trans-1,4-polychloroprene (PCP) with the glass transition temperatures Tg=170K and 226K, respectively. The measurements were performed in the temperature range of 10K to 300K.

Fig.1 shows the quasielastic neutron scattering spectra of trans-1,4-polychloroprene (PCP) measured by LAM-40 as a function of temperature. The spectra below the glass transition temperature (226K) show a broad excitation peak at 2-3 meV. This "low energy excitation" is universal for all amorphous polymers and believed to be an origin of anomalous thermal properties of amorphous materials. This problem has been discussed in detail in the previous paper [2]. The intensity of the "low energy excitation" increases with temperature and can be normalized by the Bose factor below Tg. On the other hand, the intensity begins to exceed that expected from the Bose factor near Tg in the energy range below 2 meV, accompanying a steep decrease of the elastic scattering, while the intensity above 4meV still follows the Bose factor even above Tg. This indicates onset of a new motion near Tg in the energy range below 2 meV. A similar behavior was also observed for cis-1,4-polybutadiene (PB). The excess intensity at ω=1 meV is plotted as a function of temperature for both PCP and PB in Fig.2, which appears at 120K and 200K for PB and PCP, respectively, being 50K and 30K below

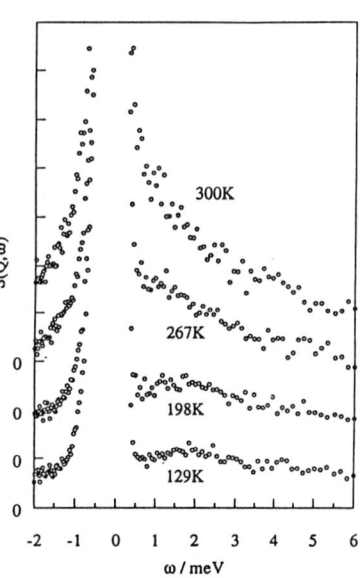

Fig.1. Quasielastic neutron scattering spectra of trans-1,4-polychloroprene (PCP) near Tg (226K) measured by LAM40.

© 1992 American Institute of Physics

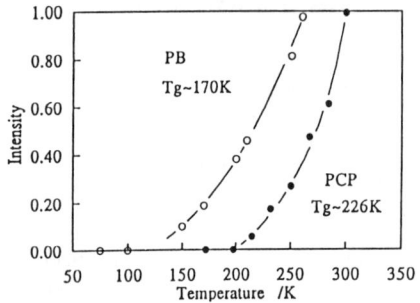

Fig.2. Excess quasielastic scattering intensity at ω=1 meV as a function of temperature for PCP(●) and PB(○).

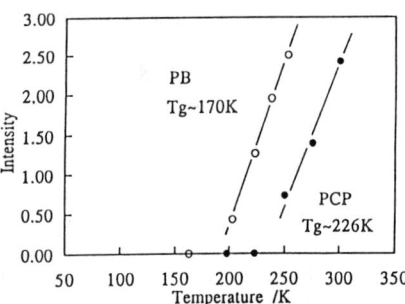

Fig.4. Quasielastic scattering intensity below 30 μeV as a function of temperature for PCP(●) and PB(○).

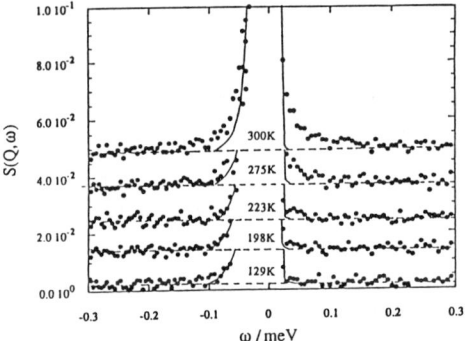

Fig.3. Quasielastic neutron scattering spectra of trans-1,4-polychloroprene (PCP) near Tg (226K) measured by LAM80ET. Dashed lines represent the contributions of the broad quasielastic component observed by LAM-40 (see Fig.1) and solid lines are the resolution function.

the corresponding Tg's. Widths of spectra of the excess scattering are about 1meV above Tg and almost independent of both temperature and length of scattering vector Q, suggesting that the new motion is very localized one.

Quasielastic neutron scattering measurements were also made in the low energy range (-0.3<ω<0.3 meV) using the high resolution spectrometer LAM-80ET. The observed spectra are shown in Fig.3 as a function of temperature. A new narrow quasielastic component is observed above Tg in the energy range below 10-20 μeV. The width of the quasielastic component increases with temperature, suggesting that the new motion is a thermally activated one. The quasielastic intensity is plotted against temperature for PB and PCP in Fig.4. These components appear at 190K and 235K for PB and PCP which are 20K and 10K higher than Tg, respectively.

There exist two modes of motions in the molten state of PB in the time range of 10^{-13}-10^{-10} s [3]; the faster and slower modes have been assigned to a damped vibrational motion and a conformational transition between trans(t) and gauche(g) states in a polymer chain, respectively. The broad (fast) quasielastic component appearing several tens K below Tg may correspond to the damped vibrational motion and the narrow (slow) component appearing only above Tg would be the trans-gauche transition, so-called primary relaxation in polymer chains.

[1] K.Inoue, T.Kanaya, Y.Kiyanagi, S.Ikeda, K.Shibata, H.Iwasa, T.Kamiyama, N.Watanabe and Y.Izumi, *Nucl. Instr. Methods*, in press.
[2] K.Inoue, T.Kanaya, S.Ikeda, K.Kaji, K.Shibata, M.Misawa and Y.Kiyanagi, *J. Chem. Phys.*, in press.
[3] T.Kanaya, K.Kaji and K.Inoue, *Macromolecules*, **24**, 1826(1991).

B. THEORY

α-RELAXATION IN SUPERCOOLED LIQUIDS

W. Götze
Physik-Department der Technischen Universität München,
D-8046 Garching, Germany
and Max-Planck-Institut für Physik und Astrophysik,
D-8000 München, Germany

L. Sjögren
Institute of Theoretical Physics, Chalmers University
of Technology, S-41296 Göteborg, Sweden

ABSTRACT

The mode coupling equations for the description of supercooled liquid dynamics exhibit cuspoid bifurcation singularities. The simplest glass transition singularity causes a liquid to glass cross over at some ideal glass transition temperature T_c. There appear α-relaxation processes whose time scale crosses over near T_c from power law to Arrhenius variation. For $T>T_c$ there holds the time-temperature superposition principle and scale universality. Below T_c scales decouple and the α-scaling law may be unvalid. Von Schweidler fractal decay is identified as reason for the stretching phenomenon. Simple schematic models can explain α-peaks quantitatively, in particular systematic deviations from the Kohlrausch law. α'-α-double peak patterns are obtained as result of self intersections of the bifurcation hypersurfaces. First principle calculations of the α-peaks for the hard sphere systems are discussed.

INTRODUCTION

Structural rearrangements of molecules become increasingly more difficult when a liquid is cooled. This leads to a slowing of relaxations and the appearance of corresponding low frequency peaks in various susceptibility spectra, the so-called α-peaks. The time scale τ_α of those processes, defined e.g. by the peak position ω_α via $\omega_\alpha \tau_\alpha = 1$, grows drastically with decreasing temperature T. If a system is quenched to sufficiently low T, if falls out of equilibrium at some temperature T_g and transforms to a glass, i.e. to a history dependent non equilibrium state of matter. The calorimetric glass transition temperature T_g is located where the characteristic probing frequency ω_p crosses the α-relaxation rate: $\omega_\alpha(T=T_g)=\omega_p$. Temperature T_g depends on the quantity probed and it varies with the test frequency ω_p[1]. An understanding of the α-peaks, i.e. an explanation of the T-variation of the scale τ_α, of the α-peak strength and, in general, of the α-peak shape is therefore an essential prerequisite for the explanation of the liquid to glass transition.

Neutron spin echo spectroscopy for $CaNO_3(CKN)$[2] and for orthoterphenyl (OT)[3], and recent light scattering experiments for CKN[4] have demonstrated, that the α-process is already fully developed within the GHz-band for T above the melting temperature

T_m. Similar results, obtained for polybutadien (PB)[5], show that this might be the case also for some polymers. There is the question: how can one understand the appearance of the α-peaks upon cooling? The quoted experiments as well as molecular dynamics studies on model systems[6] provided evidence for the existence of a cross over temperature $T_c > T_g$, contemplated first by Goldstein[7]. Near T_c the dynamics changes from one characteristic for a strongly coupled fluid to activated motion, which is characteristic for a glass. How does T_c come about and what is the full experimental signature of the T_c-cross over?

In this contribution the indicated issues shall be discussed within the mode coupling theory (MCT) for the supercooled liquid dynamics[8].

THE MODE COUPLING THEORY

The MCT deals with a set of M correlation functions of time t, $\Phi_q(t)$, or their Laplace transforms for complex frequency $z=\omega+i0$, $\Phi_q(z)$, $q=1,2,...,M$. A model for the dynamics of a system is specified by two groups of equations. The first one expresses $\Phi_q(z)$ as a Zwanzig-Mori fraction in terms of a generalized viscosity $M_q(z)$,

$$\Phi_q(z) = -1/[z - \Omega_q^2/[z + M_q(z)]] \quad , \tag{1a}$$

and the latter in terms of two other kernels $m_q(z)$, $\delta_q(z)$:

$$M_q(z) = \Omega_q^2[i\nu_q + m_q(z)]/[1 - \delta_q(z)[i\nu_q + m_q(z)]] \quad . \tag{1b}$$

The second group expresses the kernels as mode coupling functionals of the correlators and their derivatives:

$$m_q(t) = F_q(\vec{V}, \Phi_k(t)) \quad , \tag{2a}$$

$$F_q(\vec{V}, f_k) = \frac{1}{2} \Sigma_{kp} V(q,k,p) f_k f_p \quad , \tag{2b}$$

$$\delta_q(t) = \Sigma_{kp} V'(q,k,p) \Phi_k(t) \partial_t^2 \Phi_p(t) \quad . \tag{2c}$$

The coupling constants of the theory, the vertices $V(q,k,p) \geq 0$ and $V'(q,k,p) \geq 0$, are combined to mathematical control parameter vectors in some parameter spaces: $\vec{V} \in \mathbf{K}$, $\vec{V}' \in \mathbf{K}'$. The frequencies $\Omega_q > 0$ and friction coefficients $\nu_q \geq 0$ govern the short time transient motion and fix the microscopic excitation band of the model. Notice that the equations are completely regular and that they reflect no ad hoc assumption on glass transitions. The theory studies the long time properties of $\Phi_q(t)$ or the low frequency behaviour of $\Phi_q(z)$ or of the spectra for the correlations $\Phi_q''(\omega)$ and susceptibilities $\chi_q''(\omega) = \omega \Phi_q''(\omega)$ in their dependence on the vector $\vec{X} = (\vec{V}, \vec{V}') \in \mathbf{K} \times \mathbf{K}' \equiv \mathbb{R}$.

The quoted equations (1-2) can be derived[9,10] within the generalized kinetic equation approach towards simple liquid dynamics[11], exploiting Kawasaki's factorization approximation to fluctuating force correlators. In this case the correlators are the ones for density fluctuations $\rho_{\vec{q}}$: $\Phi_q(t)=\langle\rho_{\vec{q}}^*(t)\rho_{\vec{q}}\rangle/S_q$, $S_q=\langle|\rho_q|^2\rangle$ denoting the structure factor. So q is a discretizised wave vector modulus. $\Omega_q^2=q^2v^2/S_q$, with v being the thermal velocity, is the known characteristic liquid frequency and ν_q denotes the stochastic contributions to the friction kernel. The vertices V, V' can be expressed in terms of the microscopic system parameters. Hence, under the usual experimental conditions, the mathematical control parameter vector $\vec{X}$ is a function of T. Lowering T, the system moves on a path C: T→$\vec{X}$(T) through IR from the weak coupling to the strong coupling region. The kernel m_q describes the cage effect and the back flow patterns around particles which arise from the strong repulsive interaction in the densely packed system. The kernel δ_q describes particle jumps over barriers, i.e. phonon assisted hopping events.

GLASS TRANSITION SINGULARITIES

To begin with let us ignore kernel δ so that the space IR reduces to K. The latter splits into the set D_L of liquid states, the set D_G of ideal glass states, and the hypersurface D_c of glass transition singularities separating D_L and D_G. For $\vec{V}\epsilon D_L$ the correlators decay to zero for large times and the spectra are smooth functions of frequency; D_L contains the weak coupling region $\vec{V}\sim\vec{0}$. For $\vec{V}\epsilon D_G$ the dynamics is non-ergodic, the correlations do not decay but arrest at some Edwards-Anderson parameter $f_q(\vec{V})=\Phi_q(t\to\infty)>0$; D_G contains the strong coupling region $|\vec{V}|>V_0>0$ and $f_q(\vec{V})$ depends smoothly on $\vec{V}\epsilon D_G$. In this case the spectra consist of an elastic line on top of a continuum:

$$\Phi_q''(\omega) = \pi f_q(\vec{V})\delta(\omega) + \text{continuum} \quad ; \quad \vec{V} \epsilon D_G \quad . \qquad (3)$$

Thus $f_q(\vec{V})$ is the Debye-Waller factor of the arrested structure. It is obtained as solution of the set of implicit equations $f_q/(1-f_q)=F_q(\vec{V},f_k)$[9]. If there are several solutions, the largest one is $f_q(\vec{V})$. For $\vec{V}\to\vec{V}_c\epsilon D_c$ function $f_q(\vec{V})$ exhibits a bifurcation singularity; compare reference 12 for details. Driving the system through K by lowering T, the path C: T→$\vec{V}$(T) crosses D_c at some singularity $\vec{V}(T_c)=\vec{V}_c$, where an ideal liquid to glass transition occurs. The neighbourhood of V_c can be specified by some separation parameter $\sigma(\vec{V})$. It is a smooth function so that $\sigma(V)\lessgtr 0$ specifies, respectively, liquid and glass states, while $\sigma(V)=0$ is the equation for the transition hypersurface. Driving the system, e.g., by temperature changes, one can write $\sigma\propto(T_c-T)/T_c$ for T→T_c.

There are two generic possibilities for liquid to glass transitions: type A ones with $f_q(\vec{V}_c)=0$ and type B ones with $f_q^c=f_q(\vec{V}_c)>0$[13]. The latter is a Whitney fold or A_2 bifurcation and the former is a degeneracy thereof[14]. Type A transitions require linear terms in F_q, as is the case for particle motion in random fields, for phonon dynamics in mixed crystals or for polymers with some crystallinity. There are also higher order cuspoid singularities A_ℓ ($\ell\geqslant 3$), as is discussed in the accompanying paper. From now on the discussion is restricted to the B-fold.

For the full MCT including (2c) terms, one finds $\delta_q := -i\delta_q(\omega=0) > 0$ for all $\vec{V}$. The path $\vec{X}(T)$ avoids the ideal glass states ($\vec{V} \in D_G$, $\vec{V}' = \vec{0}$); in particular it does not hit glass transition singularities $\vec{X}_c = (\vec{V}_c, \vec{0})$. Hence all correlators behave regularly and do not exhibit a singularity for $T=T_c$. They describe smooth cross overs from liquid dynamics at $T>T_c$, where δ_q can often be neglected, to almost non-ergodic dynamics at $T<T_c$, where ergodicity restoring relaxation is driven by hopping events. If δ_q for $T=T_c$ is sufficiently small, however, the correlators exhibit strong anomalies for $T \sim T_c$, reflecting the existence of the singularity at $\vec{X}_c$. In case the dynamics can be evaluated for $\sigma \to 0$, $t \to \infty$, $z \to 0$ from the equation

$$\Phi_q(z)/[1 + (z + i\delta_q)\Phi_q(z)] = m_q(z) \quad . \tag{4}$$

The kernels have to be evaluated from (2). While it is important to consider the z-dependence of $m_q(z)$, kernel $\delta_q(z)$ can be approximated in leading order by its zero frequency limit. There are regions for σ, t, ω where δ_q can be dropped in (4). In case one comes back to the scenario described at the beginning of this section. The approximation $\delta_q=0$ is referred to as the idealized picture or as the simplified version of the MCT.

Idealized liquid to glass transitions have first been discussed within the MCT for particle motion in random fields, compare ref. 15 and papers quoted there. For this model mode coupling to currents (2c) as origin of ergodicity restoring mechanism was considered also[16]. Ideal type B transitions were considered first by Leutheusser[17] and Bengtzelius et al.[9].

Mazenko and collaborators have studied non-linear hydrodynamic equations via mode coupling approximations, see ref. 18 and papers quoted there. Recently an important qualitative difference between that theory and our approach has been emphasized[19]. The Das-Mazenko mechanism[18] for ergodicity restoration implies for $T<T_c$ the half width Γ_q of the α-peaks of density fluctuations to vanish proportional to the wave vector squared in the long wave length limit: $\Gamma_q^{DM} \propto q^2$ for $q \to 0$. Within our theory one gets only an irrelevant q dependence in the same limit: $\Gamma_q^{MCT} \to \Gamma_0 > 0$ for $q \to 0$. Γ_q can be measured as half width of the quasi-elastic line in the dynamical structure factor $S(q,\omega) = \Phi_q''(\omega) S_q$ or as inverse of the decay time of $\Phi_q(t)$. Experiments have been performed with photon correlation spectroscopy in the relevant regime for T close but below T_c for OT and CKN[20]. The relevant parameter q^2 was varied by factors as large as 6 without detectable change of Γ_q. It should be noted also, that the mathematical analysis of the MCT[10,13] applies also to the equations of the non-linear hydrodynamics[18]. In particular it is known, that all wave vector dependencies, and not only the $q \to 0$ limits, have to be treated properly in order to obtain the various fractal decay laws and the temperature dependence of the viscosity for $T \le T_c$ correctly. For example, the model of Leutheusser[17], which neglects all q-dependencies, describes α-relaxation by a Debye process and thus it misses the essential α-relaxation stretching phenomenon completely.

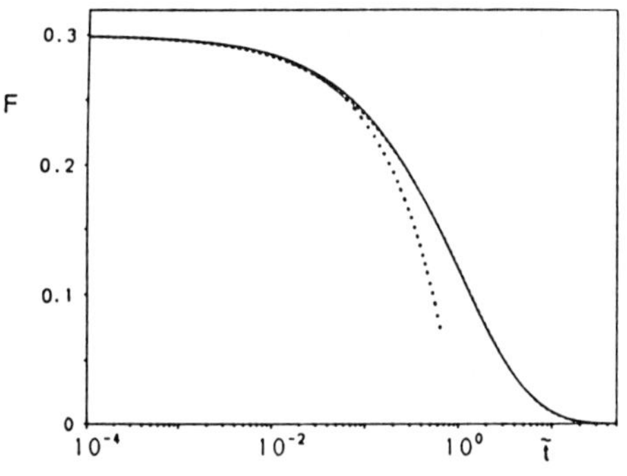

Fig. 1.

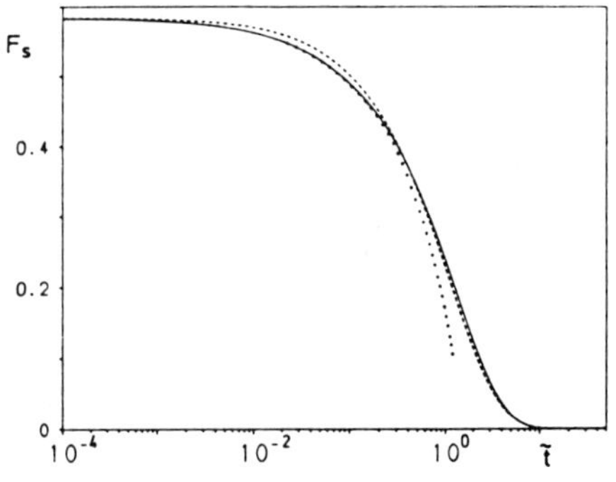

Fig. 2.

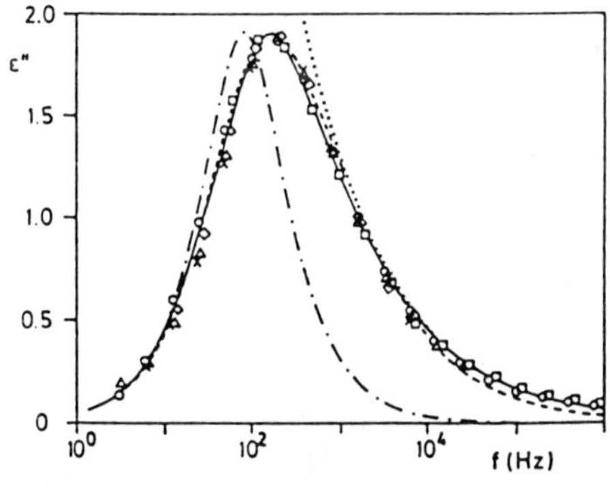

Fig. 3.

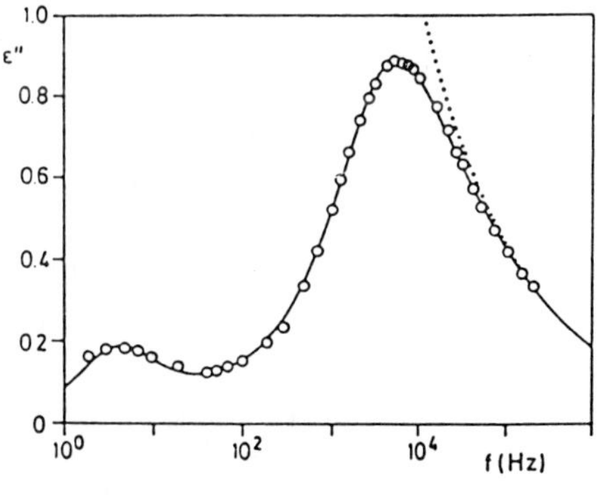

Fig. 4.

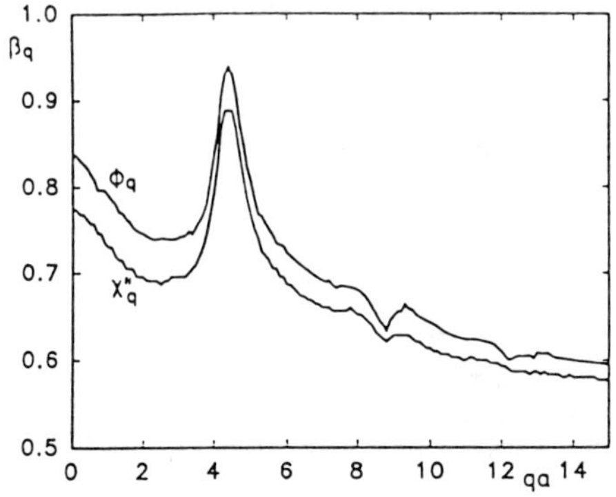

Fig. 5.

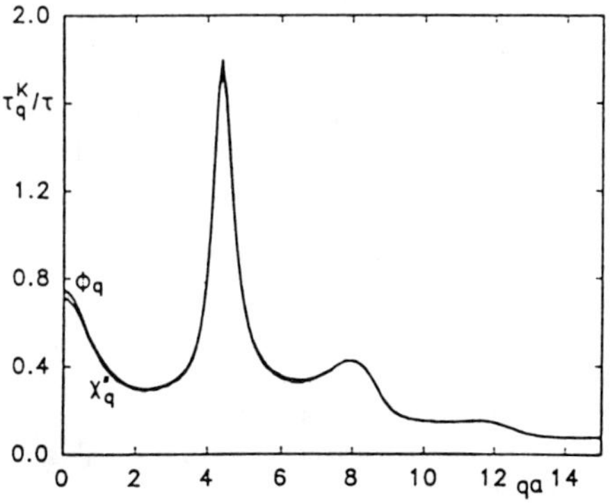

Fig. 6.

THE α-PROCESS

The dynamics of the idealized transition outside the microscopic transient region, $t_0 \ll t$, $\omega t_0 \gg 1$, is governed by two critical time scales t_σ and $\tau_\alpha = t'_\sigma$. Both scales diverge for $\sigma \to 0$ and so does their ratio: $t_\sigma \to \infty$, $t'_\sigma \to \infty$, $t'_\sigma/t_\sigma \to \infty$. The dynamics within the window $t_0 \ll t \ll t'_\sigma$ is referred to as ß-process and the one within the region $t_\sigma \ll t$ as α-process. The processes overlap on the interval $t_\sigma \ll t \ll t'_\sigma$, whose length diverges for $T-T_c \to 0$.

For the α-process one finds the scaling law:

$$\Phi_q(t) = F_q(t/\tau_\alpha) \quad , \quad T \to T_c^+ \quad . \tag{5a}$$

The sensitive T dependence enters via the one for the scale τ_α, while the master functions F_q are σ-independent in leading order. The master functions F_q can be determined as solution of the scaling equation

$$F_q(t) = m_q^c(t) - (d/dt) \int_0^t dt' \, m_q^c(t-t') F_q(t') \quad , \tag{5b}$$

where $m_q^c(t) = F_q(\vec{V}^c, F_q(t))$ is the mode coupling functional (2b) evaluated at the singularity. The scaling results hold also for the α-process of other variables Y, like dipole moments, shear etc.: $\Phi_Y(t) = F_Y(t/\tau_\alpha)$. Let us define an α-relaxation scale Γ_Y in some way, e.g. by $\Phi_Y(t=1/\Gamma_x) = \Phi_Y(t=0)/\ell$. Then one gets

$$\Gamma_Y = C_Y/\tau_\alpha \quad . \tag{6}$$

The T-independent constant C_Y depends on the variable Y and on the specific definition of Γ_Y. There is scale universality or coupling of scales in the sense, that the strongly T-dependent part of all scales Γ_Y is the same function τ_α. The scaling law (5a) was confirmed by neutron scattering data for CKN[2], OT[3] and PB[5] and the scale for density fluctuations for microscopic length scales was shown to agree with the one for shear according to (6). Light scattering data for CKN[4] are also in accord with (5a,6).

The idealized freezing signalizes itself by a power law divergency of the scale τ_α upon cooling. There is a $\sqrt{T_c-T}$ anomaly of the Debye-Waller factor upon heating as melting precursor for the ideal glass state.

$$\tau_\alpha = t_0/|\sigma|^\gamma \quad , \quad T \to T_c^+ \quad ; \quad f_q - f_q^c = h_q\sqrt{\sigma} \quad , \quad T \to T_c^- \quad . \tag{7}$$

Neutron scattering work has confirmed the predicted $\sqrt{\sigma}$-anomaly for OT[3] and PB[5]. The data followed the $\sqrt{T_c-T}$-law with the same T_c, which could be used for the power law fit to the viscosity scale. The so far published results for CKN are not conclusive in this respect.[21,22]

The initial decay of the α-process is described by the von Schweidler law

$$F_Y(\tilde{t}) = f_Y^c - h_Y B \tilde{t}^b + O(\tilde{t}^{2b}) \quad . \tag{8}$$

This implies a power law high frequency α-peak wing $\chi_Y''(\omega) \propto h_Y/(\omega\tau_\alpha)^b + O(1/(\omega\tau_\alpha)^{2b})$. The fractal exponent b, $0<b\leq 1$, is the same for all variables Y. But different transitions $\vec{V}_c$ are specified by different exponents b. Exponents b and γ are related, as shown in the accompanying paper. Notice, that the fractal dynamics is not related to any fractal structure in configuration space.

The mode coupling functional for m_q^c is given by the system's structure factor S_q at the transition, and so are all its implications like f_q^c, h_q, B, b etc. This implies, e.g., that the α-relaxation master functions $F_q(t)$ of the hard sphere system are the same for particles moving in free space according to Newton's equations of motion as for spheres diffusing in a solvent as in the case for colloidal suspensions. The utterly different microscopic dynamics of the two mentioned systems enters the microscopic time scale t_0 only, which varies smoothly with $\vec{V}$.

Activated processes as caused by $\delta_q \neq 0$ modify the quoted results as follows. The scale τ_α does not diverge for $T \to T_c$ but it crosses over to activated variation for $T \ll T_c$: $\tau_\alpha \propto \exp(E/T)^{23}$. The $\sqrt{T_c}$-T-cusp of the α-peak area f_q is smeared out. The scaling law (5a) may be unvalid for $T<T_c$ and the scales for different variables Y may decouple below T_c. For $T<T_c$ the von Schweidler exponent may become T-dependent[14]. The value of δ_q depends on the microscopic dynamics; e.g., δ_q for the hard sphere system in free space is different from that describing colloidal suspensions.

THE α-PEAK SHAPES

Suppose a variable Y like the dipole moment is composed of many small contributions Y_i: $Y = (Y_1 + \cdots + Y_N)/C_N$. One might think the different Y_i to be due to different molecular complexes. Because of (5a,8) all Y_i relax for short times like $\Phi_i = f_i(1-(\Gamma_i t/\tau_\alpha)^b)$. If the Y_i can be treated as independent random variables the limit theorem of Lévy implies for $N \to \infty$, $C_N \to \infty$ the variable Y to relax according to a stretched exponential,

$$\Phi_Y(t) = f_Y \exp{-(t\Gamma_Y/\tau_\alpha)^\beta} , \qquad (9)$$

where $\beta = b^{24}$. This result generalizes to fractal dynamics the known formula for regular time variations. In the latter case b=2 and (9) leads to the Gaussian form for inhomogeneously broadened spectral lines. The law (9), used first by Kohlrausch, is known to describe α-relaxation reasonably well[1]. Thus the MCT explains α-peak stretching qualitatively and traces it back to the von Schweidler decay process. In some rare cases the accessible part of the α-process follows (9) within the experimental uncertainties. Also the simplest meaningful schematic MCT model, using only one correlator, reproduces (9) very well[25]. Figure 1[26] shows a representative example, where the solution F of (5b), shown in full, can hardly be distinguished from a fit by (9) with $\beta=0.59$, shown in dashed.

As a rule one expects systematic deviations from the Kohlrausch law. One cannot, e.g., apply the above reasoning simultaneously to the dipolmoment, so that (9) yields a fit for the dielectric function, and to the fluctuating forces, so that (9) yields a fit for the dielectric modulus. Within the MCT variables are not independent random quantities, but non-linearly coupled correlators. Figure 2[26] reproduces as full line the second correlator F_s of an M=2 model; the first correlator for the same model is given in fig. 1. The dashed curve in fig. 2 is a Kohlrausch fit with ß=0.80, it differs systematically from the master function for $10^{-2}<t<10^{-1}$. Both correlators have the same von Schweidler part, shown for b=0.64 as dotted curves in figs. 1,2. The figures demonstrate, that different correlators at the same transition require different Kohlrausch exponents ß for the fit. The exponent ß depends on the fit interval and on the criterion for judging the fit quality. Thus ß appears as a convenient parameter for data characterization; opposed to b, exponent ß seems to have no physical or mathematical significance. The mentioned discrepancy between true master function and Kohlrausch fit shows up by the susceptibility spectra $X''(\omega)$ to lie above the Kohlrausch asymptote $1/(\omega\tau_\alpha)^\beta$ in the large frequency tail $\omega\tau_\alpha \gg 1$, since usually b<ß. This is observed for OT[27] and CKN[28] spectra; in the latter case the discrepancy between data and fit by (9) is a factor 2 and it extends over a two decade frequency interval. Figure 3[26] reproduces rescaled dielectric loss data of Ishida et al.[29] for polyvinylacetat (PVA). The von Schweidler asymptote, shown in Fig. 3 for b=0.44 in dotted, accounts for the data for the three decade interval $f>10^3$Hz. The Kohlrausch fit, shown for ß=0.58 as dashed curve, describes the upper half and the whole left wing of the α-resonance; but it fails for $f>10^4$Hz. The dashed dotted line in fig. 3 is a Debye resonance shifted to match the low frequency wing. The full curve through the data is a solution of (5b), calculated for an M=3 MCT model[26].

Some polymers exhibit a splitting of the low frequency susceptibility spectra into α'-α-peak pairs. Figure 4[26] shows this phenomenon for a dielectric loss spectrum measured by Johari for poly propylene glycol[30]. The whole spectrum follows the scaling law (5a) and therefore both peaks have to be considered as parts of one α-process[30]. Such patterns are indeed obtained as generic results of the MCT. They occur if $\vec{V}$ is close to corners of the bifurcation hypersurface, which are consequences of self crossings of parts of D_c[26]. The full curve in fig. 4 was calculated for the same M=3 schematic model mentioned above in connection with fig. 3.

The MCT explains the α-peaks as result of topological singularities. Even simple schematic models can reproduce data quantitatively. However, such schematic models cannot provide a true microscopic understanding of details, like e.g. the shown differences of the PVA and PPG loss spectra. The glass transition is a phenomenon ruled by details of the structure on microscopic length scales. A microscopic understanding of quantitative details of the α-peak shape requires an understanding of the structural details as they enter the mode coupling vertices $V(q,k,p)$ in (2b). In this sense the interpretation of the α-process within the MCT is more difficult but also much richer than the theory of the ß-process. For the latter there holds universality in several respects, as can be inferred from the accompanying paper.

THE HARD SPHERE SYSTEM

The hard sphere system (HSS) is the simplest condensed matter model. Its equilibrium functions are specified by only one control parameter, viz. the packing fraction φ. Crystallization effects render molecular dynamics studies of the supercompressed liquid difficult. However, colloidal suspensions allow for detailed experimental studies of equilibrium quantities and correlation functions for all states; compare ref. 31 for a review. An ideal liquid to glass transition was identified and the characteristic α-β-relaxation pattern was observed for the density correlations of wave vector q_0 at the peak of the structure factor S_q, compare ref. 32 and papers quoted there. For the HSS all previously mentioned MCT results for the ideal transition scenario have been calculated[9,33,34].

The theoretical value for the critical packing fraction, φ_c^{MCT}= 0.52±0.01[9,33], differs by less than 10% from the experimental value φ_c=0.560±0.005[32]. The Debye-Waller factor at the transition f_q^c was measured for three q-values[32] and was found not to differ from the prediction[9] by more than 10%; compare ref. 35 for more extensive results. The calculated HSS exponent γ=2.58[33] accounts well for the diffusivity α-scale, known from molecular dynamics studies[36].

The over all properties of the α-peaks fit to the discussion of the preceding section, but there is a variety of subtle variations of relaxation features with wave vector q of, e.g., the density correlator Φ_q[34]. For example: the Kohlrausch fit is very good for qa=0 and 20 while for qa=1.2, 3.6 and 15 the Cole-Davidson formula is a superb fit of the α-resonances; here a is the averaged interparticle distance. At the peak of the structure factor, $q_0 a$=4.4, the fit (9) differs from the true master function F_{q_0} stronger than shown in fig. 2; the data[32] are closer to F_{q_0} than to the Kohlrausch fit. Most relevant are the following two predictions. Stretching is less pronounced at the structure factor peak q_0 than off the peak. Fitting a Kohlrausch law to the master curves $F_q(\tilde{t})$ or to the susceptibility, one finds the Kohlrausch exponent ß to vary with q as shown by the upper or lower curve respectively in fig. 5[34]. The α-peak decay time $\tau_q \propto 1/\Gamma_q$ drops by about a factor 5 if q varies near q_0 as shown in fig. 6[34]. If the Pusey-van Megen experiments[32] could be extended to some range of wave vectors around q_0 and to times somewhat longer than studied so far, one could test not only the MCT predictions for scaling laws and scale variations. One could also test the relevance of our theory for predictions of structure dependent details as exhibited in figs. 5,6.

REFERENCES

1. J. Wong and C.A. Angell, *Glass: Structure by Spectroscopy* (Marcel Dekker, Basel, 1976).
2. F. Mezei, W. Knaak and B. Farago, Phys. Rev. Lett. **58**, 571 (1987) - Phys. Scr. **T19**, 363 (1987) - Europhys. Lett. **7**, 529 (1988).
3. W. Petry, E. Bartsch, F. Fujara, M. Kiebel, H. Sillescu and B. Farago, Z. Phys. **B83**, 175 (1991).
4. H.Z. Cummins, G. Li, W.H. Du and X.K. Chen, this volume.

5. D. Richter, B. Frick and B. Farago, Phys. Rev. Lett. 61, 2465 (1988) - Phys. Rev. Lett. 64, 2921 (1990).
6. J.N. Roux, J.L. Barrat and J.P. Hansen, J. Phys.: Cond. Matter 1, 7171 (1989) - Chem. Phys. 149, 197 (1990); G.F. Signorini, J.L. Barrat and M.L. Klein, J. Chem. Phys. 92, 1294 (1990).
7. M. Goldstein, J. Chem. Phys. 51, 3728 (1969).
8. L. Sjögren and W. Götze, Dynamics of Disordered Materials, eds. D. Richter et al. (Springer Verlag, Berlin, 1989), p. 18.
9. U. Bengtzelius, W. Götze and A. Sjölander, J. Phys. C: Solid State Phys. 17, 5915 (1984).
10. W. Götze and L. Sjögren, Z. Phys. B65, 415 (1987).
11. L. Sjögren, Phys. Rev. A22, 2866 (1980).
12. W. Götze, Liquids, Freezing and Glass Transition, eds. J.P. Hansen et al. (North Holland, Amsterdam, 1991), p. 287.
13. W. Götze, Z. Phys. B56, 139 (1984) - B60, 195 (1985) - Amorphous and Liquid Materials, ed. E. Lüscher et al. (Martinus Nijhoff, Dordrecht, 1987), p. 34.
14. W. Götze and L. Sjögren, J. Phys.: Cond. Matter 1, 4183 (1989).
15. W. Götze, Recent Developments in Condensed Matter, ed. J.J. Devreese (Plenum Press, New York, 1981), p. 133.
16. D. Belitz and W. Schirmacher, J. Non-Cryst. Solids, 61, 1073 (1984).
17. E. Leutheusser, Phys. Rev. A29, 2765 (1984).
18. K. Kim and G.F. Mazenko, Adv. Chem. Phys. LXXVIII, 129 (1990).
19. M. Fuchs and A. Latz, J. Chem. Phys., in print, (1991).
20. G. Fytas, C.H. Wang, D. Lilge and Th. Dorfmüller, J. Chem. Phys. 75, 4247 (1981) - E.A. Pavlatou, A.K. Rizos, G.N. Papatheodorou and G. Fytas, J. Chem. Phys. 94, 224 (1991).
21. M. Fuchs, W. Götze and A. Latz, Chem. Phys. 149, 185 (1990).
22. F. Mezei, Ber. Bunsenges. Phys. Chem., in print (1991).
23. L. Sjögren, Z. Phys. B79, 5 (1990).
24. B.V. Gnedenko and A.N. Kolmogorov, Limit Distributions for Sums of Independent Random Variables (Addison-Wesley, Reading, 1954).
25. H. DeRaedt and W. Götze, J. Phys. C: Solid State Phys. 19, 2607 (1986) - W. Götze and L. Sjögren, J. Phys. C: Solid State Phys. 20, 879 (1987).
26. M. Fuchs, W. Götze, I. Hofacker and A. Latz, J. Phys.: Cond. Matter 3, 5047 (1991).
27. G. Williams and P.J. Hains, Faraday Symp. Chem. Soc. 6, 4 (1972).
28. F.S. Howell, R.A. Bose, P.B. Maced and C.T. Moynihan, J. Chem. Phys. 78, 639 (1974).
29. Y. Ishida, M. Matsuo and K. Yamafuji, Kolloid-Z. 180, 108 (1962).
30. G.P. Johari, Polymer 27, 867 (1986).
31. P.N. Pusey, Liquids, Freezing and Glass Transition, eds. J.P. Hansen et al. (North Holland, Amsterdam 1991), p. 763.
32. P.N. Pusey and W. van Megen, Ber. Bunsenges. Phys. Chem. 94, 225 (1991) - Phys. Rev. A43, 5429 (1991).
33. J.L. Barrat, W. Götze and A. Latz, J. Phys.: Condensed Matter 1, 7163 (1989).
34. M. Fuchs, I. Hofacker and A. Latz, preprint, August 1991.
35. W. van Megen, S.M. Underwood and P.N. Pusey, Phys. Rev. Lett., in print (1991) - this volume.
36. L.V. Woodcock and C.A. Angell, Phys. Rev. Lett. 47, 1129 (1981).

β-RELAXATION NEAR GLASS TRANSITIONS

L. Sjögren
Institute of Theoretical Physics, Chalmers University of Technology,
S-412 96 Göteborg, Sweden

W. Götze
Physik Department, Technische Universität München, D-8046 Garching, Germany,
and Max-Planck Institut für Physik und Astrophysik, D-8000 München, Germany

ABSTRACT

The relaxation properties of the β-process as described within the mode-coupling theory for the dynamics of supercooled liquids is reviewed. This process is completely ruled by generic properties of underlying bifurcation singularities in the nonlinear equations of motion. The singularities belong to the cuspoid family A_k with $k \geq 2$, and lead to various scaling scenarious involving one or more parameter scaling functions depending on the order k of the relevant singularity. The theoretical predictions are compared with dielectric measurements on several polymeric systems, and in the relevant frequency region we find a quantitative agreement between theory and experimental results.

INTRODUCTION

The mode-coupling theory (MCT) on the dynamics of supercooled liquids makes very precise predictions about the frequency and temperature dependence of various correlation functions[1]. In particular the density correlation function $\phi_q(t)$ is of primary concern. Referring to the accompanying paper, we find for $\phi_q(t)$ for times $t \gg t_0$ or frequencies $\omega \ll 1/t_0$, where t_0 denotes a microscopic time, the nonlinear closed equations:

$$\frac{\phi_q(z)}{1 + [z + i\delta_q]\phi_q(z)} = m_q(z). \tag{1a}$$

Here

$$m_q(t) = \frac{1}{2} \sum_{k,p} V(q, kp)\phi_k(t)\phi_p(t). \tag{1b}$$

and δ_q is given by a similar expression with current-correlation functions. The vertex $V(q, kp) \geq 0$ is given by the structure factor S_q, and is assumed to depend regularly on external control parameters like temperature. The expression in (1b) describes collective effects arising from cooperative motions of any particle and its surrounding.

These equations describe both the α- and β-relaxation processes. The α-process describes the decay of density fluctuations on the longest time scale τ_α. The β-process exists on mesoscopic time-scales between the microscopic and α-regions. For the latter process one finds explicitly from (1)[2]

$$\phi_q(t) = f_q^c + h_q\, G(t), \tag{1c}$$

valid in the time region $t_0 \ll t \ll \tau_\alpha$. The whole relaxation pattern in the β-region is therefore described by one single function $G(t)$ only. Similar relations hold for correlation functions $\phi_{XY}(qt)$ of any variables X and Y with a nonzero overlap with density fluctuations. In particular with X and Y representing the microscopic dipole-moment we find for the dielectric function[3]

$$\epsilon(z) = f_\epsilon + h_\epsilon z G(z) \tag{2}$$

valid in the frequency region of interest here.

The function $G(t)$ satisfies the equation

$$-\delta_0/z + i\delta/z^2 + \delta_1 G(z) + zG^2(z) + (1+\delta_2)\text{LT}\left[G^2(t)\right](z)$$
$$+ [\delta_3 + \gamma_3]\,\text{LT}\left[G^3(t)\right](z) - \gamma_3 z^2 G^3(z) + \cdots = 0, \tag{3a}$$

where the Laplace transform is defined as

$$G(z) = \text{LT}\left[G(t)\right](z) = i\int_0^\infty dt\, e^{izt} G(t). \tag{3b}$$

The parameters δ_k and γ_k can be expressed as wave-vector integrals over the physical coupling constants $V(q,kp)$ in (1b), while δ is given by an integral over δ_q.

The solution to (3a) show interesting properties near various bifurcation points belonging to the cuspoid family A_k. These are defined by points in the parameter space where some coefficients δ_k vanish. This happens at some critical temperature T_c, where $\delta_0(T_c) = \delta_1(T_c) = 0$. Near such singular points there is a subtle low frequency singularity in $G(z)$, which can then be classified according to the corresponding cuspoid A_k. If $\delta_0^c = \delta_1^c = 0$ but $\delta_2^c \neq 0$ we have an A_2 or Whitney fold singularity. For parameter points where also $\delta_2^c = 0$ but $\delta_3^c \neq 0$ we have an A_3 or Whitney cusp singularity and so on[4,5].

The solution to (3) show different behaviour in the regions $|\delta_0/z| \gg |\delta/z^2|$ and $|\delta_0/z| \ll |\delta/z^2|$ respectively. In the former activated processes can be neglected. In this case the transition can be treated as an ideal one as explained in the accompanying paper. For a fixed δ, the latter region will appear for sufficiently low frequencies, or when $\delta_0 \approx 0$ near T_c. The full solution including hopping processes have been studied[6,7], but here we will only consider the ideal case with $\delta = 0$.

THE FOLD

The general cuspoid A_k has a canonical representation with $k-1$ relevant parameters $\delta_0, \ldots, \delta_{k-2}$. For the dynamical solution the cusp parameter $\mu_k = -\delta_k > 0$ also enters. All other parameters in (3a) become irrelevant for the leading order results. The Whitney fold singularity A_2 is therefore completely specified by just one separation parameter $\sigma = \delta_0$, and the exponent parameter $\lambda = 1 - \mu_2$. The function $G(t)$ is given by a conventional scaling law

$$G(t) = c_\sigma\, g_\pm(t/t_\sigma). \tag{4a}$$

The solution depends, through the correlation scale c_σ and the time scale t_σ, crucially on the distance from the transition point measured via the separation parameter σ. The masterfunctions $g_\pm$ satisfy the equation[2]

$$\mp\frac{1}{\zeta} + \zeta g_\pm^2(\zeta) + \lambda i \int_0^\infty d\tau\, e^{i\zeta\tau} g_\pm^2(\tau) = 0, \tag{4b}$$

where $\zeta = zt_\sigma$ and $\tau = t/t_\sigma$ are rescaled variables. Here $\pm$ refers to $\sigma \gtrless 0$ respectively. The scaling function depends therefore only on the exponent parameter λ, which is characteristic for any particular system. The solution of (4b) can be obtained for any value of λ [8]. We can also obtain explicit expressions for short and long rescaled times. In the former case one finds

$$g_\pm(\tau \ll 1) = 1/\tau^a. \tag{5a}$$

For long rescaled times one finds

$$g_+(\tau \gg 1) = 1/\sqrt{1-\lambda}, \tag{5b}$$

$$g_-(\tau \gg 1) = -B\tau^b, \tag{5c}$$

where $B > 0$. The exponents a and b are given by the exponent parameter λ by way of

$$\lambda = \Gamma^2(1-a)/\Gamma(1-2a) = \Gamma^2(1+b)/\Gamma(1+2b), \tag{6}$$

where $0 < a < \frac{1}{2}$ and $0 < b < 1$. The behaviour in (5c) gives the von Schweidler law

$$\phi_q(t) = f_q^c - h_q B(t/t_\sigma')^b. \tag{7}$$

This describes also the initial decay of the α-process. The new time scale $t_\sigma' = \sqrt{|\sigma|}t_\sigma$ is the predicted scale for the α-relaxation process. The von Schweidler law therefore describes the β-process for $\sigma < 0$ for long times on scale t_σ, as well as the α-process for short times on scale t_σ'. The decay in (7) can only hold for $t/t_\sigma' \ll 1$. For larger times on scale $t_\sigma' = \tau_\alpha$ the dynamics is ruled by the full equation for the α-process, as described in the accompanying paper.

The various scales are predicted to follow algebraic variations with σ. The correlation scale in (4a) has a simple square root variation

$$c_\sigma = \sqrt{|\sigma|}. \tag{8a}$$

Similarly the frequency scales $\omega_\sigma = 1/t_\sigma$ and $\omega_\sigma' = 1/\tau_\alpha$ are given by

$$\omega_\sigma = 1/t_\sigma = |\sigma|^{1/2a}/t_0, \tag{8b}$$

$$\omega_\sigma' = 1/\tau_\alpha = |\sigma|^\gamma/t_0, \qquad \gamma = 1/2a + 1/2b. \tag{8c}$$

These predicted results for the fold or A_2 singularity, can be tested most clearly for the susceptibility or the dielectric loss. The critical decay in (5a) implies a decrease of $\epsilon''(\omega) \propto \omega^a$ with decreasing frequency, while the von Schweidler law (5c) implies a subsequent increase $\epsilon''(\omega) \propto \omega^{-b}$. Hence for $T > T_c$ the β-decay process can be detected as a minimum in the dielectric loss $\epsilon''(\omega)$ located above the α-peak at some frequency $\omega_{\min}$. Such a minimum has been observed in polyethylene therephthalate[9] (PET) and in polyethylene oxybenzoate[9] (PEOB) for samples with varying degree of crystallinity. The rescaled dielectric loss $\epsilon''(\omega)/\epsilon''_{\min}$ versus $\omega/\omega_{\min}$, for a PET sample with 5% crystallinity, are shown in figure 1[10]. Here the scaling parameter $\epsilon''_{\min}$ denotes

$\epsilon''(\omega_{min})$. Clearly the experimental data for different temperatures fall on a single mastercurve around the minimum, and the frequency interval where the data-points coincide expands for decreasing temperatures. The solution to the scaling function in (4b) with $\lambda = 0.85$, giving the exponents $a = 0.25$; $b = 0.39$, is also shown as full curve. There is clearly an overall quantitative agreement between experiments and theory. Additional measurements for higher frequencies would be valuable to test the predicted behaviour. Figure 1 verifies that the experimental data satisfy a scaling law, and also that the shape of the corresponding master function agrees with the predicted one.

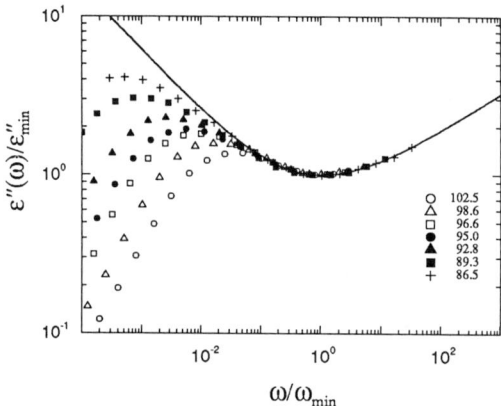

Figure 1. Rescaled dielectric loss $\epsilon''(\omega)/\epsilon''_{min}$ versus ω/ω_{min} for PET with 5% crystallinity[9]. The various symbols refer to different temperatures as indicated in the figure. Also included as the full curve is the theoretical master curve obtained from (4b) for $\lambda = 0.85$.

The two scales ϵ''_{min} and ω_{min} describe the variation of the minimum as the temperature changes. From (8) we find the predicted variation

$$\epsilon''_{min} \propto c_\sigma = |\sigma|^{1/2}, \tag{9a}$$

$$\omega_{min} \propto \omega_\sigma \propto |\sigma|^{1/2a}. \tag{9b}$$

From (8c) we also find the predicted variation of the positions of the α-peak maximum, ω_{max}

$$\omega_{max} \propto \omega'_\sigma \propto |\sigma|^\gamma. \tag{9c}$$

These results are tested in figure 2. In figure 2a we show $\epsilon''_{min}{}^2$ versus temperature T. According to (9a) the data should fall on a straight line, which is approximately the case as shown by the full line. From the intersection with the T-axis we get an estimate of the critical temperature $T_c = 84.5°C$. Similarly we show in figure 2b ω_{min}^{2a} (dots) and $\omega_{max}^{1/\gamma}$ (open circles) versus T. The data fall again on straight lines indicated by the full and dashed lines, and this supports

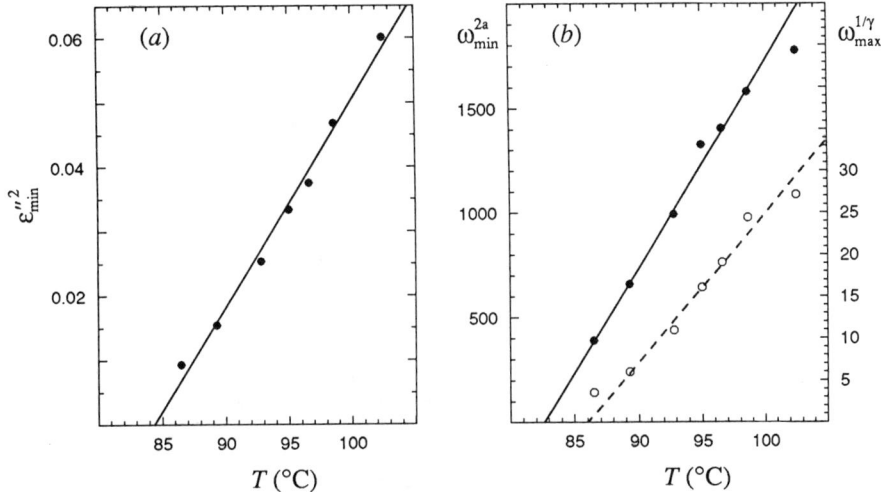

Figure 2. (a) The dots are the experimental values of $\epsilon''_{min}{}^2$ plotted versus temperature T. The straight line is the best fit through the data points, and indicate the theoretical prediction in (8a) and (9a). The intersection with the abscissa gives the critical temperature T_c. (b) Filled circles show ω^{2a}_{min} (left scale) and open circles $\omega^{1/\gamma}_{max}$ (right scale) versus temperature as obtained from experimental data. The full and dashed straight lines are the best fits through the respective data points, and gives the theoretical predictions in (8) and (9). The intersection gives again the critical temperature T_c.

the results in (9b, c). From the intersection with the T-axis we obtain the estimates $T_c = 83°C$ and $T_c = 86°C$ respectively. The three values of T_c agree within a few degrees and give the estimated critical temperature $T_c = (85\pm2)°C$. Results similar to these for PET were also found for amourphous PEOB[3].

The scaling function $g_\pm(\tau)$ can also be calculated, and the results can be directly compared with photon correlation and neutron spin echo measurements. A comparison with light scattering measurements on a hard sphere colloidal system[11] gave a quantitative agreement between experiments and theory[12]. There are also extensive molecular dynamics results on simple two component systems[13]. A similar analysis for three correlators describing relaxation with wave vector at the structure factor peak[14] produced reasonable agreement between data and MCT within the relevant intermediate time window[15]. This is shown in figure 3 for one example. The presently known molecular dynamics data[13,14] fit into the simple MCT scenario which ignores ergodicity restoring $\delta \neq 0$ contributions

THE CUSP

The relaxation scenario for the A_3 singularity or the cusp, is obtained from (3a) in the parameter region where also $\delta^c_2 = 0$ or $\lambda = 1$. To distinguish these special points with the previous ones, we will denote the corresponding critical temperatures with T_0. The solution is in this case given by the Weierstrass elliptic function[16]

$$G(t) = \rho^2 \wp\left[\ln(t/t_1); g_2, g_3\right]. \tag{10}$$

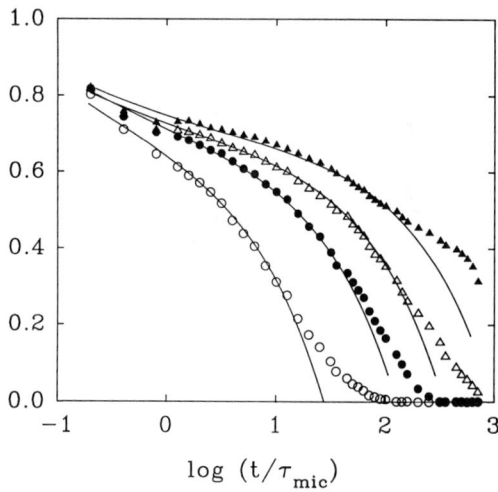

Figure 3. Relaxation data for the first tagged particle correlator of a binary mixture for the coupling constants $\Gamma = 1.38, 1.43, 1.44$ and 1.45^{14}. The solid lines are fits by (1c, 4a) with the calculated value $\lambda = 0.73^{15}$.

valid in the time region $t_1 \ll t \ll \tau_\alpha$. Here $\rho^2 = 2\pi^2/3\mu_3$ and t_1 is a microscopic time. The cusp is generically specified by two canonical separation parameters $\xi = \delta_0$ and $\eta = \delta_1$, which appear in (10) as $g_2 = 4\eta/\mu_3\rho^4$ and $g_3 = 4\xi/\mu_3\rho^6$. This gives the dielectric constant[3]:

$$\epsilon'(\omega) = f_\epsilon - \epsilon_c \wp \left[\ln(1/\omega t_1); g_2, g_3\right] = f_\epsilon - c'_\xi \wp \left[u; \pm 12(r/4)^{1/3}, \pm 4\right] \quad (11a)$$

and for the dielectric loss

$$\epsilon''(\omega) = -\frac{\pi}{2}\epsilon_c \wp' \left[\ln(1/\omega t_1); g_2, g_3\right] = -\frac{\pi}{2} c''_\xi \wp' \left[u; \pm 12(r/4)^{1/3}, \pm 4\right], \quad (11b)$$

where

$$u = \ln(1/\omega t_1)/y_\xi, \quad (11c)$$

and $\epsilon_c = h_\epsilon \rho^2$. The second relations above where obtained using the scaling property of $\wp$. The scales entering in (11) are predicted to have the following temperature dependence

$$c'_\xi = h_\epsilon |\xi/\mu_3|^{1/3} \propto |T/T_0 - 1|^{1/3}, \quad (12a)$$

$$c''_\xi = h_\epsilon/\rho|\xi/\mu_3|^{1/2} \propto |T/T_0 - 1|^{1/2} \quad (12b)$$

and

$$1/y_\xi = |\xi/\mu_3|^{1/6}/\rho \propto |T/T_0 - 1|^{1/6}. \quad (12c)$$

where T_0 denotes the location of the cusp at $\xi = \eta = 0$. To reach this point one needs in general to vary two control parameters like temperature and degree of crystallinity.

The solution in (10, 11) implies a two-parameter scaling law, where time or frequency appears as $\ln(t)$ and $\ln(\omega)$ respectively. The parameter r in (11) is defined as $r = |g_2^3/27g_3^2|$, and along the lines r =const the solution is invariant. In particular $r = 0$ and $r = \infty$ corresponds to the coordinate axis $g_2 = 0$ and $g_3 = 0$ respectively. Along these lines the Weierstrass function has relatively simple properties as well as on the line $\Delta = g_2^3 - 27g_3^2 = 0$ corresponding to $r = 1$. Here Δ denotes the discriminant and the points where $\Delta = 0$ define two fold lines which meet in a cusp.

The solution in (11) can be expanded for $u \ll 1$ and $u \gg 1$ respectively. The former case corresponds to $\omega \gg \omega_\xi = e^{-y_\xi}/t_1$ and leads to the critical behaviour

$$\epsilon'(\omega) = f_\epsilon - \epsilon_c/\left[\ln(1/\omega t_1)\right]^2 \qquad (13a)$$

and

$$\epsilon''(\omega) = \pi\epsilon_c/\left[\ln(1/\omega t_1)\right]^3 . \qquad (13b)$$

These equations are valid in the frequency interval $1/t_1 \ll \omega \ll \omega_\xi$, and here the solution becomes independent of the control parameters g_2 and g_3 or r and ξ.

For lower frequencies $\omega \ll \omega_\xi$ we find a von Schweidler region similar to the fold scenario. However, this law depends now on the sector in the $g_2 - g_3$ plane. For $r \approx 0$ we find

$$\epsilon'(\omega) = f_\epsilon - 3c'_\xi + 2c''_\xi \ln(1/\omega t_1), \qquad (14a)$$

and

$$\epsilon''(\omega) = \pi c''_\xi \qquad (14b)$$

valid in the frequency region $\omega'_\xi \ll \omega \ll \omega_\xi$, where ω'_ξ denotes the frequency where α-relaxation becomes active. Corresponding results can be found in the region $g_3 \approx 0$.

A characteristic property of the relaxation near a cusp is therefore a linear increase as $\ln \omega$ with decreasing ω for ϵ', and a constant frequency independent behaviour for ϵ'' at low frequencies. The latter property corresponds to a $1/\omega$ behaviour in the spectrum ϕ''. This $1/f$-noise or frequency independent dielectric loss for low frequencies is predicted to be followed by a rapid critical increase as frequency increases according to (13b).

These qualitative features have been observed in relaxation data for polychlorotrifluoroethylene (PCTFE) measured by Nakajima and Saito[17] and Scott et al [18]. The latter data refer to a quenched sample with 44% crystallinity with glass transition at $T_g \approx 50°C$. The observed relaxation can be attributed to the amorphous liquid region. These sets of data were also found to be close to an invariance line r =const.[3] On such lines it should be possible to rescale the data for the various temperatures so that they fall on one single master function. Such scaling plots of the data of Scott et al are shown in figures 4a and b for ϵ' and ϵ'' respectively. In the frequency region $\omega < \omega_{max}^\gamma$ all data fall on master curves, and the region where the points coincide expands when temperature is lowered. The full curves show the predicted master functions given by $\wp[u; 3, -4]$ and $-\frac{1}{2}\wp'[u; 3, -4]$ respectively corresponding to the value

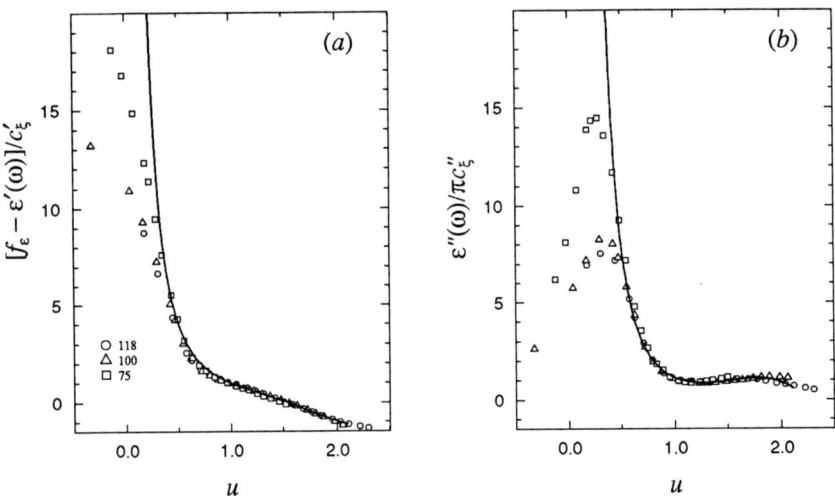

Figure 4. (a) Rescaled dielectric constant $[f_\epsilon - \epsilon'(\omega)]/c'_\xi$ for PCTFE[18] versus $u = \ln(1/\omega t_1)/y_\xi$ for the various temperatures indicated in the figure. The full curve is the predicted master curve $\wp[u; 3, -4]$ in (11a). (b) Rescaled dielectric loss $\epsilon''(\omega)/\pi c''_\xi$ versus u. The full curve is the predicted master function $-\frac{1}{2}\wp'[u; 3, -4]$ in (11b). The theoretical curves are cut off slightly below the halfperiod of $\wp$.

$r = 1/16$. The small scattering of the data points around the full curves arises because the path in parameter space do not exactly follow an invariance line $r =$const when temperature is varied.

The parameters used in the scaling plots above are shown in figures 5a and b. The theoretical predictions in (12) are shown as full curves in figure 5 with $T_0 = (62 \pm 4)°$C. Clearly more data points in the region around T_0 are necessary in order to verify or disprove the predicted functional forms. Similar results were found for the data from the annealed sample of Nakajima and Saito but with a transition at $T_0 = 19°$C[3]

The samples for PCTFE contain a crystalline component which essentially does not affect the relaxation spectra. For the 40 – 50% degree of crystallization these regions are relatively small in size and disorganized with respect to their mutual orientation. They will therefore act as static scattering centers for the relaxation of polar groups in the amourphous regions. The appearence of such static scattering centers has been treated in various contexts[19–21], and is known to give another type of glass transition singularity. The latter leads to a cusp in the static susceptibility. In the present case the static susceptibility is given by

$$\epsilon_0 - \epsilon_\infty = \frac{2}{\pi}\int_0^\infty d(\ln\omega)\,\epsilon''(\omega). \qquad (15)$$

The temperature dependence of $\epsilon_0 - \epsilon_\infty$ for PCTFE is shown in figure 6 for the data of Nakajima and Saito. Clearly there is a sharp cusp at $T = T_c = 45°$C as expected.

For the fold the scaling functions $g_\pm$ depends on one parameter λ, which will vary from system to system. Two systems which happen to have the same λ will have identical scaling functions. For the cusp, however, the scaling function

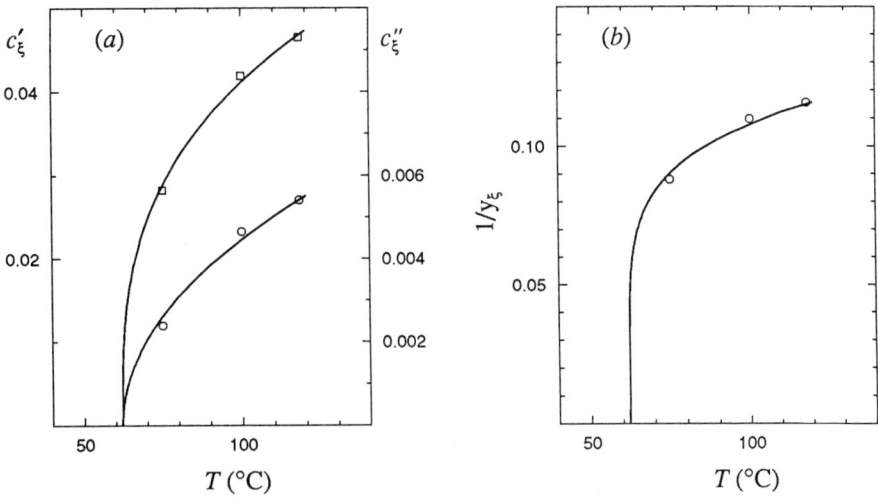

Figure 5. Scaling parameters c'_ξ (squares) and c''_ξ (circles) extracted from ϵ' and ϵ'' respectively. These parameters were used to obtain the scaling plots in figure 4. The full curves show the predicted behaviour $c'_\xi \propto |T/T_0 - 1|^{1/3}$ and $c''_\xi \propto |T/T_0 - 1|^{1/2}$ with $T_0 = 62°\text{C}$. (b) The scaling parameter $1/y_\xi$. The full curve show the predicted behaviour $1/y_\xi \propto |T/T_0 - 1|^{1/6}$ with $T_0 = 62°\text{C}$ as in (a).

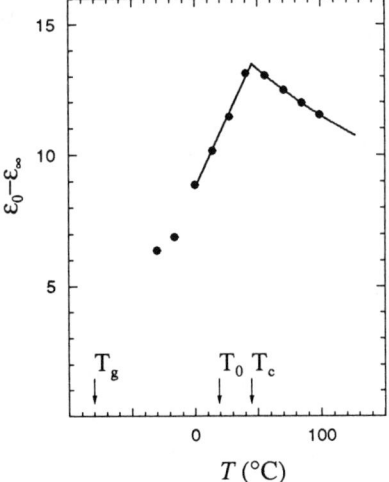

Figure 6. Static susceptibility $\epsilon_0 - \epsilon_\infty$ versus temperature for the data of Nakajima and Saito[17]. The cusp is located at $T_c = 45°\text{C}$. The full curve for $T > T_c$ show a $1/T$ law and that for $T < T_c$ a straight line. The arrows indicate also the location of the cusp temperature $T_0 = 19°\text{C}$ and the calorimetric glass transition temperature $T_g \leq -80°\text{C}$.

is universal and given by $\wp$. Any system which comes close to an A_3 singularity must therefore relax according to (10) or (11) in the β-region. Dielectric data for PET with 51% crystallinity and PEOB with 38% crystallinity were found to be close to an A_3 singularity and they could also be succesfully analysed with $(11)^3$. The constant part in ϵ'' or the $1/f$-noise in the spectrum corresponds to a $\ln(t)$ decay in $\phi(t)$. This logarithmic decay has been observed in many polymeric systems, and may be due to cusp scenarios.

REFERENCES

1. W. Götze *Liquids, Freezing and Glass Transition* ed. J. P. Hansen et al (North Holland, Amsterdam, 1991), p. 287.
2. W. Götze *Z. Phys.* B **60** 195 (1985).
3. L. Sjögren *J. Phys: Condensed Matter* **3** 5023 (1991).
4. W. Götze and R. Haussmann *Z. Phys.* B **72** 403 (1988).
5. W. Götze and L. Sjögren *J. Phys: Condensed Matter* **1** 4183 (1989).
6. W. Götze and L. Sjögren *Z. Phys.* B **65** 415 (1987),
   ——— *J. Phys. C: Solid State Phys.* **21** 3407 (1988).
7. L. Sjögren *Z. Phys.* B **79** 5 (1990).
8. W. Götze *J. Phys: Condensed Matter* **2** 8485 (1989).
9. Y. Ishida, K. Yamafuji, H. Ito and M. Takayanagi *Koll. Z. Z. Polym.* **184** 97 (1962).
10. L. Sjögren *Basic Features of the Glassy State* eds. J. Colmenero and A. Alegria (World Scientific, Singapore, 1990), p. 137.
    W. Götze and L. Sjögren *J. Non Cryst, Solids* **131-133** 153 (1991).
11. P. N. Pusey and W. van Megen *Phys. Rev. Lett.* **59** 2083 (1987),
    ——— *Ber. Bunsenges. Phys. Chem.* **94** 225 (1990),
    W. van Megen and P. N. Pusey 1991 *Phys. Rev.* A **43** 5429 (1991).
12. W. Götze and L. Sjögren *Phys. Rev.* A **43** 5442 (1991).
13. B. Bernu, Y. Hiwatari and J. P. Hansen *J. Phys. C: Solid State Phys.* **18** L371 (1985),
    B. Bernu, J. P. Hansen, Y. Hiwatari and G. Pastore *Phys. Rev.*A **36** 4891 (1987).
14. J. L. Barrat, J. N. Roux and J. P. Hansen *Chem. Phys.* **149** 179 (1990).
15. M. Fuchs, W. Götze, S. Hildebrand and A. Latz to be published (1991)
16. W. Götze and L. Sjögren *J. Phys: Condensed Matter* **1** 4203 (1989).
17. T. Nakajima and S. Saito *J. Polym. Sci.* **31** 423 (1958).
18. A. H. Scott, D. J. Scheiber, R. J. Curtis, J. I. Lauritzen and J. D. Hoffman *J. Res. Natl. Bur. Stand* A **66** 269 (1962),
    J. D. Hoffman, G. Williams and E. Passaglia *J. Polym. Sci.* C **14** 173 (1966).
19. W. Götze, E. Leutheusser and S. Yip *Phys. Rev.* A **23** 2634 (1981), **24** 1008 (1981).
20. W. Götze and L. Sjögren *J. Phys. C: Solid State Phys.* **17** 5759 (1984).
21. K. H. Michel *Z. Phys.* B **68** 259 (1987).

Stochastic dynamics in a supercooled fluid

T. Odagaki
Kyoto Institute of Technology, Kyoto 606, Japan

Y. Hiwatari
Kanazawa University, Kanazawa 920, Japan

Abstract

We discuss the stochastic motion of atoms in a supercooled fluid near the glass transition point on the basis of the trapping diffusion model. We obtain the dynamical singularities in the mean square displacement, diffusion constant, incoherent scattering function and non-Gaussian parameter. It is shown that the stochastic jump motion produces α-type relaxation.

The dynamics of atoms or ions in supercooled fluids has extensively been studied in recent years from experimental and theoretical interests, since the dynamics is believed to become singular at the glass transition point.[1] Recently, an interesting prediction was made by Götze and his coworkers[2] using the mode coupling theory: A transition from ergodic to nonergodic dynamics will take place at a certain temperature T_C somewhat higher than the glass transition temperature T_g, and above T_C two kinds of relaxation regimes, α- (or long time) and β- (or short time) relaxations, exist which crossover at a certain frequency. Recent neutron scattering experiments,[3] light scattering experiments[4] and molecular dynamics simulation[5] on $K_{0.6}Ca_{0.4}(NO_3)_{1.4}$ show dynamical characteristics consistent with this prediction.

On the other hand, molecular dynamics simulations on soft-sphere fluids[6] in the supercooled region near the glass transition have revealed that atoms make significant jump motions and there exist dynamical anomalies such as non-t-linear dependence of the mean square displacement, the residual diffusion, finite non-Gaussian parameter at time $t = \infty$ and so on.

In this paper, we focus on the stochastic jump motion of atoms and analyze the dynamical properties of supercooled fluids due to the jump motion. Figure 1 (a) shows the root square displacement of a particular atom in a softsphere binary mixture at $\Gamma = 1.6$. The effective coupling parameter Γ is defined by $\Gamma = (N/V)\sigma_{\text{eff}}^3(\epsilon/kT)^{1/4}$ for the system in which atoms interact through the pair potential $v_{\alpha\beta}(r) = \epsilon(\sigma_{\alpha\beta}/r)^{12}$, where $\sigma_{\alpha\beta} = (\sigma_\alpha + \sigma_\beta)/2$ and α and β denote the species. Here, $\sigma_{\text{eff}}^3 = \sum_\alpha \sum_\beta \sigma_{\alpha\beta}^3$, and N, V and T are the total number of

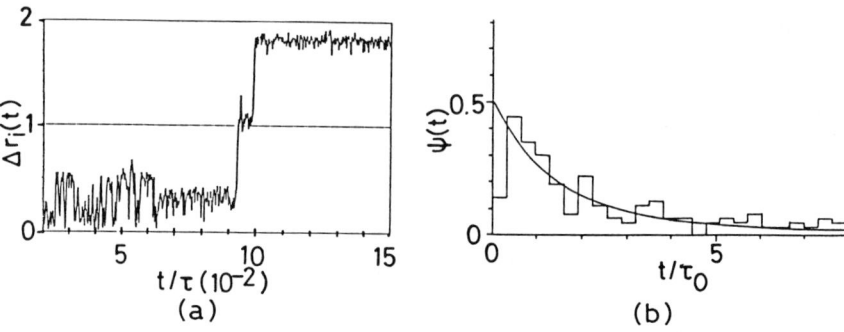

Fig. 1. (a) The mean square displacement of a particular atom observed in the molecular dynamics simulation for a supercooled binary softsphere mixture at $\Gamma = 1.6$. [From Ref. 4.] (b) The histogram shows the waiting time distribution $\psi(t)$ for a supercooled binary softsphere mixture at $\Gamma = 1.55$. The distribution function decays in time following a power law. The solid curve denotes the same quantity expected from the jump rate distribution (2) for $\rho = 0.01$. [See Ref. 9.]

particles, volume and temperature of the system, respectively. It is clearly seen that the atom performs a stochastic jump motion as well as rapid oscillatory motions. The time scale of these motions differ about two to three orders of magnitude. In order to examine long time dynamical properties, we take into acount only the jump motion of atoms. The jump motion of an atom is considered to be initiated by a certain concerted motion of a rather large number of atoms in the near neighbor, and essentially depends on their location. This kind of stochastic motion is effectively described by the trapping diffusion model,[7] where the conditional probability $P(s,t|s_0,0)$ that the atom is at s at time t when it started from s_0 at $t = 0$ obeys the master equation

$$\frac{\partial}{\partial t} P(s,t|s_0,0) = \sum_{s'} [w_{s'} P(s',t|s_0,0) - w_s P(s,t|s_0,0)]. \qquad (1)$$

The summation is taken over sites from (to) which a jump is possible. The master-equation description of incoherent dynamics has well been established by many authors.[8]

The jump rate w_s is determined by the local structure around s and is considered to have a wide distribution. In fact, the molecular dynamics simulation[9] shows that the waiting time distribution of the jump motion decays algebraically and the power decreases as the temperature is decreased. [See Fig. 1 (b).] We employ a jump rate distribution consistent with this observation. Noting that the long time behavior is dominated by small jump rates, we choose a power law form for the jump rate distribution:

$$\Phi(w_s) = \begin{cases} \frac{\rho+1}{w_0^{\rho+1}} w_s^\rho & 0 \leq w_s \leq w_0 \\ 0 & \text{otherwise} \end{cases} \qquad (2)$$

Parameter ρ is considered to be an increasing function of temperature or Γ^{-1}. We see a reasonable agreement in the waiting time distribution as shown in Fig. 1 (b).

Once $P(s, t|s_0, 0)$ is known from Eq. (1), dynamical properties can be calculated as follows.

The mean square displacement:

$$R_2(t) = \langle \sum_s (s - s_0)^2 P(s, t|s_0, 0) \rangle. \tag{3}$$

The frequency dependent diffusion constant:

$$D(\omega) = -\omega^2 \int_0^\infty R_2(t) e^{-i\omega t} dt. \tag{4}$$

The self-part of the intermediate scattering function:

$$F_S(\mathbf{q}, t) = \langle \sum_s e^{i\mathbf{q}(s-s_0)} P(s, t|s_0, 0) \rangle. \tag{5}$$

The self-part of the dynamic structure factor:

$$S_S(\mathbf{q}, \omega) = \int_{-\infty}^\infty F_S(\mathbf{q}, t) e^{-i\omega t} dt. \tag{6}$$

and so on. Here, the bracket $< \cdots >$ denotes the ensemble average.

To make further simplifications, we assume that the jump distance is sharply distributed in accordance with the recent work of molecular dynamics simulation on binary softsphere fluids[9] and that the distribution in jump distances is less important than the distribution in jump rates. We, therefore, assume that the atoms are sitting on a regular lattice (conveniently taken as the simple cubic lattice). In this phenomenological model, the physical quantities defined in Eqs. (3) $\sim$ (6) can be obtained from the ensemble average of the conditional probability $\bar{P}(s,t) \equiv < P(s,t|s_0,0) >$. It has rigorously been shown that the ensemble-averaged transition probability $\bar{P}(s,t)$ obeys a master equation with a memory effect in which no randomness is involved:

$$\frac{\partial}{\partial t} \bar{P}(s, t) = \int_0^t \sum_{s'} \Sigma_{s,s'}(t - t') \bar{P}(s', t') dt'. \tag{7}$$

The formal exact expression for the coherent jump rate $\Sigma(t)$ can be found in the literature.[10] Since we can not evaluate the formal exact expression in general, we employ the coherent medium approximation (CMA) for an approximate evaluation. The coherent medium approximation is a generalized application of the idea well-known as the molecular field approximation or the coherent potential approximation to the stochastic master equation (1). The detail of the coherent medium approximation can be found in the literature.[11]

We summarize some of the results obtained by the CMA for the long time behavior in Fig. 2.[12] We observe that (1) The mean square displacement behaves

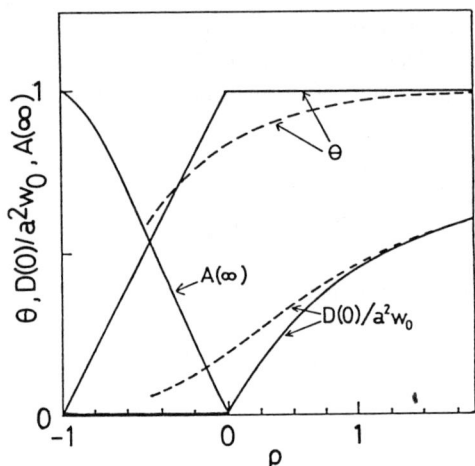

Fig. 2. The dynamical singularities expected from the trapping diffusion model. θ is the exponent for the mean square displacement. $D(0)$ is the static diffusion constant, identical to the exact result. (a is the scale of length.) Dashed curves are the corresponding quantities determined in a finite time window by appropriate time derivatives of $R_2(t)$. $A(\infty)$, the non-Gaussian parameter at $t = \infty$, plays a role of the order parameter of the transition at $\rho = 0$.

as $R_2(t) \sim t^\theta$ in the limit of $t = \infty$ with $\theta = 1$ for $\rho \geq 0$ and $0 < \theta < 1$ for $\rho < 0$. (2) The static diffusion constant $D(0)$ vanishes when $\rho \leq 0$. These dynamical behaviors indicate that the transition at $\rho = 0$ can be identified as the glass transition of the model. The observation (1) implies that the intermediate scattering function for small wave vectors and the limit of $t = \infty$ decays stretched-exponentially when $\rho < 0$.

In order to obtain further insight about the transition at $\rho = 0$, we calculate the non-Gaussian parameter which is defined as usual by

$$A(t) = \frac{3R_4(t)}{5[R_2(t)]^2} - 1, \tag{8}$$

where $R_4(t)$ is the mean quartic displacement. We find for the present model that

$$A(\infty) = \begin{cases} 0 & \text{for } \rho > 0, \\ \text{finite} & \text{for } \rho < 0. \end{cases} \tag{9}$$

Therefore, the transition at $\rho = 0$ accompanies Gaussian to non-Gaussian transition.[13] We note, however, that it is as yet not known whether this statement holds in general for glass transitions. We think that the anomaly in $A(t)$ is not necessarily equivalent to the vanishing diffusion. The mean square displacement consists of two contributions; one is due to the jump motion and the other is due

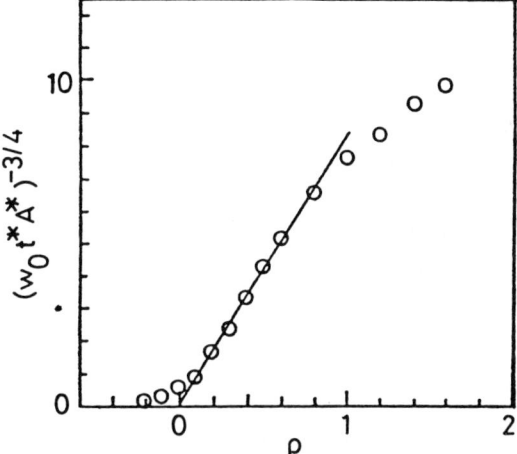

Fig. 3. The working definition of the glass transition point using the maximum of the non-Gaussian parameter. t^* denotes the time where $A(t)$ takes its maximum A^*. The product t^*A^* increases sharply at the glass transition point and thus we can determine the transition point from the plot ρ vs $[t^*A^*]^{-\nu}$ with some exponent ν.

to the rapid oscillation. When these two contributions become comparable, $A(t)$ takes its maximum. Consequently, as the jump motion becomes less frequent, it takes longer time to reach the maximum of $A(t)$. The time t^* at which $A(t)$ takes its maximum increases sharply on approaching $\rho = 0$, and therefore we can determine the transition point from t^* as shown in Fig. 3.

The dynamical characteristics obtained at $t = \infty$ or at the static limit clearly shows a significant change at $\rho = 0$. We note, however, that these dynamical properties are very sensitive to the observation time. For observations made at a finite time window, we can define, for example, the exponent θ by $d\log R_2(t)/d\log t$ which depends on the observation time. The exponent θ thus defined does not show the sharp transition as shown in Fig. 2 by the dashed curve. Furthermore, if θ is observed at a given time window, then one will see a subdiffusive (non-t-linear) behavior before the transition point is reached and the system appears to be a glass. In particular, there exist sub-anomalous terms[14] in the mean square displacement for $0 \leq \rho \leq 1$ which make it extremely difficult to reach the asymptotic region in a finite time window.

We now turn to the frequency-dependent properties observed at a finite frequency, taking the self-part of the dynamical structure factor $S_S(\mathbf{q}, \omega)$ as an example. Figure 4 shows the exponent σ as a function of frequency, where σ is defined by

$$\sigma = -\frac{d\log S_S(\mathbf{q}, \omega)}{d\log \omega}. \tag{10}$$

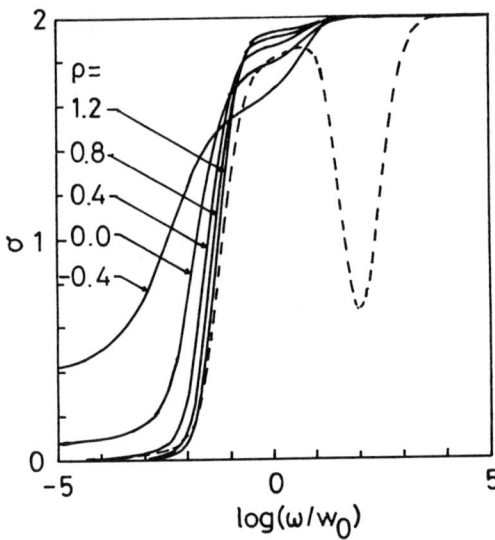

Fig. 4. The exponent σ define by $-d\log S_S(\mathbf{q},\omega)/d\log\omega$ is plotted against ω for $\rho = 1.2, 0.8, 0.4, 0.0, -0.4$. ($a^2 q^2 \sim 0.1$). The flat part for $\omega/w_0 \sim 10^{-1} - 10$ is considerd to be due to the α-relaxation. The broken line shows the exponent σ when a δ-peak in the jump rate distribution is included with equal weight when $\rho = 0.4$. In this plot we set $w_1/w_0 = 100$. For $\omega \lesssim w_0$, no significant change is seen.

Namely, when σ is constant in a frequency region, then $S_S(\mathbf{q},\omega)$ appears to be a power law function of frequency, $\sim \omega^\sigma$, in the frequency region. For large ω, we find $\sigma = 2$, which indicates that $S_S(\mathbf{q},\omega)$ decays as ω^{-2}. As we can see in Fig. 4, $S_S(\mathbf{q},\omega)$ behaves as $\sim \omega^\sigma$ with $\sigma \sim 1.9$ for $\rho > 0$ in the frequency region $0.1 \lesssim \omega/w_0 \lesssim 10$. This behavior is identical to the α-relaxation observed in certain real glasses. We can show that in this region $\sigma = 1 + \theta$ holds approximately, where θ is obtained in the time window corresponding to this frequency region.

In order to take into consideration qualitatively the higher frequency modes due to the rapid oscillation, we add a δ-peak, $\delta(w_s - w_1)$, in the jump rate distribution (2). The dashed curve in Fig. 4 represents the exponent σ for $\rho = 0.4$ when the δ-peak is included with equal weight and $w_1/w_0 = 100$. We see a significant decrease in the slope in the frequency region around $\omega \sim w_1$, which coincides with the behavior known as the β-relaxation. Inclusion of the δ-peak in the distribution, however, does not alter significantly the behavior of $S_S(\mathbf{q},\omega)$ in the lower frequency region except for a slight decrease in the exponent of the α-type relaxation regime.

In conclusion, we have presented an effective approach to understand the dynamical characteristics observed near the glass transition point on the basis of the trapping diffusion model. The present model gives results consistent with the dynamical properties observed in softsphere fluids by molecular dynamics simulation. It is believed that the hopping type motion becomes important near the glass transition point in many glass forming systems. Therefore, further analysis of such systems within the frame work of the present model will help understanding the dynamical behaviors near the glass transition.

This work was supported in part by grant-in-aids from the Ministry of Education of Japan. Data processing in this work was carried out by the Data Processing System for Perceptive Informations at Kyoto Institute of Technology.

REFERENCES

1. Y. Hiwatari, H. Miyagawa and T. Odagaki, J. Solid State Ionics, to be published, and references cited therein.

2. W. Götze, in *Liquids, Freezing and the Glass Transition*, edited by J. P. Hansen, D. Levesque, and J. Zinn-Justin, (North Holland, Amsterdam, 1991), p. 287, and and references cited therein.

3. F. Mezei, W. Knaak, and B. Farago, Phys. Rev. Lett. **59**, 1601 (1987); W. Knaak, F. Mezei, and B. Farago, Europhys. Lett. **7**, 529 (1988).

4. N. J. Tao, G. Li, and H. Z. Cummins, Phys. Rev. Lett. **66**, 1334 (1991) and preprints.

5. G. F. Signorini, J. L. Barrat, and M. L. Klein, J. Chem. Phys. **92**, 1294 (1990).

6. H. Miyagawa and Y. Hiwatari, J. Chem. Phys. **88**, 3879 (1988).

7. J. W. Haus and K. W. Kehr, Phys. Rep. **150**, 142 (1987).

8. For example, V. M. Kenkre, in *Exciton Dynamics in Molecular Crystals and Aggregates*, edited by K. G. Hohler (Springer, Berlin, 1982), p. 1.

9. H. Miyagawa and Y. Hiwatari, submitted to Phys. Rev. A.

10. M. Lax and T. Odagaki in *Macroscopic Properties of Disordered Media*, edited by R. Burridge, S. Childress, and G. Papanicolau (Springer, Berlin, 1982), p. 148.

11. T. Odagaki, J. Phys. A**20**, 6455 (1987); A. A. Ovchinnikov and K. A. Pronin, J. Phys. C**18**, 5391 (1985); J. W. Haus and K. W. Kehr, Phys. Rev. B**36**, 5639 (1987).

12. T. Odagaki and Y. Hiwatari, Phys. Rev. A**43**, 1103 (1991).

13. T. Odagaki and Y. Hiwatari, Phys. Rev. A**41**, 929 (1990).

14. T. Odagaki, Phys. Rev. B**38**, 9044 (1988).

THE GLASS TRANSITION IN ORIENTATIONAL GLASSES

D. WALTON

PHYSICS DEPT. MCMASTER UNIVERSITY, HAMILTON, ONT.

ABSTRACT

It has been possible to account for the temperature and frequency dependence of the dielectric and quadrupolar relaxations in $KBr_{1-x}KCN_x$, and the temperature dependence of T1 in glassy solid Hydrogen with a hierarchically constrained relaxation model. In particular, the model can account for the small change in the glass transition temperature with probe frequency.

Glass transitions can be understood in terms of the broadening and lengthening of the spectrum of relaxation times. The presence of structural disorder complicates the situation in canonical glasses. Orientational glasses do not suffer from this limitation, and a theory of hierarchically constrained relaxation[1] can account for many features of the glass transition in KBr-KCN alloys[2,3], and solid hydrogen.

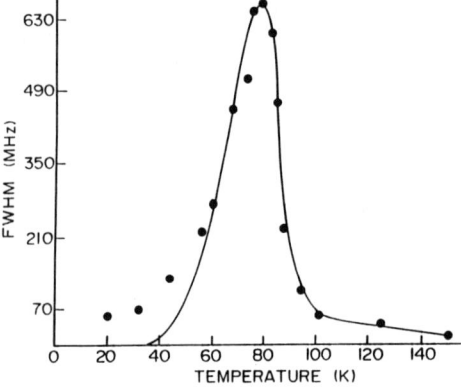

Figure 1: the linewidth of the 4 GHz transverse phonon for x=0.5, from Hu et al.[4].

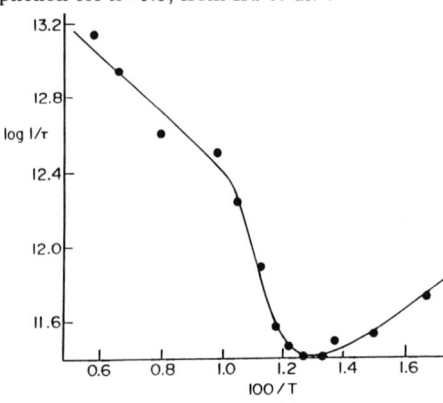

Figure 2: Logarithm of the average relaxation time obtained from the data in 1 plotted against reciprocal temperature. The Arrhenius behaviour at high temperatures breaks down near the glass transition temperature.

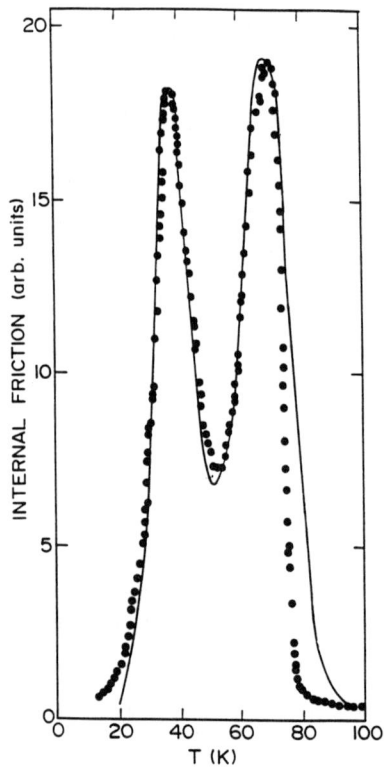

Figure 3: The temperature dependence of the relaxation in $KBr_{0.53}KCN_{0.47}$ alloys measured in a torsion pendulum at a frequency of 400 Hz, from Knorr et al.[3]. The peak at 37 K is due to dielectric relaxation. The solid lines in all three figures are calculated.

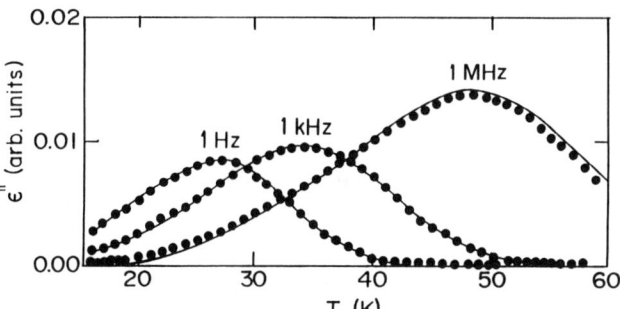

Figure 4: Dielectric relaxation as a function of temperature for various frequencies, from Ernst et al.[4]. The solid lines are calculated.

The CN ion has both an electric dipole moment, and an elastic quadrupole moment. It is generally agreed that the elastic quadrupolar interactions are responsible for a glassy phase at concentrations, x, between about 0.2 and 0.6. This manifests itself by a rapid slowing down of the quadrupolar relaxations as the temperature is lowered.

The quadrupolar relaxation results in attenuation of the acoustic phonons as shown in figs.1 and 2 and internal friction as shown in fig.3.

At temperatures below the glass transition temperature the dipoles are still able to "flip", head to tail[4,5,6], leading to strong dielectric relaxation phenomena[4]. The barriers responsible for the dielectric relaxation are provided by the elastic interaction between the CN ions, i.e. the quadrupoles. These barriers are established when the quadrupoles freeze.

What is particularly interesting is how little the temperature of the quadrupole peak, Tg, changes for a change in frequency of 7 orders of magnitude: for x=0.5 Tg is 78 K from Brillouin data[6,7]; whereas from the torsion pendulum results Tg is 70 K. The dipole peak, on the other hand, moves from about 70 K to 38 K. The data from Ernst et al.[4], reproduced in fig.4 also display the relatively much larger shift in the temperature of the dipole peak with frequency. The model to be outlined below accounts for both the quadrupolar and dielectric relaxation. Unlike previous theories[5,6], no temperature dependence of the mean and width of the distribution is introduced. Instead, it is assumed that the depth of the potential well is proportional to the fraction of nearest neighbours which are suitably oriented.

The elastic properties are conveniently described in terms of a complex elastic compliance, S, such that the velocity, v, of a sound wave of frequency ω, is determined by the real part, and it's attenuation, α, by the imaginary part[1] of S:

$$1/v - i\alpha/\omega = [\rho(ReS + iImS)]^{1/2} \qquad (1)$$

If there is a spectrum of relaxation times, it can be shown[2,3] that

$$\alpha = \rho\omega \sum_n \delta S_n \omega \tau_n / [1 + (\omega\tau_n)^2] \qquad (2)$$

where δS_n is a coupling constant.

The molecules find themselves in potential wells which have been deepened by their interaction with their neighbours. Those with fewest neighbours relax fastest. When a molecule relaxes it will reduce the potential well of it's neighbours, enabling them to relax as well. Thus the relaxation process is hierarchical in nature, with those molecules with fewest neighbours relaxing first, and those with most relaxing last. A formal theory for such a process was provided by Palmer et al.[1], and the approach presented here is closely related to it.

Let p^{-1} be the fraction of the available configurations which will result in a decreased interaction. The relaxation time of a molecule with n neighbours can be written $\tau_n = p^{\mu_n}\tau_k$

where μ_n is the number of neighbours which must be in those configurations which decrease the interaction. If the change in energy is the same for all of them, and equal to γJ, in order to remove the effect of the interaction with the neighbours, $\mu_n = n/(1-\gamma)$.

The nearest neighbours of n themselves have k neighbours where $k < n$, a similar expression can be written for τ_k, etc., leading to $\tau_n = \tau_1 p^{[(n+k+\ldots 1)/(1-\gamma)]}$

So far thermal effects have been ignored, obviously some of the neighbours will have been thermally excited, and $\tau_n = \tau_1 p^{[(nF(T,n)+kF(T,k)+\ldots)/(1-\gamma)]}$

Let

$$(nF(T,n) + kF(T,k) + \ldots)/(1-\gamma) = \beta \sum_{i=1}^{n} F(T,i) \qquad (3)$$

F(T,i) may be estimated if the occupation of the hierarchies is specified. As a first approximation we assume that the number of neighbours changes by 1. Thus, an ion with i neighbours will have i-1 of them in level i-1 (i.e. with i-1 neighbours), and one neighbour in level i+1. On the average it will find itself in a field, B(i), due to it's neighbours which will be

$$B(i) = [(i-1)F(T,i-1) + F(T,i+1)]J \qquad (4)$$

which leads to

$$F(T,i) = [2sinh(B(i)/T)][p - 2 + 2cosh(B(i)/T)]^{-1} \qquad (5)$$

Given that B(0)=0, F(T,i) may be obtained by starting with an approximation to F(T,i), and then refining it by iterating equations 14 and 13. The process converges quite rapidly, and the final result is not at all sensitive to the initial guess.

Letting $q = p^{<\beta>}$, and $\tau_1 = \tau_0 e^{J/T}$, and letting $\delta S_n = (N_n/N)C$ where N_n, is the number of CN ions with n nearest neighbours. N_n is approximated by a gaussian distribution whose width is σ. Transforming the sums to integrals, equation 5 becomes

$$S(\omega) = A \int_{1}^{12} dn e^{-[(n-12x)/\sigma]^2}/(1 - i\omega\tau_0[qe^{J/T}]\int_{1}^{n}(x)F(T,x)) \qquad (6)$$

At temperatures such that $T \gg J$ interactions are unimportant, and F(T,i)→ 0. The system then has a single relaxation time, τ_0. Thus τ_0 is given by the high temperature Brillouin results, and, from fig.2.

$$\tau(0) = 10^{(-13.5\pm.2)} exp[(240 \pm 20)/T] \qquad (7)$$

If the energy barrier is taken to be due to nearest neighbour interactions the dielectric relaxation time will be

$$\tau_d(n) = \tau_0 exp[(n-1)F(T,n-1) + F(T,n+1)]J/T \qquad (8)$$

The complex dielectric constant can be expressed as a sum of Debye equations:

$$\epsilon(\omega) = \epsilon_{inf} + (\epsilon_0 - \epsilon_{inf})\sum_{n} N(n)/(1 + i\omega\tau_d(n)) \qquad (9)$$

and the solid lines in fig.4 were calculated from the imaginary part of this equation with the same parameters used for the quadrupolar relaxation.

The adjustable parameters are then, q, J and σ, and, of course, for the KBr-CN alloys, the ratio of the dipolar to the quadrupolar coupling constant, but the latter is simply a scaling factor, and is of no interest in this calculation.

The values of these parameters used to obtain the solid lines in figs. 1, 2, 3, and 4 were, q = 12, J=103 K, σ=2.3.

The spin-lattice relaxation time in nuclear magnetic resonance, T1, is determined by the molecular relaxation time. T1 yields the average relaxation time directly. Fig.5 shows data for solid hydrogen obtained by Washburn et al.[11], and by Weinhaus and Meyer[12]. The results are strikingly similar to those shown in fig.2, obtained from the Brillouin data.

The solid lines in fig.5 were obtained with the procedure outlined above, and the following values for the parameters involved:
q=2, J=0.27 K, $\sigma=2$, and $\tau(0) = 10^{-9.3} exp[0.88/T]$.

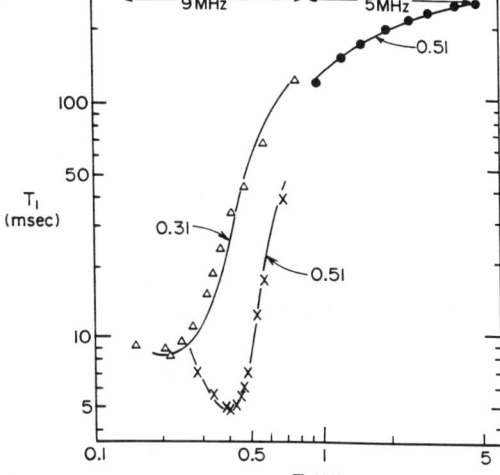

Figure 5: Temperature dependence of T1 in solid hydrogen for concentrations of 0.31 and 0.51 of ortho hydrogen. The data at 5 MHz is taken from Weinhaus and Meyer[12], and the data at 9 MHz is taken from Washburn et al.[11]. The solid lines are calculated.

Acknowledgements: This research was supported by a grant from the Natural Sciences and Engineering Research Council of Canada.

References

1) R.G.Palmer, D.L.Stein, E.Abrahams, and P.W.Anderson, Phys. Rev. Lets. 53, 958 (1984)

2) D.Walton, Phys.Rev.Lets. 65, 1599-1602 (1990)

3) D.Walton, Phys.Rev.B, in press (1991)

4) R.M.Ernst, L.Wu, S.R.Nagel, and S.Sussman, Phys.Rev. B 38, 6246 (1988)

5) N.O.Birge, Y.H.Jeong, S.R.Nagel, S.Battacharya and S.Sussman, Phys. Rev. B 30, 2306 (1984)

6) J.P.Sethna, and K.S.Chow, Phase Transitions 5, 317-340 (1985)

7) Z.Hu, D.Walton, and J.Vanderwal, Phys. Rev. B 38, 10830 (1988)

8) K.Knorr, U.G.Volkmann, and A.Loidl, Phys. Rev. Lets 57, 2544 (1986)

9) J.J.Vanderwal, Z.Hu, and Walton, D., Phys. Rev. B 33, 5782 (1986)

10) A.S.Nowick in "Physical Acoustics" v.13, edited by W.P.Mason and R.N.Thurston, 3 (1977)

11) S.Washburn, M.Calkins, and H.Meyer, J.Low Temp.Phys.53, 585 (1983)

12) F.Weinhaus, and H.Meyer, Phys.Rev.B 7, 2974 (1973)

TRANSPORT PROPERTIES IN ORDINARY AND SUPERCOOLED LIQUIDS

U.Balucani
Istituto di Elettronica Quantistica CNR, 50127 Florence, Italy

S.F.Duffy
Department of Physics of the University, Sheffield S3 7RH, UK

A.Torcini
Dipartimento di Fisica dell'Università, 50125 Florence, Italy

R.Vallauri
Dipartimento di Fisica, Università di Trento, 38050 Povo, Italy

ABSTRACT

A satisfactory microscopic theory of transport coefficients in dense fluids requires the consideration both of fast collisional events and of much slower decay mechanisms. In general, such approaches are necessarily complicated; fortunately, in ordinary and moderately supercooled liquids several physical approximations can be made to yield a considerably simpler analysis. This framework has been successfully applied to evaluate the diffusion and shear viscosity coefficients in several typical simple liquids. Further simplications are possible in the case of the diffusion coefficient, which can be expressed analytically. The practical advantages of this approach are illustrated for liquid Cs, obtaining results in good agreement with the observed data.

INTRODUCTION

Transport coefficients (TC) are known to be very important in any macroscopic treatment of the dynamics of fluids; however, their values can be predicted only by a microscopic framework based on non-equilibrium statistical mechanics. As a result, the TC are expressed in terms of "Green-Kubo (GK) relations", involving time integrals of particular time correlation functions. Clearly, the dynamical averages implicit in such integral relations are relatively insensitive to the details of microscopic motions; yet, the values of the TC are considerably affected by the possible presence of non-conventional decay mechanisms in their GK integrands. In most cases, these processes are associated with a time scale distinctly longer than the one pertinent to the "usual" decay channel, which is typically provided by essentially independent binary collisions. The classical example of this situation is the $t^{-3/2}$ long-time tail observed many years ago[1] in the velocity autocorrelation function (VACF) of a dense hard-sphere fluid; the effect has a direct impact on the value of the diffusion coefficient D, which deviates considerably from the prediction of the purely "collisional" Enskog theory. Even larger discrepancies are found at liquid densities.

A consistent theoretical framework for the entire fluid range encompassing all these kinds of dynamical events requires a judicious combination of kinetic concepts and mode-coupling methods, the latter being particularly suited to study long-lasting phenomena. Approaches in this sense have indeed been developed, and in principle provide a complete description of the dynamics of simple fluids with a minimum recourse to phenomenological assumptions (see, for example, the comprehensive review by Sjölander[2]). The only basic input required in these theories is the knowledge of the structural properties of the fluid. However, the inherent difficulty of the general problem leads to a complicated structure of the theory, and this has certainly prevented an extensive application of its results to specific cases of experimental interest. In this contribution we restrict our attention to the ordinary liquid range, and discuss how physical arguments lead to a considerable simplification of the results for the TC as well as for the underlying microdynamics. In particular, we shall consider both a single-particle case (the VACF, associated with D) and a collective problem (the stress autocorrelation function, i.e. the GK integrand of the shear viscosity coefficient η).

THE PREDICTIONS OF A BINARY MODEL

A simple fluid is usually modeled by a classical system of N particles of mass m, enclosed in the volume V and interacting through a given continuous pair potential $v(r)$. The number density $n = N/V$ and the temperature T specify the thermodynamic state of the system. As stated before, the equilibrium properties of the fluid are assumed to be known, either from a separate theory of the structure or from the data provided by real experiments and computer simulations. The GK relations for D and η can respectively be written as

$$D = (k_B T/m) \int_0^\infty dt\ \psi(t) = k_B T/mK(z=0) \qquad (1)$$

$$\eta = \int_0^\infty dt\ \eta(t) = (1/k_B TV) \int_0^\infty dt\ \langle \sigma^{xy}(0)\sigma^{xy}(t)\rangle \qquad (2)$$

In eq.(1), $\psi(t)$ is the velocity autocorrelation function of a tagged particle (VACF) $\langle v_i(0)\cdot v_i(t)\rangle$ normalized to its initial value $3k_B T/m$. Rather than $\psi(t)$, we shall find more convenient to consider its memory function $K(t)$ along with the corresponding Laplace transform $K(z)$. In eq.(2), $\eta(t)$ is the "stress autocorrelation function" (SACF) which involves the non-diagonal components of the stress tensor:

$$\sigma^{xy} = \Sigma_i\ [mv_i^x v_i^y - (1/2)\ \Sigma_{j(\neq i)}(x_{ij}y_{ij}/r_{ij})v'(r_{ij})] \qquad (3)$$

where $x_{ij} = x_i - x_j$, etc. At liquid densities the overwhelming contribution to the SACF is given by the second term in eq.(3).

A first important decay mechanism for $K(t)$ and $\eta(t)$ is provided by fast collisional events, similar to the uncorrelated binary collisions taken into account in the Enskog theory for hard spheres. In systems with a realistic $v(r)$, the predictions of such a "binary" approach for D and η can be written as[3,4]

$$D_B \approx k_B T/m\Omega_0^2 \tau_D \qquad \eta_B \approx G\tau_\eta \qquad (4)$$

In this preliminary model, the fast nature of the collisional mechanism justifies a single relaxation-time approximation for the decay of $K(t)$ and $\eta(t)$ from their initial values (the Einstein frequency Ω_0^2 and the rigidity modulus G, respectively). In turn, these times, τ_D and τ_η, can approximately be obtained from the initial decay rates of $K(t)$ and $\eta(t)$. All these quantities can be evaluated from the structural properties of the system; in particular, the order of magnitude of τ_D and τ_η in typical simple liquids is found to be $\approx 10^{-13}$ s, a figure which confirms the fast character of this decay channel. The simplicity of the predictions (4) is appealing, but unfortunately the results are found to be in severe disagreement with the observed data. Typically, in the liquid range D is overestimated and η is underestimated by $\approx 50\%$. Not surprisingly, these discrepancies are similar to the ones of the Enskog results for hard spheres at comparable packing fractions.

THE RELEVANT SLOW EVENTS AT LIQUID DENSITIES

The complete breakdown of the "binary" approach is a clear indication of the presence of decay channels associated with timescales considerably longer than the collisional mechanism. The physical origin of these slow features, as well as their form, can be found by a mode-coupling (MC) framework.[5] In particular, the latter extracts from $K(t)$ and $\eta(t)$ the portions associated with bilinear combinations of the slow dynamical variables of the system. These "modes" usually correspond to the quasi-conserved quantities typical of the hydrodynamic regime; as a result, low-wavevector modes are normally the most effective in providing a slow character to the several MC decay channels, and indeed they are responsible for the long-time tail observed in the VACF of dense gases. On the other hand, in the liquid range these tails have not been observed, and their theoretical amplitudes are too small to have any practical relevance. Then, the slow MC mechanism which we are looking for cannot involve low-wavevector modes. Beyond the hydrodynamic region all the dynamical variables have a "fast" character, with the important exception of density modes with wavevectors q near the position q_m of the main peak of the static structure factor $S(q)$. In practice, the effects of this "deGennes slowing down" are important only in the liquid region, where $S(q)$ is sharp-peaked. Consequently, at high densities the dominance of these density modes (i.e. of the "sluggish" structural relaxation) simplifies considerably the MC analysis of the channels relevant for the slow features of $K(t)$ and $\eta(t)$.

The splitting of a general phase-space memory function into the sum of "binary" and "mode-coupling" contributions can be performed by purely kinetic arguments (see e.g. Ref.2). The precise form of the MC contributions depends on "vertices" which measure the efficiency of the coupling to the pair of modes under consideration. In particular, in the liquid region the previous arguments yield a MC part of $K(t)$ involving the product $F(q_m,t)F_s(q_m,t)$, where F denotes the intermediate scattering function and F_s its self part.[3] In the collective case of the SACF, the MC portion of $\eta(t)$ involves a wavevector sum of $F^2(q,t)$, with a vertex ($\propto S'(q)$) which emphasizes the contribution of wavevectors slightly away from q_m.[4] Strictly speaking, one should evaluate the dynamic correlations $F(q,t)$ and $F_s(q,t)$ by a similar framework, i.e.

splitting their memory functions into fast and slow portions. Although this aspect is crucial near the glass transition, much simpler approximations for F and F_s are found to work very well for ordinary and moderately supercooled liquids.[3] In particular, for $q \approx q_m$ a viscoelastic model predicts that $F(q,t)/S(q) \approx \exp[-\gamma(q)t]$, where the decay rate $\gamma(q) \propto [S(q)]^{-1}$ can be expressed in terms of known structural properties. In such a way, the evaluation of $K(t)$ and $\eta(t)$ becomes relatively straightforward. A more detailed discussion is found in Refs. 3,4, where the theoretical predictions are compared with the simulation data in liquid Rb both at 318 K (just above the melting point) and at 270 K (a moderately quenched state), and in ordinary liquid Ar at 86.5 K. The comparison is quite satisfactory for all the dynamical details of $K(t)$ and $\eta(t)$, as well as for D and η.

For the diffusion coefficient, a further simplification of the theory[6] leads to the following analytic expression for the dimensionless quantity $D^* \equiv D/\sigma^2\Omega_0$ (σ is the usual length associated with $v(r)$):

$$(D^*)^{-1} = (D_0^*)^{-1} + \Gamma(C/\Omega_0^2) \exp[-2\pi B(D_0^*)]/B(D_0^*) \quad (6)$$

where $\Gamma = m\Omega_0^2\sigma^2/k_BT$, $B(D^*) = (q_m\sigma)^2 D^* + [\gamma(q_m)/\Omega_0]$ and C is the amplitude of the MC contribution, again evaluable from purely structural data. In eq.(6) the quantity D_0 follows from

$$(D_0^*)^{-1} \approx (D_B^*)^{-1} + \pi \Gamma(C/\Omega_0^2) \exp[-2\pi B(D_B^*)] \quad (7)$$

Eqs.(6),(7) follow from a perturbation procedure which "renormalizes" the diffusion coefficient from the binary result D_B to the full value D by a stepwise inclusion of MC contributions associated with progressively longer timescales. To give a specific example, let us apply this simple scheme to liquid Cs in a state specified by n = 0.0083 Å^{-3} and T = 308 K. Structural data have been obtained by neutron scattering and by computer simulation[6]: from these we deduce $\Omega_0^2 = 19.22$ ps^{-2}, τ_D 0.29 ps, yielding $D_B = 3.47 \ 10^{-5}$ cm^2/s, much larger than the actual value D = 2.35 10^{-5} cm^2/s. The MC contributions are readily evaluated from $q_m = 1.42$ Å^{-1}, $\sigma = 4.76$ Å, $S(q_m) = 2.99$. From eqs. (6),(7) we finally obtain D = 2.31 10^{-5} cm^2/s, in excellent agreement with the experimental data.

REFERENCES

1. B.J. Alder, D.M. Gass and T.E. Wainwright, J.Chem.Phys. **53**, 381 (1970).
2. A. Sjölander, in "Amorphous and Liquid Materials" ed. E.Lüscher G. Fritsch and G.Jacucci (M.Nijhoff, Dordrecht, 1987), p. 239.
3. U. Balucani, R. Vallauri, T. Gaskell and S.F. Duffy, J.Phys. Cond Matter **2**, 5015 (1990).
4. U. Balucani, Molec.Phys. **71**, 123 (1990).
5. See e.g. T. Munakata and I. Igarashi, Progr.Theor.Phys. **58**, 134 (1977); Progr.Theor.Phys. **60**, 45 (1978).
6. U. Balucani, A. Torcini and R. Vallauri (to be published).
7. T. Bödensteiner, C. Morkel, P. Müller and W. Gläser, J.Non-cryst Solids **117-118**, 116 (1990).

ACCELERATING GLASSY RELAXATION IN THE FRENKEL-KONTOROVA MODEL

Shelly Shumway
Argonne National Laboratory, Bldg 223; Argonne, IL 60439

James P. Sethna
Laboratory of Atomic and Solid State Physics
Cornell University; Ithaca, New York 14853-2501

ABSTRACT

We have developed a method for accelerating equilibration in numerical simulations of glassy models, using the Frenkel-Kontorova model for developing and testing the algorithm. This simple model is like a glass in that relaxation times diverge rapidly as the critical temperature is reached, so that no matter how slowly it is cooled, the system eventually gets stuck in some random metastable configuration. By adding "numerical enzymes," or long-range Monte Carlo moves which precisely eliminate certain barriers to relaxation, we can equilibrate rapidly to significantly lower temperatures. Our numerical method for developing these enzymes is a learning algorithm motivated by Darwinian evolution.

We begin this paper with a brief discussion of the Frenkel-Kontorova [FK] model, which is the simplest we know of that includes incommensurability and frustration in a natural way. We then discuss a general approach for equilibrating numerical models of, e.g., configurational glasses, in which the Hamiltonian is simple and not random, but relaxation time scales are so slow that the system always gets stuck in high-energy, random configurations. We find that "numerical enzymes", or complicated multiple-atom Monte Carlo moves fine-tuned to bypass the barriers to relaxation, can be of great assistance for shedding light on the underlying behavior of a model with natural dynamics too slow for effective numerical study using normal techniques. Although such enzymes may be subtle and complex, a learning algorithm motivated by principles of Darwinian evolution and natural selection can find them.

The FK model consists of a one-dimensional chain of atoms, each connected to its nearest neighbors by springs, with an externally applied sinusoidal potential:

$$H = \frac{K}{2}\sum_j (x_{j+1} - x_j - \alpha)^2 - V\sum_j \cos 2\pi x_j. \qquad (1)$$

Frustration occurs when the periodicity of the applied potential competes with the tendency of the springs to keep the atoms evenly spaced; we work in the "pinned" limit,[1] in which the spring constant K is small. We have investigated[2] the behavior on cooling from a finite temperature, at which atoms undergo thermal motion, to a zero temperature frozen chain. Any finite cooling rate γ is too fast for the system to be able to equilibrate all the way into the ground state, although cooling more slowly does result in somewhat better equilibration.

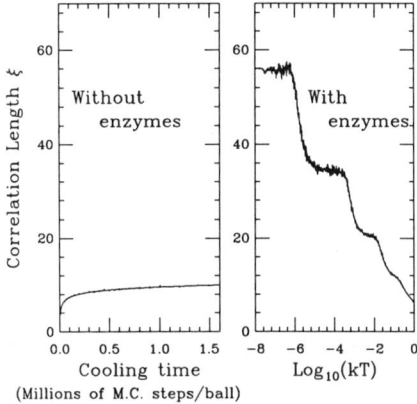

Figure 1: Dynamical Behavior of the FK Model. Without enzymes, no significant ordering develops even with very slow cooling. The addition of enzymes allows the true underlying behavior to be explored numerically in reasonable computer time. (K=1, V=3, lattice spacing = 1)

We have run simulations of this model using a standard Metropolis Monte Carlo algorithm, and have found that the correlation length ξ, which measures the average length scale of regions in local ground state configurations, diverges incredibly weakly as a function of cooling time $1/\gamma$:

$$\xi \sim \log(\log(\frac{1}{\gamma})), \qquad (2)$$

as shown in Figure 1 and discussed in detail in reference [2]. If normal simulations had been the only way to study the model, we would probably not even have suspected that the limiting behavior was actually an ordered ground state. This model is glassy in the sense that diverging time scales cause the "melt" to fall out of equilibrium at some history-dependent temperature and get stuck in one of many possible metastable configurations, generally exhibiting no long-range order. Like a glass, cooling a lot slower allows equilibration to only a slightly lower temperature.

Widom, Strandburg and Swendsen[3] encountered a similar situation in simulating a two-component Lennard-Jones [LJ] system. They, like others, found its dynamical behavior to be glassy. By adding special three-atom moves and long hops to the normal small relaxations in Monte Carlo simulations, however, they were able to equilibrate much faster and thus discover underlying quasicrystalline behavior in a system with suitable LJ parameters.

Motivated by their work, we have set out to develop a general method for finding special moves to accelerate equilibration and thereby illuminate the underlying behavior of various glassy models. We have started with the FK model because of its simplicity, but hope to be able to apply the technique to more realistic models whose true equilibrium behavior is obscured by complexity and diverging time scales in simulations.

The "numerical enzymes" we are looking for must be able to provide an accelerated mechanism for relaxation of metastable states. Calculating them directly will generally be impossible; we need to be able to find them with out knowing in advance what the high-energy configurations or corresponding low-energy relaxed states are like. We have therefore developed a learning algorithm to get the computer to systematically discover and fine-tune effective enzymes with only minimal guidance. The method works well in the FK model.

Our approach, like the genetic algorithms that have been used with great success by engineers and others,[4] is motivated by biological principles. The basic idea is that the solution to a problem is discovered through evolution of a population of some kind, in which an appropriate formula for health or fitness determines each individual member's survival and reproduction. Those that do poorly die off, and are replaced most often by the offspring of exceptionally healthy individuals. In our algorithm, we have a population of moves: each individual is represented by an n-dimensional vector telling how far to attempt to move each of n atoms. We use a modified Monte Carlo-like code in which we select one individual at random from this finite population of discrete moves for every update of the FK chain, keeping track of each one's performance in promoting equilibration. Detailed balance is preserved by multiplying the move by a random sign whenever it is used. This algorithm is not a Markov process, and therefore not genuine Monte Carlo, but it is able to discover precise and complicated enzymes that can be used to accelerate a legitimate Monte Carlo code.

The formula for health must be based on success in lowering the energy of the chain, but any reasonable implementation of this should work fine. We have chosen to multiply an individual's health by a factor $\lambda < 1$ whenever it is tried, and add an amount proportional to the energy decrease of the chain if the new configuration is accepted. The population is updated periodically, and those that are unhealthy die off. We maintain a population of constant size; whenever one dies, two moves are selected randomly with probability proportional to their health, and a new individual is created as their vector sum or difference. One parent is occasionally zero-padded so the child can move more atoms.

We begin each run at high temperature, with the moves distributed randomly from 0 to 0.3 lattice spacings in length, then let the population evolve as the FK chain slowly cools. The population quickly adjusts to an appropriate length scale for the prevailing temperature, but with an excess of long moves for hopping atoms between wells, as shown in Figure 2. Upon cooling, the thermal distribution sharpens and the enzymes become separate, no longer just an anomalously long tail to the ordinary thermal peak.

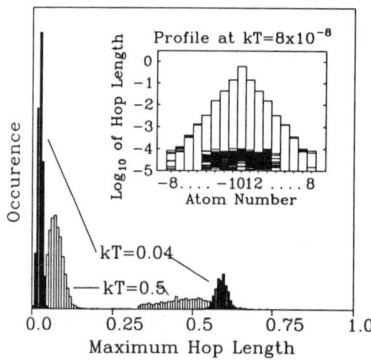

Figure 2: The population gradually develops precision, as shown by the histograms for $kT = 0.5$ and 0.04, and also complexity. By $kT = 8 \times 10^{-8}$, each enzyme moves 16 atoms: number 0 by an amount of 0.5949 lattice spacings, and neighbors by smaller amounts; thermal moves are tiny and involve 6 to 8 atoms.

By the time the chain is cold enough that the first kind of defects have equilibrated out, the population discovers that the enzymes must move at least three atoms at once, moving one a large distance across the barrier and relaxing two neighbors. This is the needed move for equilibrating out a lower energy kind of defect. At still colder temperatures, the enzymes find that they need to move more and more atoms at once. If they fail to discover some necessary improvement, they will be unable to continue equilibrating defects out as the chain cools, and will become extinct, leaving only thermal moves in the population.

This algorithm is quite robust and not particularly sensitive to details in the algorithm, although some thought must be given to ensure that it generally encourages survival of the enzymes. For example, the relative scarcity of low energy defects means that enzymes succeed infrequently at low temperatures; they simply cannot compete. We eliminate this bias by dividing the population into two classes, those that move at least one atom at least a third of a lattice spacing and those that do not, and multiplying the health of each individual by a normalization factor to promote equality between the classes. Of course, no amount of special treatment will help the enzymes if they fail to adapt to the changing environment and become unable to ever lower the energy of the chain.

If we cool slowly enough and set all the parameters to reasonable values, then long-range, fine-tuned enzymes usually develop that are capable of equilibrating out even very low energy defects. The inset to Figure 2 shows the 15-atom enzymes discovered by a run in which $\frac{K}{V} = \frac{1}{3}$. When we added this enzyme to a proper Monte Carlo code, we were able to equilibrate to a very low temperature and corresponding long correlation length, as shown in Figure 1.

In conclusion, we have found that diverging relaxation time scales prevent the FK model from equilibrating all the way to its zero-temperature ground state given finite cooling rate, and that much slower cooling results in equilibration to only slightly lower temperature. We have developed a method for accelerating relaxation in such a system in order to study the equilibrium behavior numerically. To do this, we use a modified Monte Carlo-like algorithm in which we keep track of the success of various moves, retaining and refining those that are especially useful for reducing the energy of the chain. We use the enzymes, or precisely coordinated motion of many atoms at once, that are discovered by the modified program to dramatically accelerate equilibration in a legitimate Monte Carlo code. The method should be especially useful in studying complicated models where the ground states and equilibrium properties cannot be deduced analytically.

This work was supported by Grant #NSF-DMR-88-18558-A02 from the Cornell University Materials Science Center and an NSF graduate fellowship (SLS), and computer time was provided through the Cornell–IBM Joint Study on Computing for Scientific Research.

REFERENCES

[1] The pinning transition is discussed in, e.g., S. N. Coppersmith, D. S. Fisher, *Phys. Rev. B* **28** (1983) 2566, and Braiman et. al., *Phys. Rev. Lett.* **65** (1990) 2398.

[2] S. L. Shumway and J. P. Sethna, *Phys. Rev. Lett.* **67** (1991) 995.

[3] Michael Widom, Katherine J. Strandburg, and Robert H. Swendsen, *Phys. Rev. Lett.* **58** (1987) 706.

[4] See, e.g., David E. Goldberg, *Genetic Algorithms in Search, Optimization, and Machine Learning* (Addison-Wesley, 1989).

MODE COUPLING, UNIVERSALITY AND THE GLASS TRANSITION

Gene F. Mazenko
The James Franck Institute and The Department of Physics
The University of Chicago, 5640 S. Ellis Avenue, Chicago, IL 60637

There has been a recent explosion of activity in the study of the kinetics of the liquid-glass transition. This activity has involved a beautiful interplay of experiment, theory and simulation and the progress has been remarkable. Mode coupling theory (MCT), as carefully elucidated by Götze and Sjögren, has suggested that the relaxation near the glass transition is through an elaborate sequence of different mechanisms and an intricate universal scaling picture is emerging from experiments which is very mysterious.

Let me begin with an overview of the relevant experimental[1] situation. It is fair to say that much of this experimental work was driven by theory. However, at this stage, experiment is leading theory. The experimental focus, for a wide variety of materials, has been on the behavior of two basic properties of dense fluids:

i). The temperature dependence of a characteristic relaxation time $\tau(T)$, as, for example, the shear viscosity.

(ii). The sequence of time behaviors shown by the normalized density-density time correlation function.

There are now a variety of experiments[2] probing $\tau(T)$, for simple and complex fluids, which show that for high enough temperature $(T > T_x)$, where T_x is well above the "glass transition" T_G, τ shows an algebraic behavior with temperature

$$\tau = \tau_0 (T - T_x)^{-\gamma} \tag{1}$$

with $\gamma \approx 2$. For $T < T_x$, one obtains the Vogel-Fulcher form[3]

$$\tau = \tau_0 \, exp. - \frac{A}{T - T_0} \tag{2}$$

where $T_0 \approx 0$ for "strong"[1] glass formers and $T_0 \approx T_G$ for "weak" glass formers. As shown in Fig. 1, using data from Dixon et al.,[4] the temperature T_x is a crossover temperature connecting the two behaviors given by (1) and (2).

If $\rho_{\vec{q}}(t)$ is the Fourier transform of the density in a liquid at time t, then consider the normalized density-density time correlation function

$$\phi(\vec{q}, t) = \frac{\langle \rho_{\vec{q}}(t) \rho_{-\vec{q}}(0) \rangle}{\langle \rho_{\vec{q}}(0) \rho_{-\vec{q}}(0) \rangle} \tag{3}$$

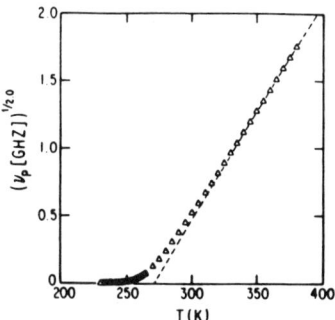
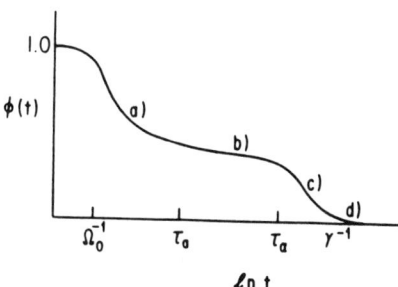

Fig. 1:
$[v_p(GHZ)]^{1/2}$ versus temperature T(K) for salol, which is taken from Ref. 4. The dotted line is a linear fit for the last 5 points of the crossover temperature is estimated to be $T_0 = 270$ K, which lies in the temperature range studied in Ref. 4.

Fig. 2:
A schematic plot of the sequence of relaxation mechanisms predicted by MCT; a) power law decay t^{-a} relaxation, b) von Schweidler relaxation $-Bt^b$, c) primary relaxation $e^{-(t/\tau)^\beta}$, d) exponential relaxation $e^{-\gamma t}$.

where the average $<>$ is over an ensemble which is stationary on the time scale of an experiment. As indicated in Fig. 2, $\phi(q,t)$, for fixed wavenumber q which is suppressed, shows a variety of behaviors as a function of time over an enormous time range (14 decades have been studied by Nagel's group).[4] The generic time behavior, shown in Fig. 2 begins at very short times, on the order of inverse phonon ($t \sim \Omega_0^{-1}$) frequencies with the rapid exponential decay of $\phi(t)$ until it enters a region ($t \sim \tau_a$) where there[5] is a power-law decay ($\phi \sim t^{-a}$). This power-law decay is followed by the so-called von Schweidler[6] regime where

$$\phi(t) \sim -Bt^b \quad . \tag{4}$$

This von Schweidler regime is really the high-frequency tail of the α or primary relaxation peak in dielectric relaxation measurements[4]. The center of the α-relaxation regime ($t \sim t_\alpha$) is well fit[4,7-15] by a stretched exponential form

$$\phi = \phi_0 \, e^{-(t/\tau)^\beta} \tag{5}$$

where β is **not** universal but depends on wavenumber, density and temperature. For even longer times, ($t \sim \gamma^{-1}$) one expects a cross-over to a final exponential decay.

This much of the experimental picture seems uncontroversial. There have been recent developments which are both exciting and mysterious. Dixon et al.[4,16] have claimed to see an exotic form of scaling in the primary relaxation region in Fig. 2 extending roughly from τ_a to γ^{-1}. Without going into detail here this has the consequence that the von Schweidler exponent b, defined by (4), and the stretching exponent β, defined by (5), are connected by the universal relation

$$\frac{1+b}{1+\beta} = \frac{3}{4} \tag{6}$$

which is independent of temperature, sample investigated, etc. In contrast, it has been claimed that scattering experiments[7-15] show "time-temperature superposition" given by (5), instead of the scaling results of Dixon et al. Of course these scattering experiments probe a much narrower time range and a much larger wavenumber regime.

There is the claim in neutron scattering experiments,[7-14] as well as some light scattering experiments, to see a qualitative change in the relaxational behavior of their systems as a function of temperature. In particular they offer evidence for an ergodic-nonergodic transition near a temperature they identify with the temperature T_x introduced in (1). Dixon et al. find no such qualitative change in the scaling properties of their results for T above, below or near T_x.

While there has been great effort to use simulational methods[17-26] to study these problems, and there has been substantial progress, it is still not possible for these studies to compete with experiments which can measure 14 time decades of an observable quantity.

Much of the experimental work discussed above has been influenced, to some extent, by the development of the mode-coupling theory (MCT) [27-35] of the glass transition. The basic assumption of the mode coupling theory is that the slowing down observed near the glass transition is governed by density fluctuations. The assumption made in the simplest MCT models is that the glass transition is insensitive to the probing wave number q and that the various wave number components can be treated independently. The Laplace transform of $\phi(t)$, has the representation

$$\phi(z) = \frac{z + id(z)}{z^2 - \Omega_0^2 + id(z)(z + i\gamma(z))} \quad . \tag{7}$$

The frequency Ω_0 in (7) corresponds to a microscopic "phonon" frequency. The renormalized viscosity $d(z)$ in (7) is coupled back to the density correlation function $\phi(t)$ via

$$d(z) = d_0 + \Omega_0^2 \int_0^\infty dt \, e^{izt} H[\phi(t)] \tag{8}$$

where d_0 is the bare viscosity governing the microscopic dynamics and $H[\phi(t)]$ is the mode coupling kernel discussed further below.

The frequency $\gamma(z)$ in (7), introduced by Das and Mazenko,[32,34] serves to cutoff the ergodic-nonergodic transition that one finds in its absence. If, as $z \to 0$, the viscosity $d(0)$ becomes very large, then (7) reduces to

$$\phi(z) = \frac{1}{z + i\gamma(0)} \tag{9}$$

and $\phi(t)$ decays to zero as $e^{-\gamma t}$. If, however $\gamma = 0$, then $\phi(z) \sim z^{-1}$ and $\phi(t)$ does not decay to zero as $t \to \infty$ and the system is not ergodic. It seems crucial, in order to obtain the very slow relaxation observed in experiment, that $\gamma(t)$ remains small for all times and γ^{-1} must be larger than any other time in the problem. This can be accomplished by requiring that short time decay processes not contribute to $\gamma(z)$. We assume here that $\gamma(0)$ can be made sufficiently small so that it can be neglected in (7) except in treating the longest time scale.

If we set $\gamma(z) = 0$, we can combine (7) and (9) to obtain the equation of motion for $\phi(z)$ in the time domain

$$\ddot{\phi}(t) + d_0 \dot{\phi}(t) + \Omega_0^2 \phi(t) + \Omega_0^2 \int_0^t ds \, H[\phi(t-s)]\dot{\phi}(s) = 0 \quad . \tag{10}$$

The original choice for the mode coupling kernel, $H = c_2\phi^2(t)$, is known as the Leutheusser[27] model and for c_2 near the value 4 one obtains the algebraic growth of η for $T > T_x$, (Fig. 1) and the power-law behavior of $\phi(q,t)$ associated with times $t \sim t_a$ in Fig. 2. One does not obtain the von Schweidler regime (labelled b in Fig. 2) or stretching in the α-relaxation regime (labelled c in Fig. 2). Instead $\beta = 1$. This picture is apparently unaltered in a qualitative sense by adding additional terms to H of the form $c_n \phi^n$ where $n > 2$. Götze, however, pointed out[29,30,6] that if one includes a linear term in H,

$$H = c_1\phi(t) + c_2\phi^2(t) \quad , \qquad (11)$$

then one **does** generate the von Schweidler region and stretching ($\beta < 1$) in the α-relaxation regime if c_1 and c_2 have values near $c_1 = 2\lambda - 1/\lambda^2$ $c_2 = 1/\lambda^2$ where $1/2 \leq \lambda < 1$. The model given by (11) leads to expression for the exponents a and b, governing sections of Fig. 2, are given in terms of the single parameter λ by

$$\frac{\Gamma^2(1-a)}{\Gamma^2(1-2a)} = \lambda = \frac{\Gamma^2(1-b)}{\Gamma^2(1-2b)} \quad . \qquad (12)$$

It can also be shown numerically[33] that (11) leads to the stretched form (5) and the stretching exponent is given by $\beta = -\ln 2/\ln(1-\lambda)$. Leutheusser's model can be derived through a series of approximations starting with kinetic theory or fluctuating nonlinear hydrodynamics. The linear term in (11) however, postulated by Götze, does **not** follow from these treatments.

Thus MCT, if one accepts the imposition of the linear term in (11), can account for the sequence of time behaviors shown in Fig. 2. This is a substantial theoretical accomplishment. One now has a theoretical model with microscopic underpinnings which does lead to stretching behavior. There are, however, a number of severe problems with this approach.

i). The origin of the parameters c_n appearing in the mode coupling kernel must be determined from a more microscopic basis. At this level they are treated, unphysically, as independent parameters and one does not know how many c_n to include and their dependence on $\vec{q}$, ρ and T.

ii). While the theory is not incompatible with the scaling of Nagel and coworkers, to obtain agreement with (6) one requires a model with (c_1, c_2, c_3 and c_4) and very specific relationships between these parameters. There is no hint, at present, of where such relationships come from. Thus this theory is not quantitatively "predictive".

iii) The theory does not naturally lead to a crossover to activated behavior for $\tau(T)$ for $T < T_x$.

Acknowledgement: This work was supported by the National Science Foundation Materials Research Laboratory at the University of Chicago.

References

1. For a review of earlier work see: C.A. Angell in **Relaxations in Complex Systems**, K. Nagi and G. Wright (eds.) U.S. Government Printing Office, Washington, D.C., 1984.
2. P. Taborek, R.N. Kleiman, and D.J. Bishop, Phys. Rev. B **34**, 1835 (1986).
3. J. Wong and C.A. Angell, **Glass: Structure by Spectroscopy** (Dekker, New York, 1976), Chapter 11.
4. P.K. Dixon, L. Wu, S.R. Nagel, B.D. Williams, and J.P. Carini, Phys. Rev. Lett. **65**, 1108 (1990).
5. N. Tao, G. Li, and M.Z. Cummins, Phys. Rev. Lett. **60**, 1334 (1990).
6. W. Götze in *Liquids, Freezing and the Glass Transition,* ed. by D. Levesque, J.P. Hansen and J. Zinn-Justin (New York, Elsevier, 1991).
7. F. Mezei, W. Knaak, and B. Farago, Phys. Rev. Lett. **58**, 571 (1987).
8. F. Mazei, W. Knaak, and B. Farago, Physica Scr.T **19**, 363 (1987).
9. D. Richter, B. Frick, and B. Farago, Phys. Rev. Lett. **61**, 2465 (1988).
10. B. Frick, D. Richter, W. Petry, and U. Buchenau, Z. Phys. B **70**, 73 (1988).
11. W. Knaak, F. Mezei, and B. Farago, Europhys. Lett. **7**, 529 (1988).
12. B. Frick, B. Farago, and D. Richter, Phys. Rev. Lett. **64**, 2921 (1990).
13. W. Petry, E. Bartsch, F. Fujara, M. Kiebel, H. Sillescu, and B. Farago, Z. Phys. B **83**, 175 (1991).
14. For a review of recent experiments (especially neutron scattering) see *Dynamics of Disordered Materials.* Ed. by D. Richter, A.J. Dianoux, W. Petry and J. Teixeira, (Berlin, Springer, 1989).
15. W. van Megen and P.N. Pusey, Phys. Rev. A **43**, 5429 (1991).
16. The most recent dielectric measurements of glass forming liquids including polymers carried out by A. Schönhal, F. Kremer, and E. Schlosser, Phys. Rev. Lett. **67**, 999 (1990) reported the breakdown of the universal scaling of Dixon et al. in the low frequency regime. This breakdown may be due to the existence of the additional peak of the imaginary part of the dielectric susceptibility in the low frequency region for these very complex systems. But is it not clear at the moment whether this very low frequency peak indeed belongs to the α-relaxation process. These experiments are in agreement with those of Dixon et al. over the frequency regime governed by the exponents β and b.
17. U. Bengtzelius, Phys. Rev. A **33**, 3433 (1985).
18. J.J. Ullo and S. Yip, Phys. Rev. Lett. **54**, 1509 (1987).
19. J.J. Ullo and S. Yip, Phys. Rev. A **39**, 5879 (1989).
20. J.N. Roux, J.L. Barrat and J.P. Hansen, J. Phys.: Cond. Mat. **1**, 7171 (1986).
21. J.J. Ullo and S. Yip, Chem. Phys. **149**, 221 (1990).
22. J.L. Barrat, J.L. Roux, and J.P. Hansen, Chem. Phys. **149**, 197 (1990).
23. G.F. Signorini, J.L. Barrat, and M.L. Klein, J. Chem. Phys. **92**, 1294 (1990).

24. M.J.D. Brakkee and S.W. deLeeuw, J. Phys. Condensed Matter, 2, 4991 (1990).
25. R. Ernst, S.R. Nagel, and G.S. Grest, Phys. Rev. A 43, 8070 (1991).
26. L. Lewis, preprint.
27. E. Leutheusser, Phys. Rev. A 29, 2765 (1984).
28. W. Götze, Z. Phys. B 56, 139 (1984).
29. U. Bengtzelius, W. Götze, and A. Sjölander, J. Phys. V. 17, 5915 (1984).
30. W. Götze, Z. Phys. B 60, 195 (1985).
31. S.P. Das, G.F. Mazenko, S. Ramaswamy, and J. Toner, Phys. Rev. Lett. 54, 118 (1985).
32. S.P. Das and G.F. Mazenko, Phys. Rev. A 34, 2265 (1986).
33. H. DeRaedt and W. Götze, J. Phys. C 19, 2607 (1986).
34. W. Götze and L. Sjögren, Z. Phys. B. 65, 415 (1987).
35. B. Kim and G.F. Mazenko, Mode Coupling, Universality and the Glass Transition, preprint.

CRYSTALLIZATION AND NON-ERGODIC MODE DYNAMICS

T. MUNAKATA

Kyoto University, Department of Applied Mathematics and Physics
Kyoto 606 Japan

ABSTRACT

To investigate dynamical aspects of crystallization, two nonlinear stochastic models are proposed based on the density functional theory. In addition to a first-order freezing transition we observe a new dynamic transition where the symmetry of the models are broken.

INTRODUCTION

During the last ten years the density functional(DF) theory has played an important role in quantitative studies of the freezing transition. The basic idea of the DF theory is to express the free energy Ω as a functional of the density field $\rho(r)$ and to calculate the difference $\Delta\Omega = \Omega[\rho_{crys.}] - \Omega[\rho_0]$ where ρ_0 and $\rho_{crys.}$ denote the density fields of a liquid and a crystalline solid, respectively with

$$\rho_{crys.}(\vec{r}) = \rho_0 \sum_G \mu_G \exp[i\vec{G}\cdot\vec{r}]. \tag{1}$$

Here $\vec{G}$ is the reciprocal lattice vector (RLV). The transition point is defined as the condition $\Delta\Omega = 0$.[1]

To investigate dynamical aspects of the transition, we propose here two stochastic models of coupled nonlinear oscillators which simulate dynamics of the order parameters(OPs) $\{\mu_G\}$. In both models we find, besides a freezing transition, a new bifurcation of a non-ergodic solution in a strong coupling region.

MODE DYNAMICS FOR CRYSTALLIZATION

(A) Ginzburg-Landau (GL) approach

Starting from the GL expansion of the DF energy we obtain, after some simplifying approximations, Ω in terms of the real OPs $\{s_i\}$ as

$$\Omega = \Sigma[s_i^2/2 + s_i^4/4] - B^*\Sigma s_i s_j s_k \quad (i < j < k \leq N). \tag{2}$$

Thus we have a system of $\mathcal{N}$ Duffing oscillators with trilinear coupling. To introduce dynamics to (2) we employ a simple relaxation model

$$ds_i/dt = -(1/\xi)\partial\Omega/\partial s_i + f_i(t) \quad (0 \leq i \leq N)$$

with the appropriate fluctuation-dissipation theorem for the random force $f_i(t)$. Properties of the model (A) are summarized as follows: 1) As $B(= B^*(N-1)(N-2)/2)$ is increased with T fixed a crystal phase appears for $B > B_f$ (if B is fixed, then for $T < T_f$). 2) For further increase of B ($B > B_{TLF} > B_f$) the system enters a two-level fluctuation state where each trajectory $S_i(t)(N = 10)$ takes two values s_+ and s_- and hops from $s_+(s_-)$ to $s_-(s_+)$ randomly keeping the number of the s_+-oscillators 7. (so 3 s_--oscillators.) 3) As B exceeds B_G the hopping rate between s_+ and s_- vanishes and we have a locked phase where symmetry among the oscilltors is broken. In this phase we have a non-zero Edward-Anderson order parameter.

(B) Time-Dependent DF approach.

A dynamical version of the DF theory yields a nonlinear Langevin diffusion equation for $\rho(r,t)$[3]. This equation can be used to derive an evolution equation for the OPs $\{\mu_G\}$. As an example, for the one component plasma(OCP) which is known to crystallize into a b. c. c. lattice, we obtain 6 couled equations for the $\{\mu_G\}$ with G denoting 12 smallest RLV. Static structure factor from the improved HNC theory is used to determine the coefficients appearing in the equations. If we set from the outset that all the μ_G are equal and real we have a liquid-crystal transition around Γ (the plasma constant)= 200. However solving the coupled equation for 6 complex OPs, we observe that for $\Gamma > 300$ the system takes a two dimensional structure probably because of the slip along the $(1,1,1)$ direction and this results in lower free energy. This transition is also related to broken-symmetry of the solution to the Langevin equation where three μ_Gs become nearly zero and the other three take nonzero values with the same amplitude.

REFERENCES

1 A. D. J. Haymet, Ann. Rev. Phys. Chem. *38*, 89(1987)
2 T. Munakata, J. Phys. Soc. Japan *60*, no. 9(1991)
3 T. Munakata, J. Non-Crystalline Solids *117/118*, 875(1990)

C. COMPUTER SIMULATION

MONTE CARLO SIMULATIONS OF THE POLYMER GLASS TRANSITION

W. Paul, Institut für Physik, Universität Mainz, D-6500 Mainz, Germany

Abstract

Simulations of glass transitions need to cope with exploding relaxation times especially in polymer systems which have an apriori slow dynamics. The longer the times your simulation method can explore, the closer you can approach the glass transition temperature. Therefore we propose the Monte Carlo study of a polymer lattice model, namely the bondfluctuation model. The infinite temperature properties of this model have been extensively studied. Adding potentials that favour long bonds and/or increase the stiffness of the chains introduces a glass transition into the model. Using a mapping to Bisphenol-A-Polycarbonate we reproduce the Vogel-Fulcher parameters experimentally found for the viscosity of BPA-PC. The microscopic structural relaxation is analyzed according to mode–coupling predictions and is well comparable to experimental results found on Polybutadiene.

INTRODUCTION

Polymers in the solid state genuinely come as amorphous materials. An understanding of the glass transition of polymers therefore is not only interesting on general physical grounds but also from the point of view of polymer design [1]. A most challenging problem in this respect is the establishment of structure property correlations for polymers. There have been a few partially successful routes to the computer modelling of real polymer glasses. But these typically are either static simulations at $T = 0K$ [2] or simulations of the dynamics of chemically realistic models that can only cover extremely short times compared to typical polymer relaxation times even at very high temperatures. Thus they cannot deal with equilibrated systems [3, 4]. We therefore decided to map a given polymer onto a lattice model on a coarse-grained scale getting rid of the fast degrees of freedom. Keeping such a coarse-grained model equilibrated is possible to much lower temperatures. As a technically relevant polymer we are studying Bisphenol-A-Polycarbonate [5, 6] and predict the Vogel–Fulcher temperature of BPA–PC oligomers.

From the theoretical point of view the most acclaimed description of the structural glass transition is the mode–coupling theory [7]. Although it has been developed for simple liquids it has been succesfully applied to experiments on polymers [8, 9]. We will redo the analysis of [8] for our model system.

THE MODEL AND IT'S MAPPING ONTO REAL POLYMERS

As a computational basis for our studies of the glass transition we are using the bond fluctuation model [10]. In this lattice model each repeat unit or monomer occupies the 2^d corners of the unit cell of a d-dimensional simple cubic lattice. The 3 dimensional version of the model is sketched in figure 1. There are 108 different possible bonds between consecutive monomers ranging in length from 2 to $\sqrt{10}$, and 87 different angles θ between consecutive bonds [11, 12]. The chain dynamics is generated by a Monte Carlo procedure where a monomer and a lattice direction for the attempted move are

146 Monte Carlo Simulations of the Polymer Glass Transition

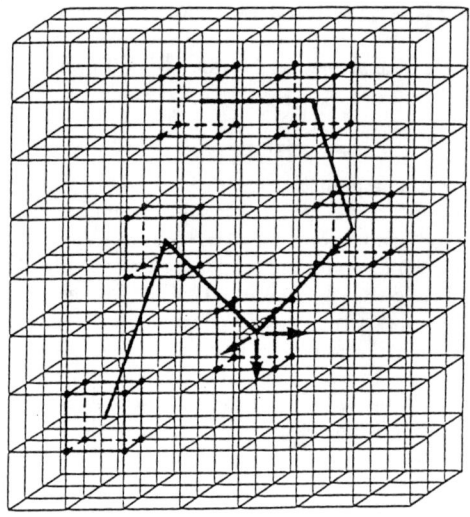

Figure 1: *The 3-dimensional version of the bondfluctuation model*

chosen at random. The move is acceptable if the two new bonds that are created lie in the allowed bond set and if the new lattice positions were empty before (self and mutually avoiding walks). Temperature comes into play by putting energies on bond lengths $E(l)$ and/or bond angles $E(\theta)$ and by using Metropolis rates for deciding whether to accept an attempted move. The model is well suited for simulations of polymer melts. The excluded volume interaction can efficiently be taken into account using lattice occupation numbers and the acceptance rate for the described moves is still high enough at melt densities because the monomers move only a fraction of the mean monomer distance. The density dependance of various static and dynamic quantities characterizing random coil polymers with Rouse dynamics was analyzed in [13, 14]. This analysis showed that a polymer volume fraction of $\Phi = 0.5$ in this model corresponds to a melt density with an excluded volume screening length of about two mean bond lengths.

A mapping between a real polymer chain (i.e. a chemically realistic polymer model) and such an abstract lattice model naturally can only be done on a mesoscopic length and time scale where a coarse-graining procedure has smeared out the differences on the microscopic scales. On the other hand the length scale has to be shorter than the statisctical segment length where everything becomes gaussian again. Our approach [5] consists in determining the centers of mass of groups of m consecutive monomers along the real chain and of n consecutive monomers along the lattice chain. These mass centers define coarse-grained chains with certain distributions of coarse-grained bond lengths $P(L)$ and bond angles $P(\Theta)$. For the chemically realistic model these are the outcome of a Monte–Carlo–RIS simulation of a single random walk chain using quantum chemically determined potentials [1, 6]. For the lattice model single non-reversal random walk chains were generated using energies

$$E(l_i) = \epsilon_a(l_i - l_0)^2 \tag{1}$$

$$E(\theta_i) = \epsilon_b \left(\cos(\wp(\theta_i)) - \cos(\theta_0)\right)^2 \tag{2}$$

Equation (2) comes about by an indexing error producing a permutation $\wp(\theta_i)$ of the energy values giving a more or less random potential for the angles. This resulted in a temperature independent stiffness of the model chains.

The mapping between the microscopic and the lattice model was done at a reference temperature $T_{ref} = 570K$. The four parameters

$$a = \frac{\epsilon_a}{T_{ref}}, \quad b = \frac{\epsilon_b}{T_{ref}}, \quad l_0, \quad \theta_0 \tag{3}$$

were determined by fitting the first two moments of the distributions $P(L)$ and $P(\Theta)$ of the lattice chains to those of the realistic ones. For $P(L)$ one needs a length scaling between lattice constants and Angstrœms. We required that our simulations were done at the mass density of BPA-PC at $T_{ref} = 570K$.

$$\rho = \frac{\Phi}{8} \frac{m}{n} \frac{\text{mass}_{BPA\text{-}monomer}}{s^3} \tag{4}$$

Fixing the polymer volume fraction at the melt value $\Phi = 0.5$ and coarse-graining $m = 1$ monomers of BPA-PC (which has a rather large repeat unit) and $n = 3$ lattice monomers we get a scale factor

$$s = 2.03 \ \frac{\text{Å}}{\text{lattice constant}} \tag{5}$$

The simulations were done with relatively short chains of $N = 20$ to be able to keep the system equilibrated to as low temperatures as possible. They were performed on the Multitransputer facility of the Condensed Matter Theory group in Mainz running 33 systems of lattice size 40^3 in parallel each containing 200 chains of length $N = 20$.

RESULTS

The simulations were started with well equilibrated athermal melts. Then the temperature was reduced in 8 steps to $T = 4560K$, $2280K$, $1520K$, $1140K$, $912K$, $760K$, $651K$, $570K$ with the reference temperature $T = 570K$ as the final one. At each temperature the chains were propagated until they had diffused for approximately one radius of gyration, which defines the Rouse time. Figure 2 shows typical curves for various mean square displacements as a function of time.

The intersection between the mean square displacement of the center of mass of a chain $g_3(t)$ and the inner monomers of the chain in the center of mass coordinate frame $g_2(t)$ allows for a convenient definition of a maximum relaxation time τ_r approximately equal to the Rouse time [14]. This time scale changes from $\tau_r = 16,500$MCS at $T = \infty$ to $\tau_r = 4,090,000$MCS at $T = 570K$, meaning a factor of 250 in the computer time needed at the two temperatures. At a final time $t_f \geq t_r$ one can perform measurements of static properties getting

statistics out of the 33 parallely treated systems. Figure 3 shows that the systems were indeed equilibrated as all polymer length scales evolve equally. Remember that the persistence length was constant, so that the end-to-end distance

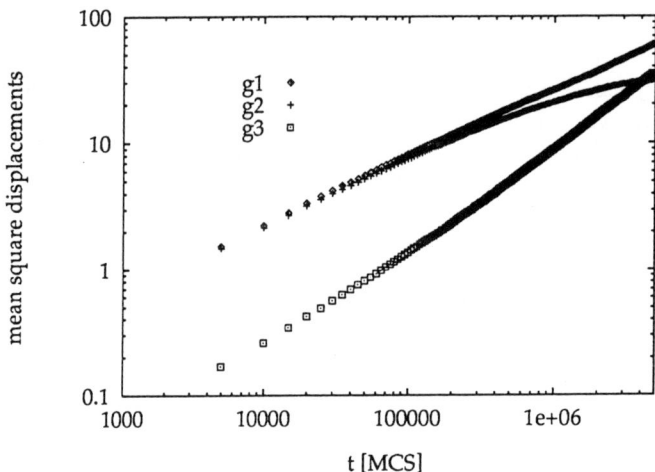

Figure 2: *Mean square displacement of inner monomers, $g_1(t)$, inner monomers in the center of mass coordinate frame, $g_2(t)$, and the center of mass of a chain, $g_3(t)$.*

$$\langle R^2 \rangle = N \langle l^2 \rangle l_p^2 \tag{6}$$

has the same temperature dependance as the mean squared bond length.

The dramatic increase in relaxation time can also be seen in the behaviour of typical microscopic timescales of the simulation [13]. One of these is the acceptance rate A of the Monte Carlo moves that displays a clear Arrhenius behaviour (see figure 4). The other one is the Rouse rate W defined from the mean square displacement of inner monomers $g_1(t)$ in figure 2.

$$g_1(t) = \langle b^2 \rangle \sqrt{Wt} \quad , \quad t \ll \tau_r \tag{7}$$

where b is the statistical segment length. The quantities A' and W' in figure 4 are defined as

$$A' = \left(\ln \left(\frac{A(\infty)}{A(T)} \right) \right)^{-1} \; ; \quad W' = \left(\ln \left(\frac{W(\infty)}{W(T)} \right) \right)^{-1} \tag{8}$$

An Arrhenius law shows up as a straight line extrapolating to $A'(0) = 0$. The Rouse rate W' that is connected with the actual mass transport in the system clearly extrapolates to a finite intersection temperature yielding a Vogel–Fulcher law. To analyse this in close connection with experiments one has to look at the viscosity of the polymers. This can be calculated from the center of mass diffusion coefficient

$$D_N = \lim_{t \to \infty} \frac{g_3(t)}{6t} \tag{9}$$

as

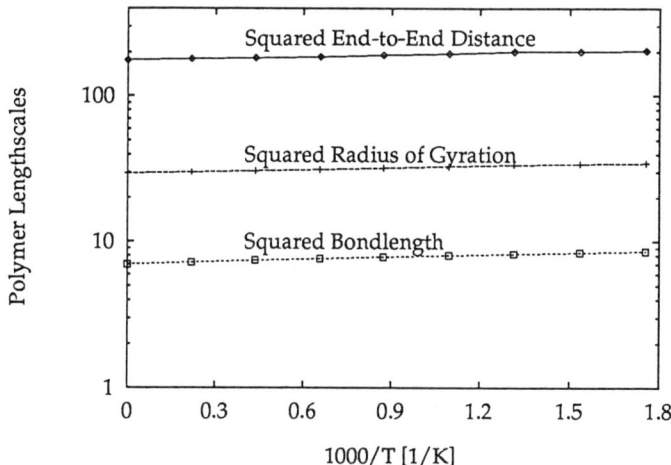

Figure 3: *Polymer lengthscales as a function of temperature.*

$$\eta = \frac{\phi}{8s^3} \frac{\langle R^2 \rangle}{36} \frac{k_B T}{N D_N} \quad \left[\text{Poise} \frac{\text{MCS}}{\text{sec}}\right] \tag{10}$$

Note that the time scale of the simulation is not yet fixed whereas the lengths can be calculated in real units. We fitted a Vogel-Fulcher law

$$\eta = \eta_\infty \exp\left(\frac{E_0}{T - T_0}\right) \tag{11}$$

to the data. The fit parameters are η_∞, E_0, T_0 and are determined as

$$T_0 = 321 \pm 27 K, E_0 = 950 \pm 113 K, \eta_\infty = 2.94 \cdot 10^{12} \left[\text{Poise} \frac{\text{MCS}}{\text{sec}}\right] \tag{12}$$

To account for chain length effects on the viscosity of polymer melts dedicated viscosity measurements of BPA-PC oligomers were done at the Bayer AG [1] giving

$$T_0 = 322 K, E_0 = 1053 K, \eta_\infty = 0.43 \, [\text{Poise}] \tag{13}$$

The agreement between Vogel temperatures and activation energies is excellent and enables us to equate the prefactors which results in a time scale for the simulation

$$1 \text{MCS} = 1.5 \cdot 10^{-13} \text{seconds} \tag{14}$$

Figure 5 shows the simulation data together with the fitted and simulated Vogel-Fulcher laws. With these data we are able to predict the glass transition temperature $T_g^{sim} = 352 K, T_g^{exp} = 356 K$ out of low viscosity data: $\eta \leq 10 Poise$. Note also that an acceptance rate of $A \approx 2\%$ at $T = 570 K$ means one accepted monomer movement every $10^{-11} sec$, which is the typical time for conformational rearrangements due to jumps in

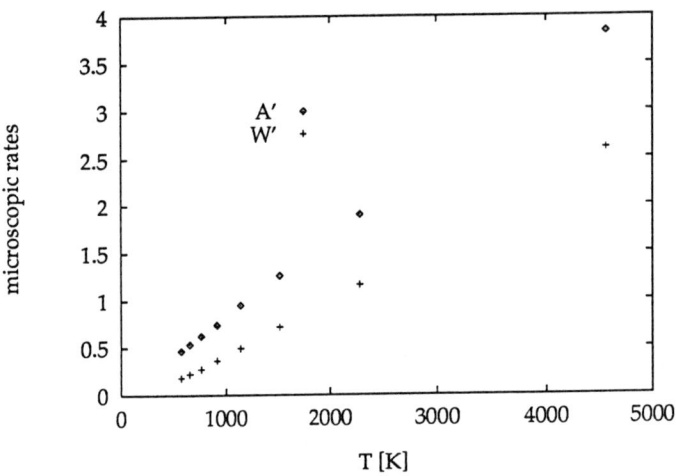

Figure 4: *Acceptance rate and Rouse rate as a function of temperature (see text).*

the torsional potentials of the $C - C$ bonds. This is the lower bound of times above which this Monte Carlo simulation can generate a realistic dynamics.

Let us now switch to more microscopic quantities. Figure 6 shows a plot of the global structure factor of the simulated lattice systems, where the abscissa is given in experimental units. It shows the genuine behaviour of an amorphous material with the first maximum at $Q = 1.48 \text{Å}^{-1}$ in roughly the right place. A detailed comparison to the experimental structure factor [15] would require an incoorporation of the monomer formfactor. To make contact with the predictions of mode–coupling theory [7] we looked at the incoherent intermediate scattering function at the structure factor maximum. For the required temperature regime we couldn't keep the system equilibrated any longer. Therefore we cooled the system down in steps of $20K$ and simulated it at each temperature for $100,000$MCS. This corresponds to a high cooling rate of $10^9 \frac{K}{\text{sec}}$. We analyzed a time window of $1,000$MCS $- 100,000$MCS corresponding to $10^{-10} - 10^{-8}$sec. This time range is comparable to Neutron-spin-echo data on polybutadien [8]. Figure 7 shows some of the simulation data together with fits of a Kohlrausch law

$$\frac{S(Q,t)}{S(Q,0)} = f \exp\left(-\left(\frac{t}{\tau}\right)^\beta\right) \tag{15}$$

Keeping the Kohlrausch exponent β free yielded good fits with exponent values around $\beta = 0.57$. Therefore we set β to that value and determined the relaxation time, τ, and the prefactor, f, the so-called non-ergodicity-parameter, with β fixed. Figure 8 shows a scaling plot of the relaxation at a few selected temperatures, where times are in units of τ and intensities in units of f. We are not able to scale our data with a macroscopic timescale extractable out of the viscosity for instance; this is due to the high cooling rate producing quenched configurations with internal temperature differing from the external one. The Kohlrausch-fit, however, yields the actual system timescale. The chosen time window is one where all the Rouse modes are important for the relaxation of

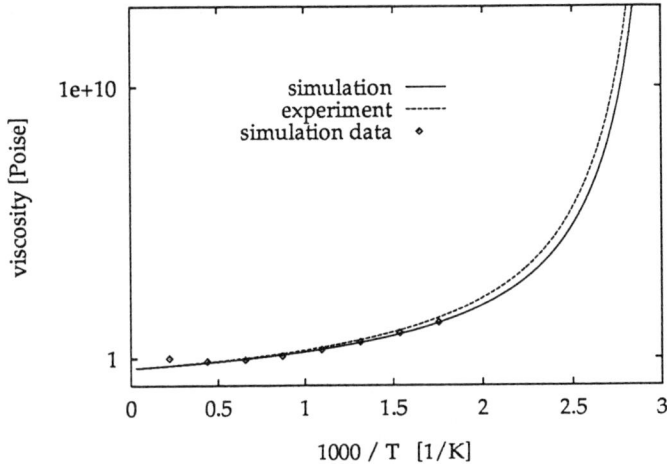

Figure 5: *Simulation data for the viscosity and Vogel–Fulcher fits to simulation and experiment*

the structure factor and therefore the Kohlrausch law is understandable from standard Rouse-dynamics as well as as a possible fit-function in mode–coupling theory. Figure 9 finally shows the non-ergodicity parameter, f, as a function of temperature. The shape of the curve is compatible with a constant for $T \geq T_c$ and a $\sqrt{T - T_c}$-dependance for $T \leq T_C$ (up to order $O(T - T_c)$) as predicted by theory. The overall results are well comparable with the experimental findings, however, it must be noted that the nice kink in figure 9 at a temperature $T_c \approx 550K$ is sensitive to the range over which you fit the data. You have to exclude larger times where the intensity has decayed to a level where the noise of your experiment or simulation becomes noticable. An inclusion of these data yielded a satisfactory Kohlrausch-fit as well but gave a smooth decay of f. The dependance of f on the chosen time window clearly has to be looked at more carefully in simulations as well as experiments because it must depend sensitively on how much of the relaxation has happened outside your measurement window. Note also that there is no sign of any singularity at all in the viscosity data that are available on equilibrated systems down to a temperature of $570K$. These data contradict any singularity at the T_c found from the non-ergodicity parameter.

CONCLUSIONS

We have presented results of a Monte Carlo simulation of a lattice model mapped to Polycarbonate. The mapping is done through a coarse-graining procedure where coarse-grained bond lengths and angles are mapped onto the lattice model. The whole procedure is still in a testing stage but nevertheless we were able to predict the Vogel–Fulcher temperature and activation energy of BPA-PC oligomers. This therefore seems a promising route into the search for structure property correlations of polymers. Equating densities and the absolute scale of the viscosity we can identify absolute length

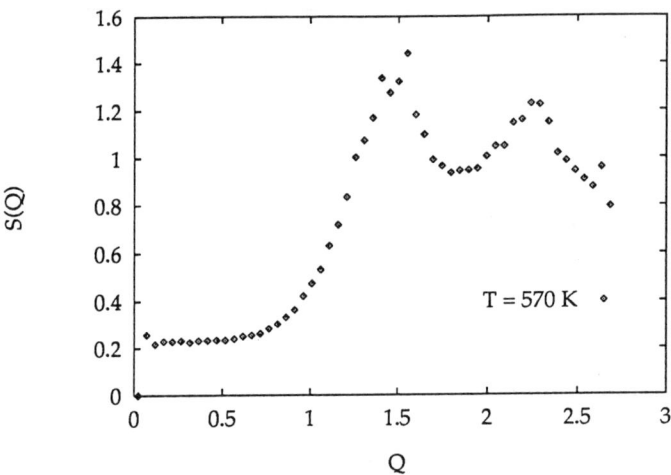

Figure 6: *Global structure factor of the simulated lattice polymer melts.*

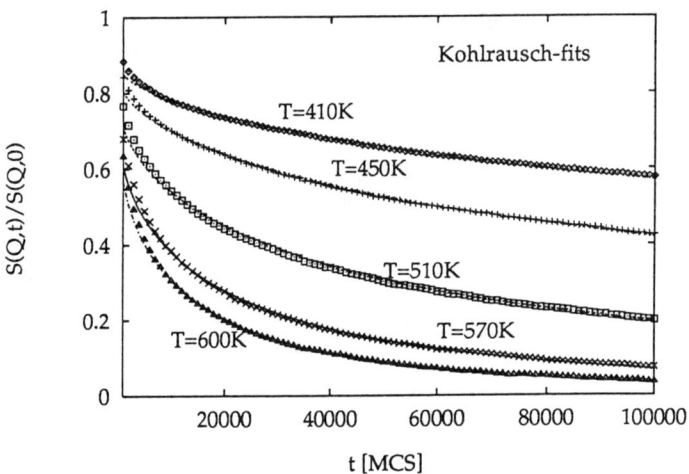

Figure 7: *Incoherent intermediate structure factor and fitted Kohlrausch-laws for selected temperatures as indicated. Data taken at the structure factor maximum* $Q = 1.48 \text{Å}^{-1}$.

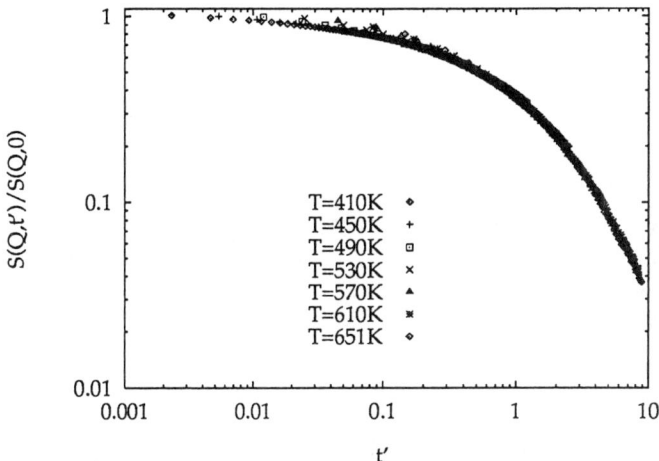

Figure 8: *Scaling plot for the incoherent structure factor at $Q = 1.48 Å^{-1}$.*

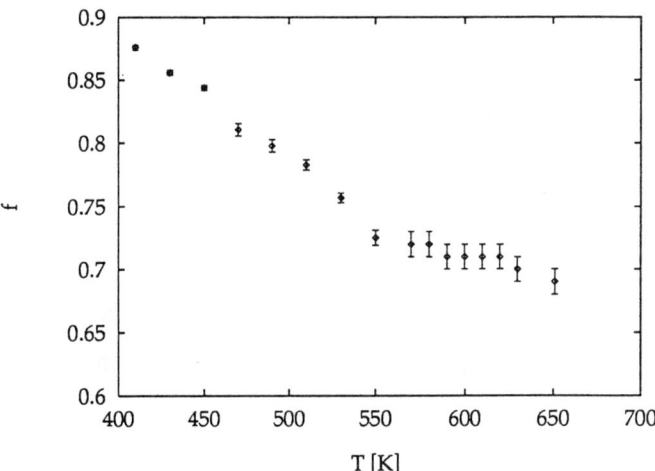

Figure 9: *Non-ergodicity parameter, f, as a function of temperature at $Q = 1.48 Å^{-1}$.*

and timescales for our simulation. The resulting global structure factor is in reasonable agreement with the experimental one. An analysis of the incoherent intermediate scattering function is closely comparable to experimental findings in the polymer glass transition regime. This kind of analysis is compatible with the predictions of mode–coupling theory, however, the viscosity data contradict the mode–coupling predictions. This contradiction remains to be resolved.

ACKNOWLEDGEMENT: It is my pleasure to thank K. Binder, K. Kremer, D. W. Heermann, J. Baschnagel, I. Batoulis and K. Sommer for a fruitful collaboration and many stimulating discussions. This work was supported by the German Ministery for Research and Technology and the Bayer AG under grant no. 03M4028.

References

[1] I. Batoulis, K. Binder, F. T. Gentile, D. W. Heermann, W. Jilge, K. Kremer, M. Laso, P. J. Ludovice, L. Morbitzer, W. Paul, B. Pittel, R. Plaetschke, R. Reuter, K. Sommer, U. W. Suter, R. Timmermann, G. Weymanns, submitted to Advanced Materials

[2] D. Theodorou, U. W. Suter, Macromolecules 19, 379 (1986) and references therein

[3] J. H. R. Clarke, D. Brown, Mol. Simul. 3, 27 (1989)

[4] D. Rigby, R. J. Roe, Macromolecules 22, 2259 (1989)

[5] W. Paul, K. Binder, K. Kremer, D. W. Heermann, Macromolecules, in press

[6] J. Baschnagel, K. Binder, W. Paul, M. Laso, U. W. Suter, I. Batoulis W. Jilge, T. Bürger, J. Chem. Phys, in press

[7] W. Götze in Liquids, Freezing an the Glass Transition, (North Holland, 1990)

[8] D. Richter, B. Frick, B. Farago, Phys. Rev. Lett. 61, 2465 (1988)

[9] B. Frick, B. Farago, D. Richter, Phys. Rev Lett. 64, 2921 (1990)

[10] I. Carmesin, K. Kremer, Macromolecules 21, 2819 (1988)

[11] H. P. Wittmann, K. Kremer, Comp. Phys. Comm. 61, 309 (1990)

[12] H. P. Deutsch, K. Binder, J. Chem. Phys, (1991)

[13] W. Paul, K. Binder, D. W. Heermann, K. Kremer, J. Phys. (Paris) II, 1, 37 (1991)

[14] W. Paul, K. Binder, D. W. Heermann, K. Kremer, J. Chem Phys, in press

[15] C. Lamers, W. Schweika, D. Richter, at International Conference on Neutron Scattering, Oxford 1991

MOLECULAR-DYNAMICS STUDY OF HIGHLY SUPERCOOLED LIQUIDS: DYNAMICAL SINGULARITIES NEAR THE LIQUID-GLASS TRANSITION

Y. Hiwatari, H. Miyagawa*, T. Muranaka and K. Uehara
Kanazawa University, Kanazawa 920, Japan

ABSTRACT

We report on the molecular-dynamics (MD) study of the binary soft-sphere fluids, focusing our attention on their dynamical properties of the supercooled fluid phase and the dynamical anomalies associated with the liquid-glass transition. The time dependent behavior of the mean-square displacement, intermediate scattering functions, both of the self (incoherent) and total correlations, also the dynamical structure factor and non-Gaussian parameter (NGP) are examined. Because of a rather short time window searched by the MD simulation, it is only an apparent dynamical behavior (e.g. on the diffusion coefficient, the long-time decay of the density autocorrelation function and so on) that we can observe through the MD simulation in the vicinity of the transition, in which the longest relaxation time far exceeds the time scale of the MD simulation. Therefore, it accompanies severe difficulties in practice with an accurate determination of the glass transition temperature of the system from the properties of the long time behavior alone. The results of the MD simulation are compared with the trapping diffusion model (TDM) recently developed by Odagaki and Hiwatari. It is concluded that our results of the MD simulation are in excellent agreement with the TDM. With this, the glass transition can be more properly determined by maxima of the NGP, which take place in an intermediate time scale comparable to that of the MD simulation.

INTRODUCTION

On approaching a glass transition, diffusivity (viscosity) decreases (increases) as low (high) as to that of a crystalline value so that the relaxation of the density (structural) fluctuations in a supercooled fluid remarkably slows down. This phenomenon is very similar to critical slow-down phenonema of a second-order phase transition. It is well known that a secon-order phase transition is driven by the critical fluctuations of infinitely long wavelength ($k \to \infty$). One may ask, then, what types of fluctuation are most essential to a glass transition, and one may be possible to discuss critical exponents relevant to the glass transition. Our main interest in this work is to discuss dynamical (anomalous) properties of supercooled fluids in the vicinity of the glass transition, if they exist, through a MD simulation on a simple model system.[1] The glass transition is often believed to be not a thermodynamic phase transition unlike a usual first- or second-order phase transition. This is true in the sense that an observation time is far shorter than a longest relaxation time of a system, which is mostly the case in highly supercooled fluids near a glass transition. This is also the case in the MD study, since a possible time window covered by it even with a supercomputer is rather very short (roughly of the order of 100–1000 pico seconds). One may also ask whether there exists an underlying phase transition relevant to observed glass

*present address: Taisho Seiyaku Co., Ltd., Ohmiya 330, Japan

tarnsition which can only be realized in the long-time limit and therefore can never be observed. This may be a purely theoretical problem, but may serve importantly as our understanding of the nature of the glass transition. This discussion is certainly beyond the work via the MD simulation and needs an approach from a theoretical point of view. In the present work, we will try to interpret our MD results with the trapping diffusion model (TDM)[2] and see to what extent the TDM becomes consistent with the MD results.

A number of works have recently been reported based on the mode coupling theory (MCT)[3] from both theoretical and experimantal sides[4]. It is, however, very problematic whether intrinsic local (short-wavelength) fluctuations like jump motions, which has been found to be particularily important in highly supercooled fluids from the works of MD simulations, can successfully be taken into consideration to the basic equations (non-linear integral-differential equations) on which the MCT is based. The idea of the MCT may be useful for modelately supercooled fluid states far above a glass transition temperature T_g, but presumably no more reliable for highly supercooled fluids near T_g, in which jump motions are dominant. Then, it is worth seeking for a different approach based on the jump motion. We will show that the TDM can be consistent with the results of the MD simulations.

This paper is organized as follows: After a brief description of the binary soft-sphere model for which our MD simulations were performed, we present the results of the MD simulation for binary soft-sphere mixtures. These includes some preliminary results of our recent MD computations over tow-million time steps, which are one order of magnitude longer than our previous MD computations.[5] The results are shown on the time dependent behavior of the mean-square displacement, density autocorrelation functions and non-Gaussian paprameter. Next, we present the results of the analyses of jump motions in binary soft-sphere fluids, which enable us to investigate the validity of a fundamental hypothesis made for the TDM. Then, we will discuss about the glass transition temperature of the present model fluid. Finally, we discuss briefly about $\alpha-$ and $\beta-$relaxations of the present model.

THE BINARY SOFT-SPHERE MODEL AND THE MD SIMULATION

As a simple model for binary alloys, we consider in our MD simulation soft-sphere mixtures composed of N_1 atoms of mass m_1 and diameter σ_1 and N_2 atoms of mass m_2 and diameter σ_2 in a volume V, which interact through the purely repulsive soft-sphere potentials:[6]

$$v_{\alpha\beta}(r) = \epsilon \left(\frac{\sigma_{\alpha\beta}}{r}\right)^{12}, \qquad (1)$$

where $1 \leq \alpha, \beta \leq 2$ are species indices, ϵ is the energy unit, and we simply assume that $\sigma_{\alpha\beta} = (\sigma_\alpha + \sigma_\beta)/2$.

According to the scaling property of the inverse power potentials and using the effective one-component approximation, it is easily shown that all reduced equilibrium properties of the soft-sphere mixtures, in excess of their ideal gas counterpart, depend only on the following coupling constant:

$$\Gamma_{\text{eff}} = n^*(T^*)^{-1/4}\left(\frac{\sigma_{\text{eff}}}{\sigma_1}\right)^3, \qquad (2)$$

$$\sigma_{\text{eff}}^3 = \sum_\alpha \sum_\beta x_\alpha x_\beta \sigma_{\alpha\beta}^3, \qquad (3)$$

where $n^* = N\sigma_1^3/V$ denotes the reduced number density with the total number of atoms $N(=N_1+N_2)$. The reduced temperature T^* equals k_BT/ϵ, $x_1 = N_1/N$ and $x_2 = 1 - x_1$. The equation of states (compressibility factor PV/Nk_BT vs Γ_{eff}) for binary soft-spheres has been found to follow the curve very close to that of the pure soft-sphere model,[6] which implies that the effective one-component approximation works very well. The freezing point of the pure soft-sphere model has been found to occur at $\Gamma_{\text{eff}}=1.15$.

In the present work, we study a binary soft-sphere mixture with the core-size ratio $\sigma_2/\sigma_1=1.2$, mass ratio $m_2/m_1=2.0$, and an equimolar system ($x_1=0.5$). Using the constant-temperature MD techniques and the seventh-order Gear algorithm together with periodic boundary conditions as usual, we have carried out MD simulations with $N=500$ particles in a cubic cell. Some runs were also made with $N = 4000$. The pair potential, Eq. (1), is cut off over the distance $r/\sigma_{\alpha\beta}=3.5$, and the number density was kept constant ($n^* = 0.8$); the temperature T^* was varied to achieve desired Γ_{eff}'s. The microscopic time scale is chosen to be

$$\tau = \left(\frac{m_1\sigma_1^2}{\epsilon}\right)^{1/2}, \qquad (4)$$

which is nearly the same order of magnitude of the Einstein period τ_E, namely a period of oscillation of atoms in an equivalent solid. This becomes of the order of 10^{-13} sec, if we used the parameters for an argon liquid in Eq. (1). The equation of motions were solved numerically with a time mesh of the order of 100th of τ, which precisely depends on the respective temperature.[5]

First, we simulated an equilibrium fluid of $\Gamma_{\text{eff}}=0.8$; then, using the configuration at the final step of this run, this fluid was quenched down to $\Gamma_{\text{eff}}=1.6$ (quenching process). The resulting cooling rate is equivalent to of the order of 10^{13} K/sec for an argon fluid, which is much faster than those achieved by experimental techniques.

Annealing MD simulation has been carried out for various different Γ_{eff}'s, starting with the final configuration of the corresponding Γ_{eff} obtained by the quenching MD simulation mentioned above.

MEAN-SQUARE DISPLACEMENT AND SELF-DIFFUSION COEFFICIENT

The time for the mean-square displacement(MSD) to reach into a t-linear asymptotic line, called as the Einstein limit behavior, becomes longer and longer, as the temperature of a supercooled fluid is decreased. This is easily seen in Fig. 1, in which the MSD for $\Gamma_{\text{eff}} = 1.5$ is shown; that is, $R_\alpha(t) =<| \mathbf{r}_i^{(\alpha)}(t) - \mathbf{r}_i^{(\alpha)}(0) |^2>$ for α species. The MSD still bends downwards, showing no t-linear bahavior in this time window. It is observed that, in $\Gamma_{\text{eff}}=1.4$, the MSD becomes t-linear in a much shorter time of the order of a few tens of τ. Referring to the data of the MSD for various supercooled fluid states all together obtained by our MD computations, it can be summarized that the MSD for $\Gamma_{\text{eff}} < 1.45$ reaches rather fast into a t-linear asymptotic limit within the time window accessible by the MD simulation. On the other hand, for larger Γ_{eff}'s the MSD never attains to a t-linear behavior within the time scales of the order of 1000τ; it is always negatively bent. Thus, the self-diffusion constant, as deduced from a least-sqaure fitting of the MSD data, more or less is always overestimated for $\Gamma_{\text{eff}} \geq 1.45$.

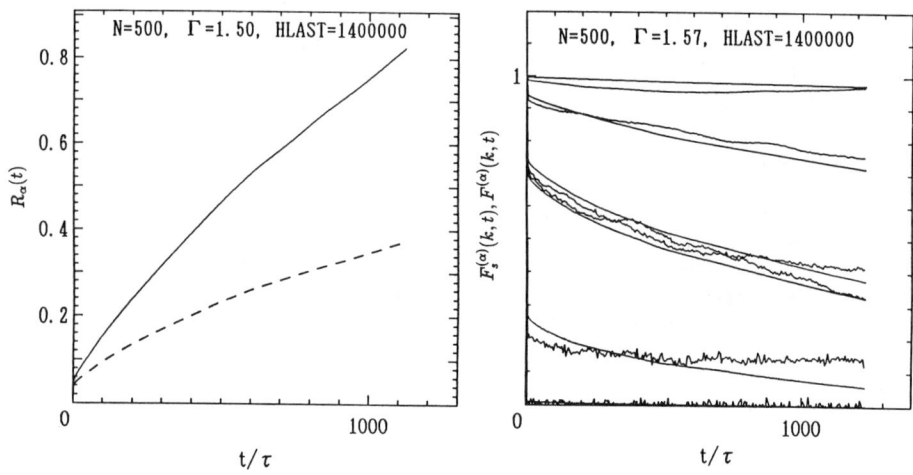

(Left): Fig. 1. MSD (in units of σ_1^2) for Γ_{eff}=1.50; solid curve is for species 1 and dashed curve for species 2.
(Right): Fig. 2. $F^{(1)}(t)$ (wavy curves) and $F_s^{(1)}(t)$ (otherwise) for $\Gamma_{\text{eff}} = 1.57$. $k^*(= k\sigma_1)$ are $k_1^*, 4k_1^*, 9k_1^*, 10k_1^*, 20k_1^*$ and $40k_1^*$ from top to bottom, where $k_1^* = 0.735 (= 2\pi\sigma_1/L)$.

DENSITY AUTOCORRELATION FUNCTION

Density autocorrelation function (DAF) $F^{(\alpha)}(k,t)$ of species α, defined below, measures how intrinsically excited density fluctuations associated with a wavelength k relaxes with an elapsed time t.

$$F^{(\alpha)}(k,t) = \frac{1}{N_\alpha} <\rho^{(\alpha)}(\mathbf{k},t)\rho^{(\alpha)}(-\mathbf{k},0)>, \quad (5)$$

$$\rho^{(\alpha)}(\mathbf{k},t) = \sum_{i=1}^{N_\alpha} \exp(i\mathbf{k}\cdot \mathbf{r}_i^{(\alpha)}(t)). \quad (6)$$

$F^{(\alpha)}(k,t)$ is sometimes called a total density autocorrealtion. The self part of it, $F_s^{(\alpha)}(k,t)$, can therefore be written as:

$$F_s^{(\alpha)}(k,t) = \frac{1}{N_\alpha} \sum_{i=1}^{N_\alpha} < \exp\{i\mathbf{k}\cdot [\mathbf{r}_i^{(\alpha)}(t) - \mathbf{r}_i^{(\alpha)}(0)]\} > \quad (7)$$

We have calculated both $F^{(\alpha)}(k,t)$ and $F_s^{(\alpha)}(k,t)$ for various Γ_{eff}'s and k's. Since the side length of the cubic simulation cell is finite, k values can not be smaller than $k_1 = 2\pi/L$ with $L = V^{1/3}$. The MD results of $F^{(\alpha)}(k,t)$ and $F_s^{(\alpha)}(k,t)$ for Γ_{eff}=1.57 and 1.60 (species 1 only) are shown in Figs. 2 and 3, respectively. It can be seen that the long time behavior of the correaltion functions shown in these two figures are remarkably different: The density autocorrelations for $\Gamma_{\text{eff}}=1.57$ reasonably tends to vanish, although the time window studied in the present MD simulation are not enough to confirm it. On the other hand, the density autocorrelation functions for $\Gamma_{\text{eff}}=1.60$ have almost no sign of a vanishing relaxation. This trend (the difference between two Γ_{eff}'s) is seen more clearly for larger k values. It is, of course, a hard question whether the larger Γ_{eff} is indeed rigorously a non-ergodic system, i.e., the correlation stays non-zero even for an infinitely long time. But this difference between these two Γ_{eff}'s is notable. The density autocorrelation functions for $\Gamma_{\text{eff}} = 1.50 - 1.57$ are similar to those of $\Gamma_{\text{eff}} = 1.57$.

We attempted to fit the MD data of $F_s^{(\alpha)}(k,t)$ in terms of the stretched exponential function of the form of the Williams-Watts law:

$$f(t) = A \exp[-(t/t_0)^\beta]. \tag{8}$$

The result of the best fitting for $k = k_1$ indicates that $\beta \simeq 1$ up to $\Gamma_{\text{eff}} = 1.40$ and that the hydrodynamic relation $t_0 = (Dk^2)^{-1}$ is nearly satisfied within an accuracy of the data for this range of Γ_{eff}. Thus, the MD simulation confirms normal relaxation (simple exponential decay and hydrodynamic relation) for the supercooled fluids of binary soft spheres up to $\Gamma_{\text{eff}} \leq 1.40$. A significant change is obtained around $\Gamma_{\text{eff}}=1.45$, above which β becomes significantly smaller than one and the hydrodynamic relation no more holds.[5] This indicates a remarkable slowing down of kinetics already to take place around $\Gamma_{\text{eff}} \simeq 1.45$. This is consistent with the behavior of the MSD mentioned above. Such drastic changes of kinetics around $\Gamma_{\text{eff}} \simeq 1.45$ may be regarded as a kinetic transition,[7] but we should note that this can not be identified as the glass transition of the system, because the diffusivity (viscosity) is still too large (small), compared with a typical value of a solid.

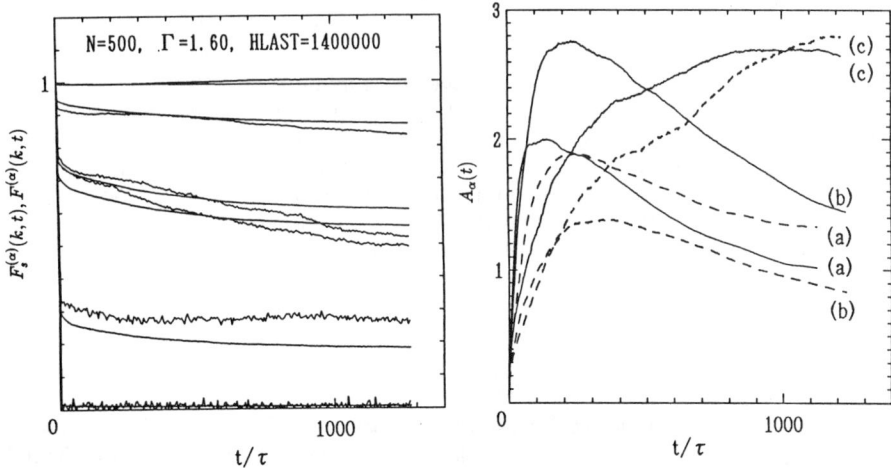

(Left): Fig. 3. Same as Fig. 2, but for $\Gamma_{\text{eff}} = 1.60$.
(Right): Fig. 4. NGP $A_\alpha(t)$ for $\Gamma_{\text{eff}} = 1.50$(a), 1.57(b) and 1.58(c). Solid curves are for species 1 and dashed curve for species 2.

NON-GAUSSIAN PARAMETER

Equation (7) can generally be written into a cumulant formula with powers of k^2:

$$F_S^{(\alpha)}(k,t) = \exp\{-\frac{1}{6}k^2 R_\alpha(t) + \frac{1}{72}k^4[R_\alpha(t)]^2 A_\alpha(t) + \mathcal{O}(k^6)\}, \qquad (9)$$

where $R_\alpha(t)$ is a MSD for species α and $A_\alpha(t)$ is so-called a non-Gaussian parameter defined as

$$A_\alpha(t) = \frac{3}{5}\frac{<|\mathbf{r}_i^{(\alpha)}(t) - \mathbf{r}_i^{(\alpha)}(0)|^4>}{[R_\alpha(t)]^2} - 1. \qquad (10)$$

$A_\alpha(t)$ obtained by the MD simulation are shown for various Γ_{eff}'s (species 1 only) in Fig. 4. It can be shown with a simple model of Einstein random walk and harmonic oscillation combined together that $A(t)$ increases monotonically from zero to a peak value at a certain time of the order of ω_0^{-1} (ω_0 is a constant jump rate) and then it decreases monotonically to zero for further increasing time.[8] Therefore, the time at which the maximum of $A(t)$ appears should be related to the time scale on which the MSD reaches a t-linear asymptotic line. This increases remarkably as Γ_{eff} increases; the maximum value of $A(t)$ also increases as Γ_{eff} increases. We will discuss below these significant properties by analyzing the maximum of $A(t)$ as a function of Γ_{eff}, which turns out to be useful to estimate the glass transition temperature.

JUMP MOTION AND COMPARISON WITH THE TRAPPING DIFFUSION MODEL

(The results of this and subsequent sections are partly based on our previous MD computations.[5]) In our previous paper,[9] we demonstrated that in supercooled fluid states near the glass transition a group of binary soft-sphere particles perform clear jump motions in a cooperative way among them. Here, we attempted to analyse jump motions more in detail, using the MD data of configurations at every time step. The analysis was made in the following way. Motions of each atom are distinguished into two types according to the length of the displacement $\Delta l(t)$ during a fixed (small enough) time period Δt between t and $t + \Delta t$: Two types are (a) $\Delta l(t) \geq \Delta_0$ and (b) $\Delta l(t) \leq \Delta_0$, where $\Delta l(t)$ is calculated by Eq. (11) and Δl_0 is a threshold parameter that is determined arbitrarily.

$$\Delta l(t) = |\bar{\mathbf{r}}_i(t + \Delta t) - \bar{\mathbf{r}}_i(t)| \quad (i = 1, 2, \ldots, N), \qquad (11)$$

$$\bar{\mathbf{r}}_i(t) = \frac{1}{2\tau_1}\int_{t-\tau_1}^{t+\tau_1} \mathbf{r}_i(t')dt'. \qquad (12)$$

In order to eliminate thermal noises properly, we took $\tau_1 \simeq$ (a few times the Einstein period). A jump motion is defined as a series of (a)-type motions, which, say, begins at time t_1 and ends at time t_2. The jump distance (JD) is therefore $|\bar{\mathbf{r}}_i(t_2) - \bar{\mathbf{r}}_i(t_1)|$ and the flying time (FT) is $t_2 - t_1$. The residence-type motion is contrarily defined as a series of (b)-type motions, which, say, begins at time t'_1 and ends at time t'_2. The travel distance (TD) is therefore $|\bar{\mathbf{r}}_i(t'_2) - \bar{\mathbf{r}}_i(t'_1)|$ and the residence time (RT) $t_2' - t_1'$. Thus, the motions of all atoms are classified into the residence-type motion and jump-type motion. The results are shown in Figs. 5 and 6 of ref. 5 for various Γ_{eff}'s. In $\Gamma_{\text{eff}}=1.0$ (normal liquid), each distribution of FT, JD and TD has a broad distribution, therefore the distinction of two types of motions are not clear. However, it is clearly seen that in supercooled fluids the histograms have a much narrower distribution, indicating that jump motions are well defined. The JD is distributed around $0.4 \sim 0.8\sigma_1$ and the FT around $1.0 \sim 2.5\tau_0$ (τ_0 is a physical time scale over which a particle travels by a mean interparticle distance, $l = (V/N)^{1/3}$, with a mean thermal velocity; this is nearly the same order of magnitude as τ in our T^* and n^*)[5]. We have found that the mean JD and FT are $0.419\sigma_1$ and $1.70\tau_0$, respectively, for $\Gamma_{\text{eff}}=1.55$. We have also made other statistical analyses about jump and resident motions: For example, for $\Gamma_{\text{eff}}=1.55$, the fraction of the number of the jump motions observed for species 1 ($N_J^{(1)}$) to the total number of jump motions observed (N_J) gives 70.2%, and the fraction of the number of the residence-type motions with a TD$\geq 0.5\sigma_\alpha$ to the total number of the residence-type motions is only 15%, for species 1. These results give us a measure of reality of such separation of two types of motions in supercooled fluid states and hence a physical meaning of the present analysis.

The RT distribution remarkably spreads out to long times as the temperature is decreased. This behavior is shown in Figs. 7 and 8 of ref. 5. These histograms are now compared with the theoretical prediction for the residence time distribution function $\Psi(t)$ of the TDM:

$$\Psi(t) = zw_0(\rho + 1)G(\rho + 2)\gamma^*(\rho + 2, zw_0 t), \tag{13}$$

where $G(x)$ is gamma function and $\gamma^*(a, x)$ Tricomi incomplete gamma function, and z and ω_0 are constant parameters. We have made least-square fits for the RT distribution obtained by MD computations with Eq. (13) by varying ρ and fixing $zw_0=0.1$ (in units of τ_0^{-1}), which works for $\Gamma_{\text{eff}} \geq 1.40$. Results of these fitting are shown in Figs. 7 and 8 of ref. 5. Best fit values of ρ for species 1 are 0.50, 0.22 and 0.045 for $\Gamma_{\text{eff}}=1.45$, 1.50 and 1.55, respectively. Similar results are also obtained for species 2. These results are consistent with the prediction of the TDM, in which the galss transition takes place at $\rho=0$.

GLASS TRANSITION

In preceding sections, we have seen that significant slowing down phenomena are observed for $\Gamma_{\text{eff}} > \sim 1.45$. In this section, we attempt to determine the glass tansition Γ_g of the binary soft-sphere model in parallel with the TDM. The TDM predicts that beyond the glass transition point the non-Gauusian parameter at an infinite time takes a non-vanishing value, that is, $A_\alpha(t \to \infty) > 0$ for $\rho < 0$. The MSD increases with time as $\sim t^\theta (0 < \theta < 1)$ for large t. Therefore, the diffusion constant, as defined by $D_\alpha = \lim_{t\to\infty} \frac{1}{6t} R_\alpha(t)$, becomes zero in glasses. The model also predicts apparent anomalous dynamical properties[10] in intermediate-time scales such as the MSD for $\rho < 1$ obeys a power-law time dependence well before the glass transition $\rho = 0$. Since $\rho = 1$ corresponds to $\Gamma_{\text{eff}} = 1.44$ of the soft-sphere model, as we will see below in Eq. (16), this prediction is again consistent with the slowing down phenomena observed in the MD simulation for $\Gamma_{\text{eff}} > 1.45$ which we have seen in preceding sections.

The task of determining the glass transition from the MD simulation results has no unambiguous solution as far as the long-time behavior is concerned. Here we will show that the glass transition can more properly be determined with the maximum of $A_\alpha(t)$. Since the maximum appears in intermediate time scales, as mentioned above, this way will, in principle, bring us no such a difficulty of the long-time behavior of the MSD or the dynamical correlations. With the TDM, it is shown that $D \sim \rho$ when ρ is small. On the other hand, we have found that the MD data of D_α for the binary soft-sphere model can be well fitted by $D \sim (\Gamma_g - \Gamma_{\text{eff}})^3$ with $\Gamma_g = 1.58$.[11] Therefore, we may expect that $\rho \sim (\Gamma_g - \Gamma_{\text{eff}})^3$. With a semi-empirical relation predicted by the TDM that $A^* \tilde{t}^* \sim \rho^{-4/3}$, which holds for $0.1 \leq \rho \leq 0.8$,[10] (A^* and $\tilde{t}^*$ are the maximum value of $A(t)$ and the corresponding time, respectively, where $\tilde{t} = z\omega_0 t$) we may expect for the binary soft-sphere model that $A_\alpha^* \tilde{t}^* \sim (\Gamma_g - \Gamma_{\text{eff}})^{-4}$. In fact, this is confirmed by a best-fit analysis of the MD data of $A_\alpha(t)$. The result is shown in Fig. 9 of ref. 5. The best-fit result obtained is

$$A_\alpha^* \tilde{t}^* \sim (\Gamma_g - \Gamma_{\text{eff}})^{-\beta'}. \tag{14}$$

where $\Gamma_g = 1.590$ and $\beta' = 3.623$ for species 1, and $\Gamma_g = 1.563$ and $\beta' = 3.322$ for species 2. Thus, we may conclude that all these results mentioned above are self consistent, taking into consideration a rather low accuracy of the MD data in the vicinity of the transition, in which the effects of the dynamical slowing down phenomena are more or less inevitable.

Detailed comparison of the residence time distribution $\Psi(t)$, Eq. (13), with those obtained by the MD simulation (see Figs. 7 and 8 of ref. 5) enables us a further check of the TDM. We can use the bahavior of $\Psi(t)$ at $t = 0$ to determine the relation between the parameter ρ and Γ_{eff} of the soft-sphere fluids. We estimate $\Psi(0)$ for each Γ_{eff} from the residence time distribution obtained by the MD simulation. Then, we obtain the corresponding $\bar{\rho}$ by the relation obtained from Eq. (13) at $t = 0$:[12]

$$\bar{\rho} = \frac{1 - 2\Psi(0)}{\Psi(0) - 1}. \tag{15}$$

Although the statistics for $\Psi(t)$ near $t = 0$ are not so high, a reasonable estimate of $\bar{\rho}$ as a function of Γ_{eff} is possible, from which we obtain that[12]

$$\bar{\rho} = 398(\Gamma_g - \Gamma_{\text{eff}})^3 \tag{16}$$

This is also consistent with the above discussions.

SUMMARIES AND DISCUSSIONS

With MD simulations of binary soft spheres, we have studied dynamical slowing down phenomena in supercooled fluids. The intermediate and long time behaviors of the MSD, DAF and NGP in highly supercooled fluid states (i.e. for $\Gamma_{\text{eff}} \geq 1.45$) remarkably differ from those of smaller Γ_{eff}'s. The drastic change of kinetics was observed around the same $\Gamma_{\text{eff}}(\simeq 1.45)$ for all these dynamical functions. This transition may be called a kinetic transition (the corresponding temperature T_c), but we note that this appears far above the glass transition temperature T_g. Physical meanings of such a transition have not yet been made clear so far. However, with the TDM, this transition can be identified as a transition, at which dynamic changes (sub-anomalous properties) begin to appear;

the intermediate- and long-time bahaviors of the MSD change from a usual t-linear behavior to anomalous t behaviors (non-t-linear terms as well as t-linear terms are involved). This transition takes place at $\rho = 1$, which corresponds to $\Gamma_{\text{eff}} \simeq 1.44$ for the present model fluid.

It is concluded that the results of the MD simulations for the binary soft-sphere model are in good agreement with the predictions of the TDM. The compariosn between the MD results and the TDM are made on several dynamical functions and Γ_{eff}'s. The glass transition of the present model was also estimated with the maximum of the NGP, which agrees, within a statistical accuracy of the MD data, with the previous estimates from other dynamical functions, based on the long time behavior. Noting that the maximum of the NGP appears in an intermediate time domain accessible by the MD simulation, the analysis of the NGP is very useful to estimate Γ_g. This should be contrasted with other estimates referring to the long-time behavior of dynamics such as diffusion constant, which encounters a serious difficulty of slowing-down phenomena in highly supercooled fluids.

The sub-anomalous properties observed for $\Gamma_{\text{eff}} \geq 1.45$ in the present model fluid essentially represent the α-relaxation, as easily be confirmed by the Fourier transform of $F^{(\alpha)}(k,t)$, the dynamic structure factor $S(k,\omega)$. α-relaxation can be characterized by $S(k,\omega) \sim \omega^{1+\beta}$, where β is a stretched-exponential exponent (see Eq. (8)).

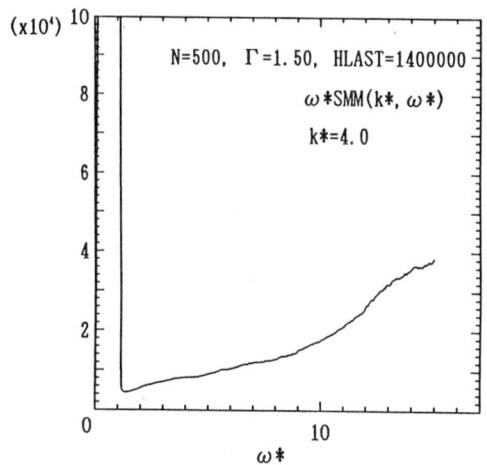

Fig. 5. $\omega^* S_{MM}(k^*, \omega^*)$ vs ω^* at $\Gamma_{\text{eff}} = 1.50$ and $k^* = 4k_1^*$. $\omega^* = \omega\tau$. $k_1^* = 2\pi\sigma_1/L$.

In Fig. 5 is shown $\omega^* S_{MM}(k^*, \omega^*)$ vs ω^* for $\Gamma_{\text{eff}} = 1.50$ and $k^* = 4k_1^*$ for the present model fluid, where $S_{MM}(k^*, \omega^*)$ is dynamical mass structure factor with $\omega^* = \omega\tau$ and $k_1^* = 2\pi\sigma_1/L$ (minimum wave number compatible with the length of the cubic simulation cell (L) and the present system size $N = 500$). The spectrum function in the range of $\omega^* > 1$ represents a β relaxation. An ω-linearly increasing behavior of the spectrum is seen in the range of $1 < \omega^* < 10$. This behavior is similar to the scaling prediction of the MCT below a critical temperature. However, the present result shows that both α- and β-relaxations are present for $T < T_c$ ($\Gamma_{\text{eff}} > \Gamma_c$, where $\Gamma_c \simeq 1.45$), which disagrees to the MCT. More detailed analyses of the spectrum function, both on α- and β-relaxations, are currently investigated.

ACKNOWLEDGMENTS

We express our sincere thanks to T. Odagaki for useful comments and discussions. This work was partly supported by a Grant-in-Aid from the Ministry of Education, Science and Culture, Japan. Computations were carried out with supercomputer systems at Kyoto University and Institute of Molecular Science.

REFERENCES

1. For a recent review, see, Y. Hiwatari, H. Miyagawa and T. Odagaki, Solid State Ionics, (in press).

2. T. Odagaki and Y. Hiwatari, Phys. Rev. A41, 929 (1990).

3. E. Leutheusser, Phys. Rev. A29, 2765 (1984); T. R. Kirkpatrick, ibid. 31, 939 (1985); W. Götze, Z. Phys. Chem. 156, S3 (1988); U. Krieger and J. Bosse, Phys. Rev. Lett. 59, 1601 (1987).

4. F. Mezei, W. Knaak and B. Farago, Phys. Rev. Lett. 58, 571 (1987); B. Frick, B. Farago and D. Richter, Phys. Rev. Lett. 64, 2921 (1990); N. J. Tao, G. Li and H. Z. Cummins, Phys. Rev. Lett. 66, 1334 (1991); N. J. Tao, G. Li and H. Z. Cummins, preprint; H. Z. Cummins, G. Li, W. M. Du, X. K. Chen, N. J. Tao and A. Sakai, preprint; G. Li, W. M. Du, X. K. Chen, H. Z. Cummins and N. J. Tao, preprint.

5. H. Miyagawa and Y. Hiwatari, Phys. Rev. A, (in press).

6. B. Bernu, J. P. Hansen and Y. Hiwatari, Phys. Rev. A36, 4891 (1987); see also ref. 5.

7. J. N. Roux, J. L. Barrat and J. P. Hansen, J. Phys.: Condes. Matter 1, 7171 (1989).

8. T. Odagaki and Y. Hiwatari, Phys. Rev. A43, 1103 (1991).

9. H. Miyagawa, Y. Hiwatari, B. Bernu and J. P. Hansen, J. Chem. Phys. 88, 3879 (1988).

10. T. Odagaki and Y. Hiwatari, (submitted to Phys. Rev. A); see also ref. 2.

11. The value of the exponent is somewhat larger than that of ref. 6. MD data of D from much longer simulations seems to be better fitted by the present formula.

12. T. Odagaki and Y. Hiwatari, J. Phys.: Condes. Matter 3, 5191 (1991).

LOSS OF ERGODICITY IN GLASSY SYSTEMS

Raymond D. Mountain
National Institute of Standards and Technology
Gaithersburg, MD 20899 USA

D. Thirumalai
Institute for Physical Science and Technology
University of Maryland
College Park, MD 20742 USA

ABSTRACT

Quantitative measures of the time required for effective ergodicity to be realized in a dynamical system are developed and applied to computer simulations of strongly supercooled Lennard-Jones fluid mixtures. The techniques employed are generally applicable to the study of ergodic-nonergodic transitions and can also be used in the design of long simulations.

INTRODUCTION

The glassy state is intimately associated with the inability of the fluid to sample adequately the phase space of the system; as such it is said to be nonergodic.[1] Nonergodic behavior is associated with barriers in phase space that inhibit the sampling of all allowed regions of phase space.[2] The argument to be expounded here is that the methods for monitoring and investigating the glass transition should be directly tied to the loss of ergodicity. Some ways to do this using computer simulations will be developed and illustrated in terms of a simple glass forming system.

The glass transition occurs when the observation time, t_{obs}, becomes less than the time, t_{ec}, required for a dynamical trajectory to adequately sample the phase space available to the system. When $t_{obs} < t_{ec}$ the ergodic hypothesis, which asserts that time averages and phase space averages are equal, breaks down and thus the glass transition is an example of an ergodic-nonergodic transition. It should be understood from the beginning that we are discussing effective ergodicity in the sense that time and ensemble averages are equivalent but not in the sense that every available point in the phase space of the system is reached by the dynamical trajectory of the particles.[3]

The major point to be stressed here is that the loss of ergodicity is a universal feature of the glassy state, independent of specific intermolecular potentials, and that properties which sample ergodicity directly should be used to explore the characteristics of the glass. We have developed some quantities which are constructed so that they explicitly determine t_{ec}, the time required for ergodic convergence to be realized in a molecular dynamics simulation. We shall apply them to the problem of characterizing the supercooled liquid and glassy states of a previously studied mixture of Lennard-Jones particles.[4]

We also want to emphasize that the methodology discussed here is quite general and can be applied to many different situations beyond those explicitly mentioned.[5,6]

EFFECTIVE ERGODIC BEHAVIOR

Effective ergodicity means that ensemble and time averages of phase space functions are equal.[2] For the microcanonical ensemble appropriate to molecular dynamics simulations, this equality can be expressed in terms of the time average of a phase space function $A(\Gamma(t))$, which is evaluated along a dynamical trajectory in phase space denoted by $\Gamma(t)$,

$$a(t_{obs}) = \frac{1}{t_{obs}} \int_0^{t_{obs}} ds\, A(\Gamma(s)) \tag{1}$$

and in terms of of the microcanonical ensemble average of A, which is evaluated over the surface in phase space with energy E,

$$\langle a \rangle = \frac{\int d\Gamma\, A(\Gamma)\delta(H(\Gamma) - E)}{\int d\Gamma\, \delta(H(\Gamma) - E)} \tag{2}$$

where $H(\Gamma)$ is the Hamiltonian of the system. Effective ergodicity means that $a(t_{obs}) = \langle a \rangle$ for observation times t_{obs} which are sufficiently long. In practice t_{obs} can be quite short and well within the range of times available for molecular dynamics simulations except for conditions which are identified with many slow degrees of freedom such as the glassy state. A major question to be addressed is that of how to determine the minimum value of t_{obs}, namely t_{ec}, needed for this equality to be realized when an independent assessment of $\langle a \rangle$ is not available. The answer to this question will be provided in the next section.

First we will expand the discussion of effective ergodicity in supercooled and glassy states by considering possible order parameters which might be of value in characterizing the glass transition. It is expected that a dynamical quantity is needed given the kinetic nature of the glass transition. Candidates should have the property that correlations which have a finite lifetime in the liquid should become very long lived as the glass transition is approached. The difficulty with this concept is that many plausible quantities have this property. For example, the non-gaussian term in the self-correlation function $F_s(q,t)$[7], the local orientational correlation function $\varphi(t)$ introduced by us[8], and the long time behavior of the dynamical structure factor $F(q,t)$[9] have been suggested as possible order parameters. In each case there is no compelling theoretical reason to focus on these quantities and the long time behavior of these quantities appears to be a consequence of the glass transition and not a guide to the origin and nature of the transition.

It should be also noted that crystals are not ergodic in the sense that not all of the energetically allowed regions of phase space are sampled in a time t_{obs}. Even so, there is no concern about the equivalence of time and ensemble averages. This is because the particles of a crystal sit at well defined sites which are crystallographically equivalent and so the different reqions of the energy surface corresponding to the crystal make essentially identical contributions to the averages. This is not the case in the glass because the sites are not equivalent and as a result, permutational symmetry is broken. A consequence of this is that the local environment of a particle in the glass is specific to that particle. This fact is an important guide to the development of ways to monitor the rate of convergence of time averages to the effective ergodic values. These methods are described in the next section.

ERGODIC CONVERGENCE TOOLS

For fluid in equilibrium, it is not possible to distinguish one particle from another on the basis of time averaged quantities associated with individual particles. For example, the average energy of a particle of a given species is the same as all other particles of that species. If it were not so, one could make a classification of particles in that way. (That is, of course, what happens when a fluid phase separates, a topic we do not consider here.) It is useful to consider this time averaged equivalence as a special sort of symmetry of the fluid, a property we call statistical symmetry.[3] The equivalent sites of a crystal enforce statistical symmetry in a crystal while the absence of equivalent sites in a glass results in the breaking of the statistical symmetry of the fluid at the glass transition. We note that the homogeneity of space is preserved even though statistical symmetry (or permutational symmetry) appears to be broken in the glassy state. The symmetry breaking associated with the glass transition is a feature which can be directly monitored in the following way.

We consider a property of a fluid which can be decomposed into parts associated with individual particles in the fluid. Examples are the energy E,

$$E_j = \frac{1}{2}(m_j v_j^2 + \sum_{k \neq j} \phi_{jk}), \qquad (3)$$

or S, the off diagonal part of the stress tensor,

$$S_j = \frac{1}{2} \sum_{k \neq j} \frac{x_{jk} y_{jk}}{r_{jk}^2} r_{jk} \frac{d\phi_{jk}}{dr_{jk}}. \qquad (4)$$

More generally we consider the case where there is a phase function F which can be written as

$$F = \sum_{j=1}^{N} F_j \qquad (5)$$

where F_j is the part of F associated with particle j. Now let

$$f_j(t) = \frac{1}{t}\int_0^t ds\, F_j(s) \qquad (6)$$

be the time average of F_j over the interval t and let $\overline{f}$ be the average of the $f_j(t)$'s. A measure of the extent to which statistical symmetry is violated, which we have called the fluctuation metric,[10,11] is defined as

$$\Omega_f(t) = \frac{1}{N}\sum_{j=1}^N [f_j(t) - \overline{f}]^2. \qquad (7)$$

This quantity has a valuable attribute which we now discuss.

For long times, $\Omega(t)/\Omega(0) \to 1/D_f t$ where the coefficient D_f is a measure of the ergodic convergence time.[10,11] This can be seen by rewriting the definition of $\Omega(t)$ in terms of the time integral expression for $f_j(t)$ and invoking the equivalence of time and ensemble averages for equilibrium states. The details of this development are to be found in Refs. 10 and 11. This result is quite general for systems in equilibrium and has been applied to other problems not associated with the glass transition. The essential feature is that the correlations in the individual F_j be of finite duration. If this is not true, as is necessarily the case for a nonergodic system, then the $1/t$ behavior is not present and $\Omega(t)$ will either remain finite for all observation times or will decay with other power laws.[12]

Experience has shown that effective ergodicity is obtained when the observation time t_{obs} satisfies $D_f t_{obs} \geq 100 = D_f t_{ec}$. D_f is obtained as the slope of $\Omega_f(0)/\Omega_f(t)$ as a function of t and usually can be reliably determined for times significantly less than t_{ec}. This means that an estimate of t_{ec} can be made from fairly short runs, a feature which can aid in the design of long simulations.

The extension of this derivation to mixtures, where the $\overline{f}$ can depend on the species is discussed in Ref. 11 in detail. In effect, one includes a species label and compares $f_j(t)$ with the average for the species group of particle j. Otherwise the analysis and discussion is unchanged.

APPLICATION TO SUPERCOOLED FLUIDS

Molecular dynamics simulations have been performed for a mixture of Lennard-Jones particles. The interparticle interactions have the form

$$\phi_{LJ}(r) = 4\epsilon[(\sigma/r)^{12} - (\sigma/r)^6], \qquad (8)$$

with a diameter σ appropriate to the specific pair of particles as described below. The mixture consists of 400 particles of type 1 with a diameter $\sigma_{11} = 0.8\sigma_{22}$ and 100 particles of type 2 with a diameter $\sigma_{22} = 1$ which is taken as the unit of length. The cross diameter $\sigma_{12} = 0.9\sigma_{22}$. All energy parameters ϵ are taken to have unit value and the mass of the type 2 particles is taken to be 1 while that of the type 1 particles is 0.9. Constant pressure simulations[13] were made for the pressure $p = 5$ for the mixture over a range of temperatures from the equilibrium liquid to the glass. As usual, time is measured in units of $\tau = \sqrt{m_2\sigma_{22}^2/\epsilon}$. Individual runs varied in duration from 50τ at high temperatures to 200τ for strongly supercooled states. The equations of motion were integrated using an iterated version of the Beeman algorithm[14] with a time step of 0.005τ. The value of the wall mass was taken to be 1. Larger values of the wall mass were examined but were found to have little influence on the results of interest here. Constant volume simulations were also performed and we conclude that the fluctuation metric coefficients are not sensitive to this aspect of the dynamics.

The fluctuation metrics for the energy and transverse stress and the corresponding coefficients D_E and D_S have been determined as functions of temperature at constant pressure for these systems. Also the self-diffusion coefficients, D_1 and D_2 have been determined using the Einstein relation for the mean square displacement as a function of time. For long times,

$$\frac{1}{N_a}\sum_{j=1}^{N_a} <[\mathbf{r}_j(t) - \mathbf{r}_j(0)]^2> \to 6D_a t. \tag{8}$$

Here the subscript a can be either 1 or 2.

The results are displayed in Figs. 1 and 2 as functions of $1/T$. The points are the computed values and the solid curves are representations of the points in terms of the Volger-Fulcher (V-F) form

$$D = A exp[-B/(T - T_0)]. \tag{9}$$

For each of the four quantities, $B = 1.5 \pm 0.05$ and $T_0 = 0.29$. The prefactors A are specific to the quantity in question. It is noteworthy that the curvature in the $1/T$ plots for D_E and D_S for the Lennard-Jones mixture starts to become pronounced near the temperature ($T = 0.6$) where Jónsson and Andersen[4] found that the local cluster motif in the fluid began to favor five-fold coordination at the expense of cubic motifs. A detailed analysis of these results can be found elsewhere.[15]

DISCUSSION

There are several observations about these results to be discussed. The first of these is that the V-F singularity temperatures T_0 are considerably lower than the observed glass transition temperatures for these systems. This is frequently the case for real glass formers as well. It means that if T_0 has any meaning as the temperature of an underlying phase transition, the transition is inaccessible physically. In particular the relevant parameter, $T_g = (T_g - T_0)/T_0$, measuring the proximity of the glass transition to T_0 is typically between 0.3–0.6. This suggests that the relevant length over which the particle motion is highly correlated can at most be of the order of $(2-3)\sigma_{22}$ for $T \approx T_g$.[3,16,17] In effect, the system is restricted to at most a "cross-over" region where physical correlation lengths are still of molecular size. Thus theoretical attempts to base the glass transition on critical phenomena type theories must address the unresolved questions of the cross-over from the critical to the non-critical regions.

As noted above, the departures of the temperature dependence of the metric coefficients D_E and D_S from the activated, Arrhenius form to the V-F form is correlated with the change in clustering noted by Jónsson and Andersen for the Lennard-Jones mixture. This is an illustration of the sensitivity of the metric to changes in the local structure of phase space. It is this sensitivity which makes the metric a useful tool for studying the structure of the supercooled and glassy states.

Finally, the metric is constructed to be sensitive to the changes in ergodic convergence rates. It is not limited to supercooled fluids. With suitable changes in the quantities monitored so that the appropriate statistical symmetry is examined, it can be usefully employed to probe the time scales for effective ergodicity to be reached in Hamiltonian systems involving large numbers of degrees of freedom. In particular these ideas have already been used to probe the relaxation of dipole moments in liquid water[6] and to probe the motion of various residues in proteins such as myoglobin and ribonuclease.[18]

REFERENCES

1. C. A. Angell, J. Phys. Chem. Solids **49**, 863 (1988).
2. R. G. Palmer, Adv. Phys. **31**, 669 (1982).
3. D. Thirumalai, R. D. Mountain, and T. R. Kirkpatrick, Phys. Rev. A **39**, 3563 (1989).
4. H. Jónsson and H. C. Andersen, Phys. Rev. Lett. **60**, 2295 (1988).
5. D. Thirumalai and R. D. Mountain, J. Stat. Phys. **57**, 789 (1989).
6. R. D. Mountain and D. Thirumalai, Comp. Phys. Comm. **62**, 352 (1991).
7. G. Pastore, B. Bernu, J. P. Hansen, and Y. Hiwatari, Phys. Rev. A **38**, 454 (1988). This is one paper in an extensive series of papers on the glass transition in soft-sphere mixtures.
8. R. D. Mountain and D. Thirumalai, J. Chem. Phys. **92**, 6116 (1990).
9. See for example, M. J. D. Brakkee and S. W. De Leeuw, J. Phys. :Condens. Matter **2**, 4991 (1990).
10. R. D. Mountain and D. Thirumalai, J. Phys. Chem. **93**, 6975 (1989).
11. D. Thirumalai and R. D. Mountain, Phys. Rev. A **42**, 4574 (1990).

12. R. D. Mountain and D. Thirumalai, (unpublished work).

13. M. Parrinello and A. Rahman, J. Appl. Phys. **52**, 7182 (1981). The original molecular dynamics constant pressure method is due to H. C. Andersen, J. Chem. Phys. **72**, 2384, (1980).

14. D. Beeman, J. Comput. Phys. **20**, 130 (1975).

15. R. D. Mountain and D. Thirumalai, Phys. Rev. Lett. (submitted).

16. G. S. Grest and S. R. Nagel, J. Phys. Chem. **91**, 4916 (1987).

17. R. M. Ernst, S. R. Nagel, and G. S. Grest, Phys. Rev. B **43**, 8070 (1991).

18. J. Straub, (private communication).

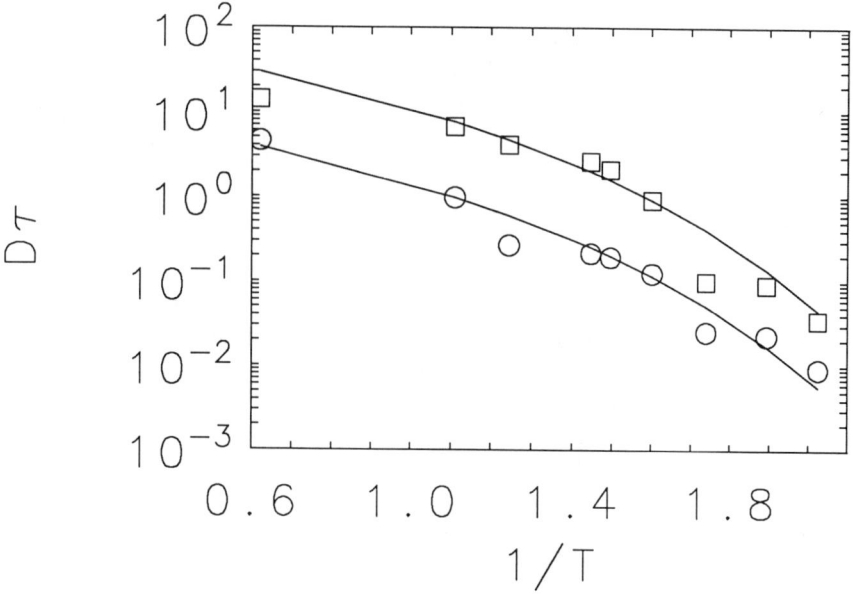

Fig. 1. The fluctutation metric coefficients D are shown as functions of $1/T$. The circles are the calculated values of D_E and the squares are the values of D_S. The solid lines are the Volger-Fulcher representations, eq.(9), fit to the points.

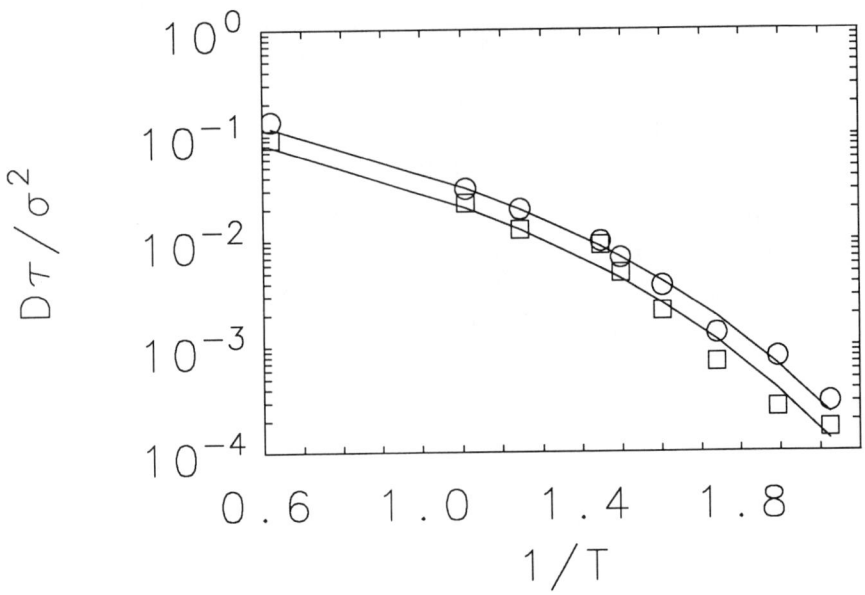

Fig. 2. The self-diffusion coefficients of the two types of particles are shown as functions of $1/T$. The circles are the calculated values for type 1 particles (the smaller size) and the squares are the values for the type 2 particles. The solid lines are the Volger-Fulcher representations, eq. (9), fit to the points.

SIMULATION OF SUPERCOOLED ATOMIC AND BROWNIAN SYSTEMS; COMPARISON WITH MODE-COUPLING THEORY.

Jean-Noël Roux*,
Laboratoire de Physique, ENSL,
allée d'Italie, 69007 Lyon, France

ABSTRACT

The main results from simulations of two different model systems, the soft-sphere alloy and a polydisperse Yukawa fluid that has been studied with both Newtonian and Brownian dynamics, in the supercooled régime, are recalled. In particular it is emphasised that the only well established qualitative change resembling a 'kinetic glass transition' is a crossover in the microscopic self-diffusion mechanism. One then briefly presents the similarities and differences with the predictions of mode-coupling theory: although the agreement is fair for intermediate relaxation times, it seems that the critical laws are never clearly seen in numerical results, and that the theory provides a poor quantitative estimate of the thermodynamic coefficients of the glass.

1. INTRODUCTION.

The recent surge of simulation work on simple liquids in the dense, supercooled régime [1] is, to a large extent, motivated by some new theoretical predictions concerning the onset of a so-called 'kinetic glass transition' *i.e.*, a qualitative change in relaxation phenomena, corresponding to a very strong slowing-down, at a certain critical temperature T_c that is somewhat above the conventional (calorimetric) glass transition temperature. This theoretical progress is due, in particular, to what is now known as the mode-coupling theory (MCT) for the glass transition. MCT was initiated by several authors [2,3,4,5], and the characteristic scaling behaviour of time-dependent correlation functions it predicts in the critical temperature range near T_c was studied, and compared to experimental results in 'fragile' glass-formers (in Angell's [6] classification), by Götze, Sjögren and their collaborators [7]. Only in model systems (hard-sphere, Lennard-Jones systems and the like), however, does MCT rely on a detailed, microscopic kinetic theory. Molecular dynamics (MD) simulations of such systems should therefore provide the most accurate and quantitative test of MCT. Of course, very long relaxation times are beyond the reach of MD investigation, which in the simplest cases is typically restricted to time scales of 10^{-9} or 10^{-8} seconds. However, the interesting viscosity régime where the 'transition' is expected to occur is $\eta \sim 1$ Pa.s (for molecular liquids), which, taking the viscosity ratio as an estimate of the relaxation time ratio, corresponds to times τ about 1000 times as large as the relaxation time in an ordinary (as opposed to supercooled) liquid. Hence, for typical molecular materials, $\tau \sim 1$ ns, and this is accessible to MD; thus comparing simulation data to theoretical predictions near T_c is conceivable.

---

* Present address: Service de chimie, point 10,
LCPC, 58 boulevard Lefèbvre, 75732 Paris cedex 15, France.

The following is a contribution to such a comparison in the case of a binary mixture of small spheres, a system for which many MD results are now available [8,9]. Results on a polydisperse Yukawa system [10] will also be quoted; this system has been simulated both as an ordinary atomic liquid, using standard MD, and as a suspension of Brownian particles, using Smoluchowski dynamics without hydrodynamic interactions. After a brief description of the systems in part 2, we shall attempt in section 3 to present the remarkable 'experimental' facts emerging from the computer studies of those model fluids in the dense, supercooled régime, in a manner that be independent on any theory of structural slowing-down. Then, in section 4, similarities and differences with MCT predictions are discussed. Part 5 is a conclusion with a few indications for future work.

2. THE SYSTEMS.

Both systems the studies of which are reported here were chosen such that crystal nucleation was rendered extremely rare and difficult, allowing thus very long simulation runs in the amorphous state.

2.1 The soft-sphere alloy.

This system has been described in many publications [8]. It consists in an equimolar binary mixture of point particles interacting with the spherically symmetric purely repulsive pair potential

$$V_{\alpha\beta}(r) = \epsilon \left(\frac{\sigma_{\alpha\beta}}{r}\right)^{12},$$

where indices $\alpha, \beta = 1, 2$ label the two particle species, $\sigma_{\alpha\beta} = \frac{\sigma_\alpha + \sigma_\beta}{2}$, and $\frac{\sigma_2}{\sigma_1} = 1.2$ for the results presented here. The mass ratio $\frac{m_2}{m_1}$ of the two particle species was set to 2, and a system of units was adopted such that $\sigma_1 = 1$, $\epsilon = 1$, $m_1 = 1$. The unit of time is thus $\tau = \frac{m_1 \sigma_1^2}{\epsilon}$. For such an inverse power potential, density and temperature play exactly the same role, static and dynamic properties of the system only depend on the coupling parameter $\Gamma = \rho \sigma_x^3 (\frac{T}{\epsilon})^{-\frac{1}{4}}$, the effective one component fluid particle diameter σ_x being defined as $\sigma_x^3 = \sum_{\alpha,\beta=1,2} x_\alpha x_\beta \sigma_{\alpha\beta}^3$, and $x_1 = x_2 = 0.5$ for the equimolar mixture. In the following $T = \epsilon$ is implicit and all measured quantities are given as functions of Γ. If one substitutes in the definition of the potential the parameters suitable for argon, one has $\tau \sim 1ps$, $\epsilon \sim 100K$. Two different cooling rates were used, corresponding to roughly $1.5 \; 10^9$ and $0.5 \; 10^9 K.s^{-1}$.

2.2 The polydisperse Yukawa system [10].

This system could be a model for a charge-stabilised colloidal suspension. The particles i and j interact via the potential:

$$v_{ij}(r) = U_0 Z_i Z_j \phi(r),$$

where Z_i is the charge of particle i, U_0 is an energy scale and

$$\phi(r) = \frac{a}{r} = exp\left(-\kappa\frac{r-a}{a}\right).$$

Particle charges are distributed according to a Schultz distribution with a polydispersity of 50%. a is the unit of length, and $\kappa = 7$, which makes the potential rather short-ranged. Density is fixed at $\rho = a^{-3}$ and temperature, expressed in units of U_0, varies. For standard MD simulations, or 'Newtonian dynamics', the unit of time is $\tau_N = \sqrt{\frac{ma^2}{U_0}}$, whereas for Brownian dynamics, in which particle are given a prescribed constant short-time diffusion coefficient D_0, it is $\tau_B = \frac{Ta^2}{U_0 D_0}$. The typical length of the coldest runs was 500 to 800 in τ_N or τ_B units.

3. 'EXPERIMENTAL' FACTS.

We summarise here the most salient phenomena that have been numerically observed in the dense, supercooled régime.

3.1 Slowing-down.

This is of course the most immediately noticeable phenomenon, witnessed e.g., by the rapid change of the transport coefficients of the soft-sphere system when $\Gamma > 1.25$: the diffusion constants $D_{1,2}$ of the two species decrease, and the shear viscosity η of the mixture increases, by a factor larger than 15 in the interval $1.25 \leq \Gamma \leq 1.43$. For larger couplings we were not able to obtain reliable values. The product ηD remains roughly constant, in agreement with the phenomenological Stokes-Einstein relationship, up to $\Gamma = 1.41$. For $\Gamma \geq 1.42$, the increase in η seems to overcome the drop in the diffusivities, but η values are already quite inaccurate. Both tagged particle and total number density correlators $F_s^{1,2}(k,t)$ and $F_{NN}(k,t)$ have been measured with k wavenumbers corresponding to the main peak in the (number-number) structure factor $S(k)$, $k \simeq k_0$, and for $k \simeq \frac{k_0}{2}$ too. As Γ increases, they exhibit slower and slower relaxation, with a relatively fast step for $t \sim \tau$, followed by a very slow one that is well fitted by a stretched exponential (or Kohlrausch function) $exp\left[-\left(\frac{t}{\tau_K}\right)^\beta\right]$. Fig. 1 shows a typical $F(k_0,t)$ versus $\log(t)$ curve. The results of the Kohlrausch fits are consistent with the choice of a Γ-independent, but wavevector-dependent, stretching exponent β: $\beta \sim 0.6$ for k_0 and $\beta \sim 0.8$ for $\frac{k_0}{2}$.

With the Yukawa system, as T gets lower than 0.15, very similar results are obtained. Moreover, there is no significant difference in the long-time part of the relaxation between both dynamics, as shown on Fig. 2.

3.2 Cross-over in the diffusion mechanism

The slowing-down phenomenon of paragraph 3.1, although fast, is gradual and does not indicate any 'transition'. However, the microscopic mechanism for the individual diffusion of particles apparently becomes qualitatively different

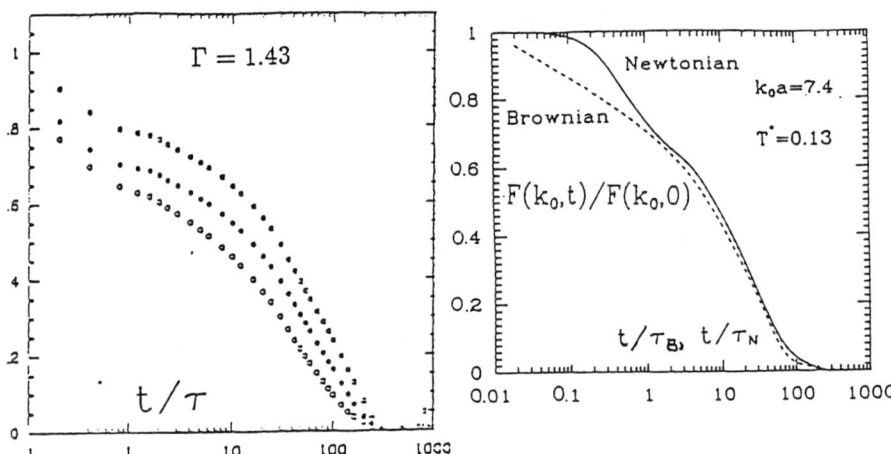

Figures 1 (left) and 2 (right). Figure 1 is a plot of the self, $F_s^{(1)}(k_0,t)$ and $F_s^{(2)}(k_0,t)$ (black and white circles), and number density, $F_{NN}(k_0,t)$ (squares, upper curve), intermediate scattering functions versus $\log(t)$ at $\Gamma = 1.44$ in the soft-sphere system. Fig. 2 represents $F_{NN}(k_0,t)$ in the polydisperse Yukawa system for both dynamics. F_{NN} functions are divided by their initial values $S_{NN}(k_0)$.

below a certain temperature. This is most obvious on a plot of the self part of the van Hove functions (the Fourier back-transform of $F_s(k,t)$) versus r at different times. For $\Gamma < 1.44$ in the soft-sphere alloy, $4\pi r^2 G_s^{(\alpha)}(r,t)$, the probability distribution for the distance r over which a particle of species α travels in time t, spreads smoothly (see fig. 3a). At higher couplings, as shown on fig. 3b, the curves possess a secondary maximum corresponding to the nearest-neighbour distance, while the first one remains at a fixed position. This means that the particles tend to stay close to a given site, and then suddenly escape by jumping to a neighbouring one, in a manner that is reminiscent of diffusion in a solid. This is confirmed by a direct examination of particle trajectories. Such a change is also apparent in $F_s(k_0,t)$ and $F_{NN}(k_0,t)$, but much less obvious in $F_{NN}(k_0/2,t)$, and it is not clear whether and to what extent the relaxation of collective quantities is really affected.

This change is about the only event one might be tempted to call a 'transition'. It is a genuine feature of equilibrium relaxation, since its existence was found to be cooling-rate independent. The only difference to be observed in the slower cooling experiment on the soft-sphere system is that the new diffusion mechanism appears a little sooner and more gradually. As one clearly sees on fig. 3b, this 'transition' is more properly termed a cross-over. In the following the corresponding thermodynamic parameters will be denoted with a star (one has $\Gamma^* \simeq 1.44$ for the soft-sphere system), and relaxation phenomena beyond the crossover will be termed 'residual'.

Once again, the Yukawa system behaves in the same way, whatever the dynamics: even in the Brownian case, when particle velocities are overdamped,

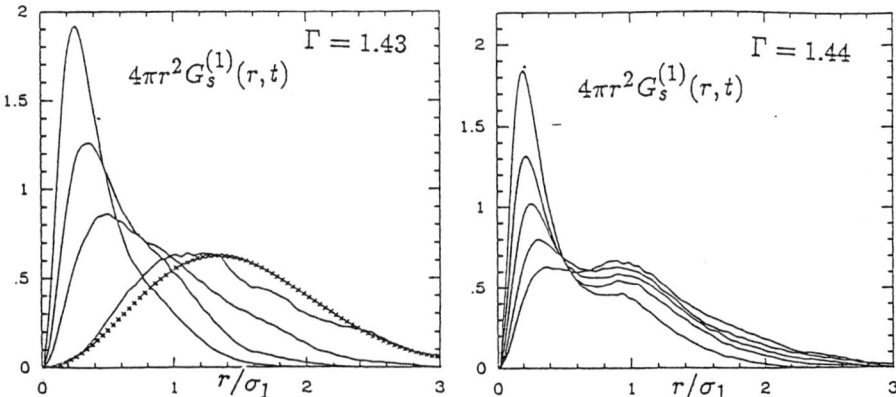

Figure 3: van Hove functions $4\pi r^2 G_s^{(1)}(r,t)$ versus r for different times for Γ just before (fig. 3a, left, $\Gamma = 1.43$, longest time 360τ, for which the hydrodynamic prediction (crosses) is approximately retrieved) and approximately at (fig. 3b, right, $\Gamma = 1.44$, $\frac{t}{\tau} = 100, 200, 300, 400, 500$) the crossover in the soft-sphere alloy.

jump diffusion takes over (although hopping events are not quite as frequent as in the Newtonian case) around $T^* \simeq 0.13$.

3.3 Tendency toward freezing.

As the soft-sphere system is compressed past Γ^*, it may suddenly freeze, without any apparent structural relaxation during the MD run. This behaviour, witnessed by the van Hove functions of fig. 4, is however not intrinsic: its reproducibility is bad, and a gentler cooling rate will shift it to higher couplings. We used a step-wise cooling program in our simulations: after each cooling step, one waits for an equilibration time τ_{eq} to elapse before recording data during a measurement time τ_m. τ_m has to be larger than the equilibrium relaxation time τ_r of the system if one wants an ergodic exploration of phase space. But, firstly, τ_{eq} must be larger than some non-equilibrium relaxation time τ_r^{ne} that depends on the magnitude of the temperature step. A likely explanation for the surprising suddenness of freezing in some MD runs (see fig. 4), is that τ_r^{ne} might be much larger than τ_r, and one has $\tau_r^{ne} \gg \tau_{eq}$, while a dramatic freezing was not expected, since one tends to set τ_{eq} and τ_m at a value somewhat larger than the extrapolated τ_r. The kind of glass one eventually makes on falling out of equilibrium depends on the interplay between those four time scales.

Such freezing events often occur close to T^*, but should not be confused with the diffusion crossover; the choice of a slower cooling rate allowed us to clearly distinguish the two phenomena in the soft-sphere system.

One may note an interesting detail on fig. 4: $G_d(r,t)$ does, in fact, change a little bit at long times: it increases close to $r = 0$, which shows that, although the pair structure is unchanged, the particle initially at $r = 0$ is replaced by another one. In other words, there is some residual jump diffusion, but no

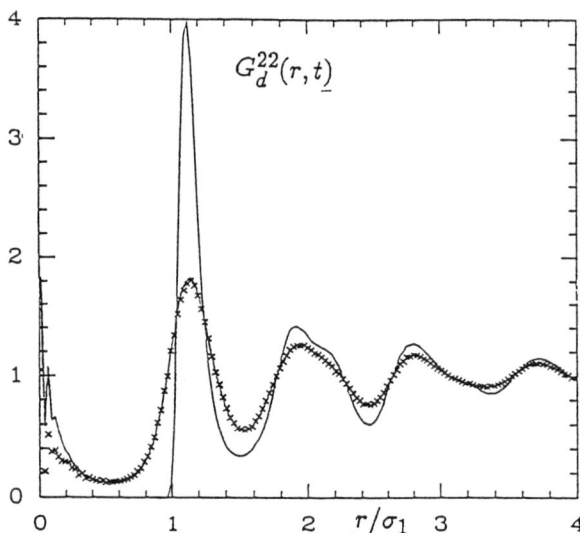

Figure 4: van Hove functions $G_d^{22}(\vec{r},t)$ versus r for $\frac{t}{\tau} = 0$ (line), 240 (crosses), 480 (line) at $\Gamma = 1.48$ and pairs of distinct particles of species 2.

residual density relaxation. This is in agreement with the observation that several particles may jump together and merely exchange their positions [11].

3.4 Onset of solid-like hydrodynamics.

(This paragraph is only concerned with atomic systems with Newtonian dynamics). By 'hydrodynamics', we mean macroscopic laws ruling the long-time and long wavelength correlations. In the soft-sphere system we have computed the total particle mass density correlator $F_m(k_1,t)$ and the transverse mass current correlator $C_t(k_1,t)$ for the smallest wavenumber k_1 compatible with the MD box, in the vicinity of Γ^*. These correlation functions (see figs. 5 and 6) exhibit oscillations the period of which is very much smaller than the relaxation time of the system. Consequently, we cannot be observing the true liquid hydrodynamic limit of the correlators. $F_m(k_1,t)$ should exhibit sound waves with a damping contribution proportional to the longitudinal viscosity, $C_t(k_1,t)$ a diffusive behaviour. But the short time processes probed on figs 5 and 6 are quite foreign to the very slow ones associated with the large value of the viscosities, and both longitudinal and transverse modes are propagative. Longitudinal waves have also been recorded at $k = 2k_1$, and the observations (cf. fig. 5) are compatible with a damping proportional to k^2, while no significant dispersion is detected. What we are in fact measuring are longitudinal and transverse waves in the solid which the system behaves like when probed on such time scales ($\frac{t}{\tau} \sim 1$), which correspond to a plateau in the relaxation (see fig. 1).

When the system is deeply supercooled, a time and wavevector domain develops [12,13], in which the correlators are characteristic of solid hydrodynamics. If C is a typical sound speed, τ_r the relaxation time, σ a molecular size, and τ_f

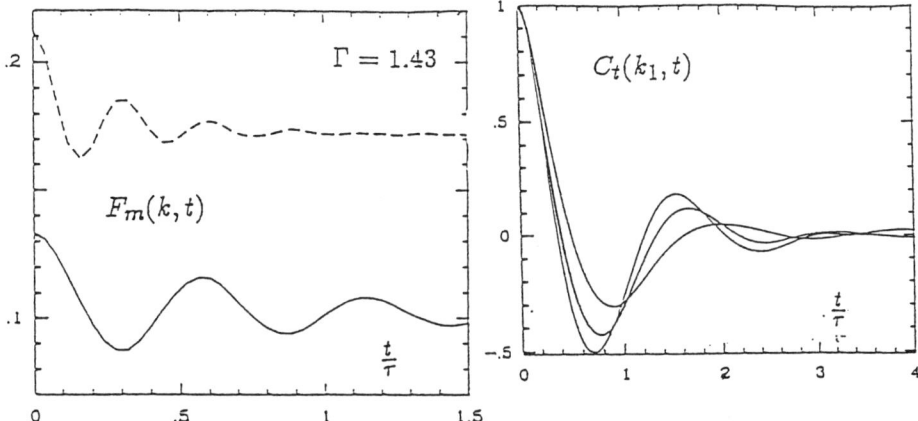

Figures 5 (left) and 6 (right). Fig. 5 shows $F_m(k,t)$ at $k = k_1$ (bottom) and $k = 2k_1$ (top). Note that in this latter case the period of the oscillations is approximately twice as short, and their decrement four times as large. Fig. 6 shows $C_t(k_1,t)$ at $k = k_1$ for $\Gamma = 1.3$, 1.38, 1.455: propagative behaviour gradually emerges.

the relaxation time for fast (vibrational) processes, this domain is defined by:
$$\frac{1}{\tau_r C} \ll k \ll \frac{2\pi}{\sigma} \text{ and } \tau_f \ll t \ll \tau_r,$$
whereas observation of the true liquid hydrodynamic régime requires:
$$k \ll \frac{1}{\tau_r C} \text{ and } t \gg \tau_r .$$
The elastic moduli of the glass, and its viscosities (in the sense of the viscosity of a solid), may be deduced from sound speeds and damping coefficients. One finds that the longitudinal modulus $B + \frac{4}{3}G$ of the glass is about 20 to 30 percent higher than the bulk modulus B of the liquid, and that its shear modulus equals 18 to $20\rho T$.

These elastic waves should not be confused with the high frequency response of ordinary liquids that is observable on much shorter time and length scales, and characterised by the high-frequency moduli B^∞ and G^∞. To wit, modulus G^∞ (an easily measurable, two-body quantity [14,13]) is much too high ($\sim 60\rho T$) to describe the transverse waves of fig. 6. High-frequency moduli are relevant when one probes the system on times t such that $t \ll \tau_f$.

One may identify the moduli B and G measured from correlators at $k = k_1$ with the static elastic moduli of the incipient glass to be made at a slightly higher coupling when one will have $\tau_m \ll \tau_r$. This interpretation is made very plausible when one measures the static elastic moduli, with appropriate fluctuation formulae, in the case of a sudden freezing (like in 3.3). Although such glasses are rather made when $\tau_{eq} \ll \tau_r^{ne}$, (and thus the non-zero large time limit of the correlators is not exactly an equilibrium plateau value), the agreement is satisfactory: one finds $G \simeq (13 \pm 5)\rho T$, and a similar compatibility for B.

These quantities are also important because of their role in liquid relaxation: in the ideal case of a sharp transition, there is a continuity between the amplitude of the diverging part of the shear viscosity in the liquid and the value of G in the glass.

4. A BRIEF COMPARISON WITH MCT.

We wish here to discuss the validity of MCT, which predicts an ideal glass transition, as a model for the diffusion crossover and relaxation close to T^*. One has to distinguish between two kinds of comparisons: qualitative ones, in which one looks for experimental evidence of the general predictions of MCT, such as scaling laws, that rely on the existence of a bifurcation in a non-linear equation but not on any particular microscopic model; and quantitative ones, in which the detailed microscopic version of MCT from kinetic theory is tested.

4.1 General predictions and scaling laws

Interpretations of simulation results according to MCT that have been attempted so far identify the crossover at T^* with the theoretically predicted transition at a temperature T_c. Residual relaxation and jump diffusion are then attributed to 'activated processes'—a convenient phrase that does not quite succeed in concealing our ignorance. The slow, Kohlrausch-like relaxation step resembles the theoretically predicted α-relaxation. But the most crucial test of MCT would be the existence of the much more specific β régime [7], another slow relaxation step involving time scales longer than the fast, vibrational ones, but shorter than the α-relaxation ones. Such times could correspond to the intermediate times plateau of correlators (see fig. 1). However, this régime was never convincingly identified, since it is always limited to a few units of τ (or ps) in simulation data, and does not extend past, say, the oscillations of the velocity autocorrelation function [10]. Thus no clear conclusion may be derived from a check of the general qualitative predictions of MCT. (However, see 15: the shape of the curve of fig. 1, near the shoulder, is well described by the theoretically calculated β-regime.)

4.2 Quantitative tests.

The microscopic version of MCT for simple systems allows, in principle, a complete computation of time-dependent correlators. Das [16] carried out such a program for the Lennard-Jones fluid, and compared his results to Ullo and Yip's [17] simulation data. In his calculations, he included a refinement of MCT [4,18] (technically, an additional coupling of the longitudinal current memory function to current correlators) that was designed to cut off the sharp transition. He thus achieved a good semi-quantitative agreement. However, this comparison was done at temperatures somewhat above T^*, outside any 'critical' or 'crossover' régime: Das's results may be considered as evidence for the validity of a mode-coupling description of liquid relaxation, but not of MCT for the glass transition.

Unfortunately, no such dynamic calculations are available for the soft-sphere and polydisperse Yukawa systems. Some interesting indications may however be gathered from the following remarks.

1– MCT would predict exactly the same transition, for the polydisperse Yukawa fluid, for both dynamics [19], which agrees with the simulations near T^*.

2- The refined version of MCT with a coupling to current modes seems unlikely to provide a good model of jump diffusion: hopping events apparently involve complicated high-order correlations [11] that are not accounted for in the theory; and they exist in the case of Brownian dynamics, while no current modes are available.

3- The critical coupling Γ_c and the 'order parameter'

$$f_k = \lim_{t \to +\infty} \frac{F(k,t)}{S(k)}$$

as predicted by MCT have been calculated [20] for the soft-sphere mixture. One has $\Gamma_c = 1.32$ while $\Gamma^* \simeq 1.44$; and the f_k versus k curve is qualitatively similar to the an experimental $\frac{F(k,t)}{S(k)}$ versus k one with an intermediate ($\frac{t}{\tau} = 10$) value of t, albeit with a significant overestimation of the tendency toward freezing.

4- From the same curve, using the formula [21,13]

$$\lim_{k \to 0} S(k) f_k = \rho T \left[\left(\frac{1}{B}\right)_{liquid} - \left(\frac{1}{B + \frac{4}{3}G}\right)_{glass} \right]$$

(valid in a one-component system, and in a mixture as well, provided there is no interdiffusion in the glass, which is the case for MCT), one deduces that the longitudinal elastic modulus $B + \frac{4}{3}G$ in the glass should be more than five times as large as the bulk modulus B of the liquid. This is in marked difference with the simulation results; the static longitudinal modulus of the glass would even be greater than its high-frequency longitudinal modulus, which is downright impossible. Such a failure may be traced back to the drastic simplifications inherent to the approximate treatment of transverse modes in MCT, which lead to a wrong identification of the memory term that might lead to a viscosity divergence [13].

5. Conclusion.

The crossover in the diffusion mechanism that is observed in simulations of soft-sphere and Yukawa systems, both for Newtonian and Brownian dynamics, is reminiscent of the breakdown of the Stokes-Einstein relationship in molecular fluids reported in laboratory experiments when the shear viscosity reaches a value of the order of $1 Pa.s$ [22]. One may try to explain it with the mode-coupling theory [23]. However, since the 'critical' aspects of MCT are hardly detectable in the simulations, and since quantitative MCT predictions remain satisfactory as long as no transition is involved, another interpretation is also possible: MCT, which reduces all correlators to combinations of one-body density autocorrelation functions, provides some reasonable approximations in some limited range of temperatures and relaxation times; then, eventually, in the crossover region, the role of higher-order correlations is more and more important, standard liquid-state kinetic theory approximations break down and a theoretical description of microscopic dynamics is beyond our current reach. MCT could, at least, describe the beginning of the slowing-down. While theoretical progress, which would require to tackle the study of high-order correlations in liquids, seems rather difficult, interesting additional information might easily be obtained from further simulation work, aiming in particular at a better knowledge of jump diffusion and residual relaxation.

REFERENCES

1. For a review, see *e.g.*, Barrat J.-L, Klein M. L. 1991 to be published in Annual Review of Physical Chemistry
2. E. Leutheusser 1984 Phys. Rev. **A29** 2765
3. U. Bengtzelius, W. Götze & A. Sjölander 1984 J. Phys. **C17** 2765
4. Das S., Mazenko G., Ramaswamy S. & Toner J. 1985 Phys. Rev. Lett. **54** 118; Das S., Mazenko G. 1986 Phy. Rev. **A34** 2265
5. Kirkpatrick T. R. 1985 Phys. Rev. **A31** 939.
6. Angell C. A. 1988 J. Phys. Chem. Solids **49** 863
7. See, *e.g.*, Götze W., in *Liquids, freezing and the glass transition*, edited by J.-P. Hansen, D. Lévesque & J. Zinn-Justin, North Holland 1991.
8. Bernu B., Hansen J.-P., Hiwatari Y. & Pastore G. 1987 Phys. Rev. **A36** 4891; Pastore G., Bernu B., Hansen J.-P. & Hiwatari Y. 1988 Phys. Rev. **A38** 454; Roux J.-N., Barrat J.-L. & Hansen J.-P. 1989 J. Phys. Condensed Matter **1** 7171; Barrat J.-L., Roux J.-N., Hansen J.-P. 1990 Chemical Physics **149** 197.
9. Mountain R., Thirumalai D. 1987 Phys. Rev. **A36** 3300.
10. Löwen H., Hansen J.-P., Roux J.-N. 1991 submitted to Phys. Rev.
11. Migayawa H., Hiwatari Y., Bernu B. & Hansen J.-P. 1988 J.Chem. Phys. **88** 3879.
12. Motorin V. I. 1987 Phys. Lett. **A122** 262.
13. Roux J.-N., articles submitted to J. Stat. Phys.
14. P. Schofield 1966, Proc. Phys. Soc. 88, 149.
15. Sjögren L., these proceedings.
16. Das S. P. 1991 Phys. Rev. **A**
17. Ullo J. J., Yip S. 1989 Phys. Rev. **A39** 5877
18. Götze W., Sjögren L. 1987 Zeitschrift für Physik **B65** 415; Sjögren L. 1990 Z. für Physik **B79** 5.
19. Szamel G. , Löwen H. 1991 preprint.
20. Barrat J.-L., Latz A. 1990 J. Phys. Condensed Matter **2** 4289.
21. See, *e.g.*, Jäckle J. 1983 J. Chem. Phys. **79** 4463, or ref. 13.
22. Rössler E. 1991, to be published in J. non-cryst. Solids
23. Barrat J.-L., Roux J.-N. 1991 J. non-cryst. Solids **131** 255.

ON THE IMPORTANCE OF KINETIC INHOMOGENEITIES IN UNDERSTANDING GLASSY DYNAMICS

Scott Butler and Peter Harrowell
Theoretical Chemistry Section, University of Sydney
Sydney N.S.W. 2006 Australia

INTRODUCTION

A (perhaps the) key question to be addressed in a theory of glassy dynamics is: how do such slow collective motions arise as a result of local particle interactions. The question can be simplified by replacing local interactions with local kinetic constraints, so neglecting much of the detail of specific interparticle potentials. Here we describe a detailed study with the aim of understanding how such local kinetic constraints couple to configurational fluctuations in a simple model to give rise to a number of features characteristic of glassy systems. Some of this work has appeared recently [1].

THE MODEL

The 2D facilitated kinetic Ising model [2] consists of spins with two states (up and down) located on the verticies of a square lattice. There is no spin–spin interaction contribution to the total energy, only an external field h which, in the ratio $h/k_B T$, determines the average concentration of up spins. Local dynamic constraints take the form of a dependence of the spin flip rate $W(\sigma_i, m_i)$ on the number m_i of up spins adjacent to the ith spin,

$$W(\sigma_i, m_i) \propto m_i(m_i - 1)\exp[(\sigma_i - 1)/k_B T] \quad (1)$$

where σ_i is the state of the ith spin. A spin must have two or more up neighbours in order to flip, a condition which becomes more difficult to satisfy as the temperature decreases and the concentration of up spins diminishes. Monte Carlo simulations of the model [3] have identified a number of features characteristic of glassy dynamics, including a non–Arrhenius temperature dependence of the spin relaxation time and long time tails in the associated spin correlation functions.

KINETIC INHOMOGENEITIES

We have concentrated on the spatial distribution of the relaxation kinetics rather than that of the spin configurations themselves. The former has the advantage that it provides a direct insight into the correlations relevant to the dynamics without requiring any assumptions as to what form these correlations might take. We find that the relaxation rates become increasingly inhomogeneous as the temperature is lowered, with small pockets of spins flipping rapidly amid 'frozen' domains. Our main result is the observation that the relaxation of these domains, the process which appears as the long time tail of the relaxation functions, is due to a mobile subset of the rapidly relaxing clusters. This picture has been quantified to the extent of an explicit evaluation of the concentration C of these mobile or active centers,

$$C = f(1) \exp\left[\frac{(1-c)^8}{8\ln(1-c)}\right] \qquad (2)$$

with $c = 1/[1+\exp(2h/k_B T)]$ and $f(1) = 6c^2(1-c)^2 + 4c^3(1-c) + c^4$. A remarkably simple power law

$$\tau \propto (1/C)^{3.8} \qquad (3)$$

is found to hold with excellent agreement for 6 orders of magnitude variation in the relaxation time τ, so that the entire non–Arrhenius temperature dependence is captured in Eq.3. With the aim of developing more general tools for the analysis of the complex structure of the relaxation kinetics we have recently completed a scaling analysis of both the spatial distribution of relaxation times and the system size dependence of the relaxation functions [4].

MEASURING COOPERATIVITY

The complex rate distributions found in the spin model present us with a variety of length scales, all possible candidates for a measure of the degree of cooperativity of the dynamics. We have examined an operational definition of such a measure based on the kinetic correlation length. This length is defined as the distance into the bulk from a free surface at which the local relaxation time has reached half its bulk value. This length was found to increase slowly with decreasing temperature. Over a temperature range which saw the relaxation time increase by 4 orders of magnitude, the kinetic correlation length increased by only a factor of three. Clearly slow dynamics does not necessarily entail large length scales.

DISCUSSION

The nature of the kinetic inhomogeneities is central to understanding the glassy dynamics of the facilitated kinetic Ising model. We believe that such inhomogeneities exist in real 'fragile' liquids and could well play a similar crucial role in determining the features of the slow relaxation kinetics in these materials.

REFERENCES

1. S.Butler and P.Harrowell, J.Chem.Phys. 95, 4454 (1991), *ibid.*, 4464 (1991)
2. G.H.Fredrickson and H.C.Andersen, Phys.Rev.Lett. 53, 1244 (1984)
3. G.H.Fredrickson and S.Brawer, J.Chem.Phys. 84, 3351 (1986)
4. M.Foley and P.Harrowell, unpublished

IONIC DYNAMICS ON THE POTENTIAL ENERGY SURFACES OF GLASSY ALKALI CHLORIDE SYSTEMS

Kenichi Kinugawa, Kohei Kadono, and Hiroshi Tanaka
Government Industrial Research Institute Osaka, Midorigaoka, Ikeda, Osaka 563, Japan

ABSTRACT

Molecular dynamics calculations are carried out for two Coulombic glasses, LiCl and LiCl-KCl-CsCl. Stillinger-Weber's inherent structures (the configurations of potential minima) which ions in glassy state undergo during 100 ps were explored for two systems. The ions in the LiCl-KCl-CsCl glass diffuse collectively via the successive potential basins whose minima are distributed along zigzag ways. On the contrary, in pure LiCl glass several ions transit to the next potential basins while the remaining ions continue to stay at almost the same potential basins.

INTRODUCTION

Dynamic processes in glass-forming materials remain somewhat unresolved despite the advances of experimental techniques[1] and the dynamical theories.[2] In particular, it is still unclear how chemical differences among substances influence the topology of the potential hypersurfaces, and even how such characteristics of the potential surfaces dominate the ionic motions in glassy state. In the present molecular dynamics study, we attempted to examine the disordered potential hypersurfaces of two ionic systems, the alkali chloride mixture (50 mol% LiCl - 40 KCl - 10 CsCl) and its reference system (pure LiCl), because of the contrastive nature of two systems; (1) the former mixture system possesses rather good glass-forming ability,[3] whereas it is difficult to obtain glassy LiCl experimentally; (2) in liquid state as well, significant decrease of Li ion mobility is observed when different alkali species are mixed together.[4]

METHODS

Molecular dynamics calculations are carried out for two systems with 64 particles in each; the glass state is achieved by cooling the liquids, and the trajectories over 100 ps are calculated at 300 K. Following Stillinger and Weber,[5] 100 configurations of potential minima (called inherent structures) which ions in glassy state undergo during 100 ps are explored by quenching the instantaneous configurations at every 1 ps. Here we concentrate on the transitions of the ions between the successive basins of the potential surfaces.

RESULTS AND DISCUSSIONS

The transition paths of each ion along the successive potential minima are plotted in Fig. 1. We can see that the ions in the mixture system diffuse via the potential basins whose minima are distributed along zigzag ways, as shown in Fig.1(a). On the other hand, the interbasin transitions in the pure LiCl system shown in Fig.1(b) are classified into two patterns; (a) Almost no displacements of the position of potential mimima are observed, so the ions continue to vibrate at nearly the same positions;

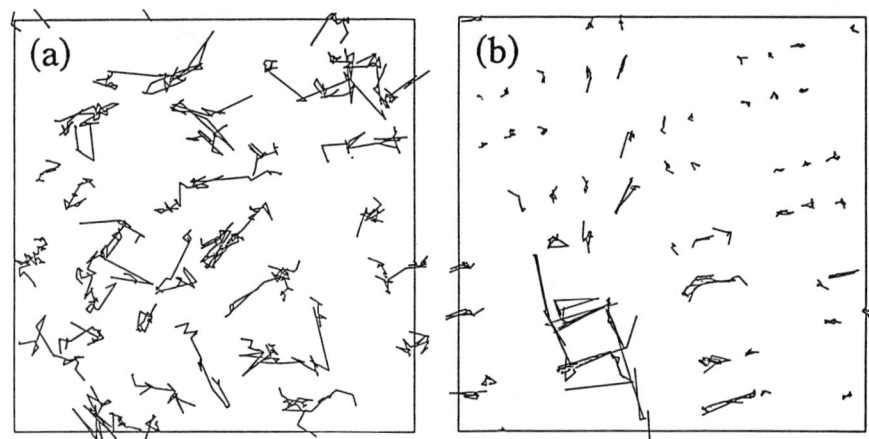

Fig.1 Displacements patterns between the inherent structures during 100 ps (every 4 ps, total 25 configurations). Only Li$^+$ and Cl$^-$ ions are drawn with projected onto yz plane. (a) LiCl-KCl-CsCl glass; (b) LiCl glass.

(b) Straight-line transitions with large displacements (typical jumping[6]) are observed. The zigzag paths in the mixture glass suggests that there are much more potential minima in this system, so that the displacements of ions are relatively short and the transition direction changes with time. In other words, the topology of the potential surfaces in the mixture glass is more complicated than in pure LiCl glass. The features of the transition paths in LiCl glass corresponds to simpler topology of the potential surface. Such differences of the distribution of the potential minima between two systems would influence the transport properties and the glass-forming ability of the materials.[7]

In order to investigate the collectivity of these transitions, we evaluated the moment ratio defined as

$$r = \langle \Delta R^4 \rangle / \langle \Delta R^2 \rangle^2,$$

where ΔR is the displacement of ion between two successive minima, and $\langle \ \rangle$ means the average over all the transitions and all the ions. For the spatially delocalized transition displacements this ratio is close to unity, while for the completely localized transition it is the order of the number of particles. The calculated moment ratio r is 5.17 for the mixture glass, and 12.47 for the LiCl glass. This shows that the interbasin transitions in the mixture system are collective and spatially delocalized, while those in LiCl are more localized phenomenon in which only several ions may participate. The delocalized and localized nature of the interbasin transition in the mixture and the LiCl glass coincides with the displacement patterns shown in Fig.1.

References
1. For example, D. Richter et al., Phys. Rev. Lett. **61**, 2465 (1988). 2. For example, E. Leutheusser, Phys. Rev. A **29**, 2765 (1984). 3. K. Kadono et al., J. Non-Cryst. Solids **122**, 214 (1990). 4. This is known as mixed-alkali effect or Chemla effect. For example, C.T. Moynihan and R.W. Laity, J. Phys. Chem. **68**, 3312 (1964). 5. F.H. Stillinger et al., Phys. Rev. A **28**, 2408 (1983). 6. H. Miyagawa et al., J. Chem. Phys. **88**, 3879 (1988). 7. F.H. Stillinger, Phys. Rev. B **41**, 2409 (1990).

A STUDY ON THE MECHANISM OF THE LIQUID-GLASS TRANSITION IN LITHIUM METASILICATE

Junko Habasaki, Isao Okada
Department of Electronic Chemistry, Tokyo Institute of Technology at Nagatsuta 4259, Yokohama 227, Japan

and Yasuaki Hiwatari
Department of Physics, Faculty of Science, Kanazawa University, Kanazawa, 920 Japan

ABSTRACT

Molecular dynamics (MD) simulation of lithium metasilicate(Li_2SiO_3) has been performed to study the structural and dynamical characteristics near the glass transition temperature.

The geometrical structure analysis reveals that the most probable numbers of "contact pairs (N)" of oxygen atoms satisfy a relation N=3V-6 (V=3 to 7) in a glassy state, where V is the co-ordination number of oxygen around the lithium. On the other hand, in a liquid state, the most probable structure is represented by N<3V-6, which means an excess 'free space' to be available among oxygen atoms. These geometrical changes cause dynamical significance near the transition point.

METHOD

Lithium metasilicate, one of the typical alkali silicates, was chosen to study the mechanism of liquid-glass transition in a coulombic liquid. MD simulation has been carried out in a similar manner as in previous papers.[1] The Gilbert-Ida type potential functions[2] with r^{-6} term were used for the simulations. The potential parameters were chosen to reproduce the si(s) function obtained by X-ray diffraction analysis.[3] The present MD simulations were performed for 6,000 steps run at 1673 K (melt) and 100,000 steps run at 700 K (glass). A time step was 4 fs.

RESULTS AND DISCUSSIONS

The system is considered to be in local equilibrium even though a rapid cooling process was employed, since the relaxation time for local rearrangement of atoms is short enough. Cooling down of the system to near the liquid-glass transition temperature, Tg, causes some changes in the short and medium range structures, while no appreciable change appears in the long range structures. Some clear peaks appear in the angular distribution function for the angle ∠Si-Si-Si, which suggests that the rotational motion of the chain structure, made up of SiO_4 units, becomes remarkably restricted near Tg. The geometrical structure in the short range was analyzed with polyhedra made of oxygen atoms around a lithium ion. This reveals that the most probable numbers of "contact pairs (N)" are at N=3V-6 (V=3 to 7) in the glassy state, where V is the co-ordination number of oxygen around the lithium. However in the liquid state, the most probable structure is N<3V-6 having 'free space' between oxygen atoms. This finding suggests that the transition can be identified as a point (temperature) where the geometrical degree of freedom substantially diminishes, that is, saturation of packing of oxygen atoms occurs at this point. These geometrical changes were confirmed as responsible for the those in dynamics near the transition point, since the structure with N=3V-6 is much more stable than structures with N<3V-6 or N>3V-6. The present analysis of the dynamics with van-Hove correlation functions reveals that, because of saturation of the packing of oxygen atoms, motions of individual ions are mostly restricted and jump motions become dominant instead in a lower temperature region. This is a similar result to that observed in a soft core system. A theory based on the 'trapping diffusion model'[4] was tested for the present system. The results are consistent with this theory, when it is assumed that the packing of oxygen atoms determines the transition in this system.

Potential parameters obtained on the basis of ab-initio MO calculations[5] also led to similar results.

REFERENCES

1   J. Habasaki, Molec. Phys. **70**, 513 (1990).
J. Habasaki, I. Okada and Y. Hiwatari, Proc. Taniguchi Int. Symposium on Molecular Dynamics Simulation, 1990, Springer-Verlag, in press.
2   Y. Ida, Phys. Earth Planet. Interiors **13**, 97 (1976).
3   Y. Waseda and H. Suito, Transactions ISIJ. **17**, 82 (1977).
4   T. Odagaki and Y. Hiwatari, Phys. Rev. **A41**, 929 (1990).
5   J. Habasaki and I. Okada, to be published.

MOLECULAR DYNAMICS STUDY ON THE GLASS TRANSITION OF LiI ABOUT DYNAMICAL SINGULARITIES

Sumiko Itoh
The Nishi-Tokyo University, Uenohara, Kitatsuru-gun, Yamanashi 409-01, Japan

Hiroh Miyagawa
Research Center Taisho Pharmaceutical Co., Ltd., Ohmiya, Saitama 330, Japan

Yasuaki Hiwatari
Kanazawa-University, Kanazawa, Ishikawa 920, Japan

ABSTRACT

The liquid-glass transition of LiI has been investigated by means of the constant-temperature and constant-pressure molecular dynamics method. Because LiI is one of the most simple example of the coulombic system and the most contrasted combination of the cation and anion among all alkaline halides (the mass ratio of I to Li is 18.3 and the ionic size ratio is 3.6), we have expected to see the ionic motion most drastically. We calculated both static and dynamic properties with microscopic analyses for ionic motions. In the vicinity of the glass transiton, the mean-square displacement shows a non-t-linear behavior, sometimes called a subdiffusive behavior, and the incoherent scattering function are well fitted by "stretched exponential" function. The non-Gaussian parameter as a function of time has a large maximum, and the product of the maximum value and the corresponding time increases sharply as approaching the glass transition. All these anomalous characteristic behavior, as well as those found for the binary soft-sphere system, are in good agreement with the prediction of the trapping diffusion theory. We discuss the validity of this theory.

Theoretical approaches for the glass transition recently developed are the mode coupling theory[1], non-linear hydro-dynamic theory[2], and the trapping diffusion theory[3]. The former two are effective to analyze in some part of the phenomenon, but not yet completely sufficient. The mode coupling theory reveals that the density-density correlation function shows the non-exponential decay. By the trapping diffusion model, which is a stochastic model based on the jump duffusion of atoms, the glass transition is expressed as the Gaussian to non-Gaussian dynamics transition.

Independent of the development of the trapping diffusion theory, we have found the fact that the ions sometimes perform large displacement in the course of the analyses of the trajectory. Therefore we classify the ionic motion into two types according to the length of the displacement $\delta l(t)$ during a fixed (small) time period δt between t and t+δt: (A) $\delta l(t) / \delta t \geq \delta l_0 / \delta t$ and (B) $\delta l(t) / \delta t < \delta l_0 / \delta t$, where $\delta l(t) = |r_i(t+\delta t) - r_i(t)|$ (i=1,2,....N). δl_0 and δt is a constant parameters which is determined in an arbitrary way. The jump motion is defined as a series of (A)-type motions, which, say, begins at time t_1 and ends at time t_2. The jump distance(JD) is therefore $|r_i(t_2) - r_i(t_1)|$ and the flying time (FT) $t_2 - t_1$. The residence motion is contrarily defined as a series of (B)-type motions which begins as time t_1' and ends at time t_2'. The traveling distance(TD) is $|r_i(t_2') - r_i(t_1')|$ and the residence time (RT) $t_2' - t_1'$.

Thus, the motion of the ions are classified into the residence-type motions and the jump-type motions. The ideal jump motion occurs when TD=0. We analysed the ionic motions with the parameter $\delta l_0 = 1.1$ Å, which is the half length of the ionic radius of the iodide ion and $\delta t = \sqrt{200[K]/T[K]}$ ps. The distribution of the RT is well fitted to the power-law distribution in agreement with the trapping diffusion theory.

We have obtained the results to support a power-law jump rate distribution proposed in the trapping diffusion theory for the simple coulombic system. Numerous anomalous behavior accompanying the glass transition are reasonably explained[4] by the theory. The glass transition temperature of LiI is estimated around 300 K ~ 360 K.

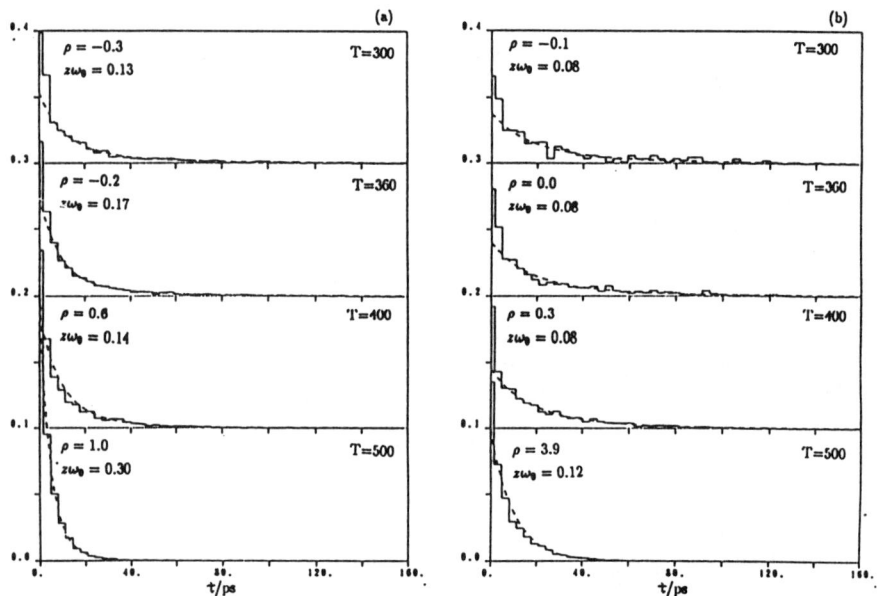

Fig. 1. Residence time distribution for the Li$^+$ ion (a) and for the I$^-$ ion (b).

1. U. Bengtzelius, W. Gotze, and A. Sjolander, J. Phys. C **17**, 5915 (1984); U. Bengtzelius, Phys. Rev. A **33**, 3433 (1986); W. Gotze and A. Sjolander, Z. Phys. B **65**, 415 (1987); W. Gotze, Z. Phys. Chem, **156**, S3 (1988); E. Leutheusser, Phys. Rev. A **29**, 2765 (1984); T. R. Kirkpatrick, *ibid*. **31**, 939 (1985); U. Krieger and J. Bosse, Phys. Rev. Lett. **59**, 1601 (1987).
2. S. P. Das, G. F. Mazenko, S. Ramaswamy, and J. J. Toner, Phys. Rev. Lett. **54**, 118 (1985); E. Siggia, Phys. Rev. A **32**, 3135 (1985). S. P. Das, G. F. Mazenko, S. Ramaswamy, and J. J. Toner, *ibid*, **32**, 3139 (1985); S. P. Das and G. F. Mazenko, Phys. Rev. A **34**, 2265 (1986); W. Gotze and L. Sjogren, Z. Phys. B **65**, 415 (1987).
3. T. Odagaki, J. Phys. A **20**, 6455 (1987); T. Odagaki, Phys. Rev. B **38**, 9044 (1988); T. Odagaki and Y. Hiwatari, Phys. Rev. A **41**, 929 (1990).
4. H. Miyagawa, Y. Hiwatari, and S. Itoh, Prog. Theor. Phys. Suppl. **103**, 47 (1991).

II. POLYMER DYNAMICS

A. EXPERIMENT

Neutron Spin Echo Investigations on the Dynamics of Dense Polymer Systems

D. Richter
Institut für Festkörperforschung
des Forschungszentrums Jülich, Germany
L.J. Fetters, J.S. Huang
Exxon Research and Engineering Co., Annandale NJ USA
B. Farago
Institut Laue-Langevin Grenoble, France
B. Ewen
MPI für Polymerforschung Mainz, Germany

Abstract

We discuss experimental results on molecular motion in polymer melts, obtained by neutron spin echo spectroscopy (NSE). We show that in the short time regime the assumption of entropic restoring forces (Rouse model) describes well the space and time dependence of the pair correlation function of one chain molecule. For longer times we observe systematic deviations from the Rouse model, revealing the presence of a well defined intermediate dynamical length scale beyond which density fluctuations within a given chain are strongly reduced. Its value is found to be in excellent agreement with the entanglement distance obtained from rheological measurements. Measurements of the temperature dependence and polymer volume fraction dependence of the entanglement distance give insight into the molecular origin of entanglement constraints.

1 Introduction

High molecular weight polymeric liquids exhibit unusual dynamic properties: Depending on the time scale of observation or temperature the same polymer may respond elastically, showing rubbery like behavior or may flow like a liquid [1]. The rubber behavior which expresses itself by the so called plateau regime in the dynamic modulus is commonly attributed to the effect of 'entanglements'. They are thought to stem from geometrical or topological constraints, mutually imposed by the interpenetrating chain molecules. Their molecular origin, however, is not well understood. The reptation theory of viscoelasticity [2-5] bases on the further assumption that the geometrical constraints can be modeled by a tube confinement surrounding a given chain. At intermediate times the polymer dynamics are restricted to a curve linear motion along the tube. The tube diameter d, thereby, may be interpreted as the distance between entanglements.

In order to observe the spatially constraint motion of chain segments in dense environment, quasielastic neutron scattering (QNS) is a unique probe. This is partly due to the fact that by hydrogen-deuterium exchange the scattering contrast can be changed at will, allowing the observation of a single chain among others. Second, since with cold neutrons the accessible momentum and energy transfer correspond to intramolecular distances and relaxation times, QNS accesses simultanously the space and time evolution of the intramelecular motion.

Here we discuss our experimental efforts to elucidate characteristics of chain motion in dense environment. We present (i) an evaluation of the short time Rouse relaxation on a melt of polyisoprene (PI); (ii) an investigation of the effect of entanglement constraints. They were studied on the alternating copolymer of poly(ethylene propylene) (PEP) and on hydrogenated polybutadiene (PEB-2) which exhibit high plateau moduli and consequently high entanglement densities; (iii) some first experiments aiming on the microscopic origin of entanglement constraints.

2 Theoretical Considerations

In the time regime of the plateau modulus the polymer melt behaves like a rubber, where the elasticity arises from the entropic forces acting on the chains between the permanent crosslinks. Consequently, at intermediate times the modulus of an entangled polymer liquid is assumed to correspond to that of a rubber network with a mesh size equivalent to the entanglement distance. For small strains the modulus of such a network is governed by entropic elasticity and is proportional to the number of elastically active chains and the temperature. Identifying the mesh of the entangled melt with the tube diameter of the reptation model, we can relate the plateau modulus G_N^0 with d [2].

$$G_N^0 = \frac{4}{5} \frac{\rho k T}{m_0} \frac{l^2}{d^2} \tag{1}$$

where $l^2 = n C_\infty l_0^2$ (the Gaussian statistical segment length), C_∞ is the characteristic ratio, l_0 the average main chain bond length, ρ the polymer density, m_0 the monomer weight and n the average bond number/monomer. Equ(1) is one of the fundamental relationships of the Doi-Edwards theory relating macroscopic viscoelastic properties to the microscopic chain confinement.

If the chains could intersect freely the chain dynamics would be described as thermal motion damped via a friction coefficient ζ. In this so called Rouse model [6], the diffusing chain segment performs a random walk on the random chain profile. This convolution of two random processes leads to a mean square segment displacement $< r^2(t) > \cong l^2(Wt)^{1/2}$ with $W = 3kT/\zeta l^2$ being the Rouse rate [7]. In Gaussian approximation the segment selfcorrelation function relates directly to the mean square segment displacement. The pair correlation function of a Rouse chain describing the interferences between the scattering from different segments is more complicated [9]. We note, however, that as S_{self} it scales with a universal 'Rouse' variable $u = Q^2\sqrt{Wl^4t}$; the $Q - t$ scaling results from the fact that besides the cut-off length scales R_g (size of the chain) and l, the model does not contain any length scale.

The presence of an intermediate dynamic length scale d changes the scaling behavior $S(Q, t)$ and causes systematic Q-dependent deviations from the Rouse scaling. In the framework of reptation de Gennes derived an explicit first order expression for $S(Q,t)$ [8]:

$$S(Q,t)/S(Q,0) = 1 - \frac{Q^2 d^2}{36} + \frac{Q^2 d^2}{36} exp(\frac{u^2}{36}) erfc(u/6) \tag{2}$$

$S(Q,t)$ describes the equilibration of density fluctuations along the tube (local reptation), neglecting any decay due to Rouse models of a spatial extend smaller than the tube diameter. The important feature of Equ(2) is the factor $Q \cdot d$ which introduces a new length scale. Due to the tube constraints, $S(Q, t)$ only partially decays to a certain Q-dependent fraction. The remaining 'elastic' part is a consequence of long living segment-segment correlation due to the tube confinement. The 'elastic' part actually decays only for times longer than a terminal time τ_d after which the chain has lost its memory of its original tube. For a quantitative analysis of scattering data originating from the crossover regime between short time Rouse motion and local reptation, it is necessary to include the initial Rouse motion which was neglected by de Gennes. The only model in the literature providing an explicit expression for the dynamic structure factor in this regime is Ronca's effective medium model [9], using an ad hoc ansatz for the time dependent friction.

The molecular origin of entanglements is not well understood but current thinking postulates them to originate mainly from the topological nature of long chain molecules as being flexible nearly one dimensional uncrossable objects. The occurance of entanglements then is governed by two length parameters, the step length of the Gaussian random walk of the chains and the lateral distance between chains determining the amount of chain contour length per volume [10-16]. Scaling models and topological calculations are brought forward.

(1) <u>General scaling model:</u> Using the chain contour length density and mutual uncrossability of chains as determining features, Graessley and Edwards [10] derived scaling relations on the basis of the experimentally known dependence of the plateau modulus on the polymer volume fraction ϕ. They find

$$d^2 \sim C_\infty^{4-a} \left(\frac{\rho\phi}{m_0}\right)^{1-a} l_0^{5-3a} \tag{3}$$

where the exponent a is given by the relation $G_N^0 \sim \phi^a (a = 2...2.3)$.

(2) <u>Packing models [11]</u> relate the occurance of an entanglement to the gradual build-up of geometrical hindrance due to the presence of the other chains. More precisely, entanglements are determined by a volume through which a certain number of other chains have to pass; or a mean number of neighbouring chain segments belonging to other non interrupted chains (chain ends do not count) is required to restrict the lateral degree of freedom. Neglecting the influence of chain ends the model yields for the entanglement distance

$$d^2 \approx \frac{1}{(\frac{\rho\phi}{m_0})^2 C_\infty^2 l_0^4} \tag{4}$$

We note that as a consequence of the packing criterion the entanglement distance increases with increasing tendency of a chain to coil - coiling diminishes the presence of other coils and the entanglement volume has to increase. Furthermore, also as result of packing, the length of an entanglement strand $N_e \sim d^2$ is inversely proportional to the bond density. Packing is a special case of general scaling for $a = 3$.

(3) <u>Binary contact models</u>. A long standing alternative scaling approach to grasp the nature of an entanglement is the assertion that an entanglement is made up by a certain fixed number of binary contacts along the chain [13,14]. This argument again is of topological nature dwelling on the non crossability of polymer chains. For the entanglement distance the model yields

$$d^2 \sim \frac{1}{(\frac{\phi\rho}{m_0})l_0} \tag{5}$$

(general scaling a=2)

Recently, Colby and Rubinstein [12] proposed an alteration of this scaling ansatz conjecturing that an entanglement was determined by a constant number of binary contact in the entanglement volume d^3 bringing thus together packing and contact models. The scaling relation for the entanglement distance then reads

$$d^2 \sim \frac{1}{C_\infty^{2/3} l_0^2 (\frac{\rho\phi}{m_0})^{4/3}} \tag{6}$$

(general scaling a=7/3)

We like to note that the apparent unability of simple scaling to give a definitive answer relates to the fact that the entanglement problem considered as a geometrical phenomenon contains

sample	$M_\omega 10^{-4}$*	M_z/M_ω**	M_ω/M_n**
h_8-PI	5.7	-	1.03
d_8-PI	5.2	-	1.03
d_{10}-PEP	8.38	1.03	1.05
d_{10}-PEP	8.22	1.02	1.05
d_{10}-PEP III	36.70	-	1.04
PEP Triblock	8.78	1.03	1.06
h_8-PEB-2	7.05	1.04	1.04
d_8-PEB-2	7.32	1.02	1.04

Table 1: Polymer molecular characteristics

two independent length scales, the step length of the random walk and the interchain distances given by the contour length density. Therefore, besides scaling arguments further assumptions as those explained above are necessary.

(4) Topological calculations [15,16] go beyond scaling in so far as founded on the mathematical concept of topological invariants geometrical constraints are actually calculated and not conjectured on the basis of scaling arguments. Using Gaussian topological invariants, very recently Iwata and Edwards [16] introduced a new quantity, the topological interaction parameter $\bar{\gamma}$ which measures the capability of a chain to entangle and may be considered as a property individual to each chain as the characteristic ratio. $\bar{\gamma}$ is mainly determined by the diameter of a polymer chain. As one would have expected polymers entangle more if the chain diameter is small or the contour length density is high. Furthermore, $\bar{\gamma}$ appears to be also related to the size of C_∞ - the larger C_∞ becomes the more a chain reaches out in the space the better it entangles. An analytic functional dependence, however, cannot be read off from their results. Finally, for the concentration dependence of the plateau modulus they calculate $G_N \sim \phi^a$ where a increases with decreasing concentration ($1.97 \leq a \leq 2.2$). For high concentrations their result agrees with $d \sim (\frac{\rho\phi}{m_0})^{-1/2}$.

3 Experiments

Given the still limited temporal resolution of neutron spin-echo ($t < 40ns$) it was essential to select thermally stable linear polymers combining high flexibility, a large plateau modulus indicating important topological constraints, and a low monomeric friction coefficient yielding high segmental mobility. For our experiments we chose polyisoprene (PI); the alternating copolymer of polyethylene propylene (PEP) which was obtained from hydrogenation of 1,4 polyisoprene and hydrogenated 1-4 polybutadiene (PEB-2) which essentially resembles polyethylene. The different polymer samples including their characteristics are listed in Table 1. The quasielastic neutron scattering experiments were performed using the neutron spin-echo (NSE) spectrometer IN11 at the Institute Laue-Langevin (ILL) in Grenoble.

3.1 Rouse Regime

Fig 1 presents the NSE spectra from PI for different momentum transfers. The solid lines represent a fit with the dynamic structure factor of the Rouse model, where we restricted the time regime of the fit to the initial part. As we may observe at short times the data are well represented by the solid lines, while at longer times deviations towards slower relaxation are

---

*light scattering
**Size exclusion chromatography

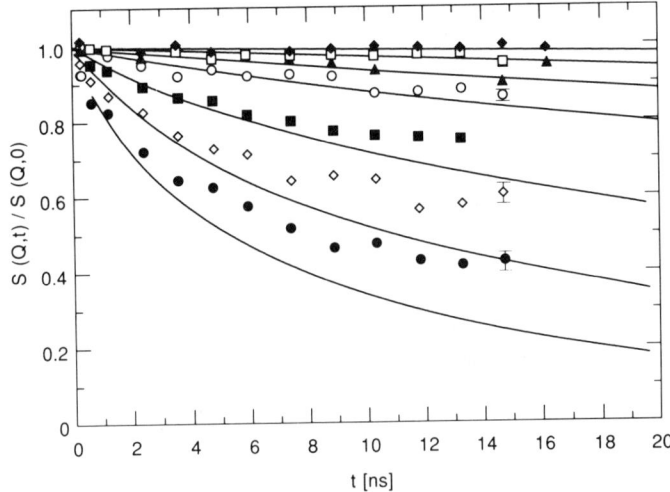

Figure 1: Dynamic structure factor as observed from PI for different momentum transfers at 468 K. ($\blacklozenge Q = 0.038\text{Å}^{-1}$; $\square Q = 0.051\text{Å}^{-1}$; $\blacktriangle Q = 0.064\text{Å}^{-1}$; $\circ Q = 0.077\text{Å}^{-1}$; $\blacksquare Q = 0.102\text{Å}^{-1}$; $\diamond Q = 0.128\text{Å}^{-1}$; $\bullet Q = 0.153\text{Å}^{-1}$). The solid lines display fits with the Rouse model to the initial decay.

Sample	T[K]	NSE d[Å]	NSE $10^9 \zeta [\frac{dyns}{cm}]$	τ_e[ns]	Rheology d[Å]	Rheology $10^9 \zeta [\frac{dyns}{cm}]$
PI 1	468		4.4±0.3	-		
PI 2	473	52±1	3.8±0.3	32	51(298K)	
PEP homopolymer	492	47.5±0.4	3.1±0.1	15	43.5±2	1.9
PEP triblock	491	47.1±0.7	2.4±0.2	15		
PEB-2	509	43.5±0.7	0.4±0.04	5	35(373K)	0.3(448K)†

Table 2: NSE results on polymer melts

obvious. In order to demonstrate further the quality of the Rouse description for the initial decay, Fig 2 presents the Q-dependence of the characteristic frequency in a double logarithmic plot. The solid line displays the $\Omega_R \sim Q^4$ law. A similar result has recently been reported for PDMS [17,18], where geometrical constraints are supposed to be even less pronounced than for PI. The insert in Fig 2 demonstrate the scaling behavior of the experimental spectra which according to the Rouse model are required to collapse to one master-curve if they are plotted in terms of the Rouse variable $u = Q^2 l^2 \sqrt{Wt}$. The solid line displays the result of a joint fit to the Rouse structure factor the only fit parameter being the Rouse rate Wl^4. An excellent agreement with the theoretical prediction is observed. The resulting value is $Wl^4 = 2.0 \pm 0.1 \cdot 10^{13} \text{Å}^4 s^{-1}$.

†from diffusion [24]

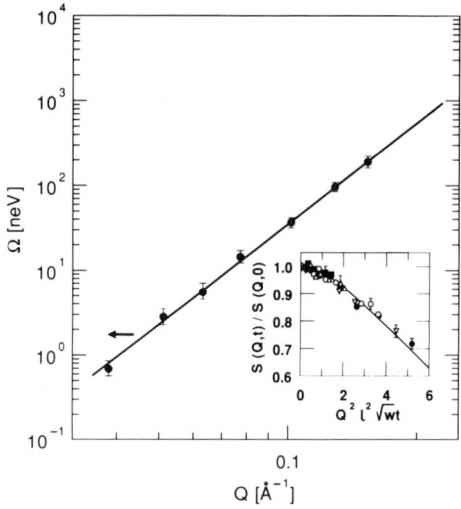

Figure 2: Characteristic frequency for the Rouse decay for the PI sample as a function of Q. The insert displays the scaling behavior of the dynamic structure factor if plotted vs the Rouse scaling variable $Q^2 l^2 \sqrt{Wt}$.

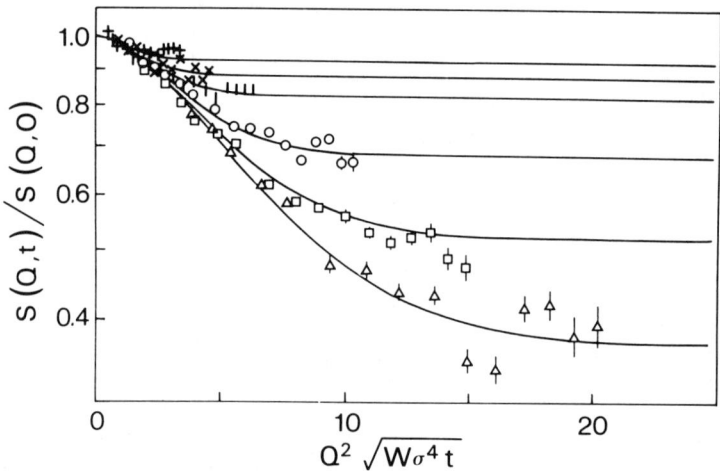

Figure 3: Rouse scaling representation of the PEP data at 492K (+ : $Q = 0.058\text{Å}^{-1}$; × : $Q = 0.068\text{Å}^{-1}$; | : $Q = 0.078\text{Å}^{-1}$; ∘ : $Q = 0.097\text{Å}^{-1}$; □ : $Q = 0.116\text{Å}^{-1}$; △ : $Q = 0.135\text{Å}^{-1}$). The solid lines are the result of a joint fit to the Ronca model.

3.2 Motion under Entanglement Constraints

The effect of entanglement constraints on the molecular dynamics in polymer melts has been investigated for all three polymers [19]. The results for all polymers are displayed in Table 2. Here we shall only remark on the data on PEP [20]. Fig 3 presents the measured dynamic structure factor for the PEP sample at $492K$ in a scaling form. The data are characterized by a common initial decay signifying the Rouse regime and a consecutive Q-dependent crossover into a plateau resulting from the presence of an intermediate dynamic length scale beyond which density fluctuations are strongly limited. The solid lines represent the result of a fit with the Ronca model. It allows a very satisfying description of the experimental data reproducing the line shape, the resulting sharp crossover and the Q-dependency of the plateau levels. In order to compare with rheological results and to avoid temperature extrapolation of existing low temperature data, we studied the plateau modulus of a high molecular weigth PEP sample ($M_\omega = 170000$) at $500^\circ C$ using a Rheometrics System 4. the value obtained is $9.3 \cdot 10^6 dyn/cm^2$. Employing Equ(1) a tube diameter or entanglement distance of 43Å is evaluated. This value compares very well with the 47Å measured by NSE on the microscopic scale.

3.3 Density and Temperature Dependence of the Entanglement Distance

The origin of entanglement constraints so far has been approached on the basis of geometrical or topological considerations relating to the nature of polymers being long flexible and uncrossable objects. As discussed above two length parameters are of importance: (i) the interchain distance which is related to the contour length density and (ii) the step length of the random walk of the chain given by the Kuhn length $l_K = C_\infty l_0$. In order to experimentally access the different theories, we varied systematically both lengths. On PEB-2 we studied the dependence on contour length density in varying the polymer volume fraction at constant temperature ($T = 509K$) diluting with the oligomer $C_{19}D_{40}$ over a wide concentration range $0.25 \leq \phi \leq 1$. The dependence on l_K was investigated changing temperature and thereby C_∞ at a given ϕ.

A detailed analysis of the spectra obtained for PEB-2 in the dilution experiment made evident that the experiments could be significantly better described in terms of the dynamic structure factor of the local reptation model (Equ(2)) - the Ronca model leads to systematic deviations both with respect to the long time behavior as well as to the Q-dependence. These deviations become evident for PEB-2 rather than for PEP because: (i) the relative dynamic range in terms of the crossover time τ_e from short time unrestricted Rouse dynamics to entanglement controlled behavior was large ($t_{max} = 3.5\tau_e$); (ii) the statistical accuracy of the result was considerably improved compared to PEP (for details see Ref 18). Because of the better accuracy of the de Gennes local reptation model all entanglement distances were evaluated in terms of this model (Table 3) - the values for d from both models differ only slightly, the maximum discrepancy of $\sim 10\%$ occuring at the lowest polymer concentration. Fig 4 displays the resulting dependence of the entanglement distance on ϕ. The solid line corresponds to $d \sim \phi^{-0.61}$ describing up to $\phi = 0.35$ the experimental data very well. At $\phi = 0.3$ the d-value is definitely larger than suggested by the $d \sim \phi^{-0.61}$ power law; at $\phi = 0.25$ an entanglement distance could be determined any more with certainty.

We also addressed the temperature dependence of the microscopic dynamics of PEB-2. The results obtained for $\phi = 1.0$ and $\phi = 0.5$ are displayed in Fig 5. They are compatible with an identical temperature coefficient for both concentrations. The solid line represents the best fit for the $\phi = 1$ data.

$$d = 23.5\ exp((1.2 \pm 0.2)10^{-3}T) \qquad (7)$$

We now compare the predictions of the various scaling models with the results obtained from PEB-2. For the concentration dependence of the entanglement distance we found $d \sim \phi^{-0.61}$. In terms of the general scaling model of Graessley and Edwards (Equ (3)) this concentration

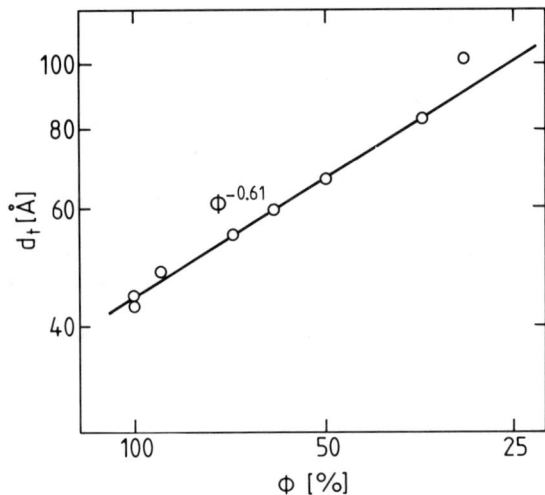

Figure 4: Double logarithmic presentation of the entanglement distance or tube diameter d for PEB-2 at 509 K as a function of polymer volume fraction ϕ.

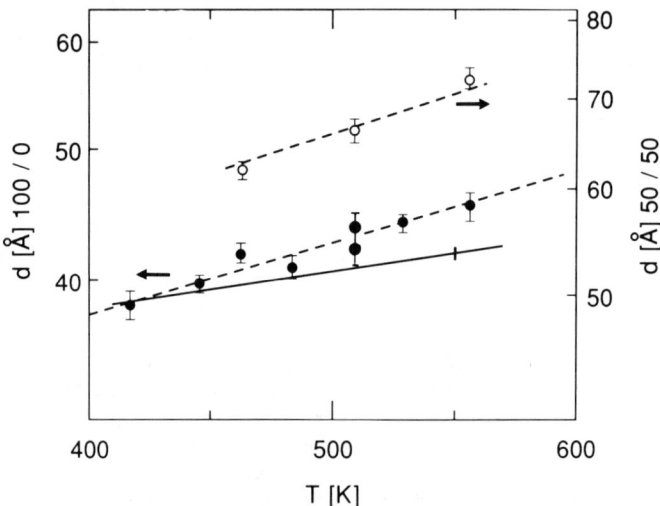

Figure 5: Temperature dependence of the entanglement distance d of PEB-2: $\circ, \phi = 1$; $\bullet, \phi = 0.5$. The dashed lines are guides to the eye. The solid line represents the scaling prediction (see text).

Polymer volume fraction φ	Wl^4 $Å^4 s^{-1} \cdot 10^{-13}$	ζ $[10^{-10} \frac{dyn\ cm}{s}]$	d [Å]
1.00	6.1±0.8	3.9±0.4	42.6±1.0
1.00	7.0±0.7	4.5±0.6	44.3±0.5
0.90	9.2±0.8	3.0±0.3	48.0±0.5
0.70	14.3±1.5	1.9±0.2	55.1±0.4
0.60	11.7±0.8	2.3±0.2	59.9±0.7
0.50	18.2±1.1	1.5±0.1	67.0±0.6
0.35	29.8±1.1	0.92±0.03	83.6±0.3
0.30	33.7±1.2	0.82±0.03	102.6±1.8
0.25	37.2±1.5	0.74±0.03	-.-

Table 3: Monomeric friction coefficients and entanglement distances in concentrated PEB-2 solutions and melts at $T = 509K$

dependence determines the scaling exponent $a = 2.22 \pm 0.04$. The exponent lies in between the binary contact model asking for a fixed amount of contacts along an entanglement strand ($a = 2$) and the Colby-Rubinst modification asking for a fixed number of binary contacts within an entanglement volume ($a = 7/3$). The packing model ($a = 3$ or $d \approx \phi^{-1}$) is clearly excluded. Our result agrees also well with recent viscoelastic results on polybutadiene solutions - PB is the mother-polymer of PEB-2 - resulting in a slightly larger value of $a = 2.3$ [22].

Having determined the scaling exponent a we now may examine the scaling prediction with respect to l_K or C_∞ respectively. The temperature dependence of C_∞ for PE has recently been studied by Boothroyd et al by SANS on PE melts [21]. In a similar T-range as our experiments they find $\frac{d \ln C_\infty}{dT} = -1.1 \pm 2 \cdot 10^{-3} K^{-1}$. Finally, using $\frac{d \ln \rho}{dT} = -0.7 \cdot 10^{-3} K^{-1}$ [23] Equ(3) predicts

$$d^2(T) \approx C_\infty^{-0.44} \rho^{-1.22} \approx exp[+(1.3 \pm 0.1)10^{-3} T] \tag{8}$$

This has to be contrasted with the experimental result of $\frac{d \ln d^2}{dT} = 2.4 \pm 0.4 \cdot 10^{-3} K^{-1}$ (Equ(7)) which is nearly three standard deviations larger. In order to visualize the discrepancy we have included the theoretical prediction of Equ(8) also in Fig 5 as a solid line. Experimental results and scaling predictions are clearly apart.

Concerning the topological calculations of Iwata and Edwards [16] a direct quantitative comparison would require extensive numerical calculations using molecular parameters of the two polymers. These calculations have not been made, leaving us with a qualitative comparison. The model predicts $G_N^0 \approx \phi^a$ with $1.97 \leq a \leq 2.2$ enveloping our value of $a = 2.22 \pm 0.04$. However, at high contour length density as realized in PEB, the exponent would be expected more close to $a = 2$. The topological interaction parameter $\bar{\gamma}$, measuring the capability of a chain to entangle, is predicted to be a monotonously increasing function of C_∞ or the entanglement distance d should monotonously decrease with increasing C_∞. On a qualitative level we observe such a behavior. In order to quantify for PEB: if we assume that the density dependence would be correctly described by the scaling model $d^2 \approx \rho^{1-a}$ then $d^2 \approx C_\infty^{-1.4}$ is found.

4 Summary

We presented neutron spin-echo investigations on the intramolecular motion in polymer melts. The regime of Rouse dynamics was analysed in the short time behavior of the dynamic structure factor from PI. The influence of entanglements was studied on PEP and PEB-2. The pertinent results may be summarized as follows: (i) The measured dynamic structure factors show that beyond a characteristic time the relaxation of the pair correlation is strongly impeded. From the dependence on the momentum transfer an associated dynamic length can be extracted. We emphasize that the existence of this dynamic length reveals itself model independently from systematic Q-dependent deviations from Rouse scaling. This microscopic length scale is found to be in excellent agreement with the entanglement distance obtained from rheological measurements assuming the reptation model. (ii) In order to access the question on the molecular origin of entanglements, in varying the temperature and polymer volume fraction, we studied the dependence of the entanglement distance on the two length parameters considered to be important for the formation of entanglements. Changing the polymer volume fraction (PEB) we altered the interchain distance and thus the contour length density. The obtained scaling exponent $a = 2.22\pm0.04$ excludes the packing model and lies in the region of the binary contact models. However, the scaling model appears not to be able to describe the dependencies on volume fraction and flexibility simultanously. Qualitatively, our results agree with the topological calculations by Iwata and Edwards [16], however, a quantitative comparison requiring important calculations has still to be done.

References

[1] J D Ferry, Viscoelastic Properties of Polymers; Wiley, New York, 1980

[2] M Doi, S F Edwards, the Theory of Polymer Dynamics, Clarendon Oxford, 1986

[3] P G de Gennes, Scaling Concepts in Polymer Physics, Cornell Univ. Press, Ithaca NY, 1979

[4] P G de Gennes, J Chem Phys 55, 572 (1971)

[5] M Doi, S F Edwards, J Chem Soc Faraday Trans 2, 74, 1789 (1978); 74, 1802 (1978); 75, 38 (1978)

[6] P E Rouse, J Chem Phys 21, 1273 (1953)

[7] P G de Gennes, Physics (Long Island City, NY) 3, 37 (1967)

[8] P G de Gennes, J Phys (Paris) 42, 735 (1981)

[9] G Ronca, J Chem Phys 79, 1031 (1983)

[10] W W Graessley, S F Edwards, Polymer 22, 1389 (1981)

[11] T Kavassalis, J Noolandi, Macromolecules 21, 2869 (1988)

[12] R H Colby, M Rubinstein, Macromolecules 23, 2753 (1990)

[13] S F Edwards, Proc Phys Soc 92, 9 (1967)

[14] P G de Gennes, Phys Lett (les Ulis Fr)35, L-133 (1974)

[15] S F Edwards, Royal Soc London A 419, 221 (1988)

[16] K Iwata, S F Edwards, J Chem Phys 90, 4567 (1989)

[17] D Richter, A Baumgärtner, K Binder, B Ewen, J B Hayter, Phys Rev Lett 47, 109 (1981); 48, 1695 (1982)

[18] D Richter, B Ewen, B Farago, T Wagner, Phy Rev Lett 62, 2140 (1989)

[19] D Richter, L J Fetters, J S Huang, B Farago, B Ewen, preprint

[20] D Richter, B Farago, L J Fetters, L S Huang, B Ewen, C Lartigue, Phys Rev Lett 64, 1389 (1990)

[21] A T Boothroyd, A R Rennie, C B Boothroyd, Europhys Lett 15, 715 (1991)

[22] R W Colby, L J Fetters, W G Funk, W W Graessley, Macromolecules 24, 3873 (1991)

[23] B Crist, J D Tanzer, W W Graessley, J Polym Science Poly Phys Ed 25, 545 (1987)

[24] D S Pearson, G Verstrate, F von Merwall, F C Schilling, Macromolecules 20, 1133 (1987)

ANALYSIS OF DNA ELECTROPHORETIC MOBILITY

B. Chu and Z.-L. Wang
Chemistry Department
State University of New York, Stony Brook, NY 11794-3400

ABSTRACT

Gel electrophoresis (GE) is an important analytical technique in biology and medicine. The technique can be improved based on a better understanding of DNA structure and dynamics in the polymer network.

Movement of fluorescence pattern after photobleaching (MOFPAP), together with its modified approaches, can measure stained DNA electrophoretic mobility in less time, permitting determinations of mobility (and probably velocity distribution) changes as a function of time during DNA chain deformations.

I. INTRODUCTION

Pulsed-field gel electrophoresis (PFG) [1] plays an important role in performing long-range restriction mapping of large DNA fragments. However, PFG could use further improvements since the measurement time is still very long in the analysis procedure. The specific aim of this work is to demonstrate that faster (by a factor of ~100) analytical techniques for identification of multiple species of stained DNA fragments in gel electrophoresis can be achieved. Furthermore, the new approach should permit measurements of time-dependent electrophoretic mobility of monodisperse DNA chains over very short time intervals. This information could play an essential role in our understanding of DNA dynamics in PFG.

II. GENERAL BACKGROUND

Fundamental studies of time-dependent DNA chain mobility, $\mu(t)$, in gels in an applied electric field by means of movements of fluorescence pattern after photobleaching (MOFPAP) [2], together with studies of DNA chain deformation by other established techniques [3-5] have been reported.

Our approach of being able to produce sharp boundaries between *photobleached* ethidium bromide (EB) stained DNA and EB stained DNA permits us to develop a new much faster analytical technique for DNA identification. The technology has potential for further developments into new analytical detectors for gel electrophoresis and can be used as a tool to study the fundamental aspects of pulsed-field gel electrophoresis of DNA.

III. DESCRIPTION OF METHODS

Let us assume that we have a DNA band (a) of width L, as shown in Fig. 1. The band has 3 DNA species (b_1, b_2 and b_3) each of which has its own mobility with $\mu_3 > \mu_2 > \mu_1$.

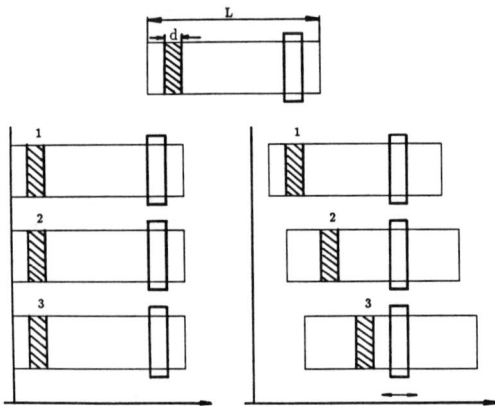

Fig. 1. Schematic illustration of simple scan approach. a: represents a DNA band of width L consisting of a superposition of 3 DNA species (b_1, b_2, b_3) each with its own electrophoretic mobility; W: window for the detector; d: denotes the region of stained DNA; ↔: suggests relative movements of window with respect to the DNA bands.

At t = 0, the 3 DNA species, as shown on the left side, are superimposed to form a single DNA band (a), as shown on the top of Fig. 1. In order to achieve an arbitrary resolution of 100, we require that the fastest species move a distance of 100 L when compared with the slowest species, which would have moved only 1 L. Thus, if L = 1 mm, and $\mu_{fastest}$ = 1 cm/hour, the run would take 10 hours when the slowest band would have moved 1 mm. Now, if we take the stained DNA band of width L and photobleach the band except for a strip of width d, the stained DNA molecules will be confined within a strip of width d and the entire band except for the strip has the dye being photobleached.

This step can be achieved by using a mask of width d which blocks the writing laser beam from photobleaching the entire DNA band of width L. At t = 0, we see a starting point as shown on the left side. It should be noted that the location of d with respect to L remains unchanged for identical DNA species during electrophoresis. However, at t = t_1, we see that species b_1 moves a distance of $v_1 t_1$ where v_1 is the velocity of species b_1; b_2 and

b_3 move a corresponding distance of $v_2 t_1$ and $v_3 t_1$, respectively, as shown schematically on the right side of Fig. 1. The three bands of width L have not been separated; but the three fluorescent bands of width d (1, 2 and 3) have been separated. Thus, we see that for *analytical* purposes, we can identify the DNA species faster by having a narrower band. The improvement in time is by a factor of L/d. If d = 0.1 mm, the measurement time for the analysis for the hypothetical example would be 1 hour instead of 10 hours. All other factors remain the same, whether the separation procedure is continuous or pulsed-field gel electrophoresis. We can see that d = 10 μm is a distinct possibility. Thus, depending on the signal-to-noise ratio (e.g. the dye content), an improvement in measurement time by a factor of 100 is within reach.

IV. EXPERIMENTAL DEMONSTRATION

The basic concept of a single-line movement of fluorescence pattern after photobleaching (MOFPAP) is illustrated in Fig. 2, where the monitor window of width d has been moved outside the DNA band of width L as shown schematically in Fig. 1. If we let the

Single Line MOFPAP

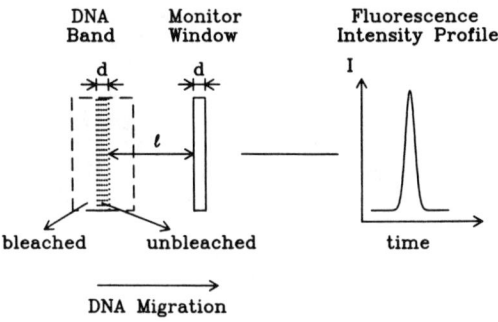

Fig. 2. Schematic representation of single-line MOFPAP.

distance between the unbleached DNA band and the monitor window be ℓ, the fluorescence intensity profile will appear as shown when the unbleached DNA band passes by the illuminated monitor window. The appearance of the fluorescence intensity profile depends on ℓ, i.e., the smaller the ℓ, the faster the peak will appear. The fluorescence intensity profile width depends on the widths of the DNA band and of the monitor window. Figure 3 shows the fluorescence intensity profile from single-line MOFPAP on a DNA sample containing two fragments of λ DNA at a total concentration of 10 μg/mL in 0.4% (w/v) agarose gel using 0.5 × TBE buffer (45

206 Analysis of DNA Electrophoretic Mobility

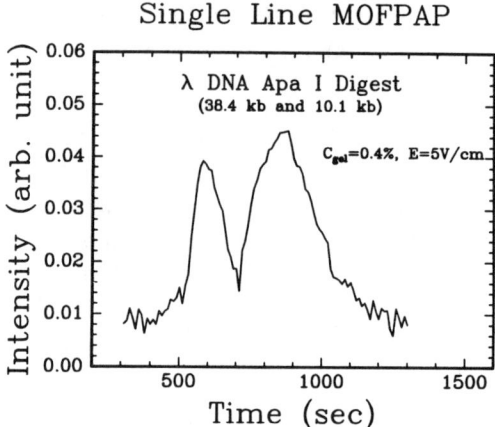

Fig. 3. Fluorescence intensity profile of λ DNA Apa I digest (38.4 kb and 10.1 kb) using single-line MOFPAP.

mM Tris base, 45 mM boric acid, 1 mM EDTA at pH = 8.0). The widths of the unbleached ethidium bromide (EB) stained DNA line and of the monitor window were 0.27 mm and 0.20 mm, respectively. The distance ℓ between the unbleached DNA line and the monitor window was ~1.5 mm. The appearance of two separate fragments could be observed in ~20 min, substantially faster than could be achieved by standard gel electrophoresis under identical conditions.

The MOFPAP approach can further be refined by means of frequency analysis yielding better signal-to-noise ratio and higher resolution. If we can detect DNA species from electrophoretic mobility differences faster by a factor of 10-100, its significance to biology and medicine is self-evident.

REFERENCES

1. Cantor, C. R., Smith, C. L. and Mathew, M. K. (1988) **Ann. Rev. Biophys. Biophys. Chem.** 17, 287, and references therein.
2. Chu, B., Wang, Z. and Wu, C. (1989) **Biopolymers**, 28, 1491; Wu, C., Wang, Z. and Chu, B. (1990) **Biopolymers**, 29, 491; Wang, Z.-L. and Chu, B. (1989) **Phys. Rev. Lett.**, 63, 2528.
3. Chu, B., Xu, R. and Wang, Z. (1988) **Biopolymers** 27, 2005; Chu, B., Wang, Z., Xu, R. and Lalande, M. (1990) **Biopolymers**, 29, 737.
4. Holzwarth, G., McKee, C.B., Steiger, S. and Crater, G. (1987) **Nucleic Acids Res.** 15, 10031.
5. Sturm, J. and Weill, G. (1989) **Phys. Rev. Lett.** 13, 1484.

THE DYNAMIC PROPERTIES OF SILICA AEROGEL: A KEY TO UNDERSTAND THE THERMAL PROPERTIES OF DENSE GLASSES?

Dorthe Posselt*, Jørgen K. Kjems

Physics dept., Risø National Laboratory, DK-4000 Roskilde, Denmark

Angelo Bernasconi, Tycho Sleator§, Hans R. Ott

Lab. für Festkörperphysik, ETH-Hönggerberg, CH-8093, Switzerland

ABSTRACT

Specific heat and thermal conductivity data for a homologous series of base-catalyzed silica aerogels in the temperature range 0.1 - 20 K are analyzed in a three regime model. The three regimes, dominated by phonons, fractons and particle modes, respectively, reflect three structural regions as found by SANS measurements. The thermal properties of aerogels are compared to those of dense amorphous quartz and important differences are found.

INTRODUCTION

Measurements of the thermal conductivity and specific heat of a wide variety of amorphous materials reveal a behaviour which to a surprising degree is independent of the chemical nature of the materials[1,2]. Silica aerogel is a material with many interesting properties, both from a scientific point of view and with respect to potential technological applications, and a rich literature exists describing both structural and dynamic properties[3]. Silica aerogel is a low density amorphous quartz material (density ρ typically 0.08 - 0.3 g/cm^3) and one would thus a priori expect the thermal properties to be closely related to those of dense amorphous silica (a-silica - $\rho \sim 2.2$ g/cm^3). However, silica aerogel, resulting from a random growth process, has a structure with a higher degree of order than found in conventional glasses. We have measured the specific heat and thermal conductivity of a series of base-catalyzed aerogels in the temperature range 0.1 - 20 K. Our data are analyzed in a model closely relating the observed behaviour to the specific aerogel structure. By comparing our data to analogous data for a-silica, we find evidence that the basic physics underlying the behaviour of the thermal properties in the investigated temperature region is of different nature for the two types of material.

EXPERIMENTAL

We have measured the specific heat and thermal conductivity of three homologous base-catalyzed silica aerogels (Airglass AB, Sweden), i.e. the samples are made using the same recipe, but with varying density: 0.145 g/cm^3 (LD), 0.190 g/cm^3 (MD) and 0.275 g/cm^3 (HD). Both oxidized and untreated samples are investigated. The method used to obtain the data is described in detail elsewhere[4]. The structure of the samples is analyzed by means of small angle neutron scattering (SANS)[5].

RESULTS

Figure 1 and 2 show the results of our measurements; only data for the highest and lowest density samples are shown, but data for the middle density sample are included in the data analysis. In figure 1, the specific heat data are plotted as C_p/T^3 vs. T on logarithmic scales. This representation emphasizes deviations from a Debye behaviour, $C_p \propto T^3$. Figure 2 shows the data for the thermal conductivity, λ.

Our SANS data shown in the insert in figure 1 are interpreted in terms of the model illustrated in figure 3. The upper part of the figure concerns the aerogel structure at different length scales, L. Three different regions are identified, separated by crossover lengths at R and

---

*Present address: Institut für Physik, Johannes-Gutenberg Universität, D-6500 Mainz, Germany
§Present address: Fakultät für Physik, Universität Konstanz, D-7750 Konstanz, Germany

208 The Dynamic Properties of Silica Aerogel

ξ. Elemental particles of diametral size R $\sim$ 40Å are aggregated into clusters as schematically shown on the top of figure 3. The size of the clusters, ξ, decreases systematically with increasing sample density with values between 200 and 350 Å. The monolithic aerogel is formed by connection of these clusters and individual clusters can be recognized in the aerogel structure. At small length scales we thus find individual particles of amorphous quartz and at large scales, the aerogel can be considered as a porous glass. The gel structure is independent of aerogel density for length scales below $\xi(\rho)$. The range between R and ξ, depending on the internal structure of the clusters, has been described using the concept of self-similarity (mass fractal scaling: M $\propto$ L^{d_f}, where d_f is the fractal dimension)[6,7]. We also apply this concept in our model ($d_f \sim 2$), however, the fractal range for all the sample densities investigated is less than a decade.

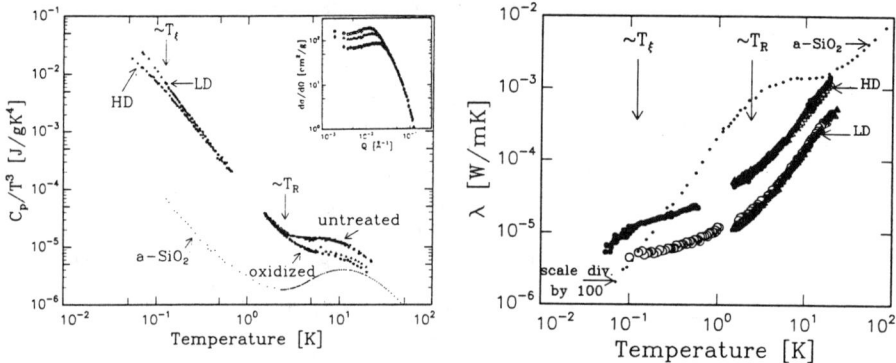

Fig. 1. Specific heat data[9]. Above 1 K, C_p for untreated samples is density independent, while for the oxidized samples, the LD data are slightly larger than the HD data. Dots: a-silica[2,11]. The insert shows SANS data, circles: LD; triangles: MD; diamonds: HD.

Fig. 2. Thermal conductivity data[9]. Filled circles: HD; open triangles: HD-ox; open circles: LD-ox; filled triangles: LD; dots: a-silica (scale divided by hundred)[2].

The dynamics of a fractal network was first described by Alexander and Orbach[8], who introduced the spectral dimension $\tilde{d}$ as the scaling of the density of states with frequency, $g(\omega) \propto \omega^{\tilde{d}-1}$. A 'dispersion relation', $\omega \propto L^{-d_f/\tilde{d}}$, binds together the length scale with the frequency scale as shown in figure 3. The characteristic crossovers in length are thus expected to be reflected in crossovers in frequency. The excitations above ω_ξ are termed fractons and are expected to be localized. The existence of an additional crossover, ω_R, is not considered in the theoretical models, but must be included for real 'fractal' systems.

We are able to analyze our C_p and λ results consistently within the above sketched model, the frequency scale being interpreted as a temperature scale through T $\sim \hbar\omega/k_B$. A detailed description of our analysis is found in ref.9. The approximate position of the crossover temperatures T_ξ and T_R are marked on figure 1 and 2. T_R is found to be independent of gel density, while T_ξ is increasing with density. Oxidation is found to influence the specific heat only in the temperature range above $T_R \sim 3$ K, consistent with this region being dominated by the dynamics of the building block particles. Oxidation removes organic groups attached to the surface of the particles and accordingly rotational degrees of freedom are removed from the excitation spectrum. In the fractal range, C_p is approximately linear and from fitting a detailed model in the range below T_R, we estimate $\tilde{d} \sim 1.1$. At the lowest temperatures measured, a deviation from the linear behaviour is seen for the sample of highest density. We interpret this to correspond to the crossing over from a regime where the excitations are localized fractons to a regime, where the fundamental excitations are phonons. A consistent behaviour is found for the thermal conductivity. λ is increasing with temperature, showing a less pronounced temperature dependence in the fractal regime. This is consistent with fractons being localized modes, thus not able to carry heat. The 'plateau' is however not a truly flat plateau, because hopping of fractons mediated via anharmonic coupling to phonons constitute an additional channel for heat conduction. This mechanism is expected to lead to a λ term linear in temperature[10] and this behaviour is confirmed by our data. For the HD sample, a marked change of the linear temperature dependence is observed at the same temperature (~ 0.13 K), where the deviation from the fractal power law was observed in the C_p data, confirming our interpretation, where the phonon-fracton crossover takes place at a frequency corresponding to this temperature. Around 3 K, the thermal conductivity changes the functional dependence on T. We observe a powerlaw, $\lambda \propto T^{1.6}$, for all samples above ~ 3 K and we interpret this as the onset of particle modes - again in consistency with the behaviour of the specific heat.

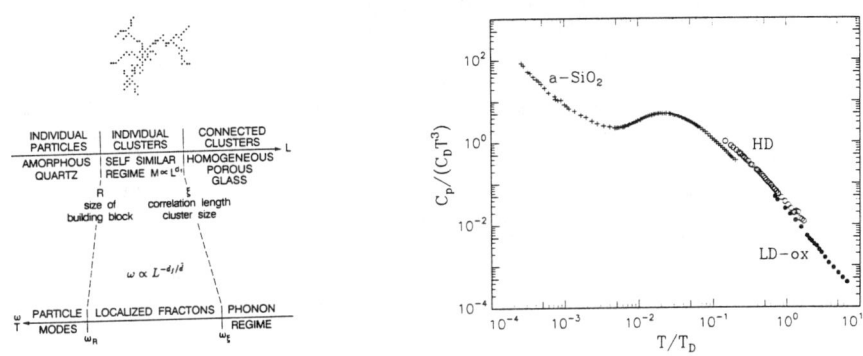

Fig.3. Schematic illustration of the model underlying our data analysis.
Fig.4. Specific heat data scaled according to the Debye model.

DISCUSSION

Also shown in figure 1 and 2 are values for C_p and λ for amorphous quartz [2,11]. Below $\sim$ 1 K, the thermal properties of amorphous materials are phenomenologically well explained in terms of two level states (TLS)[1], however the microscopic origin of these states is not clear. The existence of TLS explains the observed linear specific heat in amorphous materials and the quadratic temperature dependence of the thermal conductivity in the same temperature region. We do observe an approximately linear specific heat for aerogels around 1 K, but in the same temperature region λ is linear, i.e. the observed behaviour cannot be the 'universal' glass features. Figure 4 shows C_p data scaled with the Debye specific heat, $C_D = (2\pi/5) k_B (k_B/\hbar c)^3 / \rho$, where c the average sound velocity. The temperature axis is scaled with the Debye temperature, T_D, which is calculated to be 494 K for a-silica. It is less obvious what to choose for the aerogel Debye temperature. A scaling of the a-silica value with the aerogel sound velocity is not adequate, because the vibrating 'Debye units' are not single atoms, but individual clusters. We interpret the crossover temperature T_ξ to correspond to T_D in the aerogel case. The values for T_ξ are found by fitting a model to C_p [9]: 0.10 K (LD) and 0.37 K (HD). The scaled data in figure 4 show that it is not possible from our data to state whether excess modes are present in aerogels or not at temperatures which in terms of a Debye model correspond to the TLS regime for a-silica. The good overlap between the scaled data for HD and LD confirms the structural similarity of these samples in the fractal region, while the good overlap between the HD and a-silica data most likely is a coincidence. We also note that is has been shown that the thermal conductivity for a large variety of amorphous materials can be scaled together using material parameters[12] - it is however *not* possible to scale the aerogel λ data onto this universal curve. Altogether we must conclude that aerogels cannot be regarded as low density glasses, but exhibit phenomena in the temperature range around 1 K, which have the roots in the specific aerogel structure.

Above 1 K, the thermal properties of amorphous materials are even less well understood. The common behaviour is the occurrence of a plateau in λ and a 'bump' in C_p/T^3 (see figure 1 and 2). It has been speculated that these two features are correlated and one suggestion is that glasses exhibit a fractal connectivity at lenghtscales below $\sim 30\text{Å}$[10,13], giving rise to these distinctive features. At temperatures above ~ 1 K, we have found that the aerogel dynamics is governed by the behaviour of the individual building blocks of amorphous quartz. Disregarding the possibility of some microporosity in the aerogel building blocks, the difference to be expected between bulk and particle amorphous quartz, is the existence of a cutoff frequency in the spectrum of the latter[14]. In addition size and surface effects are expected to influence the spectrum in a region close to the cutoff[14]. This is confirmed by our C_p data, where we observe a bump in C_p/T^3 similar to the one occuring for dense amorphous quartz, but with a height depending on the detailed surface properties. However for aerogels, the bump is *not* accompanied by a plateau in λ. A λ plateau rooted in a fractal bond distribution below 30 Å would also show up in the aerogel data. There is no trace of such a plateau and we therefore find that the application of the fractal concept to dense glasses is questionable. It is found that the mean free path for phonon-like excitations in glasses (i.e. those which carry the heat) are decreasing dramatically with increasing temperature in the region of the λ plateau[15]. In aerogels, the mean free path for particle modes is limited to the size of the particles and we suggest that this is the limiting factor preventing the occurence of a true 'glass' plateau in aerogels. Below the cutoff frequency for particle modes, the excitations are of a completely different nature than found in dense glasses, i.e. the fracton modes *substitute* phonons as the fundamental excitations, whereas the two level states which are observed in dense glasses in the same temperature region, are *additional* modes, whose dynamics adds to the one governed by phonon behaviour. The linear region in the aerogel thermal conductivity is, as explained above, governed by fracton dynamics and this 'plateau' is of a completely different nature than the one observed for dense amorphous materials.

REFERENCES

1. For a review, see W.A. Phillips (ed.), Amorphous Solids, Low Temperature Properties, Springer-Verlag Berlin (1981)
2. R.C. Zeller and R.O. Pohl, Phys. Rev. B 4, 2029 (1971)
3. For a review, see E.Courtens, R. Vacher and E. Stoll, Physica D 38, 41 (1989)
4. A. Bernasconi, T. Sleator, D. Posselt and H. R. Ott, Rev.Sci.Instrum. 61, 2420 (1990)
5. D. Posselt, Thesis, Risø-M-2915 Report, Risø Nat. Lab., DK-4000 Roskilde, Denmark; D. Posselt, J.S. Pedersen and K. Mortensen, to be published in J.Non-Cryst.Solids.
6. B.B. Mandelbrot, The Fractal Geometry of Nature (Freeman, San Francisco, 1982)
7. D.W. Schaefer and K.D. Keefer, Phys.Rev.Lett. 56, 2199 (1986); R. Vacher, T. Woignier, J. Pelous and E. Courtens, Phys.Rev. B 37, 6500 (1988)
8. S. Alexander and R. Orbach, J.Phys.(Paris) 43, L-625 (1982)
9. T. Sleator, A.Bernasconi, D.Posselt, J.K. Kjems and H.R. Ott, Phys. Rev. Lett. 66, 1070 (1991); D. Posselt *et al*, Europhys. Lett. 16, 59 (1991); A. Bernasconi *et al*, submitted to Phys. Rev. B
10. A. Jagannathan, R. Orbach and O. Entin-Wohlman, Phys.Rev.B 39, 13465 (1989)
11. P. Flubacher, A.J. Leadbetter, J.A. Morrison and B.P. Stoicheff, J.Phys.Chem.Solids 12, 53 (1959)
12. J.J. Freeman and A.C. Anderson, Phys.Rev.B 34, 5684 (1986)
13. S. Alexander, C. Laermans, R.Orbach and H.M. Rosenberg, Phys.Rev.B 28, 4615 (1983)
14. D. Richter and L. Passell, Phys.Rev.B 26, 4078 (1982) and ref. therein
15. M.P. Zaitlin and A.C. Anderson, Phys.Rev.B 12, 4475 (1975)

SLOW DYNAMICS IN POLYELECTROLYTE SOLUTIONS

M. Drifford - J.P. Dalbiez - L. Belloni - O. Spalla
SCM CE Saclay 91191 Gif sur Yvette Cedex
France

ABSTRACT

Quasi elastic light scattering is used to investigate the polyelectrolyte and added salt concentration dependence of the diffusion coefficient of sodium polystyrene sulfonate (NaPSS). Two relaxation times are observed in semi dilute regime at low salt concentration. With polyvalent salts ($LaCl_3$ and $Th(NO_3)_4$) a precipitation of NaPSS is obtained at constant C_m/C_s (C_m and C_s are NaPSS and salt concentrations in mono M). The phase separation usually called "salting out phenomena" was studied and some investigations are made in single homogeneous phase.

The dynamics shows some interesting behavior of the effective diffusion coefficient as a function of NaPSS and $LaCl_3$:
- The presence of slow and fast modes at low salt concentration and the existence of a large splitting of the relaxation times.
- a q^2 dependence of the fast diffusion coefficient and a q^3 dependence of the slow mode.
- a power law $D_{fast} \approx C_m^{+0.65}$ at constant salt concentration

INTRODUCTION

Polyelectrolyte solutions are multicomponent systems made of polyions and small counterions in an aqueous solutions. One of the most obvious effect is the ionic strength dependence of the static and dynamic properties of charged macromolecules. An unexpected result of quasielastic light scattering (QLS) experiments is the existence of two modes in semi-dilute solutions, a fast one attributed to a cooperative diffusion and a slow one whose origin is not yet understood but which is believed to be associated with the presence of clusters or temporal aggregats of macromolecules. This behavior was observed in NAPSS in presence of NaCl and $CaCl_2$ [1,2,3] salts and in various other polyelectrolytes [4,5,6].

In this paper we are interested in investigating the influence of trivalent added salt concentration on the structural and dynamical properties of NaPSS solution.

MATERIALS AND METHODS

We use NAPSS of molecular weight $M_w = 7.6 \times 10^5$ g mol^{-1} and 7.0×10^4 g mol^{-1} supplied by Pressure Chemical Co. It is dissolved in deionized water and LaCl$_3$ of analytical grade (Merck) and filtered through a millipore filter of pore size 0,22 µm. The filtered solution is introduced into ligh scattering cell. All manipulations are made in an air-cleaned hood (ADS Laminaire MV 29) as to eliminate dust particles larger than 0,2 µ. Samples are centrifuged at 5000 rpm for an hour before measurements.

The QLS instrument which has been used in this study and the experimental procedure have been described previously [1]. Autocorrelation functions are measured with a Brookhaven correlator and analysed by the method of cumulants and are performed in the homodyne mode. At low salt concentration, the intensity correlation function is clearly the sum of two exponential functions where the average decay rate values Γ_i, are in the ratio of approximately 10^3 or 10^2 : 1. It is thus necessary to make two experiments on each solution using sampling times differing by a large factor. The sampling time is chosen so that the 136 channels covered 2-4 relaxation times in the observed process. In the short sampling time experiment (0.1-2 µs) the effective fast diffusion coefficient is about 10^{-6}-10^{-7} cm^{2}s^{-1}. From autocorrelation functions which are calculated with long channel times (20µs-100µs). one obtain a slow effective diffusion coefficient : $D_{eff} \approx 10^{-8}$-10^{-9} cm^{2}s. Generally, the deviation from a single exponential behaviour is large for the low-scattering solutions. The deviation factor μ^2 / Γ^2 is calculated as the mean value over the whole time range and leads to a value of 0.3-0.4.

PHASE DIAGRAM OF NaPSS AND LACl3 IN WATER

The addition of NaCl or CaCl2 salt in NaPSS solution do not lead to phase separation. Static and dynamic properties of solution are perturbed but only one homogeneous phase is observed.

With LaCl$_3$, one observe a phase separation and the formation of polyelectrolyte precipitates. Very little is known about the structure of such precipitation and the role of small ions in affecting their formation. A possible mechanism in the case of La^{3+} or Th^{4+} is the intermolecular cross linking of the polyelectrolyte molecules by salt bridges. A "zipper" model has been proposed to explain the main experimental results [8,9].

A typical set of precipitation curves is given in

Figure 1. The LaCl3 concentration is plotted versus NaPSS concentration. A simple linear relationship between C_m and C_s at the separation point provided that the precipitation phenomenon occurs when a given fraction of the monomers is associated with trivalent ions independently of polyelectrolyte concentration.
At the collapse line we obtain for LaCl3 :
C_m/C_s = 4.12
At high salt concentration $C_s \approx 0.3$ M LaCl3 we observe a redissolution of precipitates of NaPSS the solution is again optically transparent and has a low viscosity. All these results obtained on the phase diagram of NaPSS and LaCl3 or Th(NO3)4 will be discussed and analyzed in an other paper[9].

DYNAMICS IN THE SYSTEM NaPSS + LaCl3 IN DILUTE PHASE

In Fig. 1, we have investigated the line [a] in quasielastic light scattering experiments. The salt concentration is a constant $C_s = 10^{-3}$ Ml^{-1} and we have measured the diffusion coefficient as a function of NaPSS concentration from 10^{-2} Ml^{-1} to 2×10^{-1} Ml^{-1} in semi dilute regime.

The Figure 2 shows the evolution of two diffusion coefficients (Fast and Slow) at q $\approx 2.3 \times 10^{-3}$ Å^{-1} as a function of C_m. The experimental D_{fast} is increasing with the concentration power law $C_m^{+1/2}$. With the correction due to the substitution of Na$^+$ by La^{3+} we use the expression $B \equiv D_{eff} (\kappa/L_t)^{1/4}$ (κ is the inverse of Debye length and L_t is the persistence length). The effective diffusion coefficient is molar mass independent and increase with concentration according to a power law (ν = 0.65). that to close to the theoretically predicted ($\nu \approx 0.75$)[10,11].

The slow mode is decreasing as a function of C_m is agreement with the behavior observed in neutral polymers[12] but D_{slow} has no well defined power law region. Our experimental data shows the same behavior observed in free salt solution [7].

An other feature is the q^3 dependence of the slow mode, which indicates a strong contribution of internal modes in some short structured domains of polyelectrolytes due to interchain interactions..

In the Fig. 3, we have represented the behavior of the effective diffusion coefficient at constant polyelectrolyte concentration ($C_m \approx 0.1$ Ml^{-1}) as a function of C_s (line b in Figure 1).

One observed the same general behavior obtained with monovalent and divalent salts (NaCl and CaCl2)[2]. Two relaxation times are present with a large splitting.

The slow mode increases with increasing salt concentration until 5×10^{-3} Ml^{-1} very close to phase separation ($C_s \approx 7 \times 10^{-3}$ Ml^{-1}).

The fast mode is flat at very low salt

concentration and it decreases with increasing salt concentration in agreement with classical behavior of polyelectrolyte in excess of salt.

Near the phase separation, the contribution of the slow mode is much lower than that of the fast mode and one measures only the later. A pseudo drop is present and recall the ordinary-extraordinarity transition observed with NaCl and $CaCl_2$ [2].

In the redissolution phase obtained in excess of salt (Point c in Figure 1) we have measured only one mode with a large intensity. The value of the diffusion coefficient ($\approx 10^{-7} cm^{-1} s^{-1}$) is very close to those obtained for the fast mode of the phase below the line of demixtion.

CONCLUSION

QLS experiments reported on NaPSS solutions with added polyvalent salt ($LaCl_3$) has shown the presence of two relaxation times in the correlation function of the scattered intensity. A slow mode and a fast mode depending on the concentration range. The fast mode, observed in the semi-dilute and concentrated regime was attributed to the cooperative fluctuations of polyelectrolyte concentration. The slow mode can represent the contribution of labile and structured domains of polyelectrolyte with a small correlation length. The q^3 dependence of the slow time is attribuable to internal modes of some loose agregats with a length $\xi \gg q^{-1}$.

REFERENCES

1. M. Drifford, J.P. Dalbiez, J. Phys. Chem. **88** 5368 (1984)
2. M. Drifford, J.P. Dalbiez, Biopolymers 24 1501 (1985)
3. R. Koene, M. Mandel, Macromolecules **16** 220 (1983)
4. S.C. Li, W.I. Lee, J.M. Schurr Biopolymers **17** 1041 (1978)
5. A.W. Fulmer, A Julejet, A. Bensabat and V.A. Bloomfield, Biopolymers **20** 1147 (1981)
6. Xiao Li, W.F. Reed J. Chem. Phys. **94** 4568 (1991)
7. M. Drifford, J.P. Dalbiez, J. Physique Lett. **46** L311 (1985)
8. M. Drifford, J.P. Dalbiez, O. Spalla, L. Belloni, Proceeding ACS Montreal (1990)
9. M. Drifford, J.P. Dalbiez, O. Spalla, L. Belloni (to be published)
10. T. Odijk, Macromolecules **12** 688 (1979)
11. R. Koene, M. Mandel Macromolecules **16** 973 (1983)
12. W. Brown, Macromolecules **17** 66 (1984)

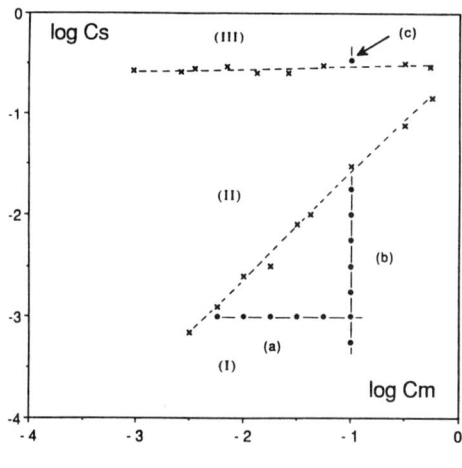

Figure 1. Phase diagram of NaPSS + LaCl$_3$ in water at 25 °C.
(I)   One phase
(II)   Two phases
(III)   One phase
(a) (b) (c) Experimental data measured by QLS.

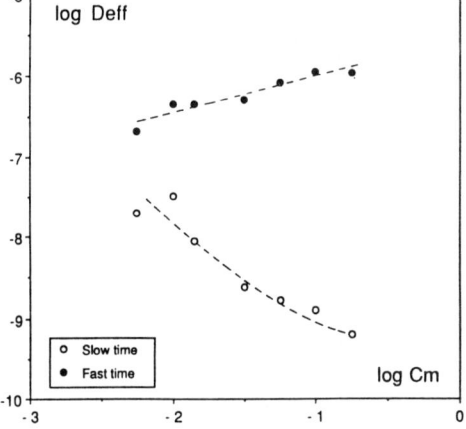

Figure 2. log D_{eff} vs log C_m for NaPSS in 0.001 M LaCl$_3$ at 25°C and q = 2.3 x 10^{-3} Å^{-1} (line **a** in figure 1).

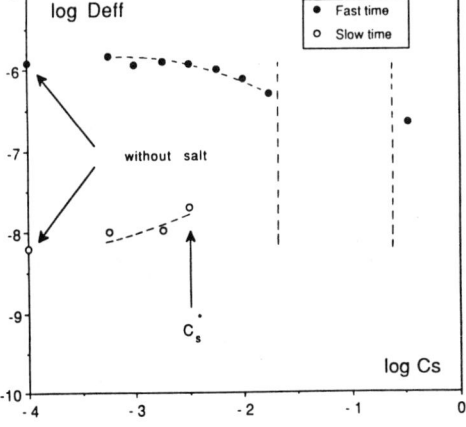

Figure 3. log D_{eff} vs log C_s for NaPSS at C_m = 0.1M at 25°C and q = 2.3 x 10^{-3} Å^{-1} (line **b** and point **c** in figure 1).

Time-Resolved Small-Angle Neutron Scattering Study
of Later Stage Spinodal Decomposition of a Polymer Blend

Hiroshi Jinnai, Hirokazu Hasegawa, and Takeji Hashimoto
Department of Polymer Chemistry, Faculty of Engineering,
Kyoto University, Kyoto 606, JAPAN

Charles C. Han
Polymers Division, National Institute of Standards and Technology,
Gaithersburg, Maryland 20899, U.S.A.

I. INTRODUCTION

Recently there have been a number of studies on the kinetics of phase separation of polymer blends in early stage spinodal decomposition (SD) and on self-assembling processes, dynamics, and pattern (morphology) in the later stage SD as a problem of non-equilibrium statistical mechanics of long chain molecules. Most of the experimental studies utilized time-resolved light scattering (LS) methods[1-5], and only a limited number of studies utilized time-resolved small-angle neutron scattering (SANS)[6-8] and small-angle X-ray scattering (SAXS)[9] methods. It has been shown, efficiently by time-resolved LS, that the growing global structure can be scaled by a single time-dependent length parameter $\Lambda_m(t)$. There still remains an important question on whether the local structure, e.g., interfacial structures including waviness of the interface and interfacial thickness, can also be scaled with $\Lambda_m(t)$. The time-resolved SANS and SAXS are useful methods to investigate this problem as they can approach much shorter length scale than LS. In this series of study we investigated the time evolution of the interfacial structure using time-resolved SANS and LS on a same blend specimen. In this report we discuss general trends on the time-change of SANS profiles in detail and the result on the time-evolution of the composition difference, $\Delta\Phi(t)$, between two coexisting domains in the later stage SD.

II. EXPERIMENTAL

Deuterated polybutadiene (DPB) (sample code: H18, $M_w=10.3\times10^4$, $M_w/M_n=1.03$) and protonated polyisoprene (HPI) (sample code: H15, $M_w=13.6\times10^4$, $M_w/M_n=1.04$) samples used in this study were synthesized by a living anionic polymerization in our laboratory. A film specimen was prepared by dissolving a binary mixture of H15/H18 with 50/50 wt%/wt% (49/51 vol%/vol%) in toluene and then casting from approximately 5wt% solution. The volume fraction of DPB of this mixture is very close to that of the critical mixture, $\phi_{C,DPB}=0.52$, according to Flory-Huggins theory. The as-cast film was thoroughly dried under vacuum at room temperature for at least 1 day. This mixture exhibited a LCST behavior and had the spinodal temperature of $T_S = 36.1°C$ which was obtained from SANS data in single-phase state. A copper heating block was used to control the specimen temperature to a desired temperature with the accuracy of $\pm 0.3°C$. Spinodal decomposition was induced by a rapid temperature jump (T-jump) from a single-phase state (23°C) to a point inside the spinodal phase boundary (40°C). Temperature of the mixture was stabilized within 2 min. Time-resolved SANS experiment was carried out at NIST Research Reactor using the focusing geometry with 11-Å neutron wavelength. For each time-slicing measurement, 3 min was needed to attain sufficient statistical accuracy for the scattering intensity.

III. RESULTS AND DISCUSSION

It is quite important to investigate growth of self-assembling structure over a wide spatial scale. This can be achieved by investigating the corresponding change of the structure factor with time over a wide range of wavenumber or scattering vector q using a combined LS and SANS technique. Here wavevector q is defined as, $q \equiv (4\pi/\lambda)\sin(\theta/2)$, where θ is the scattering angle, and λ is the wavelength of the neutron.

Figure 1 shows the time-evolution of the structure factor during early-to-late stage SD. The left and the right half of the figure correspond to the structure factors obtained by LS and SANS, respectively. At this moment there still exists a gap in q range from 10^{-2} to 7×10^{-1} nm^{-1}. Two sets of structure factors are vertically shifted for convenience but are plotted on the common logarithmic scales to compare the time-evolution of SANS profile with that of LS. It is quite impressive that the change in the structure factor at small q's observed by LS is quite large compared with that at large q's observed by SANS. The vertical broken line at $qR_g=1$ indicates the q value corresponding to $1/R_g$ where R_g is an average radius of gyration of DPB and HPI. Thus our SANS experiments are concerned with the time-change in the composition fluctuations at a length scale of R_g or at even much shorter scale in early to-late stage SD.

Figure 2 shows the SANS profiles together with the measured profile of $S(q;T=23°C)$ (broken line) before the T-jump (i.e., at 23°C) and the theoretically calculated profile $S(q;T=40°C)$ for the system decomposed into macroscopic phases 1 and 2 in equilibrium at 40°C,

$$S(q;T=40°C) = X_{1e}S_1(q) + (1-X_{1e})S_2(q). \qquad (1)$$

X_{ie} and $S_i(q)$ (i=1 or 2) are, respectively, the volume fraction and the structure factor for the thermal composition fluctuations for the phase i with the compositions of ϕ_{ij}'s (j=DPB or HPI) as determined by Flory-Huggins mean-field theory. $S_i(q)$ is given by de Gennes' RPA theory for the thermal composition fluctuations in single phase state. A similar analysis using $S(q;T=40°C)$ was performed by Bates et al.[10] on the static SANS profiles for the phase-separated mixture.

It is found in Figure 2 that time-evolution of the net scattering intensity $S(q,t)$ is different below and above the crossover wavenumber q^*. At a given q satisfying $q<q^*$ (regime I), $S(q,t)$ increases first and then decreases with time t. On the other hand, at a given q satisfying $q>q^*$ (regime III), $S(q,t)$ always decreases with t. Regime II at $q \cong q^*$ is defined as the crossover regime. The initial increase of $S(q,t)$ with t in regime I is due to the build-up of the composition fluctuations in the early stage SD. The decrease of $S(q,t)$ with t in regime I and II is due to the coarsening of the phase-separating domains in the later stage SD. As the domains grow with t, the scattering arising from the domains (*domain scattering*) shifts toward smaller q's and hence its intensity decreases. The intensity $S(q,t)$ in regime III decreases with t, implying that the local composition of A polymer (A=DPB or HPI), ϕ_A, is being biased with t from the initial state O' towards the final state F (see the inset in Figure 2), as will be discussed in detail below.

Figure 3 shows the change in the representative profiles in the (a) early, (b) intermediate, and (c) late stage SD[3]. The intermediate stage and the late stage were determined by time-resolved LS studies independently carried out for the same specimen under the same experimental conditions. In the intermediate SD the magnitude of scattering vector $q_m(t)$ at which the LS intensity becomes maximum decreases, so that the scaling exponents α and β for $q_m(t)$ and $I_m(t)$, the time-dependent maximum LS intensity, satisfy the relation $\beta > 3\alpha$ where

$$q_m(t) \sim t^{-\alpha}, \quad (2) \quad \text{and} \quad I_m(t) \sim t^{\beta}. \quad (3)$$

On the other hand, in the late stage the mean-squared fluctuations reach a constant value, and the domains grow with dynamical self-similarity, S(q,t) being scaled with a single length parameter $\Lambda_m(t)=1/q_m(t)$. Hence $\beta=3\alpha$. Observed intensity S(q,t) at $q>q^*$ systematically decreases with t in the early-to-intermediate stage [Figure 3 parts (a) and.(b)] and then finally becomes independent of time and identical to S(q;T=40°C) in the late stage SD [Figure 3 (c)].

It should be also noted that S(q,t) at a large q limit obeys q^{-2} law. This implies that S(q,t) at the large q limit arises from the local thermal composition fluctuations (defined hereafter as *internal scattering* $S_{Ti}(q)$). In fact one can easily show that $S_i(q)$ in eq.(1) for the thermal composition fluctuations in single phase state at the large q limit is given by

$$\lim_{q \to \infty} S_i(q) \equiv S_{Ti}(q) \quad (i=1 \text{ or } 2)$$
$$= [12\phi_{i,DPB}(1-\phi_{i,DPB})/<a^2/V>]q^{-2}, \quad (4)$$

where

$$<a^2/V> \equiv (a_{DPB}^2/V_{DPB})(1-\phi_{i,DPB}) + (a_{HPI}^2/V_{HPI})\phi_{i,DPB} \quad (5)$$

with a_{DPB} and V_{DPB} being Kuhn's statistical segment length and the segment volume for DPB, respectively, and a_{HPI} and V_{HPI} are the corresponding quantities for HPI. The intensity level $S_{Ti}(q)$ depends only on the volume fraction of DPB (or HPI) in single phase state or in each domain in two phase state but is essentially independent of χ and hence temperature T. Here, χ refers to Flory's interaction parameter between DPB and HPI polymers. Therefore, from eq (4), the observed intensity level of the portion of the profile which obeys q^{-2} law is expected to be due to the thermal composition fluctuations inside the two domains. The fact that this intensity level decreases with t is a consequence of increase in composition fluctuation, i.e., $\phi_{1,DPB}$ and $\phi_{2,DPB}$ both of which are initially equal to 0.49 shift to the equilibrium values of 0.3 and 0.78, respectively. The fact that this intensity level is invariant with time at the late stage SD implies that ϕ_{ij}'s (j=DPB or HPI) reach the constant equilibrium values. Again the excess scattering from S(q;T=40°C) is believed to arise from the domain scattering. Furthermore, the crossover value of q (defined as q_s) above which S(q,t) falls down to the intensity level of the internal scattering was found to shift toward smaller q. The shift of q_s can be more explicitly observed in Figure 4 where $S(q,t)q^2$ at various times were plotted as a function of q and the crossover q_s at a given time was identified by an arrow. The change of q_s with t which manifests the growth of the domain size $\Lambda_m(t)$ is plotted in Figure 5 in double logarithmic scale. Since the intensity level of the internal scattering was found to depend only on the composition in each domain, the time-change of this intensity level $S_T(q,t)$ allows one to determine the time-evolution of the difference in volume fraction of DPB between two coexisting domains, $\Delta\Phi(t) \equiv |\phi_{1,DPB}(t) - \phi_{2,DPB}(t)|$, by

$$S_T(q,t) = X_1(t)S_{T1}(q,t) + X_2(t)S_{T2}(q,t) \qquad (6)$$

with $X_1(t)+X_2(t)=1$, where $X_i(t)$ is the volume fraction of phase i at time t and $S_{Ti}(q,t)$ is the internal scattering inside phase i (i=1 or 2). $S_{Ti}(q,t)$ can be given by eq (4) by replacing equilibrium value of $\phi_{i,DPB}$ in eq.(4) with the time-dependent value of $\phi_{i,DPB}(t)$. Therefore, $\Delta\Phi(t)$ was determined by fitting SANS data to eq.(6) with one adjustable parameter, $\phi_{i,DPB}(t)$. $\Delta\Phi(t)$ is also plotted in Figure 5. The results clearly show that both $q_s(t)$ and $\Delta\Phi(t)$ undergo a characteristic change at the crossover time t_{cr} ($\cong$270 minutes) from the intermediate to the late stage SD. This t_{cr} was in very good agreement with that found independently by the time-resolved LS experiment.

REFERENCES

1) Nishi, T.,Wang, T.T. and Kwei, T.K. Macromolecules 1975,**18**,227; Hashimoto, H.,Kumaki, J. and Kawai, H. Macromolecules 1983,**16**,641; Synder, H. L.,Meakin, P. and Reich, S. Macromolecules 1983,**16**,757
2) Izumitani, T. and Hashimoto, T. J. Chem. Phys. 1985,**83**,3694; Takenaka, M.,Izumitani, T. and Hashimoto, T. Macromolecules 1987,**20**,2257
3) Hashimoto, T.,Itakura, M. and Hasegawa, H. J. Chem. Phys.1986,**85**,6118; Hashimoto, T.,Itakura, M. and Shimizu, N. J. Chem. Phys.1986,**85**,6773
4) Nose, T. Phase Transitions 1987,**8**,245; Hashimoto, T. in 'Current Topics in Polymer Science, Volume II' (Eds. R.M. Ottenbrite, L.A. Utracki and S. Inoue), Hanser, N.Y., 1987; Hashimoto, T. Phase Transitions 1988,**12**,47
5) Okada, M. and Han, C.C. J. Chem. Phys. 1986,**85**,5317; Sato, T. and Han, C.C. J. Chem. Phys.1988,**88**,2057
6) Higgins, J.S.,Fruitwala, H. and Tomlins, P.E. Macromolecules 1989,**22**,3674
7) Higgins, J.S.,Fruitwala, H. and Tomlins, P.E. British Polym. J. 1989,**21**,247
8) Schwahn, D.,Springer, T.,Hahn, K. and Streib, J. Symp. Proc. Mat. Res. Soc.: Polymer Based Molecular Composites Boston Nov.27-30,1989
9) Meier, H. and Strobl, G.R. Macromolecules 1987,**20**,649
10) Bates, F.S.,Dierker, S.B. and Wignall, G.D. Macromolecules 1986,**19**,1938

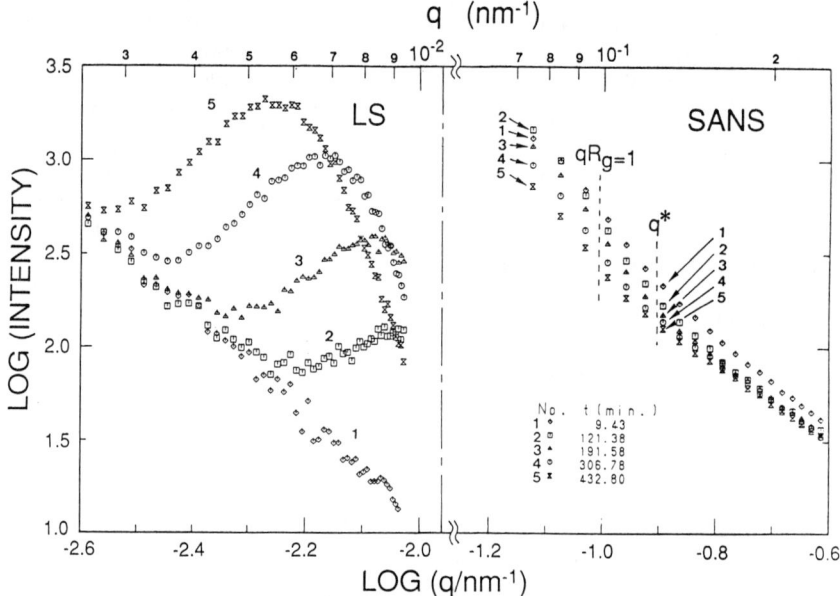

Figure 1. Time-evolution of LS and SANS profiles after the onset of the phase separation

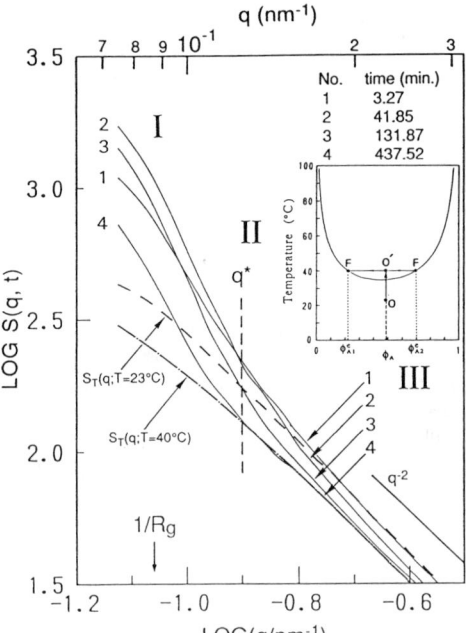

Figure 2. Time-evolution of the SANS profiles after the onset of the phase separation

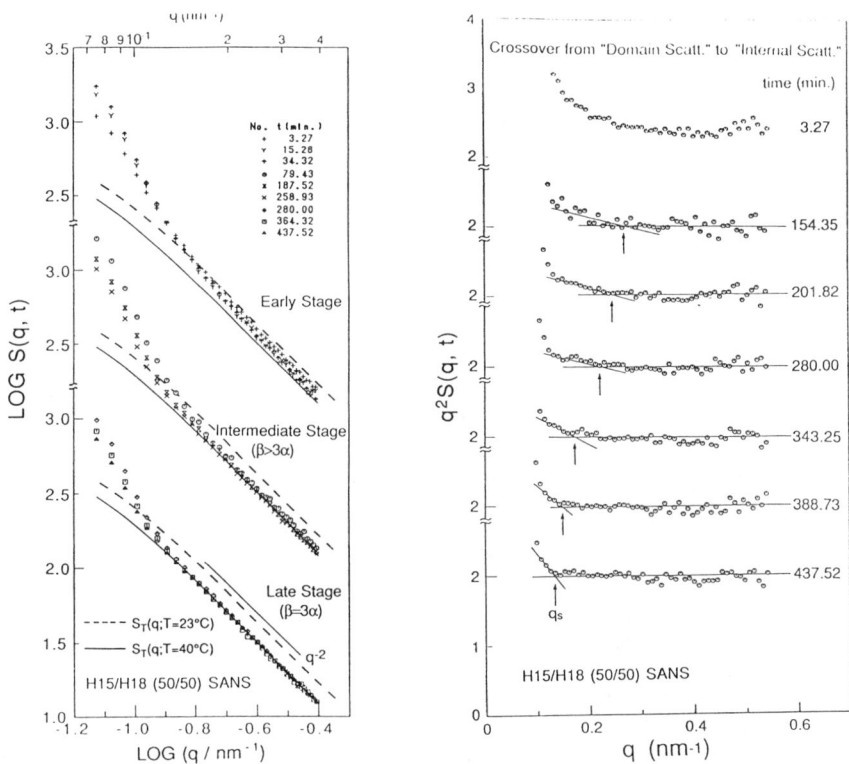

Figure 3. Time-evolution of the representative SANS profiles in the (a) early, (b) intermediate, and (c) late stage SD

Figure 4. Plot of $I(q,t)q^2$ vs time

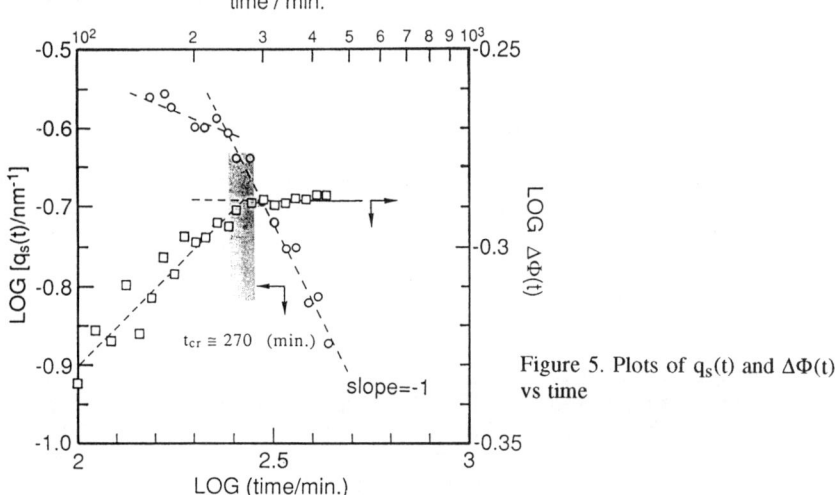

Figure 5. Plots of $q_s(t)$ and $\Delta\Phi(t)$ vs time

DIELECTRIC RELAXATION OF ISOTACTIC POLYSTYRENE IN LONG TIME REGION

K.Fukao and Y.Miyamoto
Department of Physics, College of Liberal Arts and Sciences,
Kyoto University, Kyoto 606 Japan

H.Miyaji
Department of Physics, Faculty of Science,
Kyoto University, Kyoto 606 Japan

ABSTRACT

Dielectric constant and loss were studied for the α-relaxation of isotactic polystyrene (IPS). The decay function for the relaxation was calculated directly from the dielectric loss. It was found that the dielectric relaxation becomes slower as the degree of crystallinity increases. It was also found that the relaxation undergoes a crossover from a stretched exponential decay to a power law decay in the long time region as the relaxation proceeds.

INTRODUCTION

It is known through the investigations on the dielectric relaxations of polymeric materials that the decay does not obey an ordinary Debye law $\phi(t)=\exp(-t/\tau)$, but non-Debye ones such as a stretched exponential decay, $\exp(-(t/\tau)^\beta)$ [1], or a power-law decay, t^{-n} [2]. Recently, the dielectric behavior of glass-forming materials including amorphous polymers has been studied over the wide frequency range and reported to show a deviation from that represented by a stretched exponential decay in the low and high frequency ranges[3,4]. Concerning the primary relaxation in semi-crystalline polymers, however, comparison between the dielectric relaxation behaviors and the experimental results has not been made directly by the decay functions but by the shape of dielectric loss curve $\varepsilon''(\omega)$. Polymeric materials often exhibit more than one dielectric loss peak whose tails are overlapping with each other; the frequency range available for the analysis is restricted to the frequency range where each dielectric loss is dominant. IPS exhibits a well-separated loss peak of the α-relaxation. In this paper, we investigate the dielectric relaxation behavior of the α-relaxation of IPS of various degree of crystallinities (χ); we can easily obtain a crystallized sample with various values of χ.

EXPERIMENTALS

The sample used for dielectric measurement is IPS purchased from Polymer Laboratories Ltd. The weight-averaged molecular weight Mw is 5.56×10^5 and Mw/Mn is 1.9 (Mn is the number-averaged molecular weight). Six samples were prepared; They were melted in vacuo and quenched to 175°C to crystallize isothermally. By changing the crystallization time, we obtained samples with various degree of crystallinities from χ=0% to 27%. The dielectric measurement was made by an LCR meter (HP4284A).

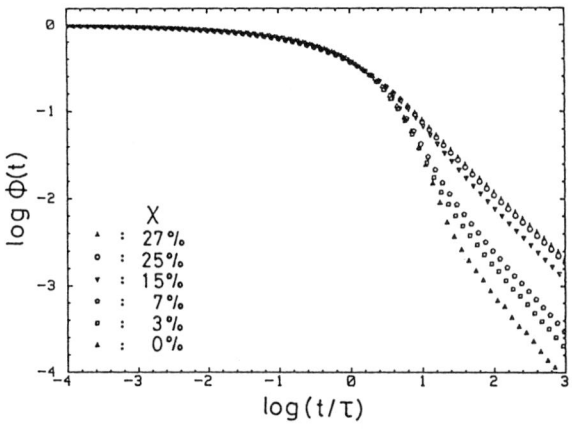

Fig.1: The dependence of the decay function on reduced time.

RESULTS AND DISCUSSION

Dielectric loss ε'' exhibits a peak whose position changes with temperature according to Vogel-Fulcher-Tamman law $\tau=\tau_0\exp(-U/(T-T_0))$. It is found that the dielectric loss curves of the same crystallinity at different temperatures can be reduced to a single master curve by normalizing them in terms of peak positions and peak values. Comparing the master curve of $\chi=0\%$ with that of $\chi=27\%$, we find that the amorphous IPS shows an asymmetric loss curve but the fully crystallized IPS shows a more symmetric one; the loss curve of $\chi=27\%$ has a shallower slope in the low frequency side than that of $\chi=0\%$. In order to elucidate the change in relaxation behavior with χ, we calculate the decay function $\phi(t)$ from the experimental values of dielectric losses. Figure 1 shows the logarithm of decay functions as a function of $\log(t/\tau)$ for various values of χ. We find that the overall behavior of relaxation can be expressed as a stretched exponential decay for all samples investigated. In the long time region, however, we can observe a crossover from a stretched exponential decay to a power-law decay as time proceeds. The exponent n of the power-law decay in the long time region decreases with increasing χ, that is, the dielectric relaxation in that region becomes a slower one as χ increases. By obtaining the decay rate $k(t)\equiv-\dot{\phi}(t)/\phi(t)$, we have investigated the more detailed relaxation behavior and found a complicated relaxation behavior in the amorpous sample[5].

REFERENCES

1. G.Williams, D.C.Watts, Trans.Faraday Soc. **66**, 80 (1970)
2. A.K.Jonscher, Nature **256**, 566 (1975)
3. P.K.Dixon, L.Wu, S.R.Nagel, B.D.Williams, J.P.Carini, Phys.Rev.Lett. **65**, 1108 (1990)
4. A.Schönhals, F.Kremer, E.Schlosser, Phys.Rev.Lett. **67**, 999 (1991)
5. K.Fukao, Y.Miyamoto, H.Miyaji, J.Phys.: Condens.Matter **3**, 5451 (1991)

ULTRASONIC AND DIELECTRIC STUDIES ON CURING PROCESS OF ALLYL-OLIGOMER

H. Okabe, H. Kanaya, K. Hara, S. Taki and K. Matsushige
Department of Applied Science, Faculty of Engineering,
Kyushu University, Hakozaki, Higashi-ku, Fukuoka 812 JAPAN

ABSTRACT

The allyl-ester is a new thermosetting resin composed of an allyl-oligomer and a curing agent. The curing time variations of imaginary parts of dielectric constant (ε'') exhibited a peak as the case of epoxy resins. The peak was detected when the sound velocities revealed considerable increases. Namely, the peak is thought to occur according to the proceed of vitrification process, suggesting the relation to α-relaxation process.

INTRODUCTION

In general, the curing process of thermosetting resins like epoxy resins are characterized by two distinct transitions, gelation and vitrification[1]. The gelation occurs at the incipient formation of an infinite network of crosslinked polymer molecules, while the vitrification involves a transformation from a liquid or rubbery state to a glassy state as a result of an increase in molecular weight.

The monitorring of curing process of the allyl-ester is of interest for material science as well as industrial application. In this study, ultrasonic and dielectric measurements were utilized to follow the thermosetting process of this resin.

EXPERIMENTAL

The specimen investigated was an allyl-oligomer(Fig. 1) and a curing agent, 1,1-bis(t-hexylperoxy) 3,3,5-trimethylcyclohexane of 1% by weight. It was poured into a special measuring cell[2-3] which had been heated to a pre-selected curing temperature. This moment was defined as zero curing time. The sound velocity and the dielectric constant were measured with an automated ultrasonic sound velocity measuring system[2] and an impedance analyzer(YHP4192A) controlled by a computer.

RESULTS AND DISCUSSION

Figure 2 shows the sound velocities as a function of curing time at various temperatures. As shown, the velocities became larger as time proceeded and saturated to certain values at the completion of the reaction. It is noticed that the sound velocity vs. curing time (in logarithmic scale) relation is monotonous at higher temperatures, while it exhibits apparent bends at lower temperatures.

These sound velocity behaviors of allyl-ester can be explained on the below basis. In curing process of thermosetting resins like epoxy resins[1], at lower curing temperatures, the crosslinking reaction proceeds normally until the molecular weight increases to the extent that the glass transition approaches the curing temperature, so the resin can remain ungelled even though it vitrifies and becomes hardened. The two straight lines respectively denote these two processes, crosslinking and vitrification. While, at higher temperatures, although the resin will not vitrify on isothermal cure, the cure can proceed to completion. In this case, the time-sound velocity curve becomes one straight line.

On the other hand, Fig. 3 shows the time variation of imaginary parts (ε'') of dielectric constant at various frequencies at 100°C. The change of ε'' in curing process of epoxy resins have been studied by a number of workers[4-7] and time-ε'' curves in Fig. 3 is similar to that of epoxy resins. The peak was detected when the substantial changes were detected at the time range where the sound velocities revealed considerable increases with time. Namely, the peak in the curing time variation of ε'' is thought to occur according to the proceed of vitrification process. This suggests that the peak is probably related to α-relaxation process.

ACKNOWLEDGEMENT

We would like to thank Kaziyama lab. (Kyushu Univ.) for making a convenient to use an impedance analyzer, and Showa Denko K.K. for supplying an allyl-oligomer and a curing agent.

REFERENCES

1. J. B. Enns, J. K. Gillham : J. Appl. Polym. Sci. 28, 2568 (1983)
2. H. Okabe, W. Takashima, S. Taki, K. Matsushige : Jpn. J. Appl. Phys. 30 Suppl.30-1 (1991) in press
3. H. Okabe, H. Kanaya, K. Hara, S. Taki, K. Matsushige : Rept. Progr. Polym. Phys. Jpn., 34 (1991) in press
4. J. Delmonte : J. Appl. Polym. Sci. 2, 108 (1959)
5. E. N. Haran, H. Gringras, D. Katz : J. Appl. Polym. Sci. 9, 3505 (1965)
6. V. Adamec : J. Polym. Sci. Part A-1 10, 1277 (1972)
7. S. Wu, S. Gedeon, R. A. Fouracre : IEEE Trans. Electrical Insulation 23, 409 (1988)

Fig.1 Constitutional formula of allyl-oligomer.

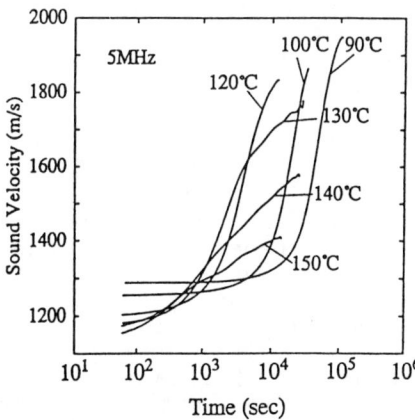

Fig.2 Sound velocity as a function of curing time at various curing temperatures.

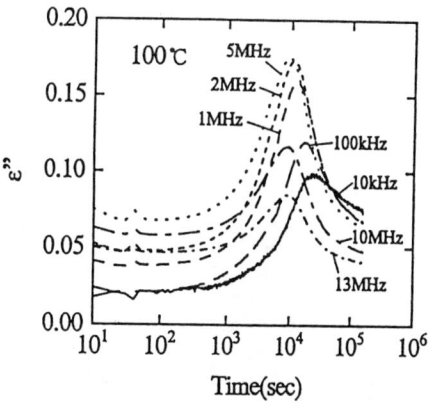

Fig.3 Time variation of ε''(imaginary part of dielectric constant) at various frequencies at the curing temperature of 100 °C.

GELATION PROCESS OF ACTOMYOSIN

H. Kanaya, K. Hara, H. Okabe and K. Matsushige,
S. Nishimuta†, M. Muguruma† and T. Fukazawa†
Department of Applied Science, Faculty of Engineering
and
† Department of Animal Science, Faculty of Agriculture,
Kyushu University, Hakozaki, Higashi-ku, Fukuoka 812 JAPAN

ABSTRACT

The heat-, and pressure-induced gelation process of actomyosin was investigated by observing the transmitted light spectra. First, pH and [NaCl] dependences of the characteristic temperature (T'_g) related to the gelation were determined at atmospheric pressure. Then, the temperature variations of the transmitted light spectra were monitorred after the temperature- and pressure-jump procedures, revealing the growth of the actomyosin assembly during these gelation processes.

INTRODUCTION

Like sausages, we usually eat protein foods as heat-processed gels. The tenderness and water holding capacity are the properties required to them, and these functions are known to be much affected by pH and salt concentration. So far, almost no investigations on the dynamical properties of the heat-induced gelation process of actomyosin, which constitutes the muscle protein, have been reported, in spite of their close interrelation to the formation mechanism. Besides, recently, the high-pressure food treatments have attracted much attention in the field of food processing technology. In these circumstances, in the present paper, we report on the influence of pH and [NaCl] on the heat-induced gelation process and on the dynamical properties of the pressure-induced one of the actomyosin, from the observation of the transmitted light spectra.

EXPERIMENTALS

The actomyosin samples, extracted from rabbit muscle by the method of SZENT-GYÖRGYI[1], were prepared in a range from 110mM to 1M by [NaCl], and from 5.0 to 7.0 by pH. The light beam through the sample was analyzed by a spectrometer (wavelength; 400~600nm) with a linear photo-diode-array (MCPD-1000, Otsuka Electronics). First, we defined the characteristic temperature (T'_g) related to the gelation temperature from the changing rate of the transmitted light intensity at atmospheric pressure, and measured pH and [NaCl] dependences of T'_g. Next, in order to characterize the dynamical features of the light scattering properties in the heating process and under high-pressure, we observed the time dependence of the transmitted light spectra during the gelation processes.

RESULTS and DISCUSSION

In the measurements, in was found that T'_g increased about 10℃ in a intermediate range of [NaCl] at each pH, and the high T'_g region seems to shift to the higher [NaCl] side according as decreasing pH, which may be caused by the formation properties of the myosin[2].

Fig. 1 shows the log-log plot of (τ d) vs. wavelength (λ) in the temperature-jump and pressure-jump experiments, respectively, where approximate linear dependence of ln(τ d) on lnλ can be seen (τ=turbidity, d=thickness of the sample). The slope of the log-log plots showed the time-dependent features, which might be correspond to the growth of the light-scattering clusters during the both gelation processes.

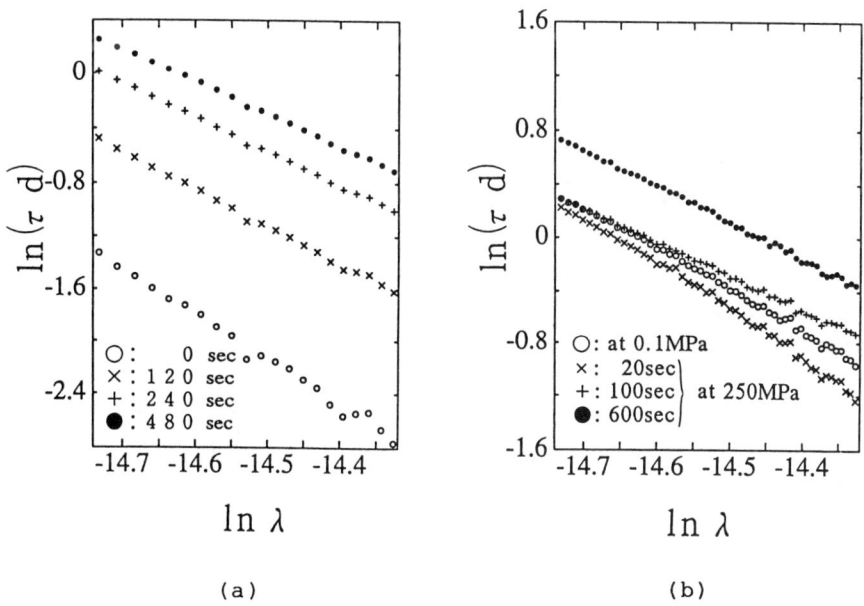

Fig .1. The plots of ln (τ d) vs. ln λ in (a) temperature-jump (from 40 to 50℃) and (b) pressure-jump (from 0 to 250MPa) measurements in 20sec (pH=6.0, [NaCl]=0.6M).

References

1) A. SZENT-GYÖRGYI, Chemistry of Muscular Contraction (Academic Press, New York 1947), p. 136.
2) T. Yasui, K. Samejima: New Food Industry 27, 130 (1985).

FREQUENCY DEPENDENCE OF THE DYNAMIC HEAT CAPACITY ACCOMPANYING THE TRANSITION PHENOMENA

Y. Saruyama
Faculty of Textile Sci., Kyoto Institute of Technology,
Matsugasaki, Sakyo-ku, Kyoto 606, Japan

ABSTRACT

Frequency dependence of the dynamic heat capacity, which is the frequency response function characterizing the temperature response to the heat absorption, was measured around the glass transition temperature of atactic polypropylene and the solid-liquid phase transition temperature of n-paraffin. It was shown that the measurement of the frequency dependence of the dynamic heat capacity is useful technique to study the dynamic properties of the glass transition and the phase transition.

INTRODUCTION

The dynamic heat capacity is defined as follows. Suppose that the sample is at the temperature just above the glass transition temperature or at the first order phase transition point. After a heat pulse to the sample at the time t=0 the temperature of the sample jumps up quickly then decays slowly accompanying the slow relaxation to the state with larger (static) heat capacity or the slow progress of the phase transition. The slow decay of the temperature after the step-like increase of the internal energy is characterized by a time response function. The frequency response function $L(\omega)$ can be defined from the time response function according to the superposition principle. The complex amplitude T of the temperature modulation is given by the equation, $T = -iJL(\omega)/\omega$ where J is the amplitude of the heat flow to the sample. The inverse of $L(\omega)$ is called dynamic heat capacity. In this study the complex amplitude T was measured by the AC calorimetry technique[1].

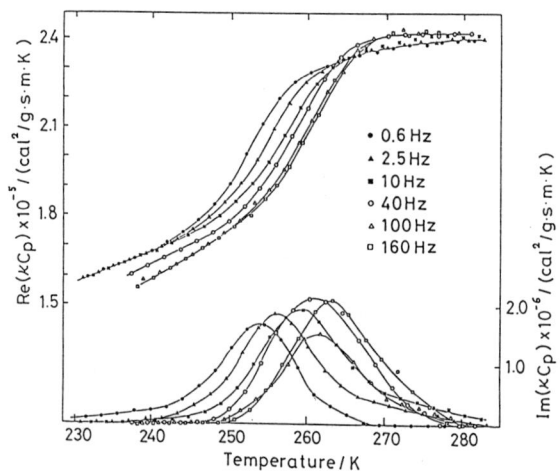

Fig. 1 Dynamic heat capacity of at-PP.

GLASS TRANSITION OF ATACTIC POLYPROPYLENE

Figure 1 shows the real part and the imaginary part of the product of the dynamic heat capacity and the thermal conductivity κ of atactic polypropylene in the temperature range just above the glass transition temperature; κ is real and almost constant in the temperature range. The step of the real part and the peak of the imaginary part moves to the higher temperature direction as the heating frequency becomes higher. This result is consistent with the result of Birge[2] about glycerol and propylene glycol. The activation energy estimated from the frequency dependence of the peak position of the imaginary part is 77[kcal/mol].

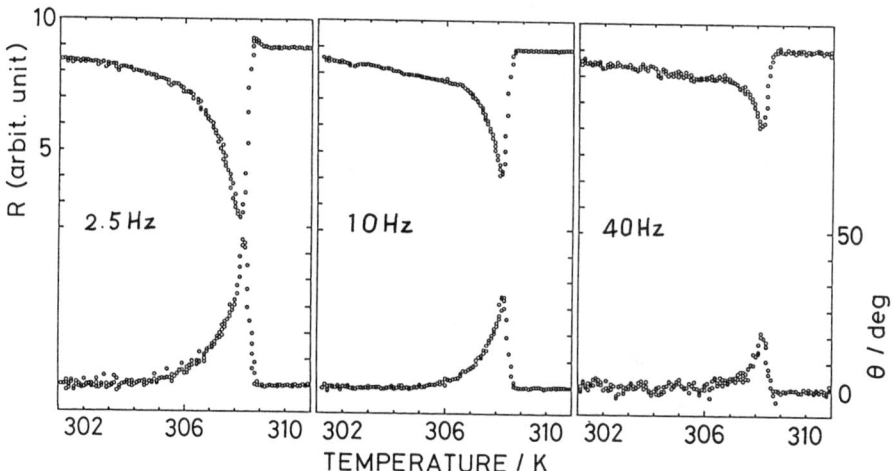

Fig. 2 Amplitude R and phase θ of $L(\omega)$.

SOLID-LIQUID TRANSITION OF n-PARAFFIN

Figure 2 shows the absolute value R and the phase θ of L of n-eicosane ($C_{20}H_{42}$) around the solid-liquid transition temperature. As the frequency becomes larger the decrease of R becomes smaller because less amount of material undergoes the phase transition. The positive phase shift is consistent with the requirement of the second law of thermodynamics. Thus the result of Figure 2 reflects the dynamic properties of the phase transition.

It was found by the detailed studies of the spectra of the temperature modulation that there is a small component of temperature modulation with the frequency twice as large as the heating frequency only in the transition temperature range.

1. I. Hatta and A. K. Ikushima, Jpn J. Appl. Phys., 20 1995 (1981)
2. N. O. Birge, Phys. Rev. B34, 1631 (1986)

STRUCTURE FORMATION OF POLY(ETHYLENE TEREPHTHALATE) DURING ANNEALING PROCESS

Masayuki Imai, Tohru Mizukami
Research Center, Toyobo Co. Ltd., Otsu, Shiga, 520-02, Japan

Toshiji Kanaya, Keisuke Kaji
Institute for Chemical Research, Kyoto University, Uji, Kyoto, 611, Japan

A new finding is reported that the long range ordered structure is formed in the induction period before the crystallization, when poly (ethylene terephthalate) (PET) is annealed just above the glass transition temperature. This ordering process can be interpreted in terms of the spinodal decomposition type phase separation phenomenon.

The annealing process of crystallizable macromolecules from the amorphous state can be divided into two stages of induction period and crystallization stage. During the induction period, the macroscopic density of the sample remains unchanged from that of the amorphous state and no local packing of the chain segments is observed[1]. In order to clarify the role of induction period in the structural formation, the ordering process of the amorphous PET was investigated using small-angle X-ray scattering (SAXS) technique.
Fig 1 shows the time evolution of the SAXS profiles after subtraction of that for amorphous state when amorphous PET (melt quenched film) is annealed at 80°C, 5°C above the glass transition temperature[2]. In this case, initial annealing time of about 120 min corresponds to the induction period and after that the crystallization begins to start. During the induction period the scattering maximum appears immediately after the initiation of annealing. As the annealng time increases the maximum position shifts toward lower Q side, where Q is the length of scattering vector. After the induction period, another peak corresponding to so-called long-period structure is observed. These results clearly show that the long range ordered structure is formed in the induction period while macroscopic density of the system does not change.
It is presumed that the formation of such ordered structure is related with the phase separation from the uniform amorphous phase to the crystallizable and uncrystallizable phases. Here we analyse the obtained data in the induction period based on the kinetics of phase separation because the macroscopic density of the system is conserved. To show the applicability of the kinetics of the phase separation, we plot the time evolution of the scattering intensity and the peak position

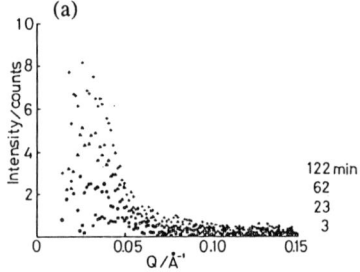

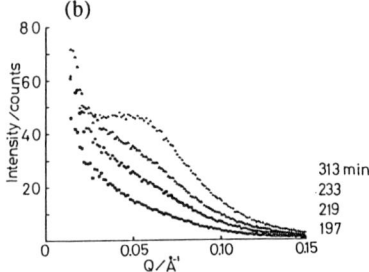

Fig. 1. SAXS profiles of annealed PET after subtraction of that of melt quenched sample. (a) is for annealing time range of 3 - 122 min (induction period) and (b) is for 197 - 313 min (crystallization stage).

Q_m as shown in Figs. 2 and 3, respectively. It should be noted from these figures that during the induction period there exist two stages of the structure formation processes. In the early stage, corresponding to the time scale up to 20 min, the intensity increases exponentially with time while the peak position does not change. These behaviour can be described by the Cahn's linearlized theory[3] for spinodal decomposition process. In the late stage, corresponding to the time scale from 30 to 120 min, the intensity increases gradually while the peak position shifts to the lower Q side. The scattering profiles in the late stage can be described in terms of the Furukawa's scaling theory[4]. According to this theory, the structure function $S(Q,t)$ is expressed by the following scaling law

$$S(Q,t) = Q_m^{-3} \tilde{S}(x) \quad (1)$$

where $x = Q/Q_m$, $\tilde{S}(x)$ is a universal function independent of time t, given by

$$\tilde{S}(x) = 3x^2/(2 + x^6) \quad (2)$$

and related to the scattering profile $I(Q,t)$ by following equation:

$$\tilde{S}(x) = Q_m^3 I(Q,t) \quad (3)$$

Fig. 4 shows the universal structure factor for the growth process of clusters fits the experimental data excellently. From these experimental results, we can conclude that the spinodal decomposition type of phase separation occurs during the induction period before crystallization.

According to the Flory's model[5], the crystallization process consists of two steps: (1) co-operative ordering of the chains in a given region into a parallel alignment without change in the intermolecular interactions, and (2) increase in intermolecular interactions made possible by the more efficient packing of the chains in the parallel state. It may be considered that the phase separation observed in the induction period corresponds to the formation of the crystallizable parallel domains from which molecular entanglements are excluded and the uncrystallizable disordered domains having a high concentration of entanglements because the entanglements prevent crystallization.

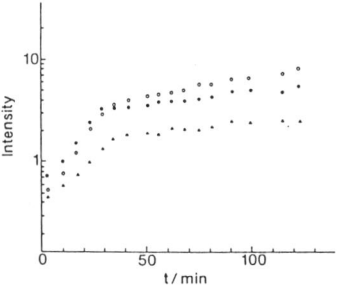

Fig. 2. Annealing time t dependence of scattering intensity as a function of Q during the induction period. The value of Q are o: 0.03, •: 0.04 and ∆: 0.05Å^{-1}.

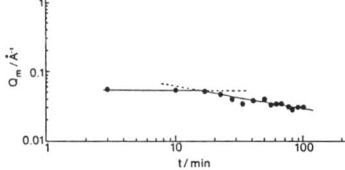

Fig. 3. Annealing time t dependence of the wavenumber Q_m at the maximum intensity.

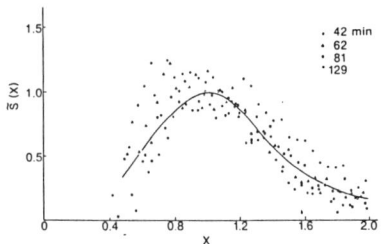

Fig. 4 Scaled structure function calculated from observed SAXS profiles. The solid curve indicates the theoretical scaled function.

References
1) Imai, M., et al. submitted to *Polymer*.
2) Imai, M., et al. submitted to *Polymer*
3) Cahn, J.W., *J. Chem. Phys.* **42**, 93 (1965).
4) Furukawa, H., *Physica* **123A**, 497 (1984).
5) Flory, P.J., *Proc.R.Soc.London* **A234**, 60 (1956).

CRITICAL BEHAVIOR OF COMPLEX SHEAR MODULUS IN CONCENTRATED POLYMER SOLUTIONS

Hajime Tanaka and Toshiaki Miura
Department of Applied Physics and Applied Mechanics, Institute of Institute of Industrial Science, University of Tokyo, Minato-ku, Tokyo 106

ABSTRACT: We demonstrate that there exists a large anomaly in both real and imaginary part of complex shear modulus near the phase-separation line of a semidilute polymer mixture, reflecting entanglement effect. The temporal change in complex shear modulus during phase separation has also been successfully observed by a real-time measurement. The behavior will be discussed on the basis of the structure-property relationship.

It is well known that there is a critical anomaly in viscosity of binary mixture near the critical point. There have been a lot of studies on the subject from both theoretical and experimental viewpoints. However, most of the studies have been limited to binary mixtures of simple liquids. In a critical binary mixture of simple liquids, a weak divergence in viscosity has been observed and also predicted by the theories. In concentrated polymer solutions, on the other hand, there is entanglement effect coming from the topological characteristics of polymers, namely one-dimensional connectivity, especially at high concentrations. The concentration fluctuation should accompany the elastic deformation and the new interesting critical dynamics may appear in the entangled system. There have been no theoretical approach and few experiments on the critical phenomena of viscoelasticity in concentrated polymer solutions.

Here we studied the critical anomaly in the complex shear modulus of a binary mixture of polystyrene (PS) and diethyl malonate (DEM) near the phase-separation line. We also performed a real-time measurement of complex shear modulus during phase separation to study the relationship between morphology and mechanical properties and the effect of interface on mechanical properties. We used a new instrument for measuring the complex shear modulus in the frequency range from 1Hz to 500Hz. The details on the instrument will be presented elsewhere[1].

Samples used were PS and DEM. The weight averaged molecular weight of PS was 355000. The sample was squeezed into a gap between two glass plates, which were attached to the piezoelectric transducers. The shear strain is excited by the one transducer and the induced stress through the sample is detected by the other transducer. The excitation is made by a piezoelectric stack, which improves the S/N ratio considerably and makes the digital signal processing possible. The ratio of the output to the input signal gives us information on the complex shear modulus $G^*(\omega)$. The frequency spectrum is obtained within a short time by applying a simultaneous multi-frequency excitation method.[1]

The appearance of the elasticity due to entanglement has been clearly observed with increasing a polymer concentration. The elasticity increases monotonously with PS concentration. Below a few wt%PS, the solution behaves as a purely viscous liquid and the imaginary part of $G^*(\omega)$, $G''(\omega)$, is expressed by $G''(\omega) = i\eta\omega$ where η is the shear viscosity. c^*, at which the chains start to overlap with each other, is roughly estimated from a simple scaling theory. The appearance of elasticity occurs around c^*. Frequency dependence of $G^*(\omega)$ can be explained by the distribution of the life time of fluctuating temporal network formed by entangled chains.

Figure 1 shows the temperature dependence of the complex shear modulus of PS/DEM(7/93) at several frequencies. The deviation from the background can be clearly observed near the phase-transition point. It becomes clear that the anomaly of viscosity can be expressed by logarithmic divergence as in the usual critical binary mixture. The deviation from the logarithmic divergence at a finite frequency may be due to relaxation behavior caused by the crossover between the measurement frequency and the characteristic time of concentration fluctuation. The anomaly of the elasticity will be discussed at the presentation.

Next we show the results of the real-time measurement of $G^*(\omega)$ during phase separation. Fig.2 shows the 3D plot indicating the temporal change in $G^*(\omega)$ spectra. The behavior is strongly dependent on both the composition and the quench depth. In particular, the composition affects the temporal change in mechanical properties seriously. The quench condition determines whether pattern appears as interconnected morphology or droplet morphology. Furthermore, in droplet morphology it is determined by the quench condition whether a polymer-rich phase appears as a sea or an island. Since the mechanical property of a polymer-rich phase is very much different from that of a solvent-rich phase, the quench condition seriously affects the mechanical behavior of phase-separated mixture. In addition to these aspects, there is another important effect: crossover phenomenon from three dimension (3D) to two dimension (2D). As the droplet grows with time, the droplet size exceeds the sample thickness and the transition from 3D to 2D occurs by wetting of the droplet to the two glasses. This transition is probably responsible for the reincrease of the shear modulus after its decrease observed in Fig.3. This is supported by an independent experiment where the droplet sizes are measured as a function of time. The crossover between the droplet size and the sample thickness coincides well with the time of the minimum of the shear modulus. The details will be discussed at the presentation. This work was partly supported by a Grant-in-Aid from the Ministry of Education, Science, and Culture, Japan.

Reference

[1] H.Tanaka et al., 1990 Ultrasonic Symp.Proc. 3, (IEEE,NY,1990) p.1325.

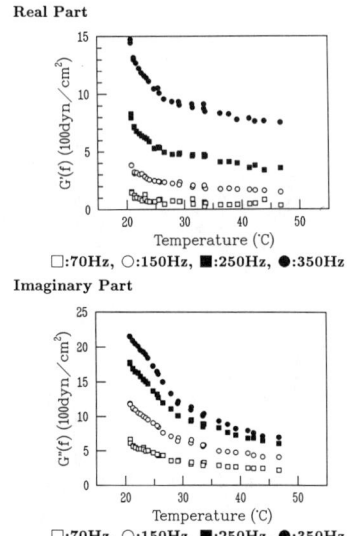

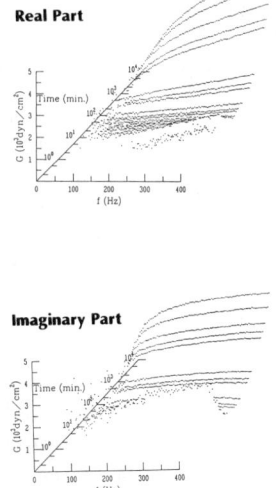

Figure 1: Temperature dependence of $G^*(\omega)$ for several frequencies at 7wt%PS.

Figure 2: Temporal change in $G^*(\omega)$ at 7wt%PS during phase separation.

ORDERING PROCESS OF LAMELLAR MICRODOMAINS FOLLOWING MORPHOLOGICAL TRANSITION FROM CYLINDRICAL MICRODOMAINS IN BLOCK COPOLYMER MELTS

Shinichi Sakurai, Kazuhiro Taie, Toshikazu Momii and Shunji Nomura

Department of Polymer Science and Engineering,
Kyoto Institute of Technology, Kyoto 606, Japan

ABSTRACT

Ordering process of lamellar microdomains of poly(styrene-*block*-butadiene-*block*-styrene) (SBS) triblock copolymers in melts has been studied using small-angle X-ray scattering (SAXS). The SAXS profiles were analyzed by means of the theoretical calculation based on the one-dimensional paracrystal model so as to characterize quantitatively how the lamellae which were formed through coalescence of cylindrical microdomains ordered. As a result of the analysis, it was found that the distribution of the interlamellar distance and the thickness distributions of polystyrene (PS) and polybutadiene (PB) lamellae decreased, whereas the interlamellar distance and the interfacial thickness did not definitely change with time.

INTRODUCTION

Recently we have found the morphological transition from cylindrical to lamellar microdomains for SBS triblock copolymers in melts when the specimen prepared by the solution cast with a selective solvent was heat-treated at above the glass transition temperature of PS.[1] In the course of the transition, coalescence of cylindrical microdomains first occurred and was followed by ordering of the lamellae which was formed through the coalescence of cylinders. In this work, the ordering process of the lamellae was studied by SAXS.

EXPERIMENTAL

The SBS specimen was obtained from Japan Synthetic Rubber Co. Ltd. (Mn = 6.31 × 10^4, Mw/Mn = 1.15 and ϕ_{PS} = 0.36). The specimens for SAXS were prepared by the solution cast, i.e., by evaporating a solvent gradually from ca. 5 wt% of polymer solutions. Methylethylketone which is selectively good for PS was used as a cast solvent. The specimens were then heat-treated at 158 °C for a given time. The SAXS measurements were performed at room temperature.

RESULTS AND DISCUSSION

Based on the results which were previously obtained by the transmission electron microscopy,[1] the change of the patterns of microdomain structures along the morphological transition is schematically represented by Fig. 1, where black and white domains are PB and PS phases, respectively. Fig. 1(A) and 1(B)

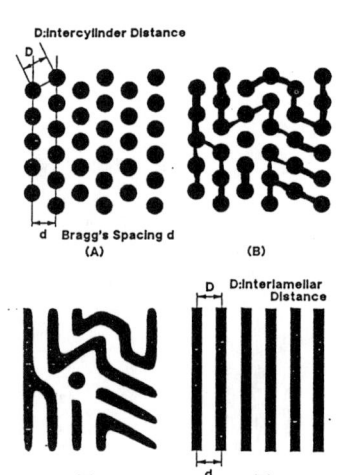

Fig. 1 Schematic representation of the change of the patterns of microdomain structures along the morphological transition.

represent, respectively, the cross-section of the nonequilibrium cylindrical morphology frozen-in due to vitrification of PS in matrix at room temperature and a transient structure induced by coalescence of cylinders when the specimen was heat-treated at 158 °C. The change from Fig. 1(A) to 1(B) appeared within 5min. The successive ordering process represented by Fig. 1(B) to 1(D) took more than 30min.

In Fig. 2 the experimental and theoretical SAXS profiles were displayed in a plot of log I(q) vs. q, where I(q) is the absolute scattered intensity and q is the magnitude of the scattering vector given by $q = (4\pi/\lambda) \sin(\theta/2)$, with λ (= 0.154 nm) and θ being, respectively, the wavelength of X-ray and the scattering angle. It is noted that the positions of the first-order peak, q_1 were found to be identical within an experimental error for all profiles. This indicates that the interlamellar distance did not definitely change through the ordering process of the lamellae. The calculation of the theoretical profiles for the lamellar microdomain structures were based on the one-dimensional paracrystal model.[2] Since several parameters characterizing the lamellar microdomain structures were used for the calculation, these parameters were evaluated when the theoretical profile well reproduced the experimental one. Best results were shown in Fig. 2. As a result of the analysis, it was found that the distribution of the interlamellar distance and the thickness distribution of PS- and PB-lamellae decreased, whereas the interfacial thickness did not change definitely with annealing time.

Fig. 3 shows scaled SAXS profiles. I_1 is the scattered intensity for the first-order peak. Here the change of profiles with time is highlighted. It is worth to mention that the reduced intensities at scattering peaks were likely time-independent. Moreover, the profiles did not change more than 30min.

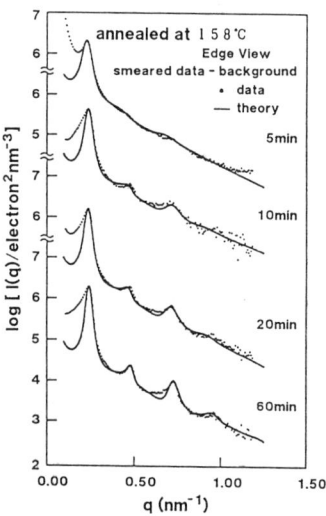

Fig. 2 Annealing time dependence of the experimental and theoretical SAXS profiles. All experimental ones were obtained at room temperature.

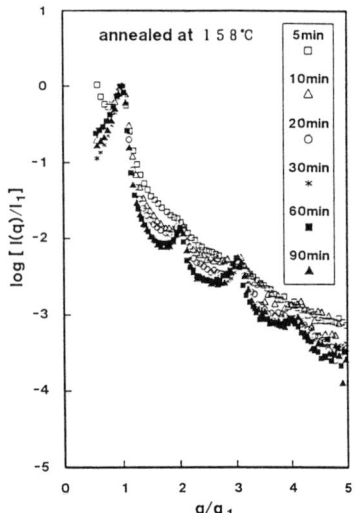

Fig. 3 Annealing time dependence of scaled SAXS profiles. The reduced scattered intensity, $I(q)/I_1$ were plotted against the reduced magnitude of the scattering vector, q/q_1.

REFERENCES
[1]. Sakurai, S.; Momii, T.; Taie, K.; Shibayama, M. and Nomura, S. *in preparation*
[2]. Sakurai, S.; Okamoto, S.; Kawamura, T. and Hashimoto, T. *J. Appl. Cryst. in press*, **1991**.

CRYSTALLIZATION OF CRYSTALLINE/AMORPHOUS POLYMER BLENDS

M. Takahashi and H. Matsuda
Shinshu University, Ueda, Nagano 386, Japan
H. Yoshida and Y. He
Tokyo Metropolitan University, Hachioji, Tokyo 192-03, Japan

ABSTRACT

Crystallization process and the phase structure formed in PEO/PMMA blends were measured by differential scanning calorimetry (DSC) and small angle X-ray scattering (SAXS). The half time of the crystallization $t_{1/2}$ increased by blending PMMA and with PMMA molecular weight increased. 2θ at which the scattered intensity showed the maximum became small by blending PMMA and decreased with PMMA molecular weight increased.

INTRODUCTION

Crystallization process and the phase structure formed in crystalline / amorphous polymer blends are largely affected by amorphous component, i.e. molecular weight, tacticity[1,2], glass transition temperature, etc. This study aims at investigating the PMMA molecular weight dependences of the crystallization process and the phase structure formed in poly(ethyleneoxide) (PEO) / poly(methylmethacrylate) (PMMA) blends by differential scanning calorimetry (DSC) and small angle X-ray scattering(SAXS).

EXPERIMANTAL

PEO(nominal molecular weight = 50,000 ± 10,000) and five kinds of atactic PMMA having different molecular weight (Mw = 22,000, 23,000, 160,000, 170,000, 770,000, Mw/Mn = 1.6, 1.4, 1.7, 1.7, 2.1) were used. Blend samples were obtained by casting from chloroform solution. The PMMA composition of the blends provided for measurements were 20 wt%. The isothermal crystallization process was measured by a Seiko DSC 200. Melting temperature (Tm) and heat of melting (ΔHm) were evaluated by heating at 10 k/min. Small angle X-ray scattering profiles were obtained by a MAC Science MXP[18] with a nickel filtered Cu-kα operating with 40 mA and 250 kV.

RESULTS AND DISCUSSION

The equilibrium melting points (Tm^o) obtained by Hoffman-Weeks Plot varied from 342.4 to 346 K. However, the systematic dependence of Tm^o on the molecular weight of PMMA was not observed, therefore, the variation of Tm^o observed here seems to be due to the experimental error. Fig.1 shows the temperature dependence of the half time of the crystallization $t_{1/2}$, which was the time for the crystallinity to become 50 %. $t_{1/2}$ largely increased with ΔT ($\equiv Tm^o - Tc$, Tc : crystallization temperature) decreased. At the constant ΔT, values of $t_{1/2}$ for blends were larger than that of

PEO, and became large with PMMA molecular weight increased. The molecular weight dependence of $t_{1/2}$ in the temperature region shown in Fig.1 was approximately $t_{1/2} \sim M^{0.3 \sim 0.6}$.

Profiles of small angle X-ray scattering for crystallized PEO and PEO/PMMA blends showed intensity peak at $2\theta = 0.2 \sim 0.3$ deg ($2\theta_{max}$). The SAXS profiles for PEO showed the higher order peak at $2\theta \sim 0.5$ deg addition with the intensity peak described above, but that for blends was not appreciable. The orderness of crystallized PEO may be higher than those for the blends. The temperature dependence of $2\theta_{max}$, at which the intensity shows the maximum, for the lowest order peak is shown in Fig.2. The values of $2\theta_{max}$ decreased with ΔT decreased. At the constant temperature, $2\theta_{max}$ for blends were smaller than that for PEO. As for the blends, $2\theta_{max}$ decreased with the molecular weight of PMMA decreased. Fig.3 shows the PMMA molecular weight dependence of $2\theta_{max}$ at Tc = 323 K. As the PMMA molecular weight increased, $2\theta_{max}$ initially increased in lower molecular weight region, and then asymptotically approached to the $2\theta_{max}$ value for PEO.

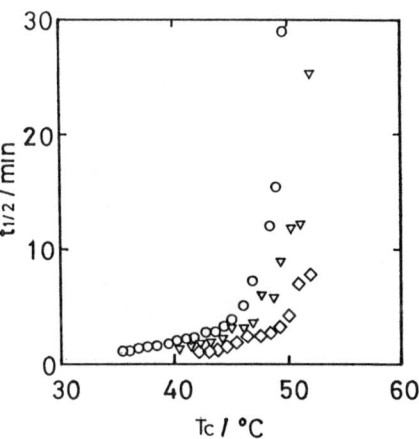

Fig. 1. Relationship between $t_{1/2}$ and Tc for PEO/PMMA blends; PMMA Mw ◇:23,000, ▽:170,000, ○:770,000.

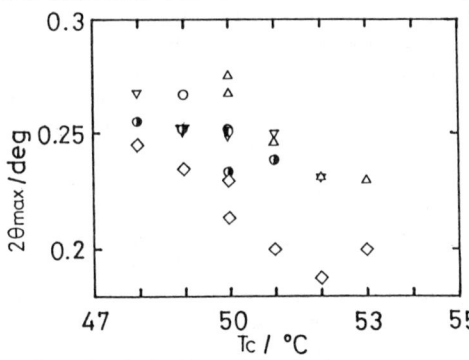

Fig. 2. Relationship between $2\theta_{max}$ and Tc for PEO/PMMA blends; PMMA Mw ●:22,000, ◇:23,000, ◐:160,000, ▽:170,000, ○:770,000.

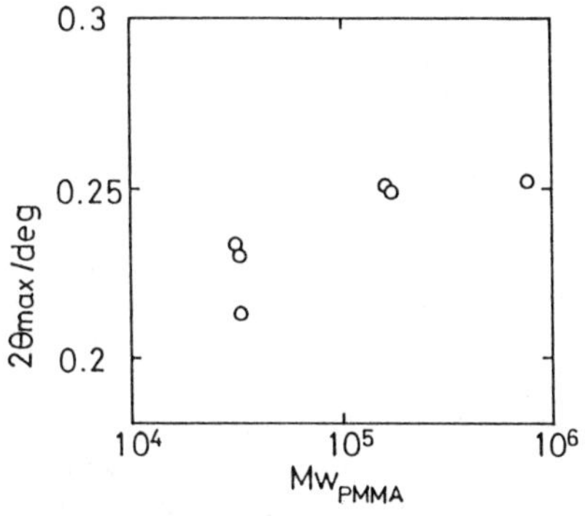

Fig.3 PMMA molecular Weight dependence $2\theta_{max}$ at T=323K.

REFERENCES

1. M. Takahashi, N. Harasawa, H. Yoshida, Kobunshi Ronbunshu, <u>47</u>, 455 (1990)
2. M. Takahashi, N. Harasawa, H. Yoshida, Dynamics and Patterns in Complex Fluids (Springer, Berlin Heidelberg New york, 1990), P. 123.

ANOMALOUS PHASE SEPARATION AND PATTERN FORMATION IN A POLYMER/WATER MIXTURE WITH A DOUBLE-WELL-SHAPED PHASE DIAGRAM

Hajime Tanaka

Department of Applied Physics and Applied Mechanics, Institute of Institute of Industrial Science, University of Tokyo, Minato-ku, Tokyo 106

ABSTRACT:We demonstrate that under deep levels of quench condition a double-well-shaped phase diagram leads to unusual pattern formations in a polymer/water mixture. In a low polymer concentration, we find an anomalous phase composed of moving droplets without coarsening. It may be the first example of *a phase stabilized dynamically and not thermodynamically*.

In the unstable region where SD occurs, decomposition and coarsening generally proceeds faster with an increase in the quench depth since the thermodynamic driving force for phase separation becomes stronger. So far there has been no exception to the above quench depth dependence of phase-separation speed. Here we present the first exception to the above general law of phase separation[1]. We have found a double-well-shaped phase diagram (DWSPD) is responsible for a dynamically stabilized phase with moving droplets. We have also found interesting, unusual phase-separated morphologies with network-like and sponge-like structures in the system.

Samples used were mixtures of poly(vinyl methyl ether)(PVME) and water. The phase diagram of the system is shown in Fig.1. The striking characteristic of the phase diagram is in its double-well shape. We define T_t as the top temperature on the binodal curve. In the region between T_t and the binodal line, usual phase separation occurs with coarsening. At low polymer concentration ϕ_p above T_t, on the other hand, an anomalous phase separation occurs as shown in Fig.2. It is anomalous since after the formation of small droplets there is no coarsening at all for a long time more than hours although the droplet density is dense enough to cause frequent collisions. Small droplets are formed immediately after the deep levels of quench above T_t and they move around rapidly by Brownian motion without coalescence. The phase diagram tells us that above T_t the system separates almost into pure-water and pure-polymer phase. The new phase seems to be metastable or stable. Here we call this phase as moving droplet phase (MDP).

Next we consider the stabilization mechanism of MDP. For the stabilization of particle, some repulsive interaction is necessary to overcome the London-van der Waals interaction. In the system there is probably no electrostatic repulsion between droplets. The stabilization mechanism may be similar to a so-called steric stabilization of polymer-adsorbed particle caused by the excluded-volume and osmotic-pressure effects. Viscous interaction is another factor which prevents direct contact between droplets.

Because a droplet is composed of mobile molecules, we need further consideration on fusion dynamics. The contact time during collision between droplets is probably too short to cause fusion of droplet, which needs interdiffusion of polymers between droplets in the entangled state. The shrinkage of a droplet by discharging water from it during the temperature jump is probably responsible for the formation of a small, dense droplet with a final equilibrium composition and the resulting rapid Brownian motion.

Furthermore, the evaporation-condensation mechanism does not work: Because of rapid droplet motion, stationary diffusion flow necessary for evaporation-condensation mechanism is not allowed since the spatial distribution of droplet is fluctuating in time much faster than the characteristic speed of diffusion. Therefore, MDP may be a stable or metastable heterogeneous phase.

MDP might be stabilized by the other mechanism: If gelation or solidification occurred only above T_t, the stabilization of MDP could be explained. However, we think at the present stage that there is little possibility of such mechanisms.

With an further increase in ϕ_p above T_t, the phase-separated structure changes from a coexistence of MDP and a phase (NP) with a unique network-like phase-separated structure (see Fig.3) to a sponge-like phase (SP) through NP. This unusual pattern formation is also caused by a DWSPD. In every case, a polymer-rich phase shrinks drastically just after the deep quench and keeps shrinking for a long period of time until the composition reaches a final equilibrium one (almost pure polymer). A coarsening process is mainly governed by the shrinkage of polymer-rich phase. This coarsening mechanism is unusual and different from that in the interconnected structure observed in usual spinodal decomposition near the symmetric composition. The process may be similar to the shrinkage of gel because of strong entanglement. Reflecting the order parameter conservation, the volume of the polymer-rich phase keeps decreasing with time. This violates the conventional assumption that after the formation of a sharp interface the system is in a local equilibrium state.

The boundary composition between MDP and NP probably corresponds to the phase inversion composition between matrix and droplet, while that between NP and SP to the symmetric composition causing the equal amount of the final two phases.

The phenomena observed here should be universal in any polymer solution with a DWSPD and may be observed even in a polymer solution with a usual phase diagram if we can perform a rapid, deep quench. It should be stressed that the large difference in mobility between two component materials is probably essential for the phenomena. The details will be published elsewhere.[2] This work was partly supported by a Grant-in-Aid from the Ministry of Education, Science, and Culture, Japan.

References

[1] H.Tanaka and T.Nishi, Jpn.J.Appl.Phys. 27, L1787 (1988).
[2] H.Tanaka, submitted.

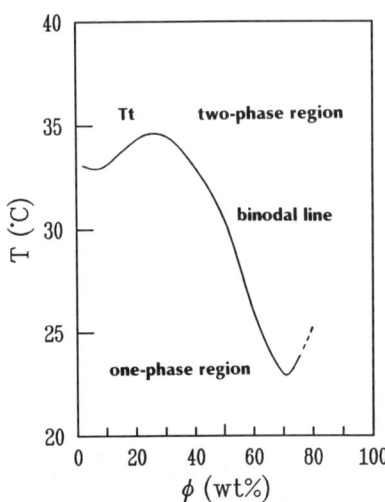

Figure 1: Phase diagram of PVME/water.

Figure 2: MDP observed at 37.3°C for $\phi_p = 5wt\%$. The bar corresponds to $20\mu m$.

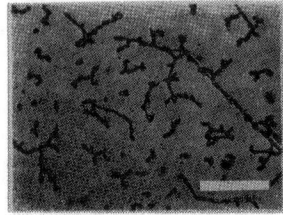

Figure 3: Coexistence of MDP and NP observed at 37.3°C for $\phi_p = 7wt\%$. The bar corresponds to $200\mu m$.

ELEMENTARY PROCESS OF COALESCENCE AMONG DROPLETS IN PHASE-SEPARATING BINARY POLYMER MIXTURES: NEW MECHANISM OF DROPLET COALESCENCE

Hajime Tanaka
Department of Applied Physics and Applied Mechanics,
Institute of Industrial Science, University of Tokyo, Minato-ku, Tokyo 106

ABSTRACT: We have found a long-range, soft collision among droplets accompanying a large deformation of droplet shape in the late-stage phase separation of a polymer mixture sandwiched by two glass plates. There are two possibilities which are responsible for the phenomenon: (1) a preferential wetting of the minority phase to glass and the resulting attractive interaction or (2) diffusion interaction through a weak diffusion field around a droplet.

In the late stage of spinodal decomposition, the morphology of phase-separated structure can be grouped into two types: interconnected structure (percolating cluster)[1] and droplet structure[2]. Here we focus our attention on the coarsening process of droplet structure. The possible coarsening mechanisms for droplet structure are (1)evaporation-condensation mechanism and (2)coalescence or collision between droplets. Binder and Stauffer proposed Smoluchowski-type kinetics for droplet coarsening. In this model a droplet is regarded as a free Brownian particle moving purely diffusively and a droplet grows by collision induced only by thermal noise acting on droplets. In any study, it has been concluded so far that the deformation of droplet or the long-range interaction is negligible and the droplet shape can always be treated as a sphere.

Here we demonstrate a direct evidence for the long-range, soft collision between droplets and the resulting deformation of droplet prior to a hard collision. We will also discuss the possible physical origins for the phenomenon.

A sample used was a mixture of polystyrene (PS) and poly(vinyl methyl ether) (PVME). The composition of the mixture was 50wt% PS. Because of the asymmetry in molecular weight, at this composition droplet pattern was observed since the initial interconnected morphology changed into the droplet morphology within a short time. The mixtures were sandwiched by two glass plates and a typical thickness of the sample was a few μm.

Figure 1 shows the collision process among droplets in the mixture. We can clearly see the deformation of droplet before hard collision, indicating that there is a long-range, attractive interaction between droplets. We measured the temporal change in the minimum distance between the interfaces of the two colliding droplets (Δl_{min}) for several quench depths. The motion of interface is found to be accelerated with a decrease in the distance. The nonlinear dependence of Δl_{min} on Δt (roughly $\Delta l_{min} = c\Delta t^{1/2}$) indicates that there is an attractive force between droplets. The motion should be determined by the competition between the attractive force and the interfacial tension.

It is known that the shape deformation of droplet is observed during the Ostwald ripening. So far this has been only an example for the experimental observation of the deformation of droplet. However, this is not the case since in our case the two interacting droplets deform to shorten the minimum distance and finally make direct, hard collision. Even though there is no difference in the curvature between two neighboring droplets, there is a long-range, attractive interaction between the droplets in our case. Furthermore, almost no droplets disappear by evaporation during the phase separation process. These facts also support the above conclusion.

There are other mechanisms which may account for the phenomenon: (1) the attractive force such as the van der Waals interaction, (2) the diffusion interaction among

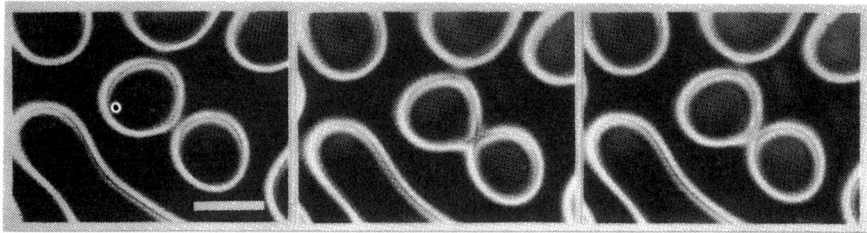

Figure 1: Collision process between droplets in PS/PVME(5/5). Photographs correspond to 1200s, 1230, and 1250s after the temperature jump to 175°C, respectively, from left to right. The bar corresponds to 20μm.

droplets, and (3) the interaction caused by a preferential wetting to the glasses. The influence of the Van der Waals forces on their relative diffusion is $\sim U/k_B T$ and estimated to be negligible for $l\rangle 1 \mu m$. Furthermore, the interfacial energy is much larger than the van der Waals interaction energy for a semimacroscopic separation.

Next we consider the effect of the interaction among droplets through diffusion. In spinodal decomposition there may remain weak diffusion flux from matrix to droplets even after the formation of a sharp interface: a local equilibrium has not been attained between droplet and matrix except for the very late stage. The diffusion flux is dependent on the size of the surrounding matrix and has a local character. This may cause the attractive interaction among the neighboring droplets and move the neighboring interfaces to shorten their distance. In this mechanism, the attractive interaction is predicted to become weaker with the phase separation time, reflecting the decrease in diffusion flux with approaching the final equilibrium. This is consistent with the experimental results on the dependence of c on t_{sp}.[3]

If an equilibrium is established between matrix and droplet, the above mechanism no longer works. In this final regime, the evaporation-condensation mechanism driven by surface tension should become important even in spinodal decomposition in the case of a viscous polymer mixture.

In this model, the coarsening of droplet structure in the late stage is probably governed firstly by interaction through diffusion, or by thermal noise (or free diffusion), and finally by surface tension. Reflecting this crossover of the mechanisms, the coarsening speed should slow down with time.

Finally we discuss the effect of wetting phenomena. In a 2D sample, droplet coalescence does not cause any energy gain associated with wetting if there is no preferential wetting. However, if there is a preferential wetting and a wetting layer, the bridging of droplets by the wetting layer may cause the interaction. In this case, the behavior should be strongly affected by whether the droplet (or the minority phase) prefers glass or not. This point is now under investigation.

Since the discussion here is not conclusive as above, we need further careful experimental and theoretical studies to clarify the mechanism and whether the interaction between droplets is originated from true 3D phenomena or from the interfacial effect.

This work was partly supported by a Grant-in-Aid from the Ministry of Education, Science, and Culture, Japan.

References
[1] H.Tanaka and T.Nishi, Phys.Rev.Lett. 59, 692 (1987).
[2] H.Tanaka et al., Phys.Rev.Lett. 65, 3136 (1990).
[3] H.Tanaka, submitted.

B. THEORY

STRESS RELAXATION IN HETEROGENEOUS POLYMERS *

T. A. Witten
James Franck Institute, University of Chicago, Chicago IL USA

ABSTRACT

When heterogeneous polymers such as diblock copolymers form a microdomain phase, an imposed strain gives rise to stress from two sources, and several mechanisms of stress relaxation. The release of stress by disentanglement is strongly influenced by the effective confinement of the junction points to the domain boundaries and by the stretching of the chains. Using accepted notions of entangled chain kinetics, it is argued that the relaxation time for sliding stress is exponential in the chainlength to the 7/9 power. A method for calculating the frequency-dependent dynamic modulus is sketched. Despite the slow relaxation implied by these mechanisms, it appears possible to create domains of high energy.

INTRODUCTION

A major puzzle of the last decade has been to understand how a mass of entangled random-walk polymers can possibly flow. The resolution of this puzzle—confirming the reptation model [1]—is an important theme of this conference. The puzzle of polymer flow is intensified when the polymers are phase-separated diblocks as in Figure 1. The polymer molecules each consist of two immiscible parts, which phase separate to form the domains shown as black and white regions. Each chain must bridge between a white domain to an adjacent black one. Since the phase separation is strong, the interface is narrow, and the black-white junction point of each chain is confined to this narrow interface. Given this confinement, the reptational motion of ordinary polymers is out of the question. The end-confined blocks cannot slither along their length for any appreciable distance.

The puzzle of how polymer domains flow is the more urgent because of the potential of domain-phase materials like that of Figure 1. Evidently these diblocks have spontaneously organized themselves into a structure of great regularity and precision. The source of energy for this regularity is easy to identify: the interfacial energy of the visible black-white interface is of the order of hundreds of thermal energies kT per repeat unit. This energy is balanced by chain elastic energy: the polymer chains in these domains are stretched. Since the energies at stake are so large, it is natural that the material will flow and relax to form the state of optimal balance between these two energies. And the

---

* Supported in part by the National Science Foundation Materials Research Laboratory under contract DMR-88-19860

246 Stress Relaxation in Heterogeneous Polymers

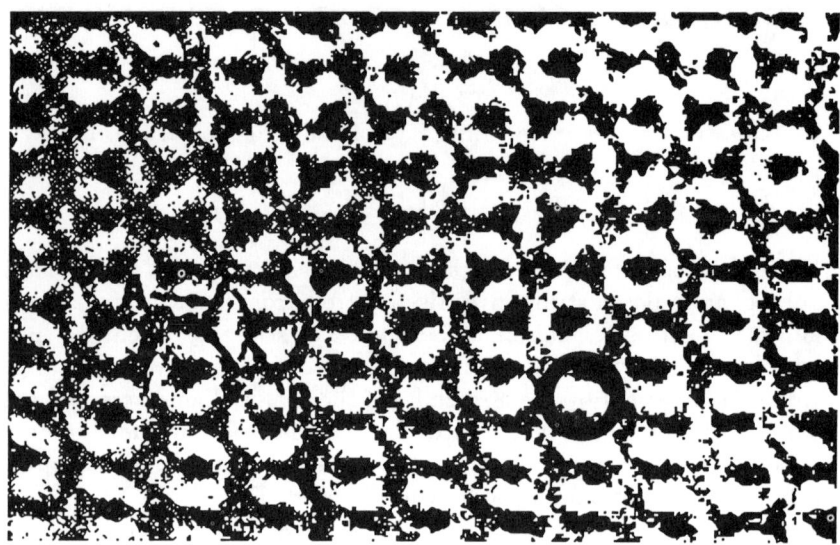

Figure 1 Transmission electron micrograph of the double-diamond or tetrapod domain structure, reproduced from Ref. 2 with the kind permission of T. Hashimoto. One block species is shown in white, the other in black. The repeat distance is several hundred Angstroms. A heavy circle encloses a region where two opposing grafted-chain layers are in contact.

energy per domain can in principle be increased indefinitely by increasing the molecular weight of the polymers [3,4]. This large driving force offers potential for making precisely controlled structures of indefinite variety, by controlling the architecture of the chains.

But the attainment of these structures requires flow: the chains must be able to relax to achieve the optimum shape. As noted above, the very formation of these domains tends to inhibit the normal mechanism of polymer flow. Because of these factors, the domain structures explored to date experimentally have used chains of limited molecular weight and limited incompatibility [5]. As molecular weight increases the materials become increasingly difficult to "process". They are too viscous to flow, so that an imposed stress does not relax over a practical time scale.

The question naturally arises, is the desired large domain energy *incompatible* with relaxation of stress? If so, the potential for higher-energy domains is severely limited. This question is the motivation for the discussion below. After exploring several mechanisms of stress relaxation, I conclude that strongly entangled blocks are inherently unprocessible. They require a relaxation time which grows exponentially with molecular weight. However, this slow relaxation may be avoided without compromising the domain energy by reducing the number of entanglements in a suitable way. The nature of the disentanglement process in stretched diblocks leads to a strong anisotropy in the relaxation rate. First I outline the energy and length scales important for stress and entanglement in a stretched-polymer domain. I survey the types of stress relaxation to be expected. I then describe a scenario for disentanglement and the accompanying stress relaxation. Finally I mention ways that entanglement may be diminished while preserving the interesting properties of the domains.

ENERGIES AND LENGTHS

Let us consider the circled section of Figure 1. The equilibrium state of the chains in such a region controls the stress in the material and its relaxation. For simplicity, we shall imagine that the opposing black-white interfaces are flat and are a distance $2h$ apart. The block chains ending at these two interfaces must fill the intervening region uniformly, so that each chain must extend over distances of order h. In order to achieve a large domain energy this h must grow much larger than the unperturbed size R of the chains. Our interest here is in the asymptotic behavior when h/R is *indefinitely* large (even though in present-day materials it is less than a factor of 2).

Generally these domains are made with mobile, flexible block chains which are long enough to be strongly entangled. If a macroscopic liquid of these chains (with the immiscible blocks removed) is subjected to a sudden strain, the entanglement constraints force the chains to distort from their random-walk configurations. From the resulting elastic energy stored, one may infer a volume per entanglement [6]. For the materials of Figure 1 this volume is of order $R_e^3 \simeq (10$ Angstroms$)^3$ and is independent of the chain length. Thus the block chains of Figure 1 have tens of entanglements apiece. As the chainlength grows, the distance between entanglements becomes arbitrarily smaller than the either R or h.

The important energy scales in this layer may readily be expressed in terms of the height h and unperturbed chain size R. Stretching a chain to lengths of order h reduces its random-walk entropy and requires work of order $(h/R)^2$. Here and below we shall express all energies in units of the thermal energy $k_B T$. This energy amounts to a unit of energy per segment of size $R_p \simeq R^2/h$. Segments smaller than this "Pincus blob" are essentially unperturbed by the stretching. In our system $R_p >> R_e$ [7]; thus at the scale of the entanglements, the chains have the same Gaussian behavior as they do in a melt. Thus we expect the density and nature of the entanglements to be altered but little by the stretching.

The chains in our circle come from two opposing interfaces; it is important to understand how much these two sets of chains entangle with one another. That is, how much do the two sets of chains interpenetrate? We shall find that the two opposing layers interpenetrate over a distance $\xi << h$; thus we must determine how ξ scales with h/R. To this end we consider a single layer of block chains in contact with a simple melt of long chains of the same species. Each block chain in this layer is held in its stretched configuration by a force arising from the other blocks. One can see that such a force must be present if one imagines inserting an additional short segment of k monomers into the layer at some height z. To make room for this segment, the nearby chains are obliged to stretch a bit further. To pay for this stretch energy one must do a work, proportional to k. Thus the stretched layer has a chemical potential field $\mu(z)$; it is this field that stretches the block chains out. This same field pushes out any unattached chains and limits penetration.

One may readily estimate the magnitude of the field μ by considering an initially unstretched block chain of size R and length N. If this chain is to stretch to a height z, it must store an elastic energy of order $(z/R)^2$, as noted above. If the chain is to be in equilibrium, this energy cost must be paid with work done by the chemical potential $\mu(z)$. This work is of order $(\mu(z) - \mu(R))N$. Thus $(\mu(z)-\mu(R))N \simeq (z/R)^2$: the chemical potential profile is parabolic [8,9,10]. Segments of unattached chain can penetrate this layer to a degree, provided that the work done against the chemical potential is of order unity (k_BT) or less. Such a segment would not be greatly distorted from its natural random-walk shape; otherwise the distortion cost alone would prohibit penetration. Thus the segment's size ξ is related to its displaced volume k by $k/N = (\xi/R)^2$. The chemical-potential cost is of order $k\mu(h-\xi) \simeq \xi h/R^2 [k/N]$. Since this pressure cost is of order unity, we find $1 \simeq \xi h/R^2 [\xi/R]^2$, so that $\xi \simeq R[h/R]^{-1/3}$. (This result applies equally well to the melt state considered here, or in the presence of any kind of solvent. [11].) Interpenetration between two opposing diblock layers like those in Figure 1 is of the same order of thickness ξ [11]. As the chainlength N grows, the interpenetration zone grows much smaller than the height h or even the unperturbed size R. Still, it grows larger in absolute size, and it thus becomes much larger than the entanglement spacing R_e.

STRESS RELAXATION MECHANISMS

We now imagine that the domain structure of Figure 1 is subjected to a step strain, thus distorting our circled region. The strain energy stored there cannot fully relax without several well-defined dynamical processes, each with its own characteristic modulus and timescale. Immediately after the initial strain, the region stores elastic energy like a rubber or homopolymer melt. The associated "plateau" modulus is of order R_e^{-3}. This stress may relax as the polymers disentangle.

In general part of the imposed strain is a "sliding" strain, tending to slide the opposing interfaces past each other. To support this kind of stress chains from one interface must exert a force on chains from the other. The important entanglements are thus those between interpenetrating chains from opposite sides. Since chains from opposite sides interpenetrate only weakly, we expect the two sides to disengage relatively rapidly, thus relaxing the sliding stress.

In general part of the initial strain is not in the sliding direction, but tends e.g. to compress the two layers. Such a strain squeezes the polymers out laterally so as to preserve their total volume. The entangled chains are stretched out laterally, owing to entanglements with their neighbors on either side. The associated modulus is of the same order as that for sliding. The resulting stress cannot relax until these chains disentangle. These entanglements are typically between adjacent chains in the same layer; they do not require the interpenetration of opposing layers. For this stress to relax the bulk of all the entanglements in the layer must relax. We expect this relaxation to be slower than for sliding stress.

Beyond this timescale the chains are free to flow along the layer like a viscous liquid. In general they will do so, in order to regain the optimum balance between interfacial energy and stretch energy. (Just as with a sheared bubble raft, the initial affine deformation leaves the interfacial energy out of balance.) The associated modulus is of the order of the interfacial energy density or the chain stretch energy density. This is much less than the density of entanglements. Relaxations of the interfacial stress were recently considered by Kawasaki and Onuki [12] and by Obukhov and Rubinstein [13]. They note the resemblence between this relaxation and that of a disordered smectic liquid crystal or spinodally decomposing liquid.

This flow can bring the domains to a state of the minimum free energy consistent with their topology. But as with a sheared bubble raft, the final state is still distorted. The associated modulus is again of the order of the density of interfacial energy. This stress cannot relax except by processes we've excluded up to now. One of these is "permeation": a black block may pass across a white domain, thus exchanging material with a disconnected region. A second is the motion of defects in the periodic structure such as dislocation lines. If such defects can migrate across the sample, even the average repeat distance can relax towards equilibrium. If the domain energy is high, all such relaxations are very strongly suppressed. The weak interfacial stress is essentially permanent.

DISENTANGLEMENT

The most distinctive of the relaxation processes sketched above is the first one: the relaxation of sliding stress. For this we must understand the disengagement of weakly interpenetrating chains from opposing layers [14]. My discussion here is a modest extension [15] of Ref. 11. We have seen that our opposing layers are rheologically heterogeneous: stress relaxes differently in different places. We must ask how the stress is transmitted when part of the stress has relaxed—that is, how is the force producing a local chain tension transmitted to the interfaces. At the most local level this force is transmitted along the chain itself. But each entanglement the chain encounters takes up part of the force and distributes it among adjacent chains. The result is that the force at distances of order R_e from the initial point is spread more-or-less uniformly over all the chains at that distance. Thus at this scale the stress in the liquid may be treated as a smoothly varying quantity. We may thus divide up our liquid into little imaginary cubes several times R_e in size, each having a well-defined dynamic modulus $G(z,\omega)$ at height z and frequency ω. We'll refer to these cubes as "stress blobs".

To determine the local modulus $G(z,\omega)$, we may imagine subjecting a single stress blob to a step-strain experiment. The stress relaxes as its initial entanglements disappear. While the stress is defined locally, the disentanglement motion is not! To release an entanglement between two chains requires one end or the other to pass through the entanglement point. These ends are typically many stress blobs away. The chains in a given stress blob have trailing segments to their free ends of various lengths. By knowing these lengths, we may infer how quickly a given stress blob relaxes.

As noted above, these trailing segments cannot pass through the blob by reptation. However, the entire chain may with some probability contract along its length by sending out little aneurisms between its entanglement contraints. This retraction is the mechanism believed to control relaxation of star polymer melts [16,6].

But retraction over a chainlength k reduces the chain entropy, and thus costs a free energy. Retraction over the whole length N means that for every N_e segment there must be a retracing segment. The loss of entropy is of order N/N_e. The free energy cost for general k is of order $N(k^2/N^2)/N_e$ [6]. Blobs at a height z have trailing segments of length $k \simeq N(z-h)/h$. This means a free energy cost $F(z) \simeq N/N_e[(z-h)/h]^2 \simeq (R/R_e)^2[(z-h)/h]^2$. This $F(z)$ amounts to an activation barrier to the disentanglement. One may express the corresponding time $\tau(z)$ in the form (mobility factor)$\times \exp(F(z))$. For $z-h \simeq h$ the time is of order (mobility factor)$\exp(N/N_e)$—the same as in a melt of star polymers.

Our formula suggests that $F(z)$ goes to zero as $z \to h$. This neglects the interpenetration of the layers. In fact, blobs in the interpenetration region have typical trailing segments of size ξ and length $k \simeq N(\xi/R)^2$. Their activation energy $F(h)$ is thus of order $N/N_e(k/N)^2 = (R/R_e)^2(R/h)^{4/3}$. Evidently this is much smaller than the energy barrier for general z. For diblocks [7] this $F(h)$ scales as $N^{7/9}$ [11]. To find the relaxation time $\tau(h)$, we must again include a mobility factor; this varies as a power of N [11].

Given the relaxation behavior of these stress blobs, one may readily find how a given overall stress relaxes. We must treat the layers as a composite material of many sublayers, each with a dynamic modulus $G(z,\omega)$. For sliding shear, the stress is transmitted in series across the layer, and the weakest element dominates. The effective *inverse* modulus $G^{-1}(\omega)$ is given by the average of the inverse moduli of the sub layers $\langle G^{-1}(z,\omega) \rangle_z$. The averaging of stronger elements with the weakest one lengthens the terminal time by a power of N, but the exponential dependence is controlled by the smallest activation energy $F(h)$. In compressional stress, the system is qualitatively like a stack of rubber sheets which are clamped together and stretched. Now the force is the sum of the layer forces, and the moduli $G(z)$ themselves average to obtain the effective modulus. The strongest moduli dominate and the relaxation time is controlled by the blobs closest to the interface. As anticipated, the relaxation is strongly anisotropic.

CONCLUSION

It appears from this discussion that disentanglement in these copolymer layers is subtle and distinctive. On the other hand, it is extremely slow whenever entanglement is substantial. This places severe limits on the use of these materials. Happily, one can sidestep these difficulties. One way is to add spacing-filling side groups to the polymer chains. This has the effect of increasing R_e without increasing R. Certainly this decreases the rubber modulus of the material. But it does not reduce the domain energy. Indeed, the energy per repeat unit increases when the side groups are added. It appears entirely possible to make domain phases much more energetic than have been made to date. And with these should come increased sensitivity to polymer architecture and new distinctive behavior.

REFERENCES

1. P. G. de Gennes, *J. Chem. Phys.* **55** 572 (1971).

2. H. Hasegawa, H. Tanaka, K. Yamazaki, and T. Hashimoto *Macromolecules*, **20** 1651 (1987).

3. D. J. Meier, *J. Polym. Sci, Part C*, **26** 81 (1969); D. J. Meier, *Polym. Prepr., Amer. Chem. Soc., Div. Polym. Chem.* **11** 400 (1970).

4. T. Hashimoto, M. Shibayama, H. Kawai, *Macromolecules*, **13** 1237 (1980); T. Hashimoto, M. Shibayama, H. Kawai, *Macromolecules*, **16** 1093 (1983); T. Hashimoto, M. Fujimura, H. Kawai; *Macromolecules*, **13** 1660 (1980).

5. Matzner, M.; Noshay, A.; and McGrath, J.E.; *Trans. Soc. Rheol.* **21** 273 (1977).

6. M. Doi and S. F. Edwards, *The Theory of Polymer Dynamics* (Clarendon Press, Oxford, 1986).

7. To infer scaling with chainlength N, we note that in diblock domains $h \sim N^{2/3} \sim R^{4/3}$ [3].

8. A. N. Semenov, *Zh. Exp. Theor. Phys.* **88** 1242 (1985), translated in *Sov. Phys. JETP* **61** 733 (1985).

9. S. T. Milner, T. A. Witten and M. E. Cates, *Macromolecules*, **21** 2610 (1988).

10. This account of the parabolic profile arose through discussion with P. Pincus and J- L Barrat.

11. T. A. Witten, L. Leibler and P. A. Pincus, *Macromolecules* **23** 824 (1990).

12. K. Kawasaki, and Akira Onuki *Phys. Rev. A* **42** 3664 (1990)

13. S. R. Obukhov and M. Rubinstein, presented at "Tethered Chains I" meeting, Brainerd MN, July 1991.

14. S. Obukhov and M. Rubinstein have recently discussed a theory of the disentanglement dynamics emphasizing aspects different from those treated here [13].

15. This picture resulted from discussions with P. Pincus and Premala Chandra.

16. Carella, J. M.; Gotro, J. T.; and Graessley, W. W.; *Macromolecules* **19** 659 (1986).

HYDRODYNAMIC, SOFT, AND RELAXATIONAL MODES IN CHIRAL LIQUID CRYSTAL SIDE-CHAIN POLYMERS

Harald Pleiner and Helmut R. Brand
FB 7 Physik, Universität Essen, D 4300 Essen 1, Germany

Abstract

We use the generalized hydrodynamic method in order to describe the macroscopic dynamics of polymeric side-chain liquid crystals. The hydrodynamic method uses only general symmetry arguments and irreversible thermodynamics to establish phenomenological dynamic equations. In the present case this method has to be generalized by taking into account not only the true hydrodynamic variables, but also some slow relaxational variables, which are slow enough to be relevant for the usual hydrodynamic frequency range.

Introduction

The true hydrodynamic variables are either due to conservation laws or due to spontaneously broken continuous symmetries.[1] They are sufficient to describe completely the dynamics of a system in the low frequency ($\omega \to 0$) and long wavelength limit ($k \to 0$). However, there are systems or situations, where this truly hydrodynamic regime is reached after a very long time, only. It is then necessary to include additional slow, but non-hydrodynamic, variables that relax on a time scale, which is slow enough to be relevant for the macroscopic dynamics. The standard example for the importance of such variables are systems near a second (or weakly first) order phase transition, where soft modes occur.[2] In addition there are less generic situations, where such slow relaxational modes are important.[3] In (isotropic) polymer melts and solutions the well-known (linear) viscoelastic behaviour is macroscopically described by a relaxing strain field,[4] which reflects the elastic properties on times shorter, and reveals the viscous liquid properties on times larger, than its relaxation time.

In nematic liquid crystalline (side-chain) polymers (PN) there are additional hydrodynamic as well as slow non-hydrodynamic variables present. The former are the orientational fluctuations of the director, which are the symmetry variables due to the broken rotational symmetry, characteristic of any nematic system. The latter include the nematic order parameter,[5] which denotes the strength or degree of nematic ordering of the side-chains. This is believed to be a slow variable in polymeric systems [6] – although in low molecular weight systems it is usually neglected (except near the isotropic phase transition or near some defects) – since any change in the nematic order (due to local reorientation of the side-chains) may affect different parts of the backbone(s) and thus the viscoelasticity of PN. Furthermore, rotations of the director relative to the local backbone orientation are possible macroscopic excitations, which have a profound influence on standard hydrodynamic experiments.[7] Not only become the nematic features "renormalized" (e.g. the viscosities rather large) and the polymeric features anisotropic (e.g. longitudinal and high frequency transverse sound), but additional macroscopic features occur (e.g. additional, anisotropic dispersion steps in the sound modes, large apparent director relaxation time, elastic response to oscillatory flow alignment) due to the coupling of the various hydrodynamic and non-hydrodynamic variables.

Throughout this paper we will restrict ourselves to side-chain systems, where it is possible to define distinct sets of variables for the liquid crystalline and polymeric features, respectively, which are coupled by the macroscopic dynamic equations only.

Chiral Systems

If a liquid crystal is chiralized, instead of phases with a homogeneous alignment of the director (e.g. nematic or smectic A, C etc.) one obtains those with a helical arrangement of the director (e.g. cholesteric, smectic C* etc.). This introduces a length scale (helical pitch $\equiv 2\pi/q_0$) and constitutes a broken translational symmetry (in the direction of the helical axis $\hat{p}$). The true hydrodynamic variable (symmetry variable) is the displacement δR along the helix axis. In the small scale picture (smaller than the pitch) the translation of the helix is described by a rotation $\delta\phi$ of the director about the helical axis, with $q_0 \delta R = \delta\phi$, while rotations of the helical axis, $\delta\hat{p}$, are related to transverse gradients of δR, i.e. $\delta\hat{p}_i = -\nabla_i^{tr} R + O(2)$, where $\hat{p}_i \nabla_i^{tr} \equiv 0$.

Contrary to the case of a real equilibrium network in solids and permanently crosslinked elastomers, in polymers and polymer liquid crystals the elasticity is due to dynamical fluctuations and entanglements of the long backbones on very small scales. This short time elasticity is described[4,7] by a (relaxing) generalized strain tensor ϵ_{ij} ($= \epsilon_{ji}$) and an antisymmetric tensor Ω_{ij} ($= -\Omega_{ji}$) characterizing rotations of the (hypothetical) polymer network. As a consequence, unlike to the case of solids and permanently crosslinked elastomers, where there exists a displacement vector $\mathbf{u}$ (denoting translations of the molecules, lattice sites, or crosslink-points away from a well defined equilibrium position), with $2\epsilon_{ij} = \nabla_j u_i + \nabla_i u_j + O(2)$ and $2\Omega_{ij} = \nabla_j u_i - \nabla_i u_j$, in polymeric systems ϵ_{ij} and Ω_{ij} are not derivable from a displacement field[8] (which does not exist anyhow). Thus, there are nine additional independent macroscopic variables, only three of which correspond to the symmetry variable $\mathbf{u}$ of solids and elastomers.

In PN there are two relative rotations possible, i.e. rotations of the director w.r.t. to the (hypothetical) network, while in helical systems there are three: one rotation about the helical axis, which is nothing else but a translation of the helix relative to the (hypothetical) network, and two rotations of the helical axis giving rise to the following harmonic contributions to the free energy density

$$2f = D_3(\delta\phi - \hat{p}_i\epsilon_{ijk}\Omega_{jk})^2 + D_1(\delta p_i - \hat{p}_j\Omega_{ij})^2 + B_1(\hat{p}_j\nabla_j R)^2 + \\ C_{ijkl}\epsilon_{ij}\epsilon_{kl} + 2D_4(\delta p_i - \hat{p}_j\Omega_{ij})\hat{p}_k\epsilon_{ik} + 2B_2\hat{p}_i\hat{p}_j\hat{p}_k\epsilon_{ij}\nabla_k R \tag{1}$$

For elastomers the first term was already discussed in ref. 9. The additional contributions involve helix compression (B_1) (as in low molecular weight materials), polymeric elasticity (C_{ijkl}) and crosscoupling terms (D_4 and B_2). Of course, all the effects involving ϵ_{ij} and Ω_{ij} cannot be observed in static experiments (since they are relaxing to zero in the long time limit), but they are manifest in high frequency measurements. The static effects described above also have their (dissipative) dynamical counterparts involving the appropriate currents (fluxes) and thermodynamic conjugate variables (forces). The full dynamical equations will be given elsewhere.

A very important feature of liquid crystals are the various linear field effects, which they support, i.e. the alignment or structure of the phase changes linearly with the applied (electric) field. In PN only the usual flexoelectric effect occurs, where inhomogeneous distortions (e.g. bend) of the director field (a rotational type of variable) are caused by an electric field (and vice versa). Analogous effects (bend of the helical axis) exist also in helical systems.[10] In addition, inhomogeneous distortions of polymeric strains and rotations are coupled to linear electrical fields. A second class of well-known linear electric effects are the piezoelectric effects in systems without an inversion center, where first order gradients of a translational type of variable are coupled to the field. Their presence was first discussed for low molecular weight systems in ref. 11 and for elastomers in ref. 9, but they also exist in helical polymeric systems. In the latter systems there are, in addition, linear electric effects connected with relative rotations and we find

$$f_E = \zeta^p \hat{p}_i \hat{p}_j E_i \nabla_j \delta\phi + \zeta^p_{ijk} q_0 E_k \epsilon_{ij} + \zeta^c q_0 E_i (\delta p_i - \hat{p}_j \Omega_{ij}) + \zeta^r q_0 \hat{p}_l E_l (\delta\phi - \hat{p}_i \epsilon_{ijk} \Omega_{jk})$$
(2)

For elastomers the two latter effects (rotations of the helical axis and of the helix about its axis, respectively, relative to the (hypothetical) network and induced by an electrical field) were coined (relative) electroclinic effect in ref. 10 and rotatoelectrical effect in ref. 9, respectively. Again, for polymeric systems the consequences of the latter three terms in eq.(2) can only be discussed within the full dynamical equations.

Smectic Systems

In chiral smectic systems there are generally two additional features compared to cholesteric liquid crystals, which have to be taken into account. First, there are layers of molecular thickness completely independent from the helix. These layers constitute a broken translational symmetry similar to the crystalline order in solids, but only in one direction. Usually this direction and the helical axis are parallel (the exception being the twisted A* phase[12], which we will not consider in the following). Layer thickness and pitch are incommensurate. Second, a helicoidal polarization accompanies the helix, because of the tilt of the director away from the helical axis. This applies to the smectic C* and the I* and F* phases, where the latter have an additional bond orientational order[13] within the layers (for the hydrodynamic consequences of the bond orientational order cf. ref. 14 and 15, some electrical effects are discussed in 16). There is no (helicoidal) polarization in the untilted C^*_M phase recently found[17], which is therefore[18,19] somehow in-between a cholesteric (especially in the electric properties) and a smectic C* phase (in the dynamical properties). In the following we will concentrate on the smectic C* case.

The helicoidal polarization does not introduce any additional dynamical mode, since (at least for hydrodynamic frequencies) it is rigidly coupled to the helix formed by the director and thus its dynamics is described by the (old) variables $\delta\phi$ and δp_i. The polarization is important for applications, when in a sufficiently strong field or, for thin films, due to surface effects the helix is unwound and a net polarization remains (in contrast to the cholesteric or untilted case, where there is none). For the dynamics the existence of the smectic layers is important, since an additional displacement variable, R_A, due to the broken translational symmetry is introduced. It describes translations of the layers along the layer axis. In contrast to the pseudo-translation of the helix (which we will call R_C now), R_A describes proper translations not connected to rotations (and is present even in non-chiral phases like the smectic A phase). Here we are interested how the polymeric elasticity couples to R_A (for the complete dynamics of the non-polymeric variables cf. ref. 11). It is obvious that a translation of the layers (defined by the position of the side-chains) affects the polymeric elasticity of the backbone, since the side-chains are physically attached to it and the backbone lies somewhere in between the layers. Thus, a relative translation between backbone and layers costs energy (similar to a relative rotation between backbone and helix, which also costs energy, cf. eq. (1)). The full dynamical equations will be given elsewhere.

References

1. D. Forster, *Hydrodynamic Fluctuations, Broken Symmetry, and Correlation Functions*, (Benjamin, Reading, Mass., 1975).
2. S.-K. Ma, *Modern Theory of Critical Phenomena*, (Benjamin, Reading, Mass. 1976).
3. H. Pleiner, in *Incommensurate Crystals, Liquid Crystals, and Quasicrystals*, eds. J.F.Scott and N.A.Clark, (Plenum New York, 1987), p.241.
4. H.R. Brand, H. Pleiner, and W. Renz, *J. Phys. France* **51**, 1065 (1990).
5. P.G. de Gennes, *The Physics of Liquid Crystals*, (Clarendon Press, Oxford, 3rd Edition, 1982).
6. H. Pleiner and H.R. Brand, *Mol.Cryst.Liq.Cryst.*, **199**, 407 (1991).
7. H. Pleiner and H.R. Brand, to be published.
8. This statement corrects Ref. 4 appropriately (compare also ref.6).
9. H.R. Brand, *Makr.Chem.Rap.Commun.* **10**, 441 (1989).
10. H.R. Brand and H. Pleiner, *Makr.Chem.Rap.Commun.* **11**, 607 (1990).
11. H.R. Brand and H. Pleiner, *J.Phys. Paris*, **45**, 563 (1984).
12. J.W. Goodby, M.A. Waugh, S.M. Stein, E. Chin, R. Pindak, and J.S. Patel, *Nature* **337**, 449 (1989).
13. D.R. Nelson and B.I. Halperin, *Phys. Rev.* **B21**, 5312 (1980).
14. H. Pleiner and H.R. Brand, *Phys. Rev.* **A29**, 911 (1984).
15. H. Pleiner, *Mol.Cryst.Liq.Cryst.* **114**, 103 (1984).
16. H.R. Brand and H. Pleiner, *Phys. Rev.* **A35**, 3122 (1987).
17. H. Leube and H. Finkelmann, *Makromol.Chem.* **192**, 1317 (1991).
18. H.R. Brand and H. Pleiner, *Makr.Chem.Rap.Commun.* **12**, ... (1991).
19. H.R. Brand and H. Pleiner, *J. Phys. II* **1**, ... (1991).

VISCOSITY AND PARTICLE-DIFFUSION OF DILUTE PARTICLE DISPERSIONS IN POLYMERS

W. Sung

Department of Physics,
Pohang Institute of Science and Technology, Pohang, 790-600, Korea

A microscopic theory for viscosity and particle-diffusion of the particle-dispersions in polymers is presented. The theory is based on the scheme of "microscopic boundary layer," which appropriately combines the statistical mechanical description of the microscopic response in particle-polymer interface with the continuum mechanical description of the collective response in the polymer background beyond the interface. While the resulting theories in large particle limit correctly recover the continuum mechanical results for intrinsic viscosity and particle diffusivity, the theories for small particles show significant deviation from the continuum results and reasonable dependence on interface structure.

Introduction

For spherical particles suspended by a low volume fraction ϕ in a polymer liquid, viscosity increase $\Delta\eta$ over the value η of the host polymer is given by Einstein [1],

$$\Delta\eta/\phi\eta = 2.5. \qquad (1)$$

The diffusivity D of the particle is given by the Stokes-Einstein relation [1],

$$D = k_B T/\zeta_H \qquad (2)$$

where $\zeta_H = 6\pi\eta R$ is the friction coefficient of the particle of radius R. The assumption behind these equations is that the particles are large enough to permit treatment of the background polymer as a continuum liquid subject to no slip boundary condition. Therefore, there are no details of the particle-polymer interface microstructure in these expressions. For particles much smaller than the characteristic lengths of polymer structure, significant deviations from these continuum results can be expected.

A simple but sensible enough to include to continuum limit as well as the interface structure, which I introduce here is the model of "microscopic boundary layer." This model was initially conceived by Hynes et al[2] and by Sung and Stell[3] for the problem of a small-particle diffusion in the background of a small-molecule fluid. According to the model, the short-range interaction between the particle and the background (fluid) within the interface is treated in microscopic detail, while the long-range interaction beyond the interface treated collectively using hydrodynamic description. The essence of the model lies in the generalized boundary condition obtained by matching these two descriptions at the microscopic boundary layer.

258 Dilute Particle Dispersions in Polymers

In adapting as described below the model mentioned above to our problems of the particles in a polymer, we shall confine ourselves to the small strain rate and the spherical particles sufficiently dilute so that the interparticle interaction may be neglected.

Microscopic boundary layer and generalized hydrodynamic description

The major frictional force on the particle is due to the elastically effective chains which remains attached to the surface of the particle and remains entangled to the background chains during the deformation and flow. An elastically effective chain exterts a frictional force on the particle

$$\mathbf{f}_s = -K_s \, \delta \mathbf{x}$$
$$= -K_s \tau_E (\mathbf{u} - \mathbf{V}) \quad (3)$$

where the $\delta \mathbf{x}$ represents the average displacement of the entanglement during the disengagement (reptation) time τ_E of the entanglement, namely, the time during which a background chain reptates through the link, $\mathbf{u}$ and $\mathbf{V}$ are the average velocity of the particle and the link (see Fig 1). The spring constant K_s is given by

$$K_s = 3k_B T / <\Delta r_s^2> \quad (4)$$

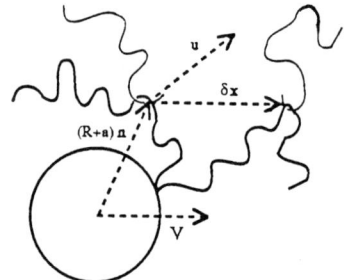

Fig.1 The chain configurations associated with the frictional force

where $\Delta \mathbf{r}_s = \mathbf{r}_s - <\mathbf{r}_s>$ and $<>$ denotes the equilibrium ensemble average. Here $\mathbf{r}_s$ is the distance of an elastically effective chain from the surface to the neighboring entanglement (see Fig 1). The average distance $<\mathbf{r}_s> = an$ naturally defines our microscopic boundary layer with width a, beyond which in the bulk the polymers are to be treated as a continuum.

Supposing that there are α elastically effective surface chains per unit surface area, we can write the frictional force per area as

$$\mathbf{F} = -\alpha K_s \tau_E (\mathbf{u} - \mathbf{V}). \quad (5)$$

In the view point of the background polymer beyond the boundary layer, the force would be regarded as the hydrodynamic stress. We can set up the relation

$$\alpha K_s \tau_E (\mathbf{u} - \mathbf{V}) = -\sigma \cdot \mathbf{n} \quad (6)$$

where σ is the stress tensor of the polymer continuum with the component

$$\sigma_{ij} = P \, \delta_{ij} - \eta(\dot{\epsilon}_{ij} + \dot{\epsilon}_{ji}). \quad (7)$$

(P : the static pressure, η, the shear viscosity $\dot{\epsilon}_{ij} = \nabla_j u_i$: the shear rate)

Provided that the velocity $\mathbf{u}$ of polymer liquid that appears in the right hand side of Eq (6) is identical to that in the left hand side, the Eq (6) is regarded as the generalized boundary condition to be satisfied at the outer-boundary ($r = R+a$) of the boundary layer. It will be clearer later that for large particle the usual hydrodynamic results of no-slip boundary condition is recovered while for a small particle significant deviations will be found.

The next task is to solve the equation (steady-state, linear)

$$\nabla \cdot \sigma = 0 \qquad (8)$$

supplemented by the condition for incompressible flow

$$\nabla \cdot \mathbf{u} = 0 \qquad (9)$$

and the boundary condition Eq (6). The solutions for $\mathbf{u(r)}, \dot{\epsilon}(\mathbf{r})$ are necessary to calculate the suspension stress and particle-diffusion as below.

Stress and viscosity of the suspension

The stress tensor Σ of the suspensions consisting of noninteracting particles in a media is given by

$$\Sigma = \Sigma^\circ + \Delta\Sigma_I + \Delta\Sigma_{II} + \Delta\Sigma_{III} \qquad (10)$$

Here $\Sigma^\circ = P\mathbf{1} - 2\eta\dot{\mathbf{E}}$ is the stress of host polymer in the absence of the particle. Here P and η are the static pressure and viscosity, $\dot{\mathbf{E}}$ is the shear rate tensor of external flow given as $2E_{ij} = \nabla_j U_i + \nabla_i U_j$

The $\Delta\Sigma_I, \Delta\Sigma_{II}, \Delta\Sigma_{III}$, the additional contributions to stress from the particles (I), polymer phases (II) and the interphase (III) has been calculated using the scheme of Batchelor[4,5] appropriately modified to include the microscopic boundary layer and the interface polymer contributions. They all involve certain hydrodynamic responses on the outer surface of the boundary layer subject to the generalized boundary condition. As the result of calculation, we obtain $\Sigma = P\mathbf{1} - 2(\eta + \Delta\eta)\dot{\mathbf{E}}$. where $\Delta\eta = \Delta\eta_I + \Delta\eta_{II} + \Delta\eta_{III}$ with

$$\Delta\eta_I = \frac{3}{2}\phi(\frac{R+a}{R})^2 \eta \frac{\eta_s - 2\eta}{\eta_s + 8\eta} \qquad (11)$$

$$\Delta\eta_{II} = \phi(\frac{R+a}{R})^3 \eta \frac{\eta_s - 2\eta}{\eta_s + 8\eta} \qquad (12)$$

$$\Delta\eta_{III} = \frac{N_p}{V}4\pi(R+a)^2 \alpha K_s (\frac{<\Delta r_s^2>}{3} + \frac{a(R+a)}{6}) \frac{10\eta}{\eta_s + 8\eta} \qquad (13)$$

In the above, N_p is the number of particle in volume V, $\eta_s \equiv \alpha K_s \tau_E (R+a)$.

Thus we have arrived at a simple analytic expression for the increase of the shear viscosity for low particle concentration in terms of interface parameters, a, η_s as well as $<\Delta r_s^2>$. In the limit of very large particle, $R/a \to \infty$, the Einstein result (Eq 1) is recovered, as it should. In the opposite limit of very small particles, the surface contribution $\Delta\eta_{III}$ is dominant. Noting that the infinitely small particle does not influence the background flow, namely, $\mathbf{u'} = 0$, we find that in that limit

$$\Delta\eta \to \nu_s \tau_E K_B T[1 + a^2/2 < \Delta r_s^2 >] \tag{14}$$

where ν_s is the density of crosslinks induced on the particles. For arbitrary sizes of particle, our result provides a highly reasonable interpolation, showing a significant deviation from the continuum result and sensitive dependence on the surface chains as the particles get small.

The particle – diffusion

The frictional coefficient ζ is determined from the drag on the particle, which is the integral of the stress over the surface area defined by $r = R + a$,

$$\int dA \ \mathbf{F}_s = -\int dA \ \sigma \cdot \mathbf{n} = -\zeta(\mathbf{V} - \mathbf{U}) \tag{15}$$

As a result of the integration, the friction coefficient is given as

$$\zeta = \zeta_E \zeta_H / (\zeta_E + \zeta_H) \tag{16}$$

where $\zeta_H = 6\pi\eta(R+a)$, the Stokes friction coefficient of the hydrodynamic radius (R + a), $\zeta_E = 4\pi(R+a)^2 \alpha K_s \tau_E$ is the friction due to the entanglement of the elastically effective surface chains. For diffusivity, we have

$$D = D_E + D_H = k_B T/\zeta_E + k_B T/\zeta_H. \tag{17}$$

This remarkable result means that the particle diffusion arises via two distinct dynamic processes : One is the disengagement (constraint release) of the surface chains from background-chain entanglement and the other one is the hydrodynamic feedback of the background. Since they occur in distinct time scales (the latter being much slower) the diffusivity appears to be the sum of their contributions. It is interesting to note that the self-diffusivity of a polymer in a melt also is given as the sum of contributions from (short-range) constraint release and reptation[6].

We see that the Stokes-Einstein result (Eq 2) is obtained in the large particle limit $(R/a \to \infty)$. For arbitrary sizes, the two dynamical processes mentioned above compete and it will be very interesting to confirm experimentally the crossover in the different size dependence, namely, $D \sim R^{-1}$ and $D \sim R^{-2}$, as predicted in our theory.

This work was supported in part by the Center for Advanced Materials Physics at POSTECH, Korea.

References

1) A. Einstein, Investigation on the theory of Brownian Movement (Dover, New York, 1956)
2) J. T. Hynes, R. Kapral, and M. Weinberg, J. Chem. Phys. 70, 1456 (1979)
3) W.Sung and G. Stell, J. Chem. Phys. 80, 3350 (1984)
4) G. K. Batchelor, J. Fluid Mech., 41, 545 (1970)
5) W. Sung and G. Stell, J. Chem. Phys. 80, 3367 (1984)
6) W. W. Graessly, Ad. in Polymer Sci, 47, 67 (1982)

DYNAMICS OF PHYSICALLY CROSSLINKED POLYMER NETWORKS

F. Tanaka* and S. F. Edwards

Cavendish Laboratory, Madingley Road, Cambridge CB3 0HE, UK

ABSTRACT

A simple transient model is introduced to describe the dynamics of physically crosslinked networks in which junctions are sufficiently weak to break and recombine in thermal fluctuations. The time-evolution equation under arbitrary macrodeformation is derived for the creation and annihilation of the junctions. Onthe basis of this equation, stress-strain relation under shear and elongational deformation is detailed. On longer time scales than the junction breakage time, the total number of effective chains decreases as time—resulting in an intranetwork flow.

INTRODUCTION

Our network model is made up of polymers of uniform molecular weight M with reversibly associating functional groups at their chain ends. We focus our attention specifically on the *unentangled regime* where M is smaller than the entanglement molecular weight M_e, so that each chain obeys Rouse dynamics modified by sticky trapping centers. During a small time interval dt of a macroscopic deformation $\hat{\lambda}(t)$, either end of a chain stretched above a critical length snaps from the junction, and relaxes to a Gaussian configuration within the Rouse relaxation time τ_R, whilst some of the dangling chains recapture junctions in their neighborhood with a constant probability (Fig.1).

Counting the variation in the number of elastically active chains from $t=0$ to t, we are led to an equation for the number $F(\mathbf{r},t)d\mathbf{r}$ of the active chains at time t whose end-to-end vectors fall onto a small region $d\mathbf{r}$:

$$F(\mathbf{r},t)d\mathbf{r} = \Theta(\mathbf{r},t;\mathbf{r}_0,0)F(\mathbf{r}_0,0) + p\int_0^t \Theta(\mathbf{r},t;\mathbf{r}',t')[n-\nu(t')]f_0(\mathbf{r}')d\mathbf{r}', \quad (1)$$

where $\Theta(\mathbf{r},t;\mathbf{r}',t') \equiv \exp[-\int_{t'}^t \beta(\mathbf{r}_{t'',t'})dt'']$ is a probability for an active chain at time t' to remain active until t (the function $\beta(\mathbf{r})$ is called *chain breakage rate*), p is the *chain recombination rate*. (We have assumed that end-to-end vectors are mapped affinely to the macrodeformation: $\mathbf{r}_{t'',t'} = \hat{\lambda}(t'') \cdot \hat{\lambda}(t')^{-1} \cdot \mathbf{r}'$ etc..) The equilibrium distribution $f_0(\mathbf{r})$ is assumed to be a Gaussian distribution with $\langle r^2 \rangle_0 = Na^2$.

---

* Present Address: Department of Physics, Faculty of General Education, Tokyo University of Agriculture and Technology, Fuchu, Tokyo 183, Japan

VISCOELASTICITY AND DYNAMICS

Viscoelastic properties and transient phenomena of the model network are studied on the basis of the fundamental time evolution equation (1). Results obtained are summarized below:

[1] The *stationary solution* is found under arbitrary macrodeformation. On the basis of the polymer statistics, a specific form of $\beta(r) = \beta_0 e^{\kappa r}$ is proposed, where β_0 and κ are functions of the temperature T and the molecular weight M of the polymer chain. Stationary viscoelastic properties are shown to exhibit an exponential dependence on T due to the activation process for the junction dissociation, differing markedly from an uncrosslinked polymer melt whose viscosity varies in a power of the temperature. Thickening and thinning conditions for shear and elongational flow are examined. It is shown that limiting behavior of the shear viscosity under high shear rate obeys the scaling law $\eta(\dot\gamma) \sim \dot\gamma^{2n/(n+1)}$ if $\beta(\mathbf{r})$ is proportional to r^n at high stretching.

[2] Linear response to oscillatory deformations is studied, and the *dynamic mechanical moduli* are obtained as functions of the frequency ω and the chain breakage rate $\beta(\mathbf{r})$. It is found that the storage modulus $G'(\omega)$ increases with the temperature, markedly differing from the uncrosslinked melts, and also that the plateau of the modulus is extended to lower ω region with increase of the life time $\tau_\times$ of the bond duration. The temperature shift factor a_T, which is necessary to carry out the frequency-temperature superposition, is proportional to $\tau_\times$, and hence, depends exponentially on the temperature. Breakdown of the Cox-Merz rule in physically crosslinked networks is suggested.

[3] Initial value problem is solved. *Stress relaxation* after a sudden macrodeformation is given is calculated. It is shown that on large time scales stress decay obeys a power law when $\beta(\mathbf{r})$ has significant r-dependence. For instance, the shear stress decays as $\sim t^{-5/n}$ if $\beta(\mathbf{r})$ is proportional to r^n at high stretching. Although the Lodge-Meissner relation still holds, time-strain separability loses its physical background. *Overshoot phenomena* in shear and normal stress, after a stationary flow is started on the initial equilibrium state, are also analyzed. It is found that the shear stress first shows a transient maximum, and then maxima of the first- and second normal stress difference follow. Larger overshoot is expected for larger values of the shear rate $\dot\gamma$, but the time at which the stress reaches its maximum is almost independent of $\dot\gamma$. The elongational stress is also obtained as a function of time.

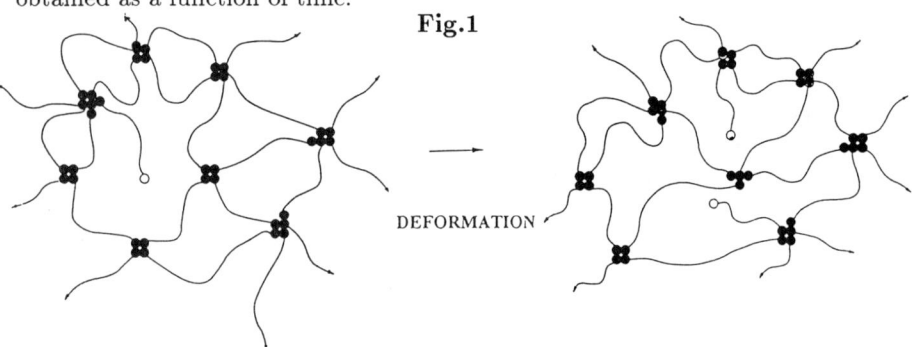

Fig.1

DEFORMATION

TOWARDS THE UNIFIED THEORY OF POLYMER SOLUTION DYNAMICS

Yasuhiro Shiwa

The Physics Laboratories, Kyushu Institute of Technology
Iizuka, Fukuoka 820, Japan

In spite of a reasonable understanding of static properties of polymer solutions, theoretical attempts at providing quantitative description of polymer solution dynamics have met considerable difficulties. A complicated interplay of excluded-volume effects and long-ranged hydrodynamic interactions among chain segments controls a slow dynamics of solutions of flexible, high-molecular-weight polymers. Moreover, the hydrodynamic and the excluded-volume interactions are gradually screened out as the concentration increases. Finally, the constraint of chain uncrossability of the connected chains is relevant to some dynamical properties.

The time-dependent Ginzburg-Landau type model we have proposed[1] is particularly convenient for assessing the above mentioned features. It reads, in appropriate units,

$$\frac{\partial \mathbf{c}_j(\tau,t)}{\partial t} = -\zeta^{-1}\frac{\delta H\{\mathbf{c}\}}{\delta \mathbf{c}_j(\tau,t)} + \int d\mathbf{r}\, \mathbf{v}(\mathbf{r},t)\delta(\mathbf{r}-\mathbf{c}_j(\tau,t)) + \mathbf{f}_j^c(\tau,t),$$

$$\frac{\partial \mathbf{v}(\mathbf{r},t)}{\partial t} = \mathsf{T}\cdot\left[\eta_e\nabla^2 \mathbf{v}(\mathbf{r},t) - \sum_{j=1}^{n}\int_0^N d\tau \frac{\delta H\{\mathbf{c}\}}{\delta \mathbf{c}_j(\tau,t)}\delta(\mathbf{r}-\mathbf{c}_j(\tau,t)) + \mathbf{f}(\mathbf{r},t)\right].$$

Here $\mathbf{c}_j(\tau,t)$ is the position vector of the monomer unit at the contour position τ ($0 \leq \tau \leq N$) of the jth chain ($j=1,2,\ldots,n$), and represents the chain conformation at time t. The solvent velocity $\mathbf{v}(\mathbf{r},t)$ at space-time point $(\mathbf{r},t)$ is assumed to satisfy the incompressibility condition, $\nabla\cdot\mathbf{v}=0$, which explains the appearance of the tensor operator, T; it selects the transverse part of the vector field it is applied to. The parameter ζ denotes the friction coefficient per segment of the chain, and η_e is the solvent viscosity. The Gaussian stochastic noises, $\mathbf{f}_j^c$ and $\mathbf{f}$, are assumed to be governed by the autocorrelation functions

$$\langle \mathbf{f}_j^c(\tau,t)\mathbf{f}_l^c(\sigma,t')\rangle = \mathsf{I}\delta_{jl}2\zeta^{-1}\delta(\tau-\sigma)\delta(t-t'),$$

$$\langle \mathbf{f}(\mathbf{r},t)\mathbf{f}(\mathbf{r}',t')\rangle = -\mathsf{I}\nabla^2 2\eta_e\delta(\mathbf{r}-\mathbf{r}')\delta(t-t'),$$

I being the unit tensor. The dynamical model so defined ensures that the system relaxes to an equilibrium state with the probability distribution $\propto$

exp $[-H\{c\} - 1/2 \int d\mathbf{r} v^2]$. The free energy functional, $H\{c\}$, associated with the polymer chain configuration $\{c\}$ is chosen to be

$$H\{c\} = \frac{1}{2}\sum_j \int d\tau [\frac{d\mathbf{c}_j(\tau)}{d\tau}]^2 + \frac{u}{2}\sum_{j,l} \int\int d\tau d\sigma \delta(\mathbf{c}_j(\tau) - \mathbf{c}_l(\sigma)) \ .$$

Here u represents the strength of the repulsive excluded-volume interaction.

The detailed predictions for various dynamical crossover behaviors (i.e., the dependence on the concentration and the degree of polymerization) of polymer solutions can be made[2,3] based on this model. Specifically,

(1) The oft-claimed puzzling discrepancy between the dynamic light scattering experiment and the renormalization-group theory for dilute solutions is resolved[4].

(2) With the help of a projection operator technique and the renormalization-group method, the full crossover behaviors of the longest relaxation time and of the viscosity of semidilute solutions are obtained[5]. In the scaling regime, the results echo the predictions of the reptation model, without resorting to a tube concept.

REFERENCES

1. Y. Shiwa, Y. Oono and P. R. Baldwin, Macromolecules **21**, 208 (1988); Y. Shiwa and Y. Oono, *ibid.* **21**, 2892 (1988).
2. Y. Shiwa and Y. Oono, Physica A **174**, 223 (1991), and references therein.
3. Y. Shiwa, Phys. Rev. Lett. **58**, 2102 (1987).
4. Y. Shiwa, J. Phys. A **24**, L579 (1991).
5. Y. Shiwa, J. Phys. (Paris) II (to be published).

DIPOLE DECAY FUNCTION AND DIELECTRIC LOSS CURVE OF POLYMERS IN DILUTE SOLUTION

Hiroshi Ogura, Riko Ozao[†] and Moyuru Ochiai

Department of Electronics, North Shore College of SONY Institute, Atugi 243, Japan

[†]Institute of Earth Science, Waseda University, Tokyo 169, Japan

ABSTRUCT

A dielectric relaxation of vinyl-type polymers in dilute solution is theoretically studied with the aid of the dipole decay function (the macroscopic time correlation function). An extension of Anderson's model for the time-dependent statistics of the Ising chain is used for the purpose. An evaluating method for the macroscopic decay function of chains with finite length is given, and results of numerical calculations under various conditions are presented.

MODEL

Several attempts to explain the observed dielectric loss curve of chain molecules in dilute solution were previouly presented. One of the most important attempts is to introduce certain kind of correlation among dipoles. The first application of the time dependent Ising model[1] to the dielectric relaxation of chain molecules was given by Anderson[2], and the influenece of the chain length was discussed using the same model by Isbister and McQuarrie[3]. However, if we intend to investigate the influence of the chain length upon dielectric relaxation propterties for vinyl-type polymers, $(CH_2-CHR)_n$, their model may be inadequate. This is the reason why the finite chain under consideration in the model calculations mentioned above consists of $N-1$ dielectrically nonactive segments but one isolated dipolar segment[2,3], while the real vinyl-type polymer chain has dipoles attatched rigidly and perpendicularly to the chain backbone for every monomeric units. In this study, we extend the model in an adequate manner so as to discuss the influence of chain length upon dielectric relaxation for the vinly-type polymers.

RESULTS

Figure 1 shows obtained macroscopic decay function $\Phi(t)$ of finite chains. Open circles indicate the results of numerical calculation obtained from the extended model mentioned above and solid curves represent the Kohlrausch-Williams-Watts function $\Phi(t)_{KWW} = \exp\{-(t/\tau_0)^\beta\}$ with adjustable parameters τ_0 and β. As seen from the figure, relaxation time τ_0 increases with increasing the nunmber of the chain segments N. This result is reasonable because the propagating time of rotational motion along the chain axis becomes longer for the large value of N. It is also seen that β decreases with increasing N. To illustrate the influence of chain length upon the dielectric loss curve, the dielectric loss sectra $\epsilon''(\omega)$ calculated from $\Phi(t)_{KWW}$ with various value of β are shown in Fig. 2. If $\beta=1$, the loss spectrum reduced to the Debye's case in which the curve is symmetric. As β decreases, the loss curve becomes more asymmetric, broader on the high frequency region. It was reported that the calculated loss spectrum from $\Phi(t)_{KWW}$ with $\beta \simeq 0.55$ agrees well with the observed asymmetric loss curve of vinyl-type polymer in dilute solution[4]. Thus, from results of Figs. 1 and 2, it can be concluded that the chain length is an important factor to reproduce the observed loss curve of vinyl-type polymers. The deviation from Debye's case for large N mentioned above may be due to the influence of cross correlation term $< \mu_i(0) \cdot \mu_j(t) >$ which arise between different dipoles μ_i and μ_j along the chain.

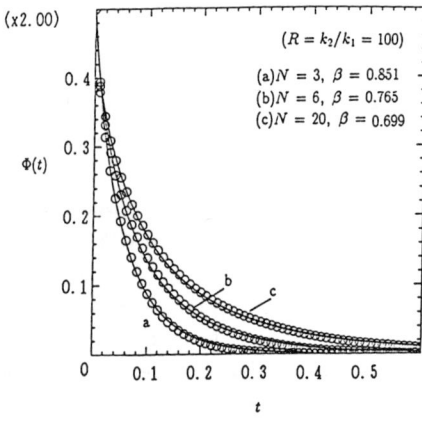
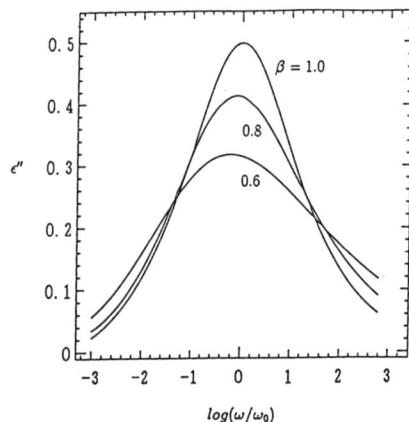

Fig. 1. Calcuclated dipole decay function $\Phi(t)$ for various values of N.

Fig. 2. Dielectric loss spectrum calculated from $\Phi_{KWW}(t)$ for various values of β.

REFERENCES

1. R. J. Glauber, J. Math. Phys., **4**, 294 (1963)
2. J. E. Anderson, J. Chem. Phys., **52**, 2821 (1970)
3. D. J. Isbister and D. A. McQuarrie, J. Chem. Phys., **60**, 1937 (1974)
4. K. Tarumi, T. Tanikawa and A. Chiba, Polym. J, **9**, 415 (1977)

NEW LANGEVIN DYNAMICS MODEL FOR POLYMER MELTS

Kyozi Kawasaki[1], Toshihiro Kawakatsu[1] and Walter Zimmermann[1,2]

[1]Department of Physics, Kyushu University 33, Fukuoka 812, Japan
[2]IFF Theorie III, Forschungszentrum Jülich, 5170 Jülich, Germany

ABSTRACT

A phenomenological model equation is proposed to describe the stochastic motion of a polymer chain in a dense medium with entanglement effects. The model involves extra friction terms which originate from the topological constraints by the other chains in the melt. The model is supposed to interpolate the Rouse and reptation regime and to generalize the respective models.

The physcis of polymer melts has been a problem of considerable interest during the recent years.[1-7] In contrast to liquids of small molecules, polymer melts display a rich and unusual viscoelastic behavior for time and length scales, where ordinary liquids are still Newtonian. The dynamics of polymer melts are typically investigated in terms of the Rouse and reptation models[1,2] or by extensive computer simulations.[5] These approaches have their natural limitations. Both models mentioned are of phenomenological character and cover only limiting cases of polymer melts and the computer simulations also have their limitations, especially in the available computation time. Facing this situation it would be desireable to develop an effective single polymer-chain description, covering in a first step at least both limiting models, which can make contact with the available computer simulations. With this in mind and based on previous works[2-4,7] we propose a phenomenological model equation of a polymer chain moving in a dense media with entanglements.

Thus we consider a single polymer chain with the polymerization index N. A chain conformation is given by a vector function $\mathbf{c}_\tau$ with τ a continuous variable between 0 and N. In our model, besides the free-energy given below, there are two types of dissipation associated with the chain. The so-called Rouse type is associated with independent movement of each test chain segment through the media. The second contribution comes from entanglement effects arising from lateral motion of a test chain segment. Starting from these general assumptions and after a series of simplifications and computation steps, which will be described in ref. 8, we obtain our model stochastic equation of motion and the accompanying fluctuation dissipation relation for the thermal noise $\mathbf{f}_\tau(t)$ in their explicit form

$$\zeta_0 \dot{\mathbf{c}}_\tau = -\zeta_1(\tau)\mathbf{n}_\tau \cdot \dot{\mathbf{c}}_\tau + \mathbf{n}_\tau \cdot \partial_\tau \zeta_2(\tau) \partial_\tau \mathbf{n}_\tau \cdot \dot{\mathbf{c}}_\tau - \frac{\delta}{\delta \mathbf{c}_\tau} H\{\mathbf{c}\} + \mathbf{f}_\tau. \quad (1)$$

$$< \mathbf{f}_\tau(t)\mathbf{f}_{\tau'}(t') > = 2T\{\zeta_0 \mathbf{1} + \zeta_1(\tau)\mathbf{n}_\tau - \mathbf{n}_\tau \cdot \partial_\tau \zeta_2(\tau)\partial_\tau \mathbf{n}_\tau\}\delta(\tau-\tau')\delta(t-t'). \quad (2)$$

where ζ_i are the different friction coefficients for the monomers. The tangent vector to the polymer chain is

$$\hat{\mathbf{t}}_\tau \equiv \partial_\tau \mathbf{c}_\tau / |\partial_\tau \mathbf{c}_\tau| \quad (3)$$

and the tensor $\mathbf{n}_\tau$

$$\mathbf{n}_\tau \equiv \mathbf{1} - \hat{\mathbf{t}}_\tau \hat{\mathbf{t}}_\tau, \tag{4}$$

projects $\dot{\mathbf{c}}_\tau$ onto the plane perpendicular to the chain tangent ($\mathbf{1}$ being the unit tensor). The free-energy functional $H\{\mathbf{c}\}$ in eq.(1) is

$$H\{\mathbf{c}\} = \frac{3T}{2b^2} \int_0^N d\tau [(\partial_\tau \mathbf{c}_\tau)^2 + \frac{\kappa}{2}(\partial_\tau^2 \mathbf{c}_\tau)^2 + ...],$$

where b is the Kuhn length and T is the Boltzmann constant times the absolute temperature. The first term in $H\{\mathbf{c}\}$ is for a Gaussian chain and the second one corresponds to the stiffness of the chain.

There are two obvious limits of (1) and (2): (i) the Rouse limit $\zeta_1(\tau) = \zeta_2(\tau) = 0$ and (ii) the reptation limit $\zeta_1(\tau) = \zeta_2(\tau) = \infty$ except for $\tau = 0$ and N (chain ends). $\zeta_2 = 0$ corresponds to a model considered in ref. 2. The model should automatically contain corrections to reptation such as constraint release processes[1] by making $\zeta_1(\tau)$ and $\zeta_2(\tau)$ finite. One may also study entanglement effects for less dense systems. The consequences of this model are currently under numerical investigation and will be reported elsewhere[8] in due course. During the further progress fixing of the free parameters in equations (1) and (2) seems also possible. Finally, also a microscopic derivation of the model equation or its generalization is highly desirable where the microscopic interpretations as well as the important N-dependences and the functional forms of $\zeta_1(\tau)$ and $\zeta_2(\tau)$ will be elucidated.

We thank K. Kremer for informative conversation. KK would like to thank hospitality he received during his stay at Forschungszentrum, Jülich where our collaboration on this work started and WZ acknowledges the financial support of the Kajima Foundation during his stay in Japan. This work is partially supported by the Scientific Research Fund of the Ministry of Education, Science and Culture of Japan.

REFERENCES

1. M. Doi and S. F. Edwards, *The Theory of Polymer Dynamics*, (Clarendon Press, Oxford, 1986).
2. C.F. Curtiss and R.B. Bird, J. Chem. Phys. **74**, 2016, (1981).
3. W. Hess, Macromolecules **19**, 1395,(1986); **20**, 2587, (1987); **21**, 2620 (1988).
4. K. S. Schweizer, J. Chem. Phys. **91**, 5802. 5822 (1989).
5. K. Kremer and G. Grest, J. Chem. Phys. **92**, 5057, (1990).
6. D. Richter, B. Farago, L.J. Fetters, J.S. Huang, B. Ewen and C. Lartigue, Phys. Rev. Lett. **64**, 1389 (1990).
7. K. Kawasaki, Mod. Phys. Lett. **B4**, 913, (1990).
8. K. Kawasaki, T. Kawakatsu and W. Zimmermann (in preparation).

C. COMPUTER SIMULATION

Relaxation Behavior of Crosslinked Polymer Melts

Kurt Kremer[1], Gary S. Grest[2], and Edgardo R. Düring[3]

(1) IFF, Forschungzentrum Jülich, 5170 Jülich, Germany
(2) Exxon Research and Engineering, CRSL, Annandale, NJ 08801, USA
(3) MPI für Polymerforschung, 6500 Mainz, Germany

Abstract

We report the first results of a recent series of simulations of randomly crosslinked polymer melts. We study the network structure and simulate the systems with two different interaction potentials. While one potential does not allow the chains to cross each other, it is allowed for the second one. By this method, we explicitly can estimate the trapping contribution to the plateau modulus. Already for an average length of an active strand of only 1/3 of the entanglement length, the modulus changes by about a third.

Introduction

There is a considerable recent interest from both theory[1] and experiment[2] in crosslinked polymer systems. Such systems, gels or rubber, are the basis of most industrial polymer products. Although these systems are subject of intense research for now more than 50 years[3], a detailed understanding is still missing. There are many parameters which influence the structural and relaxational properties of such a system. The density of crosslinks defines the average strand length $\langle N_s \rangle$. The crosslinking procedure, however, determines the distribution of links and strands. In addition, there are trapped entanglements. All these different quantities are important, however, the individual contributions are not at all understood. Here we are dealing with a chemical network. The constraints are permanent, the system is nonergodic and thus the relaxation is very slow. Experimental access is mostly very indirect. Either one measures macroscopic quantities, e.g., the plateau modulus after a shear deformation, or one uses neutron spin echo to observe microscopic displacements of monomers for only a very short time. In any case, a clear characteristization is experimentally not yet achieved.

In a simulation we can produce a variety of different systems by employing a variety of crosslinking mechanisms. The systems range from randomly crosslinked polymer melts to regular lattice structures, where the lattice bonds are replaced by polymer chains. We can also use different interaction potentials, which do or do not conserve entanglements[4]. Here we give some first results on the properties of randomly crosslinked polymer melts. The next chapter shortly describes the model. In Chapter III the structure analysis is explained, while Chapter V contains the results of the relaxation runs and a short discussion.

II Model and Methods

Here, we consider randomly crosslinked polymer melts. The crosslink density is far above the vulcanization/percolation threshold[5]. The initial configuration of the melts were generated in the same way as for a previous simulation for melt dynamics[6]. The chains are in a cubic box with periodic boundary conditions. In the standard situation with conserved entanglements all monomers interact via a pure repulsive Lennard Jones potential

$$U^{CJ}(r) = \begin{cases} 4\epsilon \left\{ \left(\frac{\sigma}{r}\right)^{12} - \left(\frac{\sigma}{r}\right)^{6} \right\} + 1/4 & r \leq r_c \\ 0 & r > r_c \end{cases} \quad (1)$$

with $r_c = 2^{1/6}\sigma$. For subsequent monomers along the chains an unharmonic attractive potential

$$U^{CH}(r) = \begin{cases} -0.5R_0^2\, k\, \ell n\left(1 - \left(\frac{r}{R_0}\right)^2\right) & r < R_0 \\ \infty & r \geq R_0 \end{cases} \quad (2)$$

is added, where $k = 30\epsilon/\sigma^2$ and $R_0 = 1.5\sigma$. For temperature $k_B T = 1.0$, and density $\rho = 0.85\sigma^{-3}$ the entanglement length is given as $N_e = 35$ and the entanglement time $\tau_e = 1800\tau$ ($\propto N_e^2$), with $\tau = \sigma(m/\epsilon)^{1/2}$. The molecular dynamics simulations were done with the particles weakly coupled to a heat bath and a background friction[6]. The integration of the equations of motion was carried out using a velocity Verlet algorithm with a time step $\Delta t = 0.0135\tau$. We studied a variety of melts, here, however, confine ourselves to a large system of $M = 400$ chains each of length $N = 50$. The longest run was 40000 τ. Here, we only study the case $p = 2$ crosslinks per chain ($p_c = 0.84$). In average each chain contains 4 crosslinked monomers cutting it into 5 pieces. The actual number is slightly smaller since in a few cases a monomer is part of two crosslinks[5,7]. To crosslink the melt, first a point in space is selected at random. Then the nearest monomer is identified as one of the crosslinking sites. This monomer is connected by a random choice to another one within a reaction radius of $r_x = 1.3\sigma$ ($= \sqrt{C_\infty}$). Nearest or second nearest neighbors along the chain are excluded to avoid artificial small intra chain loops. This procedure is repeated pN times for the same system, giving an instantaneous crosslinking. The crosslink potential is the same as the bond potential, Eq. 2. In order to be able to separate the trapping contribution in the dynamic quantities and the plateau modulus a second interaction potential for nonbonded monomers was studied. It is one of the key advantages of the simulation approach compared to experiment, that we can study the same system under a variety of different well defined conditions. The chains can now cross each other. The potential for the nonbonded monomers is replaced by

$$U^{cos}(r) = \begin{cases} 2\epsilon_a & r < r_a \\ \epsilon_a\left[\cos\frac{\pi(r - r_a)}{(r_c - r_a)} + 1\right] & r_a \leq r > r_c \\ 0 & r \geq r_c \end{cases} \quad (3)$$

r_a and ϵ_a are chosen that the chains in the uncrosslinked melt have the same persistence length (c_∞), pressure and Rouse friction as for the LJ potential. For $\rho = 0.85\sigma^{-3}$ we get $r_a = 1.0\sigma$, $\epsilon_a = 2.22\epsilon$. In Eq. 2 R_0 is increased to 1.75σ while the overall interaction of bonded monomers is reduced by a factor 0.175. With this interaction the monomers can now cross each other with an energy barrier of a few $k_B T$. Again it is important to note that the macroscopic static properties of the melts are the same for both potentials.

III Structural Properties

In order to compare the numerical results to the standard network models, we have to perform a detailed analysis of the structural properties. Using a method derived from percolation analysis we first identify the percolating network. Since the crosslink density is far above p_c, the vulcanization threshold, there is only one big cluster. A few chains are isolated "single chain clusters". In order to identify the elastically active part of the network, we first search for the backbone of the percolating network. Note that this backbone still contains 46% of the monomers. It is generally believed that this is the only part of the system, which supports zero frequency stress. It should be noted that at this stage many crosslinks, which have only two paths to the remaining network just define a bond within a strand, since the two remaining paths lead to dangling structures. This is also the reason why the average length of an active strand, i.e., the average number of bonds, which connect two crosslinks with at least three paths to the backbone of the network, is larger than the direct estimate given by p. For the average length of a strand between two crosslinks $\langle N_s \rangle$ we find $\langle N_s \rangle = 8.9\ldots$. This is

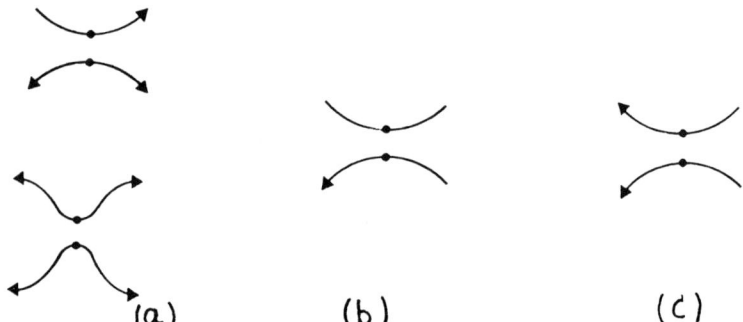

Figure 1: Illustration of the different classes of crosslinks. The lines indicate chains. The arrows mean, that a strand is connected to the backbone. Part (a) shows two typical active crosslinks. Part (b) shows a crosslink, which is in the backbone but is not counted as elastically active, while (c) shows a simple inactive crosslink. It should be noted, that these definitions are not unique, for details see ref. /7/

smaller than $N/(2p+1)$, because the chain ends have a smaller crosslinking pobability than the middle monomers. For the elasticity active strands we find $\langle N_{as} \rangle = 11$. For a more detailed discussion and an analysis of a range of crosslinking densities and chain lengths see Ref. /7/. With this information we can give the plateau-modulus as estimated from the current theories. Since our original chain is only marginally longer than N_e ($N_e = 35$) the trapping contribution caused by the conserved topology is expected to vanish[8]. Thus for the plateau modulus we expect a result between two models, which are assumed to give an upper and lower bound, namely the phantom network model and the affine deformation model. For a four-functional network these models give the plateau modulus G^0

$$G^0 = \frac{(\nu - h\mu)}{V} k_B T + T_e G_N^0 \qquad (4)$$

V is the total volume of the system. ν is the number of elastically active strands in the system while μ is the number of elastically active crosslinks. In the phantom network model $h = 0$ while $h = 1$ for the affine deformation model. $T_e G_N^0$ is the trapping contribution which should vanish, since $G_{N=50}^0 = 0$. For the present system size and $p = 2$ we get $\nu = 800 \pm 10$ and $\mu = 495 \pm 10$. Within this scheme we find $G^0 = 0.034 \pm 0.001$ for the affine model and $G^0 = 0.013 \pm 0.001$ for the phantom case respectively. Modifications of the theory typically give results in between the two limiting cases. It should be noted that G^0 actually should be slightly smaller due to the fact that not all the active crosslinks have functionality four with respect to the active backbone of the network.

IV Simulation Results

The previous chapter gave the results of a structure analysis of the network. Here we now describe the relaxational properties. One of the key questions concerns the importance of the conserved topology for networks. Using the two different potentials described in Chapter II, we can, under excluded volume conditions, explicitly test the importance of conserved entanglements. To do this we perform a standard molecular dynamics simulation of the system and investigate the relaxation functions[4]. Here we confine ourselves to the modulus of the crosslinked melt. Experimentally, the modulus is defined by the force constant of the system under a zero frequency shear in the linear regime. This, however, is numerically rather difficult to measure accurately. On the other hand, we know that the chains in the melt have random walk structure and can be well described by Rouse modes.

We can use this to calculate the elastic modulus from the relaxation of the Rouse modes. Following Doi and Edwards[9] the modulus $G(t)$ is given by

$$G(t) = \frac{\rho k_B T}{N} \sum_{p=1}^{\infty} Ap(2t) \xrightarrow{t \to \infty} \frac{\rho k_B T}{\langle N_s \rangle} \qquad (5)$$

$A_p(t) = \langle X_p(t) X_p(0) \rangle / \langle X_p^2 \rangle$ with X_p being the p-th Rouse mode of the chains. Using the reptation theory for uncrosslinked melts the above relation for $t \to \infty$ should hold for chains much larger than an entanglement length. Applied to the present simulation it means $N_s \ll N$. This, however, is not satisfied in the present case. About 46% of the monomers belong to dangling chain ends. Thus a modulus calculated directly from Eq. 5 certainly does not describe the property of the network. However, this difficulty can be avoided. The original chain structure for the network does loose its meaning, since the crosslinks add additional bonds. Thus the identification of individual chains is artificial. We can use this to circumvent the problem of short chains and huge dangling ends. In order to estimate the modulus of the network we again use Eq. ..., however, we now evaluate it for chains which were constructed not to contain any dangling part. By this we avoid the problem of the dangling ends. We performed the mode analysis for constructed chains of length $N = 100$ and $N = 50$. Within our error bars we were not able to detect a difference in the result. Fig. 2 shows the results for the time dependent modulus for both potentials for the original chains and the constructed chains. As expected the original chains give a by far too small value of the modulus. Much more important however is the result for the constructed chains. The extrapolated value of G^0 for the topology conserving Lennard Jones chains $G_{LJ}^0 = 0.040 \pm 0.002$, while the result for the cosine potential $G_{cos}^0 = 0.031 \pm 0.002$. Thus none of the classical models is capable to describe the data reasonably well. For G_{LJ}^0 the modulus is even larger than the affine deformation model predicts. G_{cos}^0 is a reasonable agreement to the affine model, however, the mean square displacements of the monomers and the crosslinks show a by far too large value in order to be explained by the standard affine model.

To conclude we give here some first results on the relaxational properties of highly crosslinked polymer networks. By employing two different monomer interaction potentials we are able to calculate the contribution of conserved topology to the network modulus explicitly. The results do not agree with any of the currently available models for rubber elasticity. Certainly, there is much more work needed in order to clarify these problems. With the use of computer simulations, however we are able to analyse both the detailed microscopic structure of the network under investigation as well as relaxation functions. Together with the different potentials available we have a unique tool to analyse polymer networks under conditions, which are by far better controlled than experimental systems. Presently, we are extending the study to longer chains and different crosslink densities.

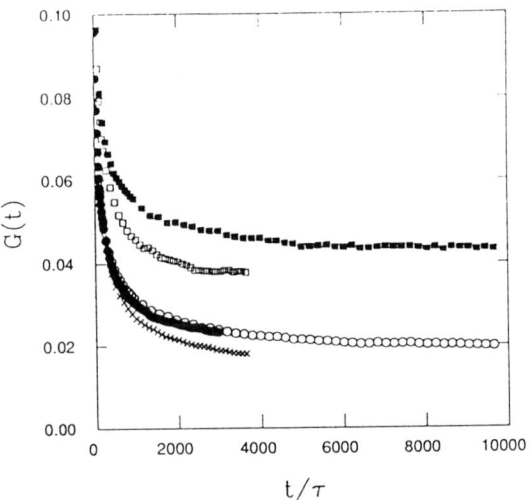

Figure 2: Time dependent modulus $G(t)$ for the original and the constructed chains for the two potentials employed. Results for the original chains are given in the lower three curves for the 400/50 system for the LJ (circles) and the cosine (crosses) potential and a smaller 50/50 system with LJ potential (dots). The upper two curves show the results for the constructed chains for the 400/50 system for the LJ (full squares) and cosine (open squares) potentials

This research was made possible by a generous grant of CPU time by the German Supercomputer Center HLRZ, Jülich. ED and KK acknowledge the support of the Sonderforschungsbereich 262 of the Deutsche Forschungsgemeinschaft DFG.

References

1. P.J. Flory, Proc. Roy. Soc. Landau A**351**, 351 (1976)
   S.F. Edwards, T.A. Vilgis, Rep. Progr. Phys. **51**, 243 (1988)

2. J.D. Ferry, Viscoelastic Properties of Polymers, Wiley NY 1990

3. L.R.G. Treolar, Theory of Rubber Elasticity, Oxford Univ. Press, Oxford (1949)

4. E. Düring, K. Kremer, G.S. Grest, Phys. Rev. Lett. 16.12.1991 and preparation

5. G.S. Grest, K. Kremer, J. Phys. (France) **51**, 2829 (1990) K. Kremer, G.S. Grest, J. Chem. Phys. **92**, 5057 (1990) erratum **94**, 4103 (1991)

6. E. Düring, K. Kremer, G.S. Grest, in preparation

7. D.S. Pearson, W.W. Graessley, Macromolecules **11**, 528 (1978)

8. M. Doi, S.F. Edwards, The Theory of Polymer Dynamics, Oxford Univ. Press, Oxford (1986)

DYNAMICAL CORRELATION FUNCTION OF FRACTAL NETWORKS : COMPUTER EXPERIMENTS

T. Nakayama and K. Yakubo
Department of Applied Physics, Hokkaido University, Sapporo 060, Japan

ABSTRACTS

We have calculated the Raman correlation function $C(q,\omega)$ for $d=2$ percolating networks. It is found that $C(q,\omega)$ can be scaled by a single wavenumber and frequency.

INTRODUCTION

Since the first observation of Raman scattering from silica gel by Boukenter et al.,[1] a variety of scattering experiments for fractal structures have been extensively carried out.[2-4] The interpretation of these scattering data, however, have been controversial. This is mainly due to the assumption that fracton wavefunctions are characterized by a smoothly averaged wavefunction, namely, the controversial conclusions on experimental data were due to the failure of this Ansatz. Computer experiments have played an important role for understanding the dynamics of fractal networks. In particular, it is useful for gaining insight into something like an "ensemble average" of a physical quantity. Examples are the calculation of the fracton wavefunction, which show that the "superlocalization" appropriate to the fracton core does not apparently extend to the outer reaches of the wavefunction.[5] Quite recently, Montagna et al.[6] calculated the Raman coupling coefficient $C_{\alpha\alpha}(\omega)$ (strain-strain correlation function) for percolating networks, and claimed that the models used so far are invalid. The aim of this paper is to report our calculations on the Raman coupling coefficient $C_{\alpha\alpha}(q,\omega)$ for large percolating networks and to clarify the dynamical properties of the correlation function.

RAMAN CORRELATION FUNCTION

The intensity $I(\mathbf{q},\omega)$ of inelastic light scattering with a frequency shift $(\omega/2\pi)$ is determined by the Raman tensor $I_{\alpha\beta}(\mathbf{q},\omega)$ defined as the Fourier transform of the correlation function for the local fluctuation of the dielectric susceptibility tensor $\delta\epsilon_{\alpha\beta}(\mathbf{r}_i,t)$. In general, this is written as

$$\delta I_{\alpha\beta,\gamma\delta}(\mathbf{q},\omega) = \frac{A}{V}\sum_{i,j}\int_{-\infty}^{\infty} dt \exp(i\omega t)\exp[-i\mathbf{q}\cdot(\mathbf{r}_i(t)-\mathbf{r}_j(0))] \times$$
$$\langle\delta\epsilon_{\alpha\beta}(\mathbf{r}_i,t)\delta\epsilon_{\gamma\delta}(\mathbf{r}_j,0)\rangle \quad (1)$$

where A is the numerical constant, V the volume of the system, and the brackets $\langle\cdots\rangle$ denote the thermal average, and $\mathbf{q}$ is the exchanged wavevector, respectively. The Greek subscripts are Cartesian components indicating the polarization vectors of incoming and outgoing light, and $\mathbf{r}_i$ denote the lattice position of the ith atom, respectively. It is standard to relate the change in the dielectric susceptibility $\delta\epsilon_{\alpha\beta}(\mathbf{r}_i)$ resulting from the lattice deformation to the elastic strain tensor $e_{\alpha\beta}(\mathbf{r}_i)$. We introduce the Raman correlation function $\tilde{C}_{\alpha\beta}(\mathbf{q},\omega)$ which describes the coupling of vibrational modes of frequency $\omega/2\pi$ to the light for a particular polarization geometry. We define this function as

$$\tilde{C}_{\alpha\beta}(\mathbf{q},\omega) = \frac{1}{V}\sum_\lambda \delta(\omega-\omega_\lambda)|e_{\alpha\beta}^\lambda(\mathbf{q})|^2 \quad, \tag{2}$$

where

$$e_{\alpha\beta}^\lambda(\mathbf{q}) = \sum_i e_{\alpha\beta}^\lambda(\mathbf{r}_i)\exp(i\mathbf{q}\cdot\mathbf{r}_i) \quad. \tag{3}$$

Here we have defined $\tilde{C}_{\alpha\beta}(\mathbf{q},\omega)$ as the average of $\tilde{C}_{\alpha\beta}(\mathbf{q},\omega_\lambda)$ over all modes λ with frequency close to ω and over some ensemble repetitions of the system. It should be emphasized that these are anisotropic random functions with fractal correlations. Describing them by an "average" eigenmode is very misleading as pointed out by Nakayama, Yakubo and Orbach,[5] and Alexander.[7] Thus, there is some ambiguity in the definition of the strains for a fractal geometry. The stress must follow the fractal geometry. We present numerical results of the Raman correlation function in two-dimensional percolating networks with large site numbers using a novel numerical method.

CALCULATED RESULTS

Consider a site-percolating networks consisting of N particles with mass m and linear springs connecting two nearest neighbor atoms. The equation of motion of the atom are

$$m\ddot{u}_i(t) = \sum_j K_{ij} u_j(t) \quad, \tag{4}$$

where u_i is the displacement of the atom on the ith site. The force constant is taken to be $K_{ij}=0$ if either sites i or j are unoccupied, and $K_{ij}=1$ otherwise. The displacement u_i has only one component. The numerical technique employed here is based on the mechanical resonance to extract pure eigenmode.[8] This method is available for investigating lattice vibrations in systems containing as many as $\sim 10^6$ particles. In particular, this algorithm has an advantage to be easily vectorized for implementation on an array-processing supercomputer. Yakubo et al.[9] has recently succeeded in finding a formula for judging the accuracy of calculated eigenmodes and eigenfrequencies for this.

The results for $C(q,\omega)(\equiv \tilde{C}(q,\omega)/\mathcal{D}(\omega))$ are shown in Figs. 1 and 2. Figure 1 shows the calculated results of the q-dependence of $C(q,\omega)$ of two-dimensional site-percolating networks at the percolation threshold $p_c(=0.593)$ for 50 frequencies ($0.005 \leq \omega \leq 0.5$). The abscissa indicates the wavenumber q/q_0, where q_0 is the wavenumber at which $C(q,\omega)$ has the largest value for each *fixed frequency*. Filled circles indicate the average value over data within narrow range close to q/q_0. We see that the results exhibit scaling property. In particular, it should be emphasized that the wavenumber dependence obeys the power law, $C(q,\omega) \propto q^{3.5\pm 0.1}$ below the characteristic wavenumber q_0 and $C(q,\omega) \propto q^{-3.5\pm 0.1}$ above q_0.

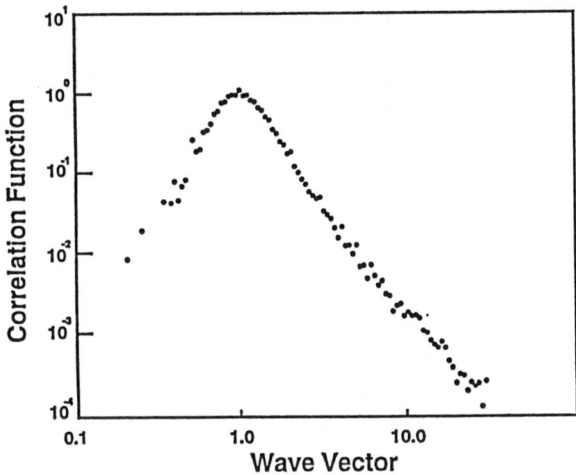

Fig. 1 The q-dependence of the Raman correlation function $C(q,\omega)$ of two-dimensional percolating networks at p_c. The correlation function is calculated by averaging over five percolating networks formed on 500 × 500 square lattices. The maximum site number of our systems is $N = 110793$ and the minimum $N = 93382$. The abscissa is scaled by the dimensionless wave number q/q_0. The value of $C(q,\omega)$ are scaled by its maximum value.

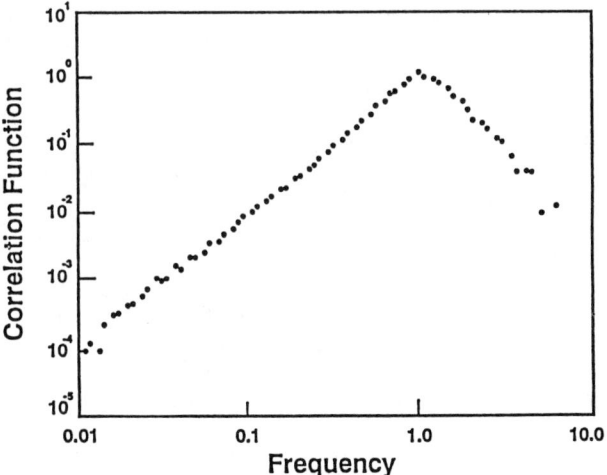

Fig. 2 The ω-dependence of the Raman correlation function $C(q,\omega)$ of two-dimensional percolating networks at p_c. The abscissa is scaled by the dimensionless frequency ω/ω_0. The value of $C(q,\omega)$ are scaled by its maximum value.

Figure 2 is the results of the ω-dependence of $C(q,\omega)$ of two-dimensional networks at p_c for 250 wavenumbers ($2\pi/250 \leq q \leq 2\pi$). The abscissa represents the frequency ω/ω_0, where ω_0 is the frequency at which $C(q,\omega)$ has the largest value for each *fixed wavenumber*. Filled circles indicate the average value over data within narrow range close to ω/ω_0. We emphasize that these results show the universal behavior scaled by the frequency ω_0. The results also indicate that the frequency dependence of $C(q,\omega)$ is expressed as $C(q,\omega) \propto \omega^{2\pm 0.1}$ in the frequency regime of $\omega \ll \omega_0$ and $C(q,\omega) \propto \omega^{-3\pm 0.1}$ for $\omega \gg \omega_0$. It is important to make clear the origin of the observed results. Strongly localized fractons are probed by plane waves in the Fourier transform and the width in $C(q,\omega)$ in q-space is ascribed by them, but ill defined q-space. Thus, the Raman coupling coefficient $C(q,\omega)$ reflects the width in q-space from the scattering of localized fracton state. Obviously, this can be mapped onto ω-space width by the dispersion relation of fractons $\Lambda(\omega) \propto \omega^{-\tilde{d}/D}$ where Λ is the localization length of a fracton, $\tilde{d}$ the fracton dimensionality, and D the fractal dimension, respectively.

CONCLUSION

To summarize, we have calculated the Raman coupling coefficient (strain-strain correlation function) for a large percolating networks. We have found that the functions $C(q,\omega)$ exhibit the universal behavior scaled by a single characteristic wavenumber and frequency.

This work was supported in part by a Grand-in-Aid of Scientific Research on Priority Areas, Computational Physics as a New Frontiers in Condensed Matter, Research of the Ministry of Education, Science and Culture, Japan. After completing this work, we have received the preprint by Alexander et al. (to be published in Phys. Rev. B) which treats analytical aspects of the subject presented here.

REFERENCES

1. A. Boukenter, B. Champagnon, E. Duval, J. Dumas, J. F. Quinson, and J. Serugetti, Phys. Rev. Lett. **57**, 2391 (1986).
2. E. Duval, G. Mariotto, O. Pilla, G. Viliani and M. Barland, Europhys. Lett. **3**, 333 (1987).
3. A. Fontana, F. Rocca, and M. P. Fontana, Phys. Rev. Lett. **58**, 503 (1987).
4. Y. Tsujimi, E. Courtens, J. Perous, and R. Vacher, Phys. Rev. Lett. **60**, 2757 (1988), and E. Courtens and R. Vacher, Proc. R. Soc. Lond. **A423**, 55 (1989).
5. T. Nakayama, K. Yakubo, and R. Orbach, J. Phys. Soc. Jpn. **58**, 1891 (1989).
6. M. Montagna, O. Pilla, G. Villiani, V. Mazzacurati, G. Ruocco, and G. Signorelli, Phys. Rev. Lett. **65**, 1136 (1990).
7. S. Alexander, Phys. Rev. **B40**, 7953 (1989).
8. K. Yakubo and T. Nakayama, Phys. Rev. **40**, 517 (1989).
9. K. Yakubo, T. Nakayama, and H. J. Maris, J. Phys. Soc. Jpn. **60** 3249 (1991).

The computer simulation of spinodal decomposition in polymer blends
- Effects of uniaxial compression during spinodal decomposition -
Mikihito Takenaka and Takeji Hashimoto
Kyoto University
Kyozi Kawasaki and Toshihiro Kawakatsu
Kyusyu Universitry

I. Introduction

The dynamics of ordering process via spinodal decomposiotn in a binary polymer mixture is nonlinear and nonequilibrium and the phase-separated structure caused by spinodal decomposition depends on the process which the system takes. Therefore we would have the unique phase-separated structure, which have not never obtained with "normal" isothermal process, by imposing the external field.

In this study, we explored the effects of a 1/2 uniaxial compression during the phase separation via spinodal decomposition with computer simulation.

II. Experimental Method

Time-dependent Ginzburg-Landau (TDGL) equation model[1] was used to investigate the dynamics of phase separation in a binary polymer mixture. In TDGL model, the binary polymer mixture is represented by a continuous field. If we define the order parameter as $\psi(\mathbf{r})$, the time evolution of $\psi(\mathbf{r})$ is given by

$$\frac{\partial \psi(\mathbf{r},t)}{\partial t} = \frac{L\nabla^2 \delta F\{\psi\}}{\delta \psi} \qquad (2.1)$$

where L is the Onsager kinetic coefficient and F is the Free energy functional given by

$$F\{\psi\} \equiv \int d\mathbf{r} \{\frac{D}{2}(\nabla \psi)^2 - \frac{c}{2}\psi^2 + \frac{u}{4}\psi^4\} \qquad (2.2)$$

where D, c and u are constant, we choose D=1.0, c=2.0 and u=2.0 in this study. We neglect the hydrodynamics interactions and the thermal noise in (2.1). We integrated (2.1) numerically on a 2-dimensional lattice with periodic boundary conditions. Our numerical 1/2-uniaxial compression is as follows. We keep the mesh size constant but change the number of lattice to compress numerically. In the parallel direction to the compression axis, we compressed the profile of the order parameter by eliminating the meshes alternatively. In the perpendicular direction to the compression axis, we expand the profile of the order parameter by interpolating the new meshes between old meshes. We imposed the 1/2-uniaxial compression both in the early stage[2] and in the intermediate stage[2]. In the early stage the time-evolution can be described by Cahn's linearized theory. And in the intermediate stage the dominant mode of fluctuations of order parameter have not reached the equilibrium value yet and Both the wavelength and the amplitude of the fluctuations grow with time. We use 128×512 meshes and 64 × 256 meshes for the early stage and the intermediate stage , respectively, at the initial state. and these systems are then deformed into 256×256 and 128×128, respectively..

III.Results
(1) 1/2 uniaxial compression at the early stage of spinodal decomposition

Fig.1 shows the time-evolution of the phase-separated structure after 1/2 uniaxial compression. As time elapsed, the diagonal mode in phase-separated structure become dominant. This Cahn's linearized theory.

(2) 1/2 uniaxial compression at the intermediate stage of spinodal decomposition

Fig.2 shows the time-evolution of the phase separated structure after the compression. In Fig.2(a), which corresponds to the snapshot just after the compression. The structure become oriented in perpendicular direction to the compression axis. However As time elapses , The structure become oriented in parallel direction to the compression axis as shown in Fig.2(d). This is because the compression shifts the wavelength of the dominant mode of the fluctuation of the order parameter to short wavelength region, where the fluctuation is unstable, and disappeared the fluctuations.

We obtained various kind of structures by uniaxial compression.

References
1. K.Kawasaki, Physica 119A (1983) 17.
2. T.Hashimoto, Phase Transitions, 12,(1988) 47.

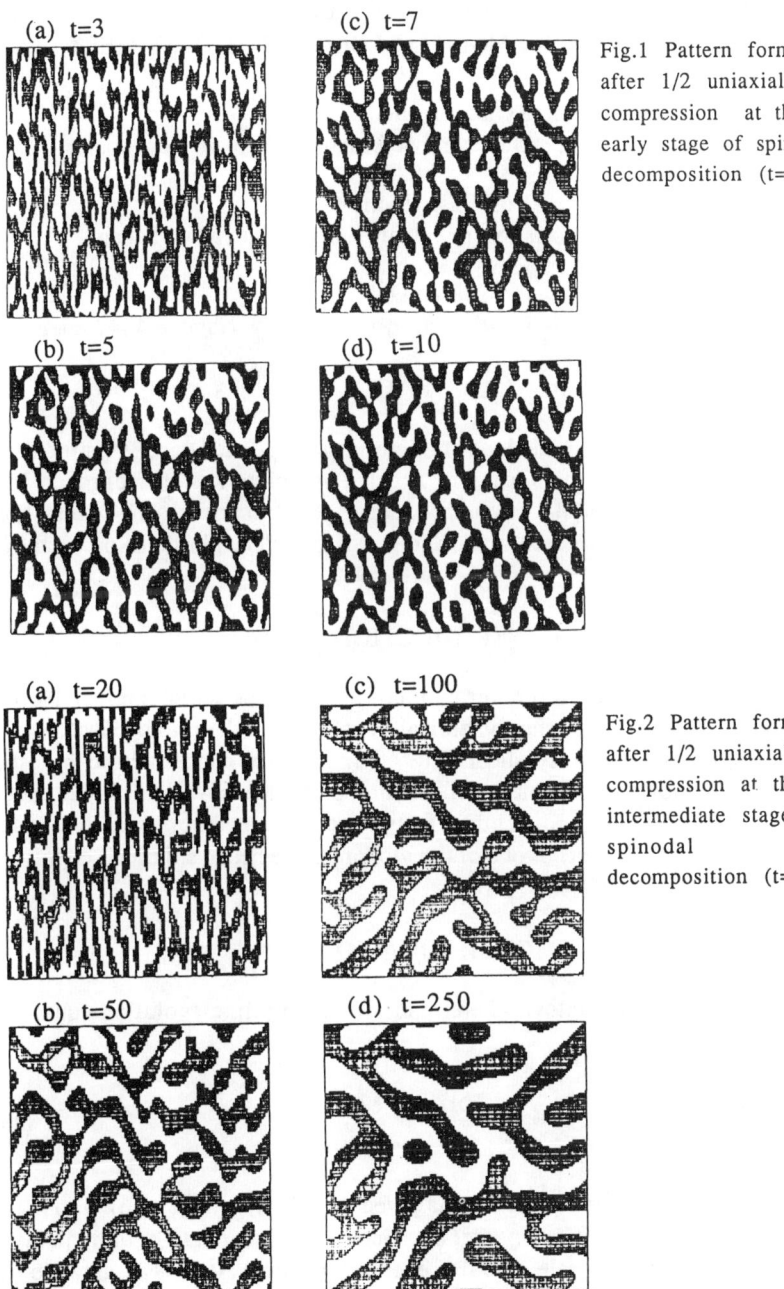

Fig.1 Pattern formation after 1/2 uniaxial compression at the early stage of spinodal decomposition (t=3).

Fig.2 Pattern formation after 1/2 uniaxial compression at the intermediate stage of spinodal decomposition (t=20).

COMPUTER SIMULATIONS OF RHEOLOGICAL RESPONSE IN A MICROPHASE SEPARATED BLOCK COPOLYMER MELT

Takao Ohta, Ayako Tetsuka
Department of Physics, Ochanomizu University, Tokyo 112 Japan

Yoshihisa Enomoto
Department of Physics, Nagoya University, Nagoya 464 Japan

Masao Doi
Department of Applied Physics, Nagoya University, Nagoya 464 Japan

ABSTRACT

We carry out computer simulations of micro-phase separated copolymer melts under an oscillatory shear flow by means of a cell dynamical system approach in two dimensions. It is found that the microphase separated structure has an extremely slow relaxation rate under a large deformation so that it exhibits a plastic response for small frequency of oscillation.

INTRODUCTION

We shall perform computer simulations of block copolymer melts in two dimensions. In order to see the dynamical properties of the micro-phase separated state, we apply an oscillatory shear flow. Our primary concern is to investigate rheological behavior due to domain deformation under the flow field. We do not take account of the possible hydrodynamic effects. Under the incompressibility condition, the local concentration $\psi(\mathbf{r}, t)$ of one of the monomer species in a diblock copolymer is the only dynamic variable.

MODEL EQUATION

The model equation employed here is the following time-evolution equation for $\psi(\mathbf{r}, t)$:

$$\frac{\partial \psi}{\partial t} + \mathbf{v}(\mathbf{r}, t) \cdot \nabla \psi = \nabla^2 \frac{\delta H\{\psi\}}{\delta \psi} \qquad (1)$$

where the external velocity field $\mathbf{v}(\mathbf{r}, t)$ is given by $\mathbf{v}(\mathbf{r}, t) = (\dot{\gamma}(t) y, 0)$ with $\dot{\gamma}(t)$ the time-dependent shear rate. The Onsager coefficient has been put to be unity in (1.1) by suitably choosing the time scale. The free energy functional $H\{\psi\}$ is assumed to be the form obtained in a previous study[1]

$$H\{\psi\} = \frac{1}{2} \int d\mathbf{r} \int d\mathbf{r}' \, u(\mathbf{r} - \mathbf{r}') \psi(\mathbf{r}, t) \psi(\mathbf{r}', t) + \int d\mathbf{r} \, W(\psi) \qquad (2)$$

In a micro-phase separated state, $W(\psi)$ is an algebraic function of ψ with two local minima. The interaction $u(r - r')$ in the bilinear term is given by $u(r - r') = D \nabla \nabla' \delta(r - r') + \alpha G(r - r')$ where D is a positive constant. $G(r - r')$ is the Green function defined through the relation $-\nabla^2 G(r - r') = \delta(r - r')$. The constant α is given by $\alpha = 3/[Nf(1 - f)]^2$ where N is the polymerization index of each copolymer chain and f is the block ratio.[1]

SIMULATIONS AND RESULTS

We perform simulations of a cell dynamical version[2] of Eq. (1). The system size LxL with L= 127 and 32 has been chosen to see a finite size effect. We have put the shear rate $\dot{\gamma}(t)$ as $\dot{\gamma}(t) = \Gamma \omega \cos \omega t$ where the frequency ω is restricted to $\omega < 10^{-2}$ for a technical reason. For the parameters f=0.4, D=1.3 and α=0.01, the disk-shaped micro-phase separated domains constitute a triangular lattice.

We have evaluated numerically the macroscopic shear stress σ_{xy} due to domain deformation, which is given in terms of the Fourier component $\psi_q(t)$ by[3]

$$\sigma_{xy} = \frac{1}{L^2} \sum_q q_x q_y [D - \frac{\alpha}{q^4}] |\psi_q(t)|^2 \tag{3}$$

The results of the simulations are as follows. (1) When the oscillating frequency ω is larger than 10^{-4}, the system for the amplitude $\Gamma = 0.4$ shows an elastic response. (2) When $\omega = 10^{-4}$, there appears a phase difference in the stress-strain relation for $\Gamma = 0.4$ and 0.8. (3) This phase difference disappears for $\Gamma = 0.1$. Thus the system exhibits a plastic response for $\omega = 10^{-4}$ and $\Gamma = 0.4$ and 0.8. This has also been confirmed by an alternative simulation of stress relaxation.

CONCLUSION

Computer simulations show that the microphase separated state has an extremely slow relaxation for large deformations such as $\Gamma = 0.4$ and 0.8, which is responsible for the plastic behavior observed for $\omega = 10^{-4}$. The details and an interpretation of the above results by a phenomenological theory and further theoretical approach to the related problem will be published in the near future.[4]

REFERENCES

1. T. Ohta and K. Kawasaki, Macromolecules, **19**, 2621 (1986) and **23**, 2413 (1990).
2. M. Bahiana and Y. Oono, Phys. Rev. **A41**, 6763 (1990).
3. K. Kawasaki and T. Ohta, Physica **A139**, 223 (1986).
4. T. Ohta, Y. Enomoto and M. Doi, to be submitted to J. Chem. Phys.

STRONG ANISOTROPY OF ORIENTATIONAL RELAXATION IN BULK AMORPHOUS POLYMER : A MOLECULAR DYNAMICS SIMULATION

H. Takeuchi
Mitsubishi Kasei Corp., 1000 Kamoshida-cho, Midori-ku, Yokohama, 227, Japan

R.J.Roe
University of Cincinnati, Cincinnati, OH 45221-0012, U.S.A.

ABSTRACT

Local chain dynamics of bulk amorphous polymer has been studied by means of molecular dynamics simulation at several temperatures including the glass transition temperature T_g. Using a realistic model mimicking amorphous phase of polyethylene, time-correlation functions of various vectors embedded in polymer segment have been evaluated. Above T_g, the result shows strong dependence of the relaxation on the direction of the vector. This strong anisotropy is then discussed in relation to effects of surrounding chains. A characteristic chain motion found below T_g, arising from highly correlated conformational transitions, is also discussed.

INTRODUCTION

The dynamics of flexible chain molecules in bulk amorphous polymer is one of the subjects of fundamental importance in polymer science. In cooperation with experimental works, computer simulations, such as molecular dynamics (MD) and Brownian dynamics simulations, have been playing an important role on providing us with microscopic information on the local chain dynamics of polymers.[1] Most of such simulations, however, have so far neglected all the nonbonded interactions, i.e., the excluded volume effect, existing in polymers, and focused on the conformational transitions (CTs) which should be one of the major factors shaping the rapid relaxation behavior. In the present work, the local chain dynamics in the bulk is thus studied using a realistic model of bulk amorphous polymer where the nonbonded interactions are rigorously taken into account, and effects of surrounding chains on the local dynamics are examined.

SIMULATION

A fully vibrational model of polymer, mimicking polyethylene, is employed with the periodic boundary condition which generates a MD cell filled with parts of several infinite length chains; details of the model are given elsewhere.[2] First, the temperature-volume relationship is evaluated by ordinary MD simulation at a constant pressure, and T_g is detected by the discontinuity in the thermal expansion coefficient. A simulation of duration up to 5ns is then performed at several temperatures. The local chain dynamics is evaluated by the time correlation functions $M_1(t)$ and $M_2(t)$ for three vectors embedded in the polymer segment (see Fig.1). At the same time, the distribution of reorientational angle is also evaluated.

RESULTS AND DISCUSSION

Table 1 shows the correlation time τ_i (i=1 or 2) corresponding to $M_i(t)$ obtained at 0.860g/cm^3 and 300K which is well above T_g (~201K). As might be expected, τ_1 for the vector perpendicular to the chain axis(c) is comparable to that of the internal rotation of chain, τ_ϕ, whereas τ_1 for the chain axis is extremely longer than τ_ϕ. This strong anisotropy is also observed for a hypothetical freely-rotating-chain system, and thus CT which occurs only in the polyethylene-like model seems not to cause this characteristic feature. Although the anisotropy of the orientational relaxation has been already pointed out for a model chain having no nonbonded interactions,[3] the present work reveals the extremely strong anisotropy in the bulk. This suggests that the surrounding chains restrict the reorientation of the chain axis. In fact, the strong anisotropy disappears as the nonbonded interactions are switched off. In addition, for a given time regime, $M_i(t)$ for the chain axis is well reproduced by the restricted rotational diffusion model [4] where a rod can undergo rotational Brownian motion only in a cone of fixed apex angle. These facts lead to the conclusion that the chain tumbles rapidly around the chain axis while the chain axis itself gradually changes its direction according to the slow structural relaxation. Consequently the surrounding chain acts as a rigid "pipe" confining the chain axis within the time scale studied here.

As the temperature is lowered through T_g, the basic feature of the orientational relaxation remains the same. Quantitative analysis is, however, difficult because of slower relaxation. Besides the anisotropy, a new feature appears in the distribution of the reorientational angle below T_g, that is, a bimodal distribution suggesting large angular jump motions. In fact, CT is found to persist even below T_g. In contrast the distribution above T_g is monomodal and similar to what is expected for a rod undergoing the rotational Brownian motion. Detail analysis on CT occurring below T_g reveals the localization of CTs and significant correlation between CTs along the chain as shown in Fig.2. The interesting odd-even effect suggests the predominance of second neighbor pair transition[1] which leaves the remainder of the chain relatively undisturbed. This cooperativity thus enables CTs to occur even in more rigid "pipe" formed by the glassy matrix.

REFERENCES

1. See for example, E.Helfand, Z.R.Wasserman and T.A.Weber, Macromolecules **13**, 526 (1980).
2. H.Takeuchi, R.J.Roe and J.E.Mark, J.Chem.Phys. **93**, 9042 (1990).
3. T.A.Weber and E.Helfand, J.Phys.Chem. **87**, 2881 (1983).
4. C.C.Wang and R.Pecora, J.Chem.Phys. **72**, 5333 (1980).

Table I Correlation times[a] for the orientational relaxation at 0.860g/cm³ and 300K.

vector	τ_1 (ps)	τ_2 (ps)	τ_1/τ_2
a	30.6±0.7	11.0±0.4	2.78±0.07
b	24.8±0.5	11.3±0.6	2.19±0.07
c	9090±3740	95.2±4.8	94.9±37.4

[a] τ_ϕ=28.3±1.7ps for the internal rotation.

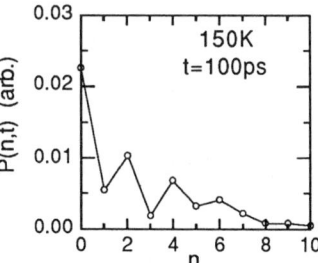

Fig.2. Correlation of CT with respect to the position n along the chain at 150K.

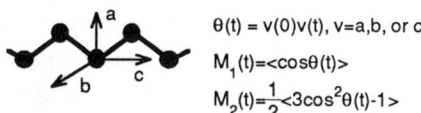

$\theta(t) = v(0)v(t)$, v=a,b, or c
$M_1(t) = \langle\cos\theta(t)\rangle$
$M_2(t) = \frac{1}{2}\langle 3\cos^2\theta(t)-1\rangle$

Fig.1. Definition of the local frame embedded in the segment.

AMPLITUDE MODE AROUND A MOVING SOLITON IN POLYACETYLENE

Makoto KUWABARA and Yoshiyuki ONO
Department of Physics, Toho University
Miyama 2 − 2 − 1, Funabashi, Chiba 274, Japan

ABSTRACT

Amplitude mode around a moving soliton in polyacetylene is analyzed by numerical simulations of its motion induced by applying an external electric field to a system with a static charged soliton within Su, Schrieffer and Heeger's(SSH) model. Its shape is visualized by numerically subtracting the Goldstone mode from phonons which excited by the moving soliton.

Linear modes around a static soliton have been studied in many papers for the SSH model[1] or its continuum version (the TLM model[2]).[3-5] For example, Terai and Ono[3] solved numerically the eigenvalue problem for the linear modes around a static soliton by using the SSH model subject to the periodic boundary condition. They have confirmed the existence of four localized modes. The purpose of this paper is to study the amplitude mode around a moving soliton.

The system is described by the modified SSH Hamiltonian,

$$H = -\sum_{n\sigma}(t_0 - \alpha y_n)[e^{i\gamma A}c_{n\sigma}^\dagger c_{n+1\sigma} + e^{-i\gamma A}c_{n+1\sigma}^\dagger c_{n\sigma}] + \frac{K}{2}\sum_n y_n^2 + \frac{M}{2}\sum_n \dot{u}_n^2,$$

where u_n is a lattice displacement of the n-th site from its equidistant lattice position, y_n defined as $y_n \equiv u_{n+1} - u_n$. The operator $c_{n\sigma}^\dagger$ and $c_{n\sigma}$ create and annihilate a π electron with spin σ at the n-th site, respectively. The nearest neighbor transfer integral of π electrons in the undimerized state is represented by t_0, and it depends on y_n through the electron-lattice coupling constant α. M is the mass of CH unit, and K the force constant due to the σ-bond. The uniform electric field is introduced by a time dependent vector potential A. The factor γ is defined as $\gamma \equiv ea/\hbar c$ with a the lattice constant, $-e$ the electronic charge and c the light velocity. We use the following parameters, t_0=2.5 eV, K=21 eV/Å^2, α=4.1 eV/Å, and a=1.22 Å.

The state with a static negatively charged soliton is obtained by solving the self-consistent equations for lattice displacements and electronic states numerically for total lattice sites $N = 99$ and total electron number $N_e = 100$ under the periodic boundary condition.[3] After the field is switched on, the time dependent wave functions are calculated from the time dependent Schrödinger equations and the lattice displacements from equations of motion by introducing discrete time, $\Delta t = 0.0025\omega_Q^{-1}$ (ω_Q the bare optical phonon frequency).[6]

The soliton is expected to propagate in a chain, exciting phonons. Since the time derivatives of the lattice displacements, $\dot{u}_n$, are proportional to the changes of the displacements, u_n, during a short time period, they can be expanded in terms of the linear modes around the soliton. The Goldstone mode leads to a translation of the soliton, so we can write its functional form as $G(n) = \partial_t \tilde{u}_n = -v_c \partial_x \tilde{u}_n$, with $\tilde{u}_n = (-1)^n u_n$, where v_c is the soliton velocity. We approximate the spatial derivative of $\tilde{u}_n$ by its spatial difference to get, $G(n) = -v_c(\tilde{u}_{n+1} - \tilde{u}_n)$. Subtracting $G(n)$ from $\dot{u}_n$, we end up with the contributions from other modes. Among them, we find the amplitude mode analogous to that for a static soliton(Fig.1).

We evaluate its frequency from the oscillation of the soliton width which is induced mainly by the amplitude mode.[3] We show the relation between the frequency and the velocity in Fig.2.

The frequency increases as the velocity deviates from zero. Although the frequency of the amplitude mode is about $0.52\omega_Q$ in the static case,[3] the present result shows a little larger value for an extrapolation $v_c \to 0$. The increment of the frequency from this extrapolation value is proportional to the square velocity in the region $v_c/v_s \leq 1$ (v_s being the sound velocity), which is consistent with the time reversal symmetry.

REFERENCES

1. W. P. Su, J. R. Schrieffer and A. J. Heeger: Phys. Rev. Lett. **42** (1979) 1689; Phys. Rev. **B22** (1980) 2099.
2. H. Takayama, Y. R. Lin-Liu and K. Maki: Phys. Rev. **B21** (1980) 2388.
3. A. Terai and Y. Ono: J. Phys. Soc. Jpn. **55** (1985) 213.
4. X. Sun, C. Wu and X. Shen: Solid State Commun. **56** (1985) 1039.
5. J. Hicks and G. Blaisdel: Phys. Rev. **B31** (1985) 919.
6. Y. Ono and A. Terai: J. Phys. Soc. Jpn. **59** (1990) 2893.

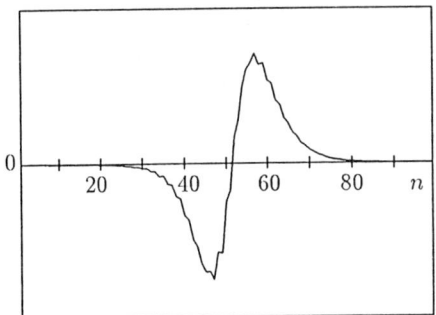

Fig.1: Shape of amplitude mode around a moving soliton. The intensity of the electric field is $0.01 E_0$, where E_0 is defined as $E_0 = \hbar\omega_Q/ea$ ($\sim 1.3 \times 10^7$V/cm). This snapshot is at 3350th time step. The abscissa indicates the site number n and the ordinate is in an arbitrary unit.

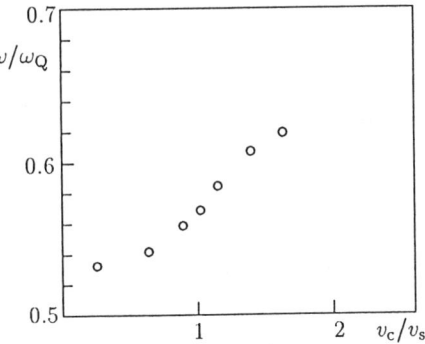

Fig.2: The velocity dependence of the amplitude mode frequency. The velocity is scaled by sound velocity $v_s (= \omega_Q a/2 = 1.53 \times 10^6$cm/sec) and the frequency by ω_Q.

DYNAMICS OF DNA IN GEL ELECTROPHORESIS

Mitsuhiro Matsumoto and Masao Doi
Department of Applied Physics, School of Engineering
Nagoya University, Nagoya 464-01, JAPAN

ABSTRACT

Brownian dynamics computer simulation was done to investigate dynamics of DNA in gel electrophoresis. Under steady electric field, long DNA molecules tend to have stretched conformation which is very different from the usual Gaussian distribution under zero field. This conformational change occurs drastically in a narrow range of field strength and chain length, and causes a "band inversion."

INTRODUCTION

Dynamics of highly charged polymers, e.g. DNA, in gel under (rather strong) electric field is now intensively studied both experimentally and theoretically because of its practical importance in biotechnology. In particular, various types of time-dependent electric field have been found effective to separate giant DNA molecules, but the basic mechanism is not still fully understood. Since this technique of gel electrophoresis with non-steady field contains many control parameters, it is essential to investigate its physical principles in order to optimize the technique and develop new methods. With this aim, we are conducting computer simulations to study various dynamic, as well as static, properties of DNA in gel electrophoresis.

SIMULATION TECHNIQUE

The model is a simple bead-spring type and not based on the reptation concept; the DNA chain consists of beads, which are connected with each other by springs. Due to external field, the chain migrates on a plane with many point obstacles which represent gel fibers. The chain is not able to cross the obstacles but is allowed only to slide on them. Each bead obeys the Langevin-type equation of motion, which contains the external force, the spring forces, and the Gaussian-type random displacement; the last one represents the kinetic fluctuations due to the surrounding solvent. For most cases reported here, the obstacles are arranged as a square lattice, but the effects of this regularity were found to be negligible, at least for this level of discussion, when we compared the results with those of an "irregular" lattice which is made by small shift of obstacles from their lattice points.

RESULTS AND DISCUSSION

We have carried out so far the simulation for several systems of different chain lengths under constant (steady) external field. We watched the conformational change of the chain during migration, and observed that (i) for most of the time, the chain is strongly stretched in the field direction; (ii) the migration of the chain is slowed down occasionally when the chain is trapped by obstacles; (iii) when this happens, the trapping is usually caused by one dominant preventing, or pinning, obstacle.

Although these observations do not appear to be consistent with the reptation model, quantitative analysis indicates that our model shows phenomena similar to those found in the previous simulation of the biased reptation model. For example, (i) the band inversion phenomenon, $i.e.$ the existence of a mobility minimum as a function of chain length, is observed. (ii) The distribution of the end-to-end vector (Fig. 1) shows a drastic change in shape around the chain length of the mobility minimum. These observations indicate that the reptation concept is also useful in the regime where the chain is strongly stretched in the direction of the field.

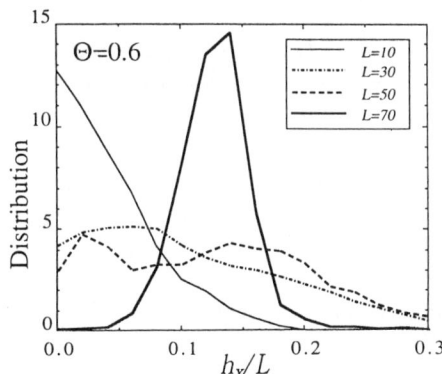

Fig. 1. Distribution of the component of end-to-end vector (h_x) along the field. The strength of the field (Θ in dimensionless unit) is moderate, roughly corresponding to 10 V/cm. The abscissa is normalized by chain length L.

We are also looking at the dynamic behavior of DNA in non-steady fields such as the field inversion method. When the field is inverted, the highly stretched chains tend to shrink, which causes the velocity overshoot phenomenon. The analysis of conformational relaxation is under way.

NEW RESULTS ON COMPUTER SIMULATION OF SPINODAL DECOMPOSITION IN BINARY FLUIDS

Tsuyoshi Koga and Kyozi Kawasaki
Department of Physics, Kyushu University 33, Fukuoka 812, Japan

ABSTRACT

The results of a computer simulation of spinodal decomposition in binary fluids are presented. These results, especially scaling function, are in good quantitative agreement with experimental results of polymer blends.

Recently, the late stage dynamics of the spinodal decomposition have been extensively studied in polymer blends,[1] from which one can obtain lots of information on the dynamical scaling law which is one of the most important properties in this subject. In such fluid systems, the domain growth in the late stage is driven by the flow induced by the surface tension of interfaces. Therefore, it is necessary to take the hydrodynamic effects into account in order to study the late stage dynamics in fluid systems.

We have already proposed a computationally efficient method, namely, the Fast Fourier Transform (FFT) method,[2] to simulate the ordering processes in binary fluids, where the dynamics is modeled by the following time-dependent Gintzburg-Landau (TDGL) equation for the order parameter $S(\vec{r})$:

$$\frac{\partial}{\partial t}S(\vec{r},t) = L\nabla^2 \mu(\vec{r}) - \nabla S(\vec{r}) \cdot \int T(\vec{r}-\vec{r}\,') \cdot \nabla' S(\vec{r}\,')\mu(\vec{r}\,')d\vec{r}\,', \quad (1)$$

where $\mu(\vec{r})$ is the chemical potential, L the Onsager kinetic coefficient and $T(\vec{r})$ the Oseen tensor. The second term on the right-hand side of Eq.(1) describes the long-range hydrodynamic interaction between the order parameter fluctuations.

We elsewhere reported about our computer simulations of spinodal decomposition in binary fluids in three dimensional cubic lattice of size 64^3 using FFT method and obtained the following results [2] : (i) The linear growth law of the characteristic length of domains in time is observed, which is peculiar to viscous fluids. (ii) The dynamical scaling law is satisfied in the time region where the hydrodynamic effects are dominant in domain growth. (iii) The scaling function is in accord with latest experimental results of polymer blends. These results were in qualitative agreement with the experimental results in polymer blends. The effects of the finite size of the system, however, affected these results, especially structure function at the small wave numbers.

To eliminate this effect, we perform another computer simulation run using a larger system which is a three-dimensional cubic lattice of size 128^3 with periodic boundary conditions. In this study, the results are averaged over two independent runs and the other parameters are the same as those in Ref.2.

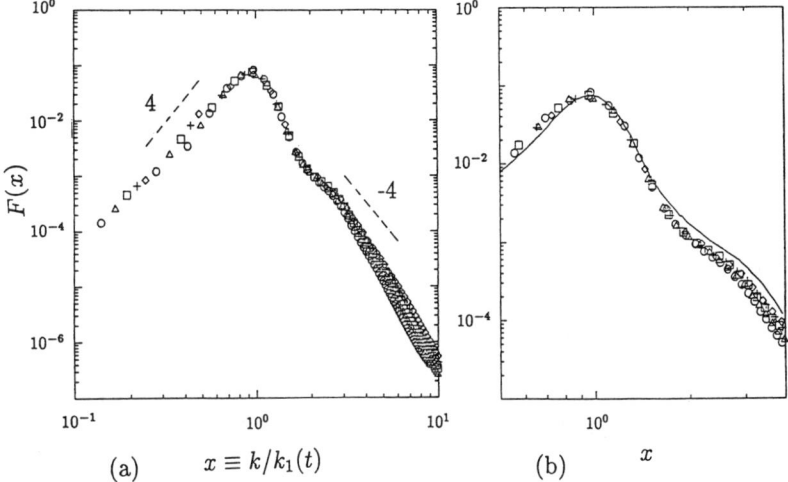

Fig.1. Plots of $F(x)$ vs x on double-logarithmic scales. The symbols $\circ, \triangle, \square, +$ and $\diamond$ correspond to $t = 400, 500, 600, 700$ and 800 time steps, respectively.

We study the scaling law for the structure function $I_k(t)$: $I_k(t) = k_1(t)^{-3} F(k/k_1(t))$, where $k_1(t)$ is the characteristic wave number defined as the first moment of $I_k(t)$ and $F(x)$ is the scaling function. The double logarithmic plots of $F(x)$ and a comparison with the experimental data are presented in Fig.1(a) and (b), respectively. The solid line in Fig.1(b) is the experimental result of polymer blends.[1] For large $x \equiv k/k_1$, $F(x)$ is still slightly time-dependent, but it approaches the experimental results, namely the form $F(x) \sim x^{-4}$, which is indicated by the dashed lines in Fig.1(a). This result corresponds to Polod's law. Except for this point, our results are in good quantitative agreement with the experimental data, especially the "shoulder" observed at $k \sim 3k_1$ and also the functional form of $F(x)$ at the small wave numbers. The details of these results will be published elsewhere.

REFERENCES

1 T. Hashimoto and M. Takenaka (preprint).
2 T. Koga and K. Kawasaki, Phys. Rev. A **44**, R817 (1991).

ACCELERATION OF SOLITON IN POLYACETYLENE BY EXTERNAL ELECTRIC FIELD – EFFECT OF SWITCHING

Yoshiyuki Ono and Makoto Kuwabara
Department of Physics, Toho University, Funabashi, Chiba 274, Japan

ABSTRACT

The time dependence of the velocity of the charged soliton in a polyacetylene subject to an elecric field is numerically studied. The velocity fluctuation is found to be quite sensitive to the switching speed of the field.

Solitons are thought to play an important role in the dynamical properties of a quasi-1D polymer t-$(CH)_x$ ($trans$-polyacetylene) which has a half-filled π-electron band.[1] A soliton in t-$(CH)_x$ is a topological defect in the dimerization pattern of the lattice due to the Peierls mechanism. Depending on the occupation number of the localized midgap state related to this defect, we have three kinds of solitons, a negatively or positively charged spinless soliton or a neutral soliton with spin. Since the distortion of the dimerization pattern and the electronic states are determined self-consistently, these solitons are complex excitations involving the electronic and lattice degrees of freedom. We are studying the dynamics of a soliton by numerical simulations.[2-4] Starting with a static soliton determined self-consistently,[5] we accelerate it by applying an external electric field. The dynamics of the soliton is determined by solving coupled equations of motion for the lattice displacements and the electronic wave functions. The advantage of this method is that it does not require any assumption on the form of a moving soliton.

In the present report we discuss the effect of switching speed of the electric field on the resulting soliton movement. The Hamiltonian used in the calculation is the so-called Su-Schrieffer-Heeger model[6] modified to include the time-dependent vector potential A giving rise to the uniform electric field $E = -\frac{1}{c}\dot{A}$;

$$H = -\sum_{n\sigma}(t_0 - \alpha y_n)[e^{i\gamma A}c_{n\sigma}^\dagger c_{n+1\sigma} + e^{-i\gamma A}c_{n+1\sigma}^\dagger c_{n\sigma}] + \frac{K}{2}\sum_n y_n^2 + \frac{M}{2}\sum_n \dot{u}_n^2, \quad (1)$$

where the bond variable y_n is related to u_n the displacement of the n-th site from its regular lattice position as $y_n = u_{n+1} - u_n$. $c_{n\sigma}^\dagger$ and $c_{n\sigma}$ are field operators of a π-electron with spin σ, t_0 the transfer integral in the undimerized state, α the electron-lattice coupling constant, M the mass of CH unit, and K the force constant due to the σ-bond. The vector potential is included through the Peierls substitution. The factor γ is defined as $\gamma \equiv ea/\hbar c$ with a the lattice constant. For explicit values of parameters we take those generally accepted for polyacetylene; t_0=2.5 eV, K=21 eV/Å^2, α=4.1 eV/Å, and a=1.22 Å. The total numbers of the lattice sites and electrons are set to be $N = 99$ and $N_e = 100$, respectively, and the periodic boundary condition is assumed. In this situation, the ground state is the one with a negatively charged soliton. The switching speed of the field is controled by changing the duration time τ which determine the time dependence of the field in the following way;[3] $\frac{1}{2}E[1 - \cos(\pi t/\tau)]$ for $0 \leq t < \tau$, E for $\tau \leq t < t_{\text{off}}$, $\frac{1}{2}E[1 + \cos(\pi(t - t_{\text{off}})/\tau))]$ for $t_{\text{off}} \leq t < t_{\text{off}} + \tau$, and 0 otherwise. As far as t_{off} is the same, the total impulse exterted to the system is the same.

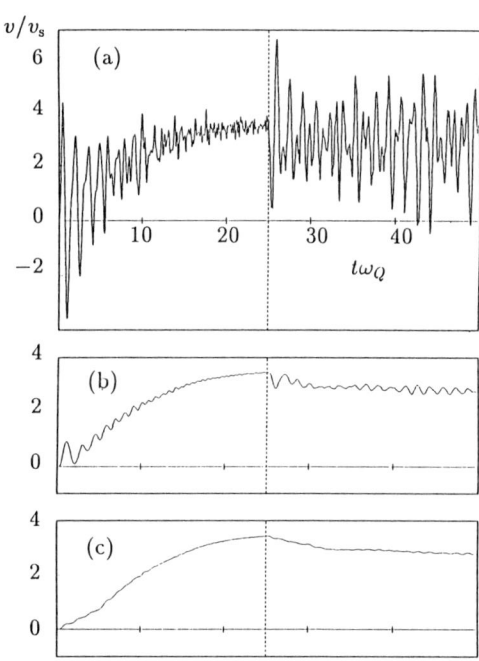

Fig. 1: Time dependence of of the soliton velocity v for different switching duration time τ; (a) 0, (b) $2\omega_Q^{-1}$ and (c) $5\omega_Q^{-1}$. v_s is the sound velocity and the vertical broken line indicates t_{off}.

In Fig. 1, we show some exampamples of the time dependence of the soliton velocity for different values of τ, where the soliton velocity was estimated from the time dependence of the center of mass of the charge distribution. The field strength E is fixed to be $0.02\hbar\omega_Q/ea$ ($\approx 2.6 \times 10^5$ V/cm), and the time is scaled by the inverse bare optical frequency ω_Q^{-1} (= $\sqrt{M/4K} \approx 0.4 \times 10^{-14}$ sec). From this figure we find that the fluctuation of the velocity is a decreasing function of the duration time. This behavior of the velocity fluctuation is one of the evidence that the soliton has an inner structure and its motion cannot be treated like the motion of a point mass. Since the inner structure has its origin in the electron-lattice coupling, it is not surprising that the velocity fluctuation gets quite small for $\tau \gg \omega_Q^{-1}$. In fact the plot of the velocity fluctuation vs. τ can be well described by a Lorentzian function $\tau_0^2/(\tau^2 + \tau_0^2)$ with $\tau_0 \approx 1.1\omega_Q^{-1}$. This behavior is common for both of the cases of switching on and off.

REFERENCES

1. A.J. Heeger, S. Kivelson, J.R. Schrieffer and W. -P. Su: Rev. Mod. Phys. **60**, 781 (1988), and references therein.
2. Y. Ono and A. Terai: J. Phys. Soc. Jpn. **59**, 2893 (1990).
3. M. Kuwabara, Y. Ono and A. Terai: J. Phys. Soc. Jpn. **60**, 1286 (1991).
4. Y. Ono, M. Kuwabara and A. Terai: J. Phys. Soc. Jpn. **60**, 3120 (1991).
5. A. Terai and Y. Ono: J. Phys. Soc. Jpn. **55** (1986) 213.
6. W.P. Su, J.R. Schrieffer and A.J. Heeger: Phys. Rev. **B22**, 2099 (1980).

III. EMULSION

A. EXPERIMENT

DYNAMIC SLOWING-DOWN IN DENSE MICROEMULSIONS NEAR THE PERCOLATION THRESHOLD

S.H. Chen
Department of Nuclear Engineering, Massachusetts Institute of Technology,
Cambridge, MA 02139, U.S.A.

F. Mallamace
Dipartimento di Fisica, Universita' di Messina, Vill. S. Agata,
I-98166 Messina, Italy.

J. Rouch
Centre de Physique Moléculaire, Optique et Hertzienne (U.R.A. n. 283 du C.N.R.S.),
Université Bordeaux I, 351 Cours de la Liberation, 33405 Talence Cedex, France.

P. Tartaglia
Dipartimento di Fisica, Universita' di Roma *La Sapienza*, P.le Aldo Moro 2,
I-00185 Roma, Italy.

ABSTRACT

We review a series of investigations of the static and dynamic properties of a three-component water-in-oil microemulsion system in which the molar ratio of water to surfactant is kept constant. This system behaves effectively like a two-component macromolecular fluid in which there are spherical, surfactant coated water droplets of macroscopic dimensions dispersed in a continuum of oil. The properties investigated include electrical conductivity, dielectric relaxation, shear viscosity and viscoelastic relaxation, static neutron and light scattering and dynamic light scattering. We focus mainly on the phenomena of the dynamic slowing-down of the dielectric relaxation and the droplet density fluctuations as the system approaches the percolation threshold from below, both in temperature and in volume fraction. A theory of static and dynamic light scattering, formulated along the lines of scattering from a system of polydisperse fractal clusters, quantitatively accounts for the dynamic slowing-down phenomenon and the non-exponential decay of the time correlation function.

INTRODUCTION

The microemulsion system, consisting of water, decane (oil) and an anionic surfactant AOT (sodium di(2-ethylhexyl) sulfosuccinate), is a model system in the sense that, in the vicinity of room temperature, the ternary phase diagram is dominated by an L_2 phase, a single phase water-in-oil microemulsion. Extensive light scattering [1] and small angle neutron scattering [2] (SANS) experiments showed conclusively that the L_2 microemulsions are made up of water droplets coated by a monolayer of surfactant molecules, much like a dense fluid containing macroscopic particles suspended in oil. The average size of the water core is a linear function of the molar ratio, X = [water] / [AOT], and the size distribution closely follows a Schultz distribution with a polydispersity of about 30% [3]. At X = 40.8 the average hydrodynamic radius of a droplet has been determined by dynamic light scattering to be 85 Å and is invariant on approaching the phase separation line and after phase separation. Thus at constant X the microemulsion behaves as a pseudo-binary system characterized by a phase diagram in T-ϕ plane at constant pressure, where ϕ is the volume fraction of water plus surfactant in the three-component mixture. This system is dilutable by oil in a continuous fashion, thus one can change the volume fraction ϕ from nearly zero up to 0.9. The T-ϕ phase diagram exhibits a rich variety of phases: there is a cloud point curve separating the one-phase and two-phase microemulsions, including a lower consolute (critical) point at 40 °C and $\phi = 0.1$ [4]. SANS experiments have established that the order parameter is the volume fraction [5]. The cloud point curve terminates abruptly at $\phi=0.4$ and from there on the lamellar phase (L_α phase) boundary appears, marking the separation of the L_α phase from the L_2 phase at lower temperatures. Within the temperature range of the one-phase region there is a percolation line, defined as the locus of the electrical percolation thresholds, cutting across from low to high volume fractions in the temperature range from 40 to 20 °C [6]. The striking electrical percolation phenomenon can be observed either by keeping the temperature constant and increasing ϕ, or by keeping ϕ fixed and varying the temperature. Thus it is a unique trajectory in the phase diagram. In Fig. 1 we display the phase diagram, where the solid symbols, marking different phase boundaries and the percolation thresholds, were determined either by electrical conductivity or by light transmission measurements [7].

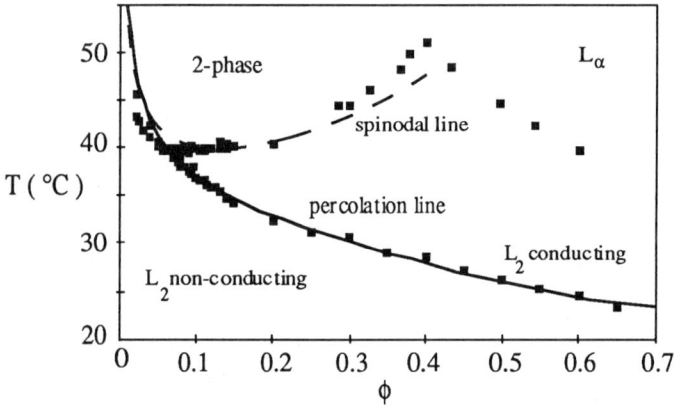

Fig. 1. The temperature vs volume fraction phase diagram of the microemulsion system.

The critical phenomena near the cloud point curve have been extensively investigated by light scattering [8,9,4] and by SANS [5]. Along the critical iso-ϕ line the correlation length ξ has been determined to obey $\xi=\xi_0 \, \varepsilon^{-\nu^*}$ and the osmotic compressibility χ obeys $\chi=\chi_0 \, \varepsilon^{-\gamma^*}$ with $\xi_0 = 13.3 \pm 1.5$ Å, $\chi_0 = 0.015 \pm 0.005$, $\nu^* = 0.70 \pm 0.04$ and $\gamma^* = 1.40 \pm 0.09$. The critical exponents are consistent with the renormalized values $\nu^* = \nu/(1-\alpha)$ and $\gamma^* = \gamma/(1-\alpha)$ where ν, γ and α have their Ising values. As was pointed out in ref.[9], the unusual feature of this critical phenomenon involving solutions of macroparticles is the existence of a crossover from the critical behavior to the single particle behavior at a finite distance from the critical point. The phenomenon has its origin in the fact that the long range correlation length in the system cannot be shorter than the size of the macroparticle for the critical phenomena to be solely dominated by one length scale. For example, by equating the long range correlation length to the average hydrodynamic diameter of the droplets we can estimate the crossover temperature to occur at 8.5 °C below the critical temperature. This crossover behavior is clearly seen in Fig. 2 both in the static and the dynamic critical phenomena. In Fig. 2 we plot the intensity and the linewidth of the scattered light observed at 90° scattering angle. The solid circles are the measurements and the lines are the standard Ornstein-Zernike plot and the mode-coupling theory results [10]. It is clearly seen from the diagram that theory and experiments deviate significantly for temperatures lower than 8 °C from T_c. Coming back to the dynamic slowing-down phenomenon, the theory of the phenomena is due to Kawasaki [11]. In the analysis of the experiment according to this theory, we found that, owing to the large dimensions of the droplets, the so-called background linewidth is unusually large, amounting to 15% of the total linewidth [12]. This comes about because of the large Debye wavenumber cutoff q_D, which enters the viscosity critical anomaly. It has a value equal to the inverse of the droplet size. The background linewidth calculated according to the theory of ref. [13] is proportional to the inverse Debye cutoff, thus proportional to the size of the droplet [4, 10]. Taking into account this correction, the resulting critical linewidth as a function of the distance to the critical point shows an excellent agreement with the mode-coupling calculation of Lo and Kawasaki [14]. It should be remarked here that this size effect on critical phenomena is universal for the systems which contain large particles, for example the micellar solutions containing non-ionic surfactants [15].

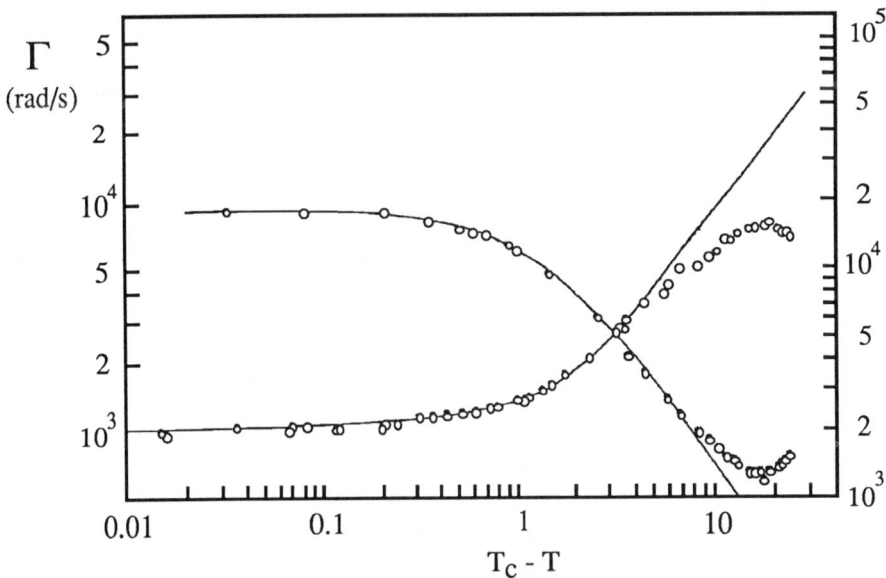

Fig. 2. The intensity and the linewidth of the scattered light in the critical region.

In the following, we shall turn our attention to denser microemulsions, near the percolation threshold, where the dynamic slowing-down can also be observed. We shall show that the static and dynamic phenomena in this regime bear a striking similarity to the critical phenomena that we described above.

PERCOLATION PHENOMENA

The thermodynamic behavior of water-in-oil microemulsions, being electrically neutral, is dominated by van der Waals attractive interaction. In particular, for AOT based microemulsions, there is evidence that an additional entropy dominated attractive interaction exists between the droplets. In fact, this effective attractive interaction, which produces the cloud point phase separation phenomena, can be shown experimentally to be temperature dependent [2, 16]. From SANS experiments near the critical point it has been deduced that the interaction strength is a quadratic function of absolute temperature [2, 17]. The attractive interaction would lead to clustering of the microemulsion droplets at elevated temperatures, depending on the volume fraction. At low volume fractions and low temperatures, where the droplets do not cluster, the conductivity of the solution, which is many order of magnitude larger than the conductivity of decane, can be shown to be due to the mobility of the positively and negatively charged droplets under the applied electric field [18, 19]. The charge fluctuations of a droplet are caused by the exchange of its surfactant anions with the surrounding pool of surfactant monomers at the critical micellar concentration of the solution [6, 7]. When a surfactant anion goes away from a droplet, leaving behind a counterion in the water core, the droplet will be positively charged by one unit; on the other hand, if one surfactant anion from the surrounding pool is incorporated into the microemulsion droplet, the droplet will be positively charged by one unit. As the volume fraction or the temperature increases, the microemulsion shows microscopic clustering of droplets. In this domain the mechanism of conduction is no longer due to the diffusion of the charged droplets, but it is mediated by hopping of certain charge carriers on the fractal clusters. This mechanism is usually called stirred or dynamical percolation, and has been developed in refs. [20,21]. The conductivity increases dramatically in this regime by six orders of magnitude, indicating the distinctly different conduction mechanism. A typical example is shown in Fig. 3.

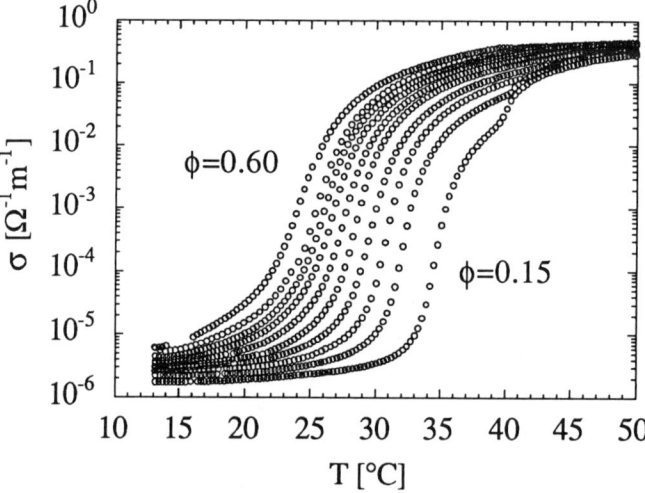

Fig. 3. The set of measured electrical conductivity data as a function of temperature at fixed volume fraction.

The open circles show the conductivity at a given ϕ as a function of temperature. The percolation point is defined as the inflection point of the sigmoidal curve. The two curves at $\phi = 0.15$ and $\phi=0.6$ show in addition discontinuities at the phase boundary of the binary phase separation and at the L_2-L_α phase boundary respectively. The conductivity near the threshold can be expressed as

$$\sigma = A\left(\frac{T_p-T}{T_p}\right)^{-s'} \quad (1)$$

coming up from below the threshold, and

$$\sigma = B\left(\frac{T-T_p}{T_p}\right)^{t} \quad (2)$$

going down from above. The exponents s' and t have been determined to be 1.2 ± 0.1 and 1.9 ± 0.1 respectively [6]. The same exponents, when T is fixed but with variable ϕ, have also been obtained [6]. In Fig. 4 we plot the scaled conductivities $T_p (\sigma/A)^{-1/s'}$ and $T_p (\sigma/B)^{1/t}$ vs $T-T_p$. We observe from this figure that, while above the threshold, all the data at different volume fractions follow the asymptotic relation of eq. (2) very well, below the threshold, the asymptotic relation of eq. (1) holds better for the higher volume fractions. The exponent s' is called the index of dynamic percolation, which is distinct from the static percolation exponent $s = 0.73$ [23]. The exponent t above percolation agrees with the static percolation index. By locating the percolation threshold from Fig. 3 we get a set of percolation loci as a function of temperature and volume fraction. These are plotted in Fig. 1. It is desirable to understand the percolation line in the phase diagram, specially in relation to the position and shape of the cloud point curve.

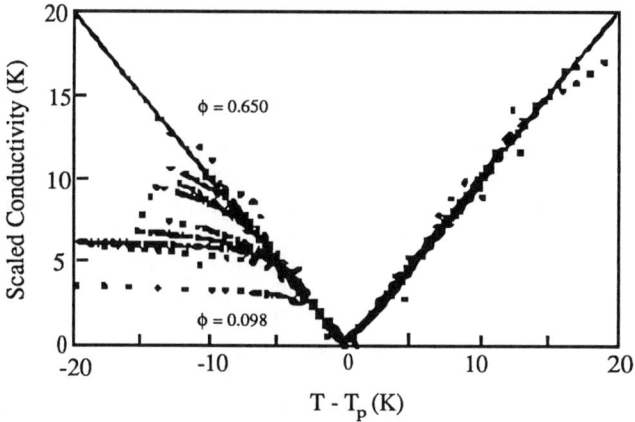

Fig. 4. The scaled conductivity as a function of the distance in temperature from the percolation threshold.

The analytical description of the percolation line can be given using a continuum percolation theory proposed some time ago by Xu and Stell [24]. They considered a system of hard spheres interacting via an attractive Yukawa pair potential and constructed the Ornstein-Zernike equation for the pair-connectdness function $h^+(r)$, introduced previously by Coniglio et al. [25]. This function is proportional to the probability of finding two particles in the same cluster, separated by a distance r. The meaning of the cluster has to be separately defined. The solution of the Ornstein-Zernike equation for the pair-connectdness function in the mean spherical approximation allows then the determination of the percolation threshold as the point where the average cluster size diverges. The counterpart of the thermal problem in the same theory allows the calculation of the spinodal line and the associated critical consolution point. A modification of this theory is needed in order to apply it to AOT-based microemulsions which, as we showed in Fig.1, have a lower consolute point with an inverted coexistence curve. As we stated earlier in the Introduction, the interaction strength of the Yukawa potential needs to be temperature-dependent in order to take into account the entropic dependence of the effective interaction. Writing the Yukawa potential outside the hard core σ_h as

$$\beta u(x) = - K e^z \exp[-zx]/x, \quad x=r/\sigma_h>1 \quad (3)$$

where $\beta=1/k_B T$, the temperature dependence of the strength parameter is then introduced as

$$K e^z = K_0 (1+K_1 \varepsilon + K_2 \varepsilon^2 + K_3 \varepsilon^3) \quad (4)$$

where $\varepsilon = (T_c-T)/T_c$. We then choose the four parameters K_0, K_1, K_2, K_3 and z, the inverse range of the potential, measured in units of the hard sphere diameter, in order to fit simultaneously the spinodal and the percolation lines. The equations that define the two curves are respectively

$$K \phi = C(z,\phi)/z^2 \quad (5)$$

and

$$K \phi = C(z,0)/(\lambda z^2) \quad (6)$$

where the function $C(z,\phi)$ is given explicitly in eq. (34) of ref. [24] and λ is a proportionality constant relating K to K^+, the strength parameter in the percolation problem, $K^+ = \lambda K$. The definition of percolation we adopt here is the so-called "percolation in probability" of Xu and Stell. It has the following meaning. In solving the Ornstein-Zernike equation for the pair-connectdness function $h^+(r)$, one assumes the direct correlation function $c^+(r)$ to have the same Yukawa form as eq.(3) outside the core, with the strength parameter K^+. Within the core one takes $h^+(r)=0$, an exact boundary condition for a hard sphere system. One can use the known critical parameters to fix two of these constants. Knowing that the critical volume fraction ϕ_c is approximately 0.1 (see Fig. 1) one chooses the range parameter z=0, corresponding to having a 1/r unscreened Yukawa potential. This gives $\phi_c = 0.1286$ close to the experimental value 0.1. Knowing that the value of T_c is approximately 313 K, λ can be calculated to be equal to 0.75. The percolation line and the spinodal curve are simultaneously fitted with the values of the three constants $K_1 = 49$, $K_2 = 957$, $K_3 = 6882$, while K_0 is fixed by the critical point parameters to have a value $K_0 = 0.888$. In Fig. 1 the solid lines are the result of such a fit. The agreement of the percolation line with the measured thresholds is rather good, considering the crudeness of the mean spherical approximation. The fact that the spinodal curve does not accurately reproduce the measured coexistence curve is due to the use of a mean spherical model which has the lowest possible value of ϕ_c at 0.1286, in slight disagreement with the experiments.

Let us turn now to the study of the relaxation phenomena involving motions of charged species at the interfaces, namely the complex permittivity as a function of frequency. Fig. 5 shows experimental data[22] for the real ε' and the imaginary part $\omega \sigma$ of the permittivity, related to the dielectric constant ε' and the conductivity σ of the sample.

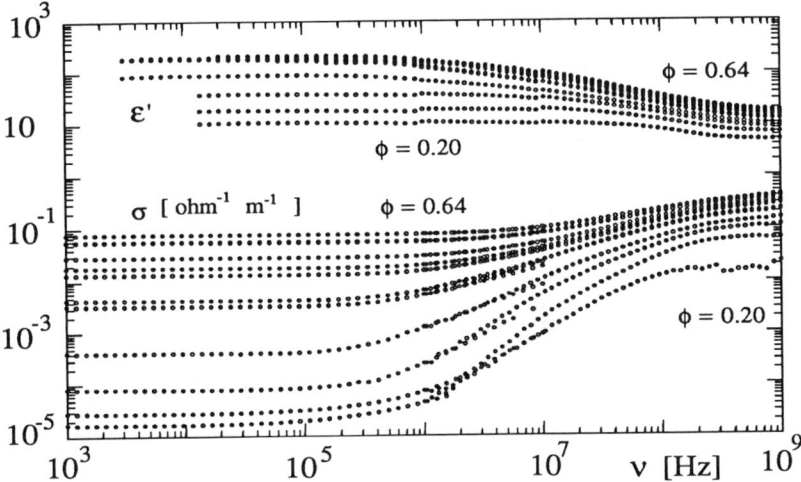

Fig. 5. The dielectric constant and the conductivity as a function of frequency for various volume fractions.

The measurements were made at a constant temperature of 23.5 °C and a series of volume fractions spanning from 0.2 to 0.64. At this temperature, the percolation threshold was determined to be ϕ_p=0.32. The frequency spans the range from 10^3 to 10^9 Hz. It is clear from the data that the microemulsion does not show any relaxation phenomena up to 10^5 Hz, and beyond that one begins to observe dispersion and absorption so that the average relaxation frequency can be estimated to be 10^7 Hz. This frequency is probably related to the motion of the charged species at the interface, e.g. to their hopping frequency from droplet to droplet belonging to the same cluster. The detailed inspection shows that the magnitude of the relaxation is much larger below the percolation threshold than above. The complete set of data for the complex permittivity can be well fitted by the Cole-Cole relaxation function given below

$$\varepsilon(\omega,\phi) = \varepsilon_\infty(\phi) + \frac{\Delta\varepsilon(\phi)}{1+[i\omega\tau(\phi)]^{1-\alpha(\phi)}} \qquad (7)$$

We extracted from the experimental data the amplitude factor $\Delta\varepsilon(\phi)$, the average relaxation time $\tau(\phi)$ and the exponent, $\alpha(\phi)$, between zero and one. The magnitude of α expresses the deviation from the simple Debye single-frequency relaxation, which takes into account the distribution of relaxation times. The Cole-Cole dispersion law can be related to the stretched exponential form of the time correlation function in the following way:

$$\frac{1}{1+[i\omega\tau(\phi)]^{1-\alpha(\phi)}} = -\int_0^\infty dt\, e^{-i\omega t} \frac{d}{dt} \exp\left[-\left(\frac{t}{\tau}\right)^\beta\right] \qquad (8)$$

Thus the Kohlrausch-Williams-Watts exponent β can be easily computed from the value of the exponent α. In Fig. 6 we show the values of the two exponents α and β as a function of the volume fraction. It is interesting to note that at the percolation threshold the values of the two exponents nearly coincide, namely α = 0.45 and β = 0.42. At this volume fraction, β has the minimum value, while α is at its maximum value, signifying the broadest distribution of relaxation times.

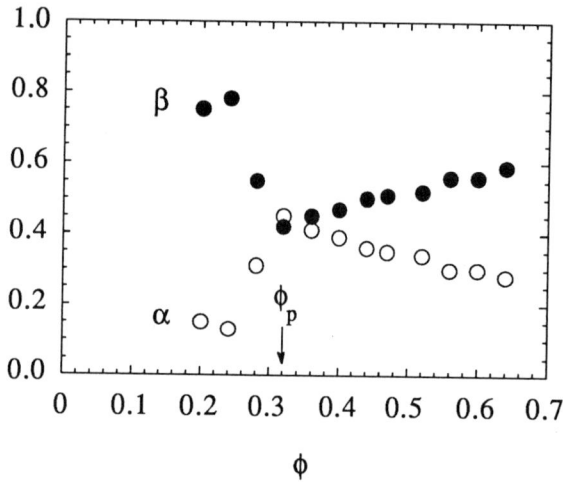

Fig. 6. The exponents α of the Cole-Cole relaxation formula and the Kohlrausch-Williams-Watts exponent β as a function of volume fraction.

Fig. 7 illustrates the volume fraction dependence of the average relaxation time τ. The figure indicates that near the percolation threshold the average relaxation time increases rapidly as a function of increasing volume fraction, but the relaxation time itself at the threshold is about 25 ns, corresponding to the average relaxation frequency of 10^7, as pointed out earlier.

Besides the rapid increase in the dielectric relaxation time or the dynamic slowing-down phenomena when approaching the percolation threshold from below, one also observes an anomaly in the transport coefficient shear viscosity. Fig. 8 illustrates a set of extensive measurements performed by Majolino et al.[26] in this microemulsion system as a function of temperature and volume fraction. A well defined maximum of the shear viscosity is observed at temperatures which decrease with increasing volume fractions. In the same figure the percolation line that we showed in Fig. 1 is traced through the curves as a dashed line. It should be noted that the line $T_p(\phi)$ crosses the viscosity curves in their steepest rising part to the peaks.

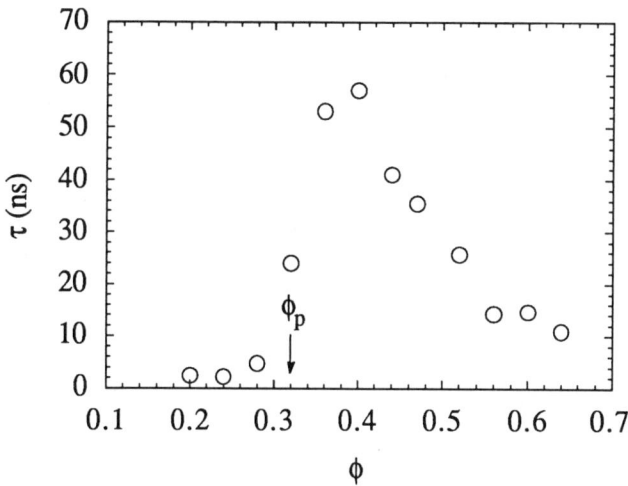

Fig. 7. Volume fraction dependence of the average dielectric relaxation time

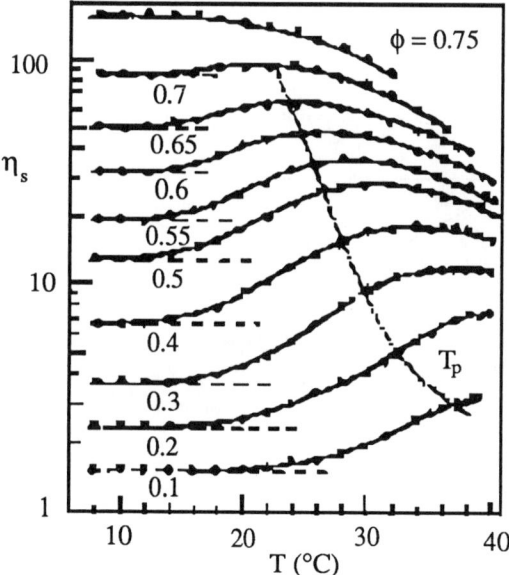

Fig. 8. The shear viscosity relative to decane as a function of temperature for various volume fractions. T_p denotes the percolation line.

The frequency dependence of sound absorption also shows a behavior similar to dielectric relaxation, i.e. a Cole-Cole relaxation [27]. We made a combined measurement of sound velocity and shear viscosity from which we could compute the static classical absorption coefficient due to the shear viscosity alone $[\alpha/f^2]_s$. The most important feature of the experimental observation is made by comparing the measured absorption coefficient $[\alpha/f^2]_{exp}$ with the classical absorption coefficient as a function of frequency. For small values of ϕ, up to 0.3, we find the measured absorption coefficient greater than the classical one, while for ϕ greater or equal to 0.3, the measured absorption coefficient is always smaller than the classical coefficient. This fact is a direct evidence of viscoelastic behavior for microemulsions of volume fractions larger than 0.3. This is an unusual property of the microemulsions when compared to the behavior of the usual molecular liquids. In order to study the relaxational properties of sound absorption we constructed the scaled quantity

$$\frac{(\alpha/f^2)_{exp} - B}{(\alpha/f^2)_s} \qquad (9)$$

where B is the experimentally determined high frequency background absorption, and assume tentatively that it is given by a Cole-Cole type relaxation

$$\frac{1}{1+\left|2\pi\tau_s f\right|^{2(1-\alpha')}} \qquad (10)$$

where the relaxation time τ_s is proportional to the shear viscosity. We plot the scaled absorption as a function of the scaled frequency $2\pi f \tau_s$ in Fig. 9 and obtain a scaling plot where all the experimentally measured points collapse, following a line of constant slope.

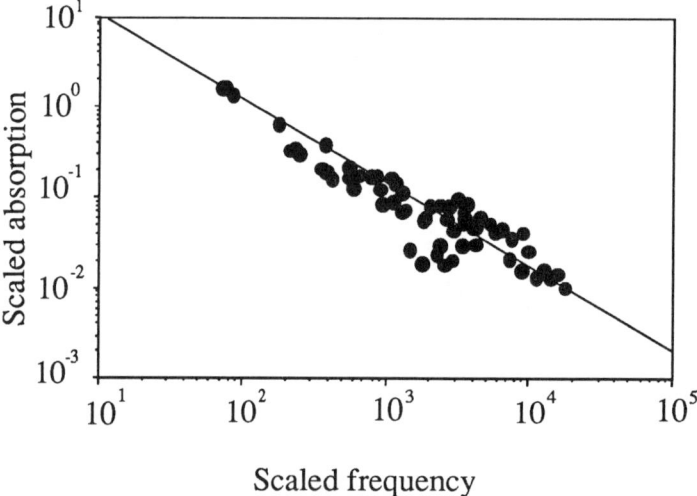

Fig. 9. The scaled sound absorption as a function of the scaled frequency, showing the high frequency Cole-Cole relaxation behavior.

This results clearly indicate that the system has a relaxation frequency which is centered at lower frequencies, less than a MHz, with respect to the ones we used in our measurements (from 10 to 300 MHz), so that we are observing only the tail of the relaxation process. The parameter α', which gives the width of the distribution of relaxation times turns out to be equal to 0.54.

STATIC AND DYNAMIC LIGHT SCATTERING

The most direct evidence of cluster formation as one approaches the percolation threshold was provided by light intensity measurements at low angles. While the SANS experiments [28] give the structure factors of dense microemulsions similar to liquids with densely packed spheres, with the first diffraction peak occurring at $2\pi/d$, where d is approximately the diameter of the droplets, the low-q intensity remains flat up to q = 0.01 Å^{-1}. However, the static light scattering experiments of Magazù et al[29] showed clearly that in the very low q range (for q less than 0.0003 Å^{-1}) the scattering intensity rises up to a Lorentzian-like peak centered around q=0. Fig. 10 shows one set of typical measurements made at 25 °C at different volume fractions. Starting from lowest volume fraction 0.05 the low angle scattering intensity increases as one approaches the critical volume fraction 0.1 and then decreases back to the original intensity at 0.2 (left hand side of Fig. 10). This is a clear indication of critical scattering. However, as the volume fraction increases from 0.2 to 0.44 and 0.48, the scattering intensity increases again (right hand side of Fig. 10). This is another clear indication of the approach to the percolation point which is at ϕ_p=0.5 for T=25 °C (see Fig. 1). It is well known from the liquid diffraction pattern studies that the zero angle peak in the structure factor indicates a strong correlation of particles due to an attractive interaction. Thus we shall present in the following a theory of static and dynamic light scattering assuming explicitly the presence of polydisperse fractal clusters near the percolation threshold.

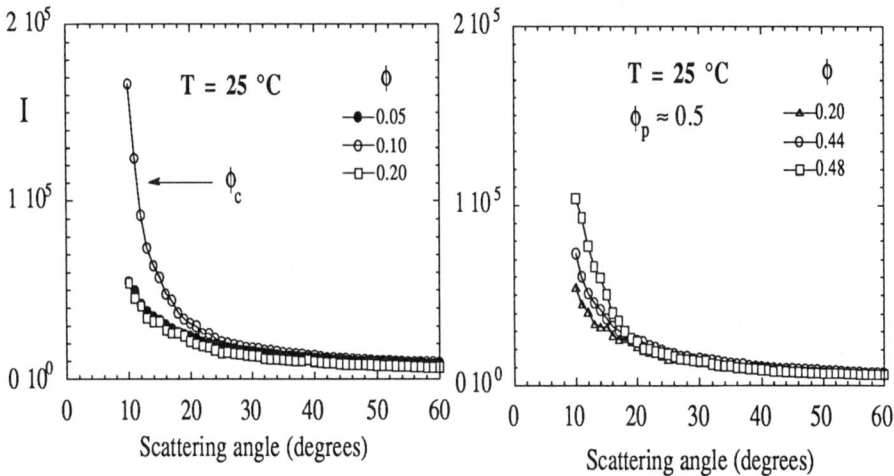

Fig. 10. The intensity of scattered light as a function of the angle at 25 °C. Left: approach to the critical volume fraction ϕ_c. Right: approach to the percolation threshold ϕ_p.

Static Light Scattering Intensity

In general the light scattering intensity I(q) from a collection of spherical particles of radius R can be written as

$$I(q) = I_0 \, P(qR) \, S(q) \qquad (11)$$

where P(qR) is the particle structure factor and S(q) is the interparticle structure factor. Since R is about 85 Å the value of qR in the light scattering range is smaller than unity and the particle structure factor can be effectively taken as unity. The interparticle structure factor can be calculated from the pair correlation function g(r) by

$$S(q) = 1 + \int_0^\infty dr \, 4\pi r^2 \, \frac{\sin(qr)}{qr} \, n \, g(r) \qquad (12)$$

where n is the number density of the particles and in the small angle region the unity can be neglected. We now make a model of the pair correlation function for a dense fluid consisting of spherical particles of radius R_1 forming polydisperse clusters of fractal dimension D. We assume the cluster size distribution function N(k) for a cluster of k particles is given by

$$N(k) \approx k^{-\tau} e^{-k/s} \qquad (13)$$

where t is the polydispersity exponent and s is related to the average number of particles in a cluster. The pair correlation function g(r) is then written as $g(r) = \langle g_k(r) \rangle_k$ where

$$n \, g_k(r) = \frac{D}{4 \pi R_1^D} \frac{e^{-\frac{r}{R_k}}}{r^{3-D}} \qquad (14)$$

and the average is over the distribution k N(k). The R_k in eq. (14) is the size of a k-cluster and can be written as $R_k = R_1 \, k^{1/D}$. Based on these assumptions, the structure factor can be calculated which has a complex expression [30]. A good approximation, in both the small and the intermediate q range turns out to be

$$S(q) \approx \left(\frac{\xi^2 / R_1^2}{1 + q^2 \xi^2} \right)^{\frac{D(3-\tau)}{2}} \qquad (15)$$

where the correlation length ξ is defined in terms of the average cluster size s as

$$\xi = R_1 \, [D(D+1)/6]^{1/2} \, s^{1/D} \qquad (16)$$

In Fig. 11 we plot the scaled intensity I(q)/I(q=0) as a function of the dimensionless parameter x = qξ. In the upper corner of the inset the actual intensity data were fitted with the expression (15) by choosing D=2.5 and τ=2.2, thus determining the correlation length ξ as a function of volume fraction near the percolation threshold. The lower inset shows that the correlation length can be well approximated by a power law divergence in the volume fraction as

$$\xi = 257 \text{ Å } (0.59 - \phi)^{-0.88} \tag{17}$$

The predicted scaled intensity according to eq. (15) is drawn by a solid line in the figure and it agrees remarkably well with the experimental data of the angular dependence of the intensity for different volume fractions at 25 °C.

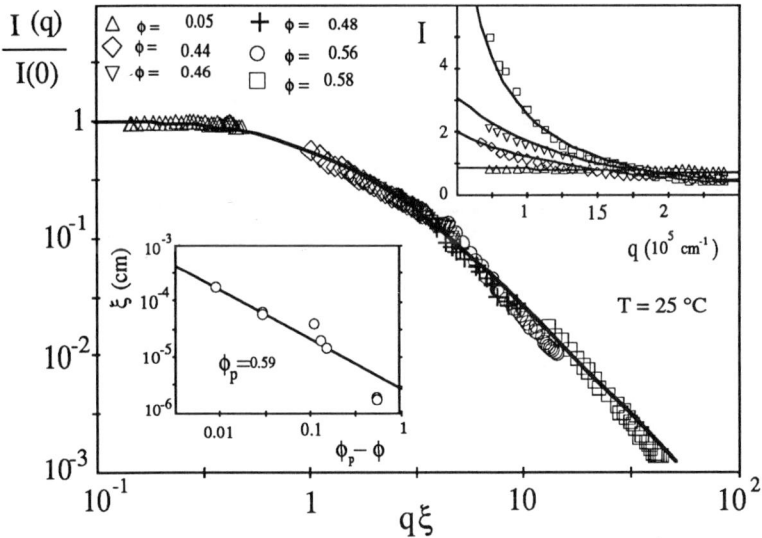

Fig. 11. The scaled scattered intensity vs the scaling variable $x = q\xi$. The symbols are experimental data at 25 °C and different volume fractions (from ref. [29]) and the solid line denotes the theoretical result $1/(1+x^2)$. Insets are explained in the text.

Dynamic Light Scattering

The time correlation function measured in conventional photon correlation spectroscopy can written as $c(q,t) = G(q,t)/G(q,0)$, where $G(q,t)$ is calculated as

$$G(q,t) = \int_1^\infty dk\, N(k)\, k\, S_k(q)\, e^{-D_k q^2 t} \tag{18}$$

where the $S_k(q)$ is the three-dimensional Fourier transform of $g_k(r)$ defined in eq. (14), its analytical form is given in ref. [31]. $D_k = D_1 k^{-1/D}$ is the diffusion coefficient of a k-cluster. The analytical form of $c(q,t)$ can be given and it is reported in ref. [30]. The most important feature of this time correlation function is that it decays initially as an exponential function $\exp[-\Gamma_c t]$ with a first cumulant Γ_c and then gradually crosses over to a stretched exponential function at long times with an asymptotic formula given by $\exp[-(\Gamma_s t)^\beta]$. The Kohlrausch-Williams-Watts exponent β has a universal value

$$\beta = D/(D+1) \qquad (19)$$

The dynamic slowing-down can be seen by plotting a scaled first cumulant $\Gamma_c^* = \Gamma_c / R_1 D_1 q^3$ as a function of the scaled variable $1/x$. This is shown in Fig. 12. Our theory, embodied in eqs. (18) and (19) predicts that far away from the percolation threshold, where $x \ll 1$, Γ_c^* goes like $1/x$, while close to the percolation threshold, where $x \gg 1$, it approaches a constant slightly below unity.

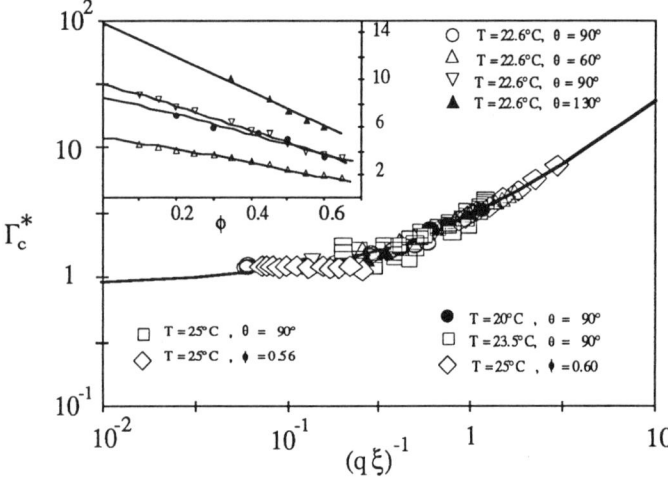

Fig. 12. The scaled first cumulant plotted as a function of the scaled variable $1/x$. The inset gives some of the actual first cumulant as a function of volume fraction, showing the dynamic slowing-down on approaching the percolation threshold.

The first cumulant data, which are the most accurately measured quantities in photon correlation spectroscopy, follow this scaling relation closely, as can be seen in Fig. 12. The analytical expression for $c(q,t)$ can also be used to fit the normalized correlation functions directly. This is shown in Fig. 13 for three volume fractions. Note that the theory uses only a single β value, 0.714, which is obtained by taking the fractal dimension as 2.5. A phenomenological single stretched exponential function can also be used to fit these measured photon correlation functions closely, but with variable stretched exponents. The stretched exponents would be close to unity for low volume fractions and decrease to 0.7 when one approaches the percolation threshold at higher volume fractions at constant temperature. The evolution of the normalized time correlation function can be best seen by plotting a double logarithmic $c(q,t)$ vs $\ln(\Gamma_c t)$. This is shown in Fig. 14. One clearly sees an evolution from an exponential to a stretched exponential with crossover occuring at $\Gamma_c t$ roughly equal to one. All the experimentally measured photon correlation functions unfortunately fall into this crossover region, thus can be fitted with a single stretched exponential with variable β. In practice this means that the universal exponent β is experimentally extremely difficult to measure accurately. In order to do so one needs to measure very accurately the photon correlation function over extremely wide time ranges, spanning five decades in time.

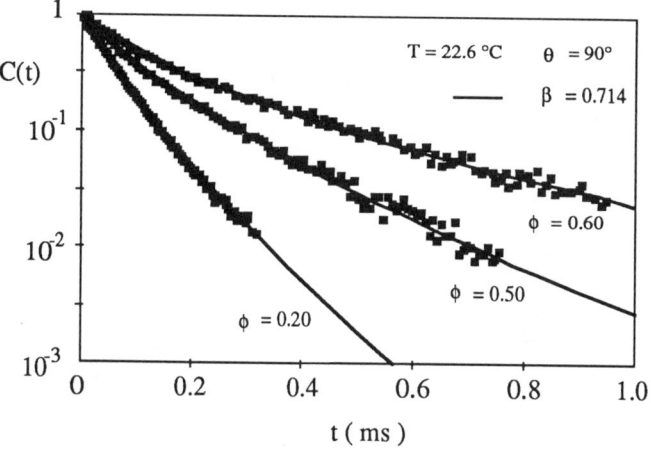

Fig. 13. Photon correlation functions at three volume fractions fitted with a universal correlation function having a single value of the stretched exponent $\beta = 0.714$.

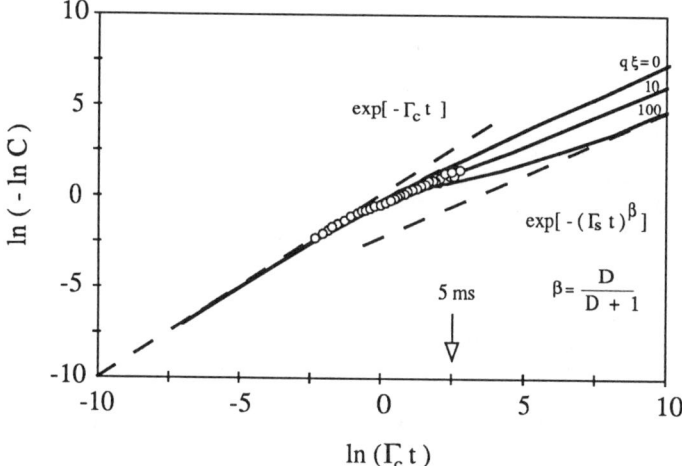

Fig. 14. Time evolution of the universal correlation function showing the evolution from exponential to stretched exponential. Open circles are experimental points.

Finally we wish to comment on the analogy of the dynamic slowing-down on approaching the percolation threshold we discuss in this paper with the well known critical slowing-down near the consolute critical point. Fig. 15 show such a comparison. The data on the left hand side figure are taken near the percolation threshold and the solid line represent our theory. The data on the right hand side figure are taken near the critical point of the same microemulsion system. The solid line represent the result of the Kawasaki mode-coupling theory. This similarity of the slow dynamics of a system with polydisperse fractal clusters and a system with critical fluctuations have been pointed out by Martin and Ackerson [32] before. The point of more than a casual interest here is that we found a system where both phenomena coexist and occur in the same phase diagram.

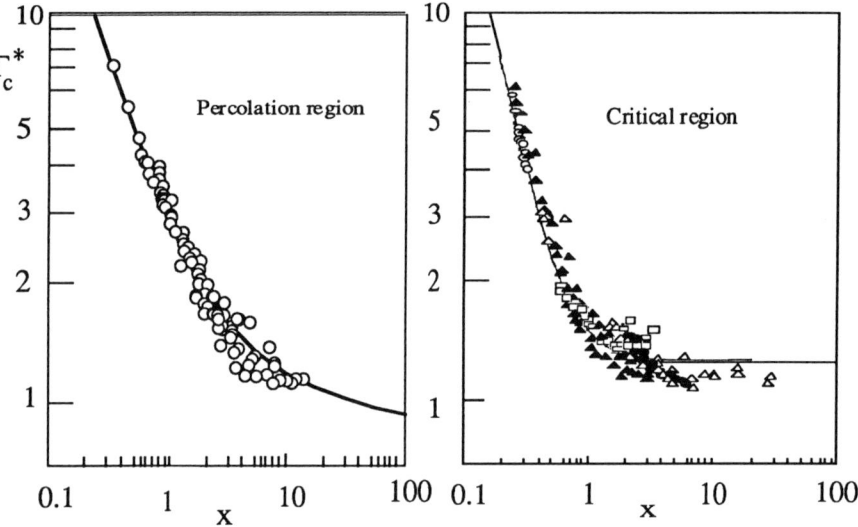

Fig. 15. Comparison of dynamic slowing-down in the percolation and the critical regime.

CONCLUSIONS

We have emphasized in this short review that water-decane-AOT microemulsions show a rich variety of interesting thermodynamic and transport properties. It is a model system in complex liquids for which many statistical mechanical models can be developed to explain its static and dynamic properties. The most striking example is the occurrence of a thermodynamic percolation phenomenon which, together with the cloud point phenomenon, dominates the entire phase diagram. The relation between the two can be explained satisfactorily by the suitably modified theory of continuum percolation due to Xu and Stell. Aside from the well-known critical phenomena associated with the cloud point curve, the system shows a number of slow dynamic relaxations of non-exponential character. These include the dielectric and ultrasonic relaxations and the droplet density correlation functions. The microscopic origin of these non-exponential relaxation phenomena can be attributed to the formation of polydisperse fractal clusters, due to the attraction between the droplets, as one approaches the percolation threshold from below. The slow dynamics is the result of the divergence of a correlation length which is in turn related to the average cluster size. The static and dynamic light scattering data can be analyzed by a theory which bears striking similarity to the well known Ornstein-Zernike theory for static and the mode-coupling theory for dynamic critical phenomena.

Acknowledgements
    The research of SHC is supported by a grant from the Division of Materials Science of the Department of Energy DE-FG01-91ER45429. The research of PT is supported by the Gruppo Nazionale di Struttura della Materia del Consiglio Nazionale delle Ricerche.

REFERENCES

1. M. Zulauf and H.F. Heicke, J. Phys. Chem. 83, 480 (1979).
2. M. Kotlarchyk, S.H. Chen, J.S. Huang, and M.W. Kim Phys. Rev. A29, 2054 (1984).
3. M. Kotlarchyk and S.H. Chen, J. Chem. Phys. 79, 2461 (1983).
4. J. Rouch, A. Safouane, P. Tartaglia and S.H. Chen, J. Chem. Phys. 90, 3756 (1990).
5. M. Kotlarchyk, S.H. Chen, J.S. Huang, Phys. Rev. A28, 508 (1983).
6. C. Cametti, P. Codastefano, P. Tartaglia, J. Rouch and S.H. Chen, Phys. Rev. Lett. 64, 1461 (1990).
7. C. Cametti, P. Codastefano, P. Tartaglia, J. Rouch and S.H. Chen, *Electrical Conductivity and Percolation Phenomena in Water-in-Oil Microemulsions*, Phys. Rev. A (submitted).
8. J.S. Huang and M.W. Kim. Phys. Rev. Lett. 47, 1452 (1981).
9. J. Rouch, A. Safouane, P. Tartaglia and S.H. Chen, J. Phys.:Condensed Matter 1, 1773 (1989).
10. J. Rouch, A. Safouane, P. Tartaglia and S.H. Chen, Prog. Colloid Polym. Sci. 79, 279 (1989).
11. K. Kawasaki, Ann. Phys., N.Y. 61, 1 (1970).
12. S.H. Chen, J. Rouch and P. Tartaglia, J. of Non-Crystalline Solids, 131, 275 (1991).
13. D.W. Oxtoby, W.M. Gelbart, J. Chem. Phys. 61, 2957 (1974).
14. S.M. Lo and K. Kawasaki, Phys. Rev. A8, 2176 (1973).
15. J. Rouch, P. Tartaglia, A. Safouane, and S.H. Chen, Phys. Rev. A40, 2013 (1989).
16. S.H. Chen, T.L. Lin and M. Kotlarchyk, in *Surfactants in Solution*, vol. 6, p.1315, edited by K.L. Mittal and P. Bothorel, Plenum Press, New York, 1986.
17. S.H. Chen, T.L. Lin and J.S. Huang, in *Physics of Complex and Supermolecular Fluids*, p. 285, edited by S.A. Safran and N.A. Clark, J. Wiley & Sons, New York, 1987.
18. H.F. Eicke, M. Borkovec and B. Das-Gupta, J. Phys. Chem. 93, 314 (1989).
19. D.G. Hall, J. Phys. Chem. 94, 429 (1990).
20. M. Lagues, J. Phys. (Paris), Lett. 40, L331 (1979).
21. G.S. Grest, I. Webman, S. Safran, and A.L.R. Bug, Phys. Rev. A33, 2842 (1986).
22. C. Cametti, P. Codastefano, A. Di Biasio, P. Tartaglia, J. Rouch and S.H. Chen, Phys. Rev. A40, 1962 (1989).
23. J.P. Clerc, G. Giraud, J.M. Laugier, and J.M. Luck, Advances in Physics 39, 191 (1990).
24. J. Xu and G. Stell, J. Chem. Phys. 89, 1101 (1988).
25. A. Coniglio, U. De Angelis, and A. Forlani, J. Phys. A10, 1123 (1977) and ibidem 10, 219 (1977).
26. D. Majolino, F. Mallamace, S. Venuto, and N. Micali, Phys. Rev. A 42, 7330 (1990).
27. C. Cametti, P. Codastefano, G. D'Arrigo, P. Tartaglia, J. Rouch, and S. H. Chen, Phys. Rev. A42, 3 421 (1990).
28. E.Y. Sheu, S.H. Chen, J.S. Huang and Y.C. Sung, Phys. Rev. A39, 5867 (1989).
29. S. Magazù, D. Majolino, G. Maisano, F. Mallamace, and N. Micali, Phys. Rev. A40, 2643 (1989).
30. P. Tartaglia, J. Rouch and S.H. Chen, *Dynamic Slowing down in Dense Percolating Microemulsions*, Phys. Rev. A (submitted).
31. S.H. Chen and J. Teixeira, Phys. Rev. Lett., 57, 2583 (1986).
32. J.E. Martin and B. J. Ackerson, Phys. Rev. A 31, 1180 (1985).

MEAN-FIELD BEHAVIOR AT PHASE SEPARATION IN 3-COMPONENT MICROEMULSION SYSTEM

Hideki Seto, Shigehiro Komura
Faculty of Integrated Arts and Sciences, Hiroshima University, Hiroshima 730, Japan

Dietmar Schwahn
Institut für Festkörperforschung des Forschungszentrums Jülich GmbH, D-5170 Jülich, Germany

Kell Mortensen
RISØ National Laboratory, DK-4000 Roskilde, Denmark

ABSTRACT

Small angle neutron scattering experiments on a phase separation process of water / decane / AOT microemulsion systems have been performed and the results could be interpreted to be the mean-field behavior although most of the earlier experiments showed that this system belongs to the 3-dimensional Ising universality class.

The critical phenomena prior to the phase separation of micelles or 'droplet structure' microemulsion systems have been investigated in the last decade,[1] and most of the results have shown that the phase separations are thought to be the 'liquid-gas' like transition and the critical phenomena can be interpreted to belong to the 3-dimensional Ising universality class.[2] While some experiments shows values close to those of 3D-Ising,[2] *i.e.*, the osmotic compressibility κ behaves like $\kappa \sim \varepsilon^{-\gamma}$ with $\gamma = 1.24$ and the correlation length $\xi \sim \varepsilon^{-\nu}$ with $\nu = 0.63$, where ε is the reduced temperature $\varepsilon \sim (T - T_S)/T_S$, T_S being spinodal temperature, other experimental results showed that the critical exponent were closer to the mean field values, *i. e.*, $\gamma = 1.0$ and $\nu = 0.5$.[3] Thus the question which universality class these systems belong to is still unclear.

From this point of view, we investigated the critical behavior of the well-known 3-component microemulsion system, water / AOT / *n*-decane mixture which forms reversed micelles and separates into two phases of different micelle concentrations with increasing temperature. The experiments were performed at the small angle neutron scattering spectrometer at the cold neutron source in RISØ National Laboratory. All the specimen were prepared to have the same water-to-AOT molar ratio as "3 / 5 / 95 composition" microemulsion with varying oil concentration following the earlier works.[2] All the diffraction profiles showed that the scattering profiles can be approximated by the Ornstein-Zernike formula in very low-Q region ($3.0 \times 10^{-3} \leq Q \leq 3.5 \times 10^{-2}$ [Å^{-1}]). A typical temperature dependence of the forward cross section $S(0)$ for $\phi = 0.1031$ is depicted in Fig. 1, where ϕ means the volume fraction of the droplets of water surrounded by surfactants. The (pseudo-) spinodal point T_S and the binodal point T_B were determined from the extrapolation to vanishing $S(0)^{-1}$ and the 'bending point' of the observed points as in Ref. 4. The proportional relation of the inverse forward scattering $S(0)^{-1}$ to the inverse temperature means that the critical exponents have the mean-field value $\gamma = 1.0$ according to Eq. 1.

$$\frac{\kappa^{-1}}{T} \propto S(0)^{-1} \propto \frac{1}{T_s} - \frac{1}{T} \propto \frac{\varepsilon}{T}, \quad (1)$$

where T is an absolute temperature and κ is the osmotic compressibility. We observe for $S(0)^{-1}$ over the whole temperature below T_B a linear power law against T^{-1} in agreement with the mean-field critical behavior for the compressibility of $\kappa \propto \varepsilon^{-1}$. Since the square of the correlation length ξ^2 must follow the temperature dependence indicated by the straight line in Fig. 2 from the requirement of the exact relation between two critical exponent $2\nu = \gamma$, mean-field behavior is observed in a limited temperature range. For temperature below $T \approx 33°C$ ($\varepsilon \geq 10^{-2}$) ξ becomes less than the diameter of the micelle spheres $D = 2\sqrt{\frac{5}{3}}R_G$ where R_G is the observed radius of gyration, and deviation from $\xi^2 \propto \varepsilon^{-1}$ occur.

From the spinodal temperature determined from $S(0)^{-1} = 0$ of various compositions ϕ, we would propose a phenomenological free energy expression assuming the van der Waals force with the Flory-Huggins parameter as in Ref. 4, and calculated the spinodal line against ϕ as follows,

$$T_s = \frac{-0.1628 + 2.480\phi - 11.51\phi^2}{1/(2\Omega\phi(1-\phi b')^2) + (-0.5389 + 7.968\phi - 37.21\phi^2) \times 10^{-3}}, \quad (2)$$

which seems to explain the observed spinodal points well. The details will appear in our coming paper.

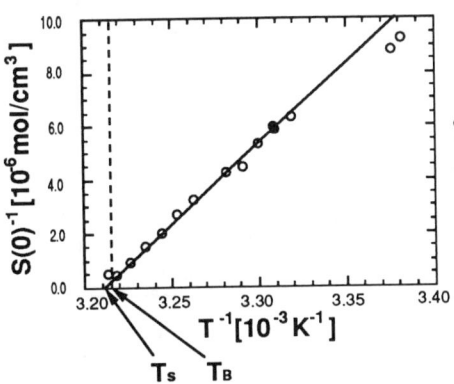

Fig. 1. The inverse of the forward scattering against the inverse of the absolute temperature for $\phi = 0.1031$ is depicted. The lines indicate the fitted straight line.

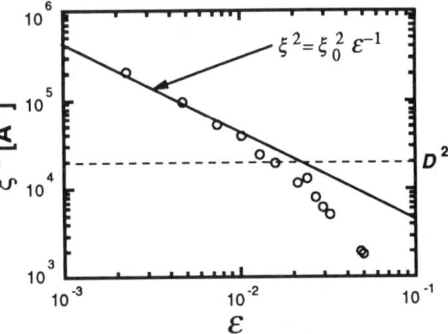

Fig. 2. Power law divergence of the correlation length for $\phi = 0.1031$. The solid line indicate the mean-field behavior fitted to some data points near the spinodal point. The dashed line means the minimum value of a distance between two droplets. The deviation from the power law was observed below the temperature where the correlation length become less than about the minimum distance between micelles.

REFERENCES

1. For a general survey, see "Micellar Solutions and Microemulsions", S. -H. Chen and R. Rajagopalan, eds., Springer-Verlag, New York, 1990.
2. J. S. Huang, M. Kotlarchyk and S. -H. Chen, in Ref. 1, p. 227.
3. M. Corti and Degiorgio, Phys. Rev. Lett. **45**, 1045, (1980).
4. D. Schwahn, K. Mortensen, T. Springer, H. Yee-Madeira and R. Thomas, J. Chem. Phys. **87**, 6078, (1987).

DYNAMIC UNIVERSALITY IN MICROEMULSION SYSTEM

Yoshifumi Harada and Mitsuru Tabuchi
Department of Applied Physics, Faculty of Engineering,
Fukui University, Bunkyo 3-9-1, Fukui 910, Japan

ABSTRACT. The low frequency sound velocities of a water-in-oil microemulsion system (AOT/water/n-decane) have been measured near a critical phase transition point together with high frequencies sound absorptions. [Here AOT is an abbreviation for sodium bis(2-ethylhexyl)sulfosuccinate.] As a result, it is shown that the excellent confirmation of dynamic scaling and dynamic universality for sound propagation have been successfully applied in the case of this microemulsion system.

1. INTRODUCTION

Recently, there is a growing awareness of the importance of critical phenomena for the description of the thermodynamic and transport properties of micellar solutions /1/, /2/ and microemulsions /3/. The microemulsion is an attractive system for the study of dynamics of the glass transition experimentally because the packing fraction of the spheres can easily be changed simply by adding more surfactant and water to the oil. A detailed knowledge of the phase transition mechanism in such systems is strongly relevant for the biochemical applications of non-ionic amphiles. In this paper, we report the observation of the low frequency sound velocity for a three component water-in-oil microemulsion (AOT/water/ n-decane) system near a critical phase transition point together with high frequencies sound absorptions.

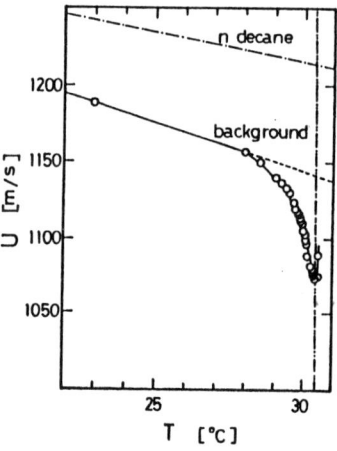

Fig. 1. Temperature dependence of sound velocity U vs Temperature T of AOT/water/n-decane system at 910 Hz.

2. EXPERIMENTS & RESULTS

Microemulsion were prepared by dissolving appropriate amounts of AOT in decane and sample tube put on a ultrasound homogenizer for emulsification. The molar ratio of water to surfactant n = [water]/[AOT] = 40.0. The critical volume fraction of droplets obtained from the coexistence curve is $\Phi_c \approx 0.06$. The sample was then allowed to age for a week before the measurements. The low frequency sound velocity were measured by means of the same LFAC(Low frequency adiabatic compressibility) methods developed by Tanaka et' al /4/.

Fig.1 shows temperature dependence of the low frequency sound velocity U at 910 Hz due to the critical slowing down. It is well known the fact that the absence of any critical temperature variation in the measured ultrasound velocity in usual critical binary mixtures is apparent contradiction to the universality occurring critical behavior of all other types of fluids. Ferrell and Bhattacharjee /5/ established quantitatively that critical behavior must exist in critical binary mixtures at sufficiently low frequencies using frequency dependent complex specific heat theory for critical sound propagation, including dynamical scaling theory and from the complex sound velocity, one obtained following linear scaling relations as:

$$\frac{U_1}{U_c} \sim \frac{g^2 U_c^2 K}{2T_c C_B^2} \left(\frac{\gamma_0}{2\pi}\right)^{\alpha_0/z_0 \nu} f^{-\alpha_0/z_0 \nu} \quad (1)$$

$$\sim \frac{g^2 U_c^2 K}{2T_c C_B^2} \ln \varepsilon + const. \quad (2)$$

Fig. 2 illustrates $-U_1/U_c$ plotted against $\ln \varepsilon$ where $\varepsilon = (T-T_c)/T_c$ is the reduced temperature. The slope of the line in Fig. 2 give $g^2 U_c^2 K/2T_c C_B^2 = -2.43 \times 10^{-2}$, which would give us the value of coupling constant g if K and C_B were known. In the case of microemulsion systems, we have confirmed, for the first time, experimentally that low frequency sound velocity of this microemulsion system (AOT/water/decane) decrease drastically as the temperature approaches the critical point.

In conclusion, from our experimental and theoretical/5/ considerations, we have shown the excellent confirmation of dynamic scaling and dynamic universality for sound propagation is also applicabe in the case of this microemulsion system.

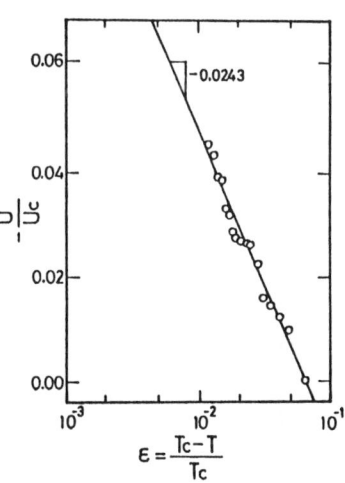

Fig. 2. Linear scaling plot of reduced low frequency sound velocity U_1/U_c vs reduced temperature ε.

REFERENCES

/1/ M. Corti and V. Degiorgio, Phys. Rev. Lett. 45, 1045,(1980), and J.Phys.Chem. 85,1442,(1981).
/2/ G. Dietler and D. S. Cannell, Phys. Rev. Lett. 60, 1852,(1988).
/3/ J. S. Huang and M. W. Kim, Phys. Rev. Lett. 47, 1462,(1981), and E.Y. Sheu, S. H. Chen, J. S. Huang and J. C. Sung, Phys. Rev. 39, 5867,(1989).
/4/ H. Tanaka, Y. Wada and H. Nakajima, Chem. Phys. 75, 37,(1983).
/5/ R. A. Ferrell and J. K. Bhattacharjee, Phys. Rev. A24, 1643,(1981).

EXPERIMENTAL AND THEORETICAL STUDIES OF CRITICAL PHENOMENA IN SUPRAMOLECULAR FLUID SYSTEMS

C CAMETTI, P. CODASTEFANO, J. ROUCH[*], AND P. TARTAGLIA
Dipartimento di Fisica, Università di Roma La Sapienza
Piazzale Aldo Moro 2, I-00185 Roma, Italy

ABSTRACT. We report on critical phenomena in multicomponent fluids consisting of three-component Water/Oil/Surfactant microemulsions and on nonionic surfactant/water binary mixtures. Due to the existence of aggregates of quasi-macroscopic sizes, static and dynamic background effects become important. A method of analysis of experimental data, involving both mode coupling equations and the linear model equation of states, is developed. All the available experimental results can be very well explained by the mode-coupling theory involving 3-d Ising critical exponents.

Sufficiently close to the critical point, all critical phenomena in usual fluid systems share universal features [1]. However for supramolecular fluids, it is quite likely that the macroscopic size of the aggregates strongly influences the critical behavior in many different ways. In that case backgrounds effects due to the scattering by individual particles (static background) or to the non-diverging terms in the transport coefficients related to a short wave length Debye cutoff q_D (dynamic background), are large [2].

In static quasi-elastic light scattering, the physical parameter that can be deduced experimentally from the total scatered intensity is the q-dependent static structure factor $S(q)$ given by the spatial Fourier transform of the static pair distribution function $g(r)$. On the other hand, as far as dynamic light scattering is concerned, one can measure the dynamic structure factor $S(q,t)$ i.e. the space Fourier transform of the particle density-density time correlation function $g(r,t)$. At constant pressure, critical phenomena in binary fluids can be theoretically accounted for in exactly the same manner as in pure system. In the case of ternary AOT/Water/Decane microemulsions it has been proven by SANS [11] that the radius of the water droplet is governed by the molar ratio w=[Water]/[AOT] which acts as a field variable. In this respect microemulsions can be considered as a binary mixture.

Mode coupling equations including backgrounds have been approximately solved by Oxtoby and Gelbart [3]. The q-dependent decay rate Γ of the order parameter (the volume fraction) is the sum of a critical part Γ_C and of a background part Γ_B respectively given by:

$$\Gamma_C(q,T) = \frac{3}{4} q^2 \frac{R k_B T}{6\pi\eta\xi} \frac{K(x)}{(q\xi)^2} \quad ; \quad \Gamma_B(q,T) = q^2 \frac{k_B T}{6\pi\eta\xi} \frac{C}{q_D\xi} (1+x^2)(q_D\xi)^\varphi \quad (1)$$

where $K(x)$ is Kawasaki's function. In formula (1), R=1.027, C is approximately equal to 0.9 and q_D is equal to the inverse of the hydrodynamic diameter of the micelle $2R_H$.

The above theoretical model allows us to treat not only the case of critical solutions following a path along the critical isochore, but also other iso-concentration lines using the so-called Schofield's linear model equation of state [4]. In the case of AOT/Water/Decane microemulsions, the long range correlation length calculated along various isochores using linear model equation of state is reported on Fig (1). For the same mixture, experimental and theoretical values of the line width including background are depicted on Fig (2). It can be seen that a good agreement between the theory and the experiment is achieved if renormalised values of the critical indices are used accordind to Fisher's predictions [5].

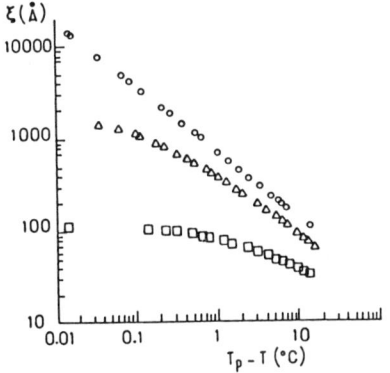

Fig. 1. Plot of the calculated value of the long range correlation length as a function of T_p-T circles: Φ=0.098 squares Φ=0.22, triangles: . Φ=0.064.

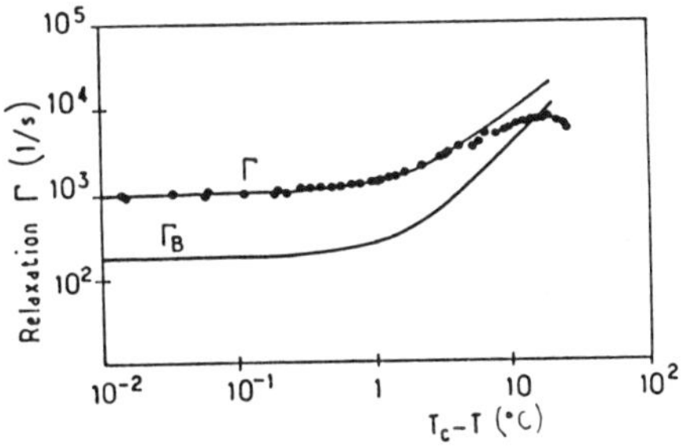

Fig 2 The relaxation rate as a function of Tc-T. The dots : experimental results. Full lines : theoretical results from mode coupling theory.

A good agreement is also observed between full mode coupling theory including backgrounds and experimental results obtained for C_iE_j-Water binary mixtures along different isochores including the most critical one [6].

*Permanent address: C P M O H , U.R.A. 283 du C.N.R.S.
Université Bordeaux I 351 Cours de la Libération 33405 Talence Cedex 5 France

References
[1] K. Kawasaki, in *Phase Transitions and Critical Phenomena*, edited by C. Domb and M.S. Grenn (Acad. Press., N.Y., 1976) Vol. 5A, p. 165.
[2] J. Rouch, A. Safouane, P. Tartaglia and S.H. Chen. J. Chem. Phys., 90, 3756, (1989).
[3] D.W. Oxtoby, and W.M. Gelbart, J.Chem. Phys. 61, 2957 (1974).
[4] P. Schofield, Phys. Rev. Lett. 22, 606 (1969).
[5] M. Fisher, Phys. Rev. Lett. 57, 1911 (1986).
[6] J. Rouch, P. Tartaglia, A. Safouane, and S.H. Chen, Phys. Rev. A40, 2013 (1989).

B. THEORY

AGGREGATION AND COALESCENCE OF WATER DROPLETS IN A DIELECTRIC LIQUID PHASE IN AN ELECTRIC FIELD

M. Yamaguchi, T. Ise and T. Katayama
Department of Chemical Engineering, Faculty of Engineering Science,
Osaka University, Toyonaka 560, Japan

ABSTRACT

When water-in-oil emulsions are placed in electric fields, the droplets line up into a lengthwise extended chain in the direction of the field, and then coalescence between the droplets proceeds. The process of the chain formation is studied theoretically and experimentally. Behavior for electrical coalescence of the droplets is discussed by using a model proposed by Harada.

THEORETICAL CONSIDERATIONS

The dipoles within two uncharged droplets(radius;R_i,R_j) in an electric field intensity(E_0) are induced. The dipole moment(μ_j) in the droplet(R_j) will be the sum of the moment(μ_{j0}) induced by the external field and that due to the additional polarization induced by the other dipole(μ_i) and is derived by the method of images as follows.

$$\mu_j = \mu_{j0} + f P_{ji} \mu_i, \quad P_{ji} = R_j^3 / z_{ji}^3 \quad (1),(2)$$

$$\mu_i = 4\pi R_i^3 \varepsilon_c \alpha E_0, \quad \alpha = (\varepsilon_d - \varepsilon_c)/(\varepsilon_d + 2\varepsilon_c) \quad (3)$$

where f is a constant depending on the conductivity of the droplet phase and ε_d and ε_c are respectively the permittivity of the droplet and continuous phase. The electric force acting on the droplet(R_j) by the droplet(R_i) is given as follows.

$$\vec{F}_j = \vec{\nabla}(\vec{\mu}_j \cdot \vec{E}_{j0}) + \vec{\nabla}(\vec{\mu}_j \cdot \vec{E}_{ji}) \quad (4)$$

The force lines around a droplet(μ_i) passing through any position (r_j,θ_j) are given as follows[1].

$$r = r_j \sin\theta \sqrt{(\cos\theta/\cos\theta_j)}/\sin\theta_j \quad (5)$$

N pieces of the droplets move on the line and line up into a lengthwise extended chain (a type of aggregation) in the direction of the field. When the process of chain formation is assumed to be controlled by the electric and viscous forces, the time(t) required for the droplets to move on the line and contact with each other can be given as follows.

$$dt = 6\pi \eta_c R_j d\vec{r}_{ji} / \vec{F}_j \quad (6)$$

where η_c is the viscosity of the continuous phase. By substituting Eqs.(1)-(5) into Eq.(6) and by integrating Eq.(6), the time is calculated numerically.

The stability between two neighboring droplets of equal radius(R_0)

in the electric field was discussed by Latham and Roxburgh[2]. They presented the field intensity in which the onset of droplet instability occurs.

$$E_0 \sqrt{(R_0/\sigma)} = \sqrt{(18/\varepsilon_c)} M(\delta) I(\delta)/\Sigma \qquad (7)$$

where $M(\delta)$ and $I(\delta)$ are functions including the eccentricity(δ) of the droplet and Σ is the enhancement factor of the field.

On the other hand, Harada[3] assumed that the chain formation of the droplets is analogous to that of molecules in the electric field and proposed a following equation.

$$E_0 \sqrt{(R_0/\sigma)} = \sqrt{(32\pi\kappa T/3\varepsilon_c \sigma \alpha^2 R_0^2)} /\sqrt{(X_0/R_0)} \qquad (8)$$

where κ and σ are Boltzmann constant and the interfacial tension, respectively, and X_0 is the initial separation between the droplets.

EXPERIMENTAL RESULTS

The validity of the present theories was investigated by using W/O emulsions (emulsifier;Span 80) and a system of a microscope and a video camera. Figure 1 shows a relation between calculated and experimental times: it is defined that an isolated droplet(R_s;μm) situated apart a droplet chain(R_i; μm; i=1∿6) is drawn up to it and comes in contact with the end of it. Figure 2 shows a relation of an applied voltage(V_0) and a distance(r_c) between centers of two droplets. A solid line represents critical values of electrical coalescence of the droplets by Eqs.(7) and (8).

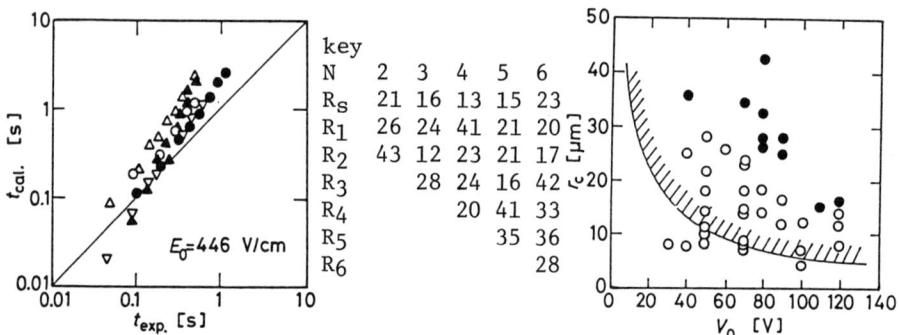

Fig. 1. Comparison of experimental lapse time with calculated one.

Fig. 2. Electrical coalescence of two droplets.

REFERENCES

1. G. Zebel, Staub, **23**, 263 (1963).
2. J. Latham and I. W. Roxburgh, Proc. Roy. Soc., **A295**, 84 (1966).
3. M. Harada, Report of Research Project, Grant-in-Aid for Co-operative Research(A), 57 (1988).

C. COMPUTER SIMULATION

PHASE SEPARATION IN BINARY MIXTURES WITH SURFACTANTS

Toshihiro Kawakatsu and Kyozi Kawasaki
Department of Physics, Kyushu University 33, Fukuoka 812, Japan

ABSTRACT

Reformulation and justifications in a microscopic level have been made on the hybrid model which has been proposed by the present authors to investigate the phase separation phenomena of binary mixtures containing surfactants. A Monte Carlo and a molecular dynamics simulation and also an analytical theory are presented for the case where the surfactant molecule is a block copolymer chain.

Systems containing amphiphilic molecules, or surfactants, are known to exhibit a variety of complex domain structures, which are related to the low surface tension of surfactant-adsorbed interfaces.[1] In order to investigate dynamics of such systems, we proposed a hybrid model of a continuous field and discrete molecules, which has been found to reproduce main features of immiscible binary mixtures containing surfactants.[2] We also investigated the early and late stage behaviors of the phase separation processes both by computer simulations and by analytical calculations.[3]

In our hybrid model, we adopted some *ad hoc* assumptions on the mesoscopic scale, for example, we treated the binary mixture and the surfactant in different manners, the former as a continuous field and the latter as discrete molecules. Thus, it is needed to justify or improve such assumptions by microscopic considerations on a more realistic model. Such attempts have been done with the use of the Monte Carlo (MC) and molecular dynamics (MD) computer simulation methods and also by a field theoretic approach, which are to be described in the following.

We performed a MC simulation on an Ising ferromagnetic spin system containing small surfactant molecules.[4] Each component of the binary mixture is represented by one of +1 and −1 Ising spin variables, and the surfactant molecule is represented by a pair of +1 and −1 Ising spins connected by a rigid bond. Phase separation processes from an initial uniform state have been simulated on a 2-dimensional square lattice with the use of the standard Metropolis algorism. We observed a deceleration of the phase separation when we add more surfactant molecules into the system, the deceleration being attributed to the impurity nature of the small surfactant molecule.

On the other hand, we have performed a MD simulation on an immiscible binary soft sphere mixture containing a large surfactant molecule, that is, an amphiphilic block copolymer chain.[5] The system is a 3-dimensional cubic box with periodic boundary conditions on each face, and contains 1000 monomers, 128 monomers of which constitute the block copolymer chain. The immiscible binary mixture is modeled by a non-additive soft spheres[6] and the block copolymer chain is composed of two subchains each of which is composed of same monomers as those of the binary mixture connected by freely rotating rigid bonds. We performed a series of simulation runs on the phase separation processes from an initial uniform mixture of the binary mixture and the block copolymer chain. We found a rapid growth of the concentration fluctuation compared with the case without the block copolymer chain. This feature originates from the fact that the block copolymer chain acts as a nucleus for the phase separation, which is due to the amphiphilic nature of the block copolymer chain.

A field theoretic approach can also be developed when the surfactant is a very long block copolymer chain.[7] We describe the binary mixture by a continuous scalar field $X(\mathbf{r})$ which is defined as the local concentration difference between the two components of the binary mixture and is assumed to obey the so-called time dependent Ginzburg Landau model as was done in the original hybrid model.[2] On the other hand, we describe the block copolymer chain by the Rouse model. Combining these two descriptions and eliminating the degrees of freedom of the polymer chain, we obtain a closed evolution equation for the field $X(\mathbf{r})$;

$$\frac{\partial X_{\mathbf{k}}(t)}{\partial t} + \int_{-\infty}^{t} K_k(t-t') \frac{\partial X_{\mathbf{k}}(t')}{\partial t'} dt' = -L_X k^2 [\frac{\delta H_R(t)}{\delta X(\mathbf{r},t)}]_{\mathbf{k}} + f_R(\mathbf{k},t), \quad (1)$$

where $X_{\mathbf{k}}$ is the Fourier transform of $X(\mathbf{r})$, K_k a memory kernel, L_X a kinetic coefficient, H_R the Hamiltonian and f_R is the thermal fluctuation. Equation (1) is an extension of the original hybrid equation of motion,[3] which predicts the same results as those of the original model in a certain limit. Equation (1) can be used to bridge the gap between a microscopic description and the hybrid description.

In conclusion, our MC and MD simulations showed that the excluded volume effect and the amphiphilic nature of surfactants decelerates and accelerates the phase separation, respectively, when the surfactant is added to immiscible binary mixtures, which is consistent with our predictions of the hybrid model.[3] Such simulations as well as the field theory mentioned above can also be used to determine the parameters of the hybrid model.

REFERENCES

1. S.A.Safran and N.Clark, eds., *Physics of Complex and Supermolecular Fluids*, (Wiley, New york, 1987),
   W.M.Gelbert, D.Roux and A.Ben-Shaul, eds., *Modern Ideas and Problems in Amphiphilic Science* (to be published).
2. T.Kawakatsu and K.Kawasaki, *Physica* **A167**, 690 (1990),
   see also T.Kawakatsu and K.Kawasaki, in *Molecular Dynamics Simulations; Proceedings of Taniguchi International Symposium*, ed. F.Yonezawa (Springer, 1991) to be published.
3. T.Kawakatsu and K.Kawasaki, *J. Colloid Interface Sci.*, **145**, 413 (1991), ibid, 420.
4. T.Kawakatsu and K.Kawasaki, *J. Colloid Interface Sci.* (to be published).
5. T.Kawakatsu and K.Kawasaki, in preparation. For a preliminary results, see Ref.7.
6. C.Hoheisel, *Phys. Rev.* **A41**, 2076 (1990).
7. T.Kawakatsu and K.Kawasaki, in *Festshrift for the 60th birthday of Prof. Szépfalusy* (World Scientific, to be published).

IV. COLLOIDAL SUSPENSION

A. EXPERIMENT

DYNAMIC LIGHT SCATTERING STUDIES OF THE GLASS TRANSITION IN COLLOIDAL SUSPENSIONS

W. van Megen, S.M. Underwood*
Department of Applied Physics,
Royal Melbourne Institute of Technology,
Melbourne, Victoria 3001, Australia

P.N. Pusey
Department of Physics
University of Edinburgh
Edinburgh, United Kingdom EH9 3JZ

ABSTRACT

This paper describes the particle dynamics, measured by dynamic light scattering (DLS), in the vicinity of the glass transition of non-aqueous suspensions of sterically stabilized colloidal spheres. The phase behavior of these suspensions, showing fluid-crystal and fluid-glass transitions, is similar to that found for a system of hard spheres. The fragility of the structure in these suspensions allows crystals to be readily "shear melted", and the slow particle dynamics means slow structural recovery and access to the metastable states.

The "glass transition" concentration is defined as that at which DLS reveals the onset of non-ergodicity by the emergence of a non-decaying component in the intermediate scattering function. This non-ergodicity is easily observed by DLS which essentially follows the stochastic evolution in time of a single spatial Fourier component of the sample's concentration fluctuations. We outline the theory of DLS in non-ergodic media which leads to a simple relationship between the ensemble-averaged intermediate scattering function and the time-averaged time correlation function of the scattered intensity.

Beyond the time scale that characterizes local or microscopic particle motions the measured decays of the intermediate scattering functions, on both sides of the glass transition at the peak of the static structure factor, closely follow the β-decay predicted by mode-coupling theory. In the glass phase the measured non-ergodicity parameters are also in agreement with mode-coupling predictions for hard-sphere atoms, with no adjustable parameters in the comparison.

---

* Present address: ICI Australia Technology Centre,
Ascot Vale, Victoria 3032, Australia

INTRODUCTION

Suspensions of spherical colloidal particles with narrow size distribution can be prepared in which the particle interaction is close to that of hard spheres[1]. Such systems show[2,3] the phase behavior, fluid, crystal and glass, predicted by computer simulation[4,5] and theory[6] for the hypothetical hard-sphere atomic system. Colloidal crystals are weak mechanically and can easily be "shear-melted", for example by shaking the sample[7]. The resulting metastable fluid states can persist for minutes or even hours before significant crystallization takes place.

In this paper we describe measurements by dynamic light scattering (DLS) of the dynamics of particle diffusion in suspensions of hard-sphere colloids in their amorphous states (equilibrium fluid, metastable fluid and glass), with emphasis on the glass transition itself. A simple picture of the glass transition of spherical particles can be given in terms of "caging". In their fluid or metastable fluid states particles are temporarily trapped in the cages formed by their instantaneous neighbors. In time, however, they can escape from their cages and ultimately diffuse through the sample. As the concentration of the particles is increased, the neighbor cages become tighter such that particle escape and long distance diffusion is increasingly hindered. At the concentration of the glass transition the particles become effectively permanently trapped by their neighbors but retain some freedom for local motions about their fixed average positions. Associated with the suppression of long distance self diffusion is the partial freezing in of density fluctuations on all spatial scales.

Given enough time, a fluid-like assembly of particles will evolve through a representative fraction of the full ensemble of all potentially available spatial configurations. Thus, in the course of a single experiment the system can explore enough of phase space that the time average inherent in a measurement of a property is effectively equivalent to its ensemble average; in other words, the system is ergodic. By contrast, a glassy assembly of particles is trapped in a restricted region of phase space, or "sub-ensemble", whose location and extent are determined respectively by the average positions of the particles and the magnitudes of their displacements about these positions. Different samples, although prepared similarly, will contain particles in different average positions and will be described by different sub-ensembles. Consequently the measured time average of a property of a particular sample in the glass phase will constitute an average over the particular sub-ensemble of configurations and will not necessarily be the same as an average over the full ensemble of allowed configurations. Glasses can therefore be regarded as non-ergodic media; the glass transition is sometimes referred to as the transition to non-ergodicity[8].

While the principles of DLS by ergodic media were established many years ago, it is only recently that the extra complications encountered when studying non-ergodic media by DLS have been fully appreciated[9]. In the next section the theory of DLS by non-ergodic media is outlined. This is followed by a section in which we describe the experiments on the colloidal fluids and glasses. Two aspects of the work are emphasized: Verification of the light scattering theory and comparison of the experimental data with the predictions of mode-coupling theory for the glass transition of hard-sphere atoms. Most of this research is already published[7,9-11], and the literature cited can be consulted for details not given in this relatively concise article.

THEORY OF DLS BY NON-ERGODIC MEDIA

We start by outlining the formal theory[9] and then give a physical interpretation in terms of fluctuating speckle patterns. For simplicity we consider an assembly of identical spherical particles. The instantaneous amplitude of the field scattered to a point in the far field by N particles in a "scattering volume" V can be written

$$E(Q,t) = \sum_{j=1}^{N} \exp[i\mathbf{Q}\cdot\mathbf{r}_j(t)], \tag{1}$$

where $\mathbf{Q}$ (magnitude Q) is the usual scattering vector and $\mathbf{r}_j(t)$ is the position of particle j at time t. We model a non-ergodic (or amorphous solid) medium by writing

$$\mathbf{r}_j(t) = \mathbf{R}_j + \Delta_j(t) \tag{2}$$

where $\mathbf{R}_j$ is the fixed average position of particle j and $\Delta_j(t)$ is its (time-dependent) displacement about this position. Then, the scattered field, Eq.(1), is expressed with the use of Eq.(2), as the sum of fluctuating, E_F, and constant, E_C, components,

$$E(Q,t) = E_F(Q,t) + E_C(Q), \tag{3a}$$

$$E_F(Q,t) = \sum_{j=1}^{N} \exp(i\mathbf{Q}\cdot\mathbf{R}_j) \{\exp[i\mathbf{Q}\cdot\Delta_j(t)] - \langle\exp[i\mathbf{Q}\cdot\Delta_j(t)]\rangle_T\}, \tag{3b}$$

$$E_C(Q) = \sum_{j=1}^{N} \exp(iQ \cdot R_j) \langle \exp[iQ \cdot \Delta_j(t)] \rangle_T, \qquad (3c)$$

where $\langle \cdots \rangle_T$ indicates a time average or, equivalently, an average over the *sub-ensemble* describing the particular scattering volume under consideration. The constant component is independent of time, but its magnitude depends on the configuration $\{R_j\}$ of the particles in the scattering volume; the variation of E_C from volume to volume, due to different configurations of the average positions, is a manifestation of the system's non-ergodicity. It can be shown[9] that the fluctuating component is a zero-mean complex Gaussian random variable whose average properties are independent of $\{R_j\}$, i.e. they are the same for all scattering volumes.

DLS measures $g_T^{(2)}(Q,\tau)$, the normalized time-averaged time correlation function of the scattered intensity $I(Q,t) = |E(Q,t)|^2$,

$$g_T^{(2)}(Q,\tau) = \langle I(Q,0)\, I(Q,\tau) \rangle_T \,/\, \langle I(Q,t) \rangle_T^2. \qquad (4)$$

After considerable manipulation, which is described in detail in Ref.9, combination of Eqs.(3) and (4) gives

$$g_T^{(2)}(Q,\tau) = 1 + (I_E/I_T)^2\, [f(Q,\tau) - f(Q,\infty)]^2$$
$$+ 2(I_E I_C/I_T^2)\, [f(Q,\tau) - f(Q,\infty)]; \qquad (5)$$

here $f(Q,\tau) = F(Q,\tau)/F(Q,0)$ is the normalized *ensemble-averaged* intermediate scattering function (ISF); the intermediate scattering function, $F(Q,\tau)$, itself is given by

$$F(Q,\tau) = (1/N) \sum_{j,k=1}^{N} \langle \exp\{iQ \cdot [r_j(0) - r_k(\tau)]\} \rangle_E, \qquad (6)$$

and $\langle \cdots \rangle_E$ indicates an ensemble average. In Eq.(5) $I_T = \langle I(Q,t) \rangle_T$ is the time-averaged total scattered intensity for the scattering volume under study, $I_C = |E_C(Q)|^2$ is the constant component and $I_E = \langle I(Q,t) \rangle_E$ is the total scattered intensity averaged over an ensemble of (different) scattering volumes. Solving the quadratic in $[f(Q,\tau) - f(Q,\infty)]$ in Eq.(5) with the use of the relationship[9]

$$I_C = I_T - I_E[1 - f(Q,\infty)] \qquad (7)$$

gives the surprisingly simple result

$$f(Q,\tau) = 1 + (I_T/I_E) \{[g_T^{(2)}(Q,\tau) - g_T^{(2)}(Q,0) + 1]^{1/2} - 1\}. \qquad (8)$$

Setting $\tau=\infty$ in Eq.(8) gives

$$f(Q,\infty) = 1 + (I_T/I_E) \{[2 - g_T^{(2)}(Q,0)]^{1/2} - 1\} \qquad (9)$$

where we have used the limit $g_T^{(2)}(Q,\infty) = 1$, which results from the total decorrelation of intensity fluctuations at long times (see Eq.(4)).

We now discuss the physical interpretation of Eqs.(8) and (9)[9]. When coherent laser light (of wavelength λ) illuminates an amorphous assembly of particles the instantaneous far-field pattern of scattered radiation at time t constitutes a random diffraction or "speckle" pattern, composed of bright and dark regions (or speckles). Each speckle subtends a solid angle $(\lambda/V^{1/3})^2$ at the scattering volume V and its intensity is $I(Q,t)=|E(Q,t)|^2$. The scattered field, $E(Q,t)$ given by Eq.(1), can be viewed as the spatial Fourier component, of wavevector Q, of fluctuations in the number density of the particles. Thus, through Eq.(4), DLS effectively follows the evolution in time (or the dynamics) of the squared amplitude of a *single* spatial Fourier component of the sample's density fluctuations. (In this sense DLS is essentially an ideal scattering technique.) Note that $F(Q,\tau)$ is simply the ensemble-averaged time correlation function of the density fluctuations in Fourier space.

In the extreme non-ergodic case where the particles are completely localized, as in a fully compressed glass at random close packing, all $\{\Delta_j(t)\}=0$. Hence in Eq.(3b) $\exp[iQ\cdot\Delta_j(t)]=1$, $E_F(Q,t)=0$ and the scattered intensity $I(Q,t)$ (or speckle pattern) is independent of time. For this case $g_T^{(2)}(Q,\tau)=1$ for all τ (Eq.(4)), and Eqs.(8) and (9) give $f(Q,\tau)=f(Q,\infty)=1$, consistent with the complete freezing in of density fluctuations. In the other extreme of an ergodic medium, the particles undergo large excursions $\{\Delta_j(t)\}$ so that $<\exp\{iQ\cdot\Delta_j(t)\}>_T=0$ and the constant component $E_C(Q)$ (Eq.(3)) of the scattered field is zero. The scattered field itself is, therefore, a zero-mean complex Gaussian variable and the speckle pattern evolves through complete fluctuations from one configuration to another. Now $g_T^{(2)}(Q,0)=2$ (from Eq.(4) and the factorization properties of such a variable); furthermore, ensemble- and time-averaged intensities are the same, $I_E=I_T$. It follows that Eq.(8) gives

$$f(Q,\tau) = [g_T^{(2)}(Q,\tau) - 1]^{1/2}, \qquad (10)$$

the well-known result for DLS by ergodic media. In addition Eq.(9) gives $f(Q,\infty)=0$, implying, as expected, complete relaxation of the density fluctuations.

Here we are interested in an arbitrary non-ergodic medium in which the particles perform limited excursions about fixed average positions. The light scattered by a particular scattering volume constitutes a speckle pattern composed of both fluctuating and non-fluctuating components. At any point in the far field the intensity undergoes restricted fluctuations about a mean value which itself varies from point to point and is determined by the average positions of the scatterers in the scattering volume. Thus the form of the measured time-averaged intensity correlation function, $g_T^{(2)}(Q,\tau)$ depends on the brightness of the actual speckle under study. For an intense speckle, for which I_C is large and $I_T > I_E$, the relative mean squared intensity fluctuation $<I^2(Q,t)>_T / <I(Q,t)>_T^2 = g_T^{(2)}(Q,0)$ is small; conversely, when $I_T < I_E$, a larger relative fluctuation is observed. However, in both cases use of Eq.(8) should give the same ensemble-averaged intermediate scattering function $f(Q,\tau)$.

The application of Eqs.(8) and (9) in an experiment requires the following quantities: First, both the time-averaged intensity I_T and the time-averaged intensity correlation function $g_T^{(2)}(Q,\tau)$ must be measured on the *actual speckle* under study. Second, the intensity I_E must be averaged over an *ensemble of speckles* having the same scattering vector Q. The first two quantities are routinely provided by a photon correlator in a single DLS measurement. The ensemble-averaged intensity must be measured separately, most conveniently by counting scattered photons while the sample is scanned through the laser beam so that a large number of different scattering volumes is illuminated, and a large number of different speckles crosses the detector.

There is a second, "brute force", method for obtaining the ensemble-averaged intermediate scattering function $f(Q,\tau)$ of a non-ergodic medium, which we employed in early work[7,10], before developing the theory outlined above. Here the intensity correlation functions $<I(Q,0)I(Q,\tau)>_T$ and average intensities are measured for a large number of different scattering volumes and are summed to provide estimates of the ensemble averages. Subsequent normalization then provides the ensemble-averaged intensity correlation function, $g_E^{(2)}(Q,\tau)$, which is related to $f(Q,\tau)$ by

$$f(Q,\tau) = [g_E^{(2)}(Q,\tau) - 1]^{1/2}; \qquad (11)$$

this relationship follows from the fact that taken *over the full ensemble* the field E(Q,t) is a zero-mean complex Gaussian variable (however, as implied above, this is not the case for the field associated with a single speckle in Eq.(3)). This brute force approach is tedious since lengthy measurements must be made for each scattering volume.

Finally we mention that, at first sight, Eq.(9) appears paradoxical since it relates the long time limit of one quantity to the zero time limit of another. The physical interpretation of Eq.(9) is that the frozen-in density fluctuations, represented by $f(Q,\infty)$, cause restriction of the intensity fluctuations, whose mean square value is $g_T^{(2)}(Q,0)$.

PARTICLE DYNAMICS AROUND THE GLASS TRANSITION

Polymer latices of polymethylmethacrylate (PMMA) particles, sterically stabilized by relatively thin (~10nm) adsorbed layers, were prepared and redispersed in a mixture of decalin and carbon disulfide whose composition was chosen so that its refractive index closely matched that of the particles. Two different latices, both with a polydispersity of about 4%, were used in the work reported here, SMU12 (particle radius 170nm) and SMU21 (radius 210nm). These suspensions show the equilibrium freezing-melting transition consistent with that of the ideal hard-sphere system[2]. This behavior allows the expression of particle concentrations as effective hard-sphere volume fractions ϕ, calculated by scaling measured weight fractions so that the observed freezing transition occurs at the hard-sphere value $\phi_F=0.494$. The agreement between the observed melting volume fraction of the colloidal crystal, $\phi_M=0.542\pm0.002$, and that of the ideal hard-sphere system, $\phi_M=0.545$, supports the utility of these suspensions as hard-sphere systems. Moreover, essentially the same equilibrium phase behavior is found for different suspensions of this type with particle diameters ranging from about 0.3 to 1.0μm[12,13].

We identify a glass transition concentration, $\phi_G\approx0.565$, where the mode of crystallization, of the shear-melted suspensions, changes from the formation of small crystals, nucleated homogeneously throughout the suspension, to the formation of large crystals, nucleated heterogeneously mainly at the cell walls and the meniscus. The former process is evident within several minutes at concentrations close to ϕ_M, but it is usually not apparent for several hours close to ϕ_G or ϕ_F. The heterogeneously nucleated crystals develop over several days and, with increasing particle concentration in excess of about 0.57, decreasing amounts of these crystals grow from the meniscus (see Ref.2 for photographs of suspensions showing crystallization by these different mechanisms). For particle concentrations $\phi>\phi_F$ DLS measurements

were performed on the metastable phases, i.e. during the period
between extensive tumbling of the samples but well before any
apparent crystallization.

The time-averaged intensity autocorrelation functions,
$g_T^{(2)}(Q,\tau)$, were measured with standard DLS equipment, the only
somewhat unusual feature being the use of a detector aperture
much smaller than one coherence area (or speckle) so that, for an
ergodic system, such as a suspension in its equilibrium fluid
phase, $g_E^{(2)}(Q,0)=1.99\pm0.01$. This setting is chosen to
approximate a point detector, for which $g_E^{(2)}(Q,0)=2$, assumed in
the derivation of Eq.(8). The other quantities appearing in the
right hand side of Eq.(8) and (9) were determined as follows.
The zero time values, $g_T^{(2)}(Q,0)$, were obtained from $g_T^{(2)}(Q,\tau)$
by cumulant analysis. The time-averaged intensity, I_T, was
determined from the total number of detections accumulated by the
photon correlator during a particular measurement. The ensemble-
averaged intensity, I_E, was measured by counting photon
detections while the sample was scanned through the laser beam at
a constant rate; during this operation at least 1000 independent
speckles crossed the detector.

The validity of the model and DLS theory for non-ergodic media,
outlined in the previous section, can be tested directly, in
principle, by comparing the ISF obtained by Eq.(8) from a single
measurement of $g_T^{(2)}(Q,\tau)$ with that obtained from a brute force
determination the ensemble average of $g_E^{(2)}(Q,\tau)$ (Eq.(11)). The
latter procedure, employed in the experiments[7,10] which preceded
the theory outlined in the previous section and discussed below,
is extremely laborious, and attaining the ISF to an accuracy of
less than a few percent requires the accumulation of data over a
very large number of independent scattering volumes (or
stationary components of the speckle). As illustrated
elsewhere[14], the validity of Eqs.(8) and (9) can be ascertained
implicitly in several ways. One of these, seen in Fig. 1, shows
that, as we have assumed in the previous section, essentially the
same values of $f(Q,\infty)$ are obtained from Eq.(9) when speckles of
widely differing intensities I_T (obtained by moving the sample
between measurements) are studied in measurements of 1000s
duration.

In Fig. 2 we show the normalized intermediate scattering
functions, $f(Q_m,\tau)$, measured at the peak, $Q=Q_m$, of the static
structure factor for suspension volume fractions that span the
freezing, melting and glass transitions. Up to a volume fraction
$\phi\approx0.55$ the suspensions, in their equilibrium or metastable fluid
phases, are ergodic so that the ISF was calculated from the
measured intensity autocorrelation function in the standard way
(Eq.(10)). The radius of the PMMA particles in these experiments
was 170nm. Note that in all cases a rapid initial ($\tau\lesssim10^{-3}$s.)
decay of $f(Q_m,\tau)$, relatively weakly dependent on the particle

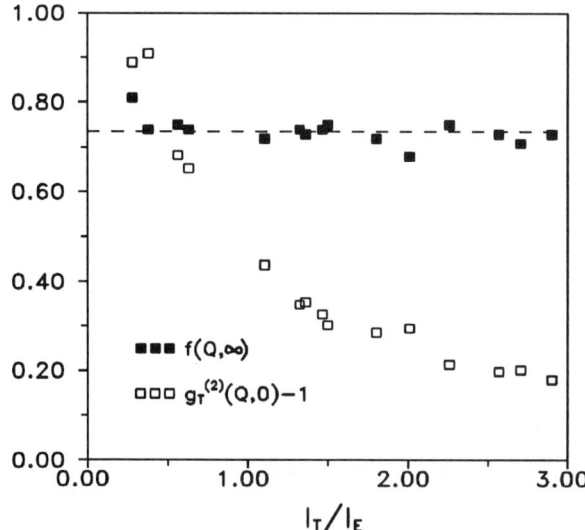

Figure 1. Verification of Eq.(9). Measured values of $g_T^{(2)}(Q,0)-1$ (open squares) and derived values of $f(Q,\infty)$ (filled squares) are plotted against I_T/I_E for a wide range of time-averaged speckle intensities, I_T. (The dashed line shows the average value for $f(Q,\infty)$.) Note the strong variation of $g_T^{(2)}(Q,0)-1$ and the essential constancy of $f(Q,\infty)$. The sample (C4) had a concentration $\phi = 0.589$ and was studied at the scattering vector $Q = 0.81 Q_m$.

concentration, is followed by a much slower decay whose characteristic time increases by a factor of about 100 for the change in concentration from 0.480 to 0.542. For $\phi=0.565$, the concentration where homogeneously nucleated crystallization was no longer observed, the glass phase is indicated by the non-decaying component in $f(Q_m,\tau)$ at long times. Qualitatively similar behavior is exhibited by the ISF's measured at scattering vectors below and above Q_m[7].

Recent applications of mode-coupling (MC) theory to dense fluids predict an ergodic to non-ergodic transition[8]. With increasing density the approach to the transition point, ϕ_G, is accompanied by the appearance of the α- and β-relaxation processes whose time scales, t_α and t_β, diverge algebraically with the separation parameter, $\sigma=(\phi-\phi_G)/\phi_G$. Here $t_\alpha \gg t_\beta \gg t_0$, where t_0 is the time

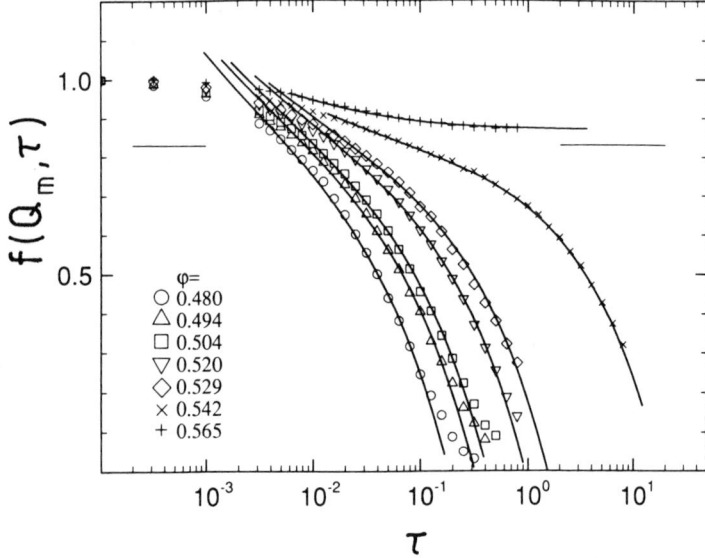

Figure 2. Normalized intermediate scattering functions, $f(Q_m,\tau)$, measured at the peak $Q=Q_m$ of the static structure factor (symbols). The solid curves are the predictions of mode-coupling theory according to Eq.(12). The horizontal lines are drawn at $f_c(Q_m) = 0.83$ (see text).

scale associated with the local particle motion as measured by the initial decay of the ISF. The α-process freezes in at the transition while the β-process persists into the non-ergodic glass phase. As discussed in these proceedings in the papers by Götze and Sjögren, in the regime close to the transition both processes allow factorization of the time and space variables. In particular, for the β-process ($t_\alpha \gg t \gg t_0$) the ISF is given by,

$$f(Q,t) = f_c(Q) + h(Q)c_\sigma g_\pm(t/t_b) \qquad (12)$$

where f_c is the non-ergodicity parameter, or the frozen in component of the structure, at the critical concentration ϕ_G, $h(Q)c_\sigma$, with $c_\sigma = |\sigma|^{1/2}$, is the amplitude of the β-process and the subscript "$\pm$" on the master function g indicates the sign of the separation parameter σ. Asymptotic expressions, such as $g_\pm(t/t_\beta \ll 1) = (t/t_\beta)^a$, which indicates that the early part of the β-process is symmetrical about the transition, are results of the theory but the detailed evaluation of the master functions and exponents requires the static structure factor as input. Such calculations have been completed for the hard-sphere system so that the predictions of MC theory can be compared quantitatively with experiments on suspensions of hard spheres[14,15].

The MC predictions shown in Fig. 2 were obtained by Götze and Sjögren[16] as follows. For the non-ergodicity parameter the experimental value, $f_c(Q_m)=0.83$, was used (see below). For $\phi<\phi_G$ the time for the measured ISF to relax to $f_c(Q_m)$ was used to obtain the scaling time t_β, whereas in the glass phase t_β was treated as a free parameter. This parameterization is necessary since this theory deals only with the relaxation of concentration fluctuations at long times and it does not provide a rigorous connection to the microscopic time scale t_0. Nevertheless, it is significant that, as may be seen in Fig. 3, the algebraic concentration dependence of t_β thus obtained is consistent with the theory. The amplitude $h_c(Q_m)c_\sigma$ was then adjusted to provide

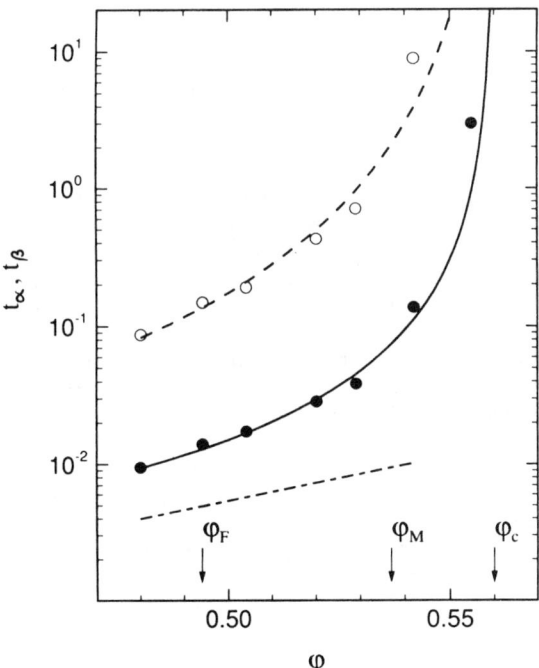

Figure 3. The two time scales t_α (open circles) and t_β (solid circles) versus ϕ. The solid and dashed lines show the predicted power laws. The dashed-dotted curve indicates the microscopic time scale t_0. (Reproduced with permission from ref.16.)

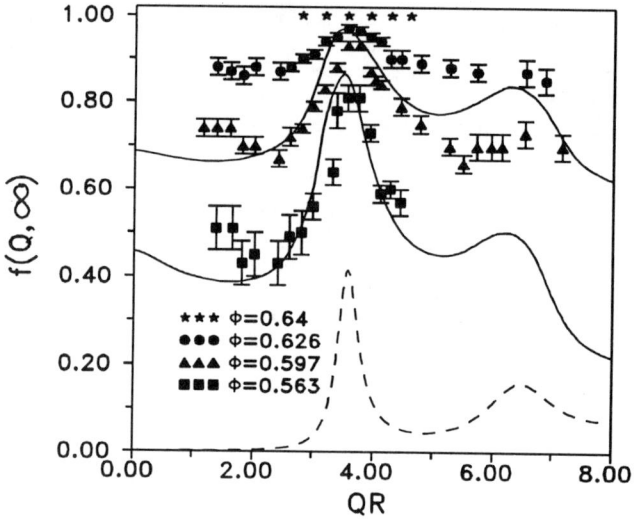

Figure 4. The non-ergodicity parameters, $f(Q,\infty)$, of hard-sphere colloidal glasses as functions of the scattering vector (R is the particle radius) for several concentrations, ϕ. The solid curves are the mode-coupling predictions[14] for hard-sphere atoms at separation parameters $\sigma = 0$ (lower curve) and $\sigma = 0.066$ (upper curve). The dashed curve is the Percus-Yevick structure factor for the hard sphere fluid at $\phi = 0.562$ reduced in magnitude by a factor of 10.

the best fit of Eq.(12) to the data. This quantity is also consistent with the theoretical square root concentration dependence[8,16]. It is evident from Fig. 2 that beyond the time scale of the local particle motion ($\tau \approx 10^{-2}$s) that the β-process closely follows the experimental decays. It has been suggested that the differences between theory and experiment for $f(Q,\tau)$ below about 0.2 are due to the α-process[16]. However, significant statistical uncertainty in the data cannot be discounted when the correlation functions have decayed to these small values.

Estimates of the scaling time of the α-relaxation have been obtained[16] from the time required for $f(Q_m,\tau)$ to decay to $f_c(Q_m)/e$. Figure 3 shows that the exponents of both t_β and t_α are consistent with MC theory and that these two times diverge on approaching the glass transition.

The non-ergodicity parameters, $f(Q,\infty)$, of colloidal glasses have been calculated[14], with the use of Eq.(9), from the zero time values of the measured time-averaged time correlation functions of the scattered intensity, $g_T^{(2)}(Q,0)$. These are shown in Fig. 4 as functions of the scattering vector for several concentrations. For this suspension of PMMA particles of radius 210nm the glass transition was visually located, as described earlier, at $\phi_G \approx 0.563$. The solid lines in this figure are the predictions of MC theory, calculated by Bengtzelius et al.[15], for the hard-sphere system at separation parameters $\sigma=0$ and $\sigma=0.066$. One can see that these agree rather well with the experimental data for $\sigma=0$ and $\sigma=0.062$, particularly in the vicinity of the main peak in the static structure factor, where experimental artifacts associated with possible incoherent and multiple scattering are small. Note however, that provided the concentrations are expressed in terms of the separation parameter σ, there are no adjustable parameters in this comparison. The increase in $f(Q,\infty)$ with increasing concentration reflects the increasingly restricted particle motions that effectively cease at random close packing ($\phi \approx 0.64$) where $f(Q,\infty) \approx 1$.

CONCLUDING REMARKS

The suspended particles used in these experiments have radii roughly 10^3 times those of atoms. The consequences of this size disparity are, (i) that any crystalline structures are mechanically very weak so that a crystallized suspension can easily be forced into a metastable phase by simply tumbling the sample and (ii) the dynamics of these suspended particles are about 12 orders of magnitude slower than those of atoms so that, with the possible exception of those states close to the melting concentration, structural recovery takes long enough to allow the study of the structure and dynamics of the metastable phases and, importantly, to allow the glass phase to be obtained. Furthermore, the equilibrium phase behavior of non-aqueous suspensions of particles with relatively thin steric barriers suggests that they constitute a close approximation to the ideal hard-sphere system.

Dynamic light scattering studies reveal that large scale particle diffusion ceases, as evidenced by the emergence of a non-decaying component in the intermediate scattering function, at approximately the same concentration as that at which homogeneously nucleated crystallization is suppressed. However, we recognize that small variations in the particle size distribution and traces of contaminants may influence the concentration where heterogeneously nucleated crystallization supersedes homogeneous nucleation.

The results in Fig. 1 along with those of ref.14 indicate that in the glass phase the particles are effectively localized over many thousands of seconds. Furthermore, as can be concluded from Fig. 2, the onset of this localization occurs over a narrow range of concentration. These features suggest that the glass transition observed in these suspensions is approximately consistent with that of a sharp transition from the ergodic fluid to ideal glass, as predicted by the simple version of mode-coupling theory[8]. Our observations along with those of the MC theory can be interpreted microscopically in terms of the cage picture mentioned in the Introduction. The microscopic time scale, t_0, which characterizes the small scale particle motion within the instantaneous neighbor cage, has only a weak concentration dependence due to the hindered response of the suspending fluid. The mesoscopic and macroscopic time scales, t_β and t_α, reflecting respectively the cage vibration and its breakdown depend strongly on the concentration, and critically so, as the glass transition is approached. The α- and β- processes are indistinguishable at lower particle concentrations but bifurcate as the glass transition is approached. At the transition the absence of the α-process indicates that the cages cannot break down so that large scale particle diffusion ceases and, as indicated by the Q-dependence of the non-ergodicity parameters, concentration fluctuations on all spatial scales are partly frozen. However, at concentrations beyond the glass transition the particles are still able to execute motions on a small scale so that cage vibration and, therefore, β-relaxation persist. It appears that all motion ceases at random close packing.

REFERENCES

1. L. Antl, J.W. Goodwin, R.D. Hill, R.H. Ottewill, S.M. Owens, S. Papworth and J.A. Waters, Colloids and Surfaces **17**, 67 (1986).
2. P.N. Pusey and W. van Megen, Nature (London) **320**, 340 (1986).
3. P.N. Pusey, W. van Megen, P. Bartlett, B.J. Ackerson, J.G. Rarity and S.M. Underwood, Phys. Rev. Lett. **63**, 2753 (1989).
4. W.G. Hoover and F.H. Ree, J. Chem. Phys. **49**, 3609 (1968).
5. L.V. Woodcock, Ann. N.Y. Acad. Sci. **37**, 274 (1981).
6. M. Baus, J. Phys.: Condens. Matter **2**, SA135 (1990).
7. W. van Megen and P.N. Pusey, Phys. Rev. A **43**, 5429 (1991).
8. W. Gotze, *Liquids, Freezing and the Glass Transition*, Eds. J.P. Hansen et al. (North Holland, Amsterdam, 1991) pg. 287.
9. P.N. Pusey and W. van Megen, Physica **A157**, 705 (1989).
10. P.N. Pusey and W. van Megen, Phys. Rev. Lett. **59**, 2083 (1987).
11. P.N. Pusey and W. van Megen, Ber. Bunsenges. Phys. Chem. **94**, 225 (1990).

12. S.E Paulin and B.J. Ackerson, Phys. Rev. Lett. **64**, 2663 (1990).
13. S.M. Underwood, W. van Megen and J.R. Taylor, to be published.
14. W. van Megen, S.M Underwood and P.N. Pusey, Phys. Rev. Lett. **67**, 1586 (1991).
15. U. Bengtzelius, W. Götze and A. Sjölander, J. Phys. C: Solid State Phys. **17**, 5915 (1984).
16. W. Götze and L. Sjögren, Phys. Rev. A **43**, 5442 (1991).

RHEOLOGICAL PROPERTIES OF SILICA SUSPENSIONS IN AQUEOUS CELLULOSE DERIVATIVES SOLUTIONS

Y. Ryo and M. Kawaguchi*
Department of Chemistry for Materials, Faculty of Engineering, Mie University, 1515 Kamihama-cho, Tsu Mie 514 JAPAN

ABSTRACT

The rheological properties of the silica suspensions in aqueous solutions of hydroxypropylmethylcellulose (HPMC) were investigated in terms of the shear stress and storage and loss moduli (G' and G'') as a function of silica content, HPMC concentration, and HPMC molecular weight by using a coaxial cylinder rheometer.

INTRODUCTION

When Fumed silica such as Aerosil is mixed with water, the resulting silica slurry is gelling due to silica aggregation.[1] It is expected that mixing aqueous polymer solutions with the silica slurry yields a silica suspension, which shows a well response to some rheological measurements. Their rheological responses should strongly depend on the silica content, the adsorbed amounts of polymer chains, and the polymer's molecular weight. Thus, an investigation of the rheological behavior of silica suspensions leads to understanding the stability of silica suspensions by adsorption of polymer chains. In this paper, shear stress, G', and G'' of the silica suspensions in aqueous solutions of hydroxypropylmethylcellulose (HPMC) were studied as functions of concentrations of silica and HPMC samples having different molecular weights by using a coaxial cylinder rheometer.

EXPERIMENTAL

Three HPMC samples (65SH-50, Mw = 107K; 65SH-400, Mw = 321K; 65SH-50000, Mw = 2670K) were kindly supplied from Shin-Estu Chemical Co., Ltd. The degree of substitution (DS) of methyl group, and molar substitution (MS) of hydroxypropyl group are determined 1.8 and 0.15 for each HPMC, respectively.

Silica suspensions were prepared by the same procedure used in the previous paper.[2] The amounts of added silicas and HPMC are expressed as weight percent in the final mixtures. The media were aqueous solutions of HPMC at concentrations of 0.5-2.5%. The silica contents in the silica suspensions were 2.5, 5.0, and 7.5%.

Steady flow measurements and oscillatory measurements were performed by using a coaxial cylinder geometry on a MR-3 Soliquid Meter (Rheology Co. Kyoto, Japan). In the former experiments the silica suspensions were subjected to preshearing at the highest shear rate of 148 s^{-1} for 5 min to destroy some structures of the silica suspensions before the measurements under the respective shear rates were carried out. The temperature of the sample chamber was maintained at 27 ± 1°C.

RESULTS AND DISCUSSION

HPMC solutions used in the dispersed media show shear-thinning. The silica suspensions showed rheopexy except for the silica suspensions in aqueous 65SH-50 solutions. Above the shear rate = 20 s^{-1} no silica suspension showed rheopexy.

From the shear rate (γ) dependence of the shear stress (σ) for the 2.5, 5.0, and 7.5% silica suspensions in aqueous 65SH-400 solutions, the log-log plots of σ against γ for the 2.5% silica suspensions can be almost fitted on the straight line with a slope of 0.8. In contrast, the 5.0 and 7.5% silica suspensions have the respective yield shear stresses. The yield stress increases with an increase in the silica content. This means that the silica suspension at the higher silica content is more strongly aggregated. Above the yield shear stress the shear stresses for the respective silica suspensions tend to increase with an increase in shear rate along with a straight line and its slope is almost constant at the same silica content, irrespective of the HPMC concentration. Moreover, such a slope decreases with an increase in silica content.

Since the Lissajous figures of the oscillations of outer and inner cylinders gave an elliptical shape for all silica suspensions studied, we did not make any correction for the values of G' and G" obtained from the Markovitz equation.[3] Both G' and G" values increase with an increase in HPMC concentration for the respective suspensions. From the frequency dependence of G' and G" values for the 2.5, 5.0, and 7.5% silica suspensions in aqueous 65SH-400 solutions, for the 2.5 and 5.0% silica suspensions there is a frequency where the G" value exceeds the G' value. In particular, for the 2.5% silica suspension the G' value is almost independent of the frequency, in contrast with the G' value the G" value above a frequency = 1.0 s^{-1} tend to increase with an increase in frequency along a straight line with the same slope as the plot of σ against γ. The 5.0% silica suspension shows a more marked frequency dependence of both G' and G" than that for the 2.5% silica suspension. The frequency where the interception between the G' and G" values for the 5.0% silica suspension is higher than that for the 2.5% silica suspension. This difference is contributed to the higher elasticity of the 5.0% silica suspension.

On the other hand, for the 7.5% silica suspensions the G' value is larger than the G" value in the entire frequency range measured. This is indicative of a viscoelastic solid behavior with weak viscous properties. The same trend is observed for the 5.0% silica suspension in 1.0% aqueous 65SH-50000 solution. Namely, such a behavior of the silica suspensions should substantially stem from the silica aggregated structure and moreover, its structure is reinforced by adsorption of HPMC.

REFERENCES

1. R. Iler, The Chemistry of Silica (John Wiley & Sons, N. Y., 1987), p. 364.
2. M. Kawaguchi, T. Ryo, and T. Hada, Langmuir, 7, 1340 (1991).
3. H. Markovitz, J. Appl. Phys., 23, 1070-1077 (1952).

OBSERVATION OF A FINITE WAVELENGTH INSTABILITY IN CRYSTALLIZATION

K. Schatzel[*] and B. J. Ackerson[**]
Department of Physics, Oklahoma State University,
Stillwater, Oklahoma 74078

ABSTRACT

The unique time and length scale properties of colloidal particle suspensions have been exploited to study homogeneous nucleation and crystal growth processes. We performed small angle light scattering experiments on sterically stabilized PMMA spheres suspended in an index matching mixture of decalin and tetralin. These particles exhibit almost pure hard sphere properties as determined from phase equilibrium and sedimentation studies. Samples of various concentrations, having either pure crystal or coexisting liquid and crystalline phases in equilibrium, are shear melted by tumbling and then left to crystallize, while we recorded the small angle scattering pattern produced by a spatially filtered and expanded laser illumination.

Qualitatively, the observations of ring shaped intensity distributions concentric with the incident beam were the most prominent feature. These rings, which are reminiscent of those produced by spinodal decomposition, generally grow in intensity and shrink in diameter as crystallization proceeds. Approximate scaling of the intensity patterns to the functional form $q^2(1+1^2)^{-\alpha}$, where q is a scaled scattering vector amplitude and with $\alpha \sim 3$, is observed for several samples. The quantitative analysis of peak intensity and ring diameter clearly reveals the existence of at least two distinct time scales during crystallization. After an initial induction time, we observe a fairly rapid and marked growth of the scattered intensity. This "fast process" is accompanied by a significant shrinking of the ring diameter in samples of lower concentrations. It is not until well into this fast process that an indication of a sharp Bragg peak can first be observed at large scattering angles. A second "slow process" typically follows the first fast one with a sudden reduction in the speed of growth by almost an order of magnitude. The slow process is again accompanied by a shrinking

---

[*]On leave from: Institut fur Angewandte Physik der Universitat, D-2300 Kiel, Germany

[**]Supported by the National Science Foundation through Grant DMR 88021301

diameter of the low angle scattering pattern. Shrinkage is by as much as a factor of 3 for highly concentrated samples just below the glass transition. The time and length scales decrease while ring diameters increase with increasing sample concentration for both "fast" and "slow" processes. The observation of ring diameter as a function of time allows absolute crystal growth rates to be calculated.

We will interpret the physical processes responsible for these different observed time scales, the reason for the finite wavelength instability, and possible connections with the glass transition.

SUPERLONG RANGE ATTRACTIVE AND REPULSIVE INTERACTIONS BETWEEN COLLOID PARTICLES

S.Yoshino
Nagoya University, Chikusa-ku, Nagoya, 464-01, Japan

ABSTRACT

Colloid particles form monolayer ordered distribution on a slide glass slip after solvent evaporated, with about 1 to 2 μm surface to surface distance. The order forms in a small drop of colloid dispersion, during solvent evaporation. In this process, colloid particles dynamically move as a results of convection of the solvent. After evaporated enough, the distribution keeps the lattice constant. The particles trap to the secondary minimum around the colloid. The distance is longer than the result from DLVO theory, and the result shows the existence of superlong range attractive and superlong range repulsive forces.

INTRODUCTION

The mutual interaction of colloid particle is one of the fundamental problem in chemical physics, physics and biophysics. Generally, the coulomb interaction between fine particles in solution is described by DLVO theory, which has been supported and believed by a lot of colloid scientists. The result from DLVO theory has no contradiction with the model for Brownian particle which has hard core potential, because the interacting distance is not so long except extremely low ionic strength. The distance form the colloid surface to the secondary potential minimum is about 20 nm in maximum[1]. The ordered phase as DLVO theory, therefore, has a short range lattice constant. The observed lattice constant of ordered phase in the case of hole volume became ordered phase or order-disorder coexistence is ten times longer than theoretical result at least. Many possibility has been proposed for the phenomena[2,3,4,5]. What kind of interaction is the essential is a important question. In my previous papers[6,7,8], therefore, reported several experimental results to show superlong range interaction between colloid particle, and a theoretical model was proposed, that is "Dynamic Electric Multipole Model for Colloid"[8,9].

In this paper it will be shown that one of the evidence for the existence of superlong range repulsive and attractive forces between colloid particles.

EXPERIMENTAL

The distribution and movement of colloid (charged polystyrene latex with 0.4 μm dispersed in aqueous solution) was observed through phase-contrast optical microscope with 40× objective lens at room temperature, and the images were recorded with video system. A slide glass slip was placed on the stage, and small amount of solution poured on the glass.

RESULT AND DISCUSSION

Colloid particles form two dimensional ordered distribution during and after evaporated solvent on slide glass slip at room temperature. Colloid particles separate about 1 to 2 µm. These distribution and the cluster size depend initial volume of solution drop, concentration of colloid and evaporation rate.

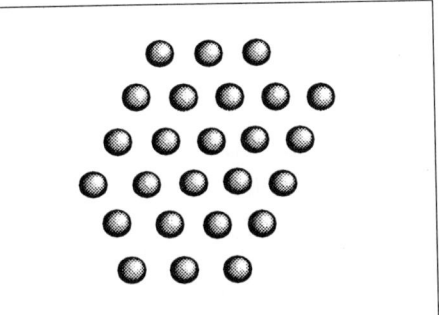

Fig.1. Typical example of two dimensional ordered distribution of colloid particles during evaporation and dried up solvent on glass slip.

During evaporation, colloid particles in solvent dynamically move with solvent convection. The drop size or the contact area with the glass become small, and the colloid concentration become high. The colloids, while, does not contact and not gather with particle surface-to-surface in solution. The particles keep the distance about 1 to 2 µm during these process and dried up. Therefore this result shows existence of a secondary minimum, consist of superlong range repulsive and attractive forces. These forces may come from the dynamical rotational movement, which was pointed out in Dynamic Electric Multipole Model.

These superlong range interactions are different phenomena from what is treated with DLVO theory based on static model.

REFERENCES

1. E. J. W. Verwey and J. Th. G. Overbeek, Theory of the Stability of Lyophobic Colloids (Elsevia, Amsterdam, 1948)
2. M. Wadachi and M. Toda, J.Phys.Soc.Japan, $\underline{32}$, 1147 (1972)
3. I. Sogami and N. Ise, J. Chem. Phys. $\underline{81}$, 6320 (1984)
4. N. Imai and S. Yoshino, J. Colloid Interface Sci. $\underline{105}$, 492 (1985)
5. P. A. Kralchevsky, N. D. Denkov, I. B. Ivanov and A. D. Nikolov, Chem. Phys. Letter $\underline{166}$, 452 (1990)
6. S. Yoshino and N. Imai, R. P. P. P. J. $\underline{28}$, 41 (1985)
7. F. Hirakawa, S. Yoshino, S. Era, K. Kuwata, M. Sogami and N. Imai, biophyis. Chem. 28, 253 (1987)
8. S. Yoshino, Ordering and Organization in Ionic Solutions (World Scientific Co. Shingapore 1988), p.449.
9. S. Yoshino, Dynamics and Patterns in Complex Fluids. (Springer-Verlag Heidelberg 1990), p.127.

B. THEORY

Transport Properties of Concentrated Colloidal Suspensions

E. G. D. Cohen
The Rockefeller University, New York, NY 10021
and
I. M. de Schepper
IRI, University of Delft, 2629 JB Delft, the Netherlands

Abstract

The macroscopic self-diffusion coefficient as well as the Newtonian and rheological viscosity coefficients of concentrated monodisperse neutral and charged colloidal suspensions consisting of spherical particles are deduced, by a simple scaling relation, from those of dense hard sphere fluids, where they have been obtained by kinetic theory. The analogy between the two dense fluid systems is based on the similarity of a basic cage-diffusion process in the two fluids. Good agreement is found with existing experiments of concentrated charged and neutral suspensions. A comparison is also made with the theory of Beenakker and Mazur for moderately concentrated colloidal suspensions.

Introduction

Already Einstein's 1905 paper on Brownian motion of a single Brownian particle[1] used the two basic ingredients that have dominated the theory of Brownian motion ever since: a hydrodynamic behavior of this particle like that of a macroscopic sphere in a continuum fluid, by using Stokes' law of friction and a probabilistic, stochastic behavior in that the particles diffuse, like the atoms in a fluid. He combined these two in his first Einstein relation:

$$D_0 = \frac{k_B T}{\zeta_0} \qquad (1)$$

Here D_0 is the (self)-diffusion coefficient of a Brownian particle in the solvent, $\zeta_0 = 3\pi\eta_0\sigma$ the Stokes friction of a Brownian particle in the solvent, where η_0 is the viscosity of the (pure) solvent with σ the diameter of the Brownian particle considered as a hard sphere, k_B Boltzmann's constant and T the temperature.

Einstein's theory only applied to infinitely dilute Brownian suspensions and many attempts have been made to generalize his work to more concentrated suspensions. Of those attempts we will discuss three, each of which emphasizes different aspects of the Brownian motion. We will confine ourselves here to the self-diffusion and the shear viscosity coefficients D_S and η, respectively, for concentrated colloidal suspensions. Before discussing the various theories, we summarize a few relevant properties of colloidal suspensions, that we use in our considerations.

1. One can distinguish two basic times in the motion of a Brownian particle: t_B, which characterizes the timescale on which a Brownian particle forgets its initial velocity

and $t_I = \sigma^2/D_0$, which characterizes the time scale on which the Brownian particles displace themselves by diffusion over a distance of the order of their size σ. Since $t_B \sim 10^{-9}$s and $t_I \sim 10^{-3}$s, typically, $t_B \ll t_I$.

2. All theories discussed here are based on a hard sphere model for the colloidal particles. While for neutral particles this is natural, for charged particles this can be based on the fact that in concentrated colloidal suspensions the static structure factor $S(k)$ exhibits a sharp maximum at $k = k^*$, which can be fitted well for k near k^* by the $S(k)$ of a dense hard sphere fluid with an appropriately chosen diameter σ[2,3]. This (effective) hard sphere diameter σ of the colloidal particles is for charged colloidal particles close to the diameter of the Debije sphere σ_D of the counterions, which surround a charged particle. Thus in this case the "effective" colloidal particles considered in $S(k)$ consist of a charged colloidal particle pit plus its surrounding Debije cloud. The same procedure can be used for concentrated neutral colloidal suspensions in which case the (effective) hard sphere diameter σ obtained from $S(k)$ for $k \approx k^*$, is close to the diameter of the colloidal particles themselves.

Transport Coefficients for Finite Concentrations

1. For *dilute* colloidal suspensions a direct and systematic extension of Einstein's work to higher concentrations has been made by Batchelor in a series of papers[4-6]. He incorporated both hydrodynamical and stochastic effects by considering the motion of two Brownian hard sphere particles in a solvent, as well as thermodynamic forces associated with the Brownian motion of the particles. His theory applied to neutral colloidal suspensions. His results, obtained as a virial or density expansion in powers of the volume fraction $\phi = \pi n \sigma^3/6$, where n is the number density of the Brownian particles, read[4,5]:

$$D_{S,s}/D_0 = 1 - 1.83\phi \tag{2}$$

$$\eta_s/\eta_0 = 1 + 2.5\phi + 5.2\phi^2 \tag{3}$$

The coefficient -1.83 of the term of $0(\phi)$ in eq.(2) decreases to -2.10, while the coefficient 5.2 of the ϕ^2 term in eq.(3) increases to 6.1, if Brownian motion effects are taken into account[6]. The coefficient of $O(\phi)$ in eq.(3), the first density correction to the viscosity of the pure solvent due to the presence of colloidal particles, was computed by Einstein[7]. The transport coefficients in eqs.(2) and (3) are short-time transport coefficients, valid for times $t_B \ll t \ll t_I$, since in the hydrodynamic treatment the spatial configurations of the particles remain approximately constant. This is indicated by a subscript s.

2. For *moderately concentrated* colloidal suspensions a completely hydrodynamic treatment of the transport properties also valid for short-times only has been given in a number of papers by Beenakker and Mazur[8-10]. They ignored the stochastic nature of the motion of the Brownian particles and treated them as if they were macroscopic hard spheres. Although their theory was applied to neutral colloidal suspensions, it applies also to charged colloidal suspensions. They developed their theory in two stages. First they solved approximately the very complicated hydrodynamical problem of the motion of many interacting hard spheres in a fluid, using a density-fluctuation expansion. A restriction to short times $t \ll t_I$ resulted from the assumption that the motion of each particle took place while all the other particles remained stationary. However, they did consider in their calculations of the mobility tensor the complicated hydrodynamic

interactions between a moving particle and its stationary neighbors induced by the displacement of the incompressible solvent. To this end they used the multipole expansion of Mazur and Van Saarloos[11] for the mobility tensor in powers of σ/R, up to $(\sigma/R)^7$, including two and three particle hydrodynamic interactions, where R is a typical distance between the Brownian particles. For moderately concentrated suspensions, σ/R would be < 1 for relevant particle configurations. Their results to $O(\phi^2)$ were[8,10]:

$$D_{S,s}/D_0 = 1 - 1.73\phi + 0.88\phi^2 \tag{4}$$

$$\eta_s/\eta_0 = 1 + 2.5\phi + 4.84\phi^2. \tag{5}$$

Eq.(5) for η_s/η_0 agrees with a calculation of Freed and Muthukumar[12]. Apart from higher multipole contributions, the neglect of Brownian motion may affect the coefficient of ϕ^2 in $D_{S,s}$.

Second, they carried out a partial resummation of their density-fluctuation expansions, in terms of correlation functions of the density fluctuations, taking into account self-correlations between the particles. This way they obtained expressions for k-dependent transport coefficients $D_s(k)/D_0$ and $\eta_s(k)/\eta_0$, which yielded expressions for $D_{S,s}/D_0$ and η_s/η_0:

$$D_s(k)/D_0 = \frac{H(k)}{S(k)} \tag{6}$$

$$D_{S,s}/D_0 = H(\infty) \tag{7}$$

$$\eta_s/\eta_0 = \lim_{k \to 0} \eta_s(k)/\eta_0. \tag{8}$$

The hydrodynamic factor $H(k)$ as well as $\eta_s(k)$ are given by complicated expressions, for which we refer to the literature[9,10]. The short-time η_s/η_0 describes experiments up to $\phi = 0.45$[13]. In fact, the results for η_s/η_0 of the resummed theory differ marginally (not more than about 5%) from that of the density-fluctuation expansion eq.(5) up to $\phi = 0.4$.

3. For *concentrated* neutral and charged colloidal suspensions with $\phi > 0.3$, expressions for $D_s(k)/D_0$ and $D_{S,s}/D_0$ as well as for the long-time *macroscopic* transport coefficients D_S/D_0 and η/η_0 have been derived by de Schepper and myself from the corresponding expressions for dense hard sphere fluids[2,3,14–16]. In this approach no distinction is made between charged and neutral colloidal suspensions and the Brownian motion in a concentrated colloidal suspension is compared with the average Newtonian motion in a dense hard sphere fluid, where the hard sphere diameter is that derived from the colloidal $S(k)$.

Explicit expressions for the scaled transport coefficients of concentrated colloidal suspensions are given below in the eqs.(9)–(12). They have been obtained from the corresponding ones for dense hard sphere fluids by relating $D_s(k)/D_0$ to $D_s^{hs}(k)/D_B$, D_S/D_0 to D_S^{hs}/D_B and $\eta(\gamma)/\eta_0$ to $\eta^{hs}(\gamma)/\eta_B$, respectively. Here D_B and η_B are the self-diffusion and viscosity coefficients, respectively, obtained from the hard sphere Boltzmann equation[17] and $\eta(\gamma)$, the rheological (shear rate γ dependent) viscosity, reduces to the Newtonian viscosity η in the limit $\gamma \to 0$. The superscript hs refers to hard sphere fluid properties. Thus it is assumed that the colloidal transport coefficients, scaled by their low-density values D_0 and η_0, can be derived from the corresponding hard sphere transport coefficients, scaled by their low-density values D_B and η_B. In fact, except for a modification discussed below for D_S/D_0, $D_s(k)/D_0$ and $\eta(\gamma)/\eta_0$ are given by the same expressions as their scaled hard sphere analogues.

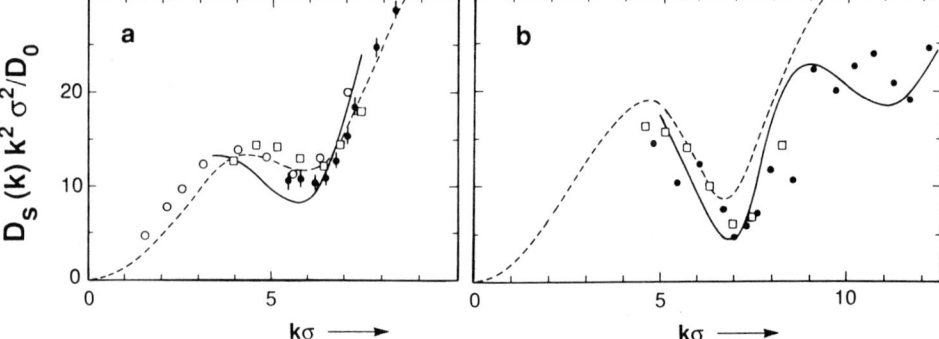

Fig. 1. $D_s(k)k^2\sigma^2/D_0$ as a function of $k\sigma$ for $\phi = 0.35$ (a) and $\phi = 0.46$ (b). Theory: 1) scaled hard spheres, eq.(9) (——-); 2) Beenakker and Mazur (ref.9) (- - - - -). Experiment: 1) neutral colloids ($\sigma = 335$nm, $\sigma^2/D_0 = 123$ms), Pusey and Van Megen (ref. 27) ($\square$ in (a) and (b)); 2) neutral colloids ($\sigma = 219$nm, $\sigma^2/D_0 = 16$ms), Fijnaut et al. (ref. 28) (o in (a)); 3) charged colloids ($\sigma = 296$nm, $\sigma^2/D_0 = 12.3$ms), Dalberg et al, (ref. 29) (• in (a)); 4) charged colloids ($\sigma = 600$nm, $\sigma^2/D_0 = 82$ms), Taylor and Ackerson (ref. 30) (• in (b)).

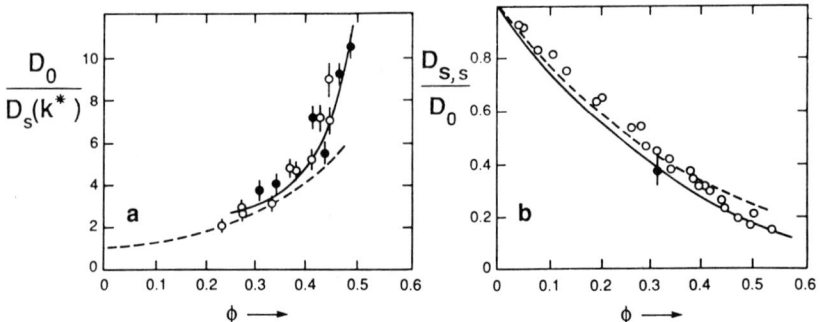

Fig. 2. (a) $D_o/D_s(k^*)$ as a function of ϕ. Experiment: 1) neutral colloids o, (ref. 31); 2) charged colloids •, (refs. 2,3). Theory: —— eq.(9), - - - - (ref.9); (b) $D_{S,s}/D_0$ as a function of ϕ. Experiment: 1) neutral o, (ref.31); 2) charged •, (ref. 32). Theory: —— eq.(10), - - - - - (ref.9).

The physics behind the eqs.(9)–(12) for concentrated colloidal suspensions is that the same physical processes take place in the colloidal suspension as in a dense hard sphere fluid, except that they occur on a much slower time scale (typically 10^9 times slower[3,14,16]). For, the typical displacement time σ^2/D_0 of a Brownian particle in a colloidal suspension is 10^9 longer than σ^2/D_B of a particle in a hard sphere fluid. Furthermore, there is a local ordering in both fluids at high concentrations, so that each particle finds itself in a cage formed by its neighbors. The basic transport coefficient, which determines all others, is $D_s^{hs}(k)$, the cage-diffusion coefficient of a hard sphere fluid – a collective short-time diffusion coefficient – which determines the diffusion rate of a hard sphere in and out of its cage. $D_s^{hs}(k)$ is derived from kinetic theory, using a linear generalized Enskog equation, which takes into account in an approximate but consistent manner the correlated motion of two neighboring hard spheres in their respective cages. The cages are represented in this equation by a mean field, characterized by $S(k)$. $D_s^{hs}(k)$ is then the lowest eigenvalue of a linear operator directly associated with this generalized Enskog equation[18]. The time scale of this cage-diffusion is of the order of $6t_0$, where the mean free time $t_0 = (m/\pi k_B T)^{1/2}/4n\sigma^2\chi$, with m, σ and n the mass, the diameter and the number density of the hard spheres, respectively. In the case of concentrated colloidal suspensions a time t_I' corresponds to t_0 for dense hard sphere fluids; these times characterize the time between successive binary collisions in the two fluid systems, respectively. As $D_s^{hs}(k)$, the expressions for D_S^{hs}/D_B and $\eta^{hs}(\gamma)/\eta_B$ are derived from kinetic theory and both contain $D_s^{hs}(k)$.

The corresponding scaled transport coefficients for concentrated charged and neutral colloidal suspensions are then the following.

For short times $t \approx 6t_I'$ and $\phi > 0.3$, the scaled cage-diffusion coefficient is given by:

$$D_s(k)/D_0 = D_s^{hs}(k)/D_B = \frac{1}{\chi} \cdot \frac{d(k)}{S(k)}, \quad (k \geq k^*) \tag{9}$$

while the scaled cage-self-diffusion coefficient $D_{S,s}$ is:

$$D_{S,s}/D_0 = D_{S,s}^{hs}/D_B = \frac{1}{\chi}. \tag{10}$$

For long times $t \gg 6t_I'$ and $\phi > 0.3$, the scaled macroscopic self-diffusion coefficient D_S is:

$$D_S/D_0 = \frac{1}{\chi} \cdot \frac{1}{1+\Sigma_{cd}^D}, \tag{11a}$$

where

$$\Sigma_{cd}^D = \frac{D_B}{6\pi n^2 \chi} \int_{6t_0}^{\infty} dt \int_0^{\infty} dk k^4 [S(k)-1]^2 e^{-D_s^{hs}(k)k^2 t} \cdot e^{-D_{S,s}^{hs} k^2 t}, \tag{11b}$$

while for $0.1 < Pe < 0.5$, the scaled macroscopic viscosity reads:

$$\eta(\gamma)/\eta_0 = \eta^{hs}(\gamma)/\eta_B \cong (\eta^{hs}/\eta_B)[1 - S(\phi)(Pe)^{1/2}], \tag{12a}$$

where the scaled Newtonian viscosity is given by:

$$\eta/\eta_0 = \eta^{hs}/\eta_B = (\eta_E/\eta_B)[1 + \Sigma_{cd}^\eta], \tag{12b}$$

with

$$\Sigma_{cd}^\eta = \frac{k_B T}{60\pi^2 \eta_E} \int_{6t_0}^{\infty} dt \int_0^{\infty} dk [kd\ln S(k)/dk]^2 e^{-2D_s^{hs}(k)k^2 t}. \tag{12c}$$

In these equations the scaled colloidal transport coefficients, on the left hand sides, are expressed in terms of hard sphere quantities only, on the right hand sides.

In eq.(9) χ is the value of the radial distribution function of the hard sphere fluid at contact, whose value varies from 1 at low densities to about 7 at $\phi \approx 0.50$[19]. It characterizes the increase in collision frequency in a dense hard sphere fluid as compared to that in a dilute hard sphere gas. $d(k) = 1/[1 - j_0(k\sigma) + 2j_2(k\sigma)]$, where $j_\ell(x)$ is the spherical Bessel function of order l. It is an oscillating function of k with values between

1/2 and 3/2[18] and characterizes the effectiveness of the collisional transfer of momentum between a particle inside a cage and the cage wall, i.e., its nearest neighbors[3]. $S(k)$ finally characterizes for $k \approx k^*$ the large amount of order in the fluid and thus the difficulty of diffusion[16]. The cage-diffusion is a short-time "micro-diffusion" on the molecular scale, characterized by $k \approx k^*$, for which eq.(9) is derived; in fact, eq.(9) also holds for $k > k^*$. Eq.(10) follows from eq.(9) by taking the limit $k \to \infty$ and using that $d(k)$ and $S(k)$ then go to unity.

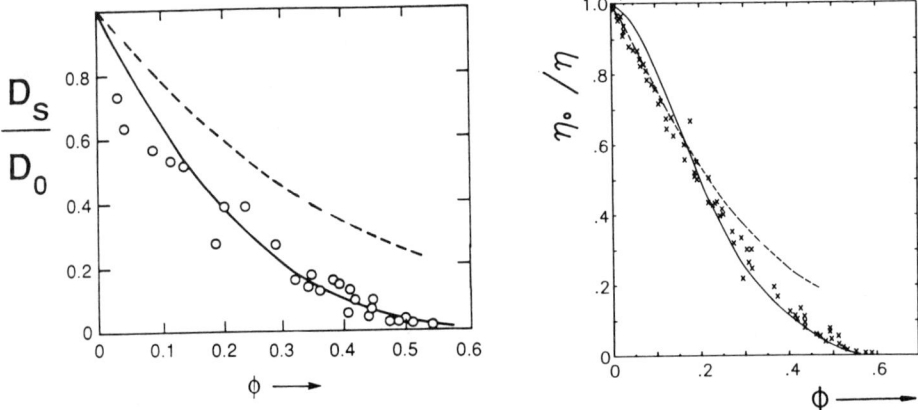

Fig. 3. D_S/D_0 as a function of ϕ. Experiment: neutral colloids (o) (ref.31). Theory: ——— eq.(11), - - - - - -, (ref.9, as in fig.2b).

Fig. 4. η_0/η as a function of ϕ. Experiment: neutral colloids (x) (ref. 33). Theory: ——— eq.(12b), - - - - -, (ref.10).

Eq.(11a) for D_S/D_0 is not identical to the corresponding one for D_S^{hs}/D_0. For, unlike in a dense hard sphere fluid, there is no momentum conservation in the (Brownian) particle system alone (in a concentrated colloidal suspension) because of the presence of the solvent. This prevents the formation of a double-vortex backflow of particles around a (tagged) particle in a concentrated colloidal suspension[41]. Or, equivalently, there is no hydrodynamic viscous mode in the colloidal particle system alone, which means that one of the modes needed for the (viscous) mode – (diffusive) mode coupling contribution to the long time tail of the velocity autocorrelation function – and hence to D_S – is absent. Consequently no long time tail contributions to D_S occur in concentrated colloidal suspensions[14,16] (see also point 3 of the Discussion). In computing therefore D_S/D_0 for concentrated suspensions from D_S^{hs}/D_B, the long-time tail related contributions to D_S^{hs}/D_B have to be omitted. Consequently D_S/D_0 and D_S^{hs}/D_B do not correspond, as is shown in fig.4 of ref.14. Eq.(11a) for D_S/D_0 has been derived by Kirkpatrick and Niewoudt[20,21]; a similar expression for D_S/D_0, has been given by Hess and Klein[22] and Medina-Noyola[23], with $d(k) = 1$. At the high densities considered here, this approximation is not important numerically. The subscript cd on Σ refers to cage-diffusion.

Eq. (12a) for $\eta(\gamma)/\eta_0$ gives the concentration dependence as well as the shear rate γ-dependence of the viscosity $\eta(\gamma)$, i.e., its rheological behavior. Eq.(12a) for $\eta^{hs}(\gamma)/\eta_B$

is for $0.1 < Pe < 0.5$, a good approximation to the complicated theoretical result for the rheological viscosity of a dense hard sphere fluid, derived by Kirkpatrick and Niewoudt on the basis of kinetic theory[20,24]. $Pe = \gamma\tau$ is the Péclet number, where the Péclet time $\tau = 3\pi\eta_B\sigma^3/4k_BT$. $\eta^{hs}(\gamma)/\eta_B$ has been computed numerically for one density ($\phi = 0.46$) by Kirkpatrick[20] (cf.fig.5a). The same behavior $\sim (Pe)^{1/2}$ of $\eta(\gamma)$ was also found numerically by Hood et al[25] for dense soft sphere fluids, where the soft spheres interact with a potential $\phi(r) = \epsilon(\sigma^{ss}/r)^{12}$, if r is the interparticle distance, ϵ a typical interaction energy and σ^{ss} the soft sphere "diameter" (with an equivalent hard sphere diameter $\sigma = 0.94\sigma^{ss}$[15]) and by Erpenbeck for a dense hard sphere fluid at the density $\phi = 0.46$[34]. Eq. (12b) gives the (Newtonian) shear viscosity, η, which was considered before in eqs.(3) and (8); η_E is the Enskog value for η in a dense hard sphere fluid[26]. Similarly as for D_S/D_0, long time tail contributions to η/η_0 have not been included. Since for $\phi > 0.3$ these contributions are negligible in hard sphere fluids, η/η_0 and η^{hs}/η_B do correspond for these densities, as is shown in fig.4.

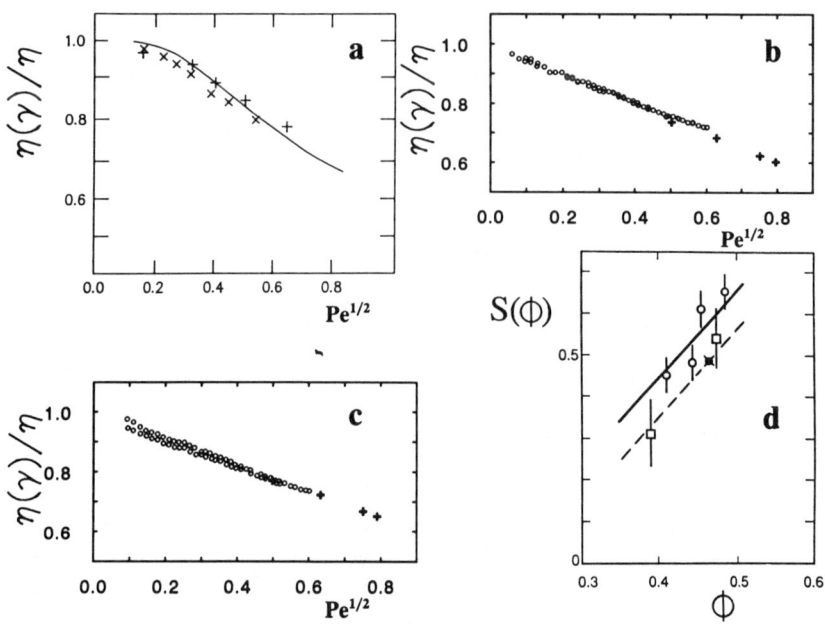

Fig. 5. (a) Reduced viscosities $\eta(\gamma)/\eta$ as a function of $Pe^{1/2}$ for $\phi = 0.46$ (a), 0.44 (b) and 0.41 (c). In (a): (—) hard sphere kinetic theory, eq.(12); (x) MD-hard spheres (ref. 34); (+) MD-soft spheres (isochoric) (ref. 25). In (b) and (c): (o) neutral colloids (ref. 35); (+) MD-soft spheres (isobaric) (ref. 25). (d) the slope $S(\phi)$ of $\eta(\gamma)/\eta_0$ as a function of ϕ, cf. eq.(12a): for neutral colloids (ref. 35) with $\sigma = 92$nm(o) and 56nm(□); for soft sphere fluids from MD (ref. 25), (—) isobaric and (- - -) isochoric; for hard sphere fluids from MD (ref. 34) (x) and from hard sphere kinetic theory (ref. 20) (•), (within error bars indistinguishable).

While the eqs.(9) and (10) are for short-time cage-(self)-diffusion coefficients, the long-time macroscopic self-diffusion coefficient D_S of eq.(11) is obtained from eq.(10) by adding the integrated effect of many successive cage-diffusions of the particles from cage to cage in the course of time[16]. This leads to the correction factor $(1+\Sigma_{cd}^D)^{-1}$. Σ_{cd}^D contains in the integrand a weight factor $\sim [S(k)-1]^2$, which is determined by the degree of order in the fluid, and for each k a mode-mode coupling contribution from a cage-diffusion mode and a cage-self-diffusion mode, characterized by their eigenvalues $D_s^{hs}(k)k^2$ and $D_{S,s}^{hs}k^2$, respectively. Σ_{cd}^D is not sensitive to the precise choice of the lower limit of the t-integral ($6t_0$) in eq.(11b) and its main contribution comes from k values $\approx k^*$. Similar considerations obtain for Σ_{cd}^η in eq.(12), except that the weight factor $\sim [dS(k)/dk]^2$ is determined by the cage deformability and that now two cage-diffusion modes are involved (cf.eq.(12c). As Σ_{cd}^D, Σ_{cd}^η is not sensitive to the lower limit of the t-integral. We remark that since we have modeled here the behavior of concentrated colloidal suspensions on that of dense hard sphere fluids, the short-time ($t \approx 6t_0$) behavior as well as the long-time macroscopic ($t \gg 6t_0$) behavior refer to dense hard sphere systems and do not always coincide with those used in the colloid literature. In fact, to conform with this literature, at least three time regimes (e.g. short $t_B \ll t \ll t_I$; long $t > t_I$ and very long (macroscopic) $t \gg t_I$) would have to be introduced, which we refrained from doing.

The eqs.(9)-(12) are compared with experiment as well as with the theory of Beenakker and Mazur in figs. 1-5. We note the following.

1) Three checks have been made of the basic cage-diffusion formula (9) for $D_s(k)$, which determines the rate of cage-diffusion and, through that, also the relaxation of all structural deformations of the fluid: be it by internal density fluctuations (D_S) or by external shearing (η)[16]:

 a. The k-dependence of $D_s(k)/D_0$ for $k \geq k^*$ is checked in fig. 1 for two values of ϕ;

 b. The ϕ-dependence of the maximum $D_0/D_s(k^*)$ of $D_0/D_s(k)$ for $k = k^*$ is given in fig. 2a;

 c. The ϕ-dependence of $D_s(k=\infty)/D_0 = D_{S,s}/D_0$ is checked in fig. 2b.

2) The ϕ-dependence of D_S/D_0 is presented in fig.3.

3) The Newtonian viscosity η/η_0 is given in fig.4.

4) Two checks have been made of the rheological behavior of $\eta(\gamma)/\eta_0$: one of its γ-dependence (figs.5a,b,c) and one of the ϕ-dependence of its slope $S(\phi)$ (fig.5d). In view of the good agreement between the theory and the soft sphere experiments in fig. 5a for $\phi = 0.46$, figs. 5b and c for $\phi = 0.44$ and 0.41, respectively, can also be expected to agree with the theory.

Comparing the eqs.(6) and (9) for $D_s(k)$, one sees that $H(k)$ corresponds to $d(k)/\chi$ (cf.fig.1 for k-values around k^*) and $H(\infty)$ to $1/\chi$, respectively.

The good agreement of the eqs.(9)-(12) with experiment gives some confidence in them and in the physics behind them. The k-dependence of the cage-diffusion coefficient $D_s(k)$ also agrees well with the results of computer simulations of Cichocki for a model of charged colloidal suspensions at $\phi = 0.3$ and $\phi = 0.4$[36].

We note that the Beenakker-Mazur theory gives better results at moderate and low concentrations $\phi < 0.3$, while the de Schepper-Cohen theory is better at high concentrations $\phi > 0.3$.

Discussion

1. All three approaches to the transport properties of colloidal suspensions discussed in the previous section have in common that they often appear to be applicable under conditions well beyond those, under which they have been derived. This same feature has been noticed for the virial expansions of the thermodynamic properties of moderately dense noble gases, which appear to agree with experiment for reduced densities $n\sigma^3 \leq$ 0.6 or $\phi \leq 0.31$, far beyond those for which they are considered to be valid[37]. This must be due to important cancellations among the higher order terms in these expansions, but the reason for this is unknown. An example in the context of this paper is Batchelor's expression (2) for $D_{S,s}/D_0$, which agrees with experiment up to $\phi \approx 0.3$[38]. Similarly Beenakker and Mazur's results give not only sensible – although not exact – values for the low density expansion coefficients of the transport coefficients, they also compare well, for η_0/η, for example, with those found experimentally up to $\phi \approx 0.4$, while the scaled hard sphere results in some cases can be extrapolated down to $\phi = 0$ (cf.figs.2b and 3), in spite of the fact that no cage is present for $\phi \leq 0.2$, since the pronounced maximum in $S(k)$ at $k \approx k^*$ virtually disappears for $\phi < 0.2$. On the other hand, Beenakker and Mazur's values for $D_s(k^*)$ are too high for $\phi > 0.4$, while the scaled hard sphere theory gives too low values for $D_s(k^*)$ for $\phi < 0.3$.

2. No direct theoretical justification is available at present for the application of the scaled hard sphere transport coefficients to those of neutral and charged colloidal suspensions and the good agreement with experiment. So far, we can only offer one consideration. For the case of a colloidal suspension and a hard sphere fluid under shear a formal correspondence has been obtained between two equations for the pair correlation function: one from the Smoluchowski equation for a colloidal suspension and the other from the Liouville equation for a hard sphere fluid, if D_B in the hard sphere equation is replaced by D_0 in the colloidal equation[16]. We note, however, that eq.(11) for the scaled D_S/D_0, derived here from D_S^{hs}/D_B for hard sphere fluids, is very similar to expressions obtained in entirely different fashions by Hess and Klein for *charged* colloidal suspensions (based on a memory kernel formalism)[22] and by Medina-Noyola for *neutral* colloidal suspensions (based on a generalized Langevin equation)[23], which also appears to support the scaling procedure used here.

3. Paul and Pusey have measured the velocity auto-correlation function of a Brownian particle at a very low concentration and found a "long-time tail'. i.e., a time dependence of this function $\sim (\nu_0 t)^{-3/2}$, where ν_0 is the kinematic viscosity of the solvent[39]. Although this time dependence is the same as that of the long-time tail of the velocity auto-correlation function of a tagged hard sphere in a hard sphere fluid, the coefficient differs. In fact, the behavior found by Paul and Pusey resembles that of a macroscopic sphere in a continuum fluid[40], rather than that of a hard sphere (atom) in a fluid, where one would have obtained a behavior $\sim [(\nu^{hs} + D^{hs})t]^{-3/2}$ [41], where ν^{hs} and D^{hs} are the kinematic viscosity and the self-diffusion coefficients, respectively, of the hard sphere fluid. In view of the discussion of eq.(11a), it is not clear to what extent the long-time tail-like behavior found by Paul and Pusey will remain with increasing concentration.

4. It would be interesting to have more experiments on concentrated charged colloidal suspensions, in order to test the validity of the eqs.(11) and (12) for the self-diffusion and viscosity coefficients, respectively, for such suspensions[14].

5. The structural relaxation by cage-diffusion in concentrated suspensions or in dense hard sphere fluids could also be applicable to many other dense fluid systems, e.g., in biology or technology, such as micelles, (micro) emulsions, coal slurries, etc.

Acknowledgement

The authors are much indebted to Ms. Sheila Rhyne for preparing this manuscript. One of them (E. G. D. C.) also gratefully acknowledges support from the U.S. Department of Energy under grant number DE-FG02-88-ER13847.

References

1. Einstein, A. Ann. d. Phys. 17, 549 (1905).
2. De Schepper, I.M., Cohen, E.G.D., Pusey, P.N. and Lekkerkerker, H.N.W., J. Phys. Condens. Matter I, 6503 (1989).
3. Pusey, P.N., Lekkerkerker, H.N.W., Cohen, E.G.D. and De Schepper, I.M., Physica 164A, 12 (1990).
4. Batchelor, G.K., J. Fluid Mech. 74, 1 (1976).
5. Batchelor, G.K. and Green, J.T., J. Fluid Mech. 56, 401 (1972).
6. Batchelor, G.K., J. Fluid Mech. 131, 155,437 (1983); 83, 97 (1977);Cichocki, B. and Felderhof, B.U., J. Chem. Phys. 89, 3705 (1988).
7. Einstein, A., Ann. d. Phys. 19, 289 (1906); 34, 591 (1911); Landau, L.D. and Lifshitz, E.M., Fluid Mechanics, Pergamon Press, (1959) p.76.
8. Beenakker, C.W.J. and Mazur, P., Phys. Lett. A 91, 290 (1982); Physica 120A, 388 (1983).
9. Id., Phys. Lett. A 98, 22 (1983); Physica 126A, 349 (1984).
10. Beenakker, C.W.J., Physica 128A, 48 (1984).
11. Mazur, P. and Van Saarloos, W., Physica 115A, 21 (1982).
12. Freed, K.F. and Muthukumar, M., J. Chem. Phys. 76, 6186, 6195 (1982).
13. Van der Werff, J.C., De Kruif, C.G., Blom, C. and Mellema, J., Phys. Rev. A 29, 795 (1989).
14. Cohen, E.G.D. and De Schepper, I.M., J. Stat. Phys. 63, 241 (1991).
15. De Schepper, I.M. and Cohen, E.G.D., Phys. Lett. A 150, 308 (1990).
16. Cohen, E.G.D.and De Schepper, I.M. in: "The Seventh International Conference on Recent Progress in Many Body Theories", C.E. Campbell and E. Krotschek, eds., Plenum (1992).
17. Chapman, S. and Cowling, T.G., "The Mathematical Theory of Non-Uniform Gases", Cambridge Univ. Press, third ed., pp. 194 and 169.
18. De Schepper, I.M., Cohen, E.G.D. and Zuilhof, M.J., Phys. Lett. A 101, 399 (1984).
19. Ree, F.H. and Hoover, W.G., J. Chem. Phys. 46, 4181 (1967).
20. Kirkpatrick, T.R., J. Non-Cryst-Solids 75, 437 (1985).
21. Kirkpatrick, T.R. and Niewoudt, J.C., Phys. Rev. A 33, 2658 (1986).
22. Hess, W. and Klein, R., Adv. in Phys. 32, 251 (1983).
23. Medina-Noyola, M. in: "Lectures on Thermodynamics and Statistical Mechanics", World Scientific (1988) p.1.
24. Kirkpatrick, T.R. and Niewoudt, J.C., Phys. Rev. A 33, 2651 (1986).
25. Hood, L.M., Evans, D.J. and Hanley, H.J.M., J. Stat. Phys. 57, 729 (1989).
26. Ref. 17, p.308.

27. Pusey, P.N. and Van Megen, B., Phys. Rev. Lett. 59, 2083 (1987).
28. Fijnaut, H.M., Pathmamanoharan, C., Nieuwenhuis, E.A. and Vrij, A., Chem. Phys. Lett.59, 351 (1978).
29. Dalberg, P.S., Boe, A., Strand, K.A. and Sikkeland, T., J. Chem Phys. 69, 5473 (1978).
30. Taylor, T.W. and Ackerson, B.J., J. Chem. Phys. 83, 2441 (1985).
31. Van Megen, B., Underwood, S.M., Ottewill, R.H., Williams, N.S.J. and Pusey, P.N., Far. Disc. Chem. Soc.83, 47 (1987).
32. Pusey, P.N. and Tough, R.J.A., in: "Dynamic Light Scattering", R. Pecora, ed., Plenum (1985) 85.
33. Van der Werff, J.C. and De Kruif, C.G., J. Rheol. 33, 421 (1989).
34. Erpenbeck, J.J., Phys. Rev. Lett. 52, 1333 (1984).
35. Van der Werff, J.C., De Kruif, C.G. and Dhont, J.K.G., Physica A 160, 205 (1989).
36. Cichocki, B., unpublished; Cichocki, B. and Hinsen, K., Physica A166, 473 (1990); Ber. Bunsenges. Phys. Chem. 94, 243 (1990).
37. Levelt, J.M.H. and Cohen, E.G.D., in: "Studies in Statistical Mechanics II", J.de Boer and G. E. Uhlenbeck, eds., North-Holland (1964) pp. 122,123.
38.Van Veluwen, A, Lekkerkerker, H.N.W., De Kruif, C.G. and Vrij, A., J. Chem. Phys. 87, 4873 (1987).
39. Paul, G.L. and Pusey, P.N., J. Phys. A 14, 3301 (1981).
40. Landau, L.D. and Lifshitz, E.M., in: "Fluid Mechanics", Pergamon Press (1959) pp. 95,96.
41. Alder, B.J. and Wainwright, T.E., Phys. Rev. A 1, 18 (1970).

BROWNIAN MOTION OF SUSPENSIONS

B.U. Felderhof
RWTH Aachen
D-5100 Aachen, Germany

ABSTRACT

We discuss diffusion of interacting Brownian particles in suspensions. A recent analysis of self-diffusion leads to a conjecture on the time-behavior of memory functions. According to the conjecture the decay of a typical memory function is characterized by a very broad spectrum of relaxation times and by a long-time tail.

INTRODUCTION

Colloidal suspensions of interacting Brownian particles consitute an interesting example of slow dynamics. On the timescale seen in light scattering experiments one observes the relatively slow change of configurations. Due to hydrodynamic and direct interactions the density time-correlation function and the mean square displacement of a selected particle show interesting dynamics characterized by a very wide spectrum of relaxation times. The dynamic behavior of a suspension on the slow timescale is described by the generalized Smoluchowski equation. We discuss general features of this equation and some exact results obtained for semi-dilute suspensions. A recent analysis of self-diffusion has suggested a conjecture for the time-behavior of typical memory functions.

SMOLUCHOWSKI EQUATION

We consider N identical spherical particles of radius a performing Brownian motion in an incompressible viscous fluid. The whole system is enclosed in a volume Ω. The particles interact via a direct pair potential and via hydrodynamic interactions. On the slow timescale seen in light scattering experiments it is sufficient to describe the system in terms of particle configurations, since the momenta quickly thermalize and assume a Maxwellian distribution. If R_i denotes the position of the i-th sphere, then the configuration of the entire suspension may be described by the $3N$-dimensional vector $X = (R_1, \cdots, R_N)$. The dynamical evolution of the configuration X is assumed to be described by a time-dependent probability distribution $P(X,t)$, which obeys the generalized Smoluchowski equation[1]. This equation reads

$$\frac{\partial P}{\partial t} = \nabla_N \cdot D \cdot [\nabla_N P + \beta (\nabla_N \Phi) P], \tag{1}$$

where the $3N \text{x} 3N$ diffusion matrix $D(X)$ incorporates the hydrodynamic interactions, and the potential $\Phi(X)$ the direct pair interactions and the wall potential. The equation describes relaxation to the time-independent equilibrium distribution

$$P_{eq}(X) = \exp\left[-\beta \Phi(X)\right] / Z(\beta), \tag{2}$$

where $\beta = 1/k_B T$ and $Z(\beta)$ is the normalization factor.

The quantity of primary interest in a light scattering experiment is the scattering function $F(q,t)$, defined by

$$F(q,t) = \lim_{\substack{N \to \infty \\ \Omega \to \infty}} \frac{1}{N} < \hat{n}(-q)\hat{n}(q,t) >, \tag{3}$$

where $\hat{n}(q)$ is the Fourier component of the density at wavevector q. We take the thermodynamic limit $N \to \infty$, $\Omega \to \infty$ at constant $n = N/\Omega$. The angle brackets indicate an equilibrium average and the time-dependence is governed by the adjoint Smoluchowski operator

$$\mathcal{L} = [\nabla_N - \beta \nabla_N \Phi] \cdot D \cdot \nabla_N \tag{4}$$

such that $\hat{n}(q,t) = [\exp \mathcal{L} t]\hat{n}(q,0)$ with $\hat{n}(q,0) = \hat{n}(q)$. The operator has real eigenfunctions with negative eigenvalues. The one-sided Fourier transform of the scattering function

$$G(q,\omega) = n \int_0^\infty e^{i\omega t} F(q,t) dt \tag{5}$$

may be written as

$$G(q,\omega) = \frac{nS(q)}{-i\omega + q^2 D(q,\omega)}, \tag{6}$$

where $S(q) = \lim < \hat{n}(-q)\hat{n}(q) > /N$ is the equilibrium structure factor. Eq. (6) defines the wavenumber- and frequency-dependent diffusion coefficient $D(q,\omega)$. It may be written as

$$D(q,\omega) = D(q,\infty) + M(q,\omega), \tag{7}$$

where $D(q,\infty)$ is the short-time limit and $M(q,\omega)$ is the memory function. The short-time diffusion coefficient is given by

$$q^2 D(q,\infty) = -\lim < \hat{n}(-q)\mathcal{L}\hat{n}(q) > /NS(q). \tag{8}$$

By use of the Mori-Zwanzig projection operator formalism one may derive a formal expression for the memory function[2]. Hence one finds the inequality[3]

$$0 \leq D(q,0) \leq D(q,\infty). \tag{9}$$

This shows that the long-time decay of the scattering function is slower than the initial decay.

Important features of the dynamics of the system are hidden in the memory function $M(q,\omega)$. Computer simulations of a suspension of hard spheres without hydrodynamic interactions have been carried out in my group by Cichocki and Hinsen[4]. These provide rather detailed information on a well-defined system for a range of volume fractions. Relatively little is known analytically, except for semi-dilute suspensions.

SEMI-DILUTE SUSPENSIONS

For semi-dilute suspensions some explicit results have been obtained. To first order in the volume fraction $\phi = (4\pi/3)na^3$ the collective diffusion coefficient may be written as

$$D(q,\omega) = D_0 \left[1 + (\lambda_C(q) + \alpha_C(q,\omega))\phi\right], \tag{10}$$

where D_0 is the single particle diffusion coefficient, the coefficient $\lambda_C(q)$ follows from $D(q,\infty)$, and the coefficient $\alpha_C(q,\omega)$ embodies dynamical effects. Both coefficients $\lambda_C(q)$ and $\alpha_C(q,\omega)$ may be evaluated from a pair problem. In abbreviated notation the coefficient $\lambda_C(q)$ may be expressed as[3]

$$\lambda_C(q) = \frac{-3}{2\pi D_0 q^2 a^3} \left(q \left| V \right| q\right), \tag{11}$$

and the coefficient $\alpha_C(q,\omega)$ may be expressed as

$$\alpha_C(q,\omega) = \frac{3}{2\pi D_0 q^2 a^3} \left(Vq \left| \frac{1}{i\omega + \mathcal{L}_r(q)} \right| Vq\right). \tag{12}$$

Here the states are defined as

$$|q\rangle = \cos\frac{q \cdot r}{2}, \qquad |Vq\rangle = V|q\rangle, \tag{13}$$

where $r = R_2 - R_1$ is the relative coordinate vector of two particles, and the perturbation operator V is given by

$$V = \mathcal{L}_r(q) - \frac{1}{2}D_0 q^2 + 2D_0 \nabla^2. \tag{14}$$

$\mathcal{L}_r(q)$ is the adjoint Smoluchowski operator in the relative coordinate space of two particles. It is given by

$$\mathcal{L}_r(q) = -Z(q,r) + [\nabla - \beta \nabla v] \cdot \mathbf{D}_r \cdot \nabla, \tag{15}$$

where $v(r)$ is the direct pair interaction, $\boldsymbol{D}_r$ is the relative diffusion tensor

$$\boldsymbol{D}_r = 2\left[\boldsymbol{D}_{11}(\boldsymbol{r}) - \boldsymbol{D}_{12}(\boldsymbol{r})\right], \tag{16}$$

and the function $Z(\boldsymbol{q},\boldsymbol{r})$ arises from the center of mass diffusion,

$$Z(\boldsymbol{q},\boldsymbol{r}) = \frac{1}{2}\boldsymbol{q} \cdot [\boldsymbol{D}_{11}(\boldsymbol{r}) + \boldsymbol{D}_{12}(\boldsymbol{r})] \cdot \boldsymbol{q}. \tag{17}$$

In Eqs. (11) and (12) we have used the scalar product

$$(A|B) = \int d\boldsymbol{r}\, g AB, \tag{18}$$

where $g(r) = \exp[-\beta v(r)]$ is the radial distribution function for a dilute system.

It is clear from Eq. (11) that the coefficient $\lambda_C(q)$ may be determined by quadratures. The coefficient has been calculated with accuracy for $q = 0$ for several models with and without hydrodynamic interactions[5,6], and some results are available[7,8] for $q \neq 0$. The calculation of the coefficient $\alpha_C(q,\omega)$ from Eq. (12) is more difficult, since it involves inversion of the operator $i\omega + \mathcal{L}_r(q)$. The calculation may be reduced to the solution of an ordinary differential equation in the radial coordinate, but this still presents a formidable problem in the general case. An explicit result is available for hard spheres without hydrodynamic interactions[9,10]. It is fairly easy to see that in general $\alpha_C(0,\omega) = 0$ at zero wavenumber. An analysis of the frequency-dependence of the corresponding coefficient $\alpha_S(0,\omega)$ for self-diffusion has suggested an approximate description, which may also be used for collective diffusion and for higher values of the density.

SELF-DIFFUSION

Self-diffusion of a selected particle is described in terms of a self-scattering function $F_S(q,t)$. Its one-sided Fourier transform

$$G_S(q,\omega) = \frac{1}{-i\omega + q^2 D_S(q,\omega)} \tag{19}$$

defines the wavenumber- and frequency-dependent self-diffusion coefficient $D_S(q,\omega)$. The mean square displacement of the selected particle is given by

$$W(t) = \frac{1}{6} < [\boldsymbol{R}_1(t) - \boldsymbol{R}_1(0)]^2 >. \tag{20}$$

Its rate of change defines a time-dependent diffusion coefficient $D_S(t)$,

$$\frac{dW}{dt} \equiv D_S(t) = D_S^S + \int_0^t M_S(0,t')dt', \tag{21}$$

where $D_S^S = D_S(0, \infty)$ is the short-time diffusion coefficient. The memory function is defined in analogy to Eq. (7) and is taken at zero wavenumber. The long-time diffusion coefficient[11] is given by

$$D_S^L = D_S^S + \int_0^\infty M_S(0,t)dt. \tag{22}$$

It is convenient to consider the relaxation function

$$\mu_S(t) = D_S(t) - D_S^L = -\int_t^\infty M_S(0,t')dt'. \tag{23}$$

From the general properties of the Smoluchowski equation it follows that its Fourier transform may be written in the form[12]

$$\mu_S(t) = \frac{1}{\pi}\int_{-\infty}^\infty F_S(v)e^{-v^2 t}dv, \tag{24}$$

where the variable v is related to frequency by

$$v = -\frac{1+i}{\sqrt{2}}\sqrt{\omega}, \tag{25}$$

with $\sqrt{\omega}$ defined with branch cut along the negative imaginary ω-axis and with $F_S(v)$ related to the Fourier transform $\hat{\mu}_S(\omega)$ by

$$F_S(v) = iv\hat{\mu}_S(-iv^2). \tag{26}$$

For a semi-dilute suspension of hard spheres without hydrodynamic interactions one finds exactly[12]

$$F_S(v) = \frac{2i\phi a^2 v}{1 + iv\sqrt{2\tau_0} - v^2\tau_0} \tag{27}$$

with $\tau_0 = a^2/D_0$. This is a meromorphic function of the complex variable v. We define $u = v^2\tau_0$ and write the expression (24) in the form

$$\mu_S(t) = \left(D_S^S - D_S^L\right)\int_0^\infty p_S(u)\,e^{-ut/\tau_0}\,du. \tag{28}$$

We have normalized such that

$$\int_0^\infty p_S(u)du = 1, \qquad \int_0^\infty \frac{1}{u}p_S(u)du = 1. \tag{29}$$

The short-time diffusion coefficient for this case is $D_S^S = D_0$ and the long-time diffusion coefficient is $D_S^L = D_0(1-2\phi)$. From Eq. (27) one finds for the distribution of relaxation rates

$$p_S(u) = \frac{1}{\pi}\frac{\sqrt{2u}}{1+u^2}. \tag{30}$$

This is a very broad spectrum behaving as $\sqrt{u}$ for $u \to 0$ and as $u^{-3/2}$ for $u \to \infty$. The corresponding relaxation function is given by

$$\mu_S(t) = A_+ \, w\left(v_+ \sqrt{t}\right) + A_- \, w\left(v_- \sqrt{t}\right), \tag{31}$$

where the function $w(z)$ is related to the error function of complex argument [13], the amplitudes are given by

$$A_\pm = (1 \pm i)\,\phi\, D_0 \tag{32}$$

and the roots $v_\pm$ by

$$v_\pm = (\pm 1 + i)/\sqrt{2\tau_0}. \tag{33}$$

For short times this yields for the diffusion coefficient

$$D_S(t) \approx D_0 \left[1 - 4\phi \sqrt{\frac{2t}{\tau_0}} + O(t)\right] \quad \text{as} \quad t \to 0. \tag{34}$$

For long times

$$D_S(t) \approx D_0 \left[1 - 2\phi + \sqrt{\frac{2}{\pi}}\phi \left(\frac{\tau_0}{t}\right)^{3/2} + O(t^{-2})\right] \quad \text{as} \quad t \to \infty. \tag{35}$$

For intermediate times the behavior is well described by a stretched exponential function.

Cichocki and I have conjectured [12] that behavior of the type (31), with modified amplitudes and roots, provides an accurate approximate description in more general cases. In order to preserve the normalization (29) we must write instead of Eq. (28)

$$\mu_S(t) = \left(D_S^S - D_S^L\right) \int_0^\infty p_S(u)\, e^{-ut/\tau_M}\, du, \tag{36}$$

with the mean relaxation time τ_M defined by

$$\tau_M = \int_0^\infty t\, M_S(0,t)\, dt \Big/ \int_0^\infty M_S(0,t)\, dt. \tag{37}$$

Our conjecture amounts to the assumption that the function $F_S(v)$ is well approximated by

$$F_S(v) \approx \left(D_S^S - D_S^L\right) \tau_M \frac{iv}{1 + i\sigma v \sqrt{\tau_M} - v^2 \tau_M}. \tag{38}$$

This is a two-pole approximation in the complex v-plane. The approximation is characterized by three parameters, the amplitude $D_S^S - D_S^L$, the timescale τ_M, and the width parameter σ. The spectral distribution corresponding to Eq. (38) is given by

$$p_S(u) \approx \frac{1}{\pi} \frac{\sigma \sqrt{u}}{1 + (\sigma^2 - 2)u + u^2}. \tag{39}$$

The distribution is sharp for $\sigma \to 0$, corresponding to purely exponential relaxation. The width of the distribution grows with increasing parameter σ.

The above conjecture has been applied successfully in an interpretation of the computer simulations of Gaylor et al. [14] and of Cichocki and Hinsen [4]. For example, in the simulation for hard spheres without hydrodynamic interactions [4] at volume fraction $\phi = 0.5$ there is an excellent fit for the parameter values $D_S^L = 0.0865 D_0$, $D_S^S = D_0$, $\tau_M = 0.0675 \tau_0$, and $\sigma = 2.5625$. Furthermore, we have supported the conjecture with exact calculations of self-diffusion in semi-dilute suspensions. If the direct pair interaction is given by a step potential and if hydrodynamic interactions are neglected, then the two-pole approximation is quite good. We have also studied hard spheres with a square well potential, again with neglect of hydrodynamic interactions. In that case we found that the two-pole approximation breaks down if the well is too deep. A four-pole approximation then provides a better fit.

COLLECTIVE DIFFUSION

In conclusion we discuss briefly the application of the above ideas to collective diffusion. One would like to find the complete wavenumber-dependent scattering function, as defined in Eq. (3). If we assume that the memory function $M(q,\omega)$ is well represented by a two-pole approximation of the type given in Eq. (38), we need to find three wavenumber-dependent parameters to characterize the scattering function. These are the difference of the short-time and long-time diffusion coefficients

$$M(q,0) = D(q,\infty) - D(q,0), \tag{40}$$

the mean relaxation time $\tau_M(q)$, and the width parameter $\sigma(q)$. Upon substitution in Eq. (6) we find that the two-pole approximation for the memory function corresponds to a four-pole approximation for the scattering function. The latter would be given by

$$F(q,t) = \sum_{j=1}^{4} A_j \, w(v_j \sqrt{t}) \tag{41}$$

with four amplitudes $\{A_j\}$ and four roots $\{v_j\}$. The roots occur in conjugate pairs. One pair, corresponding to a slow timescale, might be called the α-pair, and the other pair, corresponding to a faster timescale, might be called the β-pair. This terminology corresponds to that used in the theory of the glass transition [15]. It remains to be seen whether Eq. (41) provides a correct approximate description of the dynamic scattering function of dense suspensions.

References

1. P.N. Pusey in Liquids, Freezing and Glass Transition, edited by J.P. Hansen, D. Levesque and J. Zinn-Justin (North-Holland, Amsterdam, 1991), p. 763.

2. B.J. Ackerson, J. Chem. Phys. **69**, 684 (1978).

3. B.U. Felderhof and J. Vogel, to be published.

4. B. Cichocki and K. Hinsen, Physica **166A**, 473 (1990).

5. B. Cichocki and B.U. Felderhof, J. Chem. Phys. **89**, 1049 (1988).

6. B. Cichocki and B.U. Felderhof, J. Chem. Phys. **94**, 556 (1991).

7. H.M. Fijnaut, J. Chem. Phys. **74**, 6857 (1981).

8. W.B. Russel and A.B. Glendinning, J. Chem. Phys. **74**, 948 (1981).

9. B.U. Felderhof and R.B. Jones, Physica **122A**, 891 (1983).

10. R.B. Jones, J. Phys. **A17**, 2305 (1984).

11. B. Cichocki and B.U. Felderhof, Phys. Rev. **A42**, 6024 (1990).

12. B. Cichocki and B.U. Felderhof, to appear in Phys. Rev. **A**.

13. Handbook of Mathematical Functions, edited by M. Abramowitz and I.A. Stegun (Dover, New York, 1965).

14. K.J. Gaylor, I.K. Snook, W. van Megen, and R.O. Watts, J. Chem. Soc. Faraday Trans. 2 **76**, 1067 (1980).

15. W. Götze in Liquids, Freezing and Glass Transition, edited by J.P. Hansen, D. Levesque and J. Zinn-Justin (North-Holland, Amsterdam, 1991), p. 287.

A MODEL FOR REPULSIVE HARD SPHERES WITH SURFACE ADHESION

D. Bedeaux and G.J.M. Koper,
Dept. of Physical and Macromolecular Chemistry,
Gorlaeus Laboratories, P.O. Box 9502, 2300 RA Leiden, The Netherlands.

ABSTRACT

We have adapted Baxter's Sticky Hard Sphere (SHS) model so that it more adequately describes systems where the interparticle potential well is preceded by an energy barrier. In the original model the particles interact via a pair potential with a narrow attractive region next to a repulsive core. As such it is most suited for atomic and molecular systems at low densities. Colloidal particles, however, often exhibit long range repulsion whereas they do attract each other at shorter ranges. This motivated our effort to extend the model.

It is shown that the Percus-Yevick equation can also be solved analytically for this particular potential, provided that the region of attraction is sufficiently small. The new feature of this Repulsive Sticky Hard Sphere (RSHS) model is that the fraction of aggregated particles increases with temperature as is experimentally observed in such systems.

Structure functions from small angle X-ray studies on water/AOT/iso-octane microemulsions can be fit to those predicted by the RSHS model. The thus obtained binding enthalpy is in good agreement with earlier determinations from dielectric studies.

INTRODUCTION

Baxter's "sticky hard sphere model"[1] provides detailed structural information in terms of a radial distribution function, or equivalently of a structure function, while relatively little information is required about the interparticle potential. A model potential is used which consists of a hard core together with a rectangular well in the limiting case where the depth of the well becomes infinitely deep and its width vanishes while the product of depth and width, being proportional to the fraction of bound particles, remains constant. By construction the sticky hard sphere model is only useful for systems in which the fraction of bound particles decreases with temperature. Systems in which the fraction of bound particles *increases* with temperature cannot be described by this model. Such a system would require a potential "well" with positive well energy and this, in turn, can only be physically meaningful when the potential well is preceded by a potential barrier.

In the next section we indicate how the structure function can be calculated for such an interparticle potential. We shall use the model potential that is schematically depicted in fig. 1, where the size of the region of the well and of the barrier

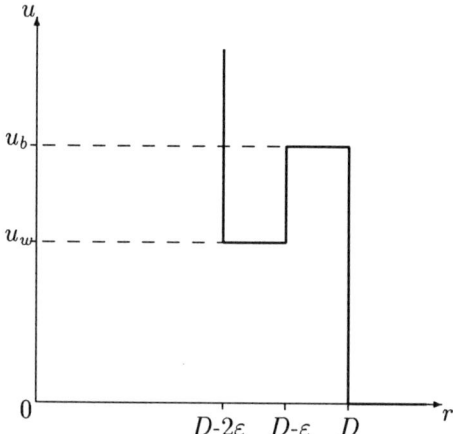
Fig. 1: Pair potential with potential barrier and positive well energy.

are both taken equal to ε. For sufficiently small values of the well width ε, compared to the core diameter D, the structure functions of Baxter's sticky hard sphere model and our repulsive sticky hard sphere model are found to be identical. Hence, the only difference with Baxter's sticky hard sphere model is that, due to the preceding energy barrier, positive well energies can be accounted for. The structure function is again determined by two parameters, namely the hard sphere diameter and the fraction of bound particles. These structure functions are, in the last section, compared to thoses obtained from small angle X-ray studies on water/AOT/iso-octane microemulsions.

METHOD OF SOLUTION

The indirect correlation function $h(r)$ is related to the radial distribution function $g(r)$ by

$$h(r) \equiv g(r) - 1. \tag{1}$$

In the region $(0, D)$ it is fully specified by the following three considerations: In the first place, the radial distribution function $g(r)$ must vanish in the core region where the potential is infinite. Secondly, the integral of the radial distribution function over the well times the particle volume fraction must give the fraction of bound particles that we shall call λ. Lastly, as the population of the barrier region becomes negligible for small ε one may also set the radial distribution function $g(r)$ equal to zero in this region. Notice that the barrier height u_b is kept constant. From these considerations we conclude that the indirect correlation function $h(r)$

in the range $(0, D)$ may be written in the form

$$h(r) = \begin{cases} -1 & \text{for} \quad 0 < r < D - 2\varepsilon \\ -1 + \frac{\lambda D}{12\varepsilon} & \text{for} \quad D - 2\varepsilon < r < D - \varepsilon \\ -1 & \text{for} \quad D - \varepsilon < r < D. \end{cases} \qquad (2)$$

The direct correlation function $c(r)$ is defined by means of the Ornstein-Zernike relation[2]

$$h(|\vec{r}|) = c(|\vec{r}|) + \rho \int d\vec{s}\ c(|\vec{s}|) h(|\vec{r} - \vec{s}|), \qquad (3)$$

with ρ the particle density. It follows from the Percus-Yevick approximation[3]

$$c(r) = \left(1 - e^{u(r)/k_B T}\right) g(r) \qquad (4)$$

with k_B Boltzmann's constant and T the absolute temperature, that for the potential given above the direct correlation function $c(r)$ vanishes beyond the range D. Consequently the Ornstein-Zernike relation (3) is sufficient to calculate the direct correlation function from the specification of the indirect correlation function (2) for $r < D$. The most convenient way to then calculate the indirect correlation function $h(r)$ for all r is to inverse Fourier transform the structure function $\hat{h}(k)$ which is related to the Fourier transformed direct correlation function $\hat{c}(k)$ by

$$\hat{h}(k) = \frac{\rho \hat{c}(k)}{1 + \rho \hat{c}(k)}; \qquad (5)$$

an expression that follows from a Fourier transformation of the Ornstein-Zernike relation, cf. eq.(3).

The calculation of the direct correlation function now proceeds in two steps[5]. Following Baxter[6] an intermediate function $Q(r)$ is calculated from the equation

$$rh(r) = -\dot{Q}(r) + 2\pi\rho \int_0^D dt\ (r-t) h(|r-t|) Q(t). \qquad (6)$$

The intermediate function $Q(r)$ is continuous everywhere and vanishes for $r \geq D$. It is given as a power series in r in each of the regions $(0, \varepsilon)$, $(\varepsilon, 2\varepsilon)$, $(2\varepsilon, D - 2\varepsilon)$, $(D - 2\varepsilon, D - \varepsilon)$, and $(D - \varepsilon, D)$, and substitution of this function in the above equation (6) while using (2) for the indirect correlation function leads to an infinite set of linear equations for the coefficients. To second order in ε it is possible to show that only a finite number of coefficients is nonvanishing, and these coefficients can then be calculated. Once the function Q is known the direct correlation function can be calculated from the following equation[5]

$$rc(r) = -\dot{Q}(r) + 2\pi\rho \int_r^D dt\ \dot{Q}(t) Q(t-r). \qquad (7)$$

$T[K]$	λ	
	$w_0 = 25$	$w_0 = 35$
298		2.3
301	3.0	2.5
304	3.3	3.0
307	4.0	3.7
311	4.3	4.3

Table 1: Experimental values of the λ.

The striking result is that for sufficiently small values of ε/D the structure functions from Baxter's sticky hard spere model and our repulsive sticky hard sphere model are found to be identical for the same volume fractions ϕ and the same fraction of bound particles λ. The only difference with Baxter's sticky hard sphere model is therefore that it also describes systems with positive well energies.

APPLICATION

We have applied the model to the small angle X-ray scattering data on water/-AOT/iso-octane microemulsions obtained by Robertus et al.[6] For various temperatures and low volume fractions they found structure functions in good agreement with Baxter's original model and they fitted the parameter λ. The data are for one volume fraction, ϕ=0.05, are tabulated in table 1. In this table w_0=[H$_2$O]/[AOT] is a parameter proportional to the droplet size. From these data it is clear that λ and therefore particle aggregation increases with temperature and thus this system cannot be completely modeled by Baxter's original sticky hard sphere model. These data motivated our extension of the model.

The parameter λ is related to the well energy u_w (see fig. 1) through

$$\lambda \sim e^{-u_w/k_B T}, \tag{8}$$

and u_w can be interpreted as the binding enthalpy per particle. Using the above data we arrive at binding enthalpies of 32 kJ/mol for the w_0=35 microemulsion and 24 kJ/mol for the w_0=25 microemulsion.

ACKNOWLEDGEMENTS

It is a pleasure to acknowledge stimulating discussions with J.G.H. Joosten and with C. Robertus.

REFERENCES

[1] R.J. Baxter, J. Chem. Phys. 49(1968)2770.
[2] L.S. Ornstein and F. Zernike, Proc. Acad. Sci. Amsterdam 17(1914)793.
[3] J.K. Percus and G.J. Yevick, Phys. Rev. 110(1958)1.
[4] G.J.M. Koper and D. Bedeaux, to be published.
[5] R.J. Baxter, Austr. J. Phys. 21(1968)563.
[6] C. Robertus, J.G.H. Joosten, and Y.K. Levine, Phys. Rev. A 42(1990)4280.

AGGREGATION KINETICS IN ELECTRO-RHEOLOGICAL FLUIDS

Howard See and Masao Doi
Department of Applied Physics, Faculty of Engineering,
Nagoya University, Furocho, Chikusa-ku, Nagoya, 464-01, Japan

ABSTRACT

We study the aggregation kinetics of Electro-Rheological Fluid particles which interact through dipoles induced by an external electric field and experience negligible thermal motion. For a system in d-dimensions, our theoretical calculations show that the average cluster size s_{av} increases with time t as $s_{av} \propto t^{\frac{d}{d+3}}$. This result shows good agreement with experiments in 3 dimensions, and with computer simulations in 2 dimensions.

EQUATION OF MOTION AND COMPUTER SIMULATION

An Electro-Rheological Fluid (ERF) is a suspension of uncharged, dielectric particles (typical size $\sim 10 \mu m$) in an insulating carrier liquid with a much lower dielectric constant. When an electric field (E) is applied, the particles form chain-like aggregates due to the interaction between the electric dipoles induced in each particle by the external field.

We will use the particles' equation of motion to study the aggregation kinetics of this system[1]. We apply the "creeping flow" approximation, and find the following force balance equation for the ith particle[2].

$$0 = \sum_{j \neq i} \mathbf{F}^{ij}_{electro} + \sum_{j \neq i} \mathbf{F}^{ij}_{repulsion} + \mathbf{F}^{i}_{flow}. \quad (1)$$

$\mathbf{F}^{ij}_{electro}$ is the induced dipole-dipole interaction force experienced by sphere i due to sphere j. Here we will use the point dipole approximation:

$$\mathbf{F}^{ij}_{electro} = -\nabla_i \left(\mathbf{p}_i \mathbf{p}_j : \nabla_i \nabla_j \frac{1}{|\mathbf{r}_j - \mathbf{r}_i|} \right), \quad (2)$$

where $\mathbf{p}_i$ is the dipole induced in sphere i. We will neglect the induced field and assume that the dipole is parallel to the applied field.

$\mathbf{F}^{ij}_{repulsion}$ is the rigid-body repulsion force between spheres i and j, which we approximate in the computer simulations by a "soft-core" potential force. $\mathbf{F}^{i}_{flow}$ is the drag force on the sphere due to its motion through the carrier liquid, which we will approximate by Stokes' equation.

In our simulations, we used a molecular dynamics approach based on eq.(1) to investigate the changes in system configuration after an electric field is applied to a random dispersion. A typical 2 dimensional configuration is shown in Fig. 1. The relationship between average cluster size s_{av} and non-dimensionalised time t^*, for various area fractions ϕ, is shown in Fig. 2. The results can be fitted by the following relationship.

$$s_{av} \propto \phi^{0.85} t^{0.39}. \quad (3)$$

We also examined the kinetics of the cluster size distribution and found agreement with the dynamic scaling theory[3], since good data collapse was obtained in plots of $s_{av}^2(t) N_s(t)$ against $s/s_{av}(t)$, where $N_s(t)$ is the number of clusters with size s at time t.

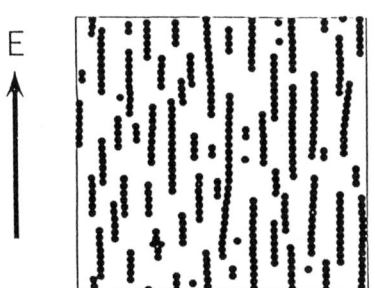

Fig.1. ERF simulation in 2 dimensions.

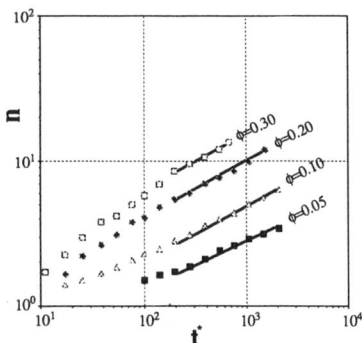

Fig.2. Graph showing the relationship between s_{av}, t^* and ϕ obtained from computer simulations.

THEORETICAL ANALYSIS

We use the "hierarchial" model of cluster growth, where we assume that at $t = 0$ all particles are separated, and that they combine to form doublets after time T_1, and then these doublets will combine to form quadruplets after a further time T_2, and so on. In order to estimate T_S we consider the time evolution of R, the average distance between nearest neighbour clusters.

Assuming that we have elongated clusters of size s, we can show that the electrostatic attraction force between clusters can be approximated by B/R^2, where B is independent of s. Since the cluster's friction coeffcient is $s(6\pi\eta_0 a)$, the time evolution of R is given by

$$s\,(6\pi\eta_0 a)\frac{dR}{dt} = -\frac{B}{R^2}. \qquad (4)$$

Assuming that after each step the aggregates are homogeneously dispersed, the average distance between the aggregates will be $a(s/\phi)^{\frac{1}{d}}$, where d is the spatial dimension of the system. Integrating eq.(4) under this condition, and finding the total time required for aggregates of size s to appear, yields the following:

$$s \approx \left(\frac{\left(\frac{\epsilon-1}{\epsilon+2}\right)^2 E^2}{\eta_0}\right)^{\frac{d}{d+3}} \phi^{\frac{3}{d+3}} t^{\frac{d}{d+3}}. \qquad (5)$$

Here ϵ is the ratio of the dielectric constant of the particles to that of the carrier liquid, and η_0 is the viscosity of the carrier liquid.

The time dependence of this result shows good agreement with our computer simulations in 2 dimensions (eq.(3)), and also with the ERF aggregation experiments of Vorob'eva et al.[4] in 3 dimensions, who found that $s_{av} \propto \phi^{0.72} t^{0.50}$. The discrepancy in the concentration dependence is thought to be due to our assumptions of a monodisperse system and isotropic aggregation.

REFERENCES

1) H. See and M. Doi, J. Phys. Soc. Jpn. **60**, 2778 (1991).
2) D.J.Klingenberg, F.van Swol and C.F.Zukoski, J.Chem.Phys. **94**, 6160 (1991).
3) T. Vicsek, *Fractal Growth Phenomena* (World Scientific 1989) p. 225.
4) T.A. Vorob'eva and I.N. Vlodavets, Kolloid. Zh. **36**, 1154 (1974).

TIME-DEPENDENT SELF-DIFFUSION COEFFICIENT OF INTERACTING BROWNIAN PARTICLES

B. Cichocki*
Institute of Theoretical Physics
Warsaw University
ul. Hoza 69
00-681 Warsaw
Poland

B.U. Felderhof
RWTH Aachen
D-5100 Aachen, Germany

ABSTRACT

We study the problem of self-diffusion for a system of interacting Brownian particles described by the generalized Smoluchowski equation. The variation in time of the mean square displacement of a selected particle may be characterized by a memory function. The Fourier transform of this function is regarded as a function of a variable v proportional to the square root of frequency. On the basis of results for simple model systems and some general properties of the Smoluchowski equation we conjecture that the transform may be analytically continued as a meromorphic function in the whole complex v-plane. In particular, near the origin it is analytic and its behavior is dominated by the nearest poles. An approximation involving a small number of poles provides an accurate description of the memory function for intermediate and long times. It turns out that numerical Brownian dynamics data are well described by a two-pole approximation.

We have studied the two-pole approximation for semi-dilute suspensions, for which analytic results may be found. We have shown that the three parameters needed to specify the approximation may be obtained from the steady-state

---

*Also at: Institute of Fundamental Technological Research, PAN, ul. Świętokrzyska 21, 00-049 Warsaw, Poland

perturbed pair distribution function. In earlier work we have developed an efficient method to calculate the perturbed distribution for general interactions. We have tested the two-pole approximation for an exactly solvable model in which the particles interact with a square step or well potential, but diffuse without hydrodynamic interaction. The approximation turns out to be accurate in the repulsive case. For strongly attractive interactions a more complicated description involving a larger number of poles is required.

PRECIPITATION IN A PARTICLE IN A COLLOIDAL SUSPENSION

Kiyoshi KISHI, Arata YOSHIDA* and Teruto YOSHIDA**
Faculty of Science, Science University of Tokyo,
Kagurazaka, Shinjuku-ku, Tokyo 162, Japan

ABSTRACT

The precipitation process of a solute is treated in a suspended submicron droplet. The precipitation needs increased concentration due to the limited volume with the effect of surface tension, when the droplet is small. The precipitation becomes increasingly difficult when the equilibrium concentration of the solute in the droplet is low.

INTRODUCTION

In accordance with Gibbs' kinetic theories[1], the critical size and the other properties of precipitations largely depend on the interactions between the precipitations and the environment[2,3]. The authors have pointed out that the presence of a liquid sheath as a slow-diffusion area markedly enhances the stability of growing precipitation[4]. This paper describes a precipitation delay in suspended small liquid particles.

THE PRECIPITATION IN A SMALL PARTICLE

When submicron liquid droplets have dispersed in a medium to form a colloidal suspension, a foreign material can be incorporate from the medium into the droplets as a solute, and consequently, the solute concentrates and then precipitates in each droplet. However, while the quantity of the solute reaches Q^*, which is the quantity in equilibrium with the critical nucleation of the precipitation, no nucleation can be triggered since the nucleation extracts the quantity and the degree of supersaturation becomes disappeared. Just before a stable critical nucleation the quantity of solute in a droplet Q_T is greater than Q^* by the quantity of critical nucleation Q^*_n.

By extending the results of Mullins and Sekerka, which were dissolved for simple solution growth with the liquid of infinity volume, the radius of the critical nucleation R^* is given for the shortest or critical time of nucleation t_c, as a positive solution of

---

* Present address : Toppan Printing Co. Ltd., Nobidome, Niiza, Saitama 352, Japan.
** Present address : R and D Division, Asahi Glass Co. Ltd., Hanezawamachi, Kanagawa-ku, Yokohama 221, Japan.

$$\frac{4}{3}\pi C(C-c_o)R'^4 - \frac{8}{3}\pi C c_o \Gamma_D R'^3 + \{C c_o V_o - (C-c_o)Q_T\}R'^2 + 2c_o \Gamma_D (Q_T + C V_o) = 0 \quad (1)$$

where C is the fixed concentration of solute in the precipitation, c_o and Γ_D are an equilibrium concentration at a flat interface and a capillary constant respectively between the precipitation and the liquid, and V_o is the constant volume of the solvent liquid in the droplet. As the minimum degree of supersaturation S for the nucleation is shown in Fig. 1, the nucleation requires a very large degree of supersaturation when the volume of the droplet and the equilibrium concentration are small. This implies that the precipitation becomes increasingly difficult. After t_c the total quantity of the charge increases and two possible values of Q^*_n are obtained for T as shown in Fig. 2. Only larger solution is stable. The hatched area shows where the degree of supersaturation remains after the nucleation.

The precipitation of solute in a suspended droplet needs increased concentration with the decrease in the diameter of the suspended droplet and with the decrease in the equilibrium concentration of the solute also.

ACKNOWLEDGEMENT

This work was supported by a Grant-in-Aid for Scientific Research on Priority Areas "Crystal Growth Mechanism in Atomic Scale" from the Ministry of Education, Science and Culture.

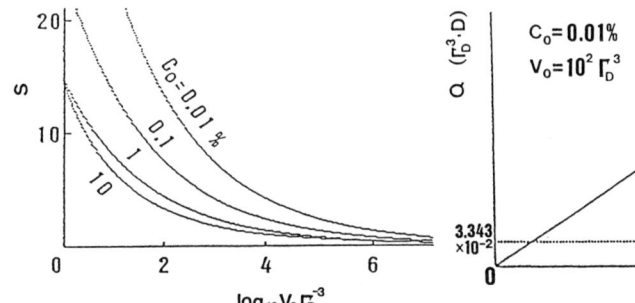

Fig. 1 The degree of supersaturation required for the nucleation.

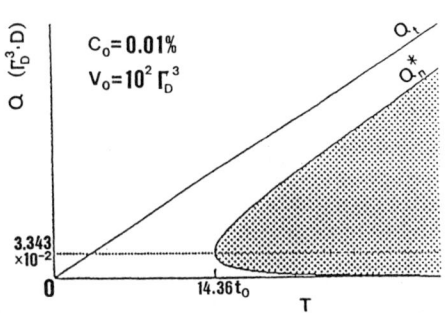

Fig. 2 The nucleation and the growth of the precipitation.

REFERENCES

1. J. W. Gibbs, The Scientific Papers of J. Willand Gibbs vol. 1 (Dover, N. Y., 1962), p. 322.
2. W. W. Mullins and R. F. Sekerka, J. Appl. Phys. 34, 323 (1963).
3. K. Nishioka, Phys. Rev. A36, 4845 (1987).
4. K. Kishi, A. Yoshida and T. Yoshida, Appl. Surface Sci. 33/34, 525 (1988).

C. COMPUTER SIMULATION

MICROSTRUCTURED FLUIDS: STRUCTURE, DIFFUSION AND RHEOLOGY OF COLLOIDAL DISPERSIONS

Thanh Phung and John F. Brady
Department of Chemical Engineering, California Institute of Technology, Pasadena, CA 91125, U.S.A.

ABSTRACT

The behavior of concentrated Brownian suspensions in a linear shear flow is studied numerically. Particle trajectories are calculated by Stokesian dynamics, a molecular-dynamics-like method that accurately represents the suspension hydrodynamics, including both many-body interactions and lubrication. The simulations are of a suspension of spheres as a function of the Péclet number, Pe, which measures the relative importance of shear and Brownian forces. Results are reported for the long-time self-diffusion coefficients and the suspension rheology and microstructure. There is a clear transition from a Brownian motion dominated regime ($Pe < 1$) to a hydrodynamically dominated regime ($Pe > 10$) with a dramatic change in the behavior of the long-time self diffusivity and the rheology. As the Péclet number is increased, the suspension first shear thins and then shear thickens at very high Péclet numbers owing to the formation of large clusters. The simulation results are shown to be in excellent agreement with the experiments of van der Werff & de Kruif[1].

INTRODUCTION

Understanding of the behavior of suspensions of particles immersed in a fluid is essential for the characterization of a variety of natural and industrial processes. The aim is to predict macroscopic transport properties (e.g. sedimentation or aggregation rates, self-diffusion coefficients, rheological behavior, etc.) from the microstructural mechanics—the hydrodynamic, colloidal, Brownian or externally imposed forces acting on the particles, as well as their distribution in space and time. In this article we describe how the dynamic simulation method know as *Stokesian Dynamics* can be used to study suspension behavior. While the applications of Stokesian dynamics are many[2], here we focus on the structure, diffusion and rheology of concentrated colloidal suspensions. First, the equations that govern both the microstructural dynamics and the macroscopic properties are presented. The essential element entering the dynamics is the hydrodynamic interactions among particles, and it will be described how the Stokesian dynamics method accurately captures both the near-field lubrication forces and the long-range many-body interactions. We shall then discuss the diffusive and rheological behavior of sheared hard-sphere suspensions, including both comparison with experiment and an interpretation of shear-thinning and -thickening behavior. It will be seen that Brownian suspensions at high solids content may order into "strings," although it will be shown that the shear thinning behavior is *not* due to this string formation.

SIMULATION METHOD

For N rigid particles suspended in an incompressible Newtonian fluid of viscosity η and density ρ, the motion of the fluid is governed by the Navier-Stokes equations, while the particle motion is described by the coupled N-body Langevin equation:

$$\mathbf{m} \cdot \frac{d\mathbf{U}}{dt} = \mathbf{F}^H + \mathbf{F}^B, \tag{1}$$

which simply states: mass × acceleration equals the sum of the forces. In (1) **m** is a generalized mass/moment of inertia matrix of dimension $6N \times 6N$, **U** is the particle translational/rotational velocity vector of dimension $6N$, and the $6N$ force/torque vectors **F** represent: 1) the hydrodynamic forces $\mathbf{F}^H$ exerted on the particles due to their motion relative to the fluid and 2) the stochastic forces $\mathbf{F}^B$ that give rise to Brownian motion. Any interparticle or external force, $\mathbf{F}^P$, could also be added to (1), but we shall only be concerned with hydrodynamic and Brownian forces in this article[2-6].

When the motion on the particle scale is such that the particle Reynolds number is small ($Re = \rho a^2 \dot\gamma/\eta \ll 1$), the hydrodynamic force exerted on the particles in a suspension undergoing a bulk linear flow is

$$\mathbf{F}^H = -\mathbf{R}_{FU} \cdot (\mathbf{U} - \mathbf{U}^\infty) + \mathbf{R}_{FE} : \mathbf{E}^\infty. \tag{2}$$

Here, $\mathbf{U}^\infty$ is the imposed flow at infinity evaluated at the particle center $\mathbf{x}_\alpha$, $\mathbf{E}^\infty(t)$ is the symmetric part of the velocity gradient tensor and is constant in space, though it may be an arbitrary function of time. $\mathbf{R}_{FU}(\mathbf{x})$ and $\mathbf{R}_{FE}(\mathbf{x})$ are the configuration dependent resistance matrices that give the hydrodynamic force/torque on the particles due to their motion relative to the fluid and due to an imposed shear flow, respectively. The vector **x** represents the generalized configuration vector specifying the location *and* orientation of all N particles. The inverse of the resistance matrix $\mathbf{R}_{FU}$ is known as the mobility matrix $\mathbf{M}\ (= \mathbf{R}_{FU}^{-1})$ and is a central element describing the hydrodynamic interactions among particles.

The stochastic or Brownian force $\mathbf{F}^B$ arises from the thermal fluctuations in the fluid and is characterized by

$$\langle \mathbf{F}^B \rangle = 0 \quad \text{and} \quad \langle \mathbf{F}^B(0)\mathbf{F}^B(t) \rangle = 2kT\mathbf{R}_{FU}\delta(t). \tag{3}$$

In (3) the angle brackets denote an ensemble average, k is Boltzmann's constant, T is the absolute temperature and $\delta(t)$ is the delta function.

The evolution equation for the particles is obtained by integrating (1) over a time step Δt that is large compared with τ, the Brownian relaxation time ($\tau = m/6\pi\eta a$), but small compared with the time over which the configuration changes. A second integration in time produces the evolution equation for the particle positions (both translational and orientational) with error of $O(\Delta t^2)$:

$$\Delta \mathbf{x} = Pe\{\mathbf{U}^\infty + \mathbf{R}_{FU}^{-1} \cdot \mathbf{R}_{FE} : \mathbf{E}^\infty\}\Delta t + \nabla \cdot \mathbf{R}_{FU}^{-1}\Delta t + \mathbf{X}(\Delta t),$$

$$\langle \mathbf{X} \rangle = 0 \quad \text{and} \quad \langle \mathbf{X}(\Delta t)\mathbf{X}(\Delta t) \rangle = 2\mathbf{R}_{FU}^{-1}\Delta t. \tag{4}$$

Here $\Delta \mathbf{x}$ is the change in particle position during the time step Δt, and $\mathbf{X}(\Delta t)$ is a random displacement due to Brownian motion that has zero mean and covariance given by the inverse of the resistance matrix. In (4), **x** has been nondimensionalized by the characteristic particle size a; the time by the diffusive time scale a^2/D_0, where $D_0 (= kT/6\pi\eta a)$ is the diffusion coefficient of a single isolated particle; and the shear forces by $6\pi\eta a^2\dot\gamma$, where $\dot\gamma = |\mathbf{E}^\infty|$ is the magnitude of the shear rate. The Péclet number, $Pe = \dot\gamma a^2/D_0 = 6\pi\eta a^3\dot\gamma/kT$, measures the relative importance of the shear and Brownian forces. Equation (4) can also be interpreted as a discretized form of the configuration-space Smoluchowski equation[2].

Equation (4) shows clearly that the suspension's behavior depends on the dimensionless parameters: Pe and ϕ, the volume fraction of particles. No restriction has been made to particles of identical size or shape; they need not be spherical, and if not, other dimensionless parameters characterizing their shape would be present.

The macroscopic properties are found from appropriate definitions and averages over particles and over time in a dynamic simulation. Here we shall be primarily interested in diffusion and rheology. Several "particle diffusivities" may be defined, and of interest is the long-time self-diffusivity $\mathbf{D}^s_\infty$, which measures the ability of a particle to wander far from its starting point. It is defined as the limit as time approaches infinity of one half of the time rate of change of the mean-square position of a particle:

$$\mathbf{D}^s_\infty = \lim_{t \to \infty} \frac{1}{2} \frac{d}{dt} \langle (\mathbf{x} - \langle \mathbf{x} \rangle)^2 \rangle. \tag{5}$$

For rheology, the bulk stress $\langle \Sigma \rangle$ is needed. This is defined as an average over the volume V containing the N particles and is given by

$$\langle \Sigma \rangle = \mathrm{IT} + 2\eta \mathbf{E}^\infty + \frac{N}{V} \{ \langle \mathbf{S}^H \rangle + \langle \mathbf{S}^B \rangle \}. \tag{6}$$

Here IT stands for an isotropic term of no interest for incompressible suspensions. In the absence of interparticle forces, the suspended particles make two contributions to the bulk stress: (a) a mechanical or contract stress transmitted by the fluid due to the shear flow, $\langle \mathbf{S}^H \rangle$; and (b) a direct contribution from the Brownian motion, $\langle \mathbf{S}^B \rangle$. Note that the contributions to the bulk stress parallel the forces in the Langevin equation (1). The particle contributions to the bulk stress are given by

$$\langle \mathbf{S}^H \rangle = -\langle \mathbf{R}_{SU} \cdot \mathbf{R}_{FU}^{-1} \cdot \mathbf{R}_{FE} - \mathbf{R}_{SE} \rangle : \mathbf{E}^\infty, \tag{7a}$$

$$\langle \mathbf{S}^B \rangle = -kT \langle \nabla \cdot (\mathbf{R}_{SU} \cdot \mathbf{R}_{FU}^{-1}) \rangle. \tag{7b}$$

In (7) $\mathbf{R}_{SU}(\mathbf{x})$ and $\mathbf{R}_{FE}(\mathbf{x})$ are configuration-dependent resistance matrices, similar to $\mathbf{R}_{FU}$ and $\mathbf{R}_{FE}$, relating the particle "stresslet" $\mathbf{S}$ to the particle velocities and to the imposed rate of strain, respectively. The stresslet is the symmetric and traceless first moment of the force distribution integrated over the particle surface.

The evolution equation (4) and the macroscopic properties (5)–(7) are the heart of the dynamic simulation. They are an exact description for N particles of arbitrary size and shape suspended in a volume V interacting through hydrodynamic and Brownian forces. Given an initial configuration, (4) is integrated in time to follow the dynamic evolution of the suspension microstructure, and (5)–(7) evaluated to determine the diffusive and rheological behavior. In addition to these macroscopic properties, a large number of particle distribution functions (e.g. $g(\mathbf{r})$, cluster sizes, etc.) can be determined.

In order to make use of the evolution equation and calculate macroscopic properties, the hydrodynamic resistance tensors $\mathbf{R}_{FU}$, $\mathbf{R}_{FE}$, etc. must be determined. A method that accurately and efficiently accounts for the near-field lubrication effects and the dominant many-body interactions has been developed[2-5]. Lubrication forces, as the name implies, result from the thin layer of viscous fluid separating particle surfaces and result in, among other effects, the relative motion of two particles approaching zero as their surfaces touch. The resistance tensors R_{FU}, R_{FE}, R_{SU} and R_{SE} can be written as part of a "grand resistance" tensor $\mathcal{R}$, which relates the force/torque (**F**) and stresslet (**S**) exerted by the fluid on the particles to the particle velocities and the rate of strain. The corresponding inverse or "grand mobility" tensor $\mathcal{M}$ is $\mathcal{M} = \mathcal{R}^{-1}$. The method proceeds by constructing an approximation to $\mathcal{M}$ by combining the Faxén laws for particle velocities with a truncated mulitpole moments representation of the particles. The grand mobility tensor, denoted $\mathcal{M}^\infty$, is then inverted to yield a far-field approximation to the grand resistance tensor. This many-body approximation to the resistance tensor lacks, however, lubrication. Lubrication would only be reproduced upon inversion of the mobility tensor if all multipole moments were included. Because of their short-range nature, lubrication forces are two-body interactions and are introduced in a pairwise additive fashion in the resistance tensor. Thus, the approximate grand resistance tensor that includes near-field lubrication and far-field many-body interactions is

$$\mathcal{R} = (\mathcal{M}^\infty)^{-1} + \mathcal{R}_{lub}, \qquad (8)$$

where $\mathcal{R}_{lub}$ stands for the near-field lubrication interactions. The grand resistance tensor is then partitioned and used in the evolution equation (4) and the macroscopic properties (5)–(7). This procedure captures both the near- and far-field physics and has given excellent results for all situations in which a comparison has been possible[2-9].

The long-range ($1/r$) nature of the hydrodynamic interactions requires care in simulating infinite suspensions, i.e. letting $N \to \infty$, $V \to \infty$, keeping $n = N/V$ fixed. A simple summation of interactions results in badly divergent expressions. While there are several alternate ways to overcome this convergence problem, the method presented by O'Brien[10] can conveniently be used in dynamic simulation. Used with periodic boundary conditions this method "renormalizes" all divergent and conditionally convergent hydrodynamic interactions and accelerates the convergence of the interactions using the Ewald summation technique.

STRUCTURE, DIFFUSION AND RHEOLOGY OF COLLOIDAL DISPERSIONS

The first studies that served to calibrated the Stokesian dynamics method with regard to the basic formulation and the treatment of the hydrodynamic interactions were for equilibrium hard-sphere microstructures. In the absence of a shear flow, a suspension of Brownian particles is distributed according to the equilibrium Boltzmann distribution: $P_N(\mathbf{x}) \sim \exp(-V(\mathbf{x})/kT)$, where $P_N(\mathbf{x})$ is the N-particle probability density and $V(\mathbf{x})$ is the N-particle interparticle potential. For the hard-sphere system the potential is zero when particles do not touch and infinite if the particles were to overlap. For this equilibrium distribution one can determine the short-time self diffusivity and the high-frequency dynamic viscosity as a function of the volume fraction. This was done in Phillips et al[11], and the results are in excellent agreement with experiment[12-14].

The simulations we would like to discuss here are for the simple shear flow of a suspension of Brownian hard spheres at a volume fraction $\phi = 0.45$. The only parameter to be varied is the Péclet number, measuring the relative importance of hydrodynamic to Brownian forces. The number of particles in the unit cell varies from 27 to 123, and the runs are for at least 50,000 times steps with a step size of 10^{-3}.

The first results presented are for the long-time self diffusivity defined in (5). Figure (1) is a log-log plot of the simulation results for the yy component of the long-time

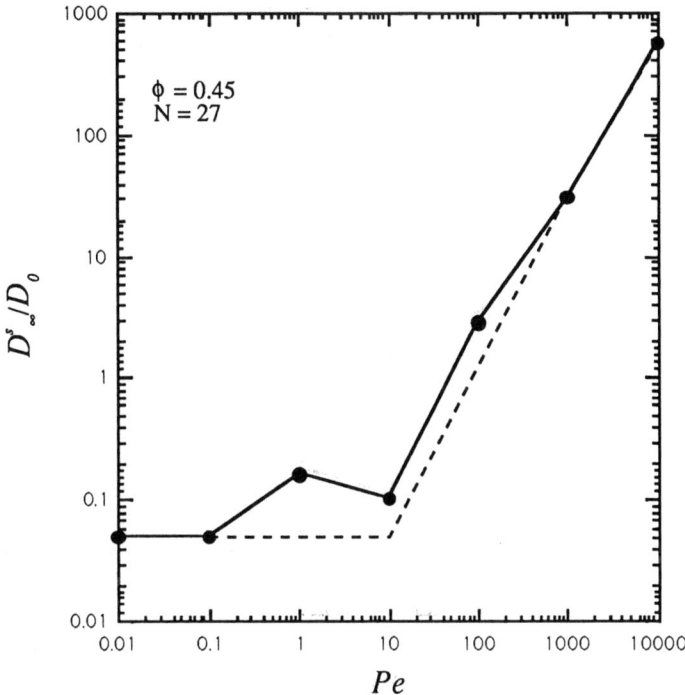

Figure 1. The long-time self diffusivity normalized by the infinite dilution diffusivity D_0 is plotted as a function of the Péclet number for a hard-sphere suspension at $\phi = 0.45$. The limiting asymptotes are: $D^s_\infty \sim 0.048$ as $Pe \to 0$ and $D^s_\infty \sim 0.055 Pe$ as $Pe \to \infty$. Note, as $Pe \to \infty$, the dimensional long-time self diffusivity scales as $\dot\gamma a^2$.

self-diffusion coefficient $\mathbf{D}^s_\infty$, nondimensionalized by D_0, as a function of Pe. (The velocity gradient of the shear flow is in the y-direction.) At small values of Pe, the system is essentially Brownian, and the dimensionless diffusion coefficient is less than unity as expected. At higher Péclet number, however, the long-time self diffusivity behaves quite differently. On purely dimensional arguments, as $Pe \to \infty$, the *dimensional* $\mathbf{D}^s_\infty$ would be expected to scale as $\dot\gamma a^2$; so $\mathbf{D}^s_\infty$, nondimensionalized by D_0, should scale with Pe as $Pe \to \infty$. Reference to Figure (1) shows that this is indeed the case. The transitional Péclet number, where the behavior changes from a Brownian motion dominated regime to a hydrodynamically dominated regime, occurs at $Pe \approx 10$.

Recent experiments by Eckstein et al.[15] and by Leighton and Acrivos[16] have demonstrated and quantified hydrodynamic dispersion of non-Brownian particles in shear flow. This hydrodynamically induced diffusive motion comes about from the deterministic chaos displayed by the nonlinear evolution equation for the particle positions (Eq. (4) as $Pe \to \infty$). Although there seems to be no question regarding the presence of hydrodynamic dispersion, the experimental measurements are not in complete agreement. For $\phi < 0.2$ both sets of experiments agree, but for higher ϕ the results of Leighton and Acrivos grow rapidly with ϕ, while on the other hand Eckstein et al. find the diffusivity

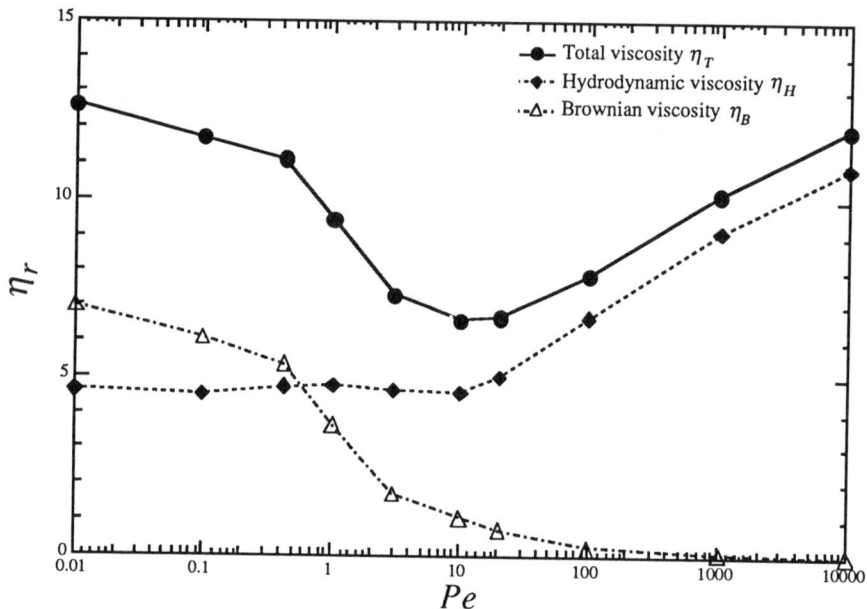

Figure 2. The relative viscosity η_r of a suspension of hard spheres at volume fraction $\phi = 0.45$ obtained by Stokesian dynamics as a function of the Péclet number: $(-\bigcirc-)$ total relative viscosity; $(-\triangle-)$ Brownian contribution η_r^B, and $(-\lozenge-)$ hydrodynamic contribution η_r^H.

to saturate at a value of approximately 0.03 for $\phi > 0.2$. Recently Leighton[17] measured anew the long-time self diffusivity and found it to saturate at a value of 0.09 for $\phi > 0.3$. Our result, 0.055, (cf. Figure (1) lies between the results of these two groups. Further, we found little change in this limiting value for a simulation at $\phi = 0.316$, and a slight decrease at $\phi = 0.49$. It should be noted, however, that we have not extrapolated our simulation results to infinite system size, which resulted in an increase in the short-time self diffusivities, and that there is a fair degree of statistical uncertainty in the simulation determination of the long-time self diffusivities owing to the very long runs required. Nevertheless, the agreement between simulation and experiment is quite reasonable.

The viscosity of a suspension is defined with the aid of (6)–(7) for the bulk stress. Specifically, for a shear flow in the xy plane the relative viscosity η_r, the viscosity of the suspension divided by the fluid viscosity, is the relation between the xy component of the bulk stress and the xy component of the bulk rate of strain. From (7)

$$\eta_r = 1 + \eta_r^H + \eta_r^B, \tag{9}$$

where η_r^H and η_r^B correspond to the contributions from the two terms in (7). In addition to the sum over particles, an average over time in a dynamic simulation is also performed. The nondimensionalization is such that for an isolated spherical particle $S_\alpha^H = 20/3\pi\eta a^3$, and $\eta_r^H = \frac{5}{2}\phi$, which gives Einstein's viscosity correction $\eta_r = 1 + \frac{5}{2}\phi$. Although the full stress is calculated in the simulation, here only the viscosity will be discussed.

In Figure (2) the evolution of the hydrodynamic and Brownian contributions to the viscosity with the Péclet number is shown. The total relative viscosity resembles

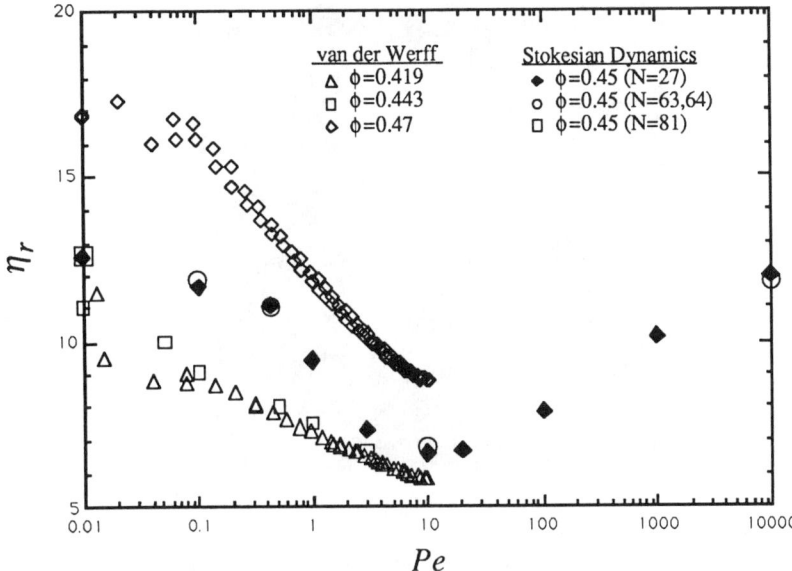

Figure 3. Comparison of the simulation results for the steady shear viscosity as a function of the Peclét number at a volume fraction $\phi = 0.45$ ($\Diamond, N = 27; \bigcirc, N = 64; \square, N = 81$) with the experiments of van der Werff and de Kruif (1989): $\Diamond, \phi = 0.47; \square, \phi = 0.443; \triangle, \phi = 0.419$.

quite strikingly experimental curves: shear thinning at low Péclet numbers ($Pe < 1$), a region where the viscosity does not vary appreciably, and finally a shear-thickening region at large Péclet number ($Pe > 10^2$). Although there is a plateau region in Pe where the total viscosity is roughly constant ($1 < Pe < 10^2$), both the hydrodynamic and Brownian components vary significantly in this region, but in such a way so as to compensate each other. For $Pe > 10$, the Brownian viscosity has essentially decayed away to zero. At low Péclet numbers, $\eta_r^B \sim \eta_r^B(0) - APe^2$, as $Pe \to 0$, where the Pe^2 scaling comes from considerations of reversing the direction of shear in a simple shear flow. Although theoretically the perturbation to the viscosity should be quadratic in Pe, our simulation results are not sufficiently accurate to check precisely this scaling. At the lowest value simulated, $Pe = 0.01$, the deformation of the equilibrium structure is so small that there is a large degree of statistical uncertainty in the Brownian contribution to the stress.

The hydrodynamic contribution to the viscosity shows a trend completely opposite that of the Brownian contribution. η_r^H remains constant for $0 < Pe < 1$, being equal to the high frequency dynamic viscosity mentioned earlier, and then continuously increases with increasing Pe. The increase in η_r^H for $Pe > 1$ is due to the formation of large clusters, whose size grows with the Péclet number.

Shown in Figure (3) is a comparison of the steady shear viscosities of Figure (2) with the experimental results of van der Werff and de Kruif[1] for silica hard-spheres. The volume fractions in the experiments are $\phi = 0.47\pm0.01, 0.443\pm0.01$ and 0.419 ± 0.01. The Péclet number range studied was $10^{-2} \leq Pe \leq 10$. The solid diamonds are Stokesian

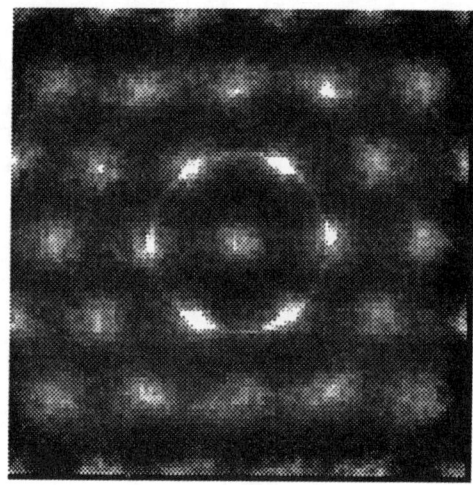

Figure 4. A plot of probability density for finding a particle at a position in the $y-z$ plane given that there is a particle at the origin, $g(y,z)$, at $Pe = 10$. There are 123 particles in the unit cell. Regions of light represent high probability and dark low.

dynamics simulations at $\phi = 0.45$ for 27 particles in the periodic cell, the open circles are for 64 particles, and the open square is for 81 particles in the unit cell. The simulation results for the different number of particles in the unit cell show little change from 27 to 81 particles. As can be seen, both the qualitative and quantitative agreement with the experiments is excellent. The major source of uncertainty in the experiments is in the determination of the volume fraction, which leads to a large uncertainty in the viscosity at high volume fraction due to the extreme sensitivity of the viscosity on volume fraction.

Both the experiments and the simulations show the suspension shear thins due to the fact that the Brownian contribution to the stress decays as the shear rate increases. The simulations show a further shear thickening region for $Pe \geq 100$. This region was not seen in the experiments of van der Werff and de Kruif because experimental limitations did not allow them to go to high enough shear rates. Shear thickening has been observed, however, in many other systems[18-20]. Shear thickening in Brownian suspensions is due to the formation of large clusters as the Péclet number increases[6]. This behavior is generic, as can be appreciated by noting that the viscosities measured for spherical particles at infinite Péclet number[21-22] are larger than the minimum in the shear thinning viscosity of the Brownian suspensions shown above in Figure (3); thus, the suspension *must* shear thicken.

Finally, we present some microstructural information from a simulation at $Pe = 10$ for 123 particles in the unit cell. The most revealing is a plot of the probability density for finding a second particle at a point in the $y-z$ plane given that there is a particle at the origin. Recall that the flow direction is x, the velocity gradient y and the vorticity z. Such a density plot is shown in Figure (4). The light areas are high probability and the dark low. This plot shows that the particles are lined up in "strings" in the flow direction, i.e. along the x-axis. We see quite clearly the hexagonal packing of the particles. This packing is such that it is easy for the particles to slide relative to one another following the imposed shear flow. (Note, one is looking back down the flow direction, so that particles coming out at you in the x-direction are situated directly above those shown in the $y - z$ plane.) This string-like formation has not been observed for the steady shear of hard spheres at this solids fraction. In oscillatory shear flow, however, this order has been seen experimentally[23-24] and it may be that the periodic boundary conditions, with particles running into their images, is much more analogous to large amplitude oscillatory flow. String formation has also been observed in nonequilibrium molecular dynamics simulations of atomic liquids[25-26], and the tendency for the particles to arrange themselves in this fashion, in which it is easiest to flow, has been used to explain the shear thinning phenomenon observed in these computer simulations. Although our Brownian suspensions appear to form this string-like order, the shear thinning we observe is *not* due to this structural arrangement; rather, the shear thinning is a direct consequence of the disappearance of the Brownian contribution to the stress, resulting from the saturation of the structure along the compressional axis[6]. Indeed, the hydrodynamic contribution to the stress is virtually unchanged from its zero Péclet number value even though the microstructure is very different.

CONCLUSIONS

In this article we have presented the Stokesian-dynamics method and applied it to study the structure, diffusive and rheological behavior of concentrated colloidal suspensions. The good agreement between simulation and experiment for the evolution of the viscosity with Péclet number shows that Stokesian dynamics is capable of excellent *quantitative* (as well as qualitative) predictive ability. Further, we have, in many cases for the first time, a detailed understanding of the mechanisms causing such behavior.

The results presented here provide only a one example of the ability of dynamic simulation to elucidate fundamental aspects of the physics of suspensions. The methods presented can be modified to include polydisperse systems, as well as systems of non-spherical particles,[8,9] and virtually any form of interparticle force is easily incorporated into the dynamic simulation[2,7,9]. The current state of the computations are such that for 27 particles in three dimensions with Brownian motion, 50,000 time steps, which represents one (ϕ, Pe) point in the simulations discussed above, requires approximately 4 hours on an IBM RISC/6000 530. For 64 particles the time requirements increase by a factor of 10. (There is a roughly N^3 scaling.) The point is that fully three dimensional simulations with 27 or 64 particles can be routinely performed on the current generation of desktop workstations. Thus, the range of problems of both a fundamental and practical nature that can be addressed through Stokesian dynamics is quite extensive and quite accessible. Finally, no comparison with theory has been presented because there is no theory for the phenomena discussed here. Clearly, there is a need for further theory, experiment and simulation of colloidal dispersions.

ACKNOWLEDGEMENTS

Portions of this work were done in collaboration with Georges Bossis and Ronald Phillips, whose help was indispensable. This work was supported in part by grants CBT-8451597, CBT-8696067, INT-8413695 and CTS-9020646 from the National Science Foundation, the Camille and Henry Dreyfus Foundation, and the San Diego Supercomputer Center.

REFERENCES

1. J.C. van der Werff and C.G. de Kruif, *J. Rheol.* **33**, 421 (1989).
2. J.F. Brady and G. Bossis, *Ann. Rev. Fluid Mech.* **20**. 111 (1988).
3. L. Durlofsky, J.F. Brady and G. Bossis, *J. Fluid Mech.* **180**, 21 (1987).
4. G. Bossis and J.F. Brady, *J. Chem. Phys.* **87**, 5437 (1987).
5. J.F. Brady, R.J. Phillips, J.C. Lester and G. Bossis, *J. Fluid Mech.* **195**, 257 (1988).
6. G. Bossis and J.F. Brady, *J. Chem. Phys.* **91**, 1866 (1989).
7. R.T. Bonnecaze and J.F. Brady, *J. Chem. Phys.* (in press 1991).
8. I.L. Claeys, Ph.D. Thesis, California Institute of Technology (1991).
9. J.F. Brady, in *Particulate Two-Phase Flow*, ed. M. Roco, Butterworths (1991).
10. R.W. O'Brien, *J. Fluid Mech.* **91**, 17 (1979).
11. R.J. Phillips, J.F. Brady and G. Bossis, *Phys. Fluids* **31**, 3462 (1988).
12. R.H. Ottewill and N. St. J. Williams, *Nature* **325**, 232 (1987).
13. P.N. Pusey and W. van Megen, *J. de Phys.* **44**, 285 (1983).
14. C.W.J. Beenakker and P. Mazur, *Physica* **126A**, 349 (1984).
15. E.C. Eckstein, D.G. Bailey and A.H. Shapiro, *J. Fluid Mech.* **79**, 191 (1977).
16. D. Leighton and A. Acrivos, *J. Fluid Mech.* **177**, 109 (1987).
17. D. Leighton, personal communication (1991).
18. R.L. Hoffman, *Trans. Soc. Rheol.* **16**, 152 (1972).
19. H.M. Laun*Prog. Trends Rheol. II*, Supplement to *Rheological Acta*, 287 (1988).
20. W.H. Boersma, J. Laven and H.N. Stein, *A.I.Ch.E.J.* **36**, 321 (1990).
21. D.J. Jeffrey and A. Acrivos, *A.I.Ch.E.J.* **22**, 417 (1976).
22. R. Pätzold, *Rheol. Acta* **19**. 322 (1980).
23. B.J. Ackerson, *J. Rheol.*, **34**(4), 553-590 (1989).
24. B.J. Ackerson, *Physica A*, **174**,15-30 (1991).
25. D.M. Heyes, *J. nonNewt. Fluid Mech.*, **27**, 47-85 (1988).
26. T. Weider, U. Stottut, W. Loose and S. Hess, *Physica A*, **174**, 1-14 (1991).

DYNAMICS OF DEFORMABLE INCOMPRESSIBLE COLLOIDAL PARTICLES IN DENSE SYSTEMS. COMPUTER SIMULATION.

T. Pakula and H. Nilgens
Max–Planck–Institut für Polymerforschung, 6500 Mainz, Germany

ABSTRACT

Dense systems of colloidal particles represented by incompressible but deformable objects on a lattice are simulated by an algorithm based on local cooperative rearrangements. Static and dynamic properties of the systems are analyzed as a function of temperature dependent deformability of particles and density of the system. Various dynamic states are observed ranging from liquid–like to glassy state. It is demonstrated that the freezing of translational diffusion can be observed as the effect of increasing density when particles are nondeformable or can result from increasing limitations of deformability at constant density. In systems with temperature dependent deformability of particle a non–arhenius slowing down with decreasing temperature is observed. A development of the simulation method for the case of three dimensional systems is presented.

INTRODUCTION

Recently, a new method of simulation of colloidal particles in dense systems has been proposed [1]. The method is based on the cooperative rearrangement algorithm applied originally to polymer chains [2]. The colloidal systems simulated have shown properties which are in many aspects characteristic for both liquids and colloids. An important parameter, the deformability of objects representing particles, has been considered. It has been shown that various structural states can be obtained by changing the density of the system with stiff particles or by changes of the particle stiffness at high density, alternatively.

Here, further developments in the simulation of such systems are briefly presented: various states of particles differing in particle deformability are related to temperature and a progress towards the simulation of three dimensional systems is reported.

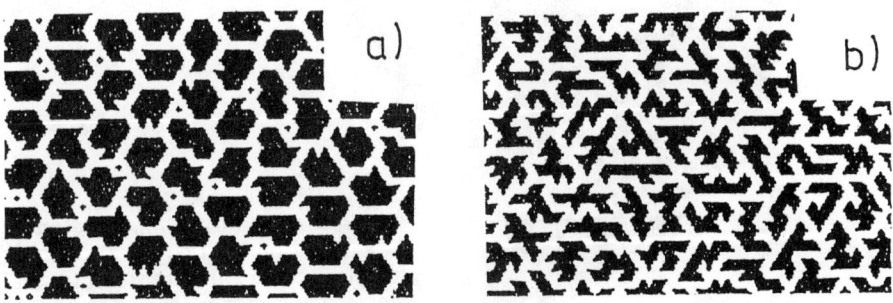

Fig.1. Illustration of systems with (a) stiff and (b) deformable particles.

ATHERMAL SYSTEMS

From the simulation of athermal systems with different densities and different deformabilities of particles, two characteristic properties have been noticed which are important for understanding of the mechanism of motion in dense colloidal and molecular systems: the existence of a kind of cage effect and the collective character of moves contributing to the diffusional steps of particles. In systems studied the following qualitatively different dynamic states have been observed: (1) only rotational and vibrational motions in the crystalline state, (2) diffusion along lines connecting defects in defected crystals, (3) collective diffusive motions in the liquid–like state (dense systems of soft particles or concentrated systems of stiff particles) and (4) noncorrelated diffusion in diluted systems. As the cage effect, a temporary entrapment of particles by nearest neighbors is interpreted. Although at high concentration, the particles retain some freedom for local motions due to incomplete volume filling or due to deformability of neighbors, they are essentially localized over long time and can escape from such entrapment only collectively with a number of other particles as demonstrated by particle trajectories [1].

TEMPERATURE DEPENDENCE

The stiffness of particles is related to temperature by assuming that various energy levels can be assigned to various discrete particle states characterized by the ratio of the membrane length to the particle surface [3]. Lower energy levels are assigned to lower deformability of particles and the simplest linear dependence between the energy and membrane length is assumed. The different states are separated by potential bariers ΔE_a for transitions to higher deformability and ΔE_b for transitions to lower deformability states.

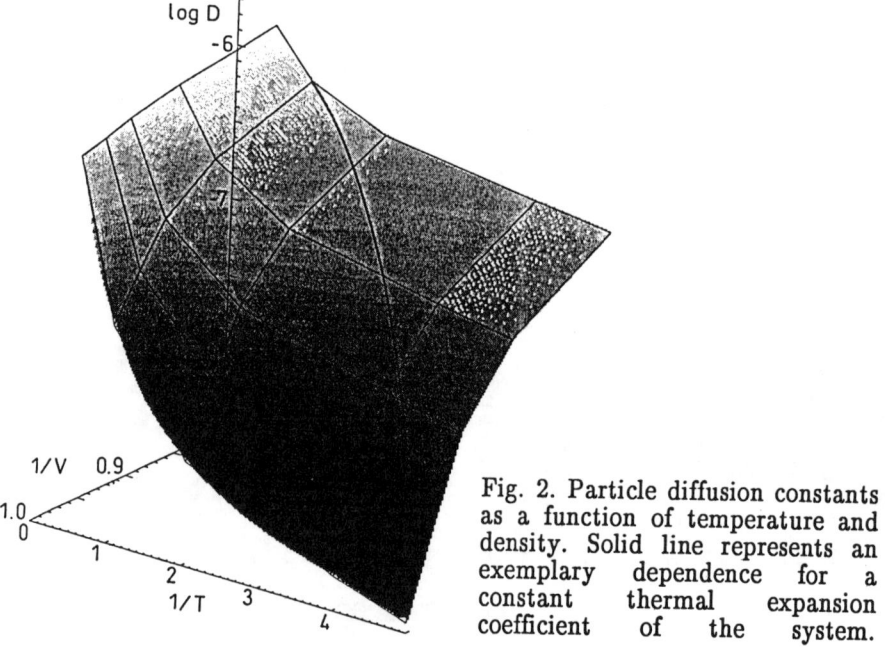

Fig. 2. Particle diffusion constants as a function of temperature and density. Solid line represents an exemplary dependence for a constant thermal expansion coefficient of the system.

According to such a model, the transition probabilities are temperature dependent, $p_a = A\exp(-\Delta E_a/kT)$ and $p_b = A\exp(-\Delta E_b/kT)$ for changes in direction of increasing and decreasing deformability, respectively. The population of various states of particles becomes in this way temperature dependent. At infinitely high temperature (the athermal case), all states are equally probable. With decreasing temperature the occupation of lower energy levels becomes more stable leading to an increase of particle stiffness. In the simulation, the following parameters of such a model have been assumed $\Delta E_a = 2kT$, $\Delta E_b = kT$ and $A = 0.5$. The equilibriated systems, generated for various particle concentrations and various temperatures, have been used as initial states for observation of the their dynamic properties. In order to characterize the mobility of particles mean square displacements of particle elements (beads) and of particle centers of mass were monitored in time. The results of particle diffusion constants as a function of temperature and density of the system are shown in Fig.2. From such a plot with coordinates $1/T$ and $\rho = 1/V$ the behavior of systems in which both the volume and the deformability of particles change with temperature can be concluded. If a constant (temperature independent) expansion coefficient is assumed (as observed normally above the glass transition) the temperature dependence of diffusion constant can be read out along the line $1/T \sim 1/V$ as schematically shown in Fig. 2 for an arbitrarily chosen case. The dependence, so obtained, shows a non–arhenius slowing down of particle mobility with decreasing temperature indicating a similarity to dependencies observed in real glass forming systems. Additionally, an effect of increasing heterogeneity of particle mobilities with decreasing temperature has been observed. This is illustrated in Fig.3 by particle trajectories. Particles with lower deformability seem to form less mobile clusters in such systems.

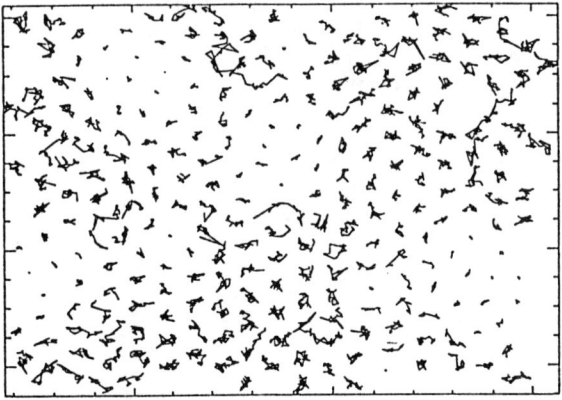

Fig. 3. Illustration of the heterogeneity of particle mobilities in a dense system at low temperature.

THREE DIMENSIONAL SYSTEMS

In the case of two dimensional systems we considered the particles as enclosed by the membrane in the form of a cyclic chain. Equivalent particles in a three dimensional space must be surrounded by a surface (membrane) being both impenetrable for the internal beads and mobile on the three dimensional lattice. Two versions of the algorithm for simulation of motion of such membranes have been developed: (1) a version for network membranes and (2) another version for liquid–like membranes. The fcc lattice was used. In the first case, the membrane is a two dimensional network with 4–functional crosslinks and the mash size of the single bond length. Such network can be moved in the three dimensional lattice space by the cooperative motion algorithm with only a slight modification considering the presence of crosslinks in the network. The membranes simulated in this way are well mobile on the fcc lattice but are relatively stiff and therefore not well suitable for objects with high curvatures as necessary for membranes enclosing particles of reasonably small sizes. In the other case of the liquid–like membranes the membrane consists of an assembly of beads joined two dimensionally with their neighbors into a continuous impenetrable surface within which no permanent bounding but only a continuity of the surface is demanded. Within the surface, the beads can change their neighbors in the way like free points mowing cooperatively on a two dimensional lattice. Fluctuations in coordination number within the membrane when locally changing shape are allowed. The beads within the membrane are enclosed by a linear chain which can change its length according to changes of the membrane boundary. The length of the membrane boundary can be reduced to zero at finite membrane surface area. In such a case the membrane forms a closed surface enclosing a part of space i.e. constituting an object which can be regarded as a deformable three dimensional particle. Details of the simulation algorithm for systems of such particles will be described elsewhere. Here, an example of the liquid–like surface simulated according to this algorithm is shown in Fig.4. by a sequence of states changing in time. The liquid–like surfaces are much more flexible than the crosslinked surfaces. In order to use such flexible surfaces as membranes for three dimensional colloidal particles, the same rules as for the two dimensional systems can be used. The stiffness of particles can be changed by changes of the ratio between the membrane surface and the volume of the enclosed part of the system. A possibility of partial permeability of such membranes can be considered as well.

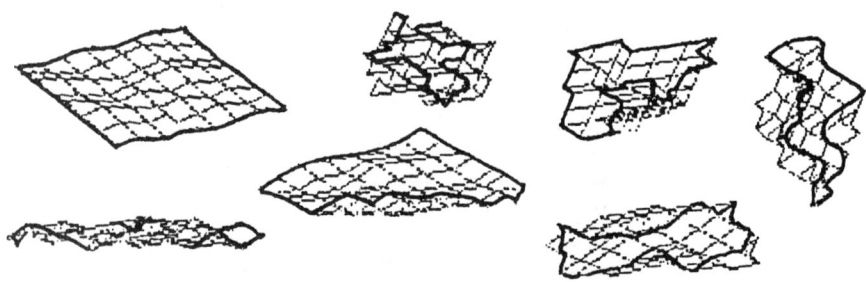

Fig. 4. An example of a membrane surface moving in the fcc lattice.

REFERENCES

1. T. Pakula, J. Chem. Phys. **94**,2104(1991); J. Non–Cryst. Solids **131**,289(1991)
2. T. Pakula, Macromolecules **20**,679(1987).
3. H. Nielgens, Diplomarbeit, University of Mainz (1991)

V. SPIN GLASS

A. EXPERIMENT

NON-EQUILIBRIUM DYNAMICS IN SPIN GLASSES

Leif E.C. Lundgren
Uppsala University, Institute of Technology, Box 534,
S-751 21 Uppsala, SWEDEN

ABSTRACT

Some recent experiments on the non-equilibrium dynamics in two- and three-dimensional spin glasses are reviewed. The experimental results are discussed within new domain (droplet) theories of the spin glass phase. It is found that the experiments convincingly verify some fundamental concepts of these theories. Especially, it is shown that if a spin glass has been 'aged' at constant temperature and afterwards subjected to a persistent temperature shift or a temperature cycling, the dynamics are governed by an interplay between the growth of spin glass domains and the overlap length, which is the maximum length scale where equilibrium spin correlations at two nearby temperatures are indistinguishable.

INTRODUCTION

Spin glasses[1] are known to exhibit slow dynamics, which is a common feature of disordered, strongly interacting systems of condensed matter, such as dielectrics, glasses and glassy polymers. In particular, it has been experimentally shown that spin glasses display ageing, which implies non-equilibrium dynamics. From extensive experiments on metallic[2-4], semiconducting[5] and insulating[6] spin glasses it has been found that the ageing phenomenon is of universal character. Ageing in spin glasses is not just caused by the existence of 'long' relaxation times, but the fundamental signatures arise from the collective nature of a system with many metastable states. Recently, domain (droplet) theories for ageing have been proposed. by Fisher and Huse[7], and by Kooper and Hilhorst[8]. In an alternative approach an ultrametric tree model has been adopted by Sibani and Hoffman[9]. Lederman et al[10] have interpreted some recent experiments within the model of an ultrametric organization of metastable states.
In this review some important experimental observations from the time dependent magnetization, ac-susceptibility and Monte Carlo simulations are reported. Both two- and three-dimensional spin glass systems are considered. The experiments are discussed within the recent domain theories. Brief reviews on the same subject have also been presented elsewhere[11,12].

EXPERIMENTAL PROBES

The spin glass relaxation occurs in a time window extending from atomic time scales up to a maximum relaxation time, which diverges on approaching the spin glass temperature. Combining different experimental techniques, e.g. neutron scattering, dynamic susceptibility and magnetic 'noise' experiments, the relaxation behaviour can be *probed* over more than 17 decades in time. Using neutron spin echo technique, the spin-spin correlation function q(t) can be observed in the time window 10^{-12}-10^{-8} sec. In dynamic susceptibility measurements, i.e. ac-susceptibility and time dependent magnetization measurements, an observation time window of about 10^{-6}-10^{4} sec. can be covered. In ac-susceptibility measurements, where $\chi(\omega) = \chi'(\omega) + i\chi''(\omega)$, the *observation time* (t) of the experiment equals $1/\omega$, where ω is the angular frequency of the applied oscillating field. In measurements of the time dependence of the zero-field-cooled (ZFC) magnetization, M(t), the observation time (t) equals the time after application of a magnetic field, H. In the regime of linear response, both ac and ZFC susceptibilities mirror the time dependence of the zero-field susceptibility, χ(t). The ZFC and ac susceptibilities can be related through:

$$\chi(t) = \chi'(\omega) \approx (1/H)M(t), \qquad t=1/\omega \qquad (1)$$

$$\chi''(\omega) \approx -(\pi/2)(1/H)(dM/d\ln t), \qquad t=1/\omega \qquad (2)$$

At equilibrium, χ(t) and q(t) are related via the fluctuation-dissipation theorem through:

$$\chi(t) = [1-q(t)]/kT \qquad (3)$$

and also, the magnetic 'noise' S(ω) relates to $\chi(\omega)$ according to:

$$S(\omega) = 4kT\chi''(\omega)/\omega \qquad (4)$$

The ageing process in spin glasses is reflected in a time-dependence of the ac-susceptibility[2,6], at constant temperature and at constant observation time, which also implies a non-stationary noise[13]. In ZFC susceptibility measurements the ageing is revealed[3] in a wait-time dependence of the relaxation before the probing field is applied.

ZFC SUSCEPTIBILITY and SPIN GLASS DOMAINS

When a spin glass is cooled in zero field to a constant temperature, the spin configuration initially attains an energetically unfavorable state - a random state. At constant temperature, the initial state changes towards an energetically most favorable spin configuration. This ageing process is revealed as a wait-time dependence in zero-field-cooled (ZFC) susceptibility measurements. In such measurements the sample is cooled in zero field, from a temperature where the spin glass is at thermodynamic equilibrium, to the measurement temperature, T_m. After a wait-time t_w, at T_m, a magnetic field is applied and the relaxation of the magnetization versus the logarithm

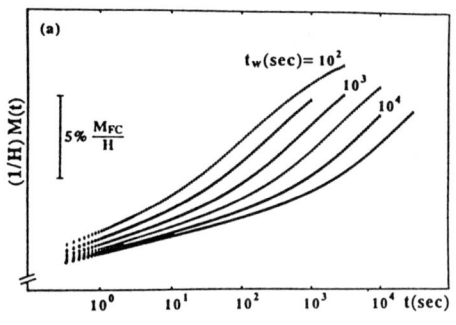
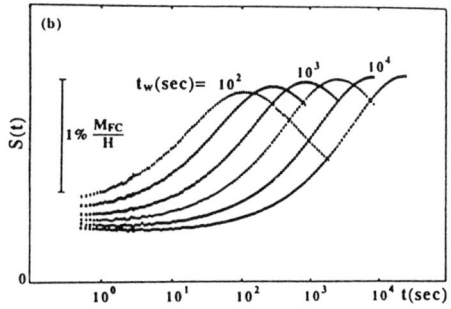

Fig.1. (a) $(1/H)M(t)$ and (b) $S(t)$ vs $\log t$ at different wait-times, tw. Three-dimensional Cu(10%Mn). $T_m/T_g=0.91$. $T_g=45.3$ K.

of time is recorded. Fig.1 shows $(1/H)M(t)$ and the corresponding relaxation rate $S(t) = (1/H)dM/d\ln t$ for different t_w for a metallic 'bulk' Cu(10%Mn) spin glass[14,15]. The figure clearly demonstrates the drastic influence of the ageing process on the observed dynamics. The salient feature of the curves is an inflection point on the $M(t)$ vs $\log t$ curves and corresponding maxima of the $S(t)$ vs $\log t$ curves at an observation time of the order of the wait-time, t_w. This specific signature implies that ageing exists for time scales corresponding to the maximum relaxation time of the system, and has the following important consequences: (i) since the relaxation times in three-dimensional (3D) systems diverge at the transition temperature T_g, the spin glass phase is inherently a non-equilibrium phase, (ii) the ageing behaviour persists through the transition temperature T_g and (iii) ageing is also an inherent property of 2D systems, which have a phase transition at zero kelvin (i.e. finite relaxation times at all T>0). This latter point is demonstrated in Fig.2, referring[16] to a 40Å Cu(13%Mn) spin glass film, which has from static non-linear and dynamic susceptibility measurements shown to display 2D behaviour. The non-equilibrium behaviour in spin glasses has recently been addressed by Fisher and Huse (FH)[7] within the droplet scaling theory

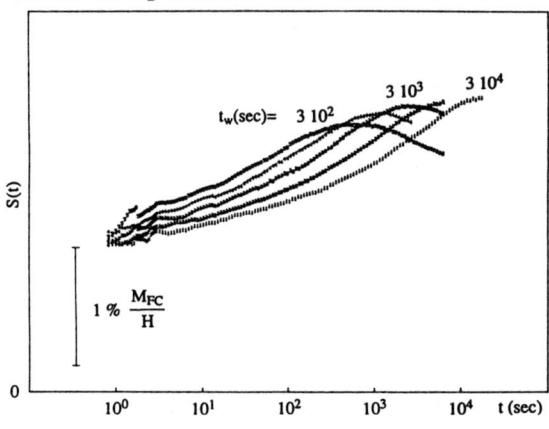

Fig.2. $S(t)$ vs $\log t$ at different wait-times, t_w. Two-dimensional 40 Å Cu(13%Mn) spin glass film. $T_m=25.8$ K. $T_g=0$ K.

(after Mattsson et al[16])

and by Koper and Hilhorst (KH)[8] in a new mesoscopic theory. A common feature of these domain (droplet) theories is that the ageing process involves the growth of domains, within which equilibrium dynamics exist. At temperatures below the spin glass freezing temperature, the size of the domains grow without limit. According to FH the characteristic size of the domains, R, grows with the age, t_a, of the system as:

$$R(t_a,T) = [T\ln(t_a/t_0)/\Delta]^{1/\psi}, \qquad (5)$$

where Δ is the free energy scale for the barriers, ψ is the barrier exponent and t_0 a microscopic time, $\approx 10^{-13}$ sec.. One important consequence of these domain theories is a direct relation between *time-scale* and *length-scale* in measurements of the time dependent susceptibility. Generally, at 'short' observation times, relaxations at 'short' length scales are probed. This *probing length-scale* (L) varies with the observation time (t) in a similar way as the domain size grows with the age of the system, i.e.,

$$L(t,T) = [T\ln(t/t_0)/\Delta]^{1/\psi}, \qquad t=t_a-t_w \qquad (6)$$

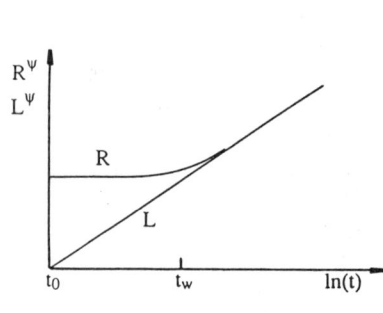

Fig.3 shows L and R versus lnt. As can be seen from the figure, when lnt<<lntw, the probing length scale is small in comparison to the domain size and *equilibrium dynamics* are probed. When lnt>>lntw, and R ≈ L, relaxation processes within 'domain walls' are probed, yielding *non-equilibrium dynamics*. When lnt ≈ lntw a *crossover* between these two relaxation regimes occurs, which is experimentally observed in an inflection point of the M(t) vs logt curves and a corresponding maximum in the S(t) vs logt curves.

The concept of growing domains may be further elucidated by studying the influence of temperature variations on the dynamics. In the following we specifically study the influence of (i) a persistent temperature shift and (ii) a temperature cycling.

Temperature shift experiments

In this experiment[14] the sample is cooled to a somewhat lower temperature ($T_m-\Delta T$) than the measurement temperature T_m. After a wait-time, the temperature is increased to T_m and the probing field is applied. Fig.4 shows S(t) versus logt curves for the same Cu(10%Mn) spin glass as above. All curves refer to the same wait-time, $t_w=10^4$ sec., and measurement temperature $T_m=0.91\ T_g$, but the

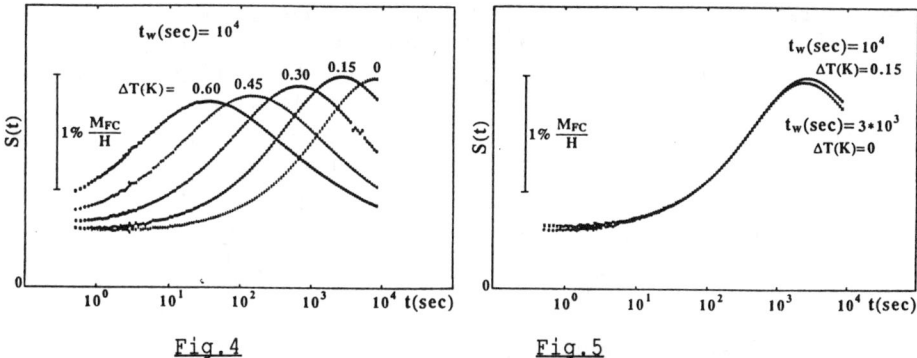

Fig.4   Fig.5

sample has been aged at different ΔT's as indicated in the figure. The characteristic signatures of the relaxation rate curves are preserved, but the location of the maximum of the S(t) versus logt curves are shifted towards shorter observation times with increasing ΔT. As can be seen, the curves in Fig.2 show similar features as the curves in Fig.4 In fact, curves with maxima occurring around the same observation time from these two different experiments are virtually identical. This is illustrated in Fig.5, where a conventional ZFC curve with $t_w=3\times10^3$ sec (from Fig.2) is plotted together with a curve with $\Delta T=0.15$ K and $t_w=10^4$ sec. (from Fig.4). This is a manifestation of a temperature dependence of the growth rate of the domains, and is schematically illustrated in Fig.6. When the sample is aged at $T_m-\Delta T$ the growth rate of the domains is lower than at T_m, and when the temperature is increased to T_m, after t_w, the size of the domain is preserved and we move from A to B in the figure. Consequently, we will probe a system that looks 'younger'; the maximum of the S(t) vs logt curve will occur at a shorter time than the original wait-time at $T_m-\Delta T$. Analogously, if the system is aged at $T_m+\Delta T$ and probed at T_m we move from C to D in the figure and the position of the maximum is pushed towards longer observation times; we consequently probe a system that looks 'older'. This effect is illustrated in Fig.7, where both positive and negative temperature shift experiments are illustrated.

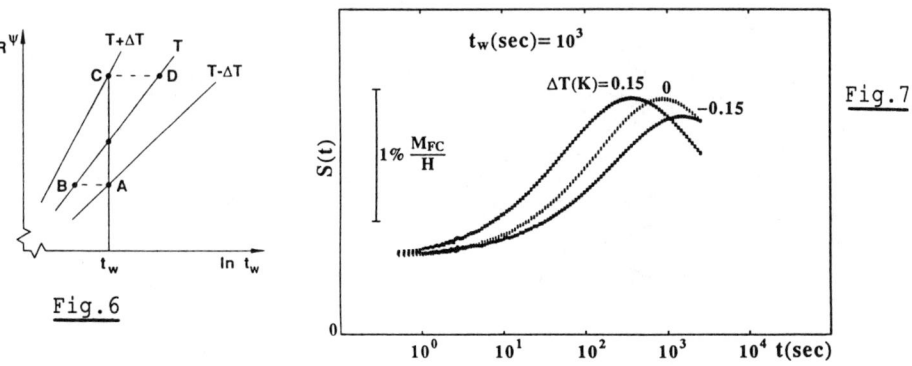

Fig.6

Fig.7

The description above is only valid if the temperature shift is small enough to yield sufficient overlap between the spin configurations at the two temperatures. As inferred by Bray and Moore[18] the overlap length, $I_{\Delta T}$, determines the maximum length scale on which similar equilibrium spin correlations exist at two different temperatures, T and T+ΔT. $I_{\Delta T}$ depends approximately on the temperature difference as $I_{\Delta T} \approx 1/\Delta T$. If the overlap length at two temperatures is shorter than the size of the spin glass domain, pronounced irreversibility effects occur if the temperature is cycled[6,11,12,19,20] between these temperatures.

Temperature cycling experiments

In this experiment on the same Cu(10%Mn) sample as above the sample is cooled to T_m, and after a wait-time of 3×10^4 sec, a positive temperature cycling (ΔT) is made, and when T_m is recovered the probing field is applied. The temperature cycling effectively takes 10-30 sec. Fig.8 shows a three-dimensional plot of S(t) vs logt for various ΔT in the the positive temperature cycling experiment. As can be seen in the figure, for ΔT<0.3 K the system is virtually unaffected by the temperature cycle, and the recorded curve corresponds to an ordinary ZFC curve at the same wait-time. However, for ΔT>0.3 K, a second pronounced maximum in S(t) at short observation times gradually develops with increasing magnitude of ΔT. The appearance of two maxima in the relaxation rate curves imply the coexistence of two characteristic domain sizes in the system, R(tw=3×10^4 sec) and R(tw$\approx$0). A schematic illustration of the effect of the temperature cycling is shown in Fig.9. When $I_{\Delta T}$>R (i.e. ΔT<0.3K) the domain are unaffected by the temperature cycling. When $I_{\Delta T}$>R (i.e. ΔT>0.3K) a fraction of the domains are broken during the temperature cycling, and after the temperature cycling the system owns two characteristic domains: one due to the overlap length, which effectively is equal to the R($t_w \approx$0) domains, and the other equal to the original wait time at T_m, equal to R(t_w=3×10^4 sec). As ΔT is

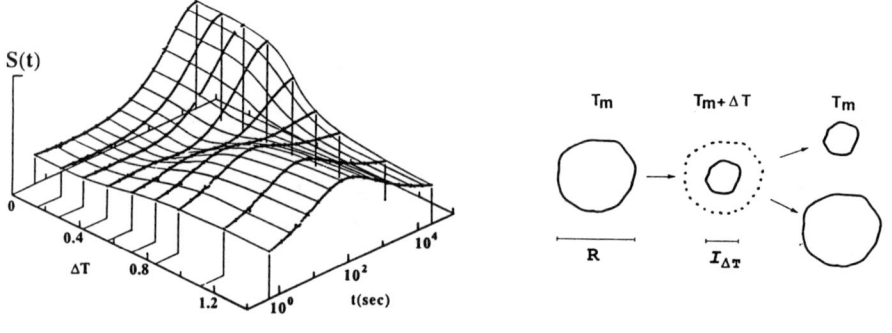

Fig.8   Fig.9

increased, the fraction of $R(t_w \approx 0)$ domains increases. This fraction (which is taken as the relative height of the $S(t)$ curves at $t_w \approx 20$ sec and $t_w = 3 \times 10^4$ sec.) is plotted versus the temperature cycling ΔT in Fig.10. As can be seen from the figure there is a rather well defined threshold value of ΔT where a finite value of k first appears. This threshold value, ΔT_0, is interpreted to correspond to the value of ΔT where the overlap length equals the original domain size.

In a negative temperature cycling experiment[20], as illustrated in Fig.11, the sample is aged 3×10^4 sec at T_m and is then subjected to a negative temperature cycle of magnitude ΔT. If the system is cooled the lower temperature and then immediately heated to T_m, the measured relaxation curve resemble an ordinary ZFC curve at the same wait-time, due to the fact that the ageing process drastically slows down when the spin glass is cooled to a lower temperature. However, if an additional wait-time (here 10^4 sec) is imposed to the system at $T_m - \Delta T$, two distinct maxima can appear in the relaxation rate curves, as shown in Fig.11.The fact that the fraction k of $R(t_w \approx 0$ sec) domains, as shown in Fig.9, has a maximum at a ΔT value of about 1-2 K depends on two competing effects: When ΔT increases the subsequent decrease of the overlap length promotes $R(t_w \approx 0)$ domains (an increase of k), whereas the subsequent slowing down of domain growth preserves $R(t_w = 3 \times 10^4$ sec) domains (a decrease of k). However, the value of ΔT_0,, where a finite k first appears is approximately the same for both positive and negative temperature cycles. A clear manifestation of the 'overlap length'.

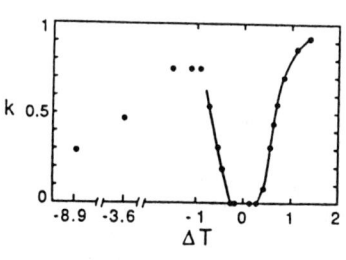

Fig.10

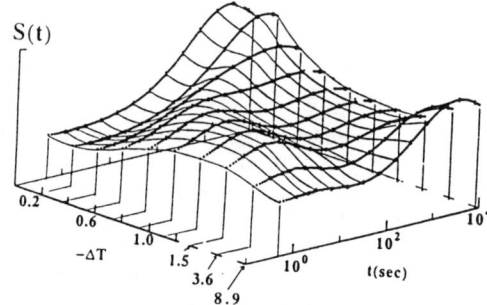

Fig.11

ac- SUSCEPTIBILITY

Oddly enough, the ageing phenomenon in spin glasses was initially discovered in a metallic Cu(4%Mn) spin glass from ac-susceptibility measurements. In these experiments[2] the ageing phenomenon is revealed as a time dependence (at constant temperature and constant frequency of the applied oscillating field) of the complex susceptibility. In the original experiments, it was observed that after a step change (increase or decrease) in temperature, and after the corresponding step change of the susceptibility, both the real (χ') and the imaginary (χ'') parts of the susceptibility slowly decreased in magnitude with time. Fig. 12 shows the time dependence of χ'' at an applied frequency of 1.7 Hz (equivalent to a constant observation time of ≈ 0.1 sec) after step changes in temperature of 1 K. For such large temperature steps the ageing process is re-started (cf. the temperature shift experiments of Fig. 4).

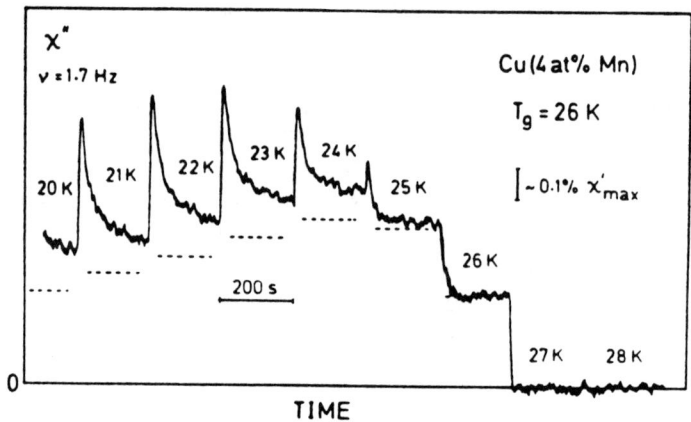

Fig.12. Time dependence of χ'' at an applied frequency of 1.7 Hz. The sample temperature is increased in discrete steps of 1 kelvin every 200 sec. The dotted lines mark equilibrium values of χ'' at respective temperature.

MONTE CARLO SIMULATIONS ON 2D, 3D MODEL SYSTEMS

Computer simulations on spin glass model systems have almost exclusively been focussed on equilibrium spin glass properties. In such simulations very time consuming equilibration processes have to be performed to find an initial state. Using a special purpose computer Ogielski[21] have studied the equilibrium dynamics of a 3D Ising model spin glass in an eight decade wide time interval. The functional form of the relaxation function from these simulations shows remarkable agreement with experimental results[22] on real spin glasses.

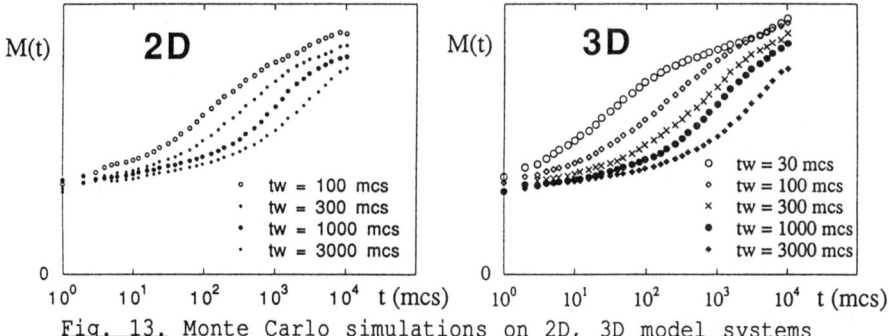

Fig. 13. Monte Carlo simulations on 2D, 3D model systems

Recently, Andersson and Svedlindh[23] have studied the non-equilibrium character of the spin glass dynamics in Monte Carlo simulations on two- and three-dimensional short range Ising spin glass systems. The simulations were performed for 2D lattices of sizes up to 200^2 and for 3D lattices of sizes up to 60^3. In analogy with ordinary ZFC magnetization measurements the system is equilibrated for various number of monte carlo steps (mcs); equivalent to various wait-times, t_w. After this equilibration time a small magnetic field is applied and the time dependence of the magnetization (M(t)) is simulated. Fig.13 shows M(t) for different equilibration times t_w for 2D and 3D systems. As can be seen from the figure the influence of ageing is quite similar in two and three dimensions, and clearly illustrates that ageing has little to do with the existence of a spin glass phase, but is merely an aspect of the dynamics. Recent experimental results by Mattsson et al[16] (see Fig.2) also show ageing behaviour in real two-dimensional spin glass systems (thin spin glass films). Both the simulations and the experiments on real systems show that aging is an inherent property of 2D spin glass systems with qualitatively similar features as in 3D systems.

Recently, Andersson et al [24] have simulated both the time dependence of the magnetization M(t) and the spin autocorrelation function q(t). They find that the fluctuation dissipation theorem (eq.(3)) holds only for short time scales (t<<tw; equilibrium dynamics) and they observe large deviations at long time scales (t>>tw; non-equilibrium dynamics). This is illustrated in Fig. 14.

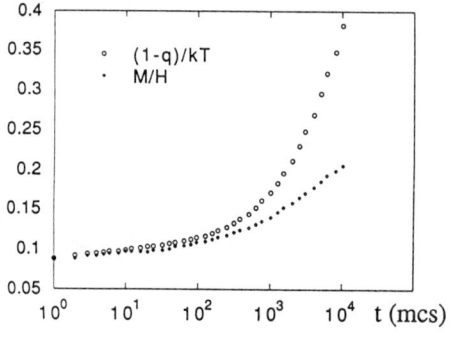

Fig. 14.

REFERENCES

1) see e.g. K. Binder and A.P. Young, Rev. Mod. Phys. $\underline{58}$, 801 (1986)
2) L. Lundgren, P. Svedlindh, O. Beckman, J. Magn. Magn. Materials $\underline{31-34}$, 1349 (1983)
3) L. Lundgren, P. Nordblad, P. Svedlindh, O. Beckman, Phys. Rev. Letters $\underline{51}$, 911 (1983)
4) P. Svedlindh, P. Granberg, P. Noedblad, L. Lundgren, H.S. Chen, Phys. Rev. $\underline{B35}$, 268 (1987)
5) P. Nordblad, P. Svedlindh, J. Ferre, M. Ayadi, J. Magn. Magn. Materials $\underline{59}$, 250 (1986)
6) Ph. Refregier, E. Vincent, J. Hammann, M. Ocio, J. Phys. France $\underline{48}$, 1533 (1987)
7) D.S. Fisher and D.A. Huse, Phys. Rev. $\underline{B38}$, 373 (1988)
8) G.J.M. Koper and H. Hilhorst, J. Phys. (France) $\underline{49}$, 429 (1988)
9) P. Sibani, K.H. Hoffman, Phys. Rev. Letters $\underline{63}$, 2853 (1989)
10) M. Lederman, R. Orbach, J.M. Hammann, M. Ocio, E. Vincent, Phys. Rev. $\underline{B44}$, 7403 (1991)
11) L. Lundgren, J. Phys. (France) $\underline{49}$, C8-1001 (1988)
12) L. Lundgren, Proc 'Relaxation in Complex Systems and Related Topics', Torino 16-20 Oct., 1989; Plenum Press (1990) p. 3.
13) M. Ocio, H. Bouchiat, P. Monod, J. Magn. Magn. Materials $\underline{54-57}$, 11 (1986)
14) P. Granberg, L. Sandlund, P. Nordblad, P. Svedlindh, L. Lundgren, Phys. Rev. $\underline{B38}$, 7097 (1988)
15) P. Granberg, L. Lundgren, P. Nordblad, J. Magn. Magn. Materials $\underline{92}$, 228 (1990)
16) J. Mattsson, P. Granberg, P. Nordblad, L. Lundgren, R. Loloee, R. Stubi, J. Bass, J.A. Cowen, (proc. ICM 91)
17) L. Sandlund, P. Svedlindh, P. Granberg, P. Nordblad, L. Lundgren J. Appl. Phys. $\underline{64}$, 5616 (1988)
18) A.J. Bray and M.A. Moore, Phys. Rev. Letters 58, 57 (1987)
19) P. Nordblad, P. Svedlindh, L. Sandlund, L. Lundgren, Phys. Lett. $\underline{A\ 120}$, 475 (1987)
20) P. Granberg, L. Lundgren, P. Nordblad, J. Magn. Magn Mater. $\underline{92}$, 228 (1990)
21) A.T. Ogielski, Phys. Rev. B32, 7384 (1985)
22) K. Gunnarsson, P. Svedlindh, P. Nordblad, L. Lundgren, H. Aruga, A. Ito
23) J.-O. Andersson, P. Svedlindh (proc. ICM 91)
24) J.-O. Andersson, J. Mattsson, P. Svedlindh (unpublished)

HIERARCHICAL ASPECTS OF THE SLOW DYNAMICS IN SPIN GLASSES

E. Vincent, J. Hammann, M. Ocio and F. Lefloch
*Service de Physique de l'Etat Condensé, C.E. Saclay,
Orme des Merisiers, 91191 Gif-sur-Yvette Cedex, France*

ABSTRACT

The dynamic properties of the spin glass phase evolve with the time elapsed after quench (*aging*). By measurements of the decay of the thermo-remanent magnetization, we have studied the effect on aging of slight temperature variations. We find that the dissymetry of the response to either heating or cooling is hard to explain in terms of an *overlap up to a characteristic distance* of the spin correlations in two states at slightly different temperatures. We interpret these experiments as a probe of a hierarchical organization of the phase space, which shows up as a continuous series of *micro-phase transitions* as the temperature is decreased from T_g. We ask the question of a possible relevance of this description to other complex systems like glassy polymers.

1. INTRODUCTION

When investigating the dynamics of the spin glass phase, the experimentalist has to face two important features: first, the observed dynamical processes extend up to as long time scales as he can wait, and second the measured properties depend on the time spent (*age*) at low temperature (non-stationarity, or *aging* effects). After having elucidated the influence of the age on isothermal measurements[1], it has become possible to consider *aging* as an observable quantity. We explain here how the non-trivial response of aging to *small temperature changes*[2,3] has brought important implications on the nature of the spin glass phase.

This paper concentrates on studies of the thermo-remanent magnetization ("TRM") relaxation in the $CdCr_{1.7}In_{0.3}S_4$ insulating spin glass (T_g=16.7 K); similar results have been obtained concerning the out-of-phase susceptibility[2]. In a TRM measurement, the sample is cooled from the paramagnetic phase down to the spin glass phase in a small field (here 20 Oe), and kept in this field at constant temperature during a waiting time t_w. After this time, the field is cut off, and the subsequent relaxation of the TRM is recorded as a function of t. Examples of results are found in Fig. 1 and 2 with **thick** lines.

The relaxations are slow, non-exponential, and depend on t_w; the larger t_w, the slower the relaxation (the spin glass is becoming *stiffer* with time). We have previously shown[1] that these isothermal *aging phenomena* can be quantitatively accounted for by considering that a given response time scales like the *age* $t_a = t_w + t$ to some power μ ($\mu \simeq 0.9$ for $0.4 \leq T/T_g \leq 0.9$).

2. SENSITIVITY OF AGING TO TEMPERATURE CHANGES

We have studied the effect on the relaxation *at a given temperature* of slight temperature variations during the waiting time[2]. The procedure is sketched in the inset of Fig. 1; after aging the system at T_0=12 K during t_w=970 min, we perform during 5 min a small heating cycle to $T_0+\Delta T$ and back. Then we wait t_w=30 min, cut the field and measure the relaxation at T_0.

The two **thick** lines in Fig. 1 correspond to the normal procedure of *isothermal* waiting during 1000 min and 30 min; they respectively illustrate the $\Delta T \to 0$ (no re-heating) and $\Delta T \to \infty$ (new quench from above T_g after 970 min) limits. The thin lines are the results for ΔT values ranging from 0.25 to 2.5 K.

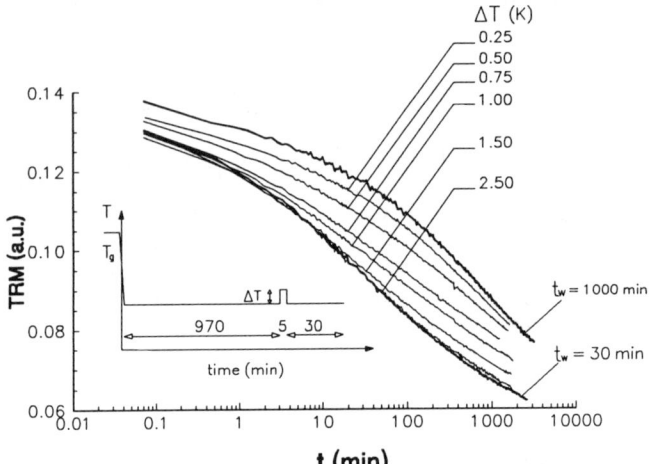

Figure 1 : Effect on the TRM relaxation at T=12 K of a slight heating cycle ΔT during the waiting time (thin lines). The procedure is sketched in the inset. The two **thick** lines are reference curves obtained after *isothermal* waiting.

The curve obtained for $\Delta T=2.5$ K is nearly identical to that measured after waiting only 30 min, except at the very right end of the experimental time window. Within this limited time window, the observed behavior mimics the limit $\Delta T \to \infty$, as if the spin glass had again been quenched from above T_g: the whole aging evolution performed during 970 min has been apparently erased.

It is worth examining how, for increasing ΔT, the shape of the 1000 min *reference curve* is progressively affected by the ΔT-perturbation. The effect on the shorter response times is strong; even for small ΔT, the curves move away from the reference. In contrast, the effect on the longer times is much weaker; the curves remain closer to the reference in the right-hand part of the graph. In brief, the effect of a small heating cycle is to "reinitialize aging at short times".

A strong sensitivity of the spin correlations to temperature variations is indeed expected within the framework of the *droplet*[4] and *domain*[5] spin glass models, in which the dynamics is that of a distribution of groups of correlated spins. After the quench from above T_g, the correlations are limited to short range, and they progressively extend towards infinite range, yielding the aging phenomena. The effect of a temperature change ΔT is to destroy the correlations extending beyond a characteristic *overlap length* $l_{\Delta T}$, which varies as an inverse power of ΔT. The behavior observed in our experiments is reminiscent of that picture, but different. Indeed, the ΔT variation has a drastic effect, but it primarily affects the short response times, i.e. the correlations at short distances, whereas in the picture of Ref. 4,5 they should *not* be affected *up to* $l_{\Delta T}$.

A quite different approach consists in attempting a phase-space description of these phenomena. It is inspired by the hierarchical structure of the equilibrium states obtained in the Parisi treatment of the mean-field problem[6]. The wide distribution of response times observed in the experiments corresponds to a rugged free-energy surface with many valleys and mountains. During aging, the system explores larger and larger regions of the phase space. If the small heating cycle does not perturb the energy landscape, it is expected to accelerate the aging evolution; on the contrary, it is found to erase (at least part of) the aging evolution, as in the case of a *new quench*. We interpret this result as due to a *modification* of the energy landscape: some valleys are merging as T increases, and backward subdivide into others as T decreases, in a hierarchical fashion.

3. COOLING OR HEATING : DISSYMETRY OF THE EFFECTS

This picture of a hierarchy of valleys for decreasing T receives further support from experiments in which a slight *cooling* cycle is performed during the waiting time. The result is very different from the case of a *heating* cycle. First, a very long time at $T_0-\Delta T$ is necessary to obtain a significant effect. The procedure therefore consists in waiting 15 min at T_0, 1000 min at $T_0-\Delta T$, and again 15 min at T_0 before cutting the field and measuring the relaxation at T_0. Second, the cooling cycle does not again yield some kind of aging reinitialization, as could be expected from the $l_{\Delta T}$ picture. For $\Delta T=1$ K, the relaxation curve falls exactly onto the reference curve of $t_w=30$ min, and is clearly distinct of a $t_w=10$ min curve : aging *after* the 1 K cooling cycle has proceeded from the state which was attained *before* (Fig. 2).

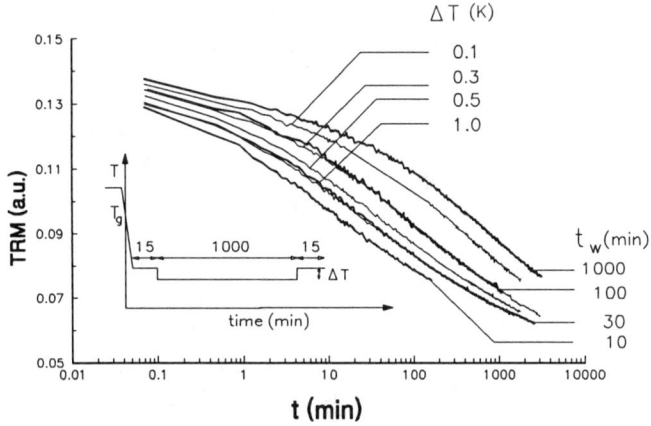

Figure 2 : Effect on the TRM relaxation at $T_0=12$ K of aging at a lower temperature (thin lines). The procedure is sketched in the inset. The **thick** lines are reference curves obtained after *isothermal* waiting at 12 K during t_w.

The dissymetry of the effects of either cooling or heating is indeed intrinsic to the hierarchical scheme proposed above; when at the start of the cooling pulse T is lowered, new aging processes are initiated among the new valleys, and this evolution is erased when T is increased back at the end of the cycle.

For intermediate ΔT values, aging at $T_0-\Delta T$ indeed affects the final relaxation; but in contrast to the case of a heating cycle, the shape of the curves remains similar to that of the reference curves. Using our scaling analysis of aging[1], we can define an *effective* waiting time t_w^{eff} at T_0 which would yield the same aging evolution as the time spent at $T_0-\Delta T$. The very rapid decrease of t_w^{eff} observed for increasing ΔT is at variance with thermally activated processes over barriers of constant height[2]. This has motivated another series of experiments, in which the growth of barriers with decreasing T has been systematically studied[7]. The results are compatible with the divergence of certain barriers at any temperature below T_g, completing the proposed hierarchical scheme towards the picture of a *continuous sequence of micro-phase transitions* starting at T_g. This picture can be compared (see Ref. 7) in more details with the predictions from the mean-field theory[6].

4. AGING IN AMORPHOUS POLYMERS

Aging effects have been observed in numerous complex systems; the case of the mechanical properties of amorphous polymers[8] presents some striking similarities with spin glasses, and the description given in Ref. 8 has inspired our scaling analysis of aging[1]. Creep tests are performed by cooling the polymer below the glass transition, waiting t_w, applying a stress and recording the strain. The slow relaxation observed is very similar to that of the TRM in spin glasses, and shows the same aging effects: the larger the time after quench, the slower the relaxation. The authors[8] consider that, due to the quench, the free volume does not reach its equilibrium value and relaxes slowly; so does the segmental mobility, and therefore the strain response to a stress solicitation.

The non-exponential relaxations in glasses have been described for a long time in terms of a potential energy surface with a large number of minima of varying depths[9]. We want to stress that, beyond that rugged energy landscape, the spin glass shows evidence for a *growth of energy barriers for decreasing T*. It would be worthy searching for such non-trivial effects in glassy systems.

In Ref. 8, the situation of storing a material at room temperature for a very long time, and then heating it to a higher measuring temperature, has been studied. The authors observe that aging which has occurred at the lower temperature is partially or completely erased. This is equivalent to our *cooling cycle* experiments; unfortunately, there are not sufficient data to conclude whether this effect remains compatible with thermal activation over *constant* barriers. To our knowledge, the inverse experiment (*heating* cycle) has not been performed; if the amorphous polymers have the same kind of hierarchical phase space as spin glasses, then a slight heating cycle (below T_g) should be enough to renew the mechanical properties of a material damaged by a long storage.

We acknowledge the contribution of Ph. Hernandez to the measurements, and fruitful discussions with C. Alba-Simoniesco and L. Leylekian.

REFERENCES

1. M. Alba, J. Hammann, M. Ocio, Ph. Refregier, *J. Appl. Phys.* **61** (1987) 3683.
2. Ph.Refregier, E. Vincent, J. Hammann, M. Ocio, *J.Phys.France* **48** (1987) 1533.
3. P. Granberg, L. Sandlund, P. Nordblad, P. Svendlindh, L. Lundgren, *Phys. Rev. B* **38** (1988) 7097.
4. D.S. Fisher and D.A. Huse, *Phys. Rev. B* **38** (1988) 373.
5. G.J.M. Koper and H.J. Hilhorst, *J. Phys. France* **49** (1988) 429.
6. M. Mézard, G. Parisi and M.A. Virasoro, *Spin Glass Theory and beyond*, World Scientific Lecture Notes in Physics Vol. **9** (Singapore, 1987).
7. M. Lederman, R. Orbach, J. Hammann, M. Ocio, E. Vincent, to appear in *Phys. Rev. B* (Oct. 91); see also the proceedings of ICM91 (Edinburgh, 1991).
8. L.C.E. Struik, *Physical Aging in Amorphous Polymers and Other Materials* (Elsevier North-Holland Inc., 1978).
9. M. Goldstein, *J. Chem. Phys.* **51** (1969) 3728; R.W. Hall and P.G. Wolynes, *J. Chem. Phys.* **86** (1987) 2943.

TIME-DEPENDENT MAGNETIC PHENOMENA IN DILUTE ANTIFERROMAGNETS $Fe_{1-x}Mg_xCl_2$ EXHIBITING RANDOM FIELD ISING MODEL AND SPIN GLASS BEHAVIOURS

Katsunori Iio, Atsunori Kitazawa* and Kazukiyo Nagata
Department of Physics, Faculty of Science,
Tokyo Institute of Technology, Oh-okayama, Meguro-ku, Tokyo 152, Japan

ABSTRACT

The time decay of thermoremanent magnetization(TRM) was studied for insulating random magnets $Fe_{1-x}Mg_xCl_2$ exhibiting both random field Ising model(RFIM) and Ising spin glass(SG) behaviours. The time dependence of TRM was found to be described over both cases with a functional form of $M_{TRM}(t) = AH_o^{\nu}[\log(t/\tau)]^{-\phi}$.

INTRODUCTION

The spin dynamics of two typical random magnets, spin glass(SG) and random field Ising model(RFIM) have attracted experimental and theoretical attention over a long term of years. Both systems have their own and in some cases common characteristics in the slow dynamics of magnetization process. However comparative and unified understanding of the mechanism governing the non-equilibrium magnetization behaviour for them have scarcely been obtained. In this note we show the results of the time decay of thermoremanent magnetization(TRM) in $Fe_{1-x}Mg_xCl_2$ observed by means of Faraday rotation.

The present mixtures, where a host $FeCl_2$ is a hexagonal layered lattice compound with ferromagnetic planes stacking antiferromagnetically along the c axis, are typical random systems with short-range Ising-like interactions. The specimen with $x < x_c = 0.5$ exhibits an ideal RFIM under the application of a magnetic field and shows reentrant spin glass behaviour near x_c, which is the site percoration threshold of the triangular net.[1,2] The specimen with $x > x_c$ displays Ising SG behaviour at low temperatures as the results of competition between the ferromagnetic nearest-neighbour, the antiferromagnetic next nearest-neighbour and the antiferromagnetic interplane interactions.

EXPERIMENTAL PROCEDURE AND RESULTS

The Faraday rotation employed in this study is a precise means for measuring the uniform magnetization of random mixtures. Because the light emitted forward after traversing specimens is irradiated exclusively from a homogeneous region of random media. The relative rotation angle of light polarization θ has been proved to be proportional to the bulk magnetization M.[3] Typical θ(t) vs. logt plots for the time dependence of TRM measured in a zero field after field cooling are shown in Fig.1 and 2 for $x < x_c$ and $x > x_c$, respectively. The measurements were performed at various waiting times (1s< t_w <3600s) under annealing field higher than 0.1kOe. The behaviour of TRM in $Fe_{1-x}Mg_xCl_2$ was not modified meaningfully for the present observing time range.

---

*Present address: Seiko Epson Corp. Ohwa, Suwa-shi, Nagano 392, Japan

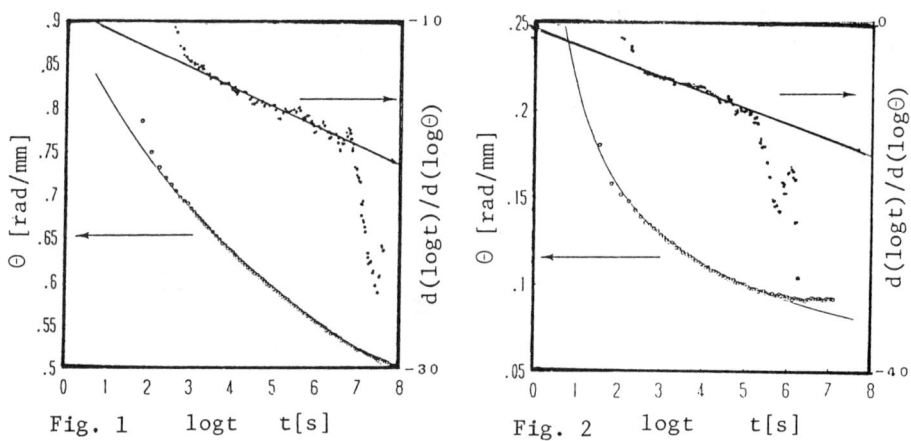

Fig. 1. $\theta(t)(\sim M_{TRM}(t))$ vs. logt for $Fe_{0.55}Mg_{0.45}Cl_2$ at $T/T_{rg}=0.63$. $\theta(t)$ can be described by the logarithmic power law with $\nu=1.34$, $\phi=0.898$, and $\tau=2.21\cdot 10^{-4}$s. The annealing field H_0 is 3kOe.

Fig. 2. $\theta(t)$ vs. logt for $Fe_{0.45}Mg_{0.55}Cl_2$ at $T/T_g=0.54$. $\theta(t)$ can be described by the logarithmic power law with $\nu=0.98$, $\phi=0.540$, and $\tau=4.39\cdot 10^{-2}$s. The field $H_0=1$kOe.

DISCUSSION

The present authors had reported previously that the decay of TRM over both concentration regions cannot be fitted to the form of a logarithmic θ_0-Slogt or a power law $\theta_0\exp(-\alpha t)$ for 1s<t<3600s.[4] Since then, cumulative experiments and analyses have been performed. The most probable time dependence at low temperatures for both RFIM and SG states can be the logarithmic power law of $M(t)\sim\theta(t)=AH_0^\nu[\log(t/\tau)]^{-\phi}$. In the above figure we have re-plotted the data as $d(\log t)/d(\log\theta)$ vs. logt, where straight lines represent the logarithmic power law. Except for short time and very long time regions, the solid lines representing the theoretical time dependence are in accord with the experimental data. The present results has revealed that $Fe_{1-x}Mg_xCl_2$ exhibits the crossover from the decay of TRM obeying the logarithmic power law intrinsic to RFIM to that to SG as x is increased.

1. D. Bertrand, A.R. Fert, M.C. Schmidt, F. Bensamka and S. Legrand: J. Phys. C15, L883(1982)
2. Po-zen Wong, S. von Molar and P. Dimon: J. Appl. Phys. 53, 7954(1982)
3. H. Yamashita, K. Iio, M. Sano, H. Masuda, H. Tanaka and K. Nagata: J. Magn. Soc. Jpn. 11, Supplement S1, 87(1987)
4. T. Kamai, K. Iio, H. Tanaka and K. Nagata, Cooperative Dynamics in Complex Physical Systems (Springer-Verlag, 1989)p.173

SMALL ANGLE NEUTRON SCATTERING STUDIES ON $Fe_{1-x}Al_x$ REENTRANT SPIN GLASS

J. Suzuki and Y. Endoh
Tohoku University, Sendai 980

M. Arai and M. Furusaka
National Laboratory for High Energy Physics, Tsukuba 305

ABSTRACT

We report small angle neutron scattering measurements on a reentrant spin glass (RSG) $Fe_{1-x}Al_x$ (x = 0.285) in a wide range of momentum transfer Q. The neutron scattering profiles below 200 K at larger Q are well described with Porod law which suggests that the system is composed from clusters with a 28 A diameter, while at smaller Q Ornstein-Zernike form is well applicable, representing macroscopic thermal spin fluctuations. The results suggest that the RSG transition is caused by cooperative freezing of the cluster spins and breaking of the ferromagnetic long range order.

INTRODUCTION

The RSG transition on the infinite range model for classical vector spins has been characterized by freezing of spin components perpendicular to the mean magnetization.[1] Experimentally it was supported by the measurement of a hyperfine field on $Au_{.81}Fe_{.19}$.[2] In this report we studied spin correlation on a single crystal $Fe_{.715}Al_{.285}$ in a wider Q range of $0.008 \leq Q \leq 0.9$ A^{-1} by utilizing pulsed cold neutrons in KEK (National Laboratory for High Energy Physics) in contrast with a previous measurement performed in $0.008 \leq Q \leq 0.065$ A^{-1}.[3]

SANS RESULTS

Figure 1 shows the SANS data in the range of $0.008 \leq Q \leq 0.9$ A^{-1} at 15 K. In the small Q region from 0.008 to 0.05 A^{-1} the scattering profile is described with the Ornstein-Zernike form; $I(Q)=Ak^2/(Q^2+k^2)$ as we discussed previously.[4] The amplitude A has a fairly sharp peak and the inverse correlation length k approaches to zero in an accuracy of the resolution of the diffractometer at the RSG transition temperature T_g (= 100 K). These features suggest that the RSG transition is due to a cooperative phenomenon. On the other hand in the large Q region the scattering profiles depend on the temperature. Figure 2 shows the SANS data at 15 K and 800 K. At 15 K the scattering function has Q^{-4} dependence (Porod law) in the Q range of $Q > 0.5$ A^{-1}, while at 800 K it holds Q^{-2} dependence which corresponds to the asymptotic form of the Ornstein-Zernike form at the large Q limit. The Q^{-4} dependence, which disappears above 200 K, represents the existence of clusters with a size of 28 A independent of the temperature. The Q^{-3} dependence in the medium Q range of $0.15 < Q < 0.45$ A^{-1} in Fig. 1 suggests a possible fractal behavior.

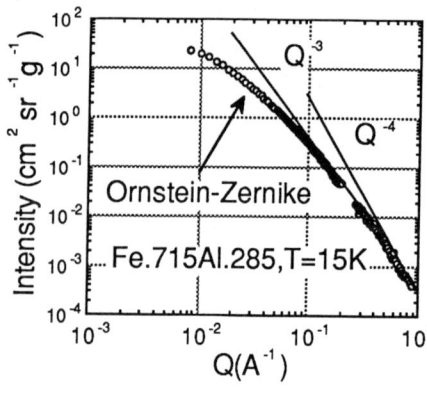 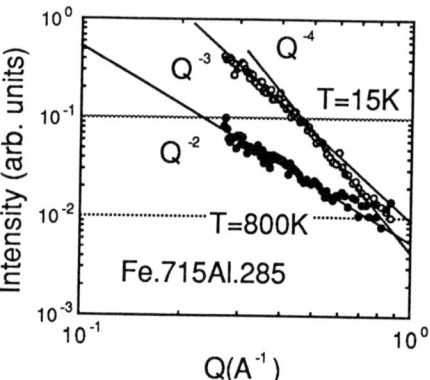

Fig.1 SANS profile in the wide Q range at 15 K.

Fig.2 SANS profiles in $Q \geq 0.25$ A^{-1} at 15 K and 800 K.

Probably this could be attributed to the segmentation of the ferromagnetic long range order. The further analysis and the experiments of the SANS data in applied magnetic fields clarified the existence of the clusters of ~34 A in a diameter, and the cluster's spins are randomly oriented each other.[5] The clusters evolve much above T_g, destroying the ferromagnetic long range order.

CONCLUSION

We have found that the RSG transition is induced by the cooperative freezing of cluster spins. This feature is very different from the scenario proposed by the mean field theory.[1] However, such behavior is very common among the realistically condensed system. Actually the very similar behavior in the Q dependence of the SANS intensity was observed in $(1-x)FeTiO_3-xFe_2O_3$ compound[6] and $Au_{1-x}Fe_x$.[7] In order to elucidate the origin of magnetic clusters, studies in the further high Q region are in progress.

REFERENCES

1. M. Gabay and G. Toulouse, Phys. Rev. Lett. **47**, 201 (1981).
2. I. A. Campbell, S. Senoussi, F. Varret, J. Teillet and A. Hamzic, Phys. Rev. Lett. **50**, 1615 (1983).
3. K. Motoya, S. M. Shapiro and Y. Muraoka, Phys. Rev. **B28**, 6183 (1983).
4. J. Suzuki, Y. Endoh, M. Arai, M. Furusaka and H. Yoshizawa, J. Phys. Soc. Jpn. **59**, 718 (1990).
5. J. Suzuki, PhD. thesis Tohoku Univ. (1990).
6. M. Arai, Y.Ishikawa and H. Takei, J. Phys. Soc. Jpn. **54**, 2279 (1985).
7. A. P. Murani, J. Appl. Phys. **49**, 1604 (1978).

B. COMPUTER SIMULATION

RELAXATION PHENOMENA OF THE FUZZY-SPIN MODEL

Seiji MIYASHITA and Tatuo KAWASAKI
Department of Physics, CLAS, Kyoto University, Kyoto 606

ABSTRACT

We study the slow relaxation phenomena of the Fuzzy spin model which is an Ising ferromagnetic model with random spin length. Distribution of the relaxation time to the equilibrium ferromagnetic state after rapid quenching from the infinite temperature is investigated by a Monte Carlo method. We study the mechanism of the slow relaxation using a cluster picture of strongly coupled spins. The dynamics of domain wall is also investigated using a one dimensional stochastic model.

Slow relaxation phenomena have been interested in the context of spin glasses, where combination of the randomness and the effect of frustration causes various interesting dynamics as well as the static properties. One of the authors (TK) has introduced a model which has only randomness but no frustration[1]. This is named as the Fuzzy-spin model and the Hamiltonian is given by

$$\mathcal{H} = -J \sum_{<ij>} S_i S_j \;, \qquad (1)$$

where the sum is taken all over the nearest neighbor pairs and J is a positive constant. The spin variables S_i can be expressed as $S_i = x_i \sigma_i$, where $\sigma_i = \pm 1$. The value of x_i is distributed randomly accoding to some distribution $F_x(x)$. Here we take the uniform distribution between 0 and 1, namely $F_x(x) = 1$ for $0 \leq x \leq 1$ and 0 for others. Relaxation phenomena in the square lattice of size $L \times L$ are studied by a Monte Carlo simulation which is equivalent to the single flip Glauber dynamics. We investigate a distribution, $P(t)$, of times which are taken before the system reaches to the equilibrium state after a rapid quenching from infinite temperature to a temperature below the critical temperature. Up to $L \leq 30$, the mean relaxation time increases exponentially with L,

$$< t > \propto \exp(aL), \qquad (2)$$

where a is a positive constant. The size-dependence of distribution is given in Fig. 1. The distribution $P(t)$ for the pure case ($x_i = 1$) has been studied[2] and it has been found that $<t> \propto L^3$. Thus, we find very slow relaxation in the Fuzzy-spin model compared with the pure case. This slow relaxation is attributed to the imhomogenety of the system. By studying snapshots of spin configurations, we conclude this dependence is due to the cluster structure of strongly coupled spins whose main frame consists of long spins. Namely, the cluster boundary tends to be located through regions with small spins. In the pure case the boundary can move freely and the motion is purely diffusive but in

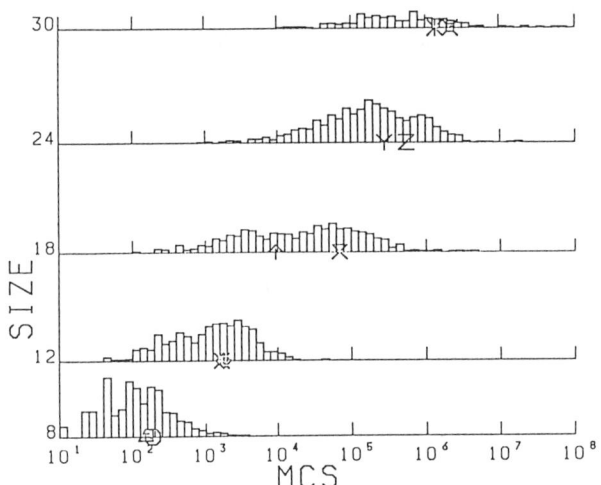

Fig. 1 Size-dependence of the distribution of the relaxation time, $P(t)$.

the present case the motion is "*massive*". Thus the dynamics becomes very slow compared with the pure case. Similar effects have been pointed out in diluted ferromagnet as a pinning effect on the domain walls.[3]

In order to study the cluster configuration, we take 100 samples of relaxation process for a configuration of x_i. In Fig. 2-1, distribution of long spins ($x_i \geq 0.9$) and small spins ($x_i \leq 0.1$) are shown. In Fig. 2-2, a typical snapshot at 100 Monte Carlo Step (MCS) is given. The bonds with small correlation will be good candidates of the cluster boundary. In Figs. 2-3 and 2-4, distributions of correlations $< S_i S_j >$ and $< \sigma_i \sigma_j >$ are given, respectively, where $< \cdots >$ denotes average over the 100 samples in the period $71 \leq \text{MCS} \leq 100$.

Here we discuss whether the exponential dependence (2) is good for larger lattices. In Fig. 2-4 we find possible positions of the cluster boundary. There we find that a typical cluster size, L_c, is about 20.

Since we do not obtain the equilibrium correlation, we can not estimate the energy gap ΔE for the overturn of the cluster. ΔE should have some distribution. If we use the smaller ΔE to define the cluster, the less boundaries survive. The orverturn of the cluster with ΔE occurs at rate of $\exp(-\Delta E/k_B T)$. For the present relaxation phenomena, the cluster distribution $P_c(L_c, \Delta E)$ should be an important quantity. If $L \leq L_c$, then the largest cluster size increases as the system size increases. Thus the exponential dependence is regarded as the time scale of jump between two stable values of magnetization, namely $M_{eq} = \pm M_0$. But for $L > L_c$, we expect more complicated mechanism. When there are only finite size clusters, the cluster effect disappears after some coase-graining procedures. Of course, the clustering causes increasement of the time scale of dynamics to the order of $\exp(aL_c)$. In this case the pinning effect is only apparent phenomena. There are, however, many other possible distributions of cluster

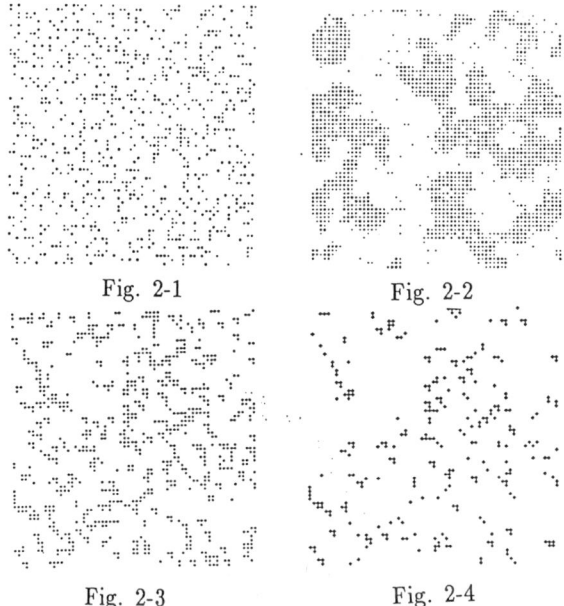

Fig. 2-1

Fig. 2-2

Fig. 2-3

Fig. 2-4

Fig. 2 Cluster structure due to the inhomogeneity: 1) distribution of long spins ($x_i \geq 0.9$ •) and small spins ($x_i \leq 0.1$ ·). 2) a metastable configuration of spins at 100MCS after quenching. Only up spins are plotted. 3) distribution of week bonds ($< S_i S_j > \leq 0.005$). 4) distribution of week bonds ($< \sigma_i \sigma_j > \leq 0.04$).

sizes. A large cluster could be divided into small ones if we use the larger ΔE. Accordingly we may have a fractal structure of clusters, which give an effective potential energy for dynamics of domain walls as is illustrated in Fig. 3. In this case we expect an essentially slower dynamics than the pure case as is discussed below. We can also consider the case where the cluster size increases with the system size. Then we will have a L-dependent expression such as eq.(2).

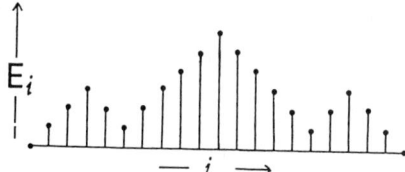

Fig. 3 Effective potential for the domain wall

In order to study dynamics of domain walls in such inhomogeneous potential, we introduce one-dimensional models with various distributions of potential energy. We use the hopping model with transition probability proportional to the weight in the cannonical distribution, namely $w_{i \to j} = \Delta t \exp(-E_i/k_B T)/(\exp(-E_i/k_B T) + \exp(-E_j/k_B T))$. Here we take $\Delta t = 0.5$. First we study the dependence of relaxation time on the shape of potential. For the purpose we

Table 1 Size and shape dependence of τ_0

System (A)

L k_BT	0.1	0.5	1.0	1.5
102	4.41×10^6	5.32×10^3	4.51×10^3	4.38×10^3
202	1.94×10^{11}	3.63×10^4	1.87×10^4	1.75×10^4
302	8.01×10^{15}	2.73×10^5	4.68×10^4	4.02×10^4

System (B)

L k_BT	0.1	0.5	1.0	1.5
102	2.68×10^6	1.06×10^4	6.53×10^3	5.61×10^3
202	5.91×10^{10}	1.26×10^5	4.12×10^4	2.98×10^4
302	1.53×10^{15}	1.02×10^6	1.56×10^5	9.18×10^4

investigate the following two models: (A) a uniform model where $E_0 = E_L = 0$ and others $E_i = 0.01(L - 2)$ and also (B) in a system where $E_i = 0.02i$ for $i \leq \frac{1}{2}L$ and $E_i = 0.01(L - 2) - 0.02(\frac{1}{2}L + 1 - i)$ for $i \geq \frac{1}{2}L + 1$. The slowest relaxation time, τ, can be obtain as $1/(1 - \lambda_2)$, where λ_2 is the second largest eigenvalue of $L \times L$ matrix for the above dynamics. Here the largest one, λ_1 is 1.0. We investigate the cases of $L = $ 102, 202 and 302 at various temperatures. In Table 1, list of τ_0 is given.

Here we find the case (B) gives slower relaxation except at very low temperatures. Since both give the relaxation times in the same order, we conclude that the shape of a single domain is not essential for the slowest relaxation which correspond to the overturn of the cluster. According to the shape illustrated in Fig. 4, we should study systems with fractal structures of E_i. The effect of the fractal structure of potential has been studied by Teitel et. al.[4] where they pointed out that there are two cases. In one case, the inhomogeneity can be renormalized and the system shows a pure diffusive nature. Namely, deviation of the position of domain wall is given by $< X^2 > \propto t$. In the other case the inhomogeneity gives an essential role and the point diffuses slower than the normal diffusion. Namely $< X^2 > \propto t^x$ with $x < 1$. This observation suggests that there are two cases in the relaxation of the Fuzzy model as well. Namely if the cluster structure is rather uniform, then the relaxation phenomena are essentially the same as the pure case, although the time scale is large. On the other hand, if the inhomogeneity is strong, then the relaxation should be changed to some slower type than the pure case. This change should depend on the form of $P_x(x)$, which will be discussed in the future.

REFERENCES

1. T. Kawasaki, Prog. Theor. Phys. **84**, 213 (1990).

2. S. Miyashita and H. Takano, Prog. Theor. Phys. **73**, 1122 (1985).

3. D. A. Huse and C. L. Henly, Phys. Rev. Lett. **54**, 2708 (1985)
   D. Chowdhury, M. Grant and J. D. Gunton, Phys. Rev. **B35**, 6792 (1987)
   H. Hayakawa, J. Phys. Soc. Jpn. **60**, 2492 (1991).

4. S. Teitel, D. Kutasov and E. Domany, Phys. Rev. **B36**, 684 (1987).

VI. OTHER RELATED TOPICS

A. EXPERIMENT

KINETICS OF ORDERING IN THE PERCOLATION MAGNET

H.Ikeda
*Booster Synchrotron Utilization Facility,
National Laboratory for High Energy Physics, Oho 1-1, Tsukuba 305, Japan*

ABSTRACT

We have observed, for the first time, the dynamics of both the ordering and disordering processes in a two-dimensional diluted Ising antiferromagnet, $Rb_2Co_{0.6}Mg_{0.4}F_4$, whose magnetic concentration was just above the percolation threshold, c_p (0.593). Time-resolved measurements by neutron scattering and magnetization experiments have provided evidence for a logarithmic power law dependence of the domain size with time.

INTRODUCTION

There are many fractal objects in nature.[1] The characteristic structure of such fractals as aerogel, aggregate, and porous media has been extensively examined in recent years by X-ray, light and neutron scattering. One of the most typical examples of fractals in the field of condensed matter physics is a percolation magnet. Upon diluting a magnetic material with nonmagnetic atoms, the phase-transition temperature decreases with decreasing proportion of magnetic atoms. At a certain magnetic concentration (percolation concentration c_p), which is determined solely by the shape of the magnetic lattice on which the magnetic atoms interact via magnetic interactions, the long-range order (LRO) disappears.[2] The connectivity of magnetic atoms at this percolation concentration takes an ideal fractal structure. If we use a neutron scattering technique, we can obtain detailed, precise information concerning the *dynamic* as well as static magnetic properties in this particular magnet with a fractal structure.

In this report, we present recent experimental results concerning both ordering and disordering kinetics in a two-dimensional (2D) diluted Ising antiferromagnet, $Rb_2Co_{0.6}Mg_{0.4}F_4$, whose magnetic concentration (0.60) is very close to and just above the percolation threshold (c_p (0.593)) of a 2D square lattice. One important parameter characterizing the structural property of a system close to the percolation threshold is the pair connectivity (ξ_G), which is defined as $\xi_G = (c - c_p)^{-\nu_G} a_0$, where c is the magnetic concentration, a_0 is the lattice constant and $\nu_G=1.36$ for the 2D square lattice. When the length-scale is less than ξ_G, the system maintains a self-similarity condition; oppositely, the system can be viewed as being homogeneous when we observe the system with a length-scale bigger than ξ_G. In the present material ($c=0.60$), ξ_G is estimated to be about $850 a_0$ using the above equation.

The ordering kinetics of a pure Ising model with a nonconserved order parameter has been extensively studied both theoretically and experimentally.[3] The order develops from the initial disordered state to the final LRO state after rapid quenching from a high-temperature paramagnetic state to an ordered state below its transition temperature. Theories as well as computer simulations and experimental observations in metallic binary alloys have proved that the temporal development of the order obeys a $t^{1/2}$ law during the late stage. However, it is now unclear how the order develops in a highly diluted magnet in which the magnetic concentration is close to the percolation threshold; the ordered clusters are therefore highly ramified.

In the following we present the first observation of ordering kinetics as well as its inverse process (disordering kinetics) in a highly diluted antiferromagnet with fractal geometry.

EXPERIMENTAL

In order to observe how a system develops in going from a disordered initial state to a final LRO state, one may cool the sample very rapidly from a high-temperature paramagnetic state to an ordered state at low temperature. In magnetic substances, however, the LRO is quickly stabilized when passing through the phase-transition temperature and, hence, it is too difficult to observe a temporal growth of the domain size on a real time scale. In the present experiments, we utilized a new idea to realize the initial disordered state; this played a key role in our successful observation. As has been extensively argued during the last decade,[4,5] the LRO in diluted Ising antiferromagnets with two dimensions is destroyed by a uniform magnetic field applied along the spin direction. Experiments have verified that the equilibrium domain size in an external field decreases with increasing field and with decreasing number of magnetic atoms. Our observation of the equilibrium domain size of $Rb_2Co_{0.6}Mg_{0.4}F_4$ in a field of 4.8 T showed only $12a_0$. This value is microscopic. Furthermore, we found that after removing the field the LRO recovers within a macroscopic time scale.[6] These facts have enabled us to observe the ordering kinetics on a real time scale. On the other hand, the LRO state realized after zero-field cooling is progressively destroyed upon applying a magnetic field, again on a macroscopic time scale. We can therefore trace the disordering process as well. In Fig.1 we show a schematic view of the ordering process ((a)→(b)→(c) in Fig.1) as well as the disordering process ((c)→(b)→(a)) in a 2D magnet close to the percolation threshold as our c=0.60 sample.

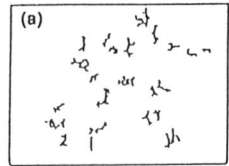

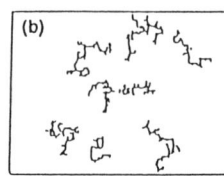

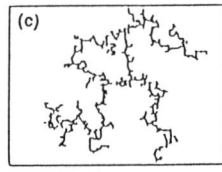

For time-resolved experiments, we performed both neutron scattering and magnetization measurements.[6,7] We mention here the possibility of observing temporal changes of the domain-size using magnetization measurements. In both pure and diluted antiferromagnets, a system having a LRO has no net magnetization at all. However, in a cluster having a finite size a ferromagnetic moment arises due to a statistical excess of the number of up spins. Simple argument gives rise to a relationship between the domain size (R) and the magnetic moment (M) as $R \sim 1/M$.[6] Therefore, the measurements of the temporal change of (M(t)) can detect the time development of the domain size (R(t)).

Fig.1. Schematic view of the time evolution of the ordering [(a)→(b)→(c)] and disordering [(c)→(b)→(a)] kinetics in a 2D percolated system.

The quality of the single crystal used in the present experiments was very high. The concentration gradient lying within the sample was estimated by measurements of the parallel susceptibility at a low field of 20 Oe. The diverging nature of the susceptibility at the phase transition is smeared due to the concentration gradient. Since the observed rounding of the transition is less than 1 K, variations in the Co^{2+} concentration are within $0.599<c<0.601$. For magnetization measurements a small piece of 0.0550g mass was cut out of the single crystal and mounted in a SQUID susceptometer. In order to directly measure the domain size as a function of time, neutron scattering experiments using a TUNS spectrometer installed at JAERI's JRR2 have been performed. We used a three-axis mode of operation with the transfer energy being fixed at zero using incident neutrons of 2.43 Å. An external field of up to 5 T in magnitude was applied in the vertical direction, so that the single crystal was mounted in a cryostat with the c-axis vertical. In order to obtain the best instrumental resolution possible, transverse scans across the (100) superlattice Bragg point were performed. The resolution of the spectrometer under these conditions was typically 0.0025 reciprocal lattice unit (rlu) half width at half maximum (1 rlu=$2\pi/a_0$=1.097 Å^{-1}).

ORDERING KINETICS

For the magnetization measurements, the crystal was cooled down to several designated temperatures below T_N (20K) while an external magnetic field was applied along the c-axis. The applied field was turned off after reaching the desired temperature and the magnetization measurement was performed in a zero field. Typical magnetization data are presented in Fig.2, where $M(t)/M(0)$ is plotted as a function of time at temperatures for 1-T cooling field conditions. The thermoremanent magnetization ($M(0)$) decreases as the temperature approaches T_N; eventually, $M(0)$ disappears near T_N. The temporal decay rate of the magnetization is clearly temperature dependent; it decreases only slightly at 2K over a period of more than 6×10^4 s, but diminishes to approximately 5% of its initial value in less than 2×10^4 s at 19K. This suggests that the growth rate of the order is governed by thermally activated magnetic fluctuations.

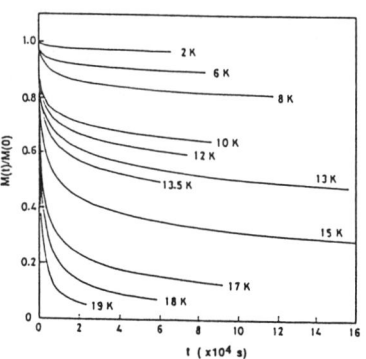

Fig.2. Time dependence of the magnetization normalized by the thermoremanent magnetization at several temperatures in the ordering process; the cooling field was 1 T in these experiments.

In order to analyze the magnetization data, we have made fits with several different functional forms, including a simple power law; in the end, however, we obtained the best results using a logarithmic power law,

$$M(t)^{-1} = A + B \, [\log_{10}(t)]^x ,$$

where A, B and x are adjustable parameters. As shown in Fig.3, fits using this functional form are excellent. We plot x as a function of temperature in Fig.4, where the results for 2-T and 3-T cooling are included. Note that x goes to 0 as T

approaches 0, indicating that the kinetics is frozen at $T=0$. This provides further support for our conjecture of thermally driven kinetics.

In order to measure the ordering kinetics directly, we have performed neutron-scattering experiments.[6] The time dependence of the line shape with the transverse scan was measured at 15K for 4.8-T field cooling of the sample and was directly compared with the magnetization measurements. The scan was repeated over a period of 15h. Each scan takes approximately 15 min. Typical line shapes at different times are shown in Fig.5. The line shape was fitted with several functions of the structure factor: a Lorentzian, a squared Lorentzian and the form $(\kappa^2+q^2)^{-1.5}$. The sum of the mean square deviations (χ^2) in the Lorentzian fitting for 62 scans is less than the other two. In Fig.6, the average domain size as determined from fits by a single Lorentzian is plotted versus $[\log_{10}(t)]^{3.5}$, the expression which best fits the magnetization measurements of this quantity. The neutron-scattering results are clearly consistent with the magnetization results. From these results we found that the observed domain size satisfies the self-similar condition of $R(t) < \xi_G$, because the pair connectivity ξ_G in the present sample is $850 a_0$, as mentioned earlier.

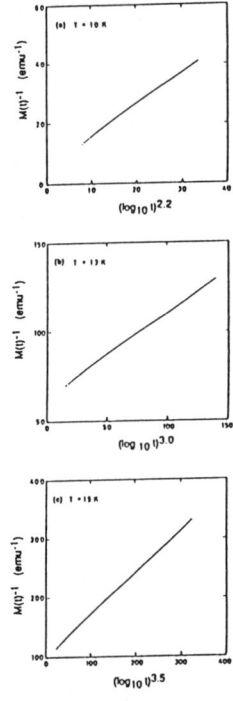

Fig.3. $M(t)$ vs $[\log_{10}(t)]^x$ at $T=10$ K (a), 13 K (b) and 15 K (c).

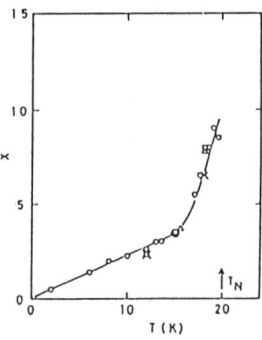

Fig.4. x vs T for data taken with external field cooling of 2 T (open circles), 2 T (triangles), and 3 T (crosses). x was found from fits of the magnetization by $[\log_{10}(t)]^x$. The neutron data point is denoted by the solid circle.

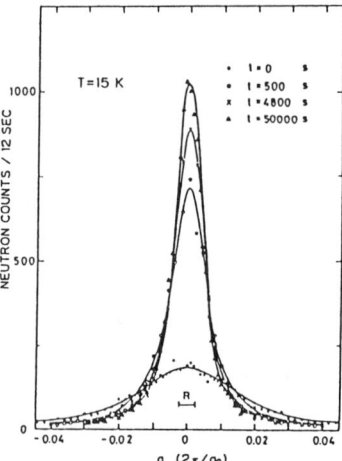

Fig.5. Temporal variation of transverse scans across the (100) superlattice position at $T=15$ K; the cooling field was 4.8 T. The lines represent fits by a single Lorentzian convolved with the instrumental resolution function. R represents the resolution width.

Although the logarithmic time dependence of the domain size ($R(t)$) has been discussed in previous theoretical works,[8-10] to the best our knowledge this is the first experiment which actually shows the log(t) power-law behavior and kinetics which exhibits freezing at $T=0$.

In a magnet with quenched impurities, the impurities act as energy barriers to domain growth; the pinning walls are therefore localized in energetically favorable positions, drastically slowing down the ordering kinetics. In particular, freezing in the percolation magnets should involve domain motion, which should be dependent on R; in other words, the energy barrier is dependent on R: $E(R) = (R-R_0)^{1/x}/F$.[11] Here, R_0 and F are only weakly temperature dependent. Since the time necessary to overcome such barriers will have an activated temperature dependence, $t \sim \tau_1 \exp[E(R)/T]$, so the growth law for the domain size in this model have a logarithmic dependence, $R(t) = R_0 + FT[\ln(t/\tau_1)]^x$.

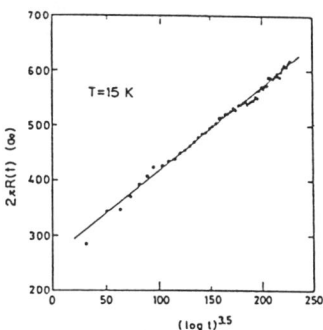

Fig.6. Domain size $R(t)$ vs $[\log_{10}(t)]^{3.5}$.

We next consider the particular situation of the percolation magnet in greater detail. The percolation magnet comprises three major characteristics regarding its structure: a one-dimensional link, a dangling end, and decoration. The dangling end does not play any important role in the ordering kinetics due to its finite size. Further, a domain boundary, or "kink", can move very quickly through the one-dimensional links, since there is no energy difference between different spin configurations with different locations of a kink in a link. Therefore, the domain motions in one-dimensional links are not relevant to the macroscopic decay rate observed in the present experiments. Different from these two, decorations are most important for the ordering kinetics. A plural number of spins connect with each other at the decorations. This gives rise to the excess energy required to run through the decorations for the domain boundary by making a plural number of Ising spins flip at the same time and, thus, increasing the correlation length. This excess energy can correspond to the activation energy mentioned above.

Our experiments show that x is dependent on the temperature and that it drastically increases as the temperature approaches T_N. We believe that although this behavior might be related to critical fluctuations, thermally controlled domain-wall motion dominates the ordering kinetics.

DISORDERING KINETICS

The disordering process was observed both by neutron scattering and magnetization measurements[7] using exactly the same crystal and instrumental configurations as in the previous observation of the ordering kinetics. Typical neutron results are shown in Fig.7, which presents the time evolution. These particular scans were made at 8.6K by applying a field of 4.8T after zero-field cooling of the sample from a temperature higher than T_N (20K). The broadening of the line-shape width and, therefore, the decrease in the peak intensity as a function of time corresponds to a decrease in the domain size (disordering process). Typical magnetization data as a function of time are presented in Fig.8 in a 1-T field. The magnetization measured under an external field (H) is expressed in terms of the sum of excess magnetization from the statistical excess of spins within the domain aligned along the field and the magnetization induced by the field (χH). After subtracting χH (which is not dependent on time) from the total magnetization, the normalized excess magnetization is plotted in the figure.

The domain size ($R(t)$) determined from both neutron scattering and magnetization measurements shows a logarithmic time dependence, and each $R(t)$ can be well fitted to the function

$$R(t) = B[\log_{10}(t)]^x + R_\infty ,$$

where B and x (negative) are adjustable parameters, and R_∞ is the equilibrium domain size, which was determined by neutron scattering after field cooling. The results of the dependence of the exponent (x) on both the temperature and strength of the magnetic field are plotted in Fig.9. We note two important facts: x is independent of both the temperature (at low temperatures of less than 10K) and the field; the absolute value of x is almost unity. On the other hand, the exponents at higher temperatures than 10K vary with the temperature. This might be again caused by critical fluctuations manifesting themselves upon approaching T_N and that the effects of thermal fluctuations play an important role in the

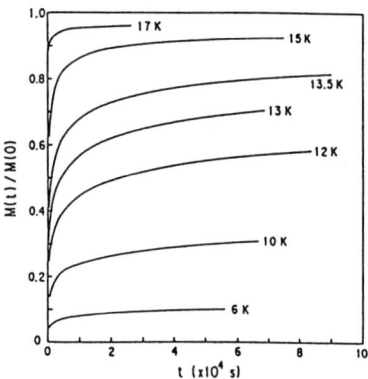

Fig.7. Time dependence of the magnetization normalized by the thermoremanent magnetization at several temperatures in the disordering process; the external field was 1 T in these experiments.

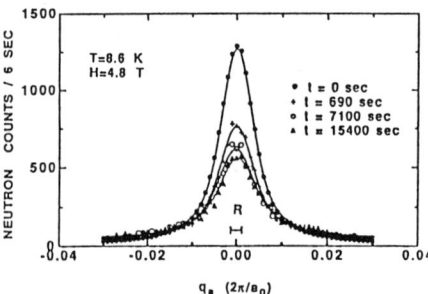

Fig.8. Temporal variation of transverse scans across the (100) superlattice position at 8.6 K. The symbols are the same as in Fig.5.

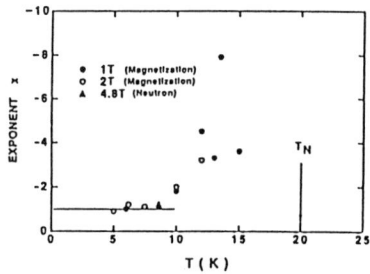

Fig.9. x vs T for the data runs taken with external field of 1 T and 2 T. x was found from fits of magnetization and neutron scattering data by $[\log_{10}(t)]^x$.

dynamics of domains at temperatures close to T_N. Strictly speaking, there is no T_N in the presence of an external field due to both random-field effects, and the an "asymptotic" singular point exists, which decreases with increasing field. This may be one reason for the sharp increase in $|x|$ above T_N.

The present experiments and analysis indicate that the disordering process should also be a thermally activated type process so as to overcome any energy barriers. The existence of energy barriers can be explained in the same way as in ordering kinetics. One notable difference from the ordering kinetics is the fact that the power of the logarithmic time dependence (x) is temperature independent at low temperatures. This result comes from the fact that disordering dynamics proceeds with time in the presence of an externally applied magnetic field. This is in remarkable contrast with the experiments concerning ordering kinetics: when the external field was switched off, the thermally controlled domain-wall motion governed the dynamics of ordering, and eventually the ordering kinetics showed a temperature-dependent power. On the other hand, in the disordering process the external field predominantly governs the dynamics. This can result in a temperature-independent power of x at low temperatures.

CONCLUDING REMARKS

In summary, we have carried out time-resolved neutron scattering and magnetization studies concerning the domain dynamics in a 2D diluted Ising antiferromagnet, $Rb_2Co_{0.6}Mg_{0.4}F_4$, whose magnetic concentration is close to the percolation threshold. Both experiments provide the first observation of a logarithmic power law below T_N An important discovery found in the measurements of the ordering kinetics is that the power of the logarithmic dependence on time is temperature dependent. We propose that thermal effects govern the ordering kinetics in magnets near the percolation threshold. This behavior must be incorporated into any theory which describes the ordering kinetics. The result concerning disordering kinetics is different from the ordering kinetics in the meaning that the power of logarithmic dependence is constant. This fact is caused by the presence of an external magnetic field which drives the dynamics.

ACKNOWLEDGMENTS

The author thanks his collaborators, especially S.Itoh, Y.Endoh and T.Suzuki for their fruitful discussions throughout the present studies. The present research was supported by a Grant-in Aid for Scientific Research from the Ministry of Education, Science and Culture.

REFERENCES

1. See for instance, *Fractals in Physics*, edited by A. Aharony and J. Feder (Elsevier-Science, Amsterdam, 1990).
2. For a review, see D. Stauffer, Phys. Rep. **54**, 1 (1979).
3. J.D.Gunton, M. San Miguel, and P. S. Sahni, in *Phase Transitions and Critical Phenomena*, edited by C. Domb and J. L. Lebowitz (Academic, New York, 1983), Vol.8, p.267.

4. For reviews of experimental results, see R. J. Birgeneau, R. A. Cowley, G. Shirane, and H. Yoshizawa, J. Stat. Phys. **34**, 817 (1984); D. P. Belanger, Phase Transitions **11**, 53 (1988).
5. For a review of the theory, see T. Nattermann and J. Villain, Phase Transitions **11**, 5 (1988).
6. H. Ikeda, Y. Endoh, and S. Itoh, Phys. Rev. Lett. **64**, 1266 (1990).
7. S. Itoh, H. Ikeda, Y. Endoh, and T. Suzuki, J. Phys. Soc. Jpn. in press.
8. G. S. Grest and D. J. Slorovitz, Phys. Rev. B **32**, 3014 (1985); D. J. Slorovitz and G. S. Grest, Phys. Rev. B **32**, 3021 (1985).
9. D. A. Huse and C. L. Henley, Phys. Rev. Lett. **54**, 2708 (1985).
10. D. Chowdhury, M. Grant, and J. D. Gunton, Phys. Rev. B **35**, 6792 (1987).
11. Z. Lai, G. F. Mazenko, and O. T. Valls, Phys. Rev. B **37**, 9481 (1988).

SIZE AND SPATIAL DISTRIBUTIONS OF PRECIPITATES IN THE OSTWALD RIPENING IN SOME ALLOYS

Tetsuo Eguchi and Kiyoshi Arita*
Department of Applied Physics, Fukuoka University, Fukuoka 814-01

Yoshitsugu Tomokiyo
Research Laboratory for High Voltage Electron Microscopy,
Kyushu University, Fukuoka 812, JAPAN

ABSTRACT

The radii and the three-dimensional coordinates of the centers of spherical precipitates were determined in Al-Li and Cu-Co alloys by the stereographic method under an electron microscope. The distribution of the radii and that of the interparticle distances to the first nearest neighbors were obtained on isothermal aging. The size distributions normalized by the average radius were consistent with any of the existing theoretical predictions without consideration of elastic interaction between precipitates. The spatial distribution can be accounted for by the assumption of an isotropic random distribution of particles of finite size. In the analysis the effect of the volume fraction of precipitates plays an essential role.

INTRODUCTION

Among a number of phenomena of slow dynamics in phase transitions in alloys, the Ostwald ripening of precipitates on isothermal aging has long been attracting attentions of many investigators. Precipitation is interesting to materials scientists for its fundamental nature in reaction kinetics, and at the same time to materials engineers for its practical importance. The Ostwald ripening in alloys proceeds much slower than in gaseous or liquid solutions, and provides us with good examples to be analyzed theoretically. In solid solutions, however, precipitation brings in certain complexities which do not usually exist in gaseous or liquid solutions; namely the directional preference resulting in the anisotropy of the shape and the distribution of precipitates on one hand, and the elastic interaction between precipitates on the other, due to the local lattice distortion around each precipitate. Thus the precipitation dynamics, or more specifically the Ostwald ripening, has long been one of the challenging important subjects of investigation for both experimental and theoretical materials scientists.

In the present work we summarize our recent investigation on the Ostwald ripening of precipitates in Cu-Co and Al-Li alloys. These alloys were chosen because they exhibit spherical precipitates up to a certain stage of aging, which can be observed directly under an electron microscope,[1] or indirectly by the method of magnetic granumetry in the case of Cu-Co.[2] We used a magnetic measurement to obtain

---

* Present Address: RD Center, Kyushu Matsushita Electric Co. Ltd., Fukuoka-812, JAPAN

the mean radius of precipitates in Cu-Co alloys, when they are too small to be observed under an electron microscope. When they are large enough the method of a stereographic analysis was used to determine the position and the radius of individual precipitates by electron microscopy. Thus we were able to obtain the size distribution and the distribution of distances to the nearest neighbors, which can be compared with implications of the existing theories. It was shown that the result of our experimental observation of the size distribution is satisfactorily consistent with any of the theoretical predictions, which neglect the effect of elastic interaction among the precipitates, and that both the scaling and $t^{1/3}$ laws hold.

EXPERIMENTAL PROCEDURE

The method of preparing the samples for the electron microscopic observation has been already explained elsewhere.[1,2] In the present work an attempt was made to renew the method of obtaining the data on the size and position of individual precipitate by the stereographic electron microscopy. For this purpose δ' (Al_3Li) particles in Al-Li alloys were imaged in a dark field with a superstructure reflection, and Co precipitates in Cu-Co alloys were imaged in a bright field with a fundamental reflection, in which a contrast was obtained due to the lattice distortion around the peripheries of spherical precipitates. For each image field the micrographs were taken as stereo pairs in order to determine the sizes and positions of individual particles in three dimensions. Examples of stereopair electron micrographs of Co precipitates in Cu-Co alloy are shown in Fig.1. We used a newly developed digitizer coupled to a personal computer, to store the data of the radii and the coordinates of precipitates in the sample. An example of the reconstruction of the image is shown in Fig.2, where the projection of all precipitates on the three coordinate planes gives us a sense of reliability of our method of image processing.

Once the data for the radii and coordinates of all the precipitates in a sample were accumulated, it is easy to obtain the size distribution and the distribution of the distance to the first nearest neighbor precipitate. From the data of radii of the precipitates which were stored, we first obtained the mean radius $\bar{r}$ for each stage of isothermal aging, and confirmed the validity of the $t^{1/3}$-rule. The standard deviation and the skewness of the distribution were also obtained automatically by the PC programming. The histogram for the size distribution was obtained in the normalized radii $\rho = r/\bar{r}$, where the increments $\Delta\rho$ were so determined that the histogram obtained seemed most reasonable as compared to the theoretical predictions.[4~6] In Fig.3 some of the histograms thus obtained are shown as compared with the theoretical distribution curves as obtained by the use of LSW[3,4] and Ardell's method.[5]

As a measure of the spatial distribution of the precipitates, we also obtained the distance R to the first nearest neighbor for each precipitate. The histogram for the nearest neighbor distance was obtained as a distribution of $\ell = R/\bar{R}$ at each stage of aging. The histograms thus obtained were compared with a simple theoretical prediction, which resembles Chandrasekhar's distribution function,[8] but takes the finite sizes and the volume fraction of precipitates into account. The result of comparison in seems to indicate that the precipitates are distributed randomly without any spatial correlation among them. The distribution function of the nearest neighbor interparticle distance we used in Fig.3 for the analysis is the following.

$$F(R)=4\pi[n/(1-\phi)]\exp[-4\pi n(R-R_c)^3/3(1-\phi)](R-R_c)^2, \quad R>R_c=\bar{r}, \quad (1)$$

which coincides with Chandrasekhar's one, when the cut-off radius R_c and the volume fraction ϕ are neglected.

RESULT

The results of our present investigation are summarized as follows. 1): Throughout the isothermal aging the $t^{1/3}$-law for the average radius of precipitates is valid in Al-Li and Cu-Co. In the latter the law observed under an electron microscope is consistent with the one obtained with a magnetic measurement. 2): Over a long period of aging the standard deviation and skewness of the size distribution change gradually to tend to their final values. The skewness is always negative, and does not change its sign in the course of long aging as some recent theory predicted.[7] 3): The observed distribution of the radius of precipitates can be fitted satisfactorily with either Ardell's or recent theoretical distribution functions[6] if the effect of the volume fraction is taken into account properly. We were not able to confirm the superiority of the recent theories[6,7] to the classical one within the accuracy of our observed distribution. 4): The spatial distribution of precipitates in the matrices seems uniform and random, as far as the distribution of the nearest neighbor interparticle distance is concerned, which is accounted for by the Chandrasekhar formula with a slight modification for the finite size and the volume fraction of precipitates. 5): As far as our observational technique is confined to use electron microscopes, the improvement of the accuracy and statistics above the present state is quite difficult. The refinement of the theory on Ostwald ripening to include, for example, the elastic interaction between precipitates does not make much sense in such circumstances.

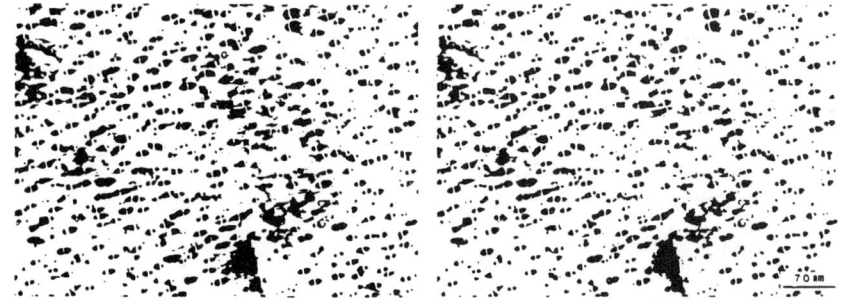

Fig.1 An example of the stereopair electron micrographs of Cu-Co

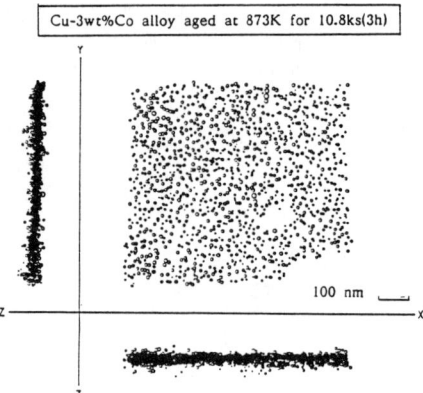

Fig.2 Reconstruction of Fig.1 into three dimensional planes

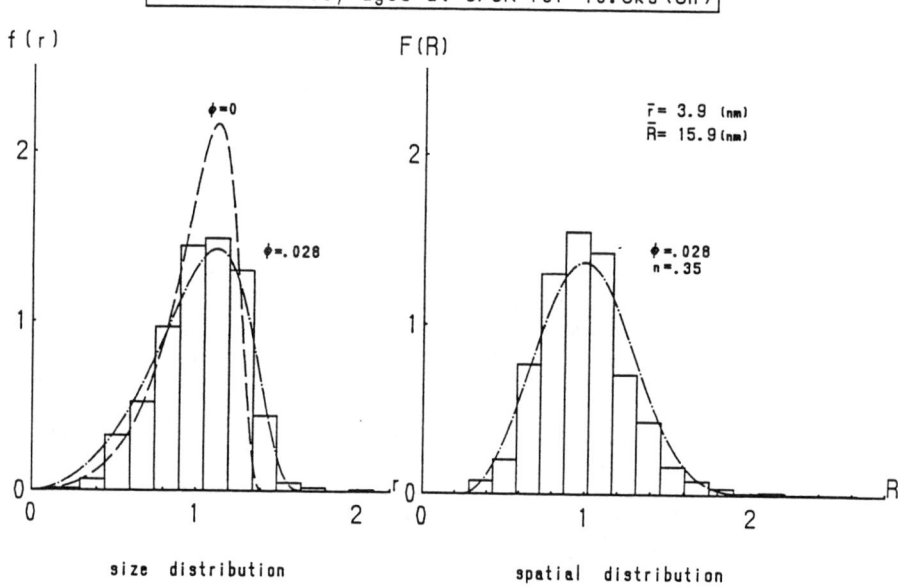

Fig.3 Examples of the distributions of radii and nearest neighbor interparticle distances of the precipitates in Fig.1

REFERENCES

1. Eguchi, Tomokiyo & Matsumura, Phase Transitions B, 8, 213 (1987)
2. Seno, Eguchi et al., Journ. Japan Inst. Metals, 45, 661 (1981)
3. Lifshitz and Slyozov, Journ. Phys. & Chem. Solids, 19, 35 (1961)
4. C. Wagner, Zeitschrift fur Elektrochemie, 65, 581 (1961)
5. A.J. Ardell, Acta Metall., 20, 61 (1972)
6. Tokuyama, Kawasaki and Enomoto, Physica A, 134, 323 (1986)
7. Kawasaki and Enomoto, Physica A, 150, 463 (1988)
8. S. Chandrasekhar, Rev. Mod. Phys., 15, 1 (1943)

NONLINEAR SLOW FLUCTUATION AND RELAXATION IN THE ORDERING OF SOME COMPLEX MAGNETIC SYSTEMS

M. Matsuura and M. Hagiwara
Department of Electronics and Information Science,
Kyoto Institute of Technology, Matsugasaki, Kyoto 606

ABSTRACT

Dynamical aspect of the intermediate state in the hierarchical spin ordering of $CoCl_2$- and $NiCl_2$-GICs (graphite intercalation compound) is investigated by examining the nonlinear magnetic response to an AC excitation field at very low frequencies (VLF) down to 10^{-3} Hz across the successive transition temperatures T_{cu} and T_{cl} ($<T_{cu}$). The singularity of the third harmonic in-phase component $M'(3\omega)$ changes the character by changing the frequency ω and the amplitude h of AC excitation field, indicating an apparent change of symmetry breaking at T_{cu} and at T_{cl}. The logarithmic divergence of $M'(3\omega)$ at T_{cu} is compared with the $1/\omega$ type slow fluctuation estimated from the linear out-of-phase component $\chi''(\omega)$ and the logarithmic slow decay of magnetization in the intermediate temperature range and discussed taking the heterogeneous lattice structure of the system into account.

INTRODUCTION

Cooperative dynamics in complex physical systems is an interesting subject from both fundamental and applied physical view points. A heterogeneous system may be further attractive, because the ordering process does not go on simultaneously and homogeneously over the whole system. The phase transition of such a system would happen successively in many steps, resulting in an intermediate or mesoscopic ordered phase without total symmetry breaking in the intermediate temperature region.

A candidate for such a system is a graphite compound intercalated by $CoCl_2$ or $NiCl_2$ (hereafter $CoCl_2$- or $NiCl_2$-GIC). In the compound, generally, the intercalated substance (intercalant) does not form a perfect two-dimensional (2D) lattice but a set of island like 2D clusters of a certain finite size (several hundred angstrom in diameter) in each layer.[1,2]

HIERARCHICAL SPIN ORDERING

For stage 2 $CoCl_2$- and $NiCl_2$-GICs, a two-step magnetic phase transition has been observed with a thermoremanent magnetization M_r between the two successive temperatures T_{cu} and T_{cl}.[3-5] In the intermediate state, a true 2D long range spin order was identified by neutron scattering experiment.[6-8]

Such a characteristic ordering process could be reasonably understood thermodynamically by taking the heterogeneous lattice structure mentioned above into account, as a hierarchical spin ordering as in the following: First at T_{cu}, the system goes into an intracluster (2D) long range order (LRO) with intercluster disorder

© 1992 American Institute of Physics

and then at T_{c1}, into the intercluster LRO, as decreasing temperature.[5,9]

Such a hierarchical spin ordering model was recently supported experimentally by examining the singularity of nonlinear susceptibility χ_2 at T_{cu} and at T_{c1}.[10] As seen in Fig. 1, the in-phase component $M'(3\omega)$ for stage 2 $CoCl_2$-GIC shows a divergent anomaly at T_{cu} and is symmetric against T_{cu}, suggesting a lack of symmetry breaking over the whole system. On the other hand $M'(3\omega)$ shows another anomaly at T_{c1} which is not apparently divergent but changes the sign or in the other word is antisymmetric against T_{c1}, which indicates a breaking of the total magnetic symmetry at the temperature. Necessarily, the state

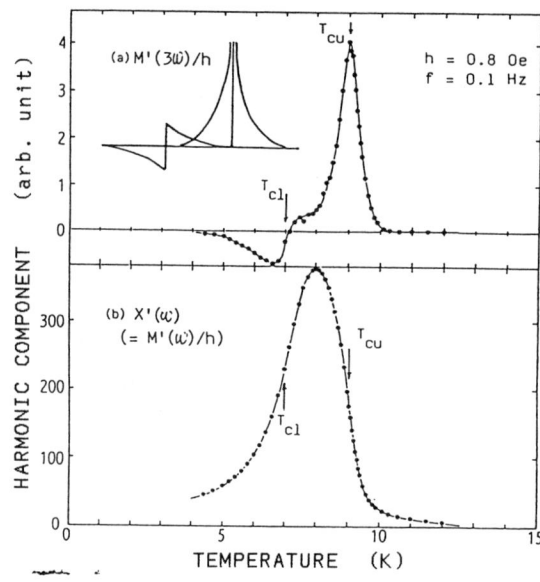

Fig. 1 Temperature dependence of (a) $M'(3\omega)/h$ and (b) linear part $\chi'(\omega)$ (= $M'(\omega)/h$) for stage 2 $CoCl_2$-GIC.

between T_{cu} and T_{c1} should be say mesoscopic from the view point of cooperativity and very attractive.

DYNAMICAL STRUCTURE OF THE INTERMEDIATE STATE

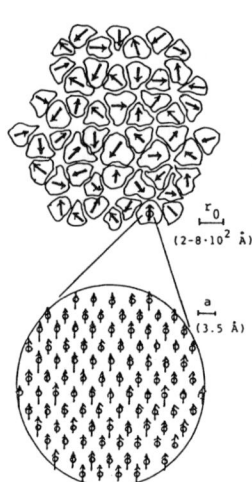

Fig.2 A schematic view of the mesoscopic pahse.

In the intermediate state between T_{cu} and T_{c1}, we can expect to find two different aspects of ordering. If we observed the system on a microscopic scale, we could find that it is already in the ordered phase. If we observed the system, however, on a larger scale than each cluster e.g. a scale of more than several hundred angstrom, we would find that it is still in the disordered phase, schematically shown in Fig. 2. Since the intercluster correlation is random both within each layer and along the interlayer direction, the magnetization of each cluster will fluctuate and change the direction to each other in a certain time scale, which is determined by individual cluster size, intercluster interaction, some pinning force and so on. The characteristic frequency ω_0, if any, of the intercluster fluctuation should be generally much smaller than that of the intracluster fluctuation.

If the frequency ω of AC excitation is

smaller than ω_o, the response signal could perfectly reflect the intercluster fluctuation. Thus, $M'(3\omega)$ shows a symmetric anomaly as seen in Fig. 1, indicating that the total symmetry of the system is not broken at T_{cu}. In the case of $\omega > \omega_o$, however, intercluster fluctuation can not follow the excitation field. Therefore, $M'(3\omega)$ does not perfectly reflect the intercluster fluctuation and a symmetry breaking phenomenon may apparently be observed.

For stage 2 $CoCl_2$-GIC, no such change of singularity has been found out in the measured frequency range up to 1 Hz. It probably indicates the situation that $\omega < \omega_o$ in the frequency range. For stage 2 $NiCl_2$-GIC, on the other hand, a change of anomaly of $M'(3\omega)$ is well expected to happen at T_{cu} by increasing ω further as seen in Fig. 3.

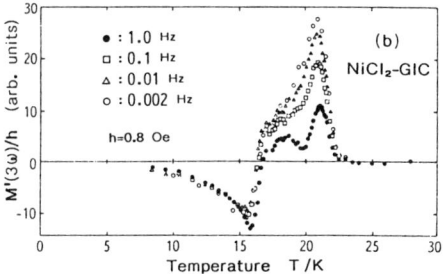

Fig. 3 Temperature dependence of $M'(3\omega)/h$ at various frequencies for stage 2 $NiCl_2$-GIC.

A similar change of dynamical anomaly of $M'(3\omega)$ is also expected to appear when the excitation field amplitude h is changed. If the characteristic pinning field H_p of each cluster moment is much larger than h, the interclustrer fluctuation could no longer be reflected on $M'(3\omega)$. In the situation, the anomaly of $M'(3\omega)$ will change, indicating an apparent symmetry breaking at T_{cu}. An example for stage 2 $CoCl_2$-GIC is given in Fig. 4. The $M'(3\omega) - T$ curve shows a divergent anomaly at T_{cu} but quite antisymmetric against T_{cu}, forming a remarkable contrast to the curve in Fig. 1. Such a dramatic change of $M'(3\omega)$ anomaly at T_{cu} tells us the fact,

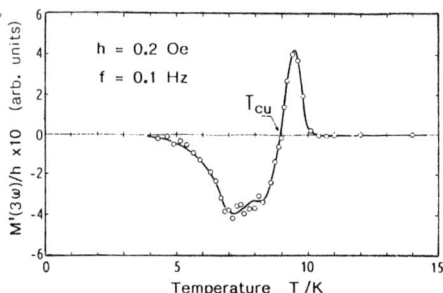

Fig. 4 Temperature dependence of $M'(3\omega)/h$ for stage 2 $CoCl_2$-GIC at small h.

$H_p > h$ in the present situation. The same change of $M'(3\omega)$ anomaly at T_{cu} is also found for stage 2 $NiCl_2$-GIC.

NONLINEAR SLOW RELAXATION

As mentioned already, the $M'(3\omega)$ anomaly at T_{cu} for stage 2 $CoCl_2$-GIC does not change qualitatively in the measured frequency range. The peak value of $M'(3\omega)$ at T_{cu}, however, increases logarithmically as decreasing ω, as shown in Fig. 5. Although the dispersion of such a nonlinear response has not been discussed, such a characteristic ω dependence reminds us of a spin glass like random magnet in which hierarchical distribution of relaxation rates could bring such nonlinear phenomena as $1/\omega$ type fluctuation spectrum, logarithmic or stretched exponential decay of thermoremanent magnetization etc..[11-14]

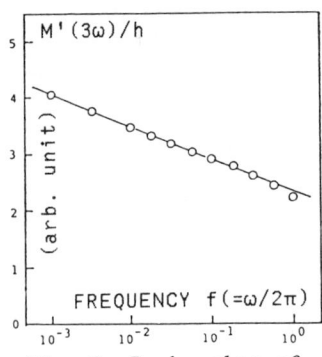

Fig. 5 Peak value of $M'(3\omega)/h$ at T_{cu} as a function of ω.

Actually, a $1/\omega$ type fluctuation has been found to exist in stage 2 $CoCl_2$-GIC by examining the linear out-of-phase susceptibility $\chi''(\omega)$ around T_{cu} in the VLF range down to about 10^{-4} Hz.[15] The thermoremanent magnetization of the compound in the intermediate temperatures between T_{cu} and T_{cl} is found to decay logarithmically long to more than 10^4 s.[16] The origin of such nonlinear slow relaxation phenomena should be attributed to the coupling scheme among the clusters. An attractive possible model is such a heterogeneous complex system in which ferromagnetic clusters are surrounded and coupled to each other by a spin glass like random media. Indeed, the existence of such a random media can be well speculated by inspecting the mechanism of staging and cluster formation in the intercalation process of GIC.

This work is supported in part by the Grant-in-Aid of Nippon Sheet Glass Foundation for Material Science.

REFERENCES

1. S. Flandrois, A.W. Hewat, C. Hauw and R.H. Bragg: Synth. Met. 7, 305 (1983).
2. M. Matsuura, Y. Murakami, K. Takeda, H. Ikeda and M. Suzuki: Synth. Met. 12, 427 (1985).
3. Y. Murakami, M. Matsuura, M. Suzuki and H. Ikeda: J. Magn. Magn. Mater. 31-34, 1171 (1983).
4. M. Suzuki, H. Ikeda, Y. Murakami, M. Matsuura, H. Suematsu, R. Nishitani and R. Yoshizaki: J. Magn. Magn. Mater. 31-34, 1173 (1983).
5. Y. Murakami and M. Matsuura: J. Phys. Soc. Jpn. 57, 1056 (1989).
6. D.G. Wiesler, M. Suzuki and H. Zabel: Phys. Rev. B36, 7051 (1987).
7. D.G. Wiesler and H. Zabel: J. Appl. Phys. 63, 3554 (1987).
8. D.G. Wiesler, H. Zabel and S.M. Shapiro: Synth. Met. 34, 505 (1989).
9. M. Matsuura: Ann. Phys. Suppl. (Paris) 11, 117 (1986).
10. M. Matsuura and M. Hagiwara: J. Phys. Soc. Jpn. 59, 3819 (1990).
11. R.V. Chamberlin, G. Mozurkewich and R. Orbach: Phys. Rev. Lett. 52, 867 (1984).
12. R.G. Palmer, D.L. Stein, E. Abrahams and P.W. Anderson: Phys. Rev. Lett. 53, 958 (1984).
13. M. Ocio, H. Bouchiat and P. Monod: J. Magn. Magn. Mater. 54-57, 11 (1986).
14. W. Reim, R.H. Koch, A.P. Malozemoff and M.B. Kechen: Phys. Rev. Lett. 57, 905 (1986).
15. M. Matsuura, Y. Endoh, T. Kataoka and Y. Murakami: J. phys. Soc. Jpn. 56, 2233 (1987).
16. Y. Murakami, M. Matsuura and T. Kataoka: Synth. Met. 12, 443 (1985).

CROSS OVER BETWEEN NON-EQUILIBRIUM RELAXATION TO EQUILIBRIUM RELAXATION IN CDW GROUND STATE

K. Biljaković
Institute of Physics of the University of Zagreb,
41001 Zagreb, P.O. Box 304, Croatia, Yugoslavia

J.C. Lasjaunias and P. Monceau
Centre de Recherches sur les Très Basses Températures,
laboratoire associé à l'Université Joseph Fourier,
C.N.R.S., BP 166, 38042 Grenoble-Cédex 9, France

ABSTRACT

We show that the relaxation of energy in some charge density wave systems indicates a cross over between a non-equilibrium state and the thermodynamical equilibrium state when the system is allowed to age. This extends work on random systems to a regime unable to be reached before, as for example in spin glasses.

INTRODUCTION

Recent experiments show that charge density wave (CDW) compounds[1] could be included in the broad class of physical systems as polymers, ionic conductors, amorphous semiconductors, spin glasses, ... showing anomalous slow relaxations in response to different excitations (mechanical stress, external electric and magnetic fields, ...). At very low temperatures (T < 1 K) there is evidence for low-energy excitations in an extra contribution to phonons in the specific heat.[2-4] In the same T range the energy relaxation shows anomalous long time response with, in addition, "aging" effects.[5]

We have studied the energy relaxation between 0.08 K and 8 K in a dilution refrigerator after cooling the sample from 300 K through the Peierls transition (210 K for TaS_3, 145 K and 59 K in $NbSe_3$ and 260 K in $(TaSe_4)_2I$) without application of any electric field. The response $\Delta T(t)$ to two different types of heat perturbations have been studied[5] : (i) to a heat pulse no longer than 1 s and (ii) to a heat flow of different durations (playing the role of a waiting time t_ω by analogy to spin glasses) up to longer than a day (~ 10^5 sec).

RESULTS

The main aim of the interpretation of the results of our experiments is to show in this paper the evolution of the distribution of relaxation time $g(\tau)$, rather than to find the functional form of the relaxation. $g(\ln\tau)$ can be obtained from the partial differentiation of the relaxation function with respect to $\ln t$: $d[\Delta T/\Delta T(0)]/d\ln t \simeq g(\ln\tau)$.[6] The logarithmic derivative of $\Delta T(t)$ drawn in Fig. 1 shows a peak in the relaxation rate at τ_p. The pronounced increase of the aging effect when T is decreased is also clearly presented, as well as the change in the shape of the relaxation rate. τ_p increases with t_ω, indicating that the system has reached its thermodynamical equilibrium when τ_p saturates at long t_ω as shown on Fig. 2.

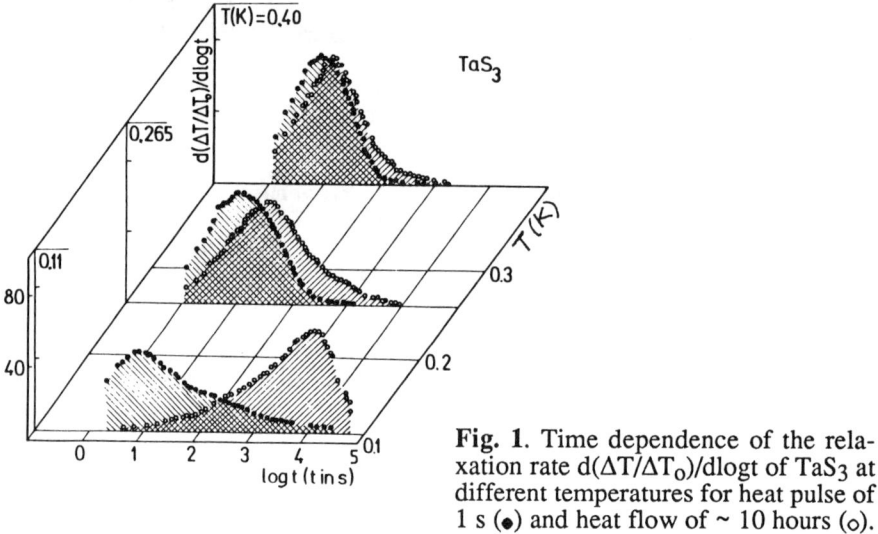

Fig. 1. Time dependence of the relaxation rate $d(\Delta T/\Delta T_0)/d\log t$ of TaS_3 at different temperatures for heat pulse of 1 s (●) and heat flow of ~ 10 hours (o).

The non-equilibrium properties reflected in the aging effect are much more pronounced in TaS_3 (Fig. 2(a)) and $(TaSe_4)_2I$ than in $NbSe_3$ (Fig. 2(b)) where they appear only below 150 mK. Thus for TaS_3 at T = 0.1 K (Fig. 2(a)) the behaviour is similar to that of spin glasses with the maximum t_ω (10 hours) which remain smaller or equal to the longest relaxation time. To our knowledge, it is the first evidence reported in glassy materials for such a *cross over between non-equilibrium and equilibrium states*.

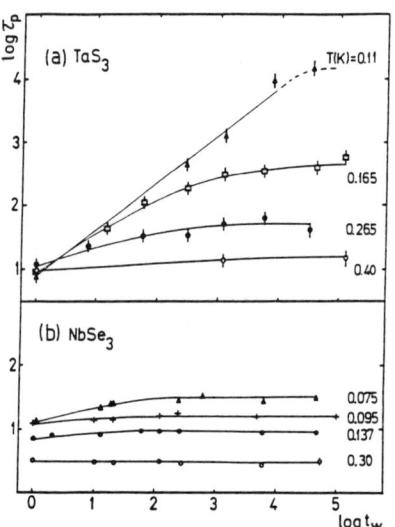

Fig. 2. Variation of the peak in the relaxation rate (Fig. 1) as a function of the waiting time t_ω at several temperatures. The cross over between non-equilibrium and equilibrium state occurs when the variation of τ_p flattens off with t_ω.

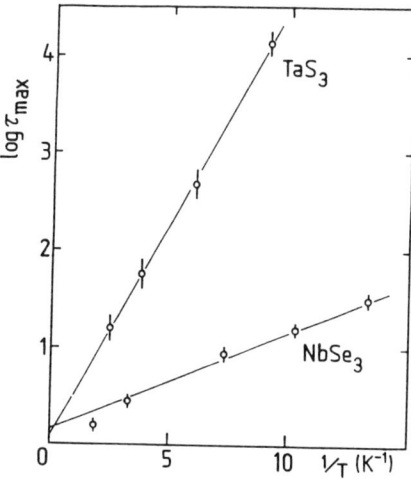

Fig. 3. Variation of τ_{max} : the saturated value of τ_p in Fig. 2, as a function of 1/T, shows activated behaviour.

Different predictions for the divergence of the relaxation time at the phase transition temperature T_c have been given by different models for disordered systems. Acceptable fits for critical slowing down : $\tau = \tau_o (1-T_c/T)^{-zv}$, would give very small T_c ($T_c \lesssim 0.01 - 0.03$ K) and huge zv (zv > 25 - 30). Alternatively, an Arrhenius law : $\tau = \tau_o \exp(W/kT)$, gives comparable fits as shown on Fig. 3 for TaS_3 and $NbSe_3$ where we have drawn the variation of τ_{max} (defined in Fig. 2) as a function of 1/T. The activation energies are W ~ 0.8 - 1 K for TaS_3 (similar value is obtained for $(TaSe_4)_2I$) and W ~ 0.25 K for $NbSe_3$.

Below 0.5 K, CDW internal degrees of freedom are still active with activation energies of ~ 1 K. We propose that defects in the CDW superstructure, phase slips in 1D, dislocation loops in 3D, are at the origin of the long time, low T, energy relaxation. One can thus think about a structure of a network of dislocations of non-uniform size similar to the ramified fractal lattice on percolation clusters with similar dynamics. The time constant of these large objects is expected to be high, which may account for the τ_o value of a few seconds in the Arrhenius behaviour (Fig. 3).

CONCLUSION

In conclusion we have shown that some CDW compounds exhibit at very low temperature long time energy relaxation, which a relaxation rate which is a function of time during which the thermal perturbation was applied. This characteristic property, in addition to the existence of low energy, extra-phonon, excitations confirm the strongly disordered nature of the CDW ground state.

REFERENCES

1. For a review, see : *Electronics Properties of Inorganic Quasi One-Dimensional Compounds*, Parts I and II, ed. by P. Monceau (D. Reidel, Dordrecht, 1985) ; *Charge Density Wave in Solids*, Modern Problems in Condensed Matter Science Series, ed. by L.P. Gor'kov and G. Grüner (Elsevier, Lausanne, 1989).
2. K.J. Dahlhauser, A.C. Anderson and G. Mozurkewich, Phys. Rev. B **34**, 4432 (1986).
3. K. Biljaković, J.C. Lasjaunias, F. Zougmoré, P. Monceau, F. Lévy, L. Bernard and R. Currat, Phys. Rev. Lett. **57**, 1907 (1986) ; J.C. Lasjaunias, K. Biljaković and P. Monceau, Physica B **165-166**, 893 (1990).
4. S.E. Brown, J.O. Willis, B. Alavi and G. Grüner, Phys. Rev. B **37**, 6551 (1988).
5. K. Biljaković, J.C. Lasjaunias, P. Monceau and F. Lévy, Phys. Rev. Lett. **62**, 1512 (1989), and Fizika (Zagreb) **21**, 135 (1989).
6. L. Lundgren, P. Svedlinh, P. Nordblad and O. Beckman, Phys. Rev. Lett. **51**, 911 (1983).

RELAXATION PROCESSES IN 2-BUTOXYETHANOL AQUEOUS SOLUTIONS NEAR THE CRITICAL REGION

G.D'Arrigo and A.Paparelli
Dipartimento di Energetica, Università di Roma, Via A.Scarpa 14, I-00161 Rome, Italy

ABSTRACT

A detailed ultrasonic investigation in 2-Butoxyethanol aqueous solutions shows that peculiar relaxation processes associated to molecular aggregates give rise to contributions to the sound absorption which are maximal in the proximity of the critical concentration and noticeably increase with decreasing temperature. However, near the critical point these effects are still so large to sensibly modify the ultrasonic attenuation characteristic trends usually found in other binary critical mixtures. We related the apparent negligible role of the critical sound attenuation near the critical point to the unusual small values of the characteristic relaxation rates of critical fluctuations.

INTRODUCTION

Simple monohydric alcohols (C_nE_0) and n-alkoxyethanols (C_nE_1) can be regarded as short-chained members of the polyoxyethylene glycol monoethers series $C_nH_{2n+1}(OCH_2CH_2)_mOH$ (usually denoted as C_nE_m) which, for large m and n, are well known nonionic surfactants. Among the homologous C_nE_1 class, butoxyethanol (C_4E_1) seems to possess the minimal length of the alkyl hydrophobic chain to form ordered aggregates. Such finding is supported by the eccentric behaviour of several physical and chemical properties as well as by the phase diagram (i.e. closed loop of miscibility in the water-rich region of concentration) which are quite similar to those observed in aqueous solutions with long-chained C_nE_m surfactants.

A direct evidence of large ordered aggregates (micelle-like structures) in C_4E_1/H_2O solutions has been recently obtained by small angle neutron scattering (SANS) [1,2] and light scattering experiments [3,4] as well as by ultrasonic spectroscopy [5].

The main purpose of this work is to analyze the influence of the molecular aggregates on the behaviour of the ultrasonic absorption in the proximity of the critical region of C_4E_1/H_2O solutions.

EXPERIMENTAL RESULTS

The low temperature portion of the miscibility loop of C_4E_1/H_2O solutions is very flat. Recent accurate determinations of the critical parameters [4,6] allow to locate the lower critical point at $T_c=49.20\pm0.06°C$ and $X_c=0.055\pm0.004$ (C_4E_1 molar fraction).

Measurements of the ultrasonic absorption coefficent (α) as a function of temperature and concentration were performed in the frequency (f) range 5-250 MHz by means of standard techniques [5]. The behaviour of the ultrasonic absorption as a function of X, T and f is shown in Fig.1-2. From these figures the following characteristic features emerge:

i) the concentration dependence of α/f^2 at different temperatures (see Fig.1) is quite similar to that found in binary critical mixtures of components of low molecular mass. However, in contrast with these systems, the absorption peaks of C_4E_1/H_2O solutions are located at concentrations X_p (depending on f and T, it is $X_p=0.03$ to 0.04)

which are different from the critical one ($X_C=0.055$) and they decrease as T increases toward T_C. At temperatures close to T_C a secondary peak appears (see Fig.1 and Ref.6);
ii) in general, the sound absorption α/f^2 at $X \approx X_C$ decreases with increasing temperature. However, at low frequencies and near T_C, α/f^2 slightly reverses this trend thus displaying the characteristic behaviour found in other binary critical mixtures;
iii) the system exhibits relaxational effects. In general, these effects are much larger at concentrations near X_P and X_C and increase as T decreases.

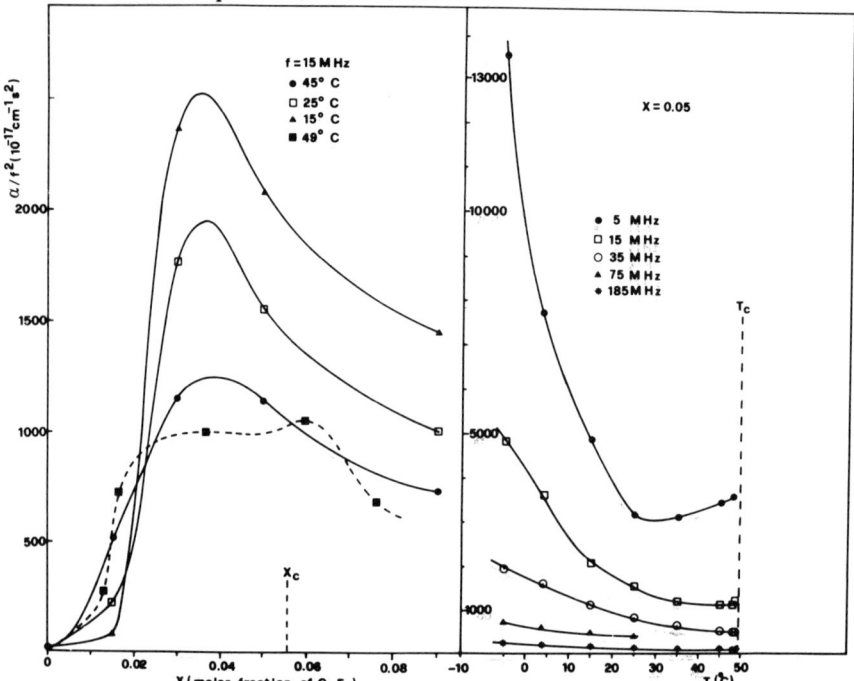

FIG.1. α/f^2 versus concentration for some temperatures (f=15 MHz; dashed line: Ref.6)

FIG.2. α/f^2 versus temperature for some frequencies (X=0.05)

DISCUSSION

The ultrasonic behaviour of binary critical mixtures of components of low molecular mass can be summarized as follows: i) the ultrasonic absorption α/f^2 as a function of concentration (at fixed T and $\omega(=2\pi f)$) displays nearly symmetrical peak values around $X=X_C$. These peaks are enhanced as ω decreases; ii) at $X=X_C$ (and fixed ω), α/f^2 noticeably increases as T approaches the critical temperature T_C. The temperature range of such increase can extend several degrees from T_C; iii) near the critical point α/f^2 exhibits large relaxational effects characterized by a distribution of relaxation times.

Some theoretical models have been proposed to explain these characteristic features which are associated to critical concentration fluctuations. Evaluations from different theoretical approachs can, in general, be expressed by [7]

$$\alpha_{\lambda,\text{crit}}(X,\omega,T) = A(X,T) F(\Omega) \qquad (1)$$

where $\alpha_{\lambda,crit}$ is the sound absorption per wavelength ($\alpha_\lambda = \alpha\lambda$), A is an amplitude and $F(\Omega)$ a scaling function of the reduced frequency $\Omega = \omega/\omega_D$. ω_D is a characteristic relaxation rate associated to concentration fluctuations and given by

$$\omega_D = 2D\xi^{-2} = \omega_0 \varepsilon^{zv} \qquad (2)$$

where D is the mutual mass diffusion coefficient, ξ the correlation length, z(=3.06) and v (=0.63) critical indices, $\varepsilon = (T-T_c)/T_c$ the reduced temperature and ω_0 a system dependent characteristic rate. There are some differences among the A's and F's obtained from different models but they, in general, well account for the main characteristic features of sound propagation in binary critical mixtures of low molecular mass.

However, in some critical mixtures with water as one component (e.g. isobutyric acid/H_2O [8] and triethylamine/H_2O [9] the critical contribution is well characterized but there are additional non critical (chemical) relaxation effects which make difficult to unambiguosly disentangle the two contributions.

Among the critical aqueous solutions, our C_4E_1/H_2O system seems to represent a special case. The temperature trend shown in Figs. 1,2 clearly indicate that the ultrasonic absorption is dominated by non critical effects which near the critical point are still so large to mask the expected critical behaviour. We attempted for separating the two contributions in order to gain some insight on the large non critical effects. For this aim we write

$$\alpha/f^2 (X,\omega,T) = (\alpha/f^2)_{crit} (X,\omega,T) + B(X,\omega,T) + B_{NS} (X,T) \qquad (3)$$

where B is a non critical contribution (background) and B_{NS} the classical Navier-Stockes term due to non relaxing shear and bulk viscosities. It turns out that B_{NS} is negligible in our system. To evaluate $(\alpha/f^2)_{crit}$ we considered the dynamical scaling theory by Ferrell-Bhattacharjie [10] where at $X = X_c$ this term can be written in the form

$$(\alpha/f^2)_{crit} = \text{const } \varepsilon^{-(\alpha^*+zv)} \, \Omega^{-\alpha^*/zv}/(1+\Omega^{1/2})^2 \qquad (4)$$

where $\alpha^* = 0.06$. In particular, Eq.(4) predicts:

$$(\alpha/f^2)_{crit}(T) \approx (1+K\varepsilon)^{-2} \quad ; \quad (\alpha/f^2)_{crit}(\omega,T_c) = \text{const } \omega^{-(1+\alpha^*/zv)} \qquad (5\text{-}6)$$

By comparing the experimental temperature dependence of α/f^2 with Eqs.(5) and (3) we find, of course, that B is temperature dependent with a trend (B≈1/T) which is opposite to $(\alpha/f^2)_{crit}(T)$. We also compared the experimental frequency dependence of α/f^2 at $X = X_c$ for $T = T_c$ and $T \neq T_c$ with the predictions of Eqs.(6) and (4), respectively. We find that these equations are inadequate to represent the observed spectra.

Rather surprisingly, however, we empirically found that α/f^2 spectra can be represented in the entire temperature range (-5 to 49 °C) and for compositions $0.03 \leq X \leq 0.09$ by a Debye two-relaxation times equation

$$\alpha/f^2 = A_1/(1+\omega^2\tau_1^2) + A_2/(1+\omega^2\tau_2^2) \qquad (7)$$

Negligible deviations from Eq.(7) are observed only very near to T_c. The fitted relaxation rates $f_i=(2\pi\tau_i)^{-1}$ depend smoothly on T and X, f_1 ranging from 38 to 73 MHz and f_2 from 3 to 16 MHz. We compared these frequencies with the characteristic relaxation rates $f_D(=\omega_D/2\pi)$ of concentration fluctuations given by Eq.(2). We evaluated f_D in two different ways. In the first case we used the value $\eta_0(= 2 \cdot 10^{-3}$ Pa s) and $\xi_0(=4.4$ Å) given in the literature [11,6] to get $f_0(=\omega_0/2\pi)=0.44$ GHz. Alternatively, we used independent D and ξ data [1-4]. The f_D values from the two evaluations were of the same order of magnitude. In particular we get: $f_D\sim 6$ kHz at 48.2°C ($\Delta T=1$°C); ~ 0.2 MHZ at 45.2 °C ($\Delta T=4.2$ °C); ~ 3 MHz at 25.2°C ($\Delta T=24.2$°C). These values are small compared to those found in other critical mixtures at the same ΔT. This finding is a consequence of the small f_0 value (~ 0.5 GHz) compared to 5-20 GHz in other systems or, alternatively, of the large value of the short range correlation length ξ_0 (=4.4 Å) compared to ~ 2Å. So we find that near the critical point the observed relaxation rates f_1 and f_2 are much larger than f_D and they do not scale with T as f_D does. Thus we must conclude that f_1 and f_2 are associated to the background rather than to the critical fluctuations. The relaxation rates f_D lie below the experimental frequency range(5 to 250 MHz): under these conditions the critical term $(\alpha/f^2)_{crit}$ in Eq.(3) should produce a small contribution to the observed absorption even near T_c. The small increase of α/f^2 as T approaches T_c at low frequencies (see Fig.2) can then be due to the high frequency tail of the relaxation spectra of critical fluctuations.

The unusual low value of ω_0 is likeky related to the existence of molecular aggregates even near the critical region as already suggested by other experiments .The presence of these aggregates can also explain the large background contribution as due to discrete structural relaxation processes associated to peculiar dynamic equilibria [5].

The ultrasonic properties of aqueous micellar solutions with nonionic surfactans C_6E_3, C_6E_5 and C_8E_6 [12] display a behaviour quite similar to that in our system, even near the critical points. This analogy support that the molecular aggregates in C_4E_1/H_2O are micelle-like.

REFERENCES

1. G.D'Arrigo and J.Teixeira,J.Chem.Soc.Faraday Trans.86,1503 (1990).
2. G.D'Arrigo,J.Teixeira,R.Giordano and F.Mallamace,J.Chem.Phys.95,2732 (1991)
3. C.Bender and R.Pecora,J.Phys.Chem. 92, 1675 (1988).
4. G.D'Arrigo,F.Mallamace,N.Micali,A.Papareli,J.Teixeira and C.Vasi,Progr.Colloid Polymer Sci. 84, (1991).
5. G.D'Arrigo et al. ,J.Chem.Phys. 91,2587(1989);Phys.Rev.A44,2578(1991).
6. C.Baaken,L.Belkoura,S.Fusenig,T.Muller-Kirschbaum and D.Woermann,Ber. Bunsenges.Phys.Chem. 94,150 (1990).
7. C.W.Garland and G.Sanchez,J.Chem.Phys. 79,3090 (1983).
8. L.Belkoura,V.Calenbuhr,T.Muller-Kirschbaum and D.Woermann,Ber.Bunsenges. Phys Chem.94 ,1471(1990).
9. C.W.Garland and C.Lai,J.Chem.Phys. 69, 1342 (1978).
10. R.A.Ferrell andJ.Bhattacharjee, Phys.Rev. B24,4095(1981); ibid.A31,1788(1985).
11. H.Hamano,T.Kawazura and N.Kuwahara,J.Chem.Phys.82,2718(1985).
12. A.Borthakur and R.Zana,J.Chem.Phys. 91, 5957 (1987).

ENHANCED EFFECT OF SALTS ON POLYMER TRANSPORT IN STRUCTURED FLOW

Hiroshi MAEDA, Takehiro MASHITA and Shigeo SASAKI
Department of Chemistry, Faculty of Science,
Kyushu University, Fukuoka 812, Japan

ABSTRACT

Transport rate of poly(vinylpyrrolidone) in aqueous dextran matrix was enhanced by the addition of CsCl or NaCl but not by LiCl. Several factors responsible for this enhancement were examined.

When an aqueous solution of dextran is layered over another aqueous solution consisting of dextran of the same concentration and poly(vinylpyrrolidone)(PVP), diffusion of PVP is expected to occur. However, a rapid transport of PVP has been observed when both dextran and PVP concentrations exceed respective critical values[1]. This rapid transport was ascribed to the occurrence of a finger-like structured flow which may be caused by rapid movement of solvent molecules driven[1].

Recently we have found that the transport rate in the presence of 2 mol/dm^3 CsCl is enhanced by about 20 times as much as that in salt-free solutions as shown in Figs. 1-2[2], where Q represents the amount of PVP transported to the upper compartment at 25°C. The experimental detail was described elsewhere[2]. Slopes of the lines in Figs. 1 and 2 give transport rates which are summarized in Table 1.

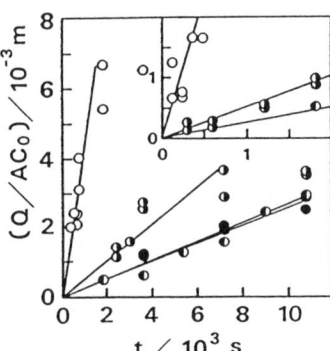
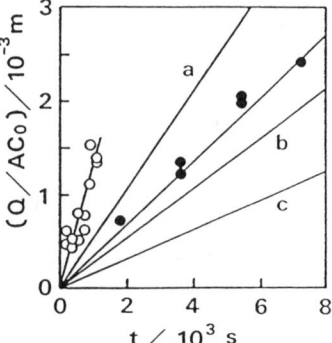

Fig.1. Transport rate in the presence of 2 M salt both in upper and lower compartments.
(○) CsCl, (◐) NaCl, (◓) LiCl and (●) no added salt. Inset: enlarged plot for initial stage($t < 2 \times 10^3$s).

Fig.2. Transport rate in the presence of CsCl at different concentrations.
(○) 0.614 M, (●) 0.2 M. Lines a,b and c represent the results with 2 M NaCl, no added salt and 2 M NaCl(upper)–2 M CsCl(lower).

The enhancement effect of salts revealed in Figs. 1 and 2 could be related to several factors such as the effect on the water structure, the polymer coil dimension and osmotic pressure. Results of viscometric measurements are shown in Table 2. Viscosity of CsCl solution was found to be much lower than the others. Decrease in viscous friction leads to enhacement of relative motion between upward and downward fingers. However, the effect of viscosity alone cannot explain the observed enhacement effect even qualitatively, as evident in the case of NaCl. Fig. 3 shows diffusion coefficients of the blobs of PVP and dextran in salt-free solutions and CsCl solutions, obtained from the fast relaxation mode in the time correlation function of scattered light intensity. The results, after corrected for differences in viscosities, indicate that the diffusion constants of blobs of PVP are smaller in 2M CsCl solutions than in salt-free solutions, while it is not the case for dextran. This fact suggests that the water-structure-breaking effect of CsCl influences mainly the interaction between water and PVP.

Table 1. Salt effect on the flow rate

Conc. (M)	No salt	CsCl	CsCl	CsCl	NaCl	LiCl
Conc. (M)	—	0.2	0.6	2.0	2.0	2.0
Rate(m/sec) ×10^7	2.7	3.4	13	45	5.2	2.6
Ratio	1	1.3	4.8	17	1.9	1

Table 2. Viscosities of salt solutions, PVP(0.1g/dl)-salt solutions and dextran(8g/dl)-salt solutions at 25 °C

	No salt	CsCl(2M)	NaCl(2M)	LiCl(2M)
η_0/η_0(No Salt)	1.0	0.77	1.13	1.21
η_{PVP}(relative)	1.0	0.72	1.10	1.21
η_{dex}(relative)	1.0	0.82	1.22	1.32

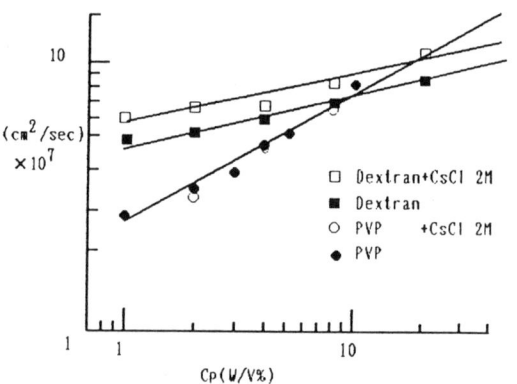

Fig. 3 Concentration dependence of diffusion coefficients of blobs of dextran and PVP

<References>
1) B.N.Preston, T.C.Laurent, W.D.Comper and G.J.Checkley, Nature, **287**, 499 (1980)
2) H.Maeda, T.Mashita and S.Sasaki, Chemistry Lett., 635 (1991)

EXPERIMENTAL STUDIES IN A PHASE-SEPARATING MIXTURE UNDER SHEAR FLOW

K. Hamano, S. Yamashita, K. Kubota, N. Kuwahara

Department of Biological and Chemical Engineering
Faculty of Technology, Gunma University, Kiryu, Japan

J. V. Sengers

Institute for Physical Science and Technology
University of Maryland, College Park, Maryland 20742

ABSTRACT

This work is concerned with critical behaviors in a nonionic micellar solution of *tetra*-ethyleneglycol *n*-decylether ($C_{10}E_4$) in water in the presence of shear flow. We find the measured viscosity can depend weakly on a shear rate S in the form $\eta \propto S^{-\omega}$ with $\omega = 0.021 \pm 0.003$. It is reported that the phase-separation processes induced by decreasing suddenly a shear rate from S to $S = 0$ approximately obey a universal function of spinodal decomposition for a fluid, in conjunction with a S-dependent temperature shift given by $[T_o(S) - T_c]/T_c = \epsilon_0 S^p$ with $\epsilon_0 = (1.3_5 \pm 0.11) \times 10^{-5}$ and $p = 0.51 \pm 0.03$. Here, $T_o(S)$ is estimated experimentally as an onset of significantly anisotropic patterns of the forward scattered light under shear flows. The viscosity behavior in a phase-separating mixtures of isobutyric acid (IB) in water is also reported.

INTRODUCTION

Near a critical point in fluids the system can include large fluctuations characterized by the long-range correlation length $\xi = \xi_0 \epsilon^{-\nu}$ with a universal value of $\nu \simeq 0.63$, and the characteristic time $\tau_\xi = 6\pi\eta\xi^3/k_B T$ of the critical fluctuations becomes extremely large, that would be an origin of shear-induced critical behaviors under shear flows. One of marked results in this region may be that strongly anisotropic light-scattering have been observed for the mixture inside the coexistence curve under shear flows. Extensive theoretical and experimental works associated with shear-induced effects have been reported recently for binary liquid mixtures and polymer mixtures.[1] Besides, amphiphilic compounds possessing in the same molecule two groups which differ greatly in their solubility tends to create micelles as a consequence of their dual nature. Phase separation into two isotropic micelle-poor and micelle-rich phases can be observed by raising temperature in a certain water-rich region, which could be tested from a critical-point universality.[2,3] In this work, we try to examine shear-induced effects for the critical mixtures $C_{10}E_4$ + water and IB + water using a rotational viscometer.

© 1992 American Institute of Physics

EXPERIMENTAL AND RESULTS

The critical mixtures $C_{10}E_4$ + water and IB + water were employed in this work. The $C_{10}E_4$ sample used here was a portion of the sample which had been employed in our previous works.[3] Measurements were performed using a rotational viscometer of a Zimm-Crothers type with aid of light-scattering observation from the mixture between two vertical coaxial clinders in the range from $S \simeq 0.6$ s^{-1} to 36 s^{-1}.

When a critical mixture is brought into the unstable region inside the coexistence curve, phase separation occurs with appearence of a characteristic ring indicating spinodal decomposition.[4] This can shrink in diameter with time. This is a general feature of a phase-separating critically binary mixture. In the presence of shear flows a significantly anisotropic scattering with sharp streak perpendicular to the flow direction was observed when a certain temperature $T_o(S)$ inside the coexistence curve was approached. We find a dimensionless shift of temperature $\epsilon_s(S) = |T_o(S) - T_c|/T_c$ obeys a power-low of $\epsilon_s(S) \propto S^p$ with $p = 0.51 \pm 0.03$ for $C_{10}E_4$ + water and $p = 0.52 \pm 0.04$ for IB + water, in good agreement with the theoretical estimation of Onuki and Kawasaki.

Non-Newtonian effects in the viscosity are examined for $C_{10}E_4$ + water in a strong shear-range characterized by $S\tau_\xi \gtrsim 1$. In this region the measured viscosity in a phase-separating mixture appeared to reach at a non-equilibrium stationary state, giving rise to $\eta \propto S^{-\omega}$ with $\omega \simeq 0.02$ independent of temperature in the region of $T_c \lesssim T \lesssim T_o(S)$. We examined tentatively the viscosity behavior for $T \gtrsim T_o(S)$, which varied approximately as $\eta \propto S^{\omega^*}$ with $\omega^* \simeq 0.02$.

The viscosity behaviors for a phase-separating mixture of IB + water have been examined by a temperature ramp and a pressure quench. Most interesting result is a significant enhancement of about 40% in the measured viscosity for the mixture with a quench depth of 0.98 bar under $S \simeq 0.2$ s^{-1}.

REFERENCES

1. See, e.g., A. Onuki and K. Kawasaki, Ann. Phys. (N.Y.) **121**, 456 (1979); A. Onuki, Phys. Rev. A **35**, 5149 (1987); T. Ohta, H. Nozaki, and M. Doi, J. Chem. Phys. **93** 2664 (1990); C. K. Chan, F. Perrot, and D. Beysens, Phys. Rev. A **43**, 1826 (1991); T. Hashimoto, T. Takebe, and S. Suehiro, J. Chem. Phys. **88**, 5874 (1988).

2. See, e.g., M. Corti, C. Minero, and V. Degiorgio, J. Phys. Chem. 88, 309 (1984).

3. K. Hamano, N. Kuwahara, K. Kubota, and I. Mitsushima, Phys. Rev. A **43**, 6881 (1991).

4. J. S. Huang, W. I. Goldburg, and A. W. Bjerkaas, Phys. Rev. Lett. **32**, 921 (1974).

PRESSURE DEPENDENCE OF THE FERROELECTRIC SOFT MODE OF KDP

T. Yagi
Res. Inst. Appl. Electricity, Hokkaido University, Sapporo 060
A. Sakai
Dept. Elect. Engineering, Muroran Inst. of Technology, Muroran 050
M. Arima
Chichibu Res. Laboratory, Showa Silicon K.K., Chichibu 369-18

ABSTRACT

This study raises a question on the displacive model of the phase transition mechanism of ferroelectric KDP, though it has been widely accepted: Raman scattering spectra of the $B_2(z)$ mode were observed under high pressure using a diamond anvil cell in the two different scattering angles. Analysis of the spectra in terms of the coupled soft phonon mechanism of the displacive model derives an unreasonable dispersion relation, where the frequency of the "soft" mode increase near Γ point. The displacive model, which is essentially accompanied by a soft phonon at the Γ point, should be reconsidered.

INTRODUCTION

The phase transition of ferroelectric KH_2PO_4 (KDP), which is one of the most typical ferroelectrics with the hydrogen bonds, has been studied intensively by many researchers since Slater proposed a simple theoretical model in 1941. In particular, the light scattering studies have been done intensively because the crystal is quite adequate to the experiment by its transparency for the visible light. The low-frequency component appeared as an intense wing centered at $\omega = 0$ has been attributed to the ferroelectric soft mode. In addition to this, the pressure dependent Raman scattering experiment revealed a gradual change of the component into an oscillatory peak with increasing hydrostatic pressure.[1] These results have been considered as a strong evidence for the displacive model of the phase transition mechanism associated with a soft phonon.

In the present study, the dispersion relation of the soft mode of KDP has been investigated by the Raman scattering under hydrostatic pressure using a diamond anvil cell at room temperature in the two geometries, 90 degree and the forward scattering geometries.

EXPERIMENT, RESULTS AND DISCUSSION

Two types of the diamond anvil cell, one is for the 90° scattering and the other for the forward scattering, were manufactured in the present study. The $B_2(z)$ spectra were observed in the two geometries, x(yx)y geometry using the 90° scattering cell and z(yx)z+δ geometry using the forward scattering cell.

The low frequency components below 300 cm^{-1} of the x(yx)y and z(yx)z+δ spectra were observed.[2] With increase of pressure, the phonon peak of the spectrum got clear indicating decrease of its damping factor. The spectra were analyzed by the coupled-phonon model which gave the spectral shape $I(\omega)$ by the soft- and hard mode frequency ω_- and ω_+. The both frequencies are given as

$$\omega_\pm = [(\omega_a^2 + \omega_b^2)/2 \pm \{[(\omega_b^2 - \omega_a^2)/2]^2 + \Delta^4\}^{1/2}]^{1/2}, \quad (1)$$

where ω_a and ω_b are the frequencies of two modes a and b coupled with each other through the coupling constant Δ. The fitting spectra gave finally the pressure dependence of ω_- and ω_+ in both geometries as shown in Fig.1. The pressure dependences of ω_- and ω_+ show a satisfactorily good agreement with the previous result as denoted by straight solid lines therein.[1]

However, ω_- in the forward scattering geometry is apparently larger than ω_- in the 90° scattering geometry as seen in the figure. The ferroelectric soft mode has its q at Γ point in the Brillouin zone. The present result, as shown in Fig.2, gives us somewhat curious dispersion relation. The q dependence of ω_+ is also contradict to the normal dispersion relation of the soft ferroelectric mode. Here we can conclude that the low-frequency component of Raman spectra is not caused by the ferroelectric soft mode. The displacive model of KDP loses experimentally its direct evidence. Another physical origin of the spectra should be discussed in order to elucidate the phase transition mechanism of KDP.

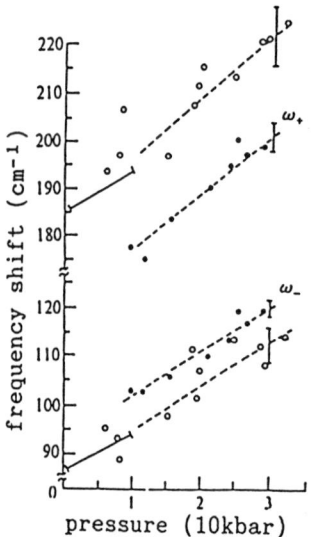
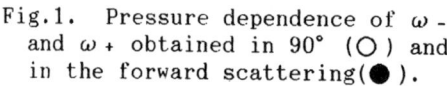
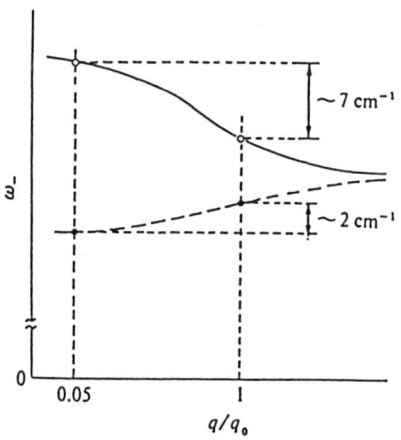

Fig.1. Pressure dependence of ω_- and ω_+ obtained in 90° (○) and in the forward scattering (●).

Fig.2. Dispersion relation of the soft modes obtained in the present study (○) and in a normal case (●).

REFERENCES

1. P.S.Peercy, Phys.Rev.Lett. <u>31</u>, 379 (1973), Phys.Rev. <u>B12</u>, 2725(1975).
2. T.Yagi, M.Arima and A.Sakai, J.Phys.Soc.Jpn. <u>59</u>, 1430(1990), Phase Transitions <u>28</u>, 63 (1990).

DIELECTRIC DISPERSION ASSOCIATED WITH THE DC-ELECTRIC-FIELD ENFORCED FERROELECTRIC PHASE TRANSITION IN ANTIFERROELECTRICS

Naohiko Yasuda
Electrical Engineering Department, Gifu University, Gifu 501-11, Japan

ABSTRACT

The strong dielectric dispersion along the b-axis associated with the field-enforced ferroelectric phase transition was observed. The slowing down of the relaxation time toward a critical electric field Ecr was observed. The frequency dependence of the complex permittivity obeys the Cole-Cole arc law. With increasing dc-electric fields, the dielectric dispersion strength increases gradually, rapidly near Ecr in the antiferroelectric phase and then decreases in the ferroelectric one.

INTRODUCTION

In many ferroelectrics and antiferroelectrics, it is known that as the transition point for either the paraelectric-ferroelectric or paraelectric-antiferroelectric phase transition is approached, the dielectric relaxation time becomes longer, i.e., a critical slowing down of the dielectric relaxation time[1], whereas little has been reported on the dielectric dispersion for the antiferroelectric-to-ferroelectric phase transition induced by a dc-electric field[2,3].

In this work, we investigated the dc-electric field dependence of the real ε_r' and imaginary ε_r'' parts of complex relative permittivity under various frequencies accompanied by the dipole reversal for the field-enforced ferroelectric phase transition in the pressure-induced antiferroelectric cesium dihydrogen phosphate(CDP) and antiferroelectric cupric formate tetrahydrate(CFT).

EXPERIMENTAL RESULTS AND DISCUSSION

The strong dielectric dispersion along the b-axis associated with the field-enforced ferroelectric phase transition is observed. The dc-electric field has a marked effect on the dielectric relaxation. The slowing down of the relaxation time τ toward a critical electric field Ecr, which causes the change from the antiferroelectric to the ferroelectric phase, is observed as seen in Fig.1. With increasing applied dc fields Ed, the dielectric dispersion strength increases gradually, rapidly near Ecr in the antiferroelectric phase, and decreases in the ferroelectric one. The increase in ε_r' at low frequencies near Ecr in the antiferroelectric phase and its decrease in the ferroelectric one with an increase in Ed are due to the nonlinear and the saturation effects on P against E, respectively, as found phenomenologically from the slope of P against E on the polarization P-electric field E double hysteresis loop[4]. Data points(ε_r', ε_r'') at the given dc field in the Cole-Cole diagram lie nearly on a circular arc, and the centers of all arcs lie on the straight line below the real axis. Thus, the frequency dependence of ε_r' and ε_r'' obey the Cole-Cole arc law[5]. The circular arc gets larger in the antiferroelectric phase and smaller in the ferroelectric one with increasing Ed. The parameter β, representing a measure of width of the distribution of dielectric relaxation times, is about 0.85 in both phases for CDP and 0.51 in both phases for CFT. This indicates the dielectric dispersion of the polydispersive[4] type. The relaxation time is of the order of 10^{-5} sec for CDP and 10^{-4} sec for CFT near Ecr in the antiferroelectric phase. It is interesting that the dielectric dispersion associated with the field-enforced ferroelectric phase transition occurs at such low frequencies, in contrast to that occurred at high frequencies in the case of a critical slowing down of the dielectric relaxation time(for example,10^{-8} sec for CDP[6]).

In many ferroelectrics and antiferroelectrics, the dielectric relaxation time has been known to be proportional to the static relative permittivity[7]. Judging from the results mentioned above for τ and the dielectric dispersion strength as a function of Ed, Mason's theory[7] may also be applicable to the dielectric dispersion associated with the field-enforced ferroelectric transition.

CONCLUSION

The strong dielectric dispersion along the b-axis associated with the field-enforced ferroelectric transition was observed. The slowing down of the relaxation time toward a critical electric field was seen.

We wish to thank J.Kawai and R.Suzuki for their help with experiments. This work was supported by a grant from Tokai Science Academy.

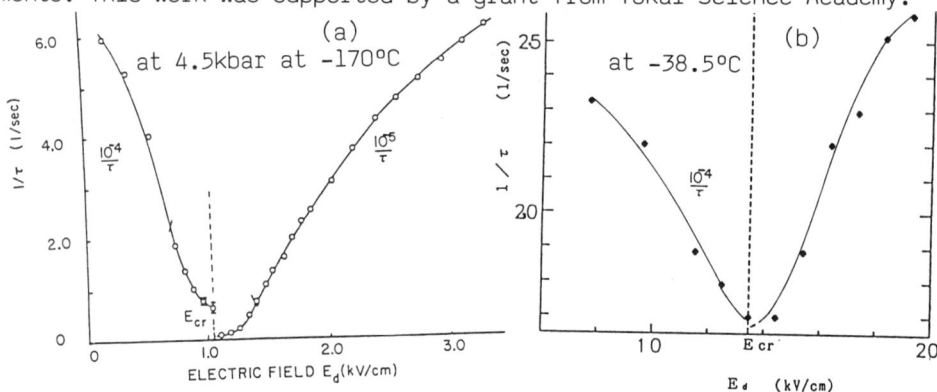

Fig.1. $1/\tau$ vs. Ed for (a) CDP and (b) CFT.

REFERENCES

1. Ferro- and Antiferroelectric Substances,Landolt-Bornstein,New Series Vol.16, edited by T.Mitsui(Springer-Verlag,Berlin 1982).
2. N.Yasuda and J.Kawai,Phys.Rev.B 42,4893(1990).
3. N.Yasuda and R.Suzuki,7th Europ.Meet.Ferro.,(Dijon,1991)P1,p.357.
4. Y.Ishibashi,J.Phys.Soc.Jpn.,56,4408(1987).
5. K.S.Cole and R.H.Cole,J.Chem.Phys.,9,841(1941).
6. K.Deguchi,E.Nakamura and F.Okaue,J.Phys.Soc.Jpn.,53,1160(1984).
7. W.P.Mason,Phys.Rev.,72,854(1947).

LIGHT SCATTERING AND TURBIDIMETRIC STUDY
OF GELLING TUNGSTIC ACID

K. Hara, H. Kanaya, H. Okabe and K. Matsushige
Department of Applied Science, Faculty of Engineering 36,
Kyushu University, Hakozaki, Higashi-ku, Fukuoka 812 JAPAN

ABSTRACT

Measurements of the time-dependences of the (static) light scattering intensity with a monochromatic light source (Ar^+ laser) and the turbidimetric ones with a multi-wavelength light source (Xe-lamp) were carried out during the gelation process of tungstic acid. In early stage of the gelation, the characteristic features due to the growing clusters were observed. In the later stage, namely around and after the gelation time, the remarkable differences with the results of the early stage were detected both in the angular dependence of the scattered light intensity and in the wavelength-dependence of the scattering cross section.

INTRODUCTION

Transition metal oxide gels, such as vanadium oxide gel, attract some attention because of various characteristic properties. Tungstic oxide gel, which belongs to the family, is known as an electrochromic material, therefore, the applications to display and printing are proposed.[1] However, the studies center mostly on the properties of the "grown gel", and the formation process have been scarcely investigated in spite of its serious influence on the products. In the present paper, at this point of view, we report on the evolution of the light scattering and turbidimetric properties of tungstic acid in the gelation process.

EXPERIMENTALS AND RESULTS

First, the time dependence of the VV-scattered light intensity at several angles was measured. In the measurements, the polarized light beam from a He-Ne laser (5mW, 632.8nm) was used as the incident light. The scattered light from the sample at each angle was detected by a photodiode.[2] During the gelation process, aniosotropic increase in the scattered light intensity was observed, and, in the later stage of the gelation process, remarkable increase in the intensity in the backward-scattering region was detected (Fig.1). The characteristic features also appeared in the 2-dimensional distribution of the forward-scattered light intensity. In the measurements, the 488nm-line of the light-power controlled Ar^+ laser was utilized as a light source. The sample cell was 1mm in thickness. In the observation, increase in the intensity of the depolarized light (VH geometry) in the diagonal directions, and that of the polarized light (VV geometry) in the direction parallel to the scattering plane was detected. Furthermore, in the measurements of the transmitted light spectra through a 10mm-thick cuvette with a multi-wavelength light source (Xe-lamp),[3] approximately linear dependence of the logarithm of the scattering cross section on ln λ was observed. The slope of the linear dependence became more gentle as time passed, and showed some anomaly in the time dependence around the gelation time.

DISCUSSION

It has been reported that, in the early stage of the gelation process of the tungstic acid, the growth of small clusters proceeds due to the condensation of the hydrates of tungsten oxide, and the gelation occurs by the crosslink of the clusters.[4] The features observed in the early stage of the gelation process in the present measurements can be well explained by the increase of the internal interference of the scattered lights from growing clusters.[2] While, the features in the later stage showed some contrasts to those in the early stage. In the later stage of the gelation, namely around and after the gelation time, the backward-scattered light intensity increased compared with that of the forward scattering, and the logarithm of the scattering cross section showed somewhat curved dependence on ln λ compared with that in the early stage. Beside, the time-dependence of the slope showed some anomaly around the gelation time. Though it seems difficult to make clear explanation without the data on the structure, it can be said that, judging from the occurrence of the anomalies around the gelation time, these features may indicate the alteration of the light scattering mechanism, related to the network formation in the later stage of the gelation process.

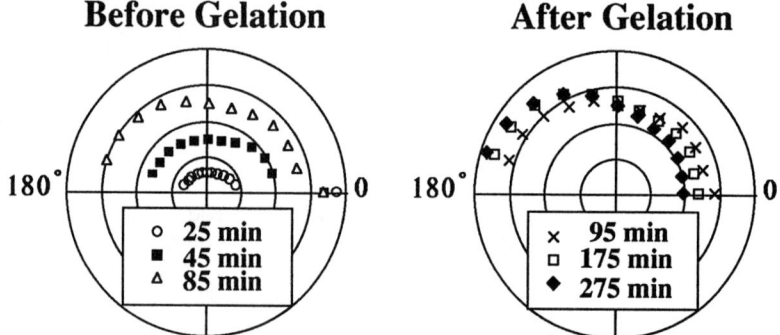

Fig.1 Polar plots of the angular dependence of the scattered light intensity. The gelation occurred at 90 minutes after the ion-exchange (examined by the tilting test).

REFERENCES

1. J. Livage and J. Lemerle, Ann. Rev. Mater. Sci. <u>12</u>, 103 (1982).
2. K. Hara and Y. Ishibashi, J. Phys. Soc. Jpn. <u>57</u>, 3838 (1988).
3. K. Hara, H. Kanaya, H. Okabe and K. Matsushige, J. Phys. Soc. Jpn. <u>60</u>, (1991) (in press).
4. J. Livage, J. Solid State Chem. <u>64</u>, 322 (1986).

X-RAY STUDIES ON STRUCTURES AND PHASE TRANSITION IN EVAPORATED FILMS OF LIQUID CRYSTALS

Y. Yoshida, T. Horiuchi and K. Matsushige
Department of Applied Science, Faculty of Engineering
Kyusyu University, Hakozaki, Higashi-ku, Fukuoka 812 Japan

ABSTRACT

The molecular orientation in thin evaporated films were investigated for the newly synthesized organic material which has a large dipole moment parallel to molecular long axis. Results of X-ray diffraction measurements revealed that this molecule forms the layer structure normal to the substrate. Further, the drastic changes in the X-ray diffraction pattern were observed during a heating process at the temperature ($80^\circ C$) where the phase transition to a liquid crystal phase (smectic A) takes place. Moreover, we elucidated molecular structures in crystalline phase and smectic A by employing X-ray diffraction and the other measurements.

INTRODUCTION

Recently we have tried to control the molecular orientation in order to prepare ferroelectric ultrathin films. In this study we used a new organic sample as an evaporating source, which is of low molecular weight and has a large dipole moment in the structural unit, expecting to get thin films with high ordering by introducing some controlling procedures during the evaporating process.

EXPERIMENTAL

The sample is a newly synthesized organic material, 5-(p-dodecyloxyphenyl)pyrazine-2-carbonitrile, called here as DOPPC, which was kindly supplied by Taniguchi laboratory of Kyusyu university, and has a large dipole moment parallel to molecular long axis. Hence this sample is expected to exhibit several interesting functional properties such as nonlinear optical effect. Further, it is liquid crystal and exhibits smectic A phase with the layer structure.

A vacuum evaporation was performed at 1×10^{-5} Pa and the substrate temperature of room temperature. The evaporation rate and thickness, which were measured by a quartz crystal oscillator, were 0.2 nm/sec and 100 nm. In this experiment we employed three kinds of substrates, namely SiO_2, Cu evaporated on SiO_2 and (001)CaF_2, which were polished in optical-flat level or freshly cleaved. Then, we examined the crystal structure and orientation in the evaporated thin films by the energy dispersive total reflection X-ray diffractometer.[1] Further, in order to investigate the phase transition behaviour in thin films, the annealing cell was mounted on the sample holder.

RESULTS & DISCUSSION

The symmetrical X-ray diffraction profile from the thin
evaporated on the SiO_2 substrate showed the existence of one sl
peak corresponding to the spacing of 1.76 nm, which implys the l
structure normal to the substrate.(Fig.1)

Moreover, the behavior of phase transition during a heat
process was studied. As shown in Fig.2, the X-ray diffract
profilesexhibited that the layer spacing changes at 80°C, and
peak appears, implying the formation of the new layer structure
the occurrence of phase transition from crystal to liquid crys
(smectic A). The value of the smectic layer spacing provides
important information to determine the molecular structure in
smectic phase. Consequently, we considered the structure such t
two molecules form the aggregates as a unit (dimer), and
molecular cores overlap in the smectic layer.(Fig.3) This model
similar to that one proposed for the structure of smectic A of
compound with cyano-biphenyl groups, named as the semibila
structure.[2] Certainly, further structural evaluation by X-ray,
IR, and others are necessary to be done, and some experiments
now in progress.

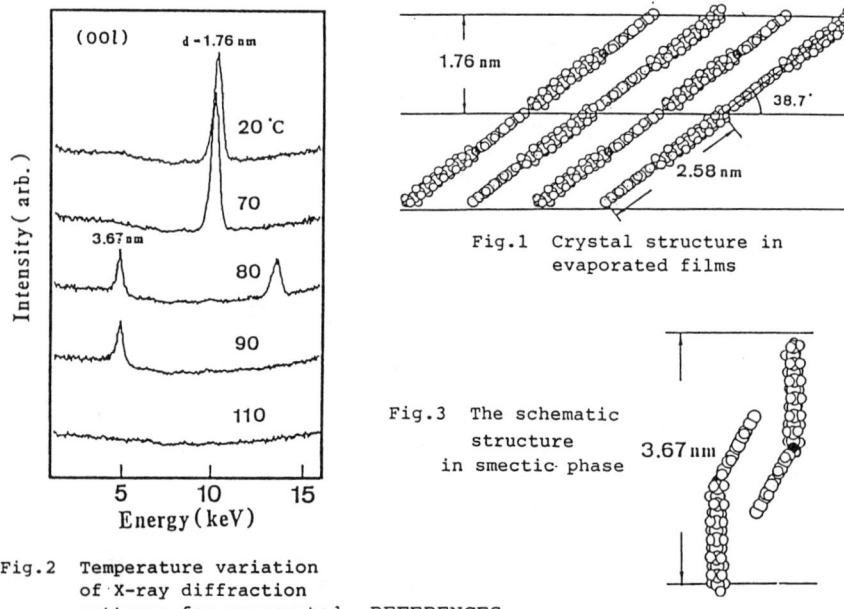

Fig.1 Crystal structure in evaporated films

Fig.3 The schematic structure in smectic phase

Fig.2 Temperature variation of X-ray diffraction patterns for evaporated films

REFERENCES

1. T. Horiuchi, K. Fukao and K. Matsushige, Jpn. J. Appl. Phys. L1839 (1987)
2. A. J. Leadbetter, J. C. Frost, J. P. Gaughan, G. W. Gray and A. Mosley, J. Physique **40**, 375 (1979)

AN EXPLOSIVE HYDRODYNAMIC FLOW INDUCED BY PHASE DIFFUSION WAVE IN THE BZ-SOLUTION LAYER

H.Miike and H.Yamamoto
Yamaguchi University, Tokiwadai 2557, Ube 755, Japan

Abstract: Hydrodynamics accompanied with chemical wave propagation is studied extensively. In an oscillatory BZ-solution layer having very long oscillation period, a spatio-temporal phase field is realized after traveling a circular wave which is triggered from a edge of the Petri dish. An explosive hydrodynamic flow is induced by a phase diffusion wave which is excited spontaneously by memory effects of the solution layer.

Introduction: Pattern formation due to the coupling between reaction-diffusion and hydrodynamic flow has recently been attracted increasing attention[1-3]. Examples are "mosaic" structures[1], chaotic or turbulent structures[2], and the deformation and irregular decomposition of chemical waves[3] observed in an uncovered layer of the unstirred Belousov-Zhabotinsky (BZ) solution. Various attempts have been made to explain the origin of the structures by the effects of convective flow caused by evaporative cooling and/or the exothermicity of the reaction[4]. Recently, one of our author and his collaborate have carried out direct and quantitative measurements with respect to the role of hydrodynamic flows in such patterns[5-7]. We introduced two-dimensional velocimetry and spectrophotometry based on microscope video imaging techniques and found that chemical wave propagations induce hydrodynamic flows. Two remarkable phenomena were detected even when the solution layer was covered with a glass plate, thus suppressing evaporation: 1) a traveling convective structures propagating with circular waves in a chemically well reduced, excitable solution layer[6], 2) an oscillatory hydrodynamic flow under the condition of periodic passage of wave trains emanating from a rotating spiral center[7]. Possible mechanisms explaining these hydrodynamic instabilities have not been well established yet.

In this report, we focus on hydrodynamics induced by a phase diffusion wave which propagates in an oscillatory Belousov-Zhabotinsky (BZ) solution layer having inhomogeneous phase distribution.

Materials and Methods: An oscillatory solution of the BZ-reaction was obtained by preparing a mixture of 48 mM NaBr, 340 mM $NaBrO_3$, 95 mM $CH_2(COOH)_2$, and 378 mM H_2SO_4. About 5 min after mixing, the catalyst and indicator ferroin (3.5 mM) was added. The solution is normally considered as an excitable medium, however, it has a long oscillation period (about 10–15 min) in a Petri dish (diameter 7 cm) at $25\pm1\ °C$. The depth of the solution layer was about 0.9 mm. For measurement of hydrodynamic flow, polystyrene particles (diameter 0.48 μm) serving as scattering centers were mixed and illuminated by He–Ne laser light (632.8 nm). Chemical activity was monitored quantitatively by CCD TV–camera with 490 nm optical band pass filter.

Experimental Results: To realize an inhomogeneous phase field, a trigger wave was excited by silver wire in the solution layer with a covered liquid/gas interface from an edge of the dish. During the wave propagation, initiation of other waves was suppressed. It takes about 5 min to reach the wave front to the counter edge of the dish. Then, after the first wave propagation, spatio-temporal phase gradient of

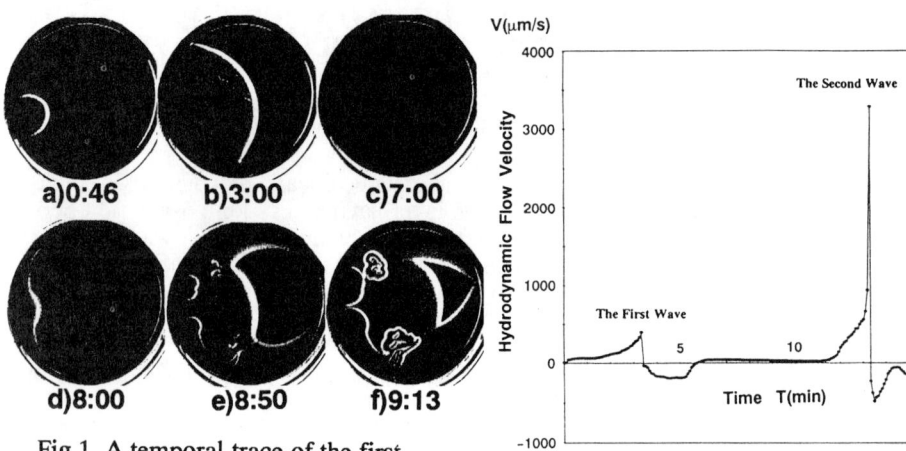

Fig.1 A temporal trace of the first wave (a–c) and the second wave (d–f) propagation.

Fig.2 Hydeodynamic flow velocity induced by trigger wave and phase diffusion wave.

the oscillatory BZ-reaction is established. About 8 min after triggering the first wave, the second wave is induced spontaneously at around the triggered position of the first wave. Memory of the first phase difference is hold and diffuses somehow, then the phase diffusion wave propagates as suggested by Bodet et al.[8]. A temporal trace of the first and the second wave propagation is shown in Fig.1(a)–(f). The characteristics of the second wave are: 1) broad wave front, 2)rapid propagation velocity, and 3)accompanying decompositions of the wave front.

On the other hand, the measurement of the hydrodynamic flow is carried out at the center of the dish during the wave propagations. In Fig.2, a time trace of the flow velocity measured near the surface of the solution is shown. In the second wave (phase diffusion wave), the flow velocity develops extremely and mostly exceeds propagation velocity of the wave. This explosive flow may bring the decomposition of the wave front.

Concluding Remarks: In summary interesting new findings are: 1) A phase diffusion wave is induced spontaneously in the BZ-solution layer after propagation of the trigger wave, 2) An explosive hydrodynamic flow accompanied with the phase diffusion wave is observed, 3) The developed flow brings decomposition of the wave front, thus brings spontaneous pattern formation of spiral waves.

The mechanisms of the hydrodynamic instability and quantitative analysis of the wave propagation are now in progress.

References
1) M.Orban, J.Am.Chem.Soc.,**102**(1980) 4311.
2) K.I.Agladge, V.I.Krinsky and A.M.Pertsov, Nature,**308**(1984) 834.
3) M.Markus, S.C.Müller, Th.Plesser and B.Hess, Biol.Cybern., **57**(1987) 187.
4) C.Vidal and P.Hanusse, Int.Rev.Phys.Chem.,**5**(1986) 1.
5) H.Miike, S.C.Müller and B.Hess, Chem.Phys.Lett., **144**(1988) 515.
6) H.Miike, S.C.Müller and B.Hess, Phys.Rev.Lett., **61**(1988) 2109.
7) H.Miike, S.C.Müller and B.Hess, Physics Letters A, **141**(1989) 25.
8) J.M.Bodet, J.Ross and C.Vidal, J.Chem.Phys.,**86**(1987) 4418.

DYNAMICAL PROPERTY OF FERROELECTRIC DOMAIN WALL NEAR THE CURIE POINT

K. Hamano, H. Sakata and J. Zhang
Department of Physics, Tokyo Institute of Technology,
O-okayama, Meguro, Tokyo 152, Japan

ABSTRACT

In almost all ferroelectrics, domain wall (DW) motion under an ac field gives rise to a low frequency dielectric dispersion of a relaxational type. We investigate theoretically the temperature change in the motion of a pinned DW, particularly near the critical point T_c, and discuss the experimental results on $NaNO_2$ and TGS.

Let $P(x,\xi)$ represent the profile of the DW displaced by ξ from its equilibrium position $x=0$. The equation for its relaxational motion is written as

$$\gamma(d\xi/dt) + \kappa\xi = E\int (\partial P(x,\xi)/\partial\xi)_{\xi=0}\, dx, \qquad (1)$$

where γ is the damping constant, κ is the pinning force constant and E is the external field. The right hand side expresses the external driving force for DW motion. From this equation the ac electric susceptibility due to DW motion is obtained as

$$\chi(\omega) = \chi_o/(1+i\omega\tau), \qquad (2)$$

where $\chi_o = [\int (\partial P/\partial\xi)_{\xi=0}\, dx]^2/\kappa$ and $\tau = \gamma/\kappa$.

We derive the expression for $P(x,\xi)$, κ and γ in the above equations on the assumption that the free energy functional of the DW is given by

$$F(P) = \int dx\left[\frac{1}{2}\left(\frac{dP}{dx}\right)^2 - \frac{\theta}{2}P^2 + \frac{g}{24}P^4 + V(x)\left(\frac{dP}{dx}\right)^2 - EP\right], \qquad (3)$$

where θ represents $(T_c - T)/T_c$ and g is a positive constant. It is well known that, if the fourth and fifth terms in the bracket are absent, the above F(P) gives the solution for a DW, $P(x) = P_s \tanh(\sqrt{2\theta}x)$, where P_s is the spontaneous polarization. The fourth term expresses the interaction energy between the DW and impurities. We assume that $V(x) = -V_o/2 < 0$ for $-\ell < x < \ell$, and 0 for $x < -\ell$ and $x > \ell$.

In principle, if F(P) is given, we should be able to calculate the contribution from DW motion to the electric susceptibility by solving the time dependent Ginzburg-Landau equation (TDGL eq.)

$$\partial P/\partial t = -L(\delta F(P)/\delta P), \qquad (4)$$

where L is the kinetic coefficient. However it cannot be done, because the assumed V(x) is discontinuous at $x = -\ell$ and ℓ. Therefore we solve the Euler equation for F(P) in the absence of the field for each of the ranges $x < -\ell$, $-\ell < x < \ell$ and $x > \ell$, and join the solutions continuously at x=±ℓ. The obtained P(x,ξ) is used to calculate $\int (\partial P/\partial \xi)_{\xi=0} dx$, κ and γ.

By calculating the energy of DW displaced by ξ from the equilibrium position, κ is found to vary as θ^3. The TDGL equation for eq. (3) can be solved rigorously if the fourth term is absent. We calculate the susceptibility X(ω) for this free DW. This X(ω) contains L. For a system of dipoles making a flip-flop motion in a double well potential with barrier ΔU as in $NaNO_2$, it is reasonable to assume L=(kT/h)exp(-ΔU/kT) where k is the Boltzmann constant and h is the Planck's constant. On the other hand, X(ω) for a free DW can be obtained from eq.(1) by putting κ to 0. By equating these two X(ω)'s, we obtain γ for a free DW. γ is found to vary as

$T^{-1}\exp(\Delta U/kT)\theta^{3/2}$. If we assume that γ for a pinned DW is approximately given by γ for a free DW, the susceptibility for a pinned DW is obtained as

$$X_o = \frac{2\sqrt{1-V_o}}{\ell(1-\sqrt{1-V_o})} \theta^{-2}, \quad (5)$$

$$\tau = \frac{\sqrt{2}h\sqrt{1-V_o} \exp(\Delta U/kT)}{3k\ell(1-\sqrt{1-V_o})T} \theta^{-3/2}. \quad (6)$$

Both X_o and τ tend to diverge toward T_c, thus the slowing down of DW motion is expected.

Figures (1) and (2) show the experimental results for $NaNO_2$, an order-disorder type ferroelectric. Here $\varepsilon_o - \varepsilon_\infty = 4\pi X_o$ is the dispersion strength and $f_o = (2\pi\tau)^{-1}$ is the dispersion frequency. These results agree qualitatively with the result of the above theory. However it is found that f_o of TGS starts to increase abruptly from just below T_o toward T_c. To understand this discrepancy, we should examine other type of pinning mechanism, the effect of emission of phonons due to DW motion, etc.

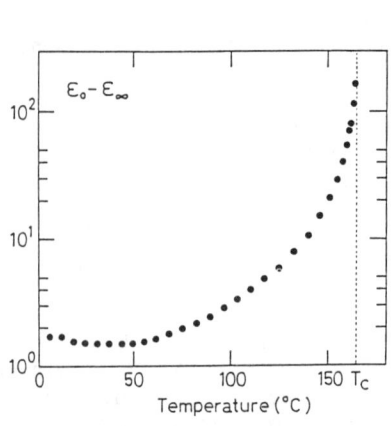

Fig.1 $\varepsilon_o - \varepsilon_\infty$ as a function of T.

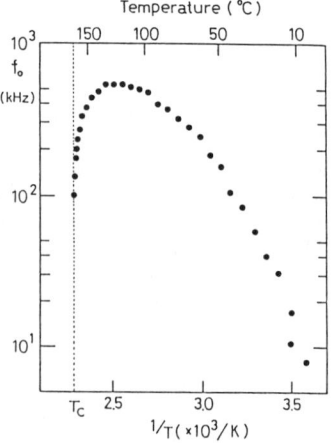

Fig.2 f_o as a function of 1/T.

OBSERVATION OF SLOW DYNAMIC PROCESS DURING THE MAIN TRANSITION IN DSPC BY TIME-DEPENDENT HEAT CAPACITY MEASUREMENT

K. Ema, H. Yao, and Y. Kawase
Department of Physics, Tokyo Institute of Technology
Oh-okayama, Meguro, Tokyo 152, Japan

ABSTRACT

The time-dependent heat capacity of DSPC has been measured around the main transition temperature to clarify the dynamic nature of the transition. It was found that an internal relaxation with a relaxation time of about a few hundreds seconds appears near the main transition temperature.

Aqueous suspensions of phosphatidylcholines such as DPPC, DMPC, and DSPC undergo three succesive phase transitions involving the structure change of the lipid bilayers. These transitions are called the main transition, the pretransition, and the sub transition from the higher temperature side. A significant experimental result is that the amount of heat capacity anomaly accompanying the main transition differs considerably for the cases when it is measured (quasi-)statically and when measured dynamically. In DPPC, for example, the transition enthalpy obtained by an adiabatic calorimeter is 38.8 kJ/mol,[1] while the integrated area of the excess heat capacity peak obtained by an AC method at 0.03 Hz is at most 6.4 kJ/mol.[2] These facts show that there exits quite a slow process during the main transition. In order to clarify the dynamic nature of the main transition, we measured the time-dependent heat capacity of DSPC using a thermal relaxation method and also an AC method.

DSPC (Avanti Polar Lipids Inc.) was dispersed in deionized distilled water and was incubated at 65 °C for 3 h. The lipid concentration was 15 wt%. About 50 mg of sample was hermetically sealed in a silver sample cell (10 mm in diameter, 1 mm in depth) using a cold-welded tin seal. A thin strain-gauge of 5 mm × 2.2 mm in area (Kyowa KFD-5-C1-11) was attached to the bottom of the cell as a heater. The temperature of the sample was measured with a small-bead thermistor of 0.3 mm in diameter (Fenwal 111 series). The sample cell was thermally linked loosely to the thermal bath made of a massive copper block.

Let us consider a case when a certain internal degree of freedom of the sample, denoted as x, relaxes to the thermal equilibrium with a relaxation time τ_{int} which is slower than the time-scale of the measurement. The heat flow $\dot{Q}$ supplied to the sample is given as

$$\dot{Q} = C_0 \frac{dT}{dt} + T \frac{\partial S}{\partial x} \frac{dx}{dt}, \qquad (1)$$

where S is the enthorpy, and C_0 is the heat capacity from degrees of freedom other than x. We can also say that C_0 is the heat capacity under constant x. An electrical equivalent circuit of the system for the present case becomes as shown in Fig.1. Here R is the thermal resistance between the sample and the bath, C' is the heat capacity associated with x, and $rC' = \tau_{int}$. We notice that the effect of the internal relaxation is identical to the so-called τ_2 effect, for which detailed calculations have already done by several workers (see for example ref.3), and the parameters can be obtained by analyzing the temperature response of the sample.

The measurements have been done both in the thermal relaxation mode ($\dot{Q}$ is a step function) and in the AC mode ($\dot{Q}$ is a square wave). Figure 2 shows the temperature dependence of τ_{int} obtained from the thermal relaxation mode measurement around the main transition temperature. We see that an internal relaxation with a relaxation time of about a few hundreds sec appears near the transition temperature. In Fig.3, closed circles show the total heat capacity $C_{total} = C_0 + C'$, and the open circles show C_0. We see that C_{total} shows a significant increase toward the transition temperature, while C_0 does not. This means that the contribution from the internal degree of freedom x becomes dominant near the transition. At this moment, the physical meaning of x is not clear, and further works including microscopic measurements such as X-ray scattering should be carried out. The results of the AC mode measurement will be reported elsewhere.

REFERENCES

1. M.Kodama, H.Hashigami and S.Seki, Biochim.Biophys.Acta 814, 300 (1985).
2. S.Imaizumi and C.W.Garland, J.Phys.Soc.Jpn. 56, 3887 (1987).
3. J.Shepherd, Rev.Sci.Instr. 56, 273 (1985).

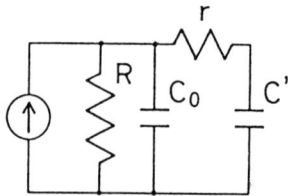

Fig.1. Electrical equivalent circuit when an internal relaxation is present.

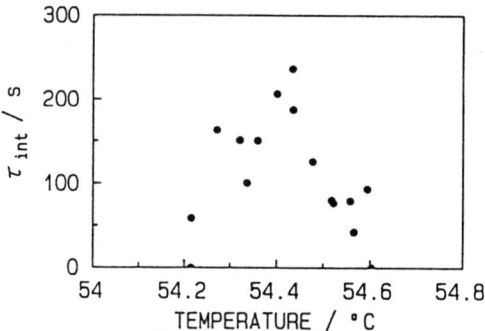

Fig.2. Temperature dependence of the internal relaxation time τ_{int}.

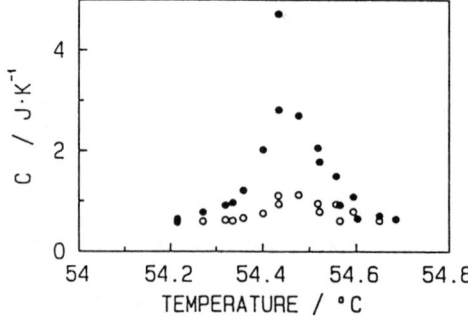

Fig.3. Temperature dependence of the heat capacity C_{total}(closed circles) and C_0(open circles).

THERMAL DIFFUSIVITY AND SURFACE TENSION OF LIQUID NITROGEN NEAR ITS CRITICAL POINT

Takeshi Shigenari, Manabu Mogi, Kohji Abe
and Masaru Suzuki*, Nobutaka Itagaki*, Akira Sato*

Department of Applied Physics and Chemistry, and Division of Natural Science*
The University of Electro-Communications, Chofu, Tokyo 182, Japan

Abstract

The critical behavior of simple liquid nitrogen N_2 were experimentally studied near the liquid-gas critical point ($T_c = 126.26K$, $P_c = 33.54$atm). The critical index of the thermal diffusivity, $\gamma - a = 0.67 \pm 0.03$ ($T > T_c$) and $\gamma' - a' = 0.71 \pm 0.03$ ($T < T_c$), was obtained by the photon correlation spectroscopy, and the the surface tension was determined as $\sigma = (31.5 \pm 0.5)|\epsilon|^{1.293 \pm 0.006}$ dyn/cm from the high precision surface-wave resonance method. Both are in agreement with the theoretical value within the experimental error.

Although nitrogen is one of the simplest molecules, the behavior near the liquid-gas critical point has not been sufficiently investigated compared to e.g.CO_2, Xe and various binary mixture of liquids. It is partly because the critical temperature of N_2 ($T_c = 126.26K$, $P_c = 33.54$atm) is much lower than others. In this work, the thermal diffusivity and the surface tension near T_c, which to our knowledge have not been reported yet, were measured by the following methods.

[1] The thermal diffusivity by the photon correlation spectroscopy.
The linewidth of the quasi-elastic scattering of light is known to be proportional to the thermal diffusivity [1]. As an alternative, we measured the correlation time of the intensity of the scattered light from an Ar-laser (514.5nm) at a nearly forward direction by a digital clipped-correlator. The obtained correlation function of the photon number $< n_c(t)n(t+\tau) >$ could be well fitted to a function $A + B\exp(-2D_T\tau)$, where $D_T = \Lambda/(\rho C_p)$ (Λ: thermal conductivity, ρ: critical density and C_p: specific heat). As shown in Fig.1., in the range of $|\epsilon| > 10^{-3}$, the critical index, $\gamma - a$ (γ and a are the indices of C_p and Λ, respectively.) was obtained as: $\gamma - a = 0.67 \pm 0.03$ ($T > T_c$) and $\gamma' - a' = 0.71 \pm 0.03$ ($T < T_c$). Within the experimental error, these values agree with those for other simple liquids (Table 1) except for SF_6 below T_c [2].

[2] The surface tension using the surface-wave resonance.
The surface tension was also measured by means of the surface-wave resonance with very high precision[3]. A block diagram of the measuring system is shown in Fig.2. The present data near the critical point were well fitted to the single power law formula.(Fig.3) The critical values of the surface tension were found to be

$\sigma = (31.5 \pm 0.5)|\epsilon|^{1.293 \pm 0.006}$ dyn/cm. The critical index is close to the theoretical value of 1.26. However, the critical amplitude is apparently different from the two-scale-factor universality. The values are consistent with the recent experiments on other liquids.

One of the authors (M.S.) would like to thank Dr.Y.Yoshino of Murata Mfg.Co.,Ltd. for the fabrication of detecting and exciting electrodes of surface-waves.

REFERENCES

1. H.G.Stanley: *Introduction to Phase Transition and Critical Phenomena* (Clarendon Press, Oxford 1971)
2. T.K.Lim, H.L.Swinney, K.H.Langley and T.A.Kachnowski: Phys.Rev.Lett.**27**, 1776 (1971).
3. M.Iino, M.Suzuki, A.J.Ikushima and Y.Okuda: Jpn.J.Appl.Phys. **23-1**,54(1983).

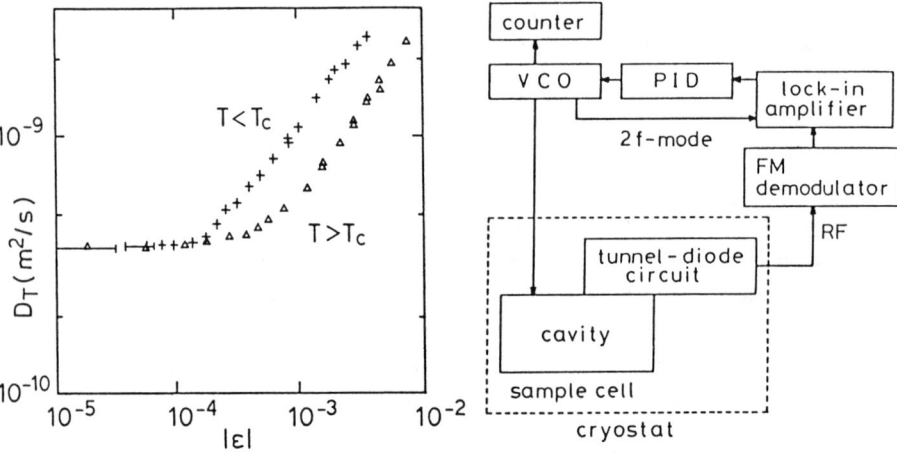

Fig.1 Thermal diffusivity

Fig.2 Block diagram of the surface-wave measuring system.

Table 1 Critical index of thermal diffusivity

	$\gamma' - a'(liq.)$	$\gamma' - a'(gas)$	$\gamma - a$
CO_2	0.72 ± 0.05	0.66 ± 0.05	0.73 ± 0.02
Xe	—	—	0.751 ± 0.004
SF_6	0.635 ± 0.003	0.632 ± 0.002	0.61 ± 0.04
N_2	—	0.67 ± 0.03	0.71 ± 0.03

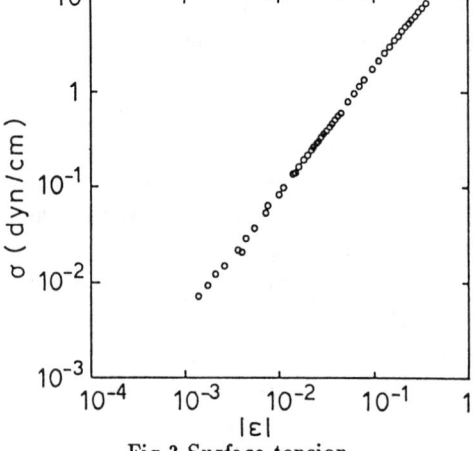

Fig.3 Surface tension

GROWTH PATTERN OF THE SURFACE OF FUNGUS *ASPERGILLUS* COLONY

Shu Matsuura
School of High-Technology for Human Welfare, Tokai University
Nisino 317, Numazu, Sizuoka, Japan

Sasuke Miyazima
Department of Technology, Chubu University
Matsumoto, Kasugai, Aichi 487, Japan

ABSTRACT

Aspergillus oryzae colonies were grown under various glucose concentrations, temperatures, and agar concentrations, and the effects on the pattern were investigated. Patterns of colony were found to vary from uniform to diffusion-limited aggregation type.

INTRODUCTION

The growth mechanism of filamentous fungi is characterized by multi-branching structures. Various textures of their colonies which are made through the irreversible diffusion process of nutrient substances and the above growth mechanism, are obtained. The main purpose of this paper is to investigate the shapes of the Aspergillus oryzae colony which show various responses to environmental conditions, i.e. temperature, nutrient, and diffusion in the medium.

SURFACE GROWTH OF THE BAND-SHAPED COLONY

The elongation of fungal filaments occurs always at their tips. The pressure difference between in and out of the hyphal cell, and the cytoplasmic flow toward the tips are considered as origins. This situation may be similar to the purely physical pattern formations such as viscous fingerings, which is essentially characterized by the diffusion-limitted aggregation (DLA).

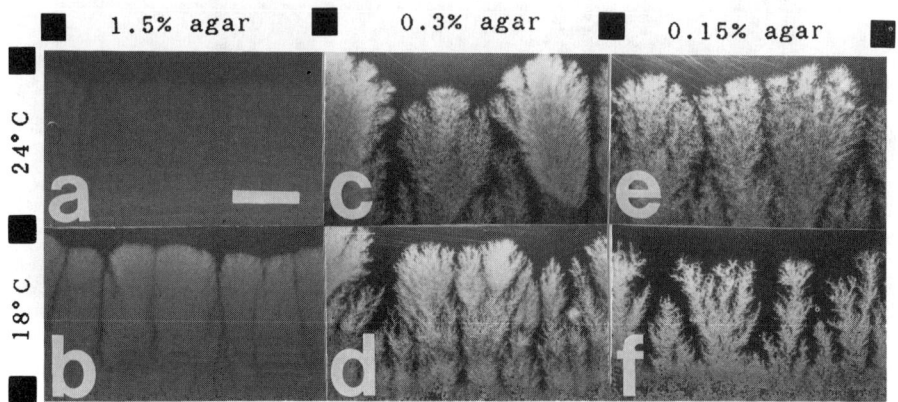

Fig.1 Fronts of Aspergillus oryzae colonies (0.1% glucose) grown under six different conditions. The white bar in Fig.1a indicates 1cm.

On the other hand, there exists the biological growth mechanism such as autotropism which regulate the spatial distribution of each filaments.

In the experiments, we inoculate a line of the fungal spores on a synthetic agar medium which contains glucose of 0.1%w (Fig.1) and 0.01%w (Fig.2, less hyphal production) in the 9cm petri dishes. To change the hyphal growth rate, colonies are incubated at 24°C and 18°C. Secondly, to change the diffusion condition of glucose and inhibitory substances produced during the growth of colony, agar concentrations are set at 1.5%w (solid), 0.3%w (very soft), and 0.15%w (semi-liquid medium). The semi-liquid agar may be also an unfavorable condition for the hyphal respiration.

As seen in Fig.1a, the surface of colony grown on the nutrient-rich solid medium remains smooth until it covers about half the medium. In the case of 18°C (Fig.1b), the fjord shaped cleavages are formed, which seems to be attributable to the inhomogeneity in spore germinations at initial. As the agar concentration is decreased (Fig.1c-f), the growth rate is significantly lowered and then the inhomogeneity is found to be amplified. Further, although the space filling efficiency is lowered in the case of 18°C incubation, rather thick branch systems are found to grow very slowly.

Under the nutrient-poor conditions (Fig.2), colonies become fairly ramified. As the agar concentration is decreased, the growing points in the surface become restricted and, as a result, the shapes of colonies become to be similar to DLA patterns (Fig.2c-f).

Extension of filaments is always observed at top fronts of the colony, where the branches are created frequently. Here, when the temperature or the agar concentration is lowered, the growth rate decreases and the inhibitory interactions seem to become dominant as a screening effect on the hyphal growth to amplify the inhomogeneity of whole patterns.

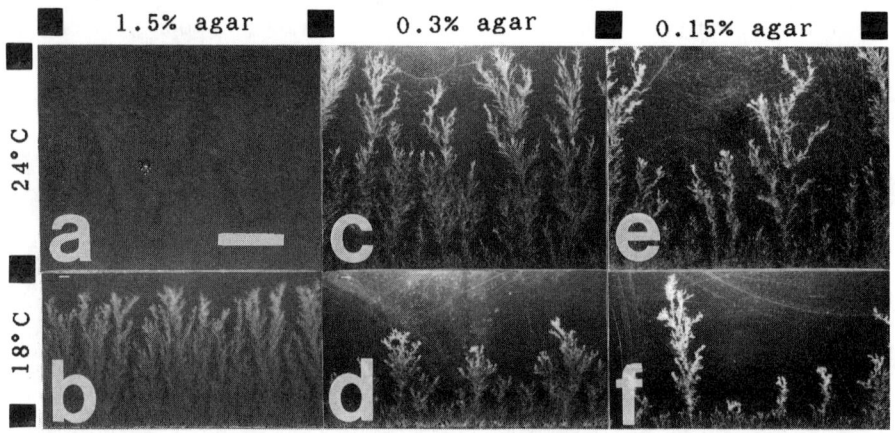

Fig.2 Fronts of Aspergillus oryzae colonies (0.01% glucose) grown under six different conditions.

SUMMARY

We found that the pattern growth of Aspergillus colony is dependent on the balance of growth speed, nutrient and diffusion conditions. Some physiological mechanism is thought to regulate the distribution of hyphal growth, subject to these physical conditions.

AN IN-SITU STUDY OF THE DYNAMICS OF OXYGEN PRECIPITATION IN SI

A. Magerl and K. D. Liss
Institut Laue Langevin, F-38042 Grenoble Cedex, France

J. R. Schneider
Hasylab at DESY, Notkestrasse 85, D-2000, Hamburg, Germany

W. Zulehner
Wacker Chemitronic GmbH, Postfach 1140 D-8263 Burghausen, Germany

ABSTRACT

The integrated intensity observed in a diffraction experiment on a large single crystal is directly related to it's defect structure. We present the first in-situ neutron and Γ ray measurements monitoring the dynamics of the build up and the annealing of strain fields at high temperatures due to the formation of SiO_2 clusters and of associated secondary defects in Czochralski grown Si crystals.

Czochralski (Cz) grown dislocation free Si crystals as widely used by semiconductor industry contains about 10^{18} oxygen atoms cm^{-3} (20ppm) [1]. The oxygen atoms in as-grown crystals are atomically dissolved in the host and the Si lattice seems to be hardly disturbed in this solid solution form. A comparison of the integrated intensities (II) measured on a Cz crystal and on a float zone crystal, which contains about a factor of 1000 less oxygen, reveals indeed the same small reflecting powers. This can be explained by the fact, that the beams are effectively diffracted only from a small volume of these perfect crystals where the extinction length is short as compared to the total thickness of the specimen. Introducing defects into the lattice will destroy perfect coherency. Different volume elements of the crystal can now reflect over an enlarged wavelength band or angular range. This will result in a pronounced increase of II for sufficiently thick crystals. Consequently, the value of II can be directly associated with properties of the defect structure.

We have performed in-situ measurements of the Si[111] Bragg peak of Cz material as a function of temperature and annealing time. Both short wavelength Γ ray and high resolution neutron diffraction have been employed. The two instruments are located at the ILL. In the first case highly collimated radiation (0.025°) from a radioactive Cs source was used. The Bragg angle is very low because of the short wavelength of 0.02 Å and, thus, these measurements are in particular sensitive to a bending of atomic lattice planes. In a second experiment, neutron backscattering in transmission geometry with extreme Bragg angles of 90° was employed. Under these conditions the measurements reflect the definition of the lattice spacing in a crystal. A more detailed comparison of the resolution functions of the two instruments is given in ref.2. The extinction lengths for both types of radiations are on the order of 10µm, considerably shorter than the thickness of the crystals, which is 1cm. We note for completeness that absorption is negligible in both cases.

Fig. 1 shows the evolution of II (part b) together with the corresponding temperature profiles (part a) for a Γ ray and a neutron diffraction experiment. The value of II has been set to one for the lowest temperature measured. This corresponds to room temperature for the Γ ray data, whereas spectra could only be taken in the neutron case from the first holding step at 1023 K. Similar temperature patterns were taken for the first 24 hours. After that point the Γ ray data have been stopped, whereas a further temperature scan was added in the neutron case.

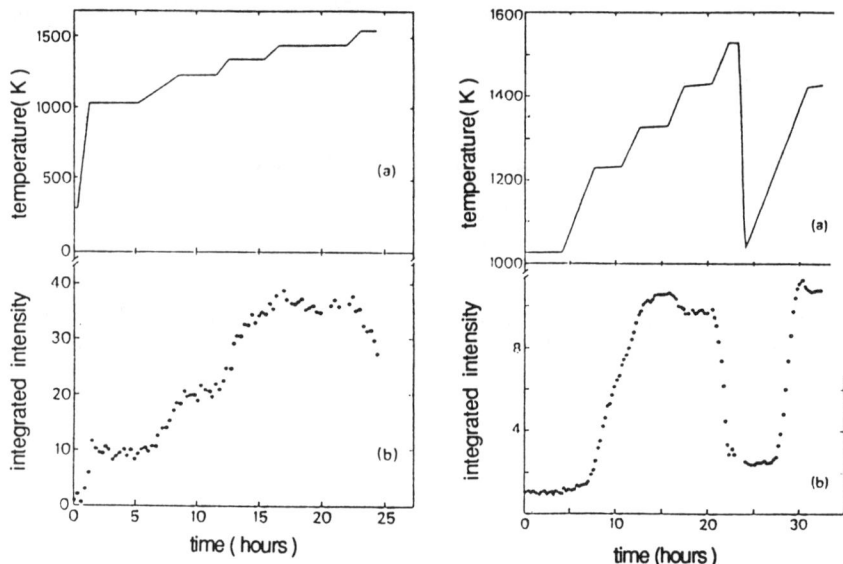

Fig. 1 Temperature profile and integrated intensity measured on a Γ ray diffractometer (left) and on a neutron backscattering diffractometer (right)

Both measurements reveal similar (although not identical) characteristics. As common salient features we note:
i) on a time scale of a few hours there seems to be no evolution of II for the first holding temperature of 1023K.
ii) the II starts to rise strongly from about 1150K.
iii) the II as particularly well visible in the right hand side of fig.1 does not come to saturation at a holding temperature of 1223K.
iv) II seems to reach an equilibrium with a time constant of about one hour at the next holding temperature of 1323K.
v) a further increase of the temperature to 1423K leads to a reduction of II. An equilibrium is rapidly reached.

In addition we note for the neutron data that an increase of the temperature to 1523K results in a drastic reduction of II. The response is immediate on our time scale. A subsequent temperature cycle reconstitutes the strained lattice with II even slightly higher than for the first temperature cycle.

In summary, the multiple facets related to the formation of oxygen precipitates in Si are observed here via the II of a Si Bragg peak. These data reveal for the first time direct access to the dynamics of the various processes involved like the formation of long needles of coesite in the hard Si lattice at lower temperatures where diffusion is very slow, or the formation of amorphous SiO_2 particles associated with the creation of numerous dislocation loops in the medium temperature range, or the annihilation and dissolution of microdefects at higher temperatures. The particular defect structures characteristic for the various temperature ranges can also distinguished from differences in the Γ ray and neutron data.

REFERENCES
1) W. Zulehner, "Oxygen-related defects and microdefects" in Landolt-Bornstein, New Series III Ed. O. Madelung, Springer Verlag 1989, 22b p391-438
2) A. Magerl, J. R. Schneider and W. Zulehner, J. Appl. Phys. 67 533(1990)

DIFFUSION LIMITED AGGREGATION IN A FLOW FIELD

K. Oota, K. Okumura, K. Maruyama and S. Miyazima
Department of Enginnering Physics, Chubu University,
Kasugai, Aichi 487, JAPAN.

INTRODUCTION

Morphorogical change of diffusion limited aggregation is investigated under a flow field in a rectangular Helle-Shaw cell with an anode lying in a flow field. Fractal dimension changes from 1.7 to 1.1 as the flow rate increases when the directions of fluid flow and ionic diffusion by the electric field are opposite. If they are in the same direction, the fractal dimension changes little.

EXPERIMENTAL APPARATUS

General shape of experimental aparatus is the similar type with Helle-Shaw cell for viscous fingering (Fig.1). The gap between two acrylic plates is less than 1 mm, and the 2-normal $ZnSO_4$ fluid is guided to flow through the narrow gap. There are several variations of arrangement of anode and cathode and the flow direction.

i) Anode lies in the gap and is parallel to the acrylic plates (forest).

ii) Anode is set up perpendicular to the acrylic plate (single tree).

On the other hand the flow direction can be considered in two different ways.

a) The directions of the fluid flow and the ionic difusion induced by the electric field are the same.

b) The directions of the above two are opposite.

Fig.1 Experimental apparatus. Main part is the central two acrylic plates with 1mm gap in which the fluid flows uniformly and the anode is set up at the centra position. The cathode (Zn-plate) is set in one of the left or right boxes.

EXPERIMENTAL RESULTS

1 i) and a)   In this case, we do not find any essential change in the DLA morphology.

2 i) and b)   The fractal dimension changes from 1.7 to 1.1 as the flow rate increases, as shown in Fig.2. The growth manner in the slow flow rate must be the same as the usual DLA growth, i.e., diffusion limited aggregation. That in the high flow rate, however, is quite different from DLA. In these regions, the ions always come from the root of each tree and stick only to the top. We found rather sharp cross-over from DLA growth to 1-dimensional growth at about 30ml/min.

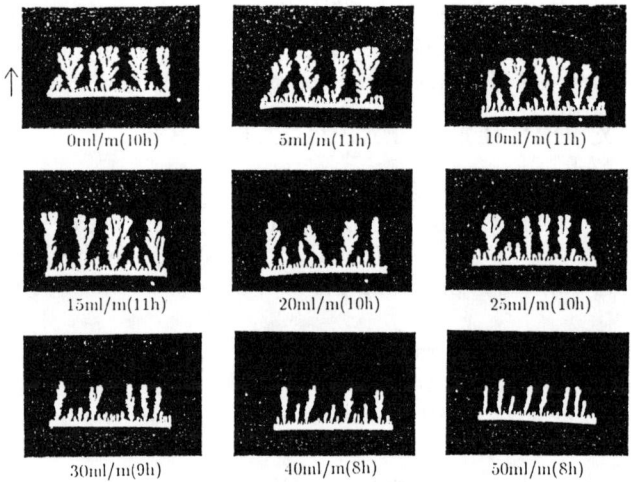

Fig.2 Morphological change of forest trees. Figures below pictures are the flow rate and time which consumed to make the aggregation.

3 ii) and a)   This case is omitted here and will be discussed elsewhere.

4 ii) and b)   The number of brances and the fan angle of each tree are changes by the fluid flow as shown in Fig.3. Two different patterns can be seen behind the anode point and front part.

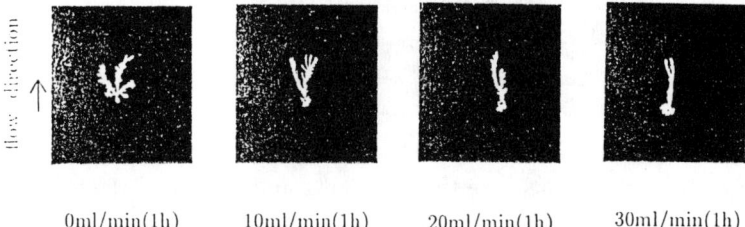

Fig.3 Morphological change of single tree. Figures below pictures are the flow rate and time which consumed to make the aggregation.

B. THEORY

TIME SCALE INVARIANCE IN TRANSPORT AND RELAXATION*

Harvey Scher
BP Research, 4440 Warrensville Center Road, Cleveland, OH 44128

The major features of transport and relaxation in disordered systems are shown to be due to the occurrence of long-tailed distributions of rate-limiting event times. The mean-time of this distribution is divergent, and hence there is no intrinsic time scale associated with the dynamic response of the system. The lack of a time scale is manifest in the film-thickness dependence of the transit time of a single carrier and in the occurence of stretched-exponential time dependence of the relaxation.

INTRODUCTION We wish to describe a simple set of ideas which have their roots in nearly paradoxical mathematical curiosities with origins from the beginning of the eighteenth century. The unifying concept behind these ideas will be the use of long-tailed probability distribution functions with infinite first moments. If the first moment existed it would define a scale. While one major thrust of physics is to find the right scale for a problem (e.g. size of an atom, mobility of an electron in a crystal), a newer thrust is to investigate problems without any characteristic scale, e.g. critical phenomena. In the work described here, we are interested in microscopic processes which do not possess a characteristic time scale. We focus on transport and relaxation in disordered solids when the mean waiting-time scale between events (electron hopping, defect movement, etc.) diverges. We call transport governed by such a long-tailed waiting-time distribution "dispersive," since many time scales coexist. We will 1) show the ubiquity of scale-invariant dynamic phenomena in a wide variety of disordered materials; 2) demonstrate that dispersive motion can account quantitatively for many of the universal aspects seen in transport and relaxation measurements. The materials which exhibit dispersive transport include amorphous semiconductors and insulators (a-Si:H, a-SiO_2, a-As_2Se_3), polymer films, molecular solid solutions, and glasses.

An analogy of the key physical idea of dispersive transport can be seen in a game of chance known as the St. Petersburg Paradox (1713)[1]. What is the game? Flip a coin. If it comes up heads, then win one coin. If the coin comes up tails then flip again until a head appears. If N tails precede a head, then win 2^N coins. This event occurs with probability 2^{-N-1}. The mean winnings are $1 \times 1/2 + 2 \times 1/4 + 4 \times 1/8 + \ldots = \infty$. The banker wants the player to ante an infinite amount of coins (the bank's expected loss); the player counters that a smaller ante is in order because his median winnings are only one coin and to win an infinite number of coins he must flip an infinite number of times, which is unreasonable. The paradox arises from trying to determine a characteristic size from a distribution which does not possess one! Winnings occur on all scales, with an order of magnitude greater winning occurring an order of magnitude less often. This will be a dominant theme in our following analysis of transport governed by waiting-time distributions with infinite means.

DISPERSIVE TRANSPORT Electronic transport in disordered systems is an easily-measured paradigm of long-tail distributions. A common example is transport via a sequence of charge-transfer steps from one localized site to another in the presence of an applied electric field. The process is called "trapping" if the transfer step involves thermal activation from the site to a conduction band, in which the charge diffuses to the next site. Tunnelling directly between localized sites is known as "hopping." Due to the disorder, the transfer time can be a random variable, which is characterized by the probability $\psi(t)dt$ that the time for an individual transfer (or event) is between t and t+dt. The accumulated sequence of these events in the charge carrier motion can be viewed as a continuous time random walk (CTRW)[2], in which each step is governed by the distribution $\psi(t)$. By specifying $\psi(t)$, and the spatial bias introduced by the electric field, one can calculate[3] the properties of a packet of charge propagating across a sample using the formalism of the CTRW.

The canonical experiment for determining electron transport is the time-of-flight measurement. A pulse of strongly absorbed light, incident on a region near the electrode, generates non-equilibrium electrons and holes. In the polarity shown in Figure [1], the electrons are swept into the near electrode, leaving a sheet of holes to transport to the far electrode; this motion gives rise to a current I(t) in the external circuit.

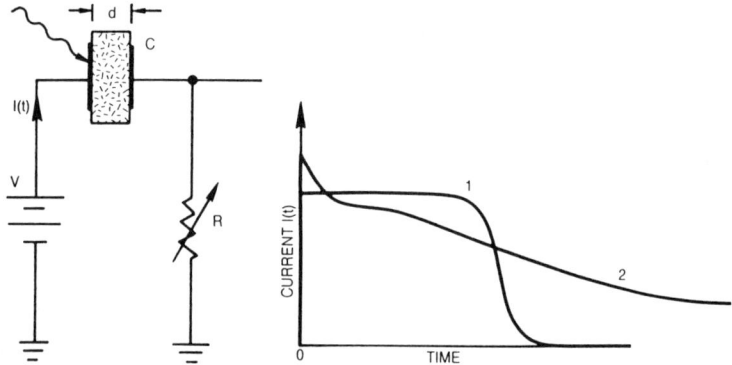

Fig. 1*. Schematic diagram for time-of-flight or transient photocurrent experiment. A light flash of duration less than t_r, the transit time, is absorbed in a depth much less than the sample thickness d. Carriers of one sign move across the sample inducing a time dependent current I(t) in the external circuit. (a) A typical I(t) current trace measured in a material with a well-defined mobility. The transit time t_r is the time for the current to drop to one half its (constant) value. (b) A highly dispersive transient photocurrent trace I(t) measured on As_2Se_3 by M. Scharfe (Ref. 4).

The expected result for the current I(t) due to normal transport is shown in curve 1 in Figure [1]. The velocity of the sheet of holes is constant; therefore the current is constant until the holes are absorbed at the electrode at which time they no longer contribute to the current. The "transition region," over which the current drops to zero, is a measure of the spread in the hole packet due to normal diffusion.

Early experiments in the late 1960s revealed rather bizarre current traces. In sharp contrast to the one shown in curve 2 in Figure [1] we show a current trace in Figure [1b] measured by M. Scharfe[4] for amorphous As_2Se_3, a material then used as a photoconductor in Xerox machines. Not only is I(t) decreasing over the entire time-of-flight (except for a small "plateau" region), but the particular shape of this decay is scale-invariant. A scale is defined by the transit time t_r, corresponding to the onset of the long tail. For a given material, $I(t)/I(t_r)$ vs. t/t_r is independent of t_r. We will discuss more recent data on a-As_2Se_3 below. In the same relative units, the shape of the "normal" current trace would depend explicitly on t_r, since the width of the region of constant current increases linearly with t_r, while the "transition region" only increases as $t_r^{1/2}$. Further, using the usual definition of the drift mobility

$$\mu_d E = d/t_r \, , \tag{1}$$

where E is the applied electric field, one observed μ_d to depend inversely on the sample thickness! These facts remained a puzzle until a fundamentally new theory clarified both the phenomenon and what the experimental current trace should be.

The theory assumes that each charge carrier independently undergoes a random walk biased by a preferred direction introduced by the applied field. The entire character of the propagating packet of charge depends on one key feature of the probability distribution $\psi(t)$. If the first two moments of $\psi(t)$ exist, the transport is normal (Figure [1]). If the first moment $<t>$ of $\psi(t)$ does not exist, the charge packet can still transit the sample; however, it exhibits unusual dispersion. For a $\psi(t)$ with an algebraic tail,

$$\psi(t) \sim t^{-1-\beta}, \qquad (2)$$

one has $<t> = \infty$ for $\beta \leq 1$ ($\beta > 0$ for $\psi(t)$ to be normalizable). The mean position of the spatially-biased time-evolving packet then varies as

$$l(t) \propto l(E) t^\beta, \qquad (3)$$

where $l(E)$ is the mean step displacement.

This sublinear variation in time is the key to all the peculiar features of I(t) discussed above. Once we have demonstrated this central point, we will discuss how the distribution in Eq. (2) arises naturally in disordered systems.

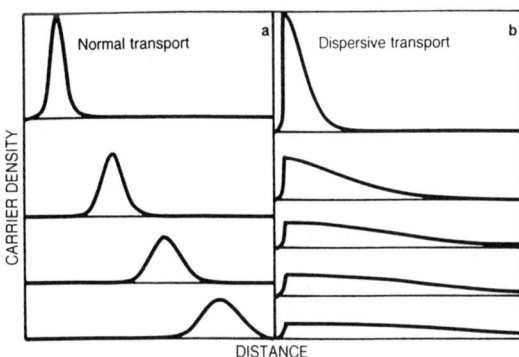

Fig. 2*. A sequence of "snapshots" of the transiting packet of charge in the sample. The normal transport in (a) corresponds to a mean packet position evolving linearly in time and gives rise to the current trace 1 in Figure [1]. In (b) the packet displacement is controlled by a large dispersion in the arrival time of the carriers at the far electrode. The mean position of the packet is a sublinear function of time (cf. text); it gives rise to the current trace 2 in Fig. [1].

In Figure [2] we contrast the packet propagation $P(l,t)$ for two types of $\psi(t)$ with the same spatial bias due to the electric field. The normal (Gaussian) transport and diffusion results from a $\psi(t)$ with finite first and second moments. The Gaussian behavior is a consequence of the well-known central limit theorem. The position of the peak of the distribution coincides with the spatial mean $l(t)$. This is not the case for $P(l,t)$ generated by the distribution in Eq. (2). The peak of this $P(l,t)$ *remains* at the initial position while with increasing t the mean is continually displaced from it. The origin of this unusual behavior is in the relatively small, but finite, occurrence of an event-time which is much larger than a typical one. As such a rare, but quite long event time can be comparable to the accumulation of typical event-times in the carrier's transit across a sample, it can have a large effect on the carrier motion. The forward "streaming" of the carriers is due mainly to those undergoing typical events. Eventually, many of these forward carriers will encounter one of the long event times. Thus, the mean position of the packet increases with time but at an ever decreasing rate. The mean is, therefore, a sublinear function of time (Eq. (3)) which leads to a current,

$$I(t) \propto dK(t)/dt \sim t^{-(1-\beta)}, \quad t<t_r \qquad (4)$$

so the current decreases even before the carriers are absorbed into the electrode. When a reasonable fraction of the carriers (~10%) reaches the electrode, the current begins to decrease at a faster rate due to the carrier loss. The detailed solution to the problem of a random walk with a bias towards an absorbing plane shows[3] a crossover to,

$$I(t) \sim t^{-(1+\beta)} \quad t>t_r . \qquad (5)$$

A double logarithmic plot of $I(t)$, corresponding to Eqs. (4,5), is simply two lines, with slopes $-(1-\beta)$ and $-(1+\beta)$, separated by a narrow transition region. Note that the sum of the slopes is -2, independent of β! An estimate for the transit time t_r, which denotes the transition region from slope $-(1-\beta)$ to $-(1+\beta)$, is easily obtained from the relation $K(t_r) \sim d$ or, using Eq. (3),

$$t_r \sim (d/K(E))^{1/\beta} \qquad (6)$$

where $K(E)$ is the measure of the bias, and where typically $K(E) \propto E$. Using Eq. (1) as a definition of mobility for this dispersive transport, one then has field- and thickness-dependent μ_d! This anomalous d and E dependence of μ_d illustrates that in a system with time-scale invariance there are no intrinsic transport coefficients. External parameters (e.g. sample thickness) constrain the dynamic response of the system, and hence these limit the "transport coefficients." The "universality" in shape of $I(t)$ is due to the variance $\sigma(t) \propto t^\beta$ and hence the ratio $\sigma(t)/K(t)$ is a constant.

We have shown that Eqs. (4,5) account for the shape of the transient current, Eq. (6), for the anomalous dependence of the defined transit time on film thickness (and also electric field), and that $\sigma(t) \propto t^\beta$ is the basis for the scale-invariance of $I(t)$. The theory, Eqs. (4-6), predicts both the shape of the transient current, $I(t)$, and the relation between this shape and the sample thickness- and field-dependence of the transit time, i.e. the β values determined from $I(t)$ and $t_r(d)$ are the same. This relationship is a hallmark of dispersive transport.

These relationships were dramatically confirmed in a careful study of the phototransients in a-As_2Se_3 by G. Pfister[4]. Shown in Figure [3a] is a double logarithmic plot of the normalized current traces in one film of a-As_2Se_3 for a range of t_r encompassing nearly 3 decades in time. The shape of $I(t)$ is scale invariant; the solid line is the theory with $\beta = 0.45$. The dependence of t_r on d, predicted by Eq. (6), was observed (we will show this feature with other transients, below). The physical significance of β can be seen in a model of transport: extensive multiple trapping, where the total time spent in traps far exceeds the total transit time in the conduction band. For an exponential distribution $\rho(\epsilon)$ of energy levels,

$$\rho(\epsilon) = \rho_0 \exp(-\epsilon/kT_0) , \qquad (7)$$

one can show[5] that Eq. (2) holds for $t > \tau$, the mean trap capture time, with

$$\beta = T/T_0 . \qquad (8)$$

Disorder in the form of a distribution of trap states, promotes a spectrum of intrinsic times ($\propto \exp(-\epsilon/kT)$ that limits the transport. In this simple example, the width of this spectrum (i.e. the relative release times) is controlled by both the temperature T and the width of the distribution T_0, and β is simply the ratio of these controlling factors. When $T > T_0$, the weighting of the release times over the entire distribution is no longer sufficient to allow the rare event (the long release time) to occur often enough to influence the accumulated typical release times. In this regime, $\beta > 1$ and $<t>$ is finite, so that the transport becomes quasi-normal although it is not truly Gaussian until $\beta > 2$.

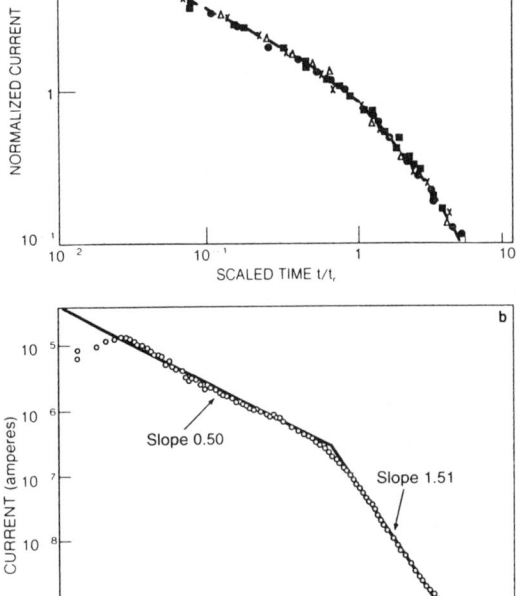

Fig. 3*. A log-log plot of the transient photocurrent measured in: (a) Amorphous As_2Se_3 (after G. Pfister, Ref. 4); the data points correspond to a superposition of transients covering nearly 3 decades of transit time. (b) Intrinsic hydrogenated amorphous silicon a-Si:H at T=160K (after T. Tiedge, Ref. 6). The current decays algebraically over the observation time of the experiment (nearly 6 decades) determined by the sample thickness, applied field, and temperature.

In Figure [3b] we show an excellent example of this mechanism of dispersive transport in a-Si:H[6]. The decay of the current is purely algebraic for nearly six decades of time. The two-slope behavior is clearly evident, and the sum of the slopes is -2.01. β is found to vary as $\beta = T/T_0$, with $T_0 \simeq 30$ meV. T_0 has been interpreted to be the width of the (exponential) conduction band-tail and is in agreement with other determinations of this width. The correlation in the scaling behavior of $t_r(E)$ has also been established.

SiO_2 is a material that exhibits hole transport characterized by a T-independent β. The transport is manifested in a measurement of the flat band voltage recovery in the oxide layer of MOS devices (Figure [4]). The details of this measurement are described in Ref. 7; it suffices to indicate that the voltage signal is proportional to $I(t)$ instead of $dI(t)/dt$ as in the I(t) measurement. In Figure [4] the value $\beta = 0.22$ of this device describes the data for 120K <T < 300K[7]. The scaling of t_r (slope $1/\beta$) with oxide thickness is in excellent agreement with the values of $\beta = 0.25$ and $\beta = 0.24$ from the voltage recovery measurements of the devices indicated in the inset for the entire (3-decade) range of t_r. The hole transport mechanism has been attributed to small polaron hopping between sites about 1 nm apart with a distribution of transfer integrals (disorder in intersite separations and/or bond angles)[7].

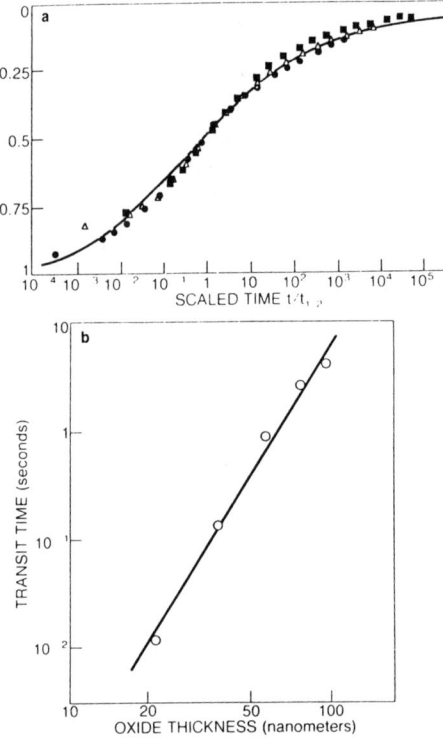

Fig. 4*. (a) A logarithmic plot of the flat-band recovery voltage of a MOS silicon device. The recovery is due to the dispersive transport of holes across the oxide layer. [After McLean et.al. Ref. 7]. (b) The dependence of the transit time on thickness of the oxide layer of a MOS silicon device. The solid lines have a slope of $1/\beta$ where β is determined from the flat-band recovery voltage of the device.

Most measurements on both hydrogenated amorphous Si and amorphous chalcogenides are consistent with $\beta \sim T$, except at low temperature. Thus excess carrier transport in amorphous semiconductors is dominated by the ubiquitous exponential band-tails and is independent of the effects of deep traps. It is highly plausible that hopping is the transport mechanism in a wide variety of polymeric systems, the exact nature of the individual charge-transfer steps has not yet been resolved.

In general, while all the materials discussed above are complex and quite varied, and though details of the transport mechanism remain to be elucidated, the near universality of these main features of their dispersive transport attest to the generality of the phenomenon.

RELAXATION LAWS The dispersive transport of a single charge (and the current it generated) was the focus of the previous section, where the scale invariance of the transient current was explored. In a natural extension of these ideas, we now consider the dispersive transport of a collection of mobile defects. We show how this can be the basis for a well-known law for relaxation in a wide variety of random systems.

Peter Debye, with his classic treatment of dielectric relaxation in fluids, set the framework for much of our intuition on relaxation. He derived the law governing how initially aligned small spherical dipolar molecules of radius R, with dipole moment $\mu(t)$, relax in a fluid of viscosity η, at temperature T, when the external electric field is removed. The relaxation from an aligned to a random configuration of dipole moments occurs because of random collisions with fluid particles. The relaxation function $\phi(t) \equiv <\mu(t)\mu(0)>/<\mu^2(0)>$ was calculated to be exponential,

$$\phi(t) = \exp(-t/\tau), \qquad (9)$$

with $\tau = 4\pi\eta R/kT$; thus only a single time scale is needed to characterize the relaxation process.

Graham Williams and David Watts[8] found (1970) empirically that the form

$$\phi(t) = \exp(-(t/\tau)^\beta), \quad 0<\beta<1 \qquad (10)$$

fit data for glassy and polymeric materials including polyethylacrylate, polymethylacrylate, and propylene oxide. An extensive survey of the role of this $\phi(t)$ in investigating the time dependent reactivity of trapped species in condensed matter has been given by Andrez Plonka[9]. In the past few years this form for $\phi(t)$ has been used to fit an ever widening variety of experimental data including mechanical, NMR, dielectric, enthalpic, volumetric, dynamic light scattering, magnetic relaxation phenomena, and reaction kinetics. It was recently shown[10] to account for the relaxation of the localized electronic structure of a-Si:H, which we will describe below. It was in a study of remnant magnetization that the expression "stretched exponential" was born to describe Eq. (10), a nomenclature which has stuck. This modern flurry of activity actually obscures the fact that the stretched exponential was introduced in 1863 to describe mechanical creep in glassy fibers[11].

Several derivations now exist[12] for systems in three dimensions involving diverse concepts such as percolation, hierarchial relaxation of constraints, and multipolar interaction transitions. We will consider a mechanism based on a reaction picture involving the *dispersive transport of defects*. Although the above ideas are dissimilar, a common mathematical structure connecting them has been discovered[13].

In the Debye model, the underlying mechanism of relaxation was fluid particles randomly hitting polar molecules. Consider now an analogy for a glass in which mobile defects hit a frozen-in dipole and instantaneously cause its relaxation. Glarum treated (1960) such a model of a defect undergoing Brownian motion in one dimension[14]. We generalize Glarum's work by allowing for a finite concentration of defects in three dimensions and, most importantly, by treating their motion as dispersive[15].

Because each defect moves independently, the probability of first encountering the frozen-in dipole is the product of N factors, where N is the number of defects in a volume V and c=N/V. For large N the product is an exponential,

$$\phi(t) = \exp(-cS(t)); \qquad (11)$$

in three dimensions,

$$S(t) \sim \begin{cases} t, & \text{for } <t> \text{ finite} \\ t^\beta, & \beta<1, \text{ for } <t> \text{ infinite} \end{cases} \qquad (12)$$

[In one dimension $S(t) \sim t^{1/2}$ and $t^{\beta/2}$ for the two cases $<t>$ finite and infinite, respectively.]

The first case occurs when the mean time $<t>$ between defect hops (jumps, etc.) is finite, and the latter case for dispersive transport where $\psi(t) \sim t^{-1-\beta}$, $\beta<1$. Within the defect-diffusion model, we obtain either Debye relaxation (for finite $<t>$) or stretched-exponential relaxation (for infinite $<t>$). In the latter case, any details which do not change the condition $<t> = \infty$ are irrelevant. Thus, an answer to the question of why the stretched exponential is so widespread is that it can be a probability limit distribution, and therefore plays a role in physics analogous to that of the Gaussian law or the Poisson law.

The defect-diffusion derivation of the stretched-exponential calls for the movement of defects, but what is a defect? In general, this is a difficult question to answer, although many possibilities have been suggested for specific materials. We will cite two distinct cases. A practical material on which we will first focus is the engineering polymer polycarbonate, a high impact thermoplastic resin. It displays ductility in the glass unlike most polymers which are brittle. Much speculation has centered on the origin of plastic flow, and a variety of solid-state NMR lineshape and relaxation experiments have provided, for the first time, structural details of molecular motions in the glass[16]. In particular, ^{13}C and deuterium lineshape measurements reveal that aromatic ring motions occur readily in the solid, with little or no disturbance of the backbone direction or orientation. A mobile carbonate bond associated with this ring motion is considered here to be the "defect" whose movement is the mechanism responsible for inducing the mechanical[17] (Fig. 5), NMR, and dielectric relaxation in this glassy polymer. Indeed, all three types of measurements find stretched exponential behavior (Eq. (10)) with $\beta = 0.15$. The low value of β may be connected to the quasi-one-dimensional motion of the mobile bond.

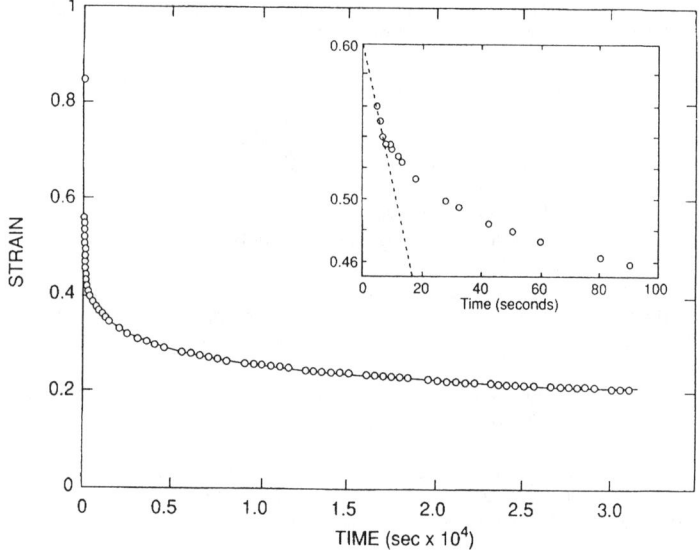

Fig. 5. Strain recovery in polycarbonate, measured using birefringence. The data are fit to a stretched exponential with $\beta = 0.15$ [Ref. 17]. The inset shows an expansion of the time region which corresponds to the nearly vertical data points on the strain vs. time plot. The dashed line is an attempt at a simple exponential fit of the data in the inset.

A recent study[10] of the relaxation of the nonequilibrium electronic and atomic structure of the material system of doped a-Si:H has provided considerable evidence for the physical mechanism we have been discussing. James Kakalios and his co-workers have examined the return to equilibrium of the T-dependent densities of dangling bond defects and donor (acceptor) states of rapidly cooled samples of n-type (p-type) a-Si:H. They probed this relaxation by monitoring the t-dependence of the band-tail states and found it to be well described by a stretched-exponential with a β varying linearly with T (the room temperature value of β for n-type a-Si:H is 0.45). A good candidate for the diffusing "defects" which account for the kinetics of the structural relaxation is the bonded hydrogen. In independent studies they have established that the hydrogen exhibits dispersive diffusion. Moreover, they have made the important observation that the $\beta(T)$

parameter determined from the measurement of H diffusion is entirely consistent with the $\beta(T)$ obtained from the stretched exponential fit of the relaxation data.

Future research is likely to involve the role of more subtle "defects" controlling the slow relaxation of glasses in the glass transition region.

*This talk is based on a review written with my co-authors Michael E. Shlesinger and John T. Bendler which appeared in **Physics Today**, January, 1991, p. 26.

REFERENCES

1. I. Todhunter, A History of the Mathematical Theory of Probability (Cambridge Univ. Press, Cambridge, UK), 1865.

2. E. W. Montroll, G. H. Weiss, J. Math. Phys. 6, 167 (1965).

3. E. W. Montroll, H. Scher, J. Stat. Phys. 9, 101 (1973). M. F. Shlesigner, J. Stat. Phys. 10, 421 (1974). H. Scher, E. W. Montroll, Phys. Rev. B 12, 2455 (1975).

4. M. E. Scharfe, Phys. Rev. B 2, 5025 (1970). G. Pfister, Phys. Rev. Lett. 33, 1474 (1974).

5. H. Scher in Proc. Seventh Int. Conf. on Amorphous and Liquid Semiconductors, Edinburgh, unpublished (1977), p. 209. G. Pfister, H. Scher, Adv. Phys. 27, 747 (1978).

6. T. Tiedje, in Semiconductors and Semimetals, V. 21C ed. by J. Pankove (Academic Press, New York) pp. 207-338 (1984).

7. H. E. Boesch, Jr., F. B. McLean, J. M. McGarrity, P. S. Winokur, IEEE Trans. Nucl. Sci. 25, 1239 (1978). F. B. McLean, H. E. Boesch Jr, T. R. Oldham, in Ionizing Radiation Effects in MOS Devices and Circuits, T. P. Ma, P. V. Dressendorfer, eds., Wiley, New York (1989), p. 87.

8. G. Williams, D. C. Watts, Trans. Faraday Soc. 66, 80 (1970).

9. A. Plonka, Time-Dependent Reactivity of Species in Condensed Matter, Lecture Notes in Chemistry 40, (Springer-Verlag, Berlin) 1986.

10. J. Kakalios, R. A. Street, W. B. Jackson, Phys. Rev. Lett. 59, 1037 (1987); Philos. Mag. B 56, 305 (1987). J. Kakalios, Hopping and Related Phenomena, World Scientific, Singapore (1990).

11. F. Kohlrausch, Pogg. Ann. Physk. 119, 352 (1863).

12. M. H. Cohen, G. S. Grest, Adv. Chem. Phys. 48, 455 (1981); R. G. Palmer, D. Stein, E. S. Abrahams, P. W. Anderson, Phys. Rev. Lett. 53, 958 (1984).

13. J. Klafter, M. F. Shlesinger, Proc. Natl. Acad. Sci. USA 83, 848 (1986).

14. S. Glarum, J. Chem. Phys. 33, 1371 (1960).

15. M. F. Shlesinger, Annu. Rev. Phys. Chem. 39, 269 (1988).

16. K. L. Li, P. T. Inglefield, A. A. Jones, J. T. Bendler, A. D. English, Macromolecules 21, 2940 (1988).

17. D. G. LeGrand, W. V. Olszewski, and J. T. Bendler, J. Polymer Sci. 25 1149 (1987).

ELECTRON TRANSPORT PROCESS INDUCED BY PROTONATION IN ALPHA-HELICAL PROTEIN

S.Ichinose and T.Minato
Nara University, 1500 Misasagi-cho, Nara 631, Japan

ABSTRACT

The simplest one-dimensional model of the protein is introduced to show that the protonated protein acts as a specific conductor of electron. In this model the protein molecule plays a catalytic role to facilitate the electron transport between the donor and acceptor molecules attached to it.

ELECTRON-PROTON TRANSFER IN PROTEIN

A great many biological phenomena are related to electron transfer from donor molecules to acceptor ones through molecular structures. Such molecular structures are often called <u>electron transport chains</u>. An experiment shows that the electron transferred covers distances of order 30-70 A [1]. It is unlikely that the electron transfer covering such large distances is realized via a simple mechanism of tunnelling. One of the possible explanations for such an electron transfer is that the transfer process is facilitated by the participation of protein molecules between the donor and acceptor molecules.

The question concerning the possibility of an electron movement along protein molecules has been discussed in the literatures [2,3]. The investigation of the real motion of an excess electron in the protein conduction band is hardly justifiable. Therefore, we need to study the effect of a protein molecule on the electron transfer between a donor and an acceptor under the assumption that an electron does not actually come to the conduction band.

The alpha-helical proteins may serve as the ideal molecular structure which transports an electron from one molecule to another. An overlapping of electron wave functions of the peptide groups, arranged in different chains, is of less importance. The movement of an electron, supplied by a donor molecule, may be considered separately in each chain. Our mechanism involves a simple, cyclic alteration in the hydrogen-bond lengths of alpha-helix protein

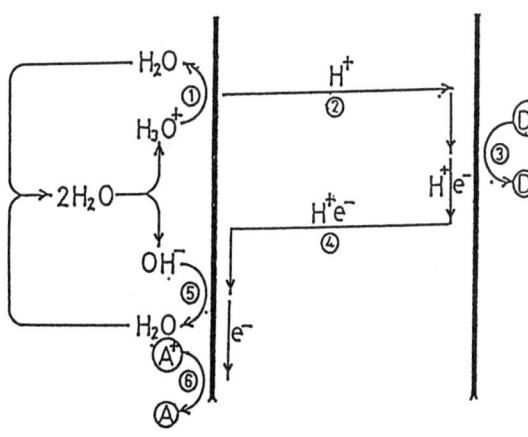

Fig.1. Basic reaction processes

accessible to a migrating proton-electron.

There are six reaction processes through which both an electron and a proton may be transferred in the space between the donor zone and the acceptor one as sketched in Fig.1. The 1st process is the protonation. Through this protonation, the energy of about 200 kcal/mol is released. The 2nd one is the proton transfer. Using that energy the proton carries itself at the end of the donor molecule. The 3rd one is the electron dopping. The donor molecule D^- takes off an excess electron and the proton impurity H^+ traps it to make a bound state like a hydrogen atom. The 4th one is the transfer of electron-proton bound state. When the chain returns to its initial state, this bound state goes back to the acceptor site. The 5th one is the deprotonation. The proton in the bound state combines with hydroxyl ion OH^- to form water molecule. The 6th one is the electron accepting. The electron taken off is captured by the acceptor molecule A^+.

Among them processes #2 and #4 are our main subjects (see Fig.2). Based on the results obtained by quantum chemistry [4], we showed that the proton can move in asymmetric double-well potentials created by the pairs of nearest-neighbour peptide groups [5]. It was also shown that the proton transferred covers distances of order 90 Å (see Fig. 2(a)). In the #4 process the transfer of the electron-proton bound state is associated with the release of energy on moving along the potential gradient. The stabilization of electron movement is very appreciable in the protein, because an electron is strongly bound to the proton motion in a chain (see Fig. 2(b),(c)).

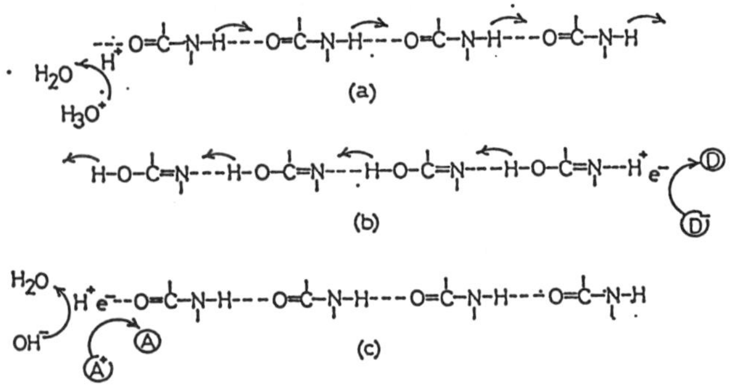

Fig.2. Electron-proton transport

In conclusion, the device presented here can function as a unidirectional electron pump or as a motor. If the electron-proton transfer finds experimental support in proteins, our model very likely plays a certain role in a number of biochemical reactions.

REFERENCES

1. B.P.Atanasov,G.B.Postnikova,Yu.Ch.Sadykov, and M.V.Volkenstein, Mol.Bil.11,537(1977).
2. V.N.Kharkyanem,E.G.Petrov, and I.I.Ukrainskii, J.theor.Biol.73,29 (1978).
3. A.S.Davydov, phys.stat.sol.(b)90,457(1978).
4. T.Minato, and S.Ichinose, unpublished.
5. S.Ichinose, and T.Minato, unpublished.

TWO-TIME FOURIER CONVOLUTION THEOREM AND ITS APPLICATIONS

H. Akama
Physics Department, Towa University, Fukuoka 815, Japan

ABSTRACT

An extension of the Fourier convolution theorem is proposed to cover nonstationary phenomena. Input functions depend on two times, namely, a fast and a slow time. Trasfer functions depend on the input and output times, so that they are no longer translationally invariant. The Fourier transforms of the input and the transfer functions, denoted by F and H, respectively, depend on the frequency ω as well as the slow time t. The first-order corrections to the usual convolution theorem consist of the Poisson brackets of F and H, calculated over the variables ω and t. The method is applied to obtain polarization for an inhomogeneous plasma.

INTRODUCTION

The aim of this paper is to extend the Fourier convolution theorem, so that it becomes valid for phenomena nonstationary in time and inhomogeneous in space.

FORMULATION

The convolution of two functions $f(t)$ and $g(t)$ is

$$h(t) = \int g(t-t')f(t')dt'. \quad (1)$$

The convolution theorem states that

$$H(\omega) = G(\omega)F(\omega), \quad (2)$$

where F, G, and H are the Fourier transforms of f, g, and h, respectively. The functions f, g, and h can be thought of as input, transfer and output functions, respectively.

The transfer function g in Eq.(1) depends on the difference between the output time t and the input time t'. The first generalization can be made by assuming that g depends on t and t'. The input-output relation (1) becomes

$$h(t) = \int g(t,t')f(t')dt'. \tag{3}$$

The second generalization is to assume that the input function f depends on a fast variable as well as on a slow variable.

$$f(t) = \frac{1}{2\pi} \int F(\omega;t)e^{-i\omega t}d\omega. \tag{4}$$

The Fourier amplitude F here is a slowly varying function of t.

For the transfer function $g(t,t')$, a change of variables $(t,t') \to (t-t', t+t')$ is introduced. If there were no $(t+t')$ dependence, g would reduce to $g(t-t')$, suitable for stationary processes. The Fourier transform of g is now designated as $G(\omega;t+t')$, where $t+t'$ is a slow variable. The Fourier transform of Eq.(3) is given by

$$H(\omega;t) = G(\omega;t)F(\omega;t) + i[G,F], \tag{5}$$

where [G,F] is the Poisson bracket of G and F calculated over the varibles ω and t. The term $i[G,F]$ is the first-order corrections to the usual convolution theorem. It is easy to extend Eq.(5) to cases where t and ω are multi-dimensional vectors.

APPLICATIONS

The polarization of an isotropic plasma is given by

$$P = \varepsilon_0 \int \chi(x,x')E(x')d^4x', \tag{6}$$

where ε_0 is the permittivity of free space, χ is the susceptibility, E is the electric field, $x = (x,y,z,t)$ and $d^4x' = dx'dy'dz'dt$. Using Eq.(5) we obtain the Fourier components of the polarization

$$P(k,\omega;x,t) = \varepsilon_0\chi E + i\varepsilon_0[\chi,E], \tag{7}$$

where $E = E(k,\omega;x,t)$ and $\chi = \chi(k,\omega;x,t)$. The Poisson bracket $[\chi,E]$ is calculated over the variables ω,t and k,x.

REFERENCES

1. H. Akama, M. Nambu and D. ter Haar, Phys. Lett. <u>13A</u>, 301 (1985).
2. H. Akama, Physica, <u>149A</u>, 631 (1988).
3. H. Akama, Theoretical and Applied Mechanics, <u>36</u>, 349 (1988).

ON ROTATING SPIRAL WAVES IN REACTION-DIFFUSION SYSTEMS

Shinji Koga

Osaka Kyoiku University, Osaka 543, Japan

Rotating spiral waves, or rotors in reaction-siffusion systems can be created in various media including the famous Belousov-Zhabotinsky reaction [1]. Characterization of this specific pattern formations in both oscillatory and excitable media [2], and the interaction of the spirals [3]-[6] have been to a large extent investigated.

In order to clarify the meaning of the phaseless point at the core, we presented the theory for the rigidly rotating spiral waves in recation-diffusion equations including the multi-armed spirals [7]. This theory is primarily based on the assumption such that the solution can be described by the radius and the phase written as

$$\alpha = \omega t + m\theta - S(r), \tag{1}$$

where ω, m, r and θ are the rotation frequency, the winding number and the polar coordinates respectively. As a result of the theory, we have found that the two quantities such as

$$R(r) = \sqrt{<\mathbf{u}_\alpha, D\mathbf{u}_\alpha>}, \tag{2}$$

$$\tilde{S}_r = S_r - \frac{<\mathbf{u}_r, D\mathbf{u}_\alpha>}{R^2}, \tag{3}$$

correspond to those of the $\lambda-\omega$ system respectively by comparing the set of equations for both cases. Here the notation $<,>$ implies the inner product averaged with respect to the phase and, $\mathbf{u}$ is the deviation part defined as $\mathbf{u} = \mathbf{X}- <\mathbf{X}>$. D is the symmetric diffusion matrix. We can actually confirm that this theory can be applied to the $\lambda - \omega$ system.

In this short note, we apply this theory to the equation written as

$$\frac{\partial}{\partial t}W = a + W - (1 + ic_2)|W|^2 W + \Delta W, \tag{4}$$

where W is the complex field, c_2 is the real parameter, and a is introduced to distort the circular limit cycle in this equation. This additional term is not very significant physically. However, as long as the parameter a is small, we expect that the stable spiral also exists for the oscillatory system exhibiting the relaxation-oscillation character slightly. We have calculated the quantities given by Eqs.(2) and (3) in the slightly distorted system $a = 0.05, 0.1, 0.15$ and $c_2 = 0.5$. As far as these quantities are concerned, we have found that the behaviors of these quantities are quite similar to those of the undistorted system. However we should note here that the the variables at the core take the values which are neither zero nor the dynamical steady state in the absense of the diffusion. The details of the computations for this concrete model are discussed elsewhere.

Acknowledgements

We thank Dr. S. Sasa and Mr. T. Mizuguchi for helpfull discussions.

References

[1] A.T.Winfree, The Geometry of Biological Time (Springer-Verlag, New York, 1980)

[2] see for the complihensive studies

Waves and Patterns in Chemical and Biological Media (H.L. Swinney and V.I. Krinsky eds.) Physica D **49** (1991)

[3] S. Rica and E. Tirapegui, Physica D **48** (1991) 396

[4] P. Pelcé and J. Sun, Physica D **48** (1991) 353

[5] H. Sakaguchi, Prog. Theor. Phys. **82** (1989) 7

[6] I.S. Aranson, L. Kramer and A. Weber, preprint

[7] S. Koga, Prog. Theor. Phys. **67** (1982) 164 and 454

SPINODAL DECOMPOSITION IN TETRAGONAL TiO_2-SnO_2 SYSTEM

S. Nambu, A. Sato, and D. A. Sagala
Central Research Laboratory, Kyocera Corporation
1-4 Yamashita-cho, Kokubu, Kagoshima 899-43, Japan

ABSTRACT

A microscopic calculation was carried out for spinodal decomposition in the tetragonal TiO_2-SnO_2 system by taking the contribution of the elastic free energy into account. Necessary elastic constants for the solid solutions and the elastic free energy of modulated structure were calculated in terms of interatomic potentials. Theoretical study and computer simulation were carried out for the dynamics of spinodal decomposition on the basis of a nonlinear diffusion equation derived from a coarse grained free energy. The time evolution of the microstructure and the effect of elastic strain on the decomposition process were investigated by Langer's approximate method and the finite difference method. It was shown that the composition fluctuation along the [001] direction develops in the first stage, the formation of interface and the grain-growth appear in the second stage, and the interface dislocations are introduced in the third stage. These results agree with the experimental observations in the tetragonal TiO_2-SnO_2 system.

INTRODUCTION

Spinodal decomposition is known to occur in the tetragonal TiO_2-SnO_2 system. Modulated structure has been experimentally observed along the [001] direction.[1] It has been shown that interlamellar surface is coherent during the early stage of the spinodal decomposition, and that, strain mismatch is accommodated by interface dislocations in the later stage. The X-ray diffraction and electron microscopy showed that there are three stages occurring in the spinodal decomposition of the TiO_2-SnO_2 system.[2] The composition fluctuations develop in the first stage which corresponds to the time evolution of satellite peaks. The formation of interface and the grain-growth appear in the second stage which corresponds to the time evolution of Bragg's peaks for separated two phases. Interface dislocations are evident to occur perpendicularly to the lamella structure in the third stage.

THE EFFECTS OF ELASTIC STRAIN

The lattice distortion for spinodal decomposition in this system discussed by Yuan and Virkar[2] is interpreted as follows: In the first stage, a rigid lattice of the solid solution shows no distortion. Only composition fluctuations occur along the [001] direction and lead to the appearance of satellite peaks in the vicinity of the (101) Bragg's peak with growing amplitudes. In the second stage, lattice parameter a does not change, but lattice parameter c shows spatial variation due to the elastic strain between separated two phases, resulting in stress fields to conserve the coherency of the lattice. Then in the third stage, the magnitude of stress field becomes larger than some critical value and interface dislocations should appear perpendicularly to lamella patterns.

Recently, Onuki[3] proposed the Ginzburg-Landau theory for systematic analysis of the elastic effect on spinodal decomposition in cubic alloys. We extend his theory to the tetragonal system to calculate the local stress field and local strain field during spinodal decomposition in the tetragonal TiO_2-SnO_2 system. A microscopic theory[4] is used for calculation of the elastic constants for ionic solid solutions. The composition dependence of the elastic constants and linear expansions of lattice constants for the solid solution system $Ti_xSn_{1-x}O_2$ is derived from interatomic potentials between constituent ions. It is complicated to solve a nonlinear diffusion equation for three dimensional system with respect to composition field and elastic fields. We assume that the composition fluctuations occur along the [001] direction and the local nonuniform deformation of lattice u(r) depends on the z-axis only. Figure 1 shows the spatial distribution of the local stress field $\sigma_{xx}=\sigma_{yy}$ ($\sigma_{zz}=0$) calculated for a local region in the second stage. The absolute value of the maximum stress in our calculation is approximately 4GPa, in contrast, the strength of the actual rutile is several hundred MPa. Consequently, the interface dislocations should be introduced in the (100) and (010) planes in the third stage. It should be noted that the slower coarsening due to the elastic long-range interactions is expected in this system as discussed by Onuki.[3] However, experimental observations will encounter a difficulty due to the appearance of interface dislocations which breaks the coherence of the lattice.

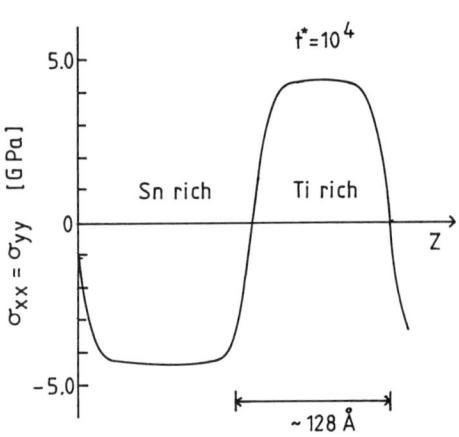

Fig.1. Snapshot of the local stress field for x=0.5, T=1000°C.

This work was supported in part by a Grant for International Joint Research Project from the NEDO, Japan.

REFERENCES

1. A. H. Schultz and V. S. Stubican, Phil. Mag. 18, 929 (1968).
2. T. C. Yuan and A. V. Virkar, J. Am. Ceram. Soc. 71, 12 (1988).
3. A. Onuki, J. Phys. Soc. Jpn. 58, 3065 (1988).
   H. Nishimori and A. Onuki, Phys. Rev. B42, 980 (1990).
4. S. Nambu and M. Oiji, J. Phys. Soc. Jpn. 59, 4366 (1990).
   S. Nambu and M. Oiji, J. Am. Ceram. Soc. 74, [8] (1991).

CONCENTRATION PROFILE OF POLYMERS NEAR A SPHERICAL SURFACE

Takashi Taniguchi, Toshihiro Kawakatsu and Kyozi Kawasaki
Department of Physics, Kyusyu University 33, Fukuoka 812, Japan

ABSTRACT

We present concentration profiles of two types of polymers (homopolymer B, copolymer A-B) near a spherical surface such as a colloidal particle or a micelle where all monomers in B-part are non penetrable to the surface whereas A-part is strongly adsorbed onto it. The concentration profile of a single copolymer (A-B) is also obtained by a numerical calculation.

Statistical properties of polymers near surfaces are important in investigating the stable morphology of the interface or for fundamental understnding of the adsorbing kinetics of polymers onto colloidal particles. We study concentration profiles of two types of polymers near a spherical surface such as a colloidal particle. The first one is a homopolymer (B) (Fig.1a.), the second one is a single block copolymer (A-B) (Fig.1b.) where all the monomers in B-part can not penetrate to the surface while A-part is strongly adsorbed onto it.

Fig.1. Schematic representations of the systems, (a) homopolymers (B) case and (b) a copolymer (A-B) case.

In such two cases, we assume the Gaussian statistic for the polymer conformation. The statistical weight or propagator of a Gaussian chain with N monomers of size a under the condition that one end is at the position $\mathbf{r}_0$ and the other at $\mathbf{r}_1$ is written as[1]

$$G_N(\mathbf{r}_0,\mathbf{r}_1) = \int_{\mathbf{R}(0)=\mathbf{r}_0}^{\mathbf{R}(N)=\mathbf{r}_1} \delta\{R(s)\} \exp(\frac{-3}{2a^2}\int_0^N \dot{\mathbf{R}}(s)^2), \quad (1)$$

where $\mathbf{R}(s)$ is the position vector of s-th monomer. When the path is forbidden to intersect the region D, the inside of the spere, we must impose a bounbary condition on the propagator. The boundary condition is expressed as $G_n(\mathbf{r}',\mathbf{r}'') = 0$ at $\mathbf{r}'$ or $\mathbf{r}'' \in D$ for $n \in [0,N]$. This boundary condition was pointed out DiMarzio[1] as a method of the proper counting of the number of conformations of a polymer near the prohibited region. In the case of Fig.1a, we calculate the concentration profile $\phi(\hat{z},\hat{b})$ which is normalized by the concentration at infinity, where $\hat{z}$ and $\hat{b}$ are the nondimensional distance from the surface and radius of the sphere that are normalized by the gyration radius ($R_g = (\frac{2Na^2}{3})^{\frac{1}{2}}$) of the polymer chain, respectively. The result is

$$\phi(\hat{z},\hat{b}) = (\frac{\hat{b}}{\hat{b}+\hat{z}})^2 \left[(\frac{\hat{z}}{\hat{b}})^2 + 2(\frac{\hat{z}}{\hat{b}})\{\operatorname{erf}(\hat{z}) - 2\hat{z}^2\operatorname{erfc}(\hat{z}) + \frac{2}{\sqrt{\pi}}\hat{z}e^{-\hat{z}^2}\} + \{2\operatorname{erf}(\hat{z}) \right.$$
$$\left. - \operatorname{erf}(2\hat{z}) + \frac{4}{\sqrt{\pi}}\hat{z}(e^{-\hat{z}^2} - e^{-4\hat{z}^2}) + 8\hat{z}^2(\operatorname{erfc}(2\hat{z}) - \frac{1}{2}\operatorname{erfc}(\hat{z}))\}\right] \quad (2)$$

As is shown in Fig.2., in the limit of the infinite radius of the sphere the concentration profile tends to the one that was derived in ref.2 as expected. We can also see that the depletion layer near the surface becomes narrower as the radius of the sphere decreases. We can estimate the width of the depletion layer from eq.(2) as follow

$$(\text{The width of the depletion layer}) \propto \left(\left(\frac{1}{\hat{b}}\right)^2 + \frac{8}{\pi}\left(\frac{1}{\hat{b}}\right) + 4 \right)^{-\frac{1}{2}} \left(\frac{2Na^2}{3}\right)^{\frac{1}{2}} \quad (3)$$

We also calculated the concentration profile of monomers of B-part of a single copolymer (A-B), where A-part is adsorbed onto the sphere. Three cases with the normalized radius $\hat{b} = \infty$ (flat case), 1 and $\frac{1}{8}$, respectively are shown in Fig.3.

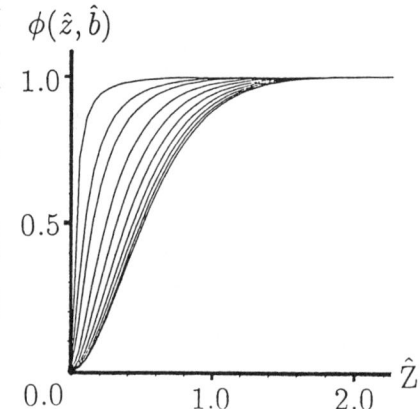

Fig.2. The concentration profiles, $\phi(\hat{z},\hat{b})$, of homopolymer (B) near the sphere with the radius $\hat{b}$=0.01, 0.05, 0.1, 0.25, 0.5, 1, 2, 4, 10 and ∞ from the left to the right, respectively.

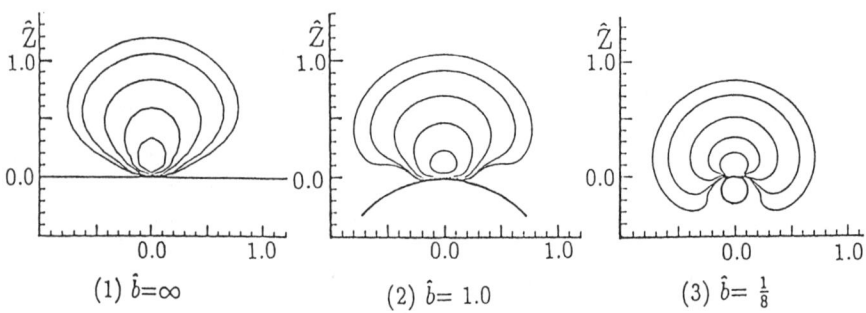

Fig.3. The concentration profiles of monomers of B-part of a single copolymer (A-B), where A-part is adsorbed onto the sphere with the radius (1) $\hat{b}=\infty$, (2) $\hat{b}=1.0$ and (3) $\hat{b}=\frac{1}{8}$, respectively. The curves are equi-concentration lines.

REFERENCE

1 E. D. DiMzarzio, J. Chem. Phys. **42** 2101 (1965).
2 C. M. Marqus and J. F. Joanny, Macromolecles **23** 268 (1990).

LONG TIME TAILS IN DIFFUSION-CONTROLLED RECOMBINATIONS

T. Ohtsuki

Department of Applied Physics, Fukui University, Fukui 910, Japan

Recently, much attention has been paid to slow asymptotic relaxation in simple diffusion-controlled recombination processes[1]. Below an upper critical dimension d_c, particle densities $\rho(t)$ exhibit nonanalytic decay in the long time limit $t \to \infty$. A typical example is A + A $\to \emptyset$ and/or A + A $\to$ A. The conventional rate equation predicts $\rho(t) \propto t^{-1} (t \to \infty)$. However, this mean-field behavior holds only above $d_c = 2$. Below it, spatial fluctuations dominate the process and causes[2]

$$\rho(t) \propto t^{-d/2} \qquad (t \to \infty), \tag{1}$$

In the present work, this behavior is investigated on the basis of a field theoretical renormalization group technique where initial conditions are taken into account explicitly[3]. Fock-space formalism[4] is used to derive the action describing the process. Renormalization is performed rigorously to all orders in a coupling constant[5]. As a result, we get an exact scaling relation for $\rho(t)$

$$\rho(t) = \rho_0 F(\rho_0 k (Dt)^{d/2}), \tag{2}$$

where ρ_0 is an initial density, k is a reaction rate, and D is a diffusion coefficient. Explicit form of scaling functions F is calculated perturbatively in two different manners. One is the usual ϵ expansion, where $\epsilon = d_c - d = 2 - d$. To first-order in ϵ, we have

$$F(X) = \frac{1}{1+X} + \left[\frac{X\{\ln(1+X) + C\}}{2(1+X)^2} - \sum_{n=2}^{\infty} K_n(-X)^n \right] \epsilon + \mathcal{O}(\epsilon^2), \tag{3}$$

where $C = \ln(8\pi) - 1$ and K_n is a numerical coefficient computed from a recurrence formula

$$K_n = \frac{1}{4n!} \sum_{m=1}^{n-1} \frac{1}{m} L_n^m,$$

$$L_n^m = \sum_{i=0}^{m} \sum_{j=0}^{n-m-2} 2(i+j)!\, {}_mC_i\, {}_{n-m-1}C_j\, L_{n-i-j-1}^{m-i} + \frac{1}{3}(m+2)(m+3)n!.$$

The other is the expansion with respect to X. The result to first-order in X is

$$F(X) = 1 - \frac{(8\pi)^{d/2-1}}{\Gamma(2-d/2)\Gamma(1+d/2)} X + \mathcal{O}(X^2). \qquad (4)$$

Note that Eq.(4) is exact to all orders in ϵ. Nonanalytic scaling behavior in the presence of source $\emptyset \to A$ is also discussed and complete scaling relations are derived.

1. R. Kopelman, Science **241**, 1620 (1988) ; V. Kuzovkov and E. Kotomin, Rep. Prog. Phys. **51**, 1479 (1988).

2. A. A. Ovchinnikov and Y. B. Zeldovich, Chem. Phys. **28**, 215 (1978) ; D. Toussaint and F. Wilczek, J. Chem. Phys. **78**, 2642 (1983).

3. T. Ohtsuki, Phys. Rev. **A43**, 6917 (1991).

4. M. Doi, J. Phys. **A9**, 1465 (1976) ; P. Grassberger and M. Scheunert, Fortschr. Phys. **28**, 547 (1980).

5. L. Peliti, J. Phys. **A19**, L365 (1986).

RELAXATION OF CRYSTAL SHAPE PROFILES NEAR THE FACET EDGE

T. Yamamoto
Department of Physics, Gunma University, Kiryu, Gunma 376

N. Akutsu
Osaka Electro-Communication University, Neyagawa, Osaka 572

Y. Akutsu
Department of Physics, Osaka University, Toyonaka, Osaka 560

ABSTRACT

The relaxation process of the crystal shape near the facet is studied. We adopt a dynamical equation which takes account of the non-analytic behavior of the vicinal-surface free-energy to analyze the relaxation process of the crystal shape near the facet. During the relaxation process, the surface gradient p behaves as $p \sim (\Delta r)^{1/2}$ (Δr: distance from the facet edge) off the facet edge and $p \sim \Delta r$ quite near the facet edge.

In recent years, behaviors of equilibrium crystal shapes at the facet edge have attracted much attention since they are analogous to phase transition phenomena. The universal behavior called the Gruber-Mullins-Pokrovsky-Talapov (GMPT) behavior at the facet edge was found theoretically.[1] However, experimental confirmation of this universal behavior has not yet been completed. Experiments both support and do not support the theoretical prediction. One possible reason for the discrepancy is the long relaxation time of the curved surface near the facet edge; if samples are not in the thermal equilibrium, the crystal shape will be different from the 'equilibrium' crystal shape showing the GMPT behavior. The aim of this report is to study the nonequilibrium effect on the crystal shape by examining the relaxation process of the vicinal surface.[2]

We consider the crystal-fluid interface in a temperature region where facets exist. We set up a Cartesian coordinate system such that the origin and x-y plane sit on one of the facets. The z axis is perpendicular to the facet. Near the facet, the equilibrium crystal shape $z = z(x, y)$ is derived from the Andreev free energy $F(p: -x, -y)$ as

$$z = \min_p F(p, :-x, -y),\tag{1}$$

where the surface gradient is denoted by p. We regard p and F in eq.(1) as the order parameter and the Landau free energy in Landau's theory of phase transitions, respectively.

We discuss the relaxation process of the crystal shape based on the time dependent Ginzburg-Landau (TDGL) equation using the Andreev free energy F. We use the Andreev free energy on the basis of the terrace-step-kink (TSK) picture of the vicinal surface. In this picture, the orientation of the steps is assumed to approach to the equilibrium value faster than the step density. Then, p ($\propto$ step density) is chosen to be dynamical variable of the TDGL equation. For small p, the TSK picture allows us to expand F (in the region $x > 0$) as

$$F(p, -x, -y) = A(x, y) p + B(y) p^3 \tag{2}$$

where the coefficients A and B are expressed by the shape of facet contour in the equilibrium state: $x = x_{f,eq}(y)$. We write a TDGL-like dynamical equation for the surface gradient $p(x, y : t)$ as

$$\partial p / \partial t = -\partial F / \partial p, \tag{3}$$

where the unit of the time t is adopted such that the prefactor of the r.h.s. is unity.

Equation (3) is easily solved. The following results are obtained. In the region moderately near the facet edge, the surface gradient p shows the GMPT behavior: $p \sim (\Delta x)^{1/2}$. The distance from the dynamical facet edge $x_f(y, t)$ is denoted by Δx : $\Delta x = x - x_f(y, t)$. However, in the region very near the facet edge, the surface gradient indicates the classical behavior: $p \sim \Delta x$. Hence, we have obtained a dynamical transmutation (GMPT→classical) of the critical behavior at the facet edge. This results gives a possible explanation for the discrepancy between theory and experiment on the facet edge critical behavior.

REFERENCES

1) C.Rottman, M.Wortis, J.C.Heyraud and J.Métois: Phys.Rev.Lett. **52** (1984) 1009.
2) T.Yamamoto, N.Akutsu and Y.Akutsu: J.Phys.Soc.Jpn.(in press).

VOID FRACTION DYNAMICS IN FLUIDIZATION

Shin-ichi Sasa
Department of Physics, Kyoto University, Kyoto 606, Japan

Hisao Hayakawa
Department of Physics, Tohoku University, Sendai 980, Japan

ABSTRACT

A nonlinear evolution equation for void fraction in fluidization is derived from a one-dimensional continuous model by applying the method of adiabatic eliminations. Based on our equation, the dynamics near the transition point from stable fluidized beds to unstable ones is investigated. Phase separations in fluidized beds are also discussed.

Fluidization refers to the phenomena which occur when a collection of solid particles is immersed in a gas fluid stream moving antipararell to gravity at velocities sufficiently large that the drag exerted on the particles exceeds their net buoyant weight and they are thus free to move.[1] Chemical engineers use fluidized beds as a means of obtaining a high rate of heat or mass transfer between the particles and the fluid. It is known that under certain conditions fluidized beds exhibits phenomena called 'bubbling' and 'slugging' although this unstable nature is not preferable from the technological point of view. Therefore, most of previous theoretical work mainly discussed the linear stability of fluidized beds.[2,3] Nonlinear analysis, however, becomes necessary to understand such interesting phenomena in unstable fluidized beds. We develop a new theory in which a nonlinear evolution equation for the void fraction ψ is obtained by eliminating adiabatically fast variables, velocity fields. This approach is useful when we are interested in long time and long distance behavior of the fluidization, because the void fraction is a conserved quantity and it evolves much slower than both of the velocity variables.

From a one-dimensional continuous model, we obtain the evolution equation for ψ in the form[4]

$$\partial_t \psi = \partial_z[-\psi V_0(\psi) + C_2(\psi)\partial_z \psi + C_3(\psi)\partial_z^2 \psi + C_4(\psi)\partial_z^3 \psi], \quad (1)$$

where z denotes a vertical position coordinate, and the ψ-dependence of V_0, C_2, C_3 and C_4 can be calculated. The negativity of $C_4(\psi)$ is required to inhibit short wavelength instability. In this case, the uniform solution $\psi = \psi_0$=const. is unstable when $C_2(\psi_0) < 0$. We note that the uniform state ψ_0 would become unstable in the range $\psi_1 < \psi_0 < \psi_2$, where ψ_1 and ψ_2 are zeros of $C_2(\psi)$.

We investigate the dynamics near the onset of instability $\psi_0 = \psi_1$ introducing a small parameter $\varepsilon = -C_2(\psi_0)$ and assuming a scaling relation $\psi = \psi_0 + \varepsilon^\delta \Psi(\varepsilon^\zeta(z + C_1(\psi_1)t), \varepsilon^\theta t)$, where $C_1(\psi) = V_0(\psi) + \psi V_0'(\psi)$ and the indices δ, ζ and θ are non-negative constants. Under this scaling assumption, the evolution equation (1) is reduced to an equation for Ψ.[4] When we start with the Batchelor model[2], for example, we can show that the equation for Ψ becomes Korteweg-de Vries equation or the Kuramoto-Sivashinsky equation[5,6] depending on the value of the particle viscosity.[4]

Phase separations between the gas phase and the solid phase is described in our model. Under the condition $C_2(\psi) < 0$, the gas phase and the solid phase may corresponds to the regime $\psi > \psi_2$ and $\psi < \psi_1$ respectively. Then, the C_2 and C_4 terms of eq.(1) is rewritten in a similar form of Cahn-Hilliard equation[7] by introducing a potential which has double minima. On the other hand, C_1 and C_3 terms express drift and dispersion effects. We first indicate the importance of these effects. Therefore, our model may exhibit new phenomena of phase separations.

From our approach, we can expect the essential development of understanding the fluidization including the pattern evolution in unstable fluidized beds.

REFERENCES

1. Fluidization, edited by J.F. Davidson, R. Clift and D. Harrison, (Academic Press, London,1985).
2. G.K. Batchelor, J.Fluid Mech. **193**, 75 (1988).
3. R. Jackson, in ref.1, p.47.
4. S. Sasa and H. Hayakawa, preprint.
5. Y. Kuramoto and T. Tsuzuki, Prog.Theor.Phys. **55**, 356 (1976).
6. G.I. Sivashinsky, Acta Astronautica 4, 1177 (1977).
7. J.W. Cahn and J.E. Hilliard, J.Chem.Phys. **28**, 258 (1958)

CALCULATION OF ROTATIONAL TUNNELING STATES OF THE METHYL GROUP IN THE HIGHER ORDER OF POTENTIAL

Yoshiaki Ozaki
Department of Chemistry, Nagoya Institute of Technology
Nagoya, Japan 466

ABSTRACT

The intensities are calculated of inelastic neutron scattering for the protons of methyl group in hindering potential. Orientationally localized states are employed as rotational bases. The Q-dependence of intensity of each peak is derived in the case of 3-fold and 6-fold potentials.

INTRODUCTION

The methyl group CH₃ of a molecule in the solid phase has rotational motion hindered by intramolecular and/or intermolecular interactions. A wide range of values of rotational energy E_t have been found directly by inelastic neutron scattering experiments. The minimum value of E_t is less than $1 \mu eV$ ($E_t/k \leq 10$ mK) while E_t can go up to 1 meV. This diversity of energy size results from the difference in strength and symmetry of the potential. When the energy is smaller than 10% of the value for free rotation, that is, than about 0.1 meV, the bottom part of energy levels makes a group and forms "rotational tunneling" levels. It corresponds to the splitting ground state of librational motion.

In general, because of the symmetry of a methyl group, the rotational potential is confined to the periodic form of $(V_n/2)\cos(n\alpha + \alpha_n)$ (n=3m), where V_n is the barrier height and α_n the phase shift in rotational angle. When the methyl group interacts significantly with the remaining part of the molecule of interest, some terms in the potential may disappear according to the symmetry of that part of molecule. Similarly, if the methyl group has the important interaction with neighbouring molecules, the potential loses several terms due to the symmetry of crystal lattice. In most cases, the first two components (n=3 and 6) are used to analyze experimental results.

© 1992 American Institute of Physics

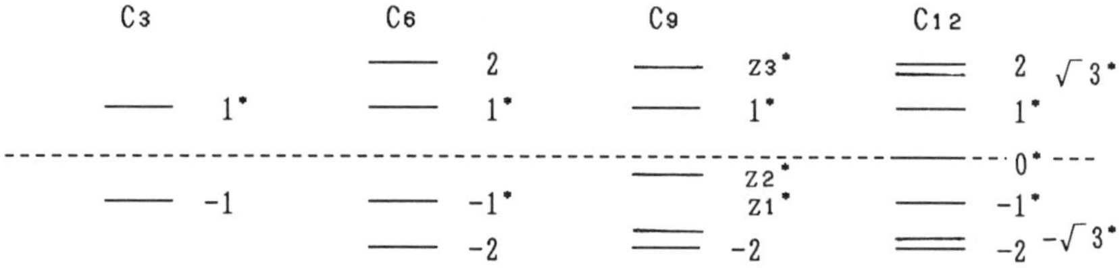

```
        C₃              C₆              C₉              C₁₂
                      ──── 2         ──── z₃*        ════ 2 √3*
        ──── 1*       ──── 1*        ──── 1*         ──── 1*
        ─────────────────────────────────── ──── z₂* ──── ──── 0* ───
                                             z₁*
        ──── -1       ──── -1*       ──── -2         ──── -1*
                      ──── -2        ══════ -2       ════ -2 -√3*
```

Fig. 1. Rotational energy levels of the localized methyl group in the n-fold (n=3, 6, 9, 12) potential. (z_1=-1.532, z_2=-0.347, z_3=1.879)

ROTATIONAL STATES AND INTENSITIES

When the potential has one of 3m-fold (m=1, 2, 3, 4, ...) symmetry, the energy levels are derived using extremely localized states (Fig. 1) and assuming the transfer is possible only to the next state. The symbol * stands for doubly degenerated states. In the case of 6-fold potential, equally separated four peaks should be observed by neutron scattering. The intensity of inelastic neutron scattering[1] is calculated by the use of similar bases. If the temperature is high enough to produce the equal population in each level, the Q-dependence of intensity[2] is expressed in a simple form. For example, the peak by transition from the ground state to the first excited state in 6-fold potential gives

$$S(0 \rightarrow 1) = (1/12)(2 - f_1 - 2f_2 + f_3) , \quad (1)$$

where f_n is $j_0(Q \cdot c_n a)$. Here, j_0 is the 0-th order of spherical Bessel function, c_1=1, c_2=√3, c_3=2, and a is H-H distance in a methyl group.

CONCLUSION

The Q-dependence has been observed in the system with four peaks.[2] All peaks have the similar behavior and therefore the agreement is not good between calculated result and observed one. First, the case of potential with 3-fold and 6-fold symmetry should be investigated. Second, the direct comparison is also necessary in spin conversion time.

REFERENCES

1. D. W. Matuschek and A. Hüller, Can. J. Chem. **66**, 495 (1988).
2. S. Ikeda et al., J. Phys. Soc. Jpn. **60**, 3340 (1991).

FLUCTUATIONS OF HYDROGEN BOND NETWORK IN LIQUID WATER

Masaki Sasai
Department of Chemistry, College of General Education
Nagoya University, Nagoya 464-01, Japan

Recently, we investigated the power spectrum of the energy fluctuation of the hydrogen bond (HB) network[1] and found that liquid water exhibits $1/f$ ($f=\omega/2\pi$) type fluctuation for $f \approx 1\text{-}1000 \text{cm}^{-1}$. We here discuss a simple cellular dynamics (CD) model to explain this long time fluctuation in liquid water.

In liquid phase, water molecules form the random HB network. The HB network is percolating throughout the system and changes its geometrical pattern with thermal vibrations. When some parts of the network become unstable and energy of the molecules in such parts exceeds the limit, the network is reorganized with rotations and large displacements of unstable molecules. Thus the unstable molecules are stabilized by distributing the excess energy to neighboring sites, which is often more than several kcal/mol per molecule.[2] Then the next accumulation of instability starts in the neighbor regions and the system never reaches to the uniform stable state. This reminds us of the dynamics in random spin systems and thus we can say liquid water is 'frustrated'. We develop the CD model of this 'frustration' dynamics and compare it to the molecular dynamics (MD) results.

We consider NxNxN cubic lattice with the periodic boundary condition and define 'energy' $V_i(t)$ at the i-th site (cell). The 'energy' at the next time step $V_i(t+\Delta t)$ is determined as

(1) If $E^- < V_i(t) + \{\eta_i(t)\}^2 < E^+$ for all i, then $V_i(t+\Delta t) = V_i(t)$, where $\eta_i(t)$ is Gaussian and white, which mimics thermal vibrations, $<\eta_i(t)\eta_j(t')> = D \delta_{ij}\delta_{tt'}$.

(2) If $V_i(t) + \{\eta_i(t)\}^2 \geq E^+$ at the i-th site, $V_i(t) \to V_i'(t) = V_i(t) - V_s$ and

$V_j(t) \to V_j'(t) + aV_s/z$, where j is the 1-st or 2-nd neighbor of i, $z=18$ is the number of neighbors, and $0 < a < 1$. Here $1-a$ is the rate of the energy dissipation.

(3) If $E^- \geq V_i(t) + \{\eta_i(t)\}^2$ at the i-th site, $V_i(t) \to V_i'(t) = V_i(t) + V_d$, where $V_d = E^- - V_i(t) - \{\eta_i(t)\}^2$.

The rule (2) represents the stabilization of high energy parts and the destabilization of surrounding molecules. The rule (2) resembles to the relaxation rule of the sand pile model of the self organized criticality[3] and thus avalanches (cascaded stabilizations and destabilizations) of various size are induced. The energy proportional to the avalanche size is lost with the rate $1-a$. Too much cooled sites, however, are heated by the rule (3). Each site is changed in parallel. When all avalanches cease,

(4) if $E^- < V_i'(t) + \{\eta_i(t)\}^2 < E^+$ for all i, then $V_i(t+\Delta t) = V_i'(t)$,

We show in Fig.1 the temporal fluctuation of the total energy $V(t) = \Sigma_i V_i(t)$ in the CD model and its power spectrum $S(\omega) = |\int dt\, V(t)\exp(i\omega t)|^2$.

When the network structures visited by the trajectory in the MD simulation are quenched to their local minima, one can get the series of inherent (quenched) structures.[2,4] Fig.2 is the total potential energy $V(t)$ of inherent structures and its power spectrum $S(\omega)$.[1] When the parameter D is decreased in the CD model, the slope of the spectrum increases and the cut off frequency (the bend in the spectrum) shifts to the lower frequency side. If we regard D as temperature, this agrees with the tendency observed in the MD data. Such similarities between the CD model and the MD simulation strongly suggests that the 'frustration' dynamics is essential in the HB network rearrangement.

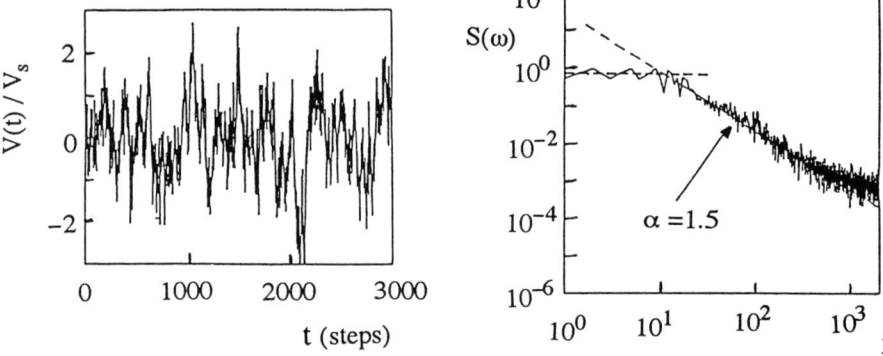

Fig.1 $V(t)$ in the CD model (left) and its power spectrum (right). N=8 cubic lattice was used. The parameters are $E^+=10$, $E^-=-11$, $V_s=20$, $D=0.9$, and $a=0.9$.

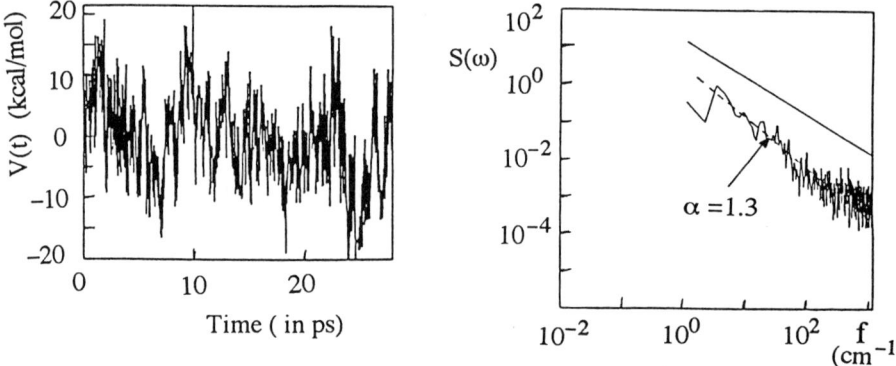

Fig.2 The total potential energy $V(t)$ of the inherent structures in the MD calculation (left) and its power spectrum (right). The system with 64 water molecules at 297K. f is in wave number (cm^{-1}) and $f=\omega/2\pi$. A solid line indicates $1/f$ and a dashed line is $1/f^\alpha$ with $\alpha=1.3$.

REFERENCES

1. M. Sasai, I. Ohmine, and R. Ramaswamy, to be published in J. Chem. Phys.
2. I. Ohmine, H. Tanaka, and P. G. Wolynes, J. Chem. Phys. 89, 5852 (1988).
3. P. Bak, C. Tang, and K. Wiesenfeld, Phys. Rev. Lett. 59, 381 (1987); Phys. Rev. A38, 364 (1988)
4. F. H. Stillinger and T. A. Weber, Phys. Rev. A25, 978 (1982); Science 225, 983 (1984).

HOLE CONDUCTION IN THE ONE-DIMENSIONAL MOLECULAR CONDUCTOR NICKEL-RICH COBALT PHTHALOCYANINE IODIDE

A. Mishima
Kanazawa Institute of Technology, Nonoichi, Ishikawa 921, Japan

ABSTRACT

A hole conduction in the one-dimensional molecular conductor, nickel-rich cobalt phthalocyanine iodide, is studied using a $d\pi$-coupled model. In this model, the d-electron spins on the metal spine couple ferromagnetically the itinerant-π-electron spins on the macrocycle. The π-band gap and the optical-absorption spectrum are computed in the mean-field approximation with the periodic-boundary condition. It is pointed out that the π holes contribute to the electric conduction in the nickel-rich cobalt phthalocyanine iodide.

INTRODUCTION

$Co_xNi_{1-x}PcI$ (CNPI) are one-dimensional (1D) alloys of the isostructural porphyrinic molecular conductors phthalocyaninatocobalt iodide, CoPcI, and phthalocyaninatonickel iodide, NiPcI.[1,2] MPcI consists of metal-over-metal stacks of MPc units surrounded by chains of iodine. In NiPcI, the metallic behavior is due to the π electrons on the Pc ring. In CoPcI, a ferromagnetic-exchange coupling between covalt and ligand is conjectured from the similarity between CoPcI and $CoTTP(SbF_6)_{1.0}$ (TPP=tetraphenylporphyrinate) with a triplet ground state.[3] We have studied the $d\pi$-ferromagnetic-exchange-coupled model for the cobalt-rich nickel phthalocyanine iodide.[4]

Here, we study the electronic and optical properties of the 1D-$d\pi$-coupled model for the nickel-rich cobalt phthalocyanine iodide at $x=1/3$. The model contains the transfer energy of d (π) electrons, $T_d(=0.15 \text{ eV})$ ($T_p(=-0.3 \text{ eV})$), the on-site Coulomb repulsion of d electrons, $U_d(=1.5 \text{ eV})$, and the ferromagnetic-exchange coupling between d and π electrons, $J(=0.027 \text{ eV})$.[3,4] We compute the energy bands, the charge and spin densities and the optical-absorption-spectral intensity by considering a single-electron excitation within the mean-field theory with the periodic-boundary condition.[4]

RESULTS AND CONCLUSION

Figure 1 shows that the gap energy of the π band increases as J increases. Thus the π-band gap occurs due to J. The π (d) band is shown by the solid (dashed) line in Fig. 2. The gap energy is about 0.013 eV at the Fermi point of the π band. Figure 3 shows the spin, S_p (S_d), and charge, C_p (C_d), densities of the π (d) electron vs. site number. The d and π spins form the spin-density-wave states with a

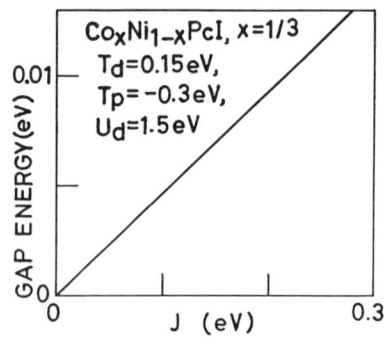

Fig. 1. Gap energy of the π band vs. $d\pi$-exchange coupling J.

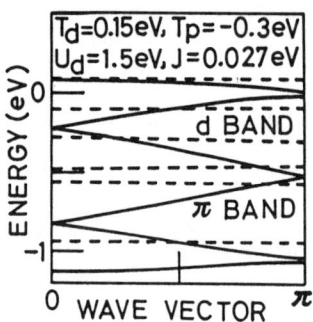

Fig. 2. Energy vs. wave vector of d and π bands.

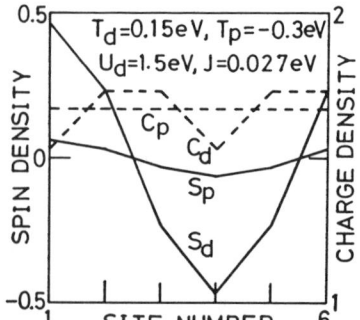

Fig. 3. Spin and charge densities vs. site number.

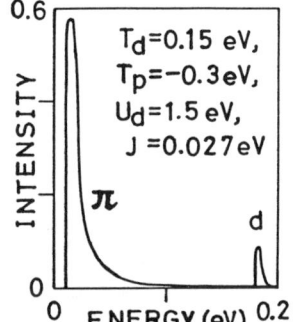

Fig. 4. Spectral intensity vs. absorption energy.

six-times period. The optical-absorption-spectral intensity vs. the absorption energy is shown in Fig. 4. The spectrum shows the asymmetric band of the π electron with a peak at 0.01 eV and with a tail to 0.2 eV, and the band of the d electron with a peak at 0.18eV.

As the carriers are the π holes at $x=1/3$ in this model, the result is consistent with the experiment of the thermoelectric power for $0 \leq x \leq 0.5$ in CNPI. Thus the π holes contribute to the electric transport and the origin of the π-band gap is J at $x=1/3$ in CNPI.

The author thanks Prof. K. Yakushi for the useful discussions.

REFERENCES

1. K. Liou, C. S. Jacobsen, and B. M. Hoffman, J. Am. Chem. Soc., in press.
2. J. Martinsen and S. M. Palmer, J. Tanaka, R. C. Greene, and B. M. Hoffman, Phys. Rev. B30, 6269 (1984).
3. K. Yakushi, H. Yamakado, T. Ida, and A. Ugawa, Solid State Commun., in press.
4. A. Mishima, Mol. Cryst. & Liq. Cryst., in press.

VARIATION PRINCIPLE FOR STOCHASTIC PROCESSES
- GENERALIZATION OF PATH PROBABILITY METHOD -

M. Kaburagi, K. Wada[A], A. Suzuki[B],
R. Kikuchi[C] and H. Sato[D]

College of Liberal Arts, Kobe Univ., Nada, Kobe 657, Japan
[A]Dep. Physics,, Hokkaido Univ., Sapporo 060, Japan
[B]Canon Inc. Research Center, Atsugi 243-01, Japan
[C]Dep. Mat. Sci. & Eng. U.C.L.A., CA. 90024-1595, U.S.A.
[D]Schl. Mat. Eng., Purdue Univ., West Lafayette IN. 47907, U.S.A.

ABSTRACT

A variation principle which leads to general expression for the joint probability in stochastic processes is discussed. A systematic way of approximation for the joint probability is proposed.

INTRODUCTION

As is well known, a great deal of development in physics has been achieved by expressing physical law in variation principle. In case of the nonequilibrium problems, the path probability method (PPM) devised by Kikuchi[1] is the one of the most successful variational approaches and has been applied to various transport phenomena[2]. The PPM, however, is applicable only to the case of Markov processes. It is, therefore, most desirable to generalize the PPM to general cases. In this paper, we derive a variation principle which leads to a general expression for the joint probability in stochastic processes. The variation principle is a natural generalization of the PPM and it provides sound bases to the PPM. A systematic way of approximation for the joint probability is proposed. The relation between the variation principle and the master equation method is examined.

VARIATION PRINCIPLE FOR STOCHASTIC PROCESSES

We start with an ensemble consisting of L identical systems. The state of the system is assumed to be designated by discrete value x. We denote the probabilities of the stochastic processes for the system as follows: nth order joint probability (the probability of finding the value $x_n, x_{n-1}, \cdots, x_1$ at respective time $t_n, t_{n-1}, \cdots, t_1$) by $P_n(x_n, t_n; x_{n-1}, t_{n-1}; \cdots; x_1, t_1)$ or abbreviated $P_n(\{n\};\{n-1\};\cdots;\{1\})$: nth order transition (or conditional) probability (the probability of finding the value x_n at time t_n under the condition that the system has the value $x_{n-1}, \cdots, x_1$ at respective time $t_{n-1}, \cdots, t_1$) by $w_n(x_n, t_n | x_{n-1}, t_{n-1}; \cdots; x_1, t_1)$ or $w_n(\{n\}|\{n-1\};\cdots;\{1\})$. Usual approach to the stochastic processes start with the "basic relation"

$$P_n(\{n\};\{n-1\};\cdots) = w_n(\{n\}|\{n-1\};\cdots) \cdot P_{n-1}(\{n-1\};\cdots). \qquad (1)$$

Before going into the variation principle, we define the "path function" $\Phi = S - E$ consisting of the "path entropy" S (logarithm of the number of ways of distributing the paths) and the "path energy" E (logarithm of the weight of the paths), where S and E are defined as

$$e^S = \prod_{\times 1 \times 2} \cdots \prod_{\times n-1} \left\{ \frac{[LP_n(\{n\};\{n-1\};\cdots)]!}{\prod_{\times n}[LP_{n-1}(\{n-1\};\cdots)]!} \right\} \quad (2)$$

$$e^{-E} = \prod_{\times 1 \times 2} \cdots \prod_{\times n} [W_n(\{n\}|\{n-1\};\cdots)]^{LP_n(\{n\};\{n-1\};\cdots)} \quad (3)$$

It is easily shown that the "basic relation" (1) can be derived by maximizing the path function $\Phi = S - E$ with respect to P_n under the condition that P_n is compatible with all the lower order joint probabilities. This is just the variation principle which leads to the "basic relation" (1). The most probable path in this variation principle is, therefore, given by the relation (1). Furthermore it is advantageous that the fluctuations around the most probable path can be treated in this variation principle by estimating the second derivative of $\Phi = S - E$ with respect to $P_n{}^3$.

SYSTEMATIC APPROXIMATIONS

In case of the cluster variation method (CVM), a systematic way of superpose approximation has been given by Morita[4]. Since the above variation principle is a natural extension of the CVM, we can employ the Morita's method by adding a time axis to the system. Let us consider the system consisting of N Ising spin variables $\{\sigma\}=x$. Then we have Nn spin variables $x_n, x_{n-1}, \cdots, x_1$ in P_n. In ref.5, the superpose approximation to this system has been given in detail for the Markov processes (n=2) where the approximation corresponds to the pair approximation with respect to the time axis. In a similar manner as in ref.5, we can expand the path entropy as

$$S = \sum_i \Gamma_1(i) + \sum_{i<j}\sum \Gamma_2(i,j) + \cdots + \Gamma_{Nn}(1,\cdots,Nn), \quad (4)$$

$$\Gamma_k(i_1,\cdots,i_k) = S_k(i_1,\cdots,i_k) - \sum S_{k-1}(j_1,\cdots,j_{k-1}) + \cdots, \quad (5)$$

$$S_k(i_1,\cdots,i_k) = Tr'[P_n(\{n\};\{n-1\};\cdots)\log\{P_n(\{n\};\{n-1\};\cdots)\}], \quad (6)$$

where Tr' denotes the trace over all spins except for i_1 to i_k spins. The "cluster entropy" S_k can be expressed in terms of the multi spin correlation functions[6]. If we neglect farther neighbor spin correlations, we obtain a close form of variation function. We can also derive the master equation for distribution function of the spin correlation as is given in ref.7. Details will be published elsewhere.

REFERNCES

1. R.Kikuchi, Progr. Theor. Phys. Suppl. **35**,1(1966)
2. R.Kikuchi et.al: Physica **123A**,227(1984); K.Wada et.al: J. Phys. Chem. Solid **46**,1195(1985); S.A.Akbar et.al:J. Am. Ceram. Soc. **70** 246(1987); J. Phys. Chem. Solid **48**,579(1987); M.Kaburagi et.al: Suppl. to Trans. JIM, **29**,199(1988); M.Asta et.al: Phys. Rev. **B44** 4907, 4914(1991)
3. T.Mohri: Acta mettall. **38**,2455(1990)
4. T.Morita: J. Phys. Soc. Jpn. **12**,753(1972)
5. K.Wada et.al: J. Stat. Phys. **53**,1081(1988)
6. M.Sanchez et.al:Phys. Rev. **B21**,216(1980)
7. T.Ishii: Monbusho joint Research Report No.01044006, 127(1991)

CALCULATION OF PHONONS IN THE AMORPHOUS STRUCTURE

H. Sato

Department of Physics, Aichi University of Education, Kariya 448, Japan

A. Ishida and M. Itoh

Department of Physics, Shimane University, Matsue 690, Japan

ABSTRACT

The roton-like excitations identified in many metallic glasses by neutron difraction experiments and computor simulations have been considered as longitudinal accoustic phonons, since they all seem to lead to the sound waves in a elastic continuum in the long wavelength limit [1]. In this paper, however, we will show that they are in fact optical phonons, by applying both analytical and computational methods to amorphous $Mg_{70}Zn_{30}$ alloy.

MODELS AND METHODS

The atomic vibration is treated by assuming there exist some "equilibrium" positions of atoms in the atomic structure. Assuming further pairwise atomic interaction, the force constant between the i-th and j-th atoms is written in the form

$$\Phi_{ss'}^{\alpha\beta}(R_i - R_j) = F_1^{ss'}(|R_i - R_j|)\cdot\delta_{\alpha\beta} + F_2^{ss'}(|R_i - R_j|)\cdot\hat{R}_{ij}^{\alpha}\hat{R}_{ij}^{\beta} , \qquad (1)$$

where $F_1^{ss'}$ and $F_2^{ss'}$ are the functions only of the atomic distance, and the indices s and s' specify the atomic species locating at R_i and R_j respectively. The dynamical matrix of the system is then obtained through the relation $D_{ij}^{\alpha\beta} = M_i^{-1/2} \Phi_{ij}^{\alpha\beta} M_j^{-1/2}$. Within this model we have attempted the calculation of the dispersion relation of binary $Mg_{70}Zn_{30}$ system by adopting the following two alternative approaches.

(1) Numerical method

We have derived the pair potentials from the perturbation theory based on the OPW pseudopotential formalism. The amorphous structure has been constructed for the system of 1000 atoms with cyclic boundary condition, by using the relaxation algorithm identical to the steepest gradient technique. The dynamical matrix is prepared and diagonalized. Eigenvectors are decomposed into the longitudinal and the transverse components, and the longitudinal spectral distributions are then calculated (Figure 1). It is possible to identify the peak positions, although the width becomes fairly broad for higher wave numbers and a single roton-like dispersion is determined (Figure 2). It is similar to those obtained by other authors experimentally or by simulations [1].

(2) Green function method

The form of the force constant (1) and the isotropy of the system imposes very strong restrictions on the form of the averaged Green function matrix. In particular, its Fourier transform takes a very simple form if the wave number k is parallel to one of the coordinate axes, say, z-axis; it is shown to be composed of the diagonal 3×3 sub-matrices defined for each pair of atomic species s and s':

$$G_k^{ss'} = \begin{pmatrix} g_{\perp}^{ss'}(k) & & \\ & g_{\perp}^{ss'}(k) & \\ & & g_{\parallel}^{ss'}(k) \end{pmatrix} . \qquad (2)$$

© 1992 American Institute of Physics

Calculation of the Green function was performed by neglecting the coupling between the Green function and its self energy part in the equation of motion. This decoupling procedure is the so-called quasi crystalline approximation (QCA). It brings no damping effect but incorporate structural information through the radial distribution functions. The equation of motion is solved for the Green function and it is expressed analytically in terms of the Fourier transforms of the force constants (1) and the radial distribution functions [2]. For these input quantities we have used those obtained in the above simulation and determined all excitations from the singularities. In contrast to the simulation study we obtained four different modes - a singlet and a doubly degenerate branches, the frequencies of which go to zero as $k \to 0$, and also a singlet and a doubly degenerate ones with a common finite frequency in the same limit. These should be properly called LA, TA, LO and TO phonons respectively. We have plotted LA and LO dispersions in the figure 2, together with that determined by the method (1) for comparison. The simulation result is seen to be rather close to the LO mode in the roton-like region.

CONCLUSIONS

From above calculations we conclude the followings:

(1) The roton-like behaviour of the observed excitations in binary metallic glasses are probably due to LO phonons: the experimental peaks, as well as those found in the simulations, are perhaps too broad to distinguish LA and LO modes.

(2) There are also TA and TO modes, and these classifications are completely general under the harmonic approximation with pairwise potentials, since the appearance of different modes is due to the hybridization between sub-matrices given in eq (2).

Figure 1   Figure 2

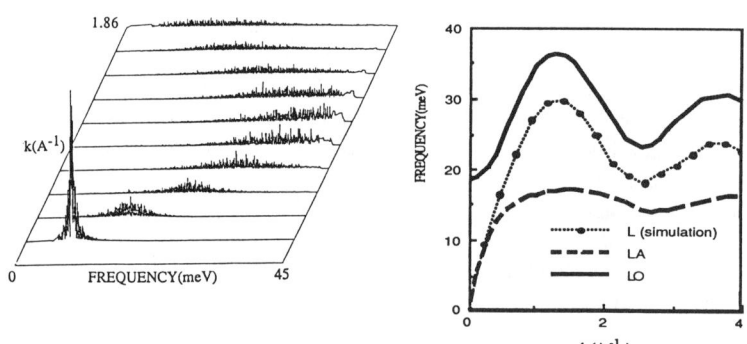

Spectral distributions of a model amorphous $Mg_{70}Zn_{30}$ alloys of 1000 atoms (Figure 1) and its phonon dispersions (Figure 2). In the roton-like region the simulation is rather close to the optical mode.

REFERENCES

1. see J. B. Suck Materials Science and Engineering (1991) A133, 40-44 and references therein.
2. Itoh et al in preparation

Long Tail Behaviors of Random Medium

Hiroaki Hara and Junji Koyama*

Department of Engineering Science, Tohoku University
Aramaki-Aoba, Aobaku, Sendai, 980 Japan
* Geophysical Institute, Tohoku University
Aramaki-Aoba, Aobaku, Sendai, 980 Japan

As an attribute of innumerable phenomena, especially in the complex systems, the concept of self similarity has been established. The self similarity is an invariant property against a scaling operation, or operations to the quantity concerned. The property is essentially important to characterize the wide variety of dissimilar phenomena[1].

Here we propose a scaling rule for the temporal behaviors of a complex system. The system is composed of many different types of elements which are grouped into clusters, and each cluster is labeled by an index i $(= 1, 2, ..., M)$. The clusters are distinguished from the elements by the dynamical response behaviors to an applied force. The clusters are assumed to be governed by the respective proper equations;

$$\frac{dX^{(i)}(t)}{dt} = -\Phi^{(i)}(t) X^{(i)}(t) + n^{(i)}(t) \tag{1}$$

$$< n^{(i)}(t_1)\, n^{(i)}(t_2) > = \sigma_i^2 \delta(t_1 - t_2) \tag{2}$$

where $X^{(i)}(t)$ is a state variable for the cluster designated by the index i, and is called a local state of the whole system. Symbols $\Phi^{(i)}(t)$ and $n^{(i)}(t)$ denote the transition probability of and the random noise to $X^{(i)}(t)$, respectively. σ_i^2 is the variance of the noise.

We assume that the cluster itself is composed of many units having a self similar structure characterized by scaling rules for $X_k^{(i)}(t)$ and $n_k^{(i)}(t)$, $(k = 0, 1, 2, ...)$ given below;

$$(a_i/b_i)^{1/2} X_{k-1}^{(i)}(b_i t) = X_k^{(i)}(t) \tag{3}$$

$$(a_i b_i)^{1/2} n_{k-1}^{(i)}(b_i t) = n_k^{(i)}(t) \tag{4}$$

where a_i and b_i are scaling factors satisfying $a_i < b_i < 1$, and are specified by σ_i^2 and $\Phi^{(i)}$. Index k stands for the scaling stage. Note that $X_0^{(i)}(t) = X^{(i)}(t)$ and $n_0^{(i)}(t) = n^{(i)}(t)$.

In order to study the dynamical behavior of the whole system, we introduce a correlation function $\tilde{C}_{XX}(\tau)$ defined by

$$\tilde{C}_{XX}(\tau) = \sum_{i=1}^{M} C_{XX}^{(i)}(\tau) \tag{5}$$

where $C_{XX}^{(i)}(\tau) \equiv\, <\tilde{X}^{(i)}(t+\tau) \tilde{X}^{(i)}(t)>$. The state $\tilde{X}^{(i)}(t)$ represents the response of i-th cluster generated by the scaling factors; the state is a sum of the scaled local states $X_k^{(i)}(t)$ $(k = 0, 1, 2, ...)$.

Repeated applications of the scaling rules to the solution of Eq. (1) lead us to evaluate $\tilde{C}_{XX}(\tau)$. The result shows a long tail behavior, $\sum w_i \tau^{-\xi_i+1}$ (w_i; weight factor: $\xi_i = \ln a_i / \ln b_i$) similar to that obtained from the Weierstrass function[2)-5)]. The parameters of ξ_i and w_i could be determined experimentally. Suppose that a set of ξ_i is arranged in an increasing order, $\xi_1 < \xi_2 ... < \xi_n < ...$, and that i-dependence of ξ_i is expressed approximately by $\xi(x) = \eta x$ ($\eta > 0$). Let $w(x)$ be a continuous weight function corresponding to w_i, and let the asymptotic behavior of $\tilde{C}_{XX}(\tau)$ be $\tau^{1-\eta}$. Then we get a functional equation for $w(x)$ in the integral form corresponding to (5);

$$w(x) = a_\eta b_\eta w(b_\eta x) \qquad (6)$$

where $a_\eta = \tau^{-\eta}$ and $b_\eta = \eta \ln \tau$. By solving (6), we obtain an asymptotic form for $w(x); \sim x^\alpha$ ($\alpha = \eta \ln \tau / \ln |\eta \ln \tau| - 1$). The scaling rule (6) is for the clusters, while the scaling rules (3) and (4) are for the units in local states. It is important that Eq.(6) manifests a nonlinear scaling rule for the complex system.

Fig. 1. Random Medium is an aggregation of clusters. Each cluster is composed of many units, which is represented by a close roop.

References

(1) H. Hara and S. Okayama; Phys. Rev. **B37**, 9504 (1987).
(2) E.W. Montoroll and M.F. Shleginger; J. Stat. Phys. **32**, 209 (1983).
(3) M.F. Shlesinger and B.D. Hughes; Physica **109A**, 115 (1981).
(4) B.J. West; Fractal Physiology and Chaos in Medicine, World Scientific, Singapore (1990).
(5) J. Koyama and H. Hara; Proc. Noise Physical System 1/f Fluctuations (1991).

CROSSOVER PHENOMENON IN SLOW PARTICLE GROWTH ON A SUBSTRATE

Michio Tokuyama
Tohwa Institute for Science, Tohwa University, Fukuoka 815, Japan

Yoshihisa Enomoto
Department of Physics, Nagoya University, Nagoya 464, Japan

A new viewpoint on the kinetics of electrochemical nucleation is presented by studying diffusive long-range interactions among hemispherical nuclei on a substrate in two theoretical aspects.

The first is to study the causal motion which is described by the single nucleus distribution function $f(R,t)$ with radius R. The distribution $f(R,t)$ is shown to obey the following Fokker-Planck type kinetic equation in the dimensionless form:[1]

$$\frac{\partial}{\partial t} f(R,t) = \lambda \frac{\partial}{\partial R} \frac{1}{R} \{ -1 + \alpha \varepsilon [1 + \beta \frac{\partial}{\partial R} \frac{1}{R}] \} f(R,t) \qquad (1)$$

with the screening term

$$\lambda(t) = \frac{1}{2} - \frac{1}{\pi} \int_0^t ds \frac{\langle R \rangle(s) \lambda(s)}{(t-s)^{1/2}}, \qquad (2)$$

and the area fraction of the hemispherical nuclei

$$\varepsilon(t) = (2Q/\pi)^{1/2} \langle R \rangle(t)^2, \qquad (3)$$

where Q is a volume fraction of the nuclei, and the brackets denote the average over $f(R,t)$. Here α and β are the parameters to be determined self-consistently by Eq.(1). Then, the function $f(R,t)$ is scaled as

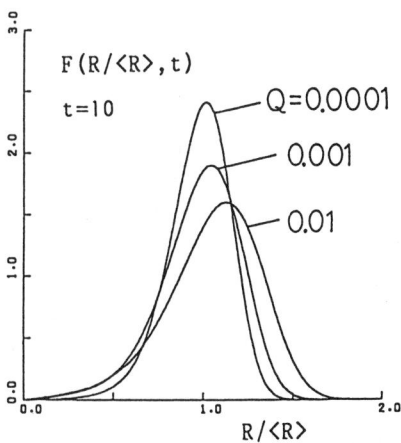

Fig.1 $F(R/\langle R \rangle)$ vs $R/\langle R \rangle$

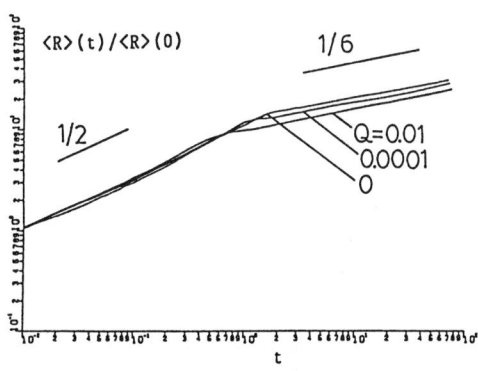

Fig.2 $\langle R \rangle$ vs t

$f(R,t) = [n/\langle R \rangle] F(R/\langle R \rangle, t)$, where n denotes the constant number density. The relative nucleus size distribution function F is experimentally observable by an electron microscope.

There are two kinds of many-body effects on diffusion-controlled nucleus growth on a substrate. The static (screening) effect of order Q^0, which is given by the first term of Eq.(1), causes the crossover of the exponent for $\langle R \rangle$;

$$\langle R \rangle \sim t^{1/2} \quad \text{for } t \langle 1, \quad \langle R \rangle \sim t^{1/6} \quad \text{for } t \rangle 1. \quad (4)$$

The dynamic (correlation) effect of order $Q^{1/2}$, which is given by the second term of Eq.(1), gives rise to a dispersion in the size of the nuclei (see Fig.1) and causes appreciable corrections to the amplitude for $\langle R \rangle$ (see Fig.2).[2]

The second is to explore the fluctuations around the causal motion. There are two types of fluctuations; initial thermal fluctuations and non-thermal fluctuations generated by the correlation effect. Although they are small as compared to the causal motion, they are still important since they are observable as a structure function $S_k(t)$ by scattering experiments. $S_k(t)$ is shown to obey a linear equation with two types of source terms which originates from initial thermal fluctuations and non-thermal fluctuations. In Fig.3 we show the time evolution of $S_k(t)$.

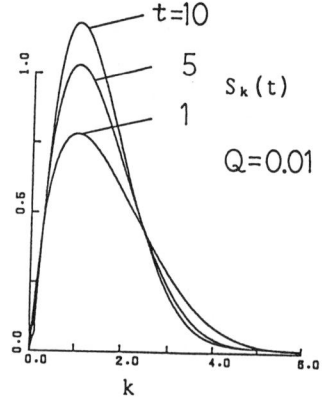

Fig.3 $S_k(t)$ vs k

$S_k(t)$ is scaled by two kinds of characteristic lengths, $\langle R \rangle$ and the screening length ℓ, as $S_k(t) \propto \langle R \rangle^6 S(k\langle R \rangle, k\ell, t)$, where $\langle R \rangle/\ell \sim Q^{1/2}$. The peak position $k_m \langle R \rangle$ of the scaled function $S(k\langle R \rangle)$ is then found to have a crossover, corresponding to that of $\langle R \rangle$;

$$k_m \langle R \rangle \sim t^{1/2} \quad \text{for } t \langle 1, \quad k_m \langle R \rangle \sim t^{1/6} \quad \text{for } t \rangle 1, \quad (5)$$

which leads to $k_m \sim t^0$. The correlation effect also causes appreciable corrections to its amplitude (see Fig.4).

In summary we stress the importance of the correlation effect on the growth of hemispherical nuclei on a substrate, which has been studied by none of authors. More experimental studies on this subject may be encouraged.

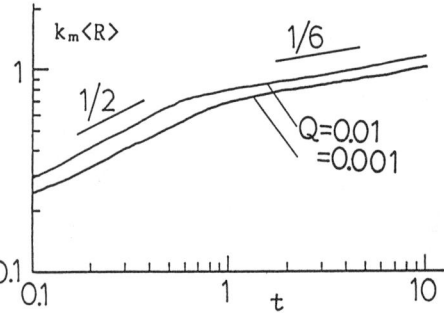

Fig.4 Peak position vs t

[1] M. Tokuyama, Physica A 169 (1990) 147.
[2] M. Tokuyama and Y. Enomoto, J. Chem. Phys. 94 (1991) 8234.

MODULATED PATTERNS AND PINNING EFFECT IN PHASE-SEPARATING ALLOYS

AKIRA ONUKI

Department of Physics, Kyoto Unviersity, Kyoto 606, Japan

HIRAKU NISHIMORI

Department of Physics, Ibaraki University, Mito 310, Japan

Domain growth in solids can be drastically influenced by elastic effects arising from the lattice misfit between the two phases. Recently we have proposed Ganzburg-Landau theory[1] and reported a series of computer simulations.[2-4]

Particularly interesting in our results is a pinning of domain growth in binary alloys due to difference $\Delta\mu$ of the shear modululs between the two phases. Namely, as the domain size R grows, the typical surface energy σR^2 of each domain becomes smaller than the elastic free energy difference $(\Delta\mu)R^3\varepsilon^2$ between the two phases, where ε is a typical strain proportional to the concentration difference Δc. Then for $R \gtrsim R_E = \sigma/(\Delta\mu)\varepsilon^2$ harder domains tend to become elastically isotropic, while softer regions are anisotropically deformed to percolate a network wrapping isolated harder regions.

Examples of 2D simulations are shown in Fig. 1 and 2 in isotropic elastic systems, where the solid regions represent the softer domains and the volume fraction of the softer component is 50 % and 30 %, respectively. We have found that the growth rate becomes extremely slow once the percolated network has been formed. Figs. 3 and 4 are examples of the pinning in cubic solids with $\Delta C_{44} \neq 0$, which closely resemble real domain structures in two-phase cubic solids. In such pinned states we have confirmed that the harder regions are almost isotropically deformed with only volume changes, while the softer regions are uniaxially deformed. The domain sizes in pinned states increases strongly with decrease of $\Delta\mu$ (or ΔC_{44} in cubic solids). The shear-deformation energy density is shown in a pinned state in Fig. 5, which is nearly zero in the harder regions.

We have thus found a new intricnisic mechanism of domain pinning without introducting impurities. Our mechanism seems to be relevant in binary alloys with large elastic misfits and also in solids undergoing martensitic transitions.

Fig. 1 Fig. 2 Fig. 3 Fig. 4

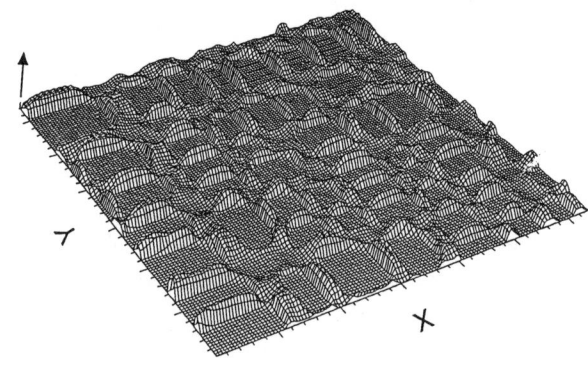

Fig. 5

REFERENCES

1) A. Onuki, J. Phys. Soc. Jpn. 58, 3065 (1989); ibid 58, 3069 (1989).
2) H. Nishimori and A. Onuki, Phys. Rev. B42, 980 (1990).
3) A. Onuki and H. Nishimori, Phys. Rev. B43, B649 (1991).
4) H. Nishimori and A. Onuki, J. Phys. Soc. Jpn. 60, 1208 (1991).

C. COMPUTER SIMULATION

SELF-AVOIDING WALKS ON A PERCOLATION CLUSTER IN FOUR DIMENSIONS

S. B. Lee and Y. J. Song
Department of Physics, Kyungpook National University, Taegu 702–701 Korea

ABSTRACT

The critical behavior and the crossover scaling of self-avoiding walks on a percolation cluster embedded in four-dimensional hypercubic lattices were investigated by Monte Carlo simulations. Results obtained appear to suggest a crossover to a new universality class at percolation threshold.

The statistics of self-avoiding walks (SAW's) on randomly diluted lattices is one of the long-standing controversial problem.[1-4] The issue is whether or not the critical behaviors of SAW's change when they are confined to a percolation cluster, and it is particularly interesting when the randomness is itself critical, i.e., near percolation threshold. Since the first appearance of this problem,[1] there have been number of works that support the opposite conclusions.[2-5]

Although there exist in the past equally unambiguous theories for the contradictory predictions, more recent analytical works appear to consistently suggest that the universality class of SAW's at p_c is different from that for $p > p_c$ where all critical exponents are now believed to remain unchanged. Unfortunately, however, there does not exist any reliable numerical work that supports these predictions. Kremer[5] has carried out Monte Carlo simulations on a diamond lattice for $p=0.55$ (where $p_c \simeq 0.428$[6]) and found the Flory exponent ν about $2/3$, where the known full lattice value is about 0.59.[7] Recently Lee and Nakanishi,[3] however, pointed out that his result was in error owing to mistakes in the data analysis and should be corrected not to be greater than 0.62, which is much closer to the full lattice value. They also presented their own Monte Carlo data for square and simple cubic lattices and obtained the Flory exponent rather similar to the full lattice results. It is thus difficult to clarify with these works alone whether or not SAW's cross over to a new universality as $p \to p_c$.

In this work, we present a clear numerical evidence for the universality of SAW's on diluted lattices in four-dimensional hypercubic lattices. We esimate from our Monte Carlo data the Flory exponent and, based upon our estimate, we discuss the crossover scaling of SAW's.

The main quantity of interest is the Flory exponent ν which characterizes the disorder average of the mean square end-to-end distances (and radii of gyration) of SAW's. The disorder average is, in principle, defined as

$$\overline{<R_N^2>} = \frac{\sum_C P(C) <R_N^2>_C}{\sum_C P(C)} \propto N^{2\nu}, \qquad (1)$$

where overbar indicates the disorder average, $<R_N^2>_C$ denotes the average of the mean square lengths on a disorder C, and $P(C)$ is the probability of having a disorder C in the percolation problem. From the theoretical point of view, the disorder average should be carried out by the following two steps. One must first carry out the configurational average over *all* possible N-step walks starting from *every* site on the cluster, weighting all walks equally. One then repeats over all possible disorder configurations, and the disorder average can subsequently be obtained. In Monte Carlo procedure, on the other hand, one is assumed to sample a certain number of probable configurations, rather than generating all possible ones. Therefore, it is crucial to develop a unbiased sampling method for the disorder average, maintaining the consistency with its analytical definition.

In the previous simulations,[3,5] certain methods were developed and employed on this problem, with attention being paid on the efficiency of the algorithm. Although those methods appear to be pertinent, there may be a valid question of whether or not each walk and each disorder were weighted properly. In order to avoid ambiguities concerning with the consistency with analytical calculation, we employ a simple, unbiased sampling method in the present work. We generate a percolation cluster on the hypercubic lattice for $p=0.202$ (where $p_c \simeq 0.197^6$) and identify an infinite cluster using the cluster labeling method.[8] We then sample a certain number of SAW's, each of which starts from the randomly chosen point on the infinite cluster. [We have attempted 10^6 walks on each cluster.] The average of the surviving walks weighted equally gives the approximate walk average on a given disorder. The disorder average was carried out over the different disorder configurations on which at least one SAW was sampled.

Data are analyzed as usual. The Monte Carlo data can be expressed in terms of the effective exponent ν_N, defined by

$$\nu_N = \frac{1}{2(N-M)} \left(N \ln R_N^2 - M \ln R_M^2 - \int_M^N \ln R_n^2 dn \right), \qquad (2)$$

and the exponent ν was obtained from the extrapolation of ν_N in the $N \to \infty$ limit. Here we have used the abbreviation R_N^2 for $< R_N^2 >$. One should, however, keep in mind that the mean square length is known to have the logarithmic correction at the marginal dimension, i.e., $d = d_c$ (=4 in the Euclidean space). If the critical behavior of SAW's on diluted lattice is similar to the ordinary SAW's, R_N^2 is expected to have the form of

$$R_N^2 \propto N^{2\nu} (\ln N)^\alpha, \qquad (3)$$

with the exponents ν and α similar to the ordinary SAW's in four dimensions ($\nu = \frac{1}{2}$ and $\alpha \simeq 0.37$).[7] If, on the other hand, SAW crosses over to a new universality at p_c as predicted by theory, one may expect that R_N^2 would be of the form similar to the ordinary SAW's ansatz because $d_c = 6$ was proposed analytically on diluted lattices, i.e.,

$$R_N^2 = AN^{2\nu}(1 + BN^{-\Delta} + CN^{-1} + \cdots), \qquad (4)$$

which leads $\nu_N = \nu + aN^{-\Delta} + bN^{-1} + \cdots$, with a and b being some constants.

Our Monte Carlo data for $p=0.202$ averaged over 200 clusters were plotted in Fig. 1 in terms of ν_N, up to 45 steps. The solid circles show the mean square end-to-end distances (R_E) and the open circles the mean square radii of gyration (R_G). As expected from the universality, the two sets of data appear to converge on the same value as $N \to \infty$. If we estimate the Flory exponent from the plot, we would get $\nu = 0.565 \pm 0.010$, which is significantly larger than the full lattice value. This indicates that the Flory exponent of SAW's crosses over to a higher value at p_c. One may, however, argue that there might be a logarithmic correction for $N \gg 1$, which makes accurate determination of ν very difficult from such a plot as Fig. 1. In order to see if the leading correction is indeed logarithmic, we have also analyzed our data according to Eq. (3): however, we were not able to observe any sign of it.

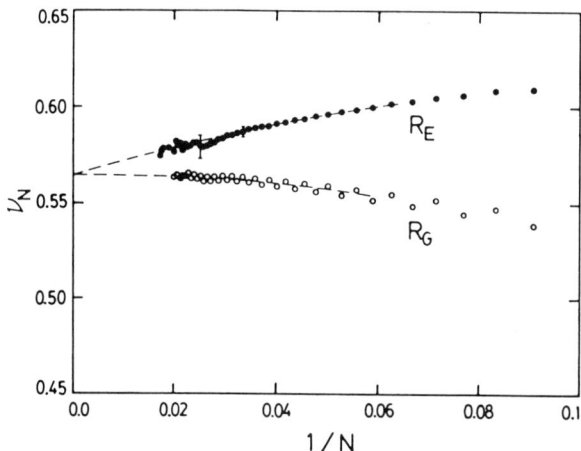

Fig. 1. The effective exponent ν_N, defined in Eq. (2), vs N^{-1} for the mean square end-to-end distance (R_E) and the radius of gyration (R_G), for $p=0.202$.

We have also studied the scaling of Monte Carlo data. Since the percolation cluster can be represented as a fractal with coherence length $\xi \sim |p - p_c|^{\nu_{perc}}$, where ν_{perc} is the percolation correlation length exponent and is known to be about 0.7 in four dimensions,[6] then one expects that SAW crosses over from a fractal lattice behavior to a Euclidean lattice behavior in the region where $N|p - p_c|^{\nu_{perc}/\nu_p} \sim 1$ holds. Thus the crossover scaling ansatz can be written as

$$<R_N^2> = N^{2\nu_p} f(N|p - p_c|^{\nu_{perc}/\nu_p}), \qquad (5)$$

with the asympototic limit of $f(x)$ given as

$$f(x) = \begin{cases} \text{const.} & \text{as } x \to 0 \\ x^{2(\nu_F - \nu_p)} & \text{as } x \to \infty, \end{cases} \qquad (6)$$

where ν_F and ν_p are the Flory exponents of SAW's on the full lattice and on the diluted lattice, respectively. Since the asymptotic behavior of SAW's is known to be observed for $N \gg 1$, the correct scaling region should be such that $N \gg 1$, $|p - p_c| \ll 1$ and $N|p - p_c|^{\nu_{perc}/\nu_p} \sim 1$. Thus the scaling relation in Eq. (6) implies that Monte Carlo data of $f(x)$ for various values of p close to p_c should converge to and follow a single curve as N increases. This type of scaling function was previously studied by Kremer[5] and more recently by Lee et al;[3] however, it is still unclear whether or not scaling indeed holds.

Shown in Fig. 2 are the data plotted in terms of $(R_N^2/N^{2\nu_p})^{1/2}$ as a function of $x \equiv N|p-p_c|^{\nu_{perc}/\nu_p}$, using $\nu_{perc} = 0.7$,[4] $p_c = 0.197$[4] and $\nu_p = 0.565$, for various values of p. For small N for each p-value, data do not appear to collapse onto a single curve. This is apparently because the walk does not reach the asymptotic region for the first several steps, as we mentioned earlier. However, as N increases, all data for $p \leq 0.26$ appear to converge onto a single curve and they overlap with the nearby data for different p's. This strongly suggests that data would collapse onto a single curve and thus scale for all p's in the correct scaling region, indicating that SAW's indeed cross over to a new universality class at p_c. For $p \geq 0.3$, data deviate slightly. We believe this to be due to that data for $p \geq 0.3$ are already outside the scaling region.

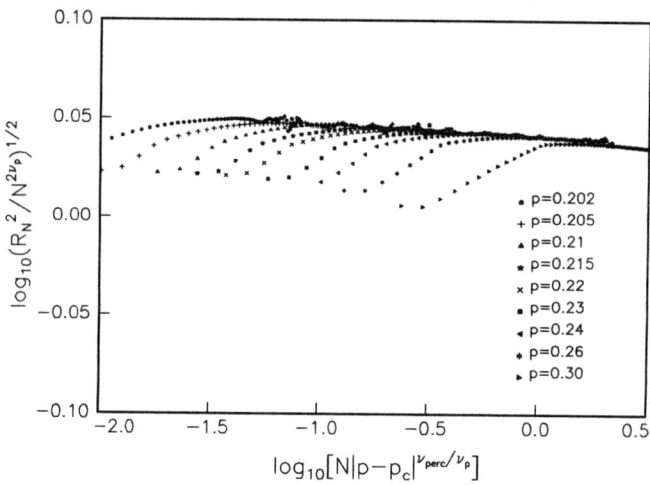

Fig. 2. The scaling function $f(x)$ in Eq. (5) plotted against the proposed scaling variable using Monte Carlo data for various values of p.

The scaling relation in Eq. (5) was also studied with $\nu_p = 0.5$. However, data do not seem to scale, nor exhibit correct asymptotic values of $f(x)$ for both extreme limits of x. This indicates that the value of ν_p used in the scaling analysis is just too small to observe the correct asymptotic behavior of $f(x)$ (not shown).

In summary, we have carried out Monte Carlo simulations for SAW's on a percolation cluster generated for p close to p_c on a hypercubic lattice in four dimensions and obtained the critical exponent ν significantly larger than the full lattice value. We have also showed that the crossover scaling holds for our estimate of ν_p, while no data collapsing was observed for the full lattice value. From these, we conclude that SAW's on diluted lattices at percolation threshold belong to the universality class different from the ordinary SAW's on Euclidean lattices, supporting the recent analytical prediction.

The authors are grateful for the partial support by Center for Theoretical Physics at Seoul National University for travel expenses to the conference. This work was supported by the Non-directed Research Fund, Korea Research Foundation, 1990.

REFERENCES

1. B. K. Chakrabarti and J. Kertesz, Z. Phys. B **44**, 221 (1981).
2. J. W. Lyklema and K. Kremer, Z. Phys. B **55**, 41 (1984).
3. S. B. Lee and H. Nakanishi, Phys. Rev. Lett. **61**, 2022 (1988); S. B. Lee, H. Nakanishi and Y. Kim, Phys. Rev. B **39**, 9561 (1989) and references therein.
4. Y. Meir and A. B. Harris, Phys. Rev. Lett. **63**, 2819 (1989).
5. K. Kremer, Z. Phys. B **45**, 149 (1981).
6. D. Staffer, Phys. Rep. **54**, 1 (1979); also see D. Stauffer, *Introduction to Percolation Theory* (Taylor and Francis, London, 1985).
7. J. C. le Guillou and J. Zinn-Justin, Phys. Rev. B **21**, 3976 (1980).
8. J. Hoshen and R. Kopelman, Phys. Rev. B **14**, 3438 (1976).

ROLE OF LOCALLY DISTRIBUTED PROPERTIES ON THE RELAXATION DYNAMICS IN DISORDERED SYSTEMS

J.L. Bocquet, Y. Limoge
Centre d'Etudes de Saclay, D.T.A./DTM/SRMP
91191 Gif sur Yvette Cedex, France

ABSTRACT

Among the great deal of models aiming to describe the origins of the "pathological" dynamical properties of disordered systems none of them starts from a realistic description of the actual properties of these systems at an atomic level. In fact we show here that this very nature has a strong influence on the global characteristics. In particular the meaning of the macroscopic properties is <u>not</u> simply linked with the pertinent parameters at an atomic level.

In strongly disordered systems, like amorphous materials, the response to a sudden variation of a state variable (temperature, electric field, pressure...) is generally much slower than in crystalline systems, displaying even frequently the "stretched exponential" form [1]. There is presently a whole range of models trying to explain this behavior. They can be classified in two classes : the first one describes the macroscopic evolution as resulting from hierarchically cooperating events acting successively [2], the second category emphasizes the role of simultaneous events occurring independently[3]. In the first case the role of the disorder consists in providing a fractal landscape of the energy surface in the phase space available to the system. In this case the dynamics of the relaxation becomes "pathological" either for the appearance of diverging activation energies or for the random walk of the relaxing systems taking place on a fractal set of accessible configurations. In this picture the disorder of the local properties (says the distribution of atomic dynamics) is of no concern. This is just the reverse in the second approach. Now the system is seen as a set of independently relaxing zones of atomic size, the total effect being given as the sum of the individual events. The main role in explaining the anomalous behavior is now given to the distribution of the local dynamics : in the case of activated process, the relaxation becomes "slow" as soon as the width of the distribution of activation energies becomes sufficiently wide. Whatever their actual pertinence in the physical effects observed, both approaches share a common drawback : in both cases are the very characteristics of the disorder in an actual physical system neglected. But it is not sufficient to describe the local disorder by a distribution of activation energies. In actual systems the disorder is twofold : the site disorder corresponding to the distribution of the energies for the system in stable positions, and the saddle disorder corresponding to the distribution of the heights of the saddles the system must overcome for jumping from a stable position to another one. We have shown recently that in strongly disordered systems the long range diffusion behavior of a tracer particle is controlled by a compensation effect between these two different disorders : in a large temperature domain a perfectly arrhenian and almost non dispersive diffusion can be recovered provided the width of both distribution is of the same order of magnitude [5].

The purpose of the present work consists in checking to what extent their interplay can affect the relaxation behavior beyond the usual treatment. We shall treat for simplicity a well known case of delayed relaxation : the Snoek effect. In this effect a population of non interacting "particles" (dipoles, defects, pairs of atoms, etc...) is distributed at random on a lattice. Under the application of some external constraint (stresses, magnetic field...) the degeneracy of the different configurations corresponding to different orientations of the "particles" with respect to the external field, is suppressed. The populations of the different classes of sites will then evolve by particle jumps, which are supposed here to be poissonian thermally activated jumps, to the new equilibrium populations according to their Boltzmann weights.

In the case of well ordered, crystalline, systems the asymmetry between the populations of the different classes of sites produces a variation with time of some physical property p(t) (the length in the Snoek effect for example). The system being well ordered, after a jump of the external constraint the populations will evolve according to a single relaxation time : the jump time τ. Therefore we get

$$p(t) = p(o) + \Delta p \, [1 - \exp(-t/\tau)] \qquad (1)$$

where Δp is the maximum amplitude of the effect.

In case of strongly disordered systems the single valued τ is generally replaced by a distribution $\pi(\tau)$ of relaxation times τ corresponding to the distribution of activation energies. If we admit that Δp doesn't vary with τ, that is with the local values of the disorder, we get therefore [6]

$$p(t) = p(o) + \Delta p \int_0^\infty [1 - \exp(-t/\tau)] \, \pi(\tau) \, d\tau \qquad (2)$$

No provision is made for handling with the different properties of site and saddle contributions. Moreover since the system has lost its translationnal symmetry a single jump of a particle is generally not sufficient to produce a maximum relaxation. This fact is not taken into account in equation (2). In our "amorphous" case the site energies of the particles and the height of the barriers they have to overcome are choosen at random following gaussian distributions, the mean values and the variances of which are respectively $\bar{\epsilon}_s, \bar{\epsilon}_c$ and σ_s, σ_c.

Given these actual distributions the relaxation time distribution is given by

$$\pi(t) = -\frac{\partial}{\partial t} <\exp(-t \sum_j w_{ij})>_{i,j} \qquad (3)$$

where w_{ij} is the activated jump frequency from site i to site j the bracket pertains to an average over the site energy distribution, i, and the z saddle ones, j, z being the coordination number of the lattice. We recognize here the well known form of the waiting time distribution for a random walk [7,4].

In a first step we have to look for the mean relaxation time $\bar{\tau}$, which is of frequent use in experimental work. In order to handle with the intricate expression (3) we introduce two approximations. The first one assumes that at high temperature all the z available jumps frequencies w_{ij} are for a given i of the same order of magnitude. Therefore one introduces $<\sum w_{ij}^{-1}>_{i,j} \sim <\sum w_{ij}>_{ij}^{-1}$ and we get easily

$$\tilde{\tau}_{H.T.} \sim \exp(+\beta(\bar{\varepsilon}_c - \bar{\varepsilon}_s))\exp(\beta^2/2(\sigma_s^2 - \sigma_c^2))$$

At low temperature on the contrary the particle will use the easiest path, which can be accounted for by admitting $\exp(-t\sum w_{ij}) = 1$ for $t < (\sum w_{ij})^{-1}$ and 0 for $t > (\sum w_{ij})^{-1}$. One gets now

$$\tilde{\tau}_{L.T.} \sim \exp(\beta(\bar{\varepsilon}_c - \bar{\varepsilon}_s))\exp(\beta^2/2(\sigma_s^2 - f\sigma_c^2))$$

In both expressions $\beta = 1/kT$ has his usual meaning, and f of the order of one is a slowly varying function of z and of β.

Exactly as for long range diffusion it is clear that with respect to the mean relaxation time the interplay of site and saddle disorders reduces the apparent disorder with respect to the mean value of the perfect system $\tilde{\tau} \sim \exp(\beta(\bar{\varepsilon}_c - \bar{\varepsilon}_s))$. Neither the mean relaxation time nor its temperature behavior are pertinent indicators of the actual level of disorder.

Turning now to the time variation p(t) we need calculate explicitly the $\pi(\tau)$ function. Using the same low temperature approximation as before we get:

* For a pure site disorder

$$\pi_s(\tau) \sim \frac{\exp}{\tau}\left(-\text{Ln}^2\left(z\tau e^{-\beta(\bar{\varepsilon}_s - \bar{\varepsilon}_c)}\right)/2\beta^2\sigma_s^2\right)$$

* For a pure saddle disorder with $\beta\sigma \geq z$

$$\pi_c(\tau) \sim \frac{\exp}{\tau}\left(-z\text{Ln}^2\left(\tau e^{-\beta(\bar{\varepsilon}_s - \bar{\varepsilon}_c)}\right)/2\beta^2\sigma_c^2\right)$$

In both cases the time distribution give rise to fractal time phenomena below a critical time. For site disorder the distribution behaves like $\pi(\tau) \sim \tau^{-1-\alpha_s}$ with $\alpha_s = \text{Ln}(z\tau e^{-\beta(\bar{\varepsilon}_s - \bar{\varepsilon}_c)})/\beta^2\sigma_s^2$ up to a critical time $\tau_c^s = \exp\left(\frac{\beta^2\sigma_s^2}{2}\right)\tilde{\tau}_s$

corresponding to $\alpha_s = 1$. Similarly for saddle disorder one gets

$\alpha_c = z\text{Ln}(\tau e^{-\beta(\bar{\varepsilon}_s - \bar{\varepsilon}_c)})/\beta^2\sigma_c^2$ and $\tau_c^c = \exp\left\{\left(\frac{\beta^2\sigma_c^2}{z}\right)\left(1 + \frac{fz}{2}\right)\right\}\tilde{\tau}_c$.

Coming back to equation (2), the relaxed fraction is easily shown to follow now

$$p(t) \sim t^{-\alpha}$$

up to the critical time where the behavior turns back to a simple behavior. The relaxation turns therefore to be extremely slow, in fact much slower than a Kohlrausch-William-Watt law, below the critical time, for it follows a power law decay. Past this critical time the diffusion becomes non dispersive [5b] and the relaxation "normal". It turns out that the situation of the site disorder is worse for a given level of disorder, since the critical time is much larger than for the saddle case. In case of an actual disorder, including site and saddle parts, the value of α can be shown to decrease with respect to both of α_s and α_c, giving rise to a more "pathological" behavior [5b]. Nevertheless the critical time value remains closer to the smaller τ_c^c than to the τ_s^c. This apparent contradiction comes from the fact that α reflects the decrease of the jump frequency with time ; starting from a higher level than in the pure site case, due to the introduction of the saddle disorder which provides for low saddles, the system has to be more dispersive, and therefore α smaller, than in either of pure site or pure saddle cases. Nevertheless the high "trapping" efficiency of the site disorder is counteracted by the more rapid dynamics involved in the saddle part, explaining now the origin of the smaller critical time. In this sense we get, as for the diffusion coefficient, a compensation effect between both kinds of disorder : the duration of the "slow" relaxation regime is much smaller in the mixed disorder case than in the pure site one.

A last remark has to be done. As is well known for a pure renewal process, the site disorder doesn't give rise to dispersive diffusion as far as the initial configuration is an equilibrium one (equation (2)) including now a Boltzmann weight pertaining to site disorder. Nevertheless on the one hand since we are checking the validity of the present approach against Monte Carlo simulations with arbitrary initial configuration this remark is of no concern. On the other hand the main result, namely the different roles of the two kinds of disorder, remains either with respect to temperature behavior or to time scale setting.

REFERENCES

1. R.H. Cole, Ann. Rev. Phys. Chem., 40, 1, (1989)
2. a) R.G. Palmer, D.L. Stein, E. Abraham and P.W. Anderson, Phys. Rev. L 53, 958, (1984)
   b) I.A. Campbell, Phys. Rev. B 33, 3587, (1986)
3. A.S. Nowick and B.S. Berry, "Anelastic relaxation in crystalline solids", Academic Press, New York (1972)
4. M.F. Schlesinger and E.W. Montroll, Proc. Nat. Acad. Sci. (USA) 81, 1280, (1984)
5. Y. Limoge and J.L. Bocquet, a) Phys. Rev. Letters, 65, 60, (1990)
   b) to be published
6. J.R. Cost, in "Nontraditional methods in diffusion", ed. G.E. Murch, H.K. Birnbaum, Met. AIME 1983, 111, (1983)
7. H. Scher and M. Lax, Phys. Rev. B, 7, 4491, (1973)

DYNAMICS OF PATTERN FORMATION OF ANTIPHASE ORDERED DOMAIN IN ALLOYS

Kenichi Shiiyama, Kazumi Horai and Tetsuo Eguchi
Department of Applied Physics, Fukuoka University,
Fukuoka 814-01, JAPAN

ABSTRACT

The kinetic analyses and computer simulations were carried out on the pattern formation in phase transitions of the alloys, in which three different types of order parameters are involved; the one is the composition, and the other two are the B2 and DO$_3$ types of order. Three different cases were investigated; one is an isotropic phase transition from A2 to A2+DO$_3$, observed in Fe$_3$Al, another is an anisotropic transition from B2 to B2+DO$_3$, observed in Fe-Si, and the third is a simple ordering from B2 to DO$_3$, also observed in Fe$_3$Al. The result of our kinetic investigation turned out consistent with the observations.

INTRODUCTION

It has been known that in some alloys the phase separation is activated by atomic ordering. For example there occurs a phase separation from A2 to A2+DO$_3$ phase, or from B2 to A2+DO$_3$ in Fe$_3$Al, and similarly from B2 to B2+DO$_3$ in Fe-Si. In these alloys two types of ordered structures, B2 and DO$_3$, are involved in the phase transition. In our previous papers[1~3] we carried out kinetic analyses and computer simulations for the processes of formations of antiphase ordered domain structures in ordering alloys. There the problem was simplified as a system with two order parameters, one for the solute concentration that is conserved, and the other the degree of order that is not conserved. Using a continuum model for the alloy, and following the TDGL (time-dependent Ginzburg-Landau) method, we were able to obtain reasonable results, which explained some essential features of the electron microscopic patterns observed in those alloys during ordering phase transitions.
In the present paper we take a more realistic standpoint by considering two types of ordered structures mentioned above. The kinetic equations representing the phase changes on isothermal aging now involve three order parameters, one is the concentration of solute atoms, which is conserved, and the other two are the degrees of B2 and DO$_3$ order, neither of which is conserved. We introduce a phenomenological anisotropy parameter, considering the elastic interaction due to the lattice distortion. We solved these equations, after appropriate scale changes, simultaneously in two-dimensional meshes, each of which consists of 200×200 points, for the three cases of particular interest. The numerical solutions obtained were then processed into images corresponding to the electron micrographs.
The first case is the isotropic phase transition from A2 disordered state into the mixed phase of A2 disordered plus DO$_3$ ordered states. The second one is the transition from B2 to B2+DO$_3$ with the

nonvanishing anisotropy parameter. In the first case the solutions started from the uniformly disordered A2 state with small fluctuations in the order parameters. They indicate a swift development of the B2 domain structure, which decomposes into two phases of A2+B2, and then to A2+DO$_3$. Such transitions in series have already been observed in Fe-Al alloy near the stoichiometric composition Fe$_3$Al.[4] In the second example of the solutions the decomposition from B2 to B2+DO$_3$ seems isotropic at first, but later becomes anisotropic with ordered domains of rectangular shape. The effect is also in conformity with the observation in Fe-Si alloy.[5]

The third case is an ordering to DO$_3$ from B2. The result of our calculation predicts a small accumulation of the solute or solvent atoms, as the case may be, along the APB (antiphase boundary) of DO$_3$ ordered domains. The effect has often been stressed by Allen[6] based on his excellent electron microscopic analysis of the ordering processes in Fe$_3$Al.[6] The result of our dynamical investigation of the pattern formation in the three representative cases, mentioned above, shows the adequateness of our interpretation of the ordering processes occurring in the Fe-based alloys.

PROCEDURE

We consider an alloy with the composition $A_{(1-U)/2}B_{(1+U)/2}$ in a continuum approximation, and introduce the degrees of order X and Y, which represent the DO$_3$ and B2 types of order, respectively. The alloy is either in the disordered state A2 (X=Y=0), or the ordered B2 (X=0, Y≠0) or DO$_3$ (X≠0, Y≠0) state. The phase transitions from A2 to B2 and from B2 to DO$_3$ are both of the second order, and the phase separation is possible only in the ordered states. In order to satisfy the above conditions we assume the following expression for the free energy density of the uniform system:

$$f(U,X,Y;T) = A(T) + B(T)U^2 - B(T)X_1(T)^2 X^2/2 - B(T)X_0(T)^2 Y^2/2 \\ - B(T)C(T)X^2Y^2/2 + B(T)U^2X^2/2 + B(T)U^2Y^2/2 \\ + B(T)X_2(T)^2 X^4/4 + B(T)X_3(T)^2 Y^4/4, \quad (1)$$

where $A(T)$, $B(T)$, $C(T)$, $X_i(T)$ (i=0,1,2,3) are all positive parameters depending on the temperature. The equilibrium values for U, X and Y are obtained by the conditions $\partial f/\partial U = \mu$ (the chemical potential) and $\partial f/\partial X = \partial f/\partial Y = 0$. If we choose an appropriate set of the values for the parameters C(T) and X(T), we can obtain an equilibrium phase diagram that contains a mixed phase field of either A2+B2, or A2+DO$_3$, or B2+DO$_3$. For the following analysis the condition $0 < X_1(T) < X_0(T) < 1$ must hold, and then $X_0(T)$ gives the phase boundary between the phase fields A2 and B2, and $X_3(T)$ the one between A2 and A2+B2, in the equilibrium phase diagram.

When the system is not uniform in the course of a phase transition, then the thermodynamical potential is the sum of the bulk free energy and the interface energy as shown by the following expression:

$$F[\{u(r,t),x(r,t),y(r,t);T\}] = \int \{f(u,x,y;T) + H(T)(\nabla u)^2/2 \\ + K(T)(\nabla x)^2/2 + L(T)(\nabla y)^2/2\}d^3r, \quad (2)$$

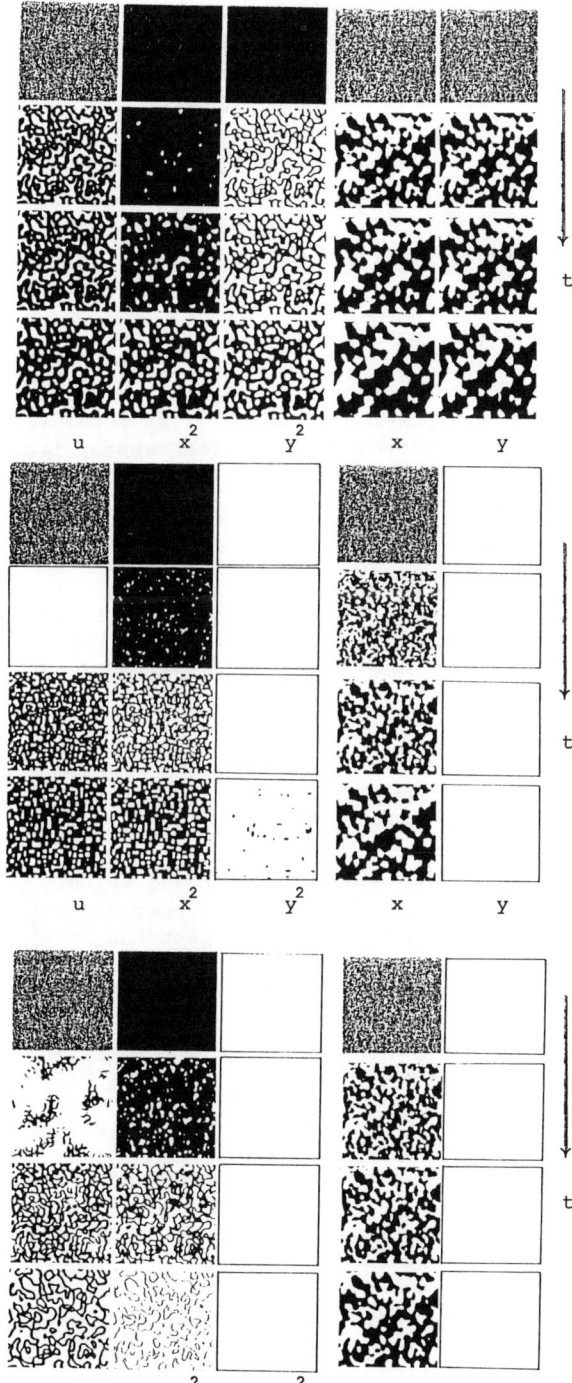

Fig.1 Simulation pattern for the transition from A2 to A2+DO$_3$ via B2 and A2+B2. Compare with Fig.8 in Reference 4.

Fig.2 Simulation pattern for the anisotropic transition from B2 to B2+DO$_3$. Compare with Fig.5 in Reference 5.

Fig.3 Simulation pattern for the ordering from B2 to DO$_3$. Note the deviation of solute concentration along APBs from its average value. This has been pointed out by Allen as the "solute drag effect."[6]

where the order parameters depending on the space-time coordinates are shown in the corresponding lower-case letters. Here $H(T)$, $K(T)$ and $L(T)$ are the interface energy densities per unit length along the varying order parameters. We can obtain in a straightforward way the equations of motion for the three order parameters, as follows:

$$\partial u/\partial t = P(T)\nabla^4(\delta F/\delta u) = -P(T)H(T)\nabla^4 u + 2P(T)B(T)\nabla^2(2u+ux^2+uy^2), \quad (3)$$
$$\partial x/\partial t = -Q(T)(\delta F/\delta x) = Q(T)B(T)x\{X_1(T)^2 + Cy^2 - u^2 - X_2(T)^2 x^2\}$$
$$+ Q(T)K(T)\nabla^2 x, \quad (4)$$
$$\partial y/\partial t = -R(T)(\delta F/\delta y) = R(T)B(T)y\{X_0(T)^2 + Cx^2 - u^2 - X_3(T)^2 y^2\}$$
$$+ R(T)L(T)\nabla^2 y, \quad (5)$$

where $P(T)$, $Q(T)$ and $R(T)$ are the reaction rates depending upon the temperature. We also take the elastic effect[7] into account by introducing the tensor natures of the coefficients, along the same line as done in the previous paper.[3] Thus the final equations of motion to be solved numerically are, after appropriate scale changes:

$$\partial u/\partial t = -\nabla^4 u + 2\gamma \nabla'^4 u + \nabla^2(2u+ux^2+uy^2), \quad (6)$$
$$\partial x/\partial t = \alpha_1 x(X_1^2 + Cy^2 - u^2 - X_2^2 x^2) + \beta_1 \nabla^2 x, \quad (7)$$
$$\partial y/\partial t = \alpha_2 y(X_0^2 + Cx^2 - u^2 - X_3^2 y^2) + \beta_2 \nabla^2 y, \quad (8)$$

where temperature dependence of the parameters has been suppressed. In the above equations α_1, α_2, β_1, and β_2 are certain positive parameters, and γ the anisotropy coefficient, which is either positive or negative. $\nabla'^4 \equiv \partial^4/\partial\eta^2\partial\zeta^2 + \partial^4/\partial\zeta^2\partial\xi^2 + \partial^4/\partial\xi^2\partial\eta^2$.

RESULT

We made numerical analyses for the three cases of interest, as mentioned in INTRODUCTION. The results of the analyses are represented in Figs.1∿3, where the time-evolution of the patterns of u, x^2, y^2, x and y are shown in black and white. In the figures the regions either u<U, x^2>.1, y^2>.1, y>0 or x>0 are all shown in white and otherwise in black. Thus the patterns of x^2 and y^2 correspond to the electron microscopic dark field images taken with the DO_3 and B2 superstructure reflections, and those of u to the segregation of solute atoms. In Fig.1 we notice a series of phase transitions A2→ B2→ A2+B2→ A2+DO₃, and in Fig.2 an anisotropic phase separation is activated by an isotropic ordering. Fig.3 shows that a small amount of deviation of the concentration along APB. All of those phenomena have been observed in Fe-based alloys, Fe₃Al or Fe-Si.[4,5]

REFERENCES

1. Ninomiya, Eguchi and Kanemoto, Phase Transitions B **28**, 125 (1990).
2. Shiiyama, Kanemoto, Ninomiya and Eguchi, Proc. IWCMS, 103 (1990).
3. Shiiyama, Ninomiya and Eguchi, "Res. Pattern", ed. Takagi (1991).
4. Oki, Matsumura and Eguchi, Phase Transitions B **10**, 257 (1987).
5. Matsumura, Oyama and Oki, Mat. Trans. JIM **30**, 659 (1989).
6. Allen and Krzanowski, Int. Conf. Solute-Defect Inter., 400 (1985).
7. J. W. Cahn, Acta Metall. **10**, 179 (1962)

LOGARITHMICALLY SLOW COARSENING IN NONRANDOMLY FRUSTRATED MODELS

Joel D. Shore, James P. Sethna, Mark Holzer, and Veit Elser
Physics Department, Cornell University, Ithaca, New York 14853

ABSTRACT

We study the growth ("coarsening") of domains following a quench in an Ising model with weak next–nearest–neighbor antiferromagnetic (AFM) bonds and single–spin–flip dynamics. The AFM bonds introduce free energy barriers to coarsening and thus greatly slow the dynamics. In three dimensions, simple physical arguments suggest that the barriers are proportional to the characteristic length scale $L(t)$ for quenches below the corner rounding transition temperature T_{CR}. This should lead to $L(t) \sim \log(t)$ at long times t. Monte Carlo simulations provide strong support for this claim.

We also predict logarithmic growth in a purely two–dimensional tiling model, which can be thought of as describing a single interface in our three–dimensional model viewed from the [111] direction. Here, the slow coarsening dynamics should persist all the way up to the order–disorder transition (at T_{CR}). However, if the model is cooled slowly at a rate Γ, the final length scale should have power–law, not logarithmic, dependence on $1/\Gamma$. Simulations support both of these claims.

INTRODUCTION

When a system is quenched from high temperatures to a temperature below the order-disorder transition, domains form and coarsen. Of particular interest is how the characteristic length scale $L(t)$ grows with time t at long times.

Historically, there have been some theoretical predictions that certain systems without randomness in their Hamiltonians would show logarithmically slow coarsening at long times.[1] For a while, such claims could not be disproved since the numerical evidence was ambiguous due to long time transients and finite–size effects. However, large Monte Carlo simulations, bolstered by more careful theoretical arguments, eventually showed that the long time growth in these models obeys the naively–expected power laws: $L(t) \sim t^n$ with $n = 1/3$ or $1/2$ (depending on whether the dynamics does or does not conserve the order parameter, respectively).[2] Indeed, the only models known to exhibit logarithmic domain growth are those which contain randomness explicitly in their Hamiltonians, such as the random–field Ising model and spin glasses.

In light of these results, there seems to be a growing belief that, for nonrandom systems quenched to nonzero temperature, the $n = 1/3$ and $n = 1/2$ power law behavior is universal (i.e., independent of the details of the Hamiltonian), and even independent of the dimensionality. Motivated by the slow dynamics present in glasses,[3] we have been looking for counter examples, i.e., models without randomness which display logarithmically slow ordering dynamics.[4]

ARGUMENT FOR LOGARITHMICALLY SLOW GROWTH

Consider the nearest–neighbor Ising ferromagnet on a square or cubic lattice in $d = 2$ or 3 dimensions, with frustration added by introducing weak next–nearest–neighbor (NNN) antiferromagnetic (AFM) bonds. The Hamiltonian is

$$H = -J_1 \sum_{NN} s_i s_j + J_2 \sum_{NNN} s_i s_j \ , \tag{1}$$

where $s_i = \pm 1$. The first sum is over all nearest–neighbor (NN) bonds while the second is over all NNN bonds. We have chosen our sign convention so that both J_1 and J_2 are positive when

© 1992 American Institute of Physics

the NN bonds are ferromagnetic and the NNN bonds are antiferromagnetic. We will require that $J_1/J_2 > 2(d-1)$ so that the ground state for this model is ferromagnetic. We will study this Hamiltonian under single-spin-flip (*i.e.*, nonconserved) dynamics.

The NNN AFM bonds introduce free energy barriers to coarsening and thus greatly slow the dynamics (freezing the system completely at $T = 0$). In two dimensions, these barriers are independent of the characteristic length scale $L(t)$, and thus $L(t) \sim t^{1/2}$ at long times.[4] Let us now study what happens in three dimensions by considering the time to shrink a cubic domain of, say, up spins in a larger sea of down spins (see inset of Fig. 1). The energy barrier to flip a corner spin (black cube) is $12J_2$. Once a corner flips, the neighboring spins along an edge (white cube) can flip in turn, but there is an energy barrier of $4J_2$ for each to flip. The barrier to flip the spins along an entire edge is then $E = 4J_2(L+1)$, where L is the linear size of the domain. The time t to do this is given by activation over this barrier and is thus exponential in L:

$$t = \tau_0 \, e^{4(L+1)J_2/T} . \qquad (2)$$

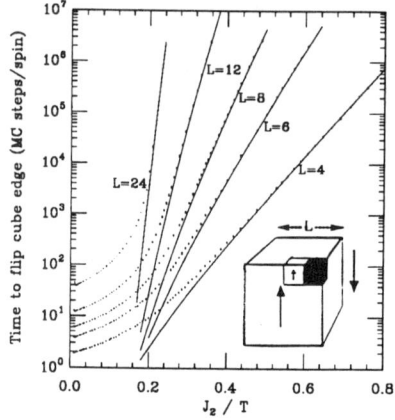

Fig 1. Arrhenius plot of the time to flip all the spins along the edge of a cubic domain of size L (shown in inset). Curves are the theoretical forms derived from a low temperature expansion, as discussed in the text.

Naively inverting this equation to solve for the size of the smallest structure which we expect to remain in a coarsening system at time t, we find

$$L(t) \sim \frac{T}{4J_2} \log(t/\tau_0) . \qquad (3)$$

This gives the expected result[5] that energy barriers which diverge with the characteristic length scale $L(t)$ should lead to logarithmically slow coarsening.

Of course, the above discussion is only valid in the limit $T \to 0$. What happens at nonzero temperatures where we must consider not energy barriers but, rather, *free* energy barriers? Fig. 1 shows Monte Carlo simulation results for the average time t to flip all the spins along the edge of a cubic domain. We see that the slope on this Arrhenius plot increases with domain size L, thus confirming our prediction of an activation barrier which grows with L.

Furthermore, if we write

$$t = \tau_0(T) e^{F_B(L,T)/T} , \qquad (4)$$

we can perform a low temperature expansion for both $\tau_0(T)$ and $F_B(L,T)$.[4] The curves in Fig. 1 show the resulting prediction, which has no free parameters and is in excellent agreement with the simulation results at low temperatures.

We expect our argument for logarithmically slow coarsening to break down when the free energy barrier per unit length (to flip the spins along a cube edge) goes to zero. This occurs at the corner rounding temperature T_{CR}, which has previously been studied in the context of equilibrium crystal shapes.[6] In the limit $J_1/J_2 \to \infty$, T_{CR} can be calculated exactly[7,4] and yields $T_{CR} \approx 7.11 J_2$.

SIMULATIONS OF THE COARSENING PROCESS

There is still a large gap in our argument: Although we have identified a special configuration in which there are energy barriers that scale with the length scale L, we have not shown that during the process of coarsening the system will necessarily find itself in configurations in which it will have to cross these barriers in order to coarsen further. It is conceivable that the system could find a path through configuration space which goes around these barriers. To construct a proof that the barriers must be crossed is very difficult since it requires a detailed understanding of the spin configurations which form in a quench. Instead, we turn to numerical simulations of the coarsening process in order to test our conjecture.

Fig. 2 shows the growth of $L(t)$ following a quench from infinite temperature (a random spin configuration) to a final temperature T. Since this is a log–log plot, power law behavior would give a straight line. We see that as the ratio of T/J_2 is decreased, the coarsening slows dramatically. Furthermore, at temperatures below T_{CR} (for $T/J_2 = 2$, 3, and 4), the Monte Carlo data show some downward curvature on this log–log plot at late times. [By contrast, for $J_2 = 0$ and $T/J_2 = 8$, there is no downward curvature until finite–size effects lead to a sharp leveling off of $L(t)$ once it is approximately 1/3 the system size.] This suggests that the growth is becoming slower than a power law. In fact, if we replot this on a log–normal plot we find that, while there is considerable upward curvature at early times, the last one to two decades of data are quite straight and thus in reasonably good agreement with logarithmic growth of $L(t)$.[4]

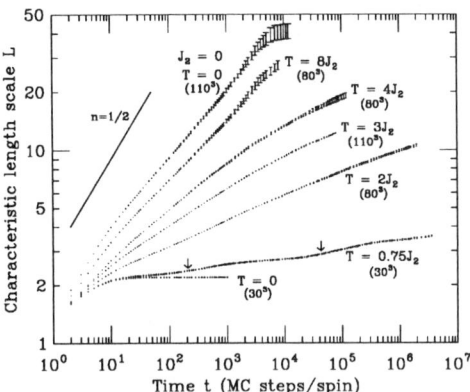

Fig 2. Growth of $L(t)$ following a quench from infinite temperature to a final temperature T. Numbers in parentheses give system sizes.

THE TILING MODEL

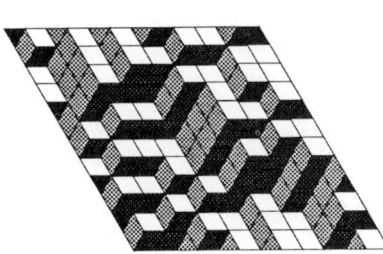

Fig. 3. A sample configuration for the tiling model.

We now briefly discuss a closely related model which is also expected to show logarithmic coarsening. This is a two–dimensional model for a single interface in the three–dimensional model as viewed from the [111] direction. If we require that the configurations of this interface have no bubbles or overhangs (when viewed from this direction), then we obtain the so-called "[111]-restricted solid-on-solid (RSOS) model" for our three-dimensional model.[7] The RSOS restriction corresponds to taking the limit $J_1/J_2 \to \infty$ in the 3-d model. Any configuration in the RSOS model (of which an example is shown in Fig. 3) can be represented as a tiling of the plane by 60° rhombi of three different orientations. (The model also has a third representation as an Ising spin system on a triangular lattice.[7])

The order–disorder transition in this model occurs at T_{CR}. Above T_{CR}, the interface is rough (i.e., the tiles intermingle); below T_{CR}, the interface forms a sharp corner (i.e., the tiles phase separate). When the system is quenched from infinite temperature to $T \leq T_{CR}$, we expect that the interface will coarsen under the dynamics, which consists of adding or removing cubes subject to the RSOS restriction. Since this dynamics conserves the order parameter in this model, the naively-expected behavior would be $L(t) \sim t^{1/3}$. However, the mechanism by which the interface coarsens involves activation over precisely the same sort of barriers which grow with $L(t)$ as in the three-dimensional model. Thus, the same arguments we made for logarithmically slow coarsening in the three-dimensional model should apply here as well. Simulations of the coarsening process once again lend support to this claim.

Unlike in the three-dimensional model, here the ordering temperature and the temperature at which the dynamics becomes slow coincide. Thus, we might hope that this system would be glassy, *i.e.*, that it would have great difficultly ordering even when cooled slowly at a rate Γ. Specifically, we'd want the final $(T=0)$ value of L to depend only logarithmically on the time $1/\Gamma$ spent cooling. We have simulated slow cooling in this model and find that this does not appear to be the case. Furthermore, more careful arguments suggest that we should expect $L(T=0) \sim \Gamma^{-1/4}$ in the limit $\Gamma \to 0$, which is in reasonably good agreement with the simulation results.[4] The reason why the dependence is a power law and not a logarithm is because the free energy barrier goes continuously to zero at T_{CR}, and thus there is a region of temperature just below T_{CR} where the barriers are small and the system can still coarsen quite rapidly.

CONCLUSIONS

We have discussed two closely related models in which we conjecture that the growth of the domains should be only logarithmic in time following a quench. Simulations lend strong support to this conjecture. However, if cooled slowly at a rate Γ, these models are not expected to order sluggishly: the final domain size has a power law, rather than a logarithmic, dependence on $1/\Gamma$.

ACKNOWLEDGEMENTS

We thank David Huse, Peter Nightingale, Jennifer Hodgdon, and David DiVincenzo for helpful discussions. This work was supported in part by NSF Grant No. DMR 88-15685 and computing facilities were provided in part by the Cornell-IBM Joint Study on Computing for Scientific Research.

REFERENCES

1. S. A. Safran, Phys. Rev. Lett. **46**, 1581 (1981); G. F. Mazenko, O. T. Valls, and F. C. Zhang, Phys. Rev. B **31**, 4453 (1985).
2. D. A. Huse, Phys. Rev. B **34**, 7845 (1986); J. Viñals and M. Grant, Phys. Rev. B **36**, 7036 (1987); G. S. Grest, M. P. Anderson, and D. J. Srolovitz, Phys. Rev. B **38**, 4752 (1988); J. G. Amar, F. E. Sullivan, and R. D. Mountain, Phys. Rev. B **37**, 196 (1988); C. Roland and M. Grant, Phys. Rev. B **39**, 11971 (1989).
3. For more details, see J. P. Sethna, J. D. Shore, and M. Huang, to appear in Phys. Rev. B (Sept. 1991); and references therein.
4. More details on the work presented here can be found in J. D. Shore, Ph.D. thesis, Cornell University (1992). An earlier report on some of the work is given in J. D. Shore and J. P. Sethna, Phys. Rev. B **43**, 3782 (1991).
5. Z. W. Lai, G. F. Mazenko, and O. T. Valls, Phys. Rev. B **37**, 9481 (1988).
6. C. Rottman and M. Wortis, Phys. Rev. B **29**, 328 (1984).
7. A.-C. Shi and M. Wortis, Phys. Rev. B **37**, 7793 (1988); and references therein.

A DLA MODEL OF INTERACTIVE PARTICLES

Masahiro Nakagawa and Koichiro Kobayashi
Department of Electrical Engineering, Faculty of Engineering, Nagaoka University of Technology, Kamitomioka 1603-1, Nagaoka, Niigata 940-21, Japan

MODEL

A simple model is devised as an extension of the diffusion-limited aggregation (DLA) model[1,2] taking account of a local drift force, whose direction and strength depend on the local structure of a growing cluster, as well as of the thermal fluctuation for the Brownian motion. Such a local drift force may be regarded as a particle-particle interaction modeling on the van der Waals (relatively short-range attraction as $\propto 1/r^6$) interaction or the Coulomb (relatively long-range repulsion as $\propto 1/r$) one. The moving particle was displaced according to the fllowing relation.

$$\Delta x = \text{sgn}\{(1-\alpha)f_x + \beta\alpha d_x\}, \quad (1a)$$
$$\Delta y = \text{sgn}\{(1-\alpha)f_y + \beta\alpha d_y\}, \quad (1b)$$

where Δx and Δy are the relative displacement components along x and y axes on the square lattice, respectively, sgn(x) is a function such that sgn(x<0)=-1, sgn(0)=0, and sgn(x>0)=+1, f_x and f_y are the fluctuation forces range over [-1/2,+1/2], d_x and d_y are the drift forces, α is the drift parameter representing the ratio between (f_x,f_y) and (d_x,d_y) and ranges over [0,1], and β is set to +1 or -1 for the attractive and the repulsive interactions, respectively. The magnitudes of d_x and d_y are restricted to 1/2 and defined by

$$d_x = \{\sum_{i,j} W_{x\,i,j}\delta(i+x,j+y)\}/\{2\sum_{i,j} W_{i,j}\delta(i+x,j+y)\}, \quad (2a)$$
$$d_y = \{\sum_{i,j} W_{y\,i,j}\delta(i+x,j+y)\}/\{2\sum_{i,j} W_{i,j}\delta(i+x,j+y)\}, \quad (2b)$$

where the summations over i and j, which are relative coordinate with respect to the moving particle, are restricted within the window, or the interaction range, $\delta(p,q)$ is the local density function such that $\delta(p,q)=1$ on an occupied site otherwise $\delta(p,q)=0$, and three matrices, $W_{x\,i,j}$, $W_{y\,i,j}$ and $W_{i,j}$ are defined by

$$W_{x\,i,j} = \text{sgn}(i)W_{i,j}, \quad (3a)$$
$$W_{y\,i,j} = \text{sgn}(j)W_{i,j}, \quad (3b)$$
$$W_{i,j} = \begin{cases} 1 & (\text{within the window}) \\ 0 & (\text{otherwise}) \end{cases}. \quad (3c)$$

Here the increase of the area of the window W_{ij} just corresponds to a long-range interaction, whereas the decrease of it does to a short-range one. In the present study the window function W_{ij} was assumed to be a diamond shape with the diagonal length w_{eff}.

RESULTS

In the following results, the total number of the particles of each cluster N was set to 10^4. The fractal dimension D was evaluated by the scaling property of the integrated correlation function, $C(r)$, such that $C(r) \propto r^D$.

In Figs.1 and 2, the dependences of D on α and w_{eff} are shown for $\beta=+1$ (attractive). From these one may confirm that the attractive interaction makes the cluster more dendritic and reduces D. Then, in Fig. 3 and 4, those for $\beta=-1$ (repulsive) are summarised. On the contrary to the previous case, one may conclude that the repulsive interaction makes the cluster more compact and increases D[3].

References
1) T. A. Witten AND L. M. Sander, Phys. Rev. Lett. **47**, 1400(1981): Phys. Rev. **B27**,5686(1983).
2) T. Vicsek, **Fractal Growth Phenomena**(World Scientific, 1989).
3) M. Nakagawa AND K. Kobayashi, **Chaos, Solitons** and **Fractals** (in press).

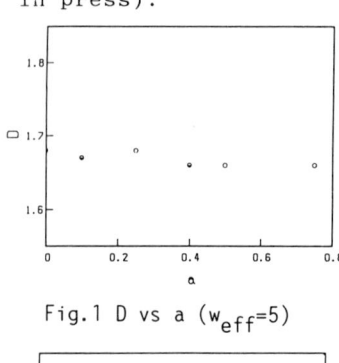

Fig.1 D vs a (w_{eff}=5)

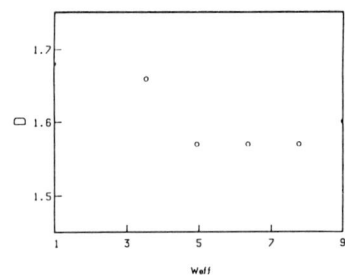

Fig.2 D vs w_{eff} (a=0.5)

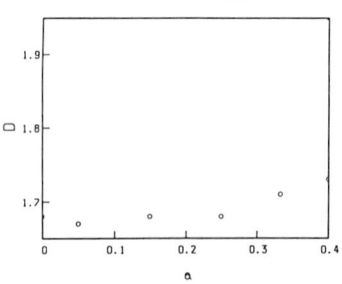

Fig 3 D vs a (w_{eff}=5)

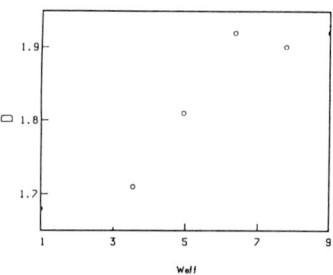

Fig.4 D vs w_{eff} (a=0.5)

FLOW-INDUCED FIRST ORDER TRANSITION OF THE AGGREGATION IN A DIFFUSION FIELD

Yukio Saito§, Makio Uwaha†‡and Susumu Seki‡
§Department of Physics, Keio University, Yokohama 223, Japan
†Institut Laue-Langevin, B.P.156X 38042 Grenoble Cedex, France
‡Institute for Materials Research, Tohoku University, Sendai 980, Japan

ABSTRACT

Effect of a uniform flow on the dynamics and the structure of growing patterns in a diffusion field is studied theoretically, and the resulting predictions are confirmed by the Monte Carlo simulation of a lattice gas system.

In an aggregation growth such as solidification from a gas with a density n_g (or similarly from a solution), the local density of diffusing gas $n(\vec{x}, t)$ obeys the time-dependent diffusion equation: $\partial n / \partial t = \Delta n$. The static and low density-limit, $n_g \to 0$, is the Laplace equation, $0 = \Delta n$, and an adhesive growth therein produces the diffusion-limited aggregation(DLA).[1] With the use of a computer model of this DLA, a great progress has been achieved in the study of diffusion growth. The aggregate is found fractal without a characteristic length, since the Laplace equation does not involve any length scale. The fractal structure of the aggregate is characterized by the fractal dimension D, which is smaller than the spatial dimension d.

We studied previously the aggregation growth from a time-dependent diffusion field, which is simulated by a lattice gas model with a finite gas density.[2] Our model is a natural generalization of the DLA model and, in the limit of high gas density, $n_g \to 1$, it becomes another well-studied growth model, the Eden model.[3] In a unidirectional diffusion growth, the spatial variation of the gas density is characterized by a diffusion length $\xi \sim 1/V$, where V is the growth velocity. Introduction of this characteristic length changes the structure of our aggregate from that of the DLA drastically. It is a DLA fractal up to the length scale ξ, whereas it is uniform and compact for scales larger than ξ. This characteristic length ξ is determined by the fractal dimension of the DLA as $\xi \approx n_g^{-\nu}$, where $\nu = 1/(d - D)$. This relation is an analogue of that between the correlation length and the temperature difference from the critical point value in a second order phase transition. Consequently, dynamics of the growth is determined by the fractal structure of the aggregate, and the growth velocity changes in a power of the gas density as $V \sim n_g^\nu$. Monte Carlo simulation confirmed this power law behavior as is shown in Fig.1, and for the fractal dimension the value $D = 1.71$ is obtained, which agrees with the previously obtained value.[4]

Homogeneous flow of the gas with a drift velocity U also introduces a characteristic length in the system via the drift term, $U \cdot \nabla n$, in the diffusion equation.

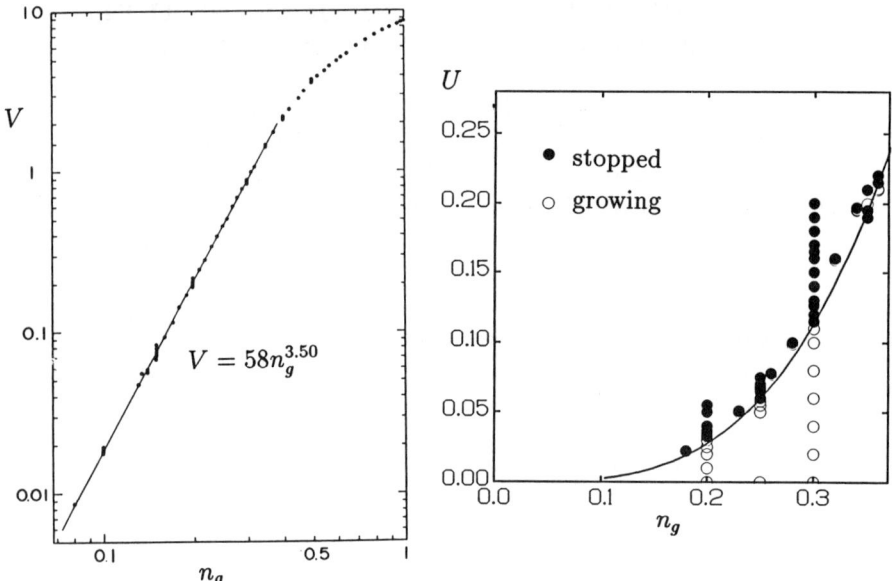

Fig.1 Growth velocity V versus gas density n_g without drift flow, $U = 0$.

Fig.2 Phase diagram in the space of flow velocity U and gas density n_g.

We studied coupled effect of the drift flow and the finite gas density in the aggregation growth, and found a dynamical phase transition of the first order.[5] For a drift velocity smaller than a critical value U_c, the aggregate grows steadily, whereas for $U > U_c$ the aggregate stops growing. At $U = U_c$, the growth velocity and other physical quantities remain finite. The ratio of the critical drift velocity U_c to the growth velocity V_0 without flow takes a constant value, determined only by the fractal dimension of the DLA. Our Monte Carlo simulation confirms this theoretical prediction, and in two dimensions the value of fractal dimension $D = 1.70$ is obtained by fitting the phase boundary between the steadily growing phase and the non-growing phase.(Fig.2) The value is quite close to $D = 1.71$ obtained previously. If the flow is perpendicular to the growth direction, it does not alter the growth velocity, although it affects the growth direction.

Part of this work is supproted by the Grant-in-Aid for Scientific Research on Priority Areas "Crystal Growth Mechanism in Atomic Scale", No. 03243101.

REFERENCES

1 T. A. Witten and L. M. Sander, Phys. Rev. B **27**, 5686 (1983).
2 M. Uwaha and Y. Saito, J. Phys. Soc. Jpn. **57**, 3285 (1988).
3 M. Eden, in *Proc. 4-th Berkley Symp. on Math. Statistics and Probability*, Vol.4, Ed. F. Neyman (Berkley, 1961).
4 P. Meakin, Phys. Rev. A **33**, 3371 (1986).
5 S. Seki, M. Uwaha and Y. Saito, Europhys. Lett. **14**, 397 (1991).

METASTABILITY AND TRANSFER-MATRIX FINITE-RANGE SCALING [1]

P. A. Rikvold* ◊,[2], B. M. Gorman†‡, and M. A. Novotny†
* Tohwa Institute for Science, Tohwa University, Fukuoka 815, JAPAN
◊ Department of Physics, Kyushu University 33, Fukuoka 812, JAPAN
† Department of Physics, and Center for Materials Research and Technology, and
‡ Supercomputer Computations Research Institute,
Florida State University, Tallahassee, FL 32306, U.S.A.

Abstract. Metastable free energies obtained from numerical transfer-matrix calculations yield scaling results in reasonable agreement with field-theoretical predictions.

The status and domain of validity of Langer's proportionality relation between the inverse lifetime of a metastable state and the imaginary part of the analytic continuation into that state of the equilibrium free-energy density, $\mathrm{Im} G_{\mathrm{ms}}$ [1], have been studied intensively[1,2]. Here we report progress with a new method for obtaining G_{ms} from transfer-matrix calculations[3].

We consider cylinder-shaped Ising systems of size $N \times \infty$, where N is the number of spins in a $(d-1)$-dimensional subsystem, whose equilibrium partition function and probability distributions are obtained from the dominant eigenvalue of the transfer matrix $\mathbf{T}$, λ_1, and its eigenvector, $|1\rangle$. We seek a spatially homogeneous, stationary, *constrained* joint probability distribution[4], $P_\alpha(X_l, Y_m) = \langle \alpha | X \rangle \langle X | (\lambda_\alpha^{-1} \mathbf{T}_\alpha)^{|m-l|} | Y \rangle \langle Y | \alpha \rangle$, where X_l and Y_m are the spin configurations of the l-th and m-th subsystems, and the constraint is identified with the α-th eigenspace of $\mathbf{T}$. The matrix $\mathbf{T}_\alpha$ is taken to commute with $\mathbf{T}$, with eigenvalues given by $\mathbf{T}_\alpha | \beta \rangle = f_\beta(\alpha) | \beta \rangle$. Reasonable regularity conditions on $P_\alpha(X_l, Y_m)$ (including convergence as $|m-l| \to \infty$) yield the conditions, $0 < f_\beta(\alpha) < f_\alpha(\alpha) = \lambda_\alpha$ [5]. However, since $|1\rangle$ is the only positive eigenvector of $\mathbf{T}$, $P_\alpha(X_l, Y_m)$ is *not* generally nonnegative, nor is $\mathbf{T}_\alpha$ uniquely determined by these conditions. We propose that $\mathbf{T}_\alpha$ should correspond to an extremum of a *constrained* free-energy density, $G_\alpha = U_\alpha - H M_\alpha - T S_\alpha$, where H is the field, U_α and M_α straightforward generalizations of the equilibrium internal energy and magnetization, and the *generalized entropy* is defined as $S_\alpha = -N^{-1} \sum_{X,Y} \langle \alpha | X \rangle \left[\langle X | \lambda_\alpha^{-1} \mathbf{T}_\alpha | Y \rangle \mathrm{Ln} \langle X | \lambda_\alpha^{-1} \mathbf{T}_\alpha | Y \rangle \right] \langle Y | \alpha \rangle$, where

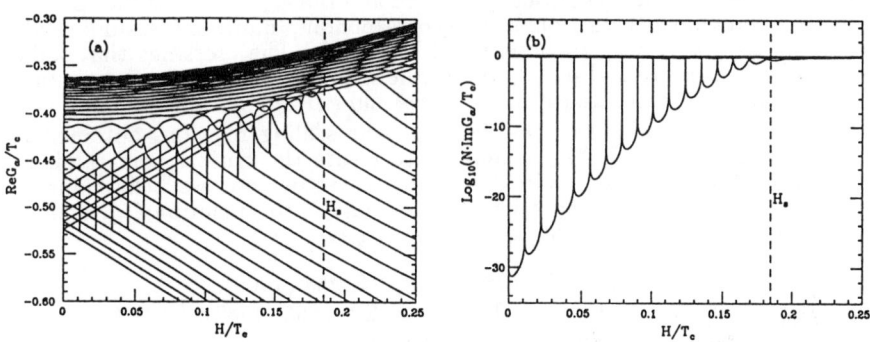

Figure 1: $\mathrm{Re} G_\alpha$ (a) and $\mathrm{Im} G_\alpha$ (b) for the Q1DI model with $N=56$ at $T=0.6 T_c$.

---

[1] Supported by FSU supercomputer time, SCRI (DOE Contract DE-FC05-85ER25000), MARTECH, NSF Grant DMR-9013107, PRF/ACS, the State of Florida, and Tohwa University.

[2] Permanent addresses: † and ‡ at Florida State University.

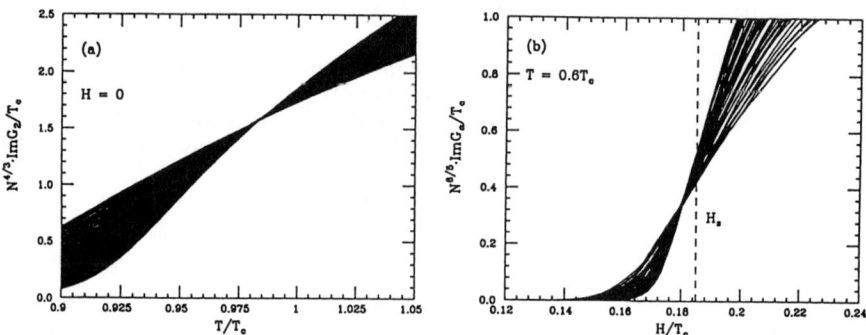

Figure 2: Scaling plots of the metastable $\text{Im}G_\alpha$ near T_c (a) and H_s (b), $N=35\text{--}84$.

$\text{Ln} z = \ln |z| + i\text{Arg} z$ is the principal branch of the complex logarithm. Here the analytic continuation does not enter via a complex field or a truncated summation over cluster sizes, as in field theory[1], but via the constraint which reduces the weights of the eigenstates with $|\lambda_\beta|>|\lambda_\alpha|$, through the logarithms of the resulting negative elements in $\mathbf{T}_\alpha$. Rather than performing the functional extremization proposed above, we obtained the numerical results shown here using the form $\mathbf{T}_\alpha = \lambda_\alpha \big\{ \sum_{|\lambda_\beta|>|\lambda_\alpha|} |\beta\rangle \frac{\lambda_\alpha}{\lambda_\beta} \langle\beta| + |\alpha\rangle\langle\alpha| + \sum_{|\lambda_\beta|<|\lambda_\alpha|} |\beta\rangle \frac{\lambda_\beta}{\lambda_\alpha} \langle\beta| \big\}$, applied to a specific Ising model with interactions J/N of range N (the Quasi-One-Dimensional Ising, or Q1DI, model[5]), which approaches mean-field behavior as $N\to\infty$ [5,6].

Numerical results for G_α in the Q1DI model below T_c are shown vs H in Fig. 1. In (a) the lowest branch of $\text{Re}G_\alpha$ corresponds to equilibrium, and the lowest branch with opposite magnetization (slope) is the metastable one. In addition to its magnetization opposite to the field, the metastable branch is uniquely identified by its extremely small imaginary part, shown in (b). The vertical line indicates the mean-field spinodal field H_s. For $|H|\leq H_s$, $N\text{Im}G_{\text{ms}}$ vanishes as $N\to\infty$.

Field-theoretical results for $\text{Im}G_{\text{ms}}$ yield the following finite-range scaling relations: $\text{Im}G_{\text{ms}}\sim N^{-4/3}$ at the critical point[1,6,7], and $\text{Im}G_{\text{ms}}\sim N^{-6/5}$ at the spinodal[6,8] (ignoring logarithmic corrections). The corresponding numerical scaling relations for our metastable $\text{Im}G_\alpha$ are illustrated in Fig. 2. The crossings that indicate scaling with the theoretical scaling powers lie within 2–3% of the exact mean-field T_c (a) and H_s (b). Accurate estimates of the numerically obtained scaling powers will be reported elsewhere[6]. The fact that the scaling relations obtained from field theory, applied to $\text{Im}G_\alpha$, enable us to locate the critical point and, more importantly, the mean-field spinodal line quite accurately, we believe indicates that our metastable G_α is equivalent to the analytic continuation G_{ms} defined in field theory[1,7], at least to within scaling corrections.

1. J. S. Langer, Ann. Phys. **41**, 108 (1967); **54**, 258 (1969).
2. B. Gaveau and L. S. Schulman, J. Phys. A **20**, 2865 (1987); Lett. Math. Phys. **18**, 201 (1989), and references cited therein.
3. P. A. Rikvold, Prog. Theor. Phys. Suppl. **99**, 95 (1989).
4. O. Penrose and J. L. Lebowitz, J. Stat. Phys. **3**, 211 (1971).
5. M. A. Novotny, W. Klein, and P. A. Rikvold, Phys. Rev. B **33** 7729 (1986).
6. B. M. Gorman, P. A. Rikvold, and M. A. Novotny, Phys. Rev. A (to appear).
7. N. J. Günther, D. A. Nicole, and D. J. Wallace, J. Phys. A **13**, 1755 (1980).
8. C. Unger and W. Klein, Phys. Rev. B **29**, 2698 (1984).

RHEOLOGICAL PROPERTIES OF FOAMS: COMPUTER SIMULATION OF VERTEX MODEL FOR TWO-DIMENSIONAL RANDOM CELLULAR STRUCTURES

Tohru Okuzono, Kyozi Kawasaki
Department of Physics, Kyushu University 33, Fukuoka 812, Japan

Tatsuzo Nagai
Physics Department, Kyushu Kyoritsu University, Kitakyushu 807, Japan

ABSTRACT

Rheological properties of foams which have two-dimensinal random cellular structures are studied by computer simulations of vertex models. The simulations are performed under simple shear flow for small shear rates. Through these computer simulations, existence of the yield stress and the shear rate dependence of shear stress are investigated.

Despite its practical importance, our understanding of the rheology of foams or concentrated emulsions has not yet reached a satisfiable level because of the complex nature due to cellular structure[1]. Although our vertex models[2-3] were originally constructed to simulate a cellular pattern growth, they are also applicable to this problem if we accept some assumptions and simplifications.

In our two-dimensional models, a random cellular system consists of a set of straight edges and a set of vertices at which three edges meet. For the i-th vertex whose position and velocity are $\mathbf{r}_i$ and $\mathbf{v}_i$, respectively, the equation of motion takes the form:

$$\frac{\partial}{\partial \mathbf{v}_i}\mathcal{R} + \frac{\partial}{\partial \mathbf{r}_i}\mathcal{F} = 0 \qquad (\ i=1,2,\cdots\ ) \qquad (1)$$

Here, $\mathcal{R}$ is the dissipation function and $\mathcal{F}$ is the interface free energy. Time evolution of the system is determined by Eq.(1) supplemented by the topological changes of the structure, that is, **T1** and **T2** processes.[4] For the detailed description of this model, see Ref. 2-3.

Now we consider a system with a large volume fraction of dispersed phase which is assumed to be inviscid and incompressible. In such a case, the effect of viscous flow of the liquid inside the films near Plateau borders is important. Schwartz & Princen[5] calculated the dissipation due to this viscous flow in the small capirally number region. In our model we can incorporate their result into the dissipation function $\mathcal{R}_P$. (We don't write it down here explicitly.) In Eq.(1), now, $\mathcal{R}$ is replaced by $\mathcal{R}_D + \mathcal{R}_P$, where $\mathcal{R}_D$ arises from diffusion across cell boundaries. In actual systems, the dynamics associated with $\mathcal{R}_D$ is very slow compared with that associated with $\mathcal{R}_P$. For our purpose we neglect the diffusion process and impose constraints on the system that the area of every cell is constant.

When we apply a simple shear flow $\mathbf{u}(\mathbf{r})$ defined as $\dot{\gamma} y \mathbf{e}_x$ at the point $\mathbf{r} = (x, y)$, where $\dot{\gamma}$ is a simple shear rate and $\mathbf{e}_x$ is an unit vector parallel to x-axis, we should replace $\mathbf{v}_i$ in the equations of motion with $\mathbf{v}_i - \mathbf{u}(\mathbf{r}_i)$. We carry out the computer simulations in this case for several values of $\dot{\gamma}$. The system consists of 459 cells with periodic boundary conditions. The results are presented in Fig. 1 and Fig. 2. In the Fig. 1, we plot the shear stress $\tau(t; \dot{\gamma})$ versus shear strain $\dot{\gamma} t$ for the various values of $\dot{\gamma}$. A recent experiment[6] on oil-water emulsion systems suggest that the steady state shear stress $\tau(\dot{\gamma})$ obeys the power law : $\tau(\dot{\gamma}) - \tau_0 \propto \dot{\gamma}^{1/2}$, where τ_0 is an yield stress. Fig. 2 shows a relation between $\tau(\dot{\gamma})$ and $\dot{\gamma}^{1/2}$, where the values of $\tau(\dot{\gamma})$ are obtained by averaging over the time interval $4.0 < \dot{\gamma} t < 5.0$ for each $\dot{\gamma}$. This result seems to be in agreement with the experimental data, although the statistics is rather poor. Further investigations are now under way.

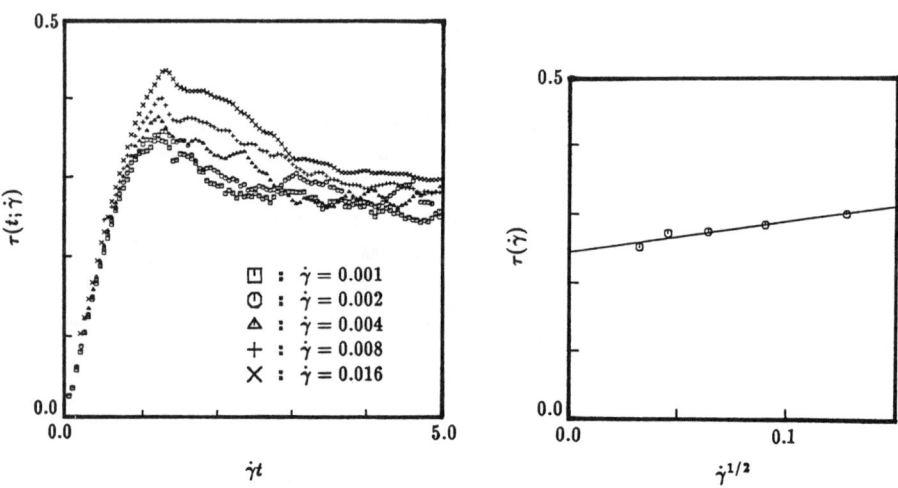

Fig. 1 Stress vs. strain.   Fig. 2 Shear rate dependence of stress

REFERENCES

1. A.M.Kraynik, Ann.Rev.Fluid Mech. **20**, 325 (1988).
2. K.Kawasaki, T.Nagai and K.Nakashima, Phil.Mag. **B60**, 399 (1989).
3. K.Nakashima, T.Nagai and K.Kawasaki, J.Stat.Phys. **57**, 759 (1989).
4. D.Weaire and N.Rivier, Contemp.Phys. **25**, 59 (1984).
5. L.W.Schwartz and H.M.Princen, J.Colloid Interface Sci.**118**, 201 (1987).
6. H.M.Princen and A.D.Kiss, J.Colloid Interface Sci.**128**, 176 (1989).

OSTWALD RIPENING IN OPEN SYSTEMS

Akio Nakahara, Toshihiro Kawakatsu and Kyozi Kawasaki
Department of Physics, Kyushu University 33, Fukuoka 812, Japan

ABSTRACT

New scaling laws are obtained for the Ostwald ripening in a open cell where the total amount of both the solute and the precipitates flows out into the neighboring cell.

We investigate the Ostwald ripening of precipitates in open systems which consist of two homogeneous cells, specified by an index i, coupled with each other by the diffusion of the solute. For such systems, the continuity equation for the size distribution function of droplets of radius r in cell i at time t, $f_i(r,t)$, is expressed as[1-4]

$$\frac{\partial}{\partial t} f_i(r,t) + \frac{\partial}{\partial r}[v_i(r,t) f_i(r,t)] = 0, \qquad (1)$$

where the growth rate $v_i(r,t)$ is given by

$$v_i(r,t) = \frac{1}{r}\left[\sigma_i(t) - \frac{1}{r}\right], \qquad (2)$$

and $\sigma_i(t)$ is a supersaturation of the solution in cell i. We consider the situation where mass transport takes place due to the difference in the supersaturation between the two adjacent cells. Thus, the mass conservation holds in the whole system.[4]

Analytical treatments based on dynamical scaling assumptions and numerical simulations shows the existence of new scaling laws for the Ostwald ripening in such open systems.[5] Since the difference in the supersaturation between two cells becomes asymptotically negligible, the supersaturation in each cell asymptotically obeys the same Lifshitz-Slyozov[1]-Wagner[2] (LSW) scaling law for a closed system, i.e., $\sigma_i(t) \simeq (3/2)^{2/3} t^{-1/3}$. The mass conservation in the whole system also leads to the result that the total amount of both the solute and the precipitates in at least one of these cells approaches a finite value for large t. For that cell, physical quantities asymptotically show the LSW scaling behaviors,

$$n_i(t) \sim t^{-1}, \qquad (3)$$

$$f_i(r,t) \sim t^{-4/3} F^{\text{LSW}}(r/t^{1/3}), \tag{4}$$

where $n_i(t)$ is the total number of droplets in cell i and $F^{\text{LSW}}(z)$ is the LSW scaling function. On the other hand, when the total amount in the cell flows out into the other cell, the number of droplets, $n_i(t)$, and the size distribution function of droplets, $f_i(r,t)$, asymptotically obey new scaling laws, which are characterized by a parameter x_i and are different from the LSW predictions;

$$n_i(t) \sim t^{-(1+x_i)}, \tag{5}$$

$$f_i(r,t) \sim t^{-(4/3+x_i)} F^{x_i}(r/t^{1/3}). \tag{6}$$

Here, $F^x(z)$ is a scaling function for the open subsystem and depends on the parameter x. In the limit $x \to 0$, $F^x(z)$ coincides with the LSW scaling function $F^{\text{LSW}}(z)$, and the scaling behavior (5), (6) reduces to (3), (4). We also find that the asymptotic long time behavior of these physical quantities, i.e., the value of x_i, is highly dependent on the initial conditions.

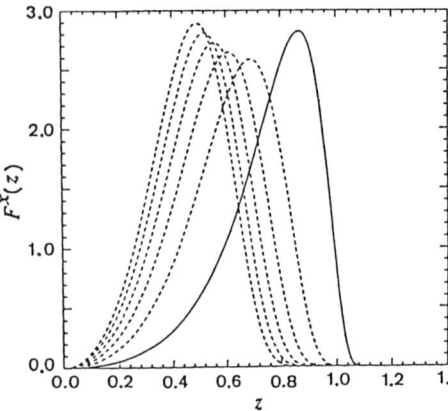

Fig.1. The scaling functions are shown as functions of the scaled radius z. The solid line represents the LSW scaling function $F^{\text{LSW}}(z)$ and dotted lines represent the scaling function for the open subsystem $F^x(z)$. The parameter x in $F^x(z)$ is, from the larger z side to the smaller z side, $x = 2, 4, 6, 8$ and 10, respectively.

REFERENCES

1. I.M.Lifshitz and V.V.Slyozov, J.Phys.Chem.Solids **19**, 35 (1961).
2. C.Wagner, Z.Elektrochem. **65**, 581 (1961).
3. J.A.Marqusee and John Ross, J.Chem.Phys.**79**, 373 (1983).
4. G.Venzl, Phys.Rev.**A31**, 3431 (1985).
5. A.Nakahara, T.Kawakatsu and K.Kawasaki, J.Chem.Phys.**95** (1991).

RELAXATION DYNAMICS OF PHASONS IN QUASICRYSTALS

Y. Ishii
*Department of Material Science, Himeji Institute of Technology
Kamigouri-cho, Akou-gun, Hyogo 678-12, Japan*

ABSTRACT

In order to understand characteristic feature of relaxation dynamics of the phasons in quasicrystals, we investigate two-dimensional Penrose tiling by a computer simulation. A power law behaviour is found in relaxation of the quenched phason fluctuations at low temperatures.

The discovery of an icosahedral phase of aluminum-based alloys stimulates studies of a novel phase of matter called *quasicrystal*[1]. Quasicrystal is an incommensurate solid where dynamics at low temperatures is described by the hydrodynamical modes due to the incommensurability (*phasons*) as well as those due to the broken translational symmetry (*phonons*). It turns out that structural modulations caused by phasons in quasicrystals can describe crystal–quasicrystal transformations[2]. It is also known that the phason disorder is frozen in some quasicrystalline samples. Therefore relaxation of the phason degrees of freedom is crucial in the formation of quasicrystalline order.

In the two-dimensional Penrose tiling, the phason fluctuation causes a local flip of a three-fold vertex and the arrow matching rule, which describes the specific ways of quasicrystalline ordering, is violated there (Fig.1). Then the relaxation of the phason degrees of freedom is realized by migration of an arrow mismatch (a local defect) along a string, which is usually called a worm. In a system with many defects, the migration of one defect is sometimes prohibited by the other ones because a string, along which a flipped vertex migrates, is terminated at a crossing point with the other strings called a decapod. Then the relaxation of the frozen-in phason degrees of freedom takes place by resolving entangled strings and might be extremely slow[3].

Phason relaxation on the two-dimensional Penrose tiling is investigated by a computer simulation. First we generate an ideal Penrose tiling and randomize it by flipping three-fold vertices. This randomization introduces a large amplitude of the phason fluctuation. Then the system is quenched and relaxed to its equilibrium according to a

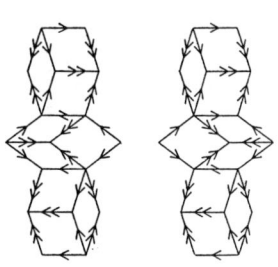 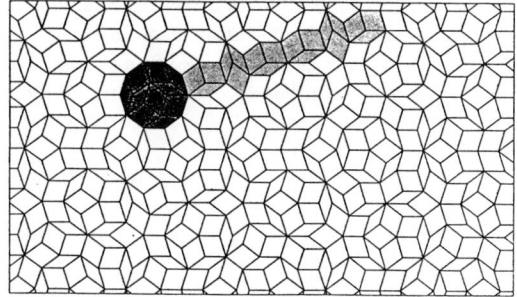

Fig.1 : Migration of a defect by a flip of the three-fold vertex (left) and a worm and a decapod (right).

© 1992 American Institute of Physics

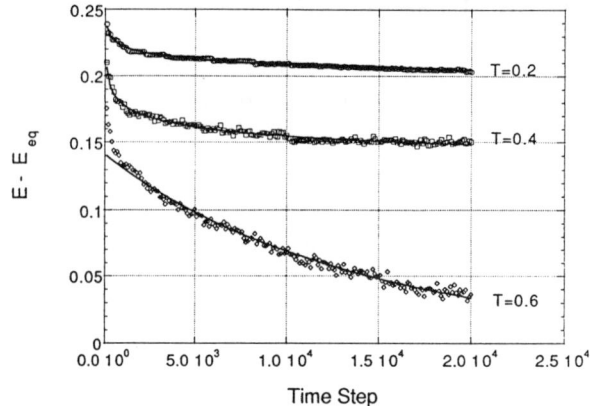

Fig.2: Relaxation of energy due to the arrow mismatch.

following algorithm: (1) Choose a vertex. If it is a three-fold one, then (2) a trial flip is made. (3) The trial flip is accepted with a probability proportional to $\exp(-\Delta E/k_B T)$ where ΔE is the change in energy due to the trial flip. One time step consists of N procedures (1)-(3) where N is the total number of tiles.

There are two types of arrow mismatches in the Penrose tiling, i.e., single and double arrow mismatches. We find that the system is easily relaxed to its equilibrium after quenching if the double arrow mismatch (DAM) is prohibited. If DAM is allowed, on the other hand, DAM defects are frozen and hardly relaxed at low temperatures. Figure 2 illustrates the relaxation of energy due to the matching rule violation where temperature is normalized to the energy cost per arrow mismatch. The relaxation curve fits to a power law

$$E - E_{eq} \propto t^{-\nu}, \qquad (1)$$

at lower temperatures ($\nu = 0.028$ at $T = 0.2$ and $\nu = 0.063$ at $T = 0.4$) while it fits to an exponential law

$$E - E_{eq} \propto \exp(-\gamma t), \qquad (2)$$

at $T = 0.6$ ($\gamma = 7.2 \times 10^{-5}$). At even higher temperatures, the system can be relaxed to its equilibrium rapidly.

A crossover from a power law relaxation to an exponential one is certainly interesting. At present, physical implication of the crossover is not clear. We should check also whether the above results are due to a specific topological nature of the two-dimensional Penrose tiling or universal in a certain family of the quasiperiodic systems. Detailed analysis is in progress.

REFERENCES

[1] P. J. Steinhardt and S. Ostlund, *The Physics of Quasicrystals*, (World Scientific, Singapore, 1987).
[2] For example, Y. Ishii, Phil. Mag. Lett. **62**, 393 (1990).
[3] L. D. Gronlund, D. C. Wright, J. P. Sethna and D. S. Rokhsar, Phys. Rev. B**42**, 8517 (1990).

A NUMERICAL MODELING OF CONVECTIVE MOTIONS IN GRANULAR MATERIALS

Y-h. Taguchi
Department of Physics, Tokyo Institute of Technology, Oh-okayama, Meguro-ku, Tokyo 152, Japan

ABSTRACT

We propose a new modeling of convective motions in powder which Faraday has observed in 1831 for the first time. In this model, powder is treat as a set of elastic particles with viscostic interaction. In the two dimensional modeling, convective motions are induced by the instability. This instability turns out to be caused by elasticity.

First we report the phenomena which are the subject of this paper. The experimental setup is as follows. First a shallow dish is prepared and filed with small spheres. The typical diameter of the dish is 10 cm, and spheres has the diameter of a little bit smaller than 1 mm. Initial depth of the layer of spheres are about a few centimeter. When the strong vertical vibration is applied to the shallow dish, powder reveal an amazing bahaviour. After accelation amplitude exceeds a critical value which is always slightly larger than that of gravity, a heap is construted spontaniously. At the same time, the convective motion starts; upwards at the center, downwards beside wall. This phenomenon has been firstly observed by Faraday at 1831[1]. Since then, many researches have been done and especially some experiments have been done and cleared out some points[2]. Although theoratical approachs[3] are carried out, no one has not yet succeed in reproduing neither convetion nor a appearance of heap. In this study, we can show how the convection start using numeriacal modeling.

Our model is a set of spheres having elasticity and viscosity. The equation of motion of each sphere is

$$\ddot{\mathbf{x}}_i = -\sum_{j=1}^{N} \theta(a-\mid \mathbf{x}_i - \mathbf{x}_j \mid) \left(k(\mathbf{x}_i - \mathbf{x}_j - a\frac{\mathbf{x}_i - \mathbf{x}_j}{\mid \mathbf{x}_i - \mathbf{x}_j \mid}) + \eta(\mathbf{v}_i - \mathbf{v}_j) \right) - \mathbf{g},$$

where $\theta(x)$ is a step function, N is a total number of spheres, $\mathbf{x}_i$ is the position vector of the i-th shpere. k and η are the elastic constant and the viscosity coefficient respectivly. a is a diameter of a sphere, $\mathbf{v}$ is velocity. If two spheres colide with each other in front, they have effective coefficient of restitution $e = \exp(-\eta\pi/\omega)$ and collision time $\Delta t = \pi/\omega$ and $\omega = \sqrt{2k - \eta^2}$. $\mathbf{g}$ is a glavity accelation. Here we should mention that each sphere does not always correspond to each particle of powder. It is a sort of 'quasi-particle' which is used in the simulation of DLA[4].

First trial is two dimensional simulation. $N = 100, e = 0.5, \Delta t = 0.5, g = 1.0, a = 2.0$ The horizontal size of a 'dish' is 30. Therefore initial depth is $6.0 \sim 7.0$.

The vibration of the bottom obeys the function of $b\cos(\omega_0 t)$, where t is time. Here $b = 1.1$ and the period $T = 2\pi/\omega_0$ is 6.0.

Figure 1. shows the flow line after some time interval. We can easily see

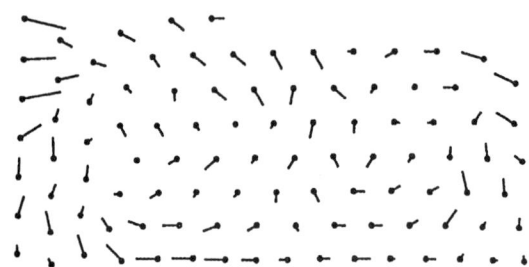

Figure 1. Flow lines. The dots indicates the origin of the velocity vector.

convection. However the flow is downward at the center which is opposite to that of experimants. This is because there is no friction between wall and spheres. If we introduce the friction, the convection is reversed. The mechanizm of the convection is as flows. At the beginning of the simulation, powder has clear layered structure. When the bottom rizes up, the shperes are pressed to the bottom and packed very strongly. And the bottom move down, the packed sphere are released. However the inner stress is not only vertical but also horizontal. Therefore each particle has some tendency to move horizontaly. If there is some anisotolopy in packing, Each layer of spheres can have some wrinkles. This causes relative motion and results in convection.

In summary, we propose a new modeling in powder which exhibit the convection. However the heap is not observed in our model. Perhaps three dimensional simulation must be done because the slope in two dimension is very unstable.

References

1 M. Faraday, Phil. Trans. R. Soc. Lond. *52*, 299 (1831)

2 P.Evesqu and J.Pajchnbach, Phys. Rev. Lett.*61*, 44 (1989)
  C.Laroche, S.Douady and S.Fauve, J. Phys. (Paris)*50*, 699 (1989)
  H.M.Jeager,Chu-heng Liu and S.R.Nagel, Phys. Rev. Lett.*62*, 40 (1989)
  P.Evesqu,E,Szmatula and J.P.Denis, Europhys. Lett.*12*, 623 (1990)

3 A.Mehta and G.C.Barker, Phys. Rev. Lett.*67*, 394 (1991) and references therein

4 P.Meakin, Phase Transitions and Critical Phenomena *12* (Academic Press, New York), p335

GROWTH MECHANISM OF HOMOGENEOUS DIFFUSION-LIMITED AGGREGATION

Shonosuke Ohta
Department of Physics, College of General Education,
Kyushu University, Ropponmatsu, Fukuoka, 810 Japan

ABSTRACT

Growth features of the aggregates including surface diffusion processes are studied by the Monte Carlo simulations on two-dimensional square lattice. Anisotropy of dendritic pattern changes from <10> direction to <11> one depending on the surface diffusion mechanism. Homogeneous diffusion-limited aggregation (DLA) grows in the crossover domain between two anisotropies. Results indicate that the random tip-splitting mechanism of the hottest tips is necessary to hold the homogeneity and self-similarity of DLA pattern.

INTRODUCTION

Homogeneous diffusion-limited aggregation (DLA) has been experimentally observed in the system of dendritic crystallization[1]. Dendrite is discussed as the anisotropic pattern formation growing in the diffusion field under the boudary condition of surface kinetics[2]. Thus it is attractive problem what growth mechanism of the hottest tips can control whether or not the pattern grows homogeneous.

SIMULATION

On the kinetics of surface particles surface diffusion and the effects of surface potential and noise are assumed, and also on the pattern growth kink and surface nucleation growths, as is well known in crystal growth, are modeled. The simulation procedures on two-dimensional square lattice are as follows: A random-walker released from a distant boundary is moving on the square lattice, just as in the ordinary DLA model[3]. After it reaches an unoccupied nearest-neighbor (NN) site of the cluster particles, call surface site, further random-walk along the surface site is performed. Hopping site of the walker is chosen from among the NN and next-nearest-neighbor (NNN) surface sites. Here, taking into account different potential walls for two directions, the ratio γ of the hopping probability for a NNN surface site to that for a NN one is assumed. Such a random-walk is repeated until the walker visits a kink site or until the trial of surface walk becomes τ times, where the kink site with low potential energy is assumed as where the two NN sites neighboring each other and a NNN site between them are occupied by the cluster. The number of walkers stuck to the surface site is counted by its own counter. If its count reaches a given number m then the surface site becomes part of the cluster and the new surface sites are born. The considerable growth on the flat surface without kink, which is recognized the surface nucleation arising from the increase of surface concentration in real crystal growth, is reflected by the finish of surface diffusion up to τ times. The value 1/m is called noise para-

© 1992 American Institute of Physics

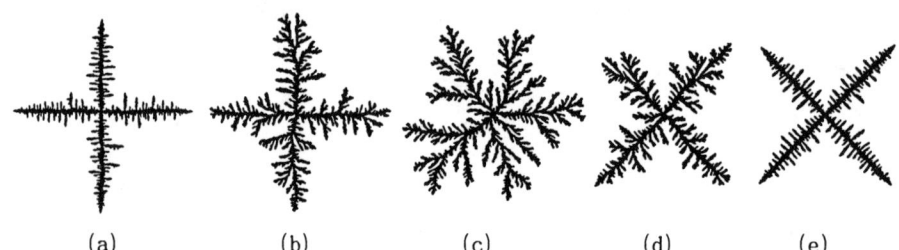

(a) (b) (c) (d) (e)

Fig.1 Simulated patterns for fixed m=5 and τ=10 with radius of 600 lattice units. The hopping parameters are γ=(a)0.05, (b)0.175, (c)0.23, (d)0.3 and (e)0.6.

meter[4].   Another random-walker is released from a distant boundary and the procedure iterated.

RESULTS

Typical examples of the simulated patterns for $\underline{m}$=5 and τ=10 are shown in Fig.1.   Dendritic pattern changes its anisotropy from <10> direction (Fig.1a) to <11> one (Fig.1e) depending on the hopping parameter γ in the surface random walk. In case of small γ, the surface diffusion is limited within a plane surface and the tip front has rough surface composed by kinks.   Crystallization of the particles adsorbed in the tip front derives the pattern growth toward <10> direction.   While in case of large γ, such adsorbed particles can penetrate into the screened surface near the tip due to the diffusion process transfer to NNN site, and then smooth surfaces develope through the growth at kink sites.   This effect leads to such a tip as facets form in its both sides.   As a result of the nucleation and kink growth on both faceted surfaces, the tip grows in the <11> direction.   Hence, we can understand that the dominant process of the tip growth is frontal (lateral) type under the condition of $\gamma < \gamma_c$ ($\gamma > \gamma_c$), where γ_c is the hopping parameter at the crossover domain between <10> and <11> anisotropies.   From the measurement of pattern anisotropy we obtained γ_c=0.233±0.005.   And in this crossover domain (Fig.1c) a homogeneous DLA pattern grows with the fractal dimension d_f=1.716±0.003 which coincides with that of off-lattice DLA. In case of low noise for m=100, however, the crossover pattern yields mixed state which contains two anisotropies.   Results indicate that the growth mechanism of the hottest tips of DLA is the random tip-splitting process arising from the noise under the crossover condition of the frontal and lateral growths, and that such a randomly ramifing process holds the homogeneity and self-similarity of DLA pattern.

REFERENCES

1. H. Honjo, S. Ohta and M. Matsushita, J. Phys. Soc. Jpn. <u>55</u>, 2487 (1986).
2. J. S. Langer, Rev. Mod. Phys. <u>52</u>, 1 (1980).
3. T. A. Witten and L. M. Sander, Phys. Rev. Lett. <u>47</u>, 1400 (1981).
4. J. Nittman and H. E. Stanley, Nature <u>321</u>, 663 (1986).

STRUCTURAL CHARACTERIZATION OF MOLTEN CALCIUM CHLORIDE BY MOLECULAR DYNAMICS SIMULATION

Norimasa Umesaki
Material Physics Department, Government Industrial Research Institute, osaka
1-8-31, Midorigaoka, Ikeda, Osaka 563, Japan

ABSTRACT

A molecular dynamics study has been carried out of calcium chloride crystal and melt. We have empirically determined the potential parameters for Ca^{2+} and Cl^- by performing MD runs of $CaCl_2$ crystal at room temperature. The computer-generated structure of molten $CaCl_2$ seemed to be realistic in comparison with our X-ray diffraction data.

INTRODUCTION

In recent years, a molecular dynamics (MD) method is increasingly being used to study physical properties of computer-simulated solids and liquids. Unfortunately, most of the calculated systems have been simple ionic liquids such as molten alkali chlorides. However, complex ionic liquids such as molten alkaline-earth chlorides are of interest from many points of view. The most remarkable difference in physical properties between alkali chlorides and alkaline-earth chlorides is the volume expansion $\Delta V_m/V_m^s$ on melting. It is well known that alkali chlorides have rather large $\Delta V_m/V_m^s$ Values, i.e., LiCl (26.2%), NaCl (25.0%) and KCl (17.3%), whereas alkaline-earth chlorides have much smaller than values, i.e., $CaCl_2$ (0.9%), $SrCl_2$ (4.2%) and $BaCl_2$ (3.5%)[1]. This property of alkaline-earth chlorides may be attributed to the small change of coordination numbers and the minimal structural disordering on melting. However, the structures of molten alkaline-earth chlorides such as $CaCl_2$ and the mixed systems of them have hardly been studied and are not yet clear.

The purpose of this study is initially to determine the suitable potential parameters of $CaCl_2$ crystal and melt for MD calculation, and the to seek the characteristics of the atomic scale structure of molten $CaCl_2$ by an MD simulation with the obtained potentials.

MD SIMULATION

It is usually assumed in MD that the interaction between the particles in the system pairwise additive. We use the suitable Busing-type pair potential function. The MD runs were carried out numerically solved by the Newtonian equations of motion of interacting particles using Verlet's algorithm.

RESULTS AND DISCUSSION

For any successful MD calculation, we have empirically determined the potential parameters for Ca^{2+} and Cl^- by performing MD runs of $CaCl_2$ crystal at room temperature. A distorted rutile structure (space group: P_{nnm}) of $CaCl_2$ crystal[2] was consequently reproduced by our MD simulation. The lattice parameters of MD-

© 1992 American Institute of Physics 561

simulated $CaCl_2$ crystal changed by less than one percent to become equal, and the cation/anion coordinates became equal in a similar way.

The MD calculation for molten $CaCl_2$ assumed a cube as a basic cell. The edge length of the basic cell was calculated from the observed value of density of the molten $CaCl_2$ at 1073K[3]. The number of particles within the basic cell was 360 (Ca:120;Cl:240) or 600 (Ca:200;Cl:400). Structure of molten $CaCl_2$ at 1073K was represented by our MD simulation. The MD-simulated equilibrium melt structure was analyzed in terms of pair radial distribution functions $g_{ij}(r)$ for comparison with the corresponding total radial distribution function $G(r)$ obtained from our X-ray results[4]. Fig. 1 shows the $g_{ij}(r)$ curves and the distribution of the coordination numbers $N_{ij}(r)$ of the pairs Ca-Cl, Cl-Cl and Ca-Ca in the MD-calculated molten $CaCl_2$. As shown in this figure, the structure information obtained from the MD runs is in good agreement with our X-ray results[4]. The MD-simulated coordination number $N_{Ca/Cl}$ (=5.6) of the nearest-neighbor Ca-Cl pair is close to the $N_{Ca/Cl}$ (=5.8) estimated from the volume expansion of $CaCl_2$ on melting[1]. Moreover the MD-calculated distances and coordination numbers of the nearest-neighbor pairs Ca-Cl and Cl-Cl are nearly equal to those found in the crystalline form[2]. Since $CaCl_2$ has a rutile structure, Ca^{2+} cations are centered at slightly distorted $CaCl_6$ clusters composed of six Cl^- ions with Ca-Cl distance of 2.7~2.76Å. It is, therefore, believed that the distances and coordination numbers of the nearest-neighbor ionic pairs scarcely change on melting.

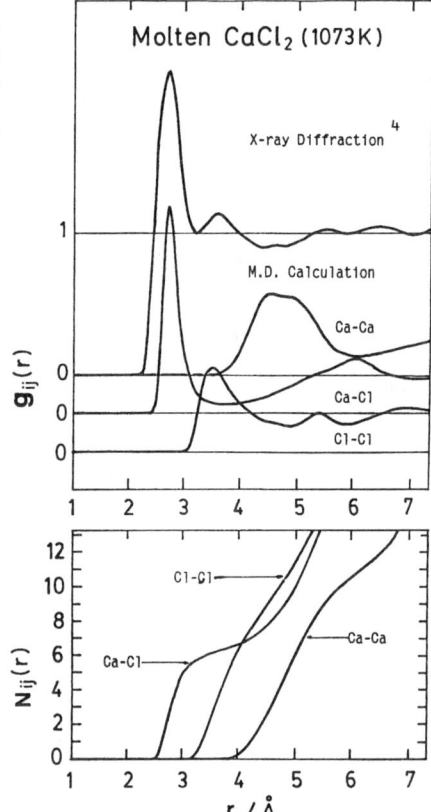

Fig. 1 A comparison of the pair correlation functions $g_{ij}(r)$ and the distribution of coordination numbers $N_{ij}(r)$ obtained from our MD simulation with the total X-ray correlation function $G(r)$ for molten $CaCl_2$ at 1073K.

REFERENCES

1. H. Schinke and F. Sauerwald, Z. Anorg. Allg. Chem., <u>287</u>, 832 (1956).
2. Ralph W. G. Wyckoff, Crystal Structures, Vol. 1, Second Edition, Interscience Publisher, 1963, p.252-253.
3. G. Brautigam, H. H. Emons and H. Vogt, Z. Chem., <u>10</u>, 344 (1970).
4. N. Iwamoto, N. Umesaki, T. Asahina and M. Kosaka, High Temp. Sci., <u>23</u>, 1 (1987).

SLOW RELAXATION PROCESSES IN THE FERROMAGNETIC STATE

Hiizu Nakanishi
Department of Physics, Faculty of Science and Technology,
Keio University, Yokohama 223, Japan

ABSTRACT We study the relaxational modes in the Glauber dynamics and their contribution to the spin auto-correlation function by means of numerical diagonalization. Differences in the dynamics between the disordered and the ordered phase are discussed.

INTRODUCTION

Effects of a symmetry breaking on dynamics of a system has been attracting much interest. Existence of several physically equivalent equilibrium states in the broken symmetry phase gives the system a rich dynamical structure.

In the disordered phase above the critical temperature, size of typical equilibrium fluctuations is order of the correlation length and the time correlations of all the physical quantities will decay exponentially within the correlation time associated with these fluctuations.

On the other hand, in the ordered phase, in addition to these fluctuations there exist the long-lived droplet fluctuations, which is a direct consequence of the symmetry breaking. The effects of these droplet fluctuations on the spin auto-correlation function have been investigated by several authors[1-6] and it is argued that they might cause stretched exponetial decay of the spin auto-correlation function $C(t) \sim \exp(-at^\beta)$ ($0 < \beta < 1$) for the long time limit. Although this type of slow relaxation has been known very well, it is really unexpected that it is found even in a regular system.

We study, in this report, the equilibrium fluctuations of the Glauber model and their effects on the spin auto-correlation function below and above the ferromagnetic phase transition temperature using the numerical diagonalization technique for the small two-dimensional square lattice system(up to 20 spin system).

FORMULATION

The Glauber dynamics can be described by the master equation for the probability distribution function of the spin configuration $P(\{S\}, t)$ as

$$\frac{\partial P(\{S\}, t)}{\partial t} = -\sum_{\{S'\}} \Lambda(\{S\}, \{S'\}) P(\{S'\}, t),$$

where Λ is a time evolution matrix of $2^N \times 2^N$ dimension for a N spin system. The spin auto-correlation function in an equilibrium state can be decomposed by the eigenvectors of Λ as

$$C(t) = \sum_{k,n} |\mu_{k,n}|^2 \exp(-\lambda_{k,n} t), \qquad \mu_{k,n} = \sum_{\{S\}} \phi_{k,n}(\{S\}) S_0 \exp(-E(\{S\})/T),$$

where $\phi_{k,n}$ is the n-th left eigenvector of Λ with the wave number k, $\lambda_{k,n}$ is the corresponding eigenvalue, and T is temperature. $E(\{S\})$ is system energy for the configura-

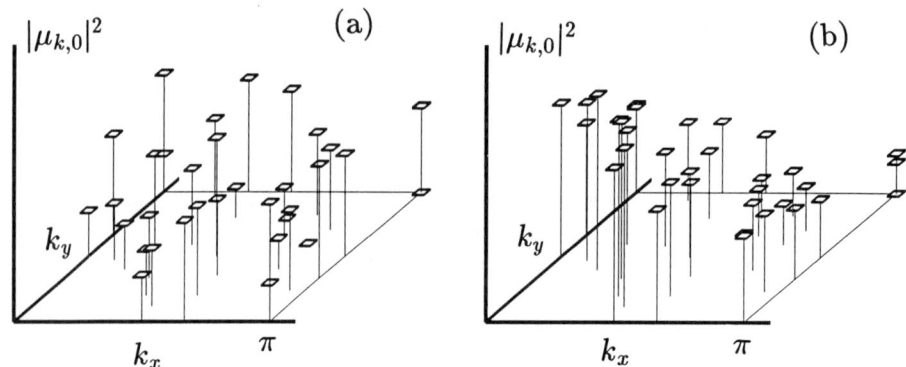

Figure: $|\mu_{k,0}|^2$ v.s. (k_x, k_y). All the data for N=16, 17, 18, and 20 system are plotted on a single graph for J/T=0.6 (a) and J/T=0.3 (b). For the critical temperature J/T_c=0.441.

tion $\{S\}$ and is given by $E(\{S\}) = -\sum_{i,j} JS_iS_j$, where J is exchange energy and the summation is over the nearest neighbor pairs.

For the slowest mode $\mu_{k,0}$, Takano et al [2] assumed $|\mu_{k,0}|^2 \sim \exp(-ck^{-(d-1)})$ for small k in the ordered phase and derived $\beta = (d-1)/(d+1)$ with d being spatial dimensionality, which contradicts Huse and Fisher's results[1], who expect $\beta = 1/2$ for $d = 2$ and $\beta = 1$ for $d \geq 3$.

METHOD AND RESULTS

Although the dimension of Λ is very large even for the small system, we can calculate the slowest mode $\phi_{k,0}$ and $\lambda_{k,0}$ for a given k for $N \leq 20$ by means of Lanczos method and the conjugate gradient method using the spin up-down symmetry and the translational symmetry. The systems calculated are square system on the square lattice with the twisted boundary condition with N =16, 17, 18, and 20. Examples of the results for $|\mu_{k,0}|^2$ v.s. (k_x, k_y) are shown in Figure for above ($J/T = 0.6$) and below ($J/T = 0.3$) the critical temperature.

It can be seen from Figure that $|\mu_{k,0}|^2$ is suppressed for small k in the ordered phase compared with those for the disordered phase. This agrees qualitatively with the assumption used by Takano et al[2].

REFERENCES

1. D.A.Huse and D.S.Fisher, Phys.Rev.B **35**,6841(1987).
2. H.Takano, H.Nakanishi, and S.Miyashita, Phys.Rev.B **37**,3716(1988).
3. A.T.Ogielski, Phys.Rev.B **36**,7315(1987).
4. C.Tang, H.Nakanishi, and J.S.Langer, Phys.Rev.B **40**,995(1989).
5. H.Nakanishi, Phys.Rev.B **42**,1997(1990).
6. I.S.Graham, and M.Grant, unpublished(1990).

FORCED TWO-DIMENSIONAL PATTERNS IN ANISOTROPIC CONVECTIVE SYSTEMS

Atsushi Ogawa, Walter Zimmermann[A], Kyozi Kawasaki
and Toshihiro Kawakatsu

Department of Physics, Kyushu University 33, Fukuoka 812, Japan
[A]IFF Theorie III, Forschungszentrum Jülich, 5170 Jülich, Germany

ABSTRACT

In convective systems the first emerging pattern are often straight convection rolls. Bringing the wavelength of these rolls into competition with the wavelength of a spatial modulation of the external stress, one can investigate in convective systems also commensurate/incommensurate transitions. We here report that by a spatially modulated external stress also new patterns like undulated convection rolls or skew-varicose type patterns can be forced.

When in Rayleigh-Bénard convection or in electroconvection, for example, the external stress like the temperature difference or the voltage difference across the fluid layer is increased above some critical strength convection starts in the form of straight rolls.[1] Nematic liquid crystals have in addition an intrinsic anisotropy (director $\hat{n}$) which can be fixed in planarly aligned nematic layer by appropriate boundary conditions along a well defined direction (here the x-direction; $\hat{n} \parallel \hat{x}$). The straight periodic convection rolls are usually normally oriented to these prefered direction, which means that the wavevector $\vec{k}_c$ characterizing them is parallel to the director: $\vec{k}_c \parallel \hat{n}$. However, depending on the parameters the rolls can be also oblique with respect to $\hat{n}$, whereas two oblique orientations are then degenerate: $\vec{k}_c = (q_c, \pm p_c)$. By changing external parameters, e.g. the frequency of the external voltage in electroconvection or a magnetic field in Rayleigh-Bénard convection in nematics, one can also induce a continuous transition from oblique to normally orientated rolls.[2,3]

In our study the case in the vicinity of the normal/oblique transition(which is also called Lifshitz Point) is considered. Here we introduce the wave vector $\vec{k}$ describing the modulation of the external stress. If $\vec{k}$ is parallel to $\vec{k}_c$, one observes commensurate/incommensurate transition of the convection pattern[4], like in solid state physics. Here we fix $\vec{k}$ parallel to $\vec{n}_0$ ($\vec{k} = (k,0)$). Physical quantity such as the flow field near the threshold can be described by the following expression: $\vec{u}(x,y,z,t) = A(x,y,t)e^{iq_c x}\vec{U}_0(z) + c.c.$, where $A(x,y,t)$ can capture slow spatial modulations and the slow dynamics of the periodic convection rolls. $\vec{U}_0$ describes the variation of the physical quantities across the fluid layer. By separating in a systematic way fast and slowly varying properties of the pattern one can derive generalized Ginzburg-Landau equations(GGLE) for the amplitude $A(x,y,t)$. The GGLE turn out to be very appropriate to consider the resonance case $k = n(q_c + Q)$ (Q:misfit wavenumber, n=1,2,3,4).[5] The explicit form of the GGLE near the normal/oblique transition is[6]

$$\partial_t A = (\partial_x{}^2 - iZ\partial_x \partial_y{}^2 + W\partial_y{}^2 - \partial_y{}^4 + \mu - |A|^2)A + \alpha e^{inQx}A^{*(n-1)}, \quad (1)$$

© 1992 American Institute of Physics

where W depends on the external parameters and for $W < 0$ one has slightly oblique rolls when $\alpha = 0$, μ measures the distance from the threshold, α is proportional to the external modulation amplitude and Z depends on the system and the material(usually one has $0 < Z < 2$). We can notice that eq.(1) can also be derived from a potential.

In the oblique rolls range for $k = 2(q_c+Q)$ and $\alpha \neq 0$, one can expect above threshold a rectangular pattern[7] which is followed with increasing μ by a stable skew-varicose-like pattern as sketched in Fig.1. At the resonance condition $k = 3(q_c + Q)$ ($W < 0$), eq.(1) exhibits also a stable undulated solution[7] as shown in Fig.2. That a pattern like in Fig.1 and undulated convection rolls can exist stably has been shown for the first time in ref.7. Should such solutions become unstable for some values of Q, μ and α, we can also expect topologically new types of transitions from regular rectangles and undulations to their higher order commensurate or incommensurate patterns.

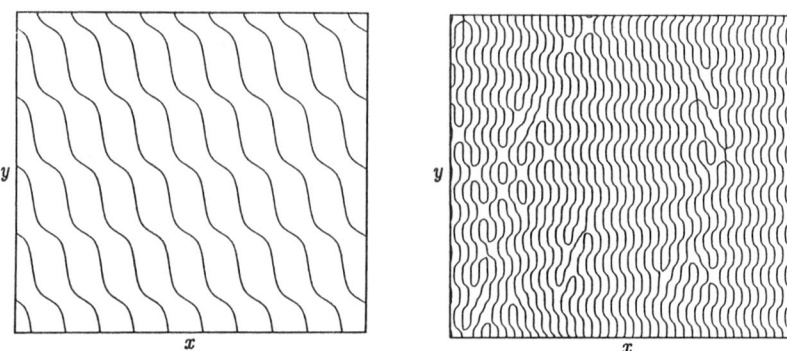

Fig.1.(*left*) Stable skew-varicose type pattern ($k = 2q_c$). **Fig.2.**(*right*) A roll pattern at the late stage of a simulation. Zero contours of $Re\{A(x,y)e^{iq_c x}\}$ are drawn ($n = 3, W = -1.5, Q = 0.1, Z = 0.8, \mu = 1.0, \alpha = 0.9$).

W.Zimmermann acknowledges the financial support of the Kajima Foundation during his stay in Japan.

REFERENCES

1. F. H. Busse, in *Hydrodynamic Instabilities and the Transition to Turbulence*, 2nd ed., edited by H. L. Swinney and J. P. Gollub(Springer 1986); W. Zimmermann, MRS Bulletin Jan. (1991).
2. R. Ribotta, A. Joets and Lin Lei, Phys. Rev. Lett. 56, 1595, (1986)
3. W. Zimmermann and L. Kramer, Phys. Rev. Lett. 55, 402 (1985); E. Bodenschatz, W. Zimmermann and L. Kramer, J. Physique. 49, 1875 (1988); W. Zimmermann, Ph.D. Thesis, Bayreuth (1987).
4. M. Lowe, J. Gollub and T. Lubensky, Phys. Rev. Lett. 51, 786 (1983)
5. P. Coullet, Phys. Rev. Lett. 56, 724 (1986).
6. A. Ogawa, K. Kawasaki, W. Zimmermann and T. Kawakatsu, in *Dynamics and Patterns in Complex Fluids*, edited by A. Onuki and K. Kawasaki (Springer 1990), p 78.
7. W. Zimmermann, A. Ogawa, S. Kai, K. Kawasaki and T. Kawakatsu (submitted for publication).

Monte Carlo Study of Vesicles

Shigeyuki Komura
Department of Applied Physics, Faculty of Science,
Tokyo Institute of Technology, Ohokayama, Meguro-ku,
Tokyo 152, Japan

Artur Baumgärtner
Institut für Festkörperforschung, Forschungszentrum Jülich,
D-5170 Jülich, Fed.Rep.Germany

ABSTRACT

Model of self-avoiding polymerized vesicles are investigated by Monte Carlo simulations. Flaccid polymerized vesicles are *not* crumpled, while deflated polymerized vesicles exhibit fully collapsed configuration.

1. FLACCID POLYMERIZED VESICLES

Details of the simulation technique have been already reported in the previous papers.[1,2]

The mean squared radius of gyration, $\langle R^2 \rangle$, and the mean volume, $\langle V \rangle$, of vesicles are expected to scale with the exponents ν_R and ν_V as

$$\langle R^2 \rangle \sim N^{\nu_R}, \quad \langle V \rangle \sim N^{3\nu_V/2}, \tag{1}$$

respectively, where N is the number of monomers constituting the vesicle. Our Monte Carlo results for $\langle R^2 \rangle$ and $\langle V \rangle$ are $\nu_R = 0.95 \pm 0.05$ and $3\nu_V/2 = 1.48 \pm 0.04$. The fact that the exponents exhibit their upper limiting value ($\nu \leq 1$) implies that flaccid polymerized vesicles are not crumpled. The absence of the crumpled phase can be explained by the large bending rigidity which is induced due to the short range repulsive interactions between adjacent hard spheres representing monomers.

2. DEFLATED POLYMERIZED VESICLES

Polymerized vesicles under negative pressure difference $\Delta p = p_{\text{in}} - p_{\text{out}}$ between inside and outside have been also investigated by using the "stress ensemble." The shapes of vesicles are expected to deviate from flaccid configurations in accordance with the cross-over scaling forms,

$$\langle R^2 \rangle \approx N^\nu X(x), \quad \langle V \rangle \approx N^{3\nu/2} Y(x), \tag{2}$$

where $\nu \approx 1.0$ and $x = \bar{p} N^{\varphi\nu/2}$ is the scaled pressure variable with $\bar{p} \equiv \Delta p a^3 / k_B T$ (a is the radius of a sphere). The crossover exponent $\varphi = 4.40 \pm 0.20$ is determined from several attempts to obtain optimal overlap of the curves for all values of N. In Fig.1, $X = \langle R^2 \rangle / N^{\nu_R}$ and $Y = \langle V \rangle / N^{3\nu_V/2}$ are plotted as a function of $x = \bar{p} N^{\varphi\nu/2}$. One observes power-law behavior of the scaling functions for $|x| > 10^4$, namely,

$$X(x) \approx \frac{X_-}{|x|^\rho}, \qquad Y(x) \approx \frac{Y_-}{|x|^\tau}, \qquad (3)$$

with $\rho = 0.140 \pm 0.007$ and $\tau = 0.185 \pm 0.008$. These results imply that $\langle R^2 \rangle \sim N^{\nu_R^-}$ and $\langle V \rangle \sim N^{3\nu_V^-/2}$ with $\nu_R^- = \nu_R(1 - \varphi\rho/2) = 0.66 \pm 0.08$ and $3\nu_V^-/2 = (3\nu_V/2)(1 - \varphi\tau/3) = 1.08 \pm 0.08$.

Our result for ν_R^- is very close to the lower limiting value for the exponent $\nu = 2/3$, corresponding to the "fully collapsed" configuration. This compact structure is also observed by the exponent for the volume, $3\nu_V^-/2$, since $\langle V \rangle \sim N a^3$ is expected for this configuration.

Results for inflated vesicles ($\bar{p} > 0$) will be published elsewhere.[2]

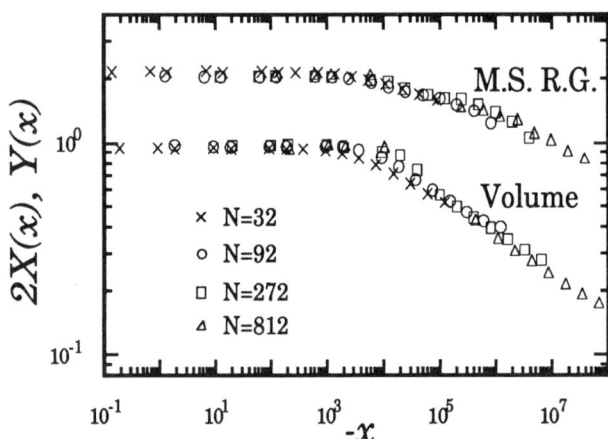

Fig.1 : Scaling plots of the mean square radius of gyration and the volume for the deflated ($\bar{p} < 0$) polymerized vesicles. Here $x = \bar{p} N^{\varphi\nu/2}$, $X = \langle R^2 \rangle / N^{\nu_R}$ and $Y = \langle V \rangle / N^{3\nu_V/2}$ with $\varphi = 4.40$, $\nu_R = 0.95$ and $\nu_V = 0.99$ are used. $X(x)$ is shifted to avoid the overlap of two curves.

REFERENCES

1. A. Baumgärtner and J. -S. Ho, *Phys. Rev. A.* **41** 5747 1990.
2. S. Komura and A. Baumgärtner, *Phys. Rev. A.* **44** 3511 1991.

DOMAIN GROWTH IN QUENCHED RANDOM IMPURITIES

Hisao Hayakawa and Toshiya Iwai
Department of Physics, Tohoku University, Sendai 980, Japan

ABSTRACT

Domain growth in a system with quenched random impurities for the conserved scalar order parameter is investigated using a cell-dynamics method. We find that the growth rate of domains are reduced due to the pinning by impurities.

Domain growth in a system with impurities is quite different from that for pure systems. Interfaces between domains are pinned by impurities and the rate of domain growth is reduced. Although Huse and Henley[1] predicted the logarithmic growth of the characteristic length $l(t)$ as $l(t) \sim (T\log t)^4$ with temperature T for two dimensional systems in the late stage of the ordering dynamics, the Monte Carlo simulations only found the existence of a cross over from a power law growth to a more slow growth law of the characteristic length.

Hayakawa[2] performed a numerical simulation for a two dimensional system with impurities for the nonconserved scalar order parameter based on the cell-dynamics method (CDS)[3]. He assumed that a system is divided into two kinds of cells, pure cells and impurities, where the order parameter $S_{\vec{n}}$ is only defined in pure cells and $S_{\vec{n}} = 0$ at impurities. He found that the effect of the thermal noise which is irrelevant in pure systems is important in the late stage of domain growth. For systems without the noise domain configurations are frozen after finite time steps. In his analysis the free energy

$$F = \sum_{\vec{n}} \sum_{\vec{m}} K(\vec{n},\vec{m})(S_{\vec{m}} - S_{\vec{n}})^2 + V\{S_{\vec{n}}\} \quad (1)$$

was used by assuming that the coupling between pure cells and impurities satisfies $K(\vec{n},\vec{m}) = 0$, and the functional form of $V\{S_{\vec{n}}\}$ in pure cells is independent of the configuration of impurities. In reality, the potential form should be modified due to the existence of impurities. Then, we adopt the potential form $V\{S_{\vec{n}}\} = -\tau_{\vec{n}}(T)S_{\vec{n}}^2 + S_{\vec{n}}^4$ in a suitable unit. The effects of impurities appear in $\tau_{\vec{n}}(T) = [(T/T_0) - \Theta(\vec{n})]/[(T/T_0) - 1]$, where $\Theta(\vec{n})$ and T_0 are the number of nearest-neighbor pure cells of $\vec{n}$ divided by 4 and the mean-field critical temperature, respectively.

We investigate a system with the conserved scalar order parameter. The time evolution of the order parameter field is given by

$$S_{\vec{n}}(t+1) = S_{\vec{n}}(t) - D\Delta_{imp}^2 S_{\vec{n}} - \Delta_{imp} f[S_{\vec{n}}(t)] + B\nabla \cdot \vec{\eta}_{\vec{n}}(t), \quad (2)$$

where $D = 0.5$ and $B = 0.3\sqrt{2T/T_0}$. The Laplacian Δ_{imp} includes the effect of impurities[2]. We use a uniform random number (uniformly distributed between -1 and 1) for each component of the vector noise $\vec{\eta}_{\vec{n}}(t)$

© 1992 American Institute of Physics

with a boundary condition satisfying the conservation of $S_{\vec{n}}$. The map function in (2) is given by

$$f[S_{\vec{n}}] = \frac{A^{\tau_{\vec{n}}} S_{\vec{n}}}{[1 + (1/\tau_{\vec{n}}) S_{\vec{n}}^2 (A^{2\tau_{\vec{n}}} - 1)]^{1/2}} \quad (3)$$

with $A = 1.3$, which is obtained from a single cell dynamics[3] with the help of (1). Note that our model is reduced to that of Ref.3 in the absence of impurities.

We have a set of two parameters, the concentration of impurities c and the depth of quench T/T_0. For a set of parameters we perform 49 runs up to $t = 30000$ for a simulation, and changing sets of parameters as $c = 0.1$ and 0.2, and $T/T_0 = 0$, $1/3$ and $2/3$. Our system has the periodic boundary condition and its size is 128×128 in the cell unit. The results of our simulation can be summarized as follows: (i) The characteristic length does not obey $l(t) \sim t^{1/3}$ but obeys a more slow growth law (Fig.1). It is difficult to find the growth law reported in Ref.1. (ii) The scattering function is well satisfied a dynamical scaling law. A profile of the order parameter for the volume fraction 0.5 is displayed in Fig.2, where interfaces are pinned by impurities.

The details of our results including those for the nonconserved case will be reported elsewhere.

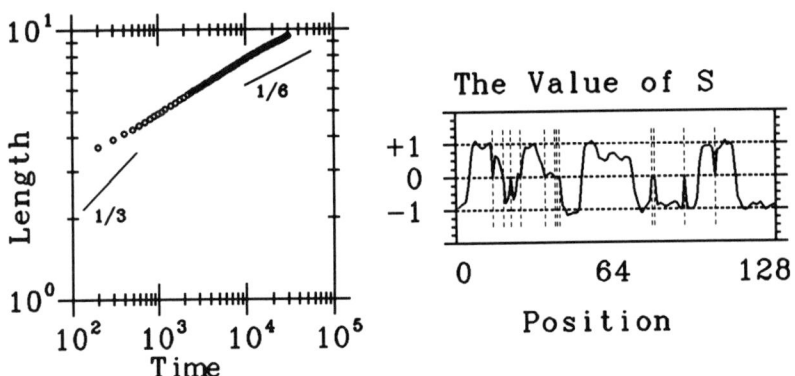

Fig.1: *(left)* A log-log plot of $l(t)$ for $(c, T/T_0) = (0.1, 1/3)$, where $l(t)$ seems to obey $l(t) \sim t^{1/6}$.

Fig.2: *(right)* The value of the order parameter along a line at $t = 30000$, where vertical dashed lines represent the positions of impurities.

REFERENCES

1. D.A.Huse and C.L.Henley, Phys.Rev.Lett. **54**, 2708 (1985).
2. H.Hayakawa, J.Phys.Soc.Jpn. **60**, 2492 (1991).
3. Y.Oono and S.Puri, Phys.Rev.A **38**, 434 (1988).

Clustering Motion in Conservative Coupled Map Systems

Tetsuro KONISHI

Dept. of Phys., School of Science, Nagoya Univ., Nagoya, 464-01, JAPAN

E-mail address (JUNET) : c42636a@nucc.cc.nagoya-u.ac.jp

and

Kunihiko KANEKO

Dept. of Pure and Applied Sciences, College of Arts and Sciences

Univ. of Tokyo, Komaba, Meguro-ku, Tokyo, 153, JAPAN

E-mail address (JUNET) : chaos@tansei.cc.u-tokyo.ac.jp

Abstract

Clustering motion of particles is found in Hamiltonian dynamics of symplectic coupled map systems. Particles assemble and move with strong correlation. The motion is chaotic but is distinguishable from random chaotic motion. Lyapunov analysis distinguishes global instability and local fluctuations. Cluster motions have finite lifetime. It has a fractal geometric structure in the phase space, and trapped to ruins of KAM and islands.

Long time behavior of chaos in conservative systems is usually considered to be well described by uniformly random state, i.e., thermal equilibrium. Structure and order are thought to be of dissipative systems, where the ordered states are mostly represented as attractors. All the above is a limited viewpoint. In this paper we discuss a new kind of cluster-like dynamical order in Hamiltonian systems which is at the same time chaotic.

Order formation in Hamiltonian systems can be an emergent property of the systems, as are seen in astrophysics (globular stellar clusters), microclusters, and in many other fields. Even pattern formation in dissipative systems should be originally described in a Hamitonian system, when we start from a molecular level description.

In the present paper we introduce a very simple model giving rise to order formation within Hamiltonian dynamics by using a symplectic version of a coupled map lattice.

In our model we have N particles on a unit circle. State of each particle is specified by its phase (position) $2\pi \cdot x_i$ and its conjugate momentum p_i. Temporal evolution is defined as;

$$p'_i = p_i + \frac{K}{2\pi\sqrt{N-1}} \sum_{j=1}^{N} \sin 2\pi(x_j - x_i), \quad K > 0, \quad x'_i = x_i + p'_i \text{ mod } 1, \quad i = 1, 2, \cdots, N. \quad (1)$$

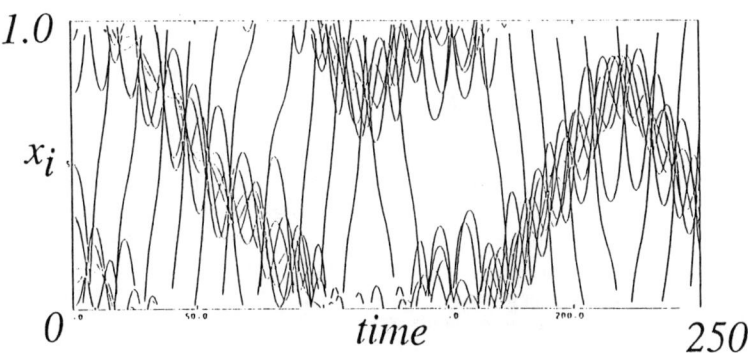

Figure 1: clustered motion

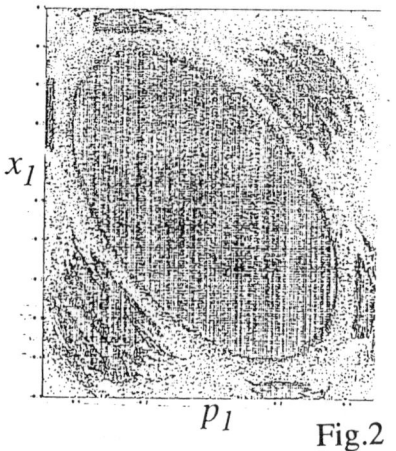

Fig.2

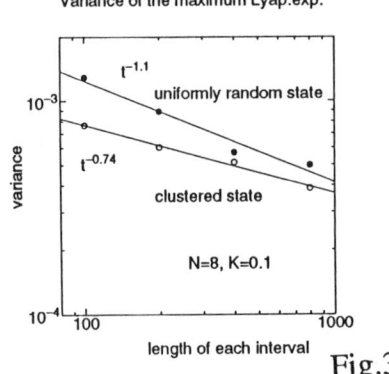

Fig.3

Since $K > 0$, the interaction term $\dfrac{K}{2\pi\sqrt{N-1}}\sin 2\pi(x_j - x_i)$ between two particles i and j is *attractive*. The model preserves symplectic structure, so that it is volume preserving.

Fig. 1 shows a typical example of clustering motion of the model (1). Particles initially distributed over a unit circle (with almost zero momenta) gradually get together to form a macro cluster. Some particles may not participate in cluster formation and wander around the cluster. These particles increase the entropy of spatial configuration and serve to stabilize the clustered state. With the same parameters, the model shows uniformly random behavior, just by changing initial condition. By emitting and absorbing particles, the cluster wonders around several states, then collapses, then comes back, and so on.

By measuring Lyapunov spectra, we can see that both clustered and random states are chaos. Thus we have two distinct chaotic seas coexisting in the phase space. Also we note that the Lyapunov spectra differ mostly at the $N - 1$'th exponent. This degree of freedom corresponds to the macroscopic instability of the cluster motion (when it exists), as we can see from their Lyapunov vectors.

The clustered state seems to be a motion trapped to the remnants of KAM tori and islands. This is verified by i) looking at the phase space slice (Fig.2) and ii) examining fluctuation property of Lyapunov exponents.(Fig.3)

In Fig.2, 2-dimensional slice of phase space of a 4-particle system is shown. Dots indicate points where the states make cluster. We see that they have fractal structure around the origin, which reminds one of island structure. Also, in the clustered state, the variance of a Lyapunov exponent measured at finite time T decays as a fractional power $(\Delta\lambda(T))^2 \propto T^{-\alpha}$, $0 < \alpha < 1$, which indicates the existence of long time correlation in the clustered state.(See Fig.3)

In this paper we have shown the formation of structure (which we call cluster) by chaos in *conservative* systems. This clustering is a remarkable novel feature in Hamiltonian systems distinguishable from uniform thermalization. The clustered states are chaotic and show crossover to uniform chaos, which is a chaos–chaos crossover. Theoretical study on chaos–chaos crossover process for high-dimensional systems is important. Origin of stability of clustered states can be found in the intricate structure of high-dimensional phase space. Clustering process will hopefully found in other physical systems, such as gravitational systems, microclusters of atoms, and so on. Chaos will give a new light on their study, in addition to traditional view based on barrier structures of potential energy landscapes.

Reference T. Konishi and K. Kakeko, preprint.

GLASSY ENTRAINMENT IN A LARGE POPULATION OF LIMIT-CYCLE OSCILLATORS WITH RANDOM AND FRUSTRATED INTERACTIONS

Hiroaki Daido
Kyushu Institute of Technology, Kitakyushu 804, Japan

ABSTRACT

A model of a large population of coupled limit-cycle oscillators is numerically shown to exhibit a transition to a peculiar phase of entrainment with glass-like behaviors.

Many of far from equilibrium systems can be regarded as a large assembly of coupled limit-cycle oscillators, e.g. chemical reactors, diverse living organisms such as intestines and hearts, and so on. One of important phenomena exhibited by them is mutual entrainment which occurs for coupling strength greater than a threshold. Although quite a few studies have been done on it using simplified models[1], nothing is known as to the behavior of coupled-oscillator systems with interactions endowed randomness as well as frustration, both of which should be more or less unavoidable in nature. Let us consider the following model[2]: for $j=1,\ldots,N$

$$d\theta_j/dt = \Omega_j + (2\pi)^{-1}\sum_{i=1}^{N} J_{ij}\sin 2\pi(\theta_i - \theta_j) , \quad (1)$$

where θ_j is the phase of the jth oscillator, Ω_j its natural frequency, and $J_{ij}=J_{ji}$ are independent random constants obeying an identical distribution as follows:

$$P(J_{ij}) = (2\pi J^2/N)^{-1/2}\exp(-NJ_{ij}^2/2J^2) . \quad (2)$$

In simulations described below, Ω_j were chosen so as to

Fig.1. Behavior of the averaged distribution of LF's.

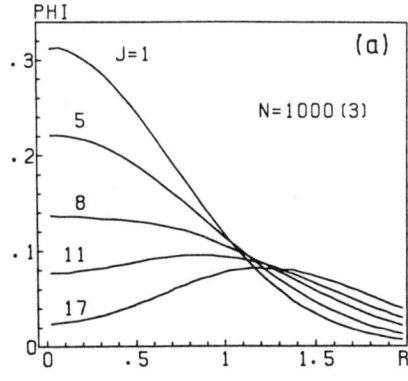
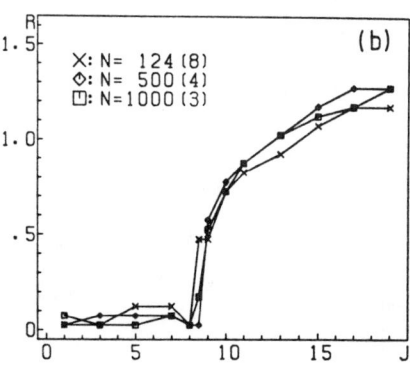

obey a Gaussian with variance 1 and mean 0 over the population. As J is increased, the distribution for temporal variation in a local field(LF), $p_j \equiv \Sigma_i J_{ij} \exp(2\pi\theta_i)$, changes from a Gaussian-like shape to a volcano-like one at $J=J_c \simeq 8$. See Fig.1 where Φ is the distribution function averaged over all j(and some samples of J_{ij} whose number is shown in parentheses), and $R \equiv |p|$. Figure 1(b) shows the behavior of the peak point of Φ which is an order parameter of the transition. For $J>J_c$ average frequencies(time average of $d\theta_j/dt$) were found to have a distribution with a sharp peak near 0(conveniently chosen as the average of natural frequencies), suggesting existence of a cluster of mutually "entrained" oscillators. It was confirmed that this "entrainment" is frequency entrainment, but NOT phase locking with the property as $<(\delta\theta_{ij}(t+\tau)-\delta\theta_{ij}(t))^2>$ $\propto \tau$ for τ large, where $< >$ stands for a time average, and $\delta\theta_{ij} \equiv \theta_i - \theta_j$. In other words this new type of entrainment is something between pure entrainment and nonentrainment. Each entrained oscillator performs a diffusive motion without drift, being analogous to the atomic motion below a glass transition. This "freezing" seems to cause a change in the manner of the relaxation of $Z \equiv \Sigma_j \exp(2\pi\theta_j)$ from the type as $\exp(-\Gamma t)$ to the one as $t^{-\alpha}$(Fig.2). The latter nonexponential decay is also reminiscent of typical glassy systems, tempting one to expect existence of an OSCILLATOR GLASS[2].

REFERENCES

1. A.T.Winfree, The Geometry of Biological Time(Springer, N.Y., 1980);Y.Kuramoto, Chemical Oscillations, Waves and Turbulence(Springer,Berlin,1984).
2. H.Daido, Prog.Theor.Phys.77(1987),622 and preprint.

Fig.2. Relaxation of Z averaged over ten samples.

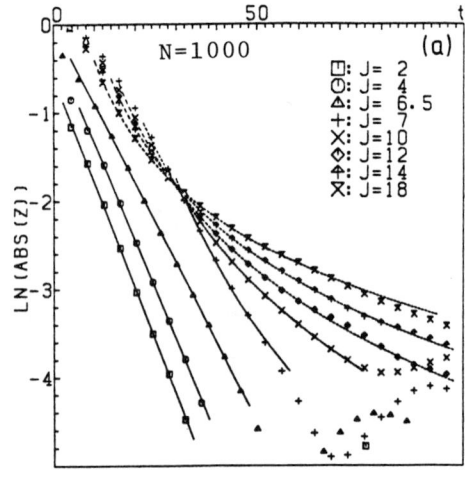
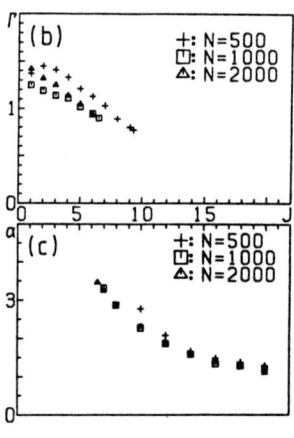

STRUCTURAL RELAXATION PROCESS IN A TWO-DIMENSIONAL XY CLOCK MODEL

Shigehiro Komura, Hideki Kobayashi, Satoru Ueno and Takayoshi Takeda

Faculty of Integrated Arts and Sciences, Hiroshima University, Hiroshima 730, Japan

ABSTRACT

The structural relaxation process has been examined by a Monte Carlo simulation for a two-dimensional XY clock model that has two transition points, a Kosterlitz-Thouless transition at T_{KT} and a long range order transition at T_C ($T_C < T_{KT}$). The system was quenched from the middle temperature phase between T_C and T_{KT} to low temperature phase below T_C. The structure relaxation rate at low temperature phase is slower for the initial middle temperature state that has been created directly from the high temperature phase above T_{KT} than that created from the low temperature phase below T_C, both middle temperature phases being kept for the time insufficient for equilibration.

INTRODUCTION

A Monte Carlo simulation was performed on a two-dimensional XY model described by the Hamiltonian,

$$H = -J \sum_{i<j} s_i \cdot s_j, \qquad (1)$$

where the spin variable s_i, whose magnitude is one, takes one of 6-clock directions on a hexagonal lattice. Each spin has ferromagnetic exchange interactions ($J > 0$) with its 6 nearest neighbours. The Hamiltonian is motivated to model a lipid-water lamellar system, in which there are two transition points, a main transition at T_M and a pretransition at T_P ($T_P < T_M$). Each lipid molecule is free to rotate on its axis above T_M, but below T_M it stops to rotate and tilts toward one of the six directions to its neighbours. Between T_M and T_P the domains of the unidirectional tilts of molecules are spatially limited and there is no long range order. Below T_P such unidirectional domains develop into a long range order.

In fact the model Hamiltonian (1) reveals such two transitions,[1,2] a Kosterlitz-Thouless transion T_{KT} corresponding to T_M and a long range order phase transition T_C corresponding to T_P. The ordering process of the low temperature phase below T_C of the system, that is quenched-in from the middle temperature phase between T_C and T_{KT} created in two different ways, is examined by this model.

RESULTS & DISCUSSIONS

We have calculated the average energy of the system per lattice, $E = <H>/N$ by usual Metropolis rule for various temperatures T, where N is the number of lattices. From the two peaks of the specific heat $C(T) = \Delta E / k_B \Delta T$ we could determine the two transition temperatures at $k_B T_C / J = 0.95$ and $k_B T_{KT} / J = 1.7$ for a system size N = 32 x 32 (Fig.1) after 50000 Monte Carlo Steps (MCS), which is thought to be enough for a thermal equilibrium.

Now we create the middle temperature phases at $k_B T / J = 1.2$ between T_C and T_{KT} in two different ways; in one way we realize the middle temperature by heating the system from below T_C and in another by cooling the system from above T_{KT} and wait for 1000 MCS which is insufficient for equilibration. We then quench the respective two middle temperature phases to a low temperature phases below T_C and persue how they reach an equilibrium. An example of the results for quenching to $k_B T / J = 0.3$ is shown in Fig.2, in which the E / J is plotted as a function of time for the first 1000 MCS. It is evident that the quenching from the middle phase created from above T_{KT} is slower to equilibrate than that from below T_C. Such tendency is shown for every quenched-in temperature below T_C from the middle temperature in which the samples are kept for the time less than 1×10^4 MCS which is insufficient for equilibration. The same results were obtained also for the relaxation process of the average magnetization per lattice M.

The results are consistent to what we expected. In our experimental observation on the lipid-water lamellar system the middle temperature phases between T_M and T_P were prepared in two different ways, one cooling from above T_M and another heating from below T_P. The structural relaxation of the quenched-in system below T_P was observed to be slower for the one cooled from above T_M than that heated from below T_P [3]. Such an agreement shows an applicability of our model to the lipid systems. Such a calculation might present a clue for understanding the structural difference between the stable and metastable rippled phases that are believed to exist between T_P and T_M; an existence of two middle temperature phases that have almost equal energy but different topologies.

REFERENCES

1) J. Tobochnik, *Phys. Rev.* **B26** (1982), 6201-6207.
2) Y. Ueno, G. Sun and I. Ono, *J. Phys. Soc. Jpn.* **58** (1989), 1162-1181.
3) S. Ueno, T. Takeda, S. Komura and H. Seto, to be published in *J. Appl. Cryst.*

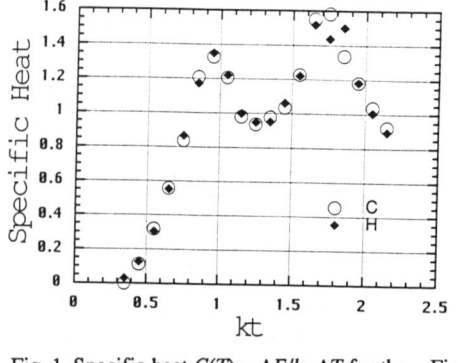
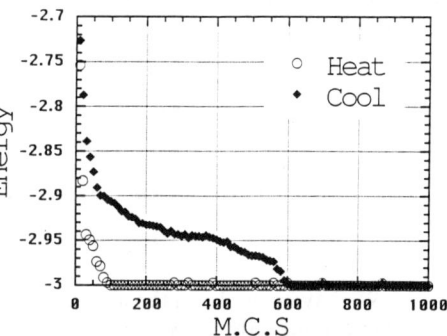

Fig. 1. Specific heat $C(T) = \Delta E/k_B \Delta T$ for the two-dimensional XY clock model. C stands for samples in cooling process and H for heating process.

Fig. 2. Relaxation process of the two dimensional XY clock system quenched-in to $k_B T/J = 0.3$ from $k_B T/J = 1.2$ created in two different ways. "Cool" stands for middle temperature phase cooled from above T_{KT} and "Heat" for that heated from below T_C.

DYNAMICS OF RANDOM CELLULAR STRUCTURES IN THREE DIMENSIONS

S. Ohta and T. Nagai
Physics Department, Kyushu Kyoritsu University
Kitakyushu 807, Japan

K. Kawasaki
Department of Physics, Faculty of Science, Kyushu University 33
Fukuoka 812, Japan

ABSTRACT

Computer simulations of evolving random cellular structures found in soap froths and grain aggregates have been carried out by employing the three-dimensional vertex model. Scaling properties of the cellular structures have been examined. A new relation for correlation of the number of faces per cell has been found.

THREE-DIMENSIONAL VERTEX MODEL AND COMPUTER SIMULATION

Recent studies for random cellular structures concluded that such systems exhibit universal properties, i.e. power law growth and dynamical scaling.[1-5] Those however were limited mostly to two dimensions. Our purpose is to investigate the scaling properties characteristic of three dimensions.

We extend the two-dimensional vertex model[4] to three dimensions. Three-dimensional random cellular structures are regarded as assemblies of vertices connected to each other by straight edges. Four edges meet at each vertex. We assume that a vertex is driven by the thermodynamic forces trying to reduce the areas of 6 triangular interfaces which touch the vertex. Extending the method used in two dimensions then gives the following equation for the velocity $\vec{v}_i$ of the i-th vertex located at $\vec{r}_i$:

$$\sum_\tau^{(i)} D_\tau \cdot \vec{v}_i = \sum_\tau^{(i)} \sigma \vec{n}_\tau \times (\vec{r}_j - \vec{r}_k), \qquad D_\tau \equiv (1/6)\, \eta\, A_\tau\, \vec{n}_\tau\, \vec{n}_\tau \qquad (1)$$

The symbol $\tau = (i, j, k)$ denotes each of 6 triangular interfaces with the common vertex i and $\vec{n}_\tau$ is the unit normal vector to τ chosen in such a way that the vertices i, j and k are arranged in a clockwise manner when viewed in the direction of $\vec{n}_\tau$. The quantity σ is the interfacial energy density. The tensor D_τ is the friction coefficient of τ where η and A_τ are the friction constant and the area of τ, respectively. The topological change of the system is described by the two types of elementary processes, i.e. recombination and tetrahedron annihilation (see Ref. 5 for details).

We have carried out computer simulation by employing Eq.(1) with the elementary processes and starting with the pattern of 1,000 Voronoi cells under the periodic boundary condition. Figure 1 is a snapshot of evolving patterns at t=1(dimensionless time). The system exhibits the dynamical scaling after t~1 where the average radius of cell grows as $t^{0.5}$ which is the same as in two dimensions. We show the distribution of the number of faces of a cell f_n in the scaling regime by the broken line in Fig. 2 where n is the number of faces of a cell. We have for $\bar{n}$, the average n in the scaling regime, the value $\bar{n}=13.46\pm0.07$, while $\bar{n}=15.56$ at t=0. The average number m(n) of faces of cells surrounding a cell with n faces is shown by the squares in Fig. 2 and can be expressed by the form

$$m(n)=13.31+25.1/n\pm0.01. \qquad (2)$$

This demonstrates existence of a topological correlation that a cell with small n is surrounded by cells with large n and vice versa. This new relation corresponds to the two-dimensional Aboav-Weaire law.

REFERENCES

1. J. A. Glazier, M. P. Anderson and G. S. Grest, Phil. Mag. **B62**, 615(1990).
2. J. Stavans, Phys. Rev. **A42**, 5049(1990).
3. M. P. Anderson, G. S. Grest and D. J. Srolovitz, Phil. Mag. **B59**, 293(1989).
4. K. Nakashima, T. Nagai and K. Kawasaki, J. Stat. Phys. **57**, 759 (1989).
5. T. Nagai, S. Ohta, K. Kawasaki and T. Okuzono, Phase Transitions **28**, 177(1990).

FIGURES

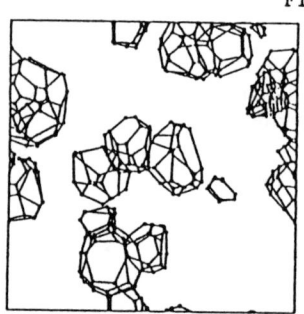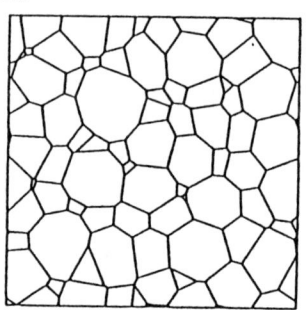

Fig. 1. Snapshot of evolving cellular structures. (a)Projections of cells sampled out of 430 cells onto a plane. (b)Pattern in a cross-section.

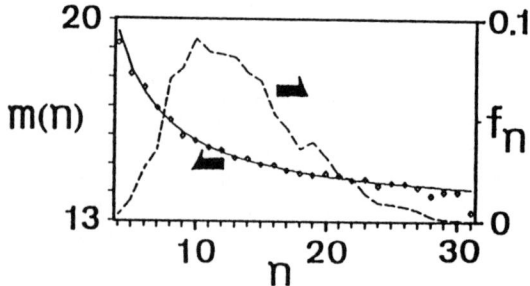

Fig. 2. Topological correlation (solid line) and distribution of the number of faces of a cell (broken line).

SINTERING OF CRYSTALLINE SOLIDS : NEW MODELIZATION TECHNICS

A. PAVLOVITCH and G. MARTIN
DTA/DTM/SRMP, CE- Saclay 91191 Gif-sur-Yvette Cedex, France

The sintering path of a powder agregate at given temperature and pressure is known to depend on the diffusion mechanism (bulk, grain boundary or surface diffusion) and on the distribution and efficiency of defect sources and sinks [1]. While bulk and grain boundary diffusion induce densification, surface diffusion leads to the spheroïdisation of the pores. To our knowledge no numerical model is now available to simulate the time evolution of the morphology of a powder compact under the influence of both diffusion mechanisms.

The proposed model is based on the technique recently introduced by Kawasaki [2] : since sintering is diffusion controled, the dynamics is such that the gain of free energy per unit time equals the power dissipated by diffusion :

$$\frac{\partial R}{\partial \mathbf{v}} + \frac{\partial F}{\partial \mathbf{r}} = 0 \qquad (1)$$

where $\mathbf{v}$ and $\mathbf{r}$ are the set of velocities and positions of the points of the system, F the free energy :

$$F = \int \sigma \, dS \qquad (2)$$

with σ the surface free energy density and dS the elementary area;
R is the dissipation function given by :

$$R = \int \mathbf{x} \cdot \mathbf{j} \, d\Omega \qquad (3)$$

with $\mathbf{j}$ the diffusion flux, $\mathbf{x}$ the thermodynamic force and $d\Omega$ the elementary volume; the integration is performed on the total volume of the system.
Equation (1) is to be solved with the constraint of fixed volume of the material.
We studied two limiting cases for which eq. 3 can be reduced to a surface integral.

1/- <u>Bulk diffusion</u> in the presence of a uniform distribution of defects source and sinks :

$$R = \frac{kT}{\kappa D_V C_V \Omega_V^2} \int v_n^2 \, dS \qquad (4)$$

where Ω_V is the vacancy volume, kT the thermal energy, κ^2 the sink strength, D_V and C_V the bulk diffusion coefficient and concentration of vacancies and v_n the normal velocity of the surface.

2/ <u>Surface diffusion</u> :

$$R = \frac{kT}{D_S C_S} \int j_s^2 \, dS \qquad (5a)$$

where D_S, C_S and j_s are respectively the diffusion coefficient, concentration and flux of vacancies on the surface; finaly,

$$v_n = -\Omega_V \, \text{div}(j_s) \qquad (5b)$$

After discretisation of the outer surface, equations 4 or 5 are solved numerically in two dimensions for cases 1 and 2 and in three dimensions in the case of bulk diffusion only. Figure 1 represents the sintering of two disks by bulk diffusion and Figure 2 is the firtst stage of the sintering by surface diffusion.

References :
[1] I. Gaal, *Defect and Diffusion Forum* <u>66-69</u>, 1227(1989).
[2] K. Kawasaki, T. Nagai and K. Nakashima, *Philos. Mag B*<u>60</u>, 399(1989) and references cited therein.

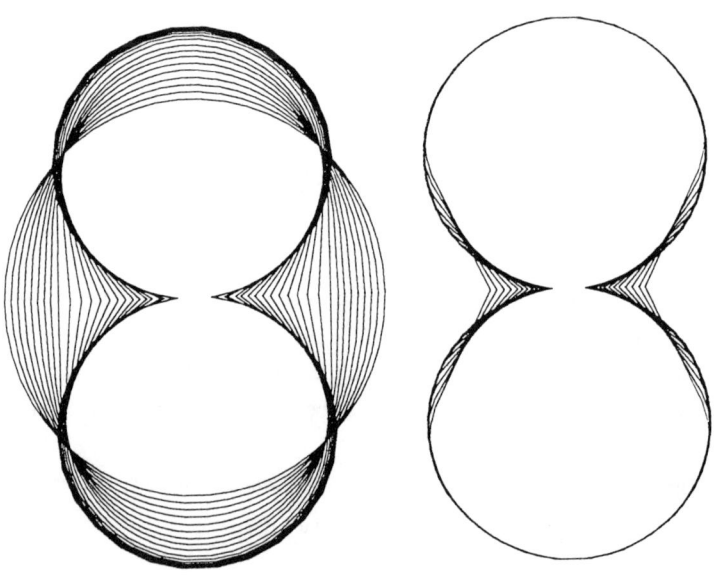

Figure 1.   Figure 2.

VII. AFTERWORDS

GOLD SEAL AND KYUSHU DYNASTY: UNSOLVED MYSTERIES OF ANCIENT JAPAN

Masayasu Kamimura
Department of Physics, Kyushu University, Fukuoka, Japan

ABSTRACT

A central dogma of the Japanese history is that, *throughout the history, there was absolutely no other dynasty or kingdom than the dynasty of the Imperial House of Japan (Yamato dynasty)* which continues to exist since some 1500~2000 years ago. Although much inconsistency has been noted between this dogma and accumulated records of Chinese historical documents on ancient Japan, almost all the Japanese historians have not taken such inconsistency seriously and even attributed it to mistakes or confusion of the Chinese historians who wrote the documents. But, one historian, Takehiko Furuta, who fully believes in the Chinese records, analyzed them recently in a very logical and even scientific manner and reached a surprising conclusion that there was a dynasty in northern Kyushu (Kyushu dynasty) and it was always this dynasty that appeared in the Chinese old records until 648 A.D., whereas the Yamato dynasty which started as a branch of the Kyushu dynasty finally overcame the Kyushu dynasty around 700 A.D. This new theory seems to be suppoted by many archaeological discoveries in and around the Fukuoka area, one of the most notable examples of which is the Gold Seal that appears in the Symposium poster. These remind us of the famous story that Heinrich Scliemann who believed in Homer's *The Iliad* as real finally succeeded in excavating the ruin of Troy.

1. INTRODUCTION

The city of Fukuoka, the symposium site, is located in the north part of Kyushu island, which is close to Korea and China (Fig.1). The northern Kyushu had the advantage, in ancient times, of absorbing the developed culture from Korea and China much earlier than other places in Japan. Thus, the Fukuoka area is one of the most important in ancient Japanese history, and many archaeological discoveries have been reported in this area. I have been deeply interested in the ancient history of this city of Fukuoka and the surrounding area though I am not a historian but a physicist studying theoretical nuclear physics.

At the beginning of my talk, let me remind you of the famous story which recounts that Heinrich Schliemann completely believed in Homer's epic poem *The Iliad* (~800 B.C.) as a real description of the Troy War (~1270 B.C.), and that he finally succeeded in excavating the ruins of Troy in 1872 at the

---

Talk given at the symposium banquet.

same place that Homer described. My talk will be somewhat analogous to this story (Fig.2). Namely, a Japanese historian, Takehiko Furuta, who believes in the Chinese historical records on ancient Japan, arrived at a surprising and revolutionary conclusion concerning the ancient history of Japan and the special importance of the Fukuoka area. The aim of my talk is to explain to you this new theory of Furuta's (Ref.1). Furuta is a professor of history at Showa College of Pharmaceutical Science in Tokyo.

Fig. 1. Location of Fukuoka, Kyushu island and Yamato area.

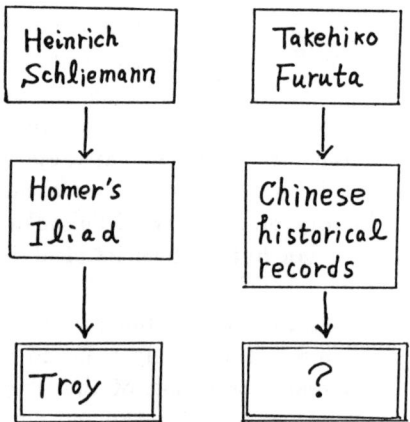

Fig. 2.

One of the most important central dogmas of the history of Japan is that, *throughout Japanese history, there was absolutely no other dynasty or kingdom than the dynasty of the Imperial House of Japan* which continues to exist since some 1500~2000 years ago. The ancient dynasty of the Imperial House is called *the Yamato dynasty*, because it was located in the Yamato area, the present-day Kyoto-Nara-Osaka region (Fig.1). This dogma is deeply believed by almost all Japanese historians, and all Japanese learn it in school, so it is a matter of common knowledge in Japan. Please remember this dogma, because the main aim of my talk is to deny it according to Professor Furuta's new theory.

The ancient times before the 6th century are, unfortunately, not reliably described in Japanese historical documents because their oldest editions were written in the 8th century by the Yamato dynasty. But, very fortunately, many contemporary records of ancient Japan were given in Chinese official historical documents, because diplomatic missions were often sent from Japan to ancient Chinese dynasties from the 2nd century B.C. through the 9th century A.D..

Though it has been noticed that there is some inconsistency between the Chinese records and the afore-mentioned dogma, the inconsistency has not been taken to be serious. All the historians who believe in this dogma consider that such disagreement is due to mistakes or confusion by the Chinese historians who wrote the records.

However, the historian, Professor Furuta, who believes in the Chinese records, recently investigated them in a very logical and even scientific manner. He pointed out that the Japanese missions that were sent in the 8th and 9th centuries were really sent by the Yamato dynasty, but that all the missions that were sent from the 2nd century B.C. through the 7th century A.D. do not seem to have been sent by the Yamato dynasty, but rather by a different dynasty which represented Japan in that period. So, if we believe the old Chinese records, what will happen to the history of ancient Japan?

2. THE GOLD SEAL AND HISTORY OF THE LATER HAN DYNASTY

Let us start by discussing the Gold Seal. Official historians of ancient Chinese dynasties recorded not only diplomatic exchanges between China and Japan, but also information on the society of ancient Japan. Therefore, we may expect archaeological discoveries which show exact coincidence or agreement with the old Chinese records. We have many examples of such exact coincidence.

One of the most notable examples is the following. The historical documents of the Later Han dynasty mentions that *in the year 57 A.D. an Emperor, Kuang-Wu, granted a gold seal to the king of the Wa state. Wa* was the name of Japan in ancient times.

It was a great surprise when, about 1700 years after, this seal was accidentally discovered here in Fukuoka on Shika island by a farmer. By examining the gold seal from the viewpoints of size, material, design and choice of Chinese characters, it became clear that it was genuine. A big picture of the Gold Seal is used on the symposium poster. The seal impression is seen in Fig.3. This Gold Seal is one of the Japan's National treasures of the first class, and it is now on display in the Fukuoka City Museum.

Fig.3. Impression of the Gold Seal (2.3cmx2.3cm).
It reads *King of Japan of China.*

Since the Gold Seal was found in Fukuoka, it is most reasonable to assume that the center of the Wa state was in or around the Fukuoka area. It is known that ancient Chinese emperors often granted gold, silver and bronze seals to rulers in the surrounding subject states. Professor Furuta investigated all the seals so far discovered in China and the surrounding countries, and he reached the conclusion that gold seals were granted exclusively to the representative king of a state and were never given to 2nd-rank or 3rd-rank rulers or local kings. The latter were given silver or bronze seals, just like in the Olympic games. There is no exception to this rule.

Therefore, there seems to be no doubt that the king who received this Gold Seal here in Fukuoka was recognized as the representative king of the whole of Japan in the 1st century. This conclusion seems reasonable, but it is seriously contradictory to the afore-mentioned dogma of Japanese history that, throughout the history of Japan, there was absolutely no other dynasty than the Yamato dynasty.

No historian who believes this dogma has been able to find any reason to explain why the Gold Seal was discovered in Fukuoka, if it was really granted to a king of the Yamato dynasty. One historian suggested desperately that when the Yamato dynasty mission's ship was passing by the Shika Island at Fukuoka on its way back from China, the Gold Seal accidentally dropped over board. This is of course nonsense. The Gold Seal and the corresponding record in the Chinese historical documents must be regarded as the first signature of the presence of the Kyushu dynasty.

3. THE HISTORY OF THREE DYNASTIES

The next Chinese reference to Japan can be found in the History of Three Dynasties written by the official historian Ch'en-shou around 280 A.D.. Since the Three Dynasties existed in the 3rd century (220-265 A.D.), dividing China into three parts, this is a contemporary history concerning China and the surrounding countries. Therefore, descriptions in this history are considered to be accurate and reliable.

The History of Three Dynasties is composed of three chapters: a chapter on the Wei dynasty, a chapter on the Wu dynasty and a chapter on the Shu-Han dynasty. The chapter on the Wei dynasty includes a section on the Wa state, that is Japan. Since the Wa state often sent missions to the Wei dynasty, and vice versa, the section on the Wa state describes vividly this society in the 3rd century.

According to this Chinese record, the Wa state was composed of about 30 city-states and was ruled by Queen Himiko. Queen Himiko often sent diplomatic missions to Wei dynasty of China since 239 A.D.. She was granted another gold seal, but unfortunately it is not discovered yet. This Queen Himiko is one of the most popular persons in Japanese history.

So, where was the Queen's capital, which was the center of Japan in the 3rd century? If you believe the historical dogma on the Imperial House of Japan, which I mentioned before, then Queen Himiko should be one of the ancestors of the Imperial Family, and the location of her capital should be in the Yamato area. But, how does the Chinese record describe its location? Let me discuss this point.

The Chinese Wei dynasty sent their return missions several times to the Queen Himiko's capital and obtained a lot of first-hand informations on Japan. The Chinese record describes how to reach the queen's capital. It is interesting and goes as follows: The Chinese missions came through Korea and crossed the sea, landing at Karatsu (Fig.4). They proceeded along the route shown in the map until reaching the west entrance to Fukuoka city. Up until this point the record is very precise and clear, and poses no problem. However, it is a big problem that, from this point, the description in the Chinese record becomes rather ambiguous to the readers (except Professor Furuta who is considered to have solved this problem).

By the way, some of you may notice now that the *true* purpose of yesterday's excursion to Karatsu was just for you to experience the route of the ancient Chinese missions from Karatsu to Fukuoka.

So, how did the Chinese mission reach the Queen's capital from the west entrance to Fukuoka? Here, the theories on this problem are devided into two types. One type of theories says that the queen's capital must be somewhere in northern Kyushu. This is reasonable, and many sites have so far been proposed as candidates. The other type insists of course that the queen's capital should be definitely in the Yamato area, because the dogma on the Yamato dynasty should be true. However, if you believe this theory, you are forced to assume that there are a lot of mistakes and errors in the Chinese records, and you have to modify those records in order for the Chinese mission to reach the Yamato area which is some 500km away from Kyushu. I cannot agree with this theory.

On the contrary, Professor Furuta believes completely in the Chinese records

Fig. 4. The route of the Chinese missions to the Queen Himiko's capital.

Fig. 5. Location of the sites where Wa silks and Chinese silks were excavated. No other site in all of Japan. One big circle is for the Chinese silks.

Fig. 6. Number of excavated Han's bronze mirrors.
The dashed lines are the borders of the present-day prefectures.

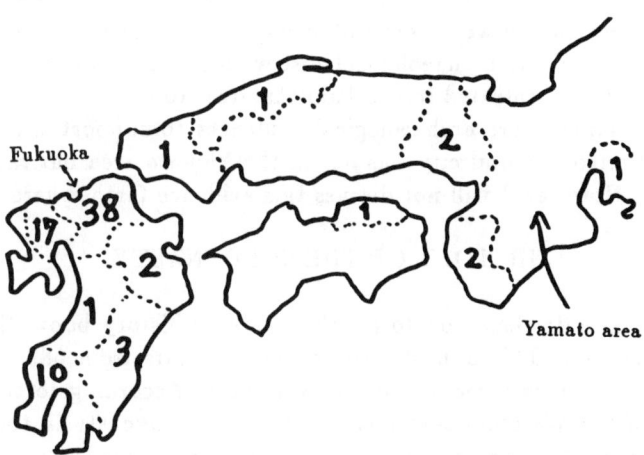

Fig. 7. Number of excavated iron weapons of the 3rd century A.D. and before.

and investigated, in a very logical and scientific manner, not only the section on the Wa state itself but also the whole book of the History of Three Dynasties, checking the Ch'en-shou's rule, throughout the book, on the description of distance and direction of travels; it is rather surprizing to know that Furuta is the first person who made such type of analysis with the whole book referred to. He found that the Ch'en-shou's manner of description on travels is always strict and consistent throughout the History of Three Dynasties. Furuta finally concluded that the location of Queen Himiko's capital city is nowhere but in the Fukuoka area. I am sorry that I have no time to introduce you his analysis of this problem, but let me show you some archaeological evidence to support his conclusion.

According to the Chinese records, a Chinese Emperor granted some silk products to Queen Himiko, and in turn Queen Himiko presented some Wa silk products as tribute to the Chinese Emperor. In all of Japan, silk products from Himiko's period and before have been found only in Fukuoka and the surrounding area as shown in Fig.5. Among these, Chinese silk products from this period have been discovered only at the single site in Fukuoka.

Another example. According to the Chinese records, the same Later Han Emperor gave Queen Himiko one hundred bronze mirrors. As shown in Fig.6, the majority of Han's bronze mirrors have been excavated in the Fukuoka area.

Still another example. As far as the weapons of soldiers in the 3rd century are concerned, bronze weapons were popular. However, the Chinese record says that iron weapons, which were new and stronger, were already used in the Wa state. As seen in Fig.7, archaeological discoveries of iron weapons from this period have been concentrated in the Fukuoka area, too.

There are many more archaeological evidences to support the conclusion that Queen Himiko's capital city was not in the Yamato area but rather in the Fukuoka area. However, I will not discuss this evidence further here.

4. HISTORY OF THE SUI DYNASTY

Next, let me introduce you to another Chinese history book, the History of the Sui Dynasty. This dynasty existed for 30 years from 589 A.D.. The Chinese record includes a section on the Wa state. According to this section, an Emperor of the Wa state sent missions to the Sui dynasty in the year 600 A.D.. The Chinese historian recognizes that this Wa state is the same one that was granted the Gold Seal by the Han Dynasty Emperor and also the same one that was ruled by Queen Himiko in the 3rd century.

The Chinese records describe the Emperor of the Wa state as follows: *"The Japanese Emperor's family name is Ame, his given name is Tarishihoko, and his wife's name is Kimi. He has 600-700 ladies in his harem."* If you believe that the mission was actually sent by the Yamato dynasty, then this statement is very astonishing and unbelievable, because, at that very time, the Yamato dynasty was ruled by *Empress* Suiko. It is therefore clear that Empress Suiko who is female can not be the Emperor Tarishihoko Ame who is male. Also, the Chinese record says that the Emperor of the Wa state has the family name, Ame. However, it is well known that the Japanese Imperial Family has had no family name since ancient times.

According to the Sui dynasty record, once a diplomatic mission from the Sui dynasty was sent to the capital city of the Emperor Tarishihoko. According to the travel record of the Chinese ambassador, the location of the capital city is estimated to be in the Fukuoka area. Another decisive piece of evidence for the capital city being near Fukuoka is that, near the capital city, the Chinese ambassador observed an actively erupting volcano. The name of this volcano is recorded as Mt. Aso, which is located here in Kyushu (Fig.8) and is some 500 km far away from the Yamato area, in and around which there is no volcano at all.

Fig. 8. Emperor Tarishihoko's Kyushu dynasty and Mt. Aso.

All these Chinese records seem to prove that the Emperor Tarishihoko's dynasty was definitely located in northern Kyushu. In the other words, it is not the Yamato dynasty but a Kyushu dynasty. Professor Furuta considers it is to be so, and I agree with him. However, almost all other historians maintain that those Chinese records are nonsense and contain a lot of misunderstandings about the *true* Japan, because any dynasty other than the Imperial House of Japan should not exist in Japanese history. I cannot agree with them.

5. HISTORY OF THE T'ANG DYNASTY

Let me enter the History of the Early T'ang Dynasty. The T'ang dynasty existed for about 300 years from 618 A.D.. The historical document of this dynasty has a section on the Wa state as usualy. Very interestingly, however, in addition to this section, there is another section on a *Nippon* state which is also in Japan. This Nippon was a newcomer to the T'ang dynasty, and Chinese historians of the T'ang period recognized that Japan was composed of two different states, Wa and Nippon.

The diplomatic missions that arrived in China before 648 A.D. were listed under the section on the Wa state, whereas the ones that came after 703 A.D. were listed under the section on Nippon. It is very likely that the Wa state lost its power during this period (probably due to a complete defeat in a big naval battle at a Korean shore on 663 A.D. against a combined fleet of the T'ang dynasty and a state in Korea) and ceased sending of missions to China, while Nippon began to represent Japan in place of the Wa state.

However, at the beginning, Nippon was a newcomer to the T'ang dynasty. When the Chinese officials met the mission from Nippon for the first time, their first impression was recorded as follows: *"The mission members are almost all arrogant. They do not answer our questions with facts. Thus, we are suspicious of these people."* This is actually written in the Chinese document. Nippon were just new visitors to China. Later, however, Nippon and the T'ang dynasty became very intimate, and Nippon imported not only the advanced Chinese culture, but also the political and social systems of the T'ang dynasty. It is obvious that this Nippon is just the Yamato dynasty, because the records of the section on Nippon coincide nicely with the Japanese earliest history book, which was written in the 8th century by the Yamato dynasty.

Then, what was the Wa state which was described in the section on Wa, and what became of it? In the book of Early T'ang History, it is explicitly recorded that *"Nippon was originally a small state, but afterwards absorbed the area of the Wa state."*

Therefore, Professor Furuta concludes that this Wa state was nothing but the Kyushu dynasty, which continued to send delegations to China during several hundred years before 648 A.D.. However, almost all historians of ancient Japan still maintain their opinion that the Wa state and the Nippon state are the same entity, namely the Yamato dynasty, and that the existence of the two independent sections on Wa and Nippon must be due to mistakes or confusion by the Chinese official historian who wrote the document. I cannot agree with this opinion.

6. CONCLUDING REMARKS

Let me summarize my talk by the diagram of Fig.9. Namely, Heinrich Schliemann, who believed in Homer's *Iliad*, succeeded in excavating the ruin of Troy. Takehiko Furuta, who believes in the Chinese official records on the ancient Japan, arrived at the conclusion of the existence of the Kyushu dynasty in the Fukuoka area in ancient times. However, the academic community of Japanese historians does not want to accept Furuta's theory. This is understandable because if they accept it once, then, logically speaking, their authorized theory of ancient Japan is totally destroyed. It is very strange, however, that no member of that community has tried to deny or criticize Furuta's theory explicitly by writing papers or books. Japanese ancient historians continue to ignore his theory and have kept silent during these 15 years since Furuta first published his entire theory in three books. I guess that, even if they wanted to destroy Furuta's theory, they could not do so because the theory is very strongly supported by the Chinese historical records.

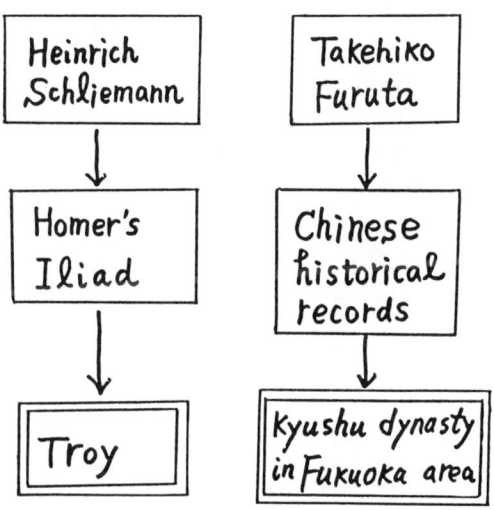

Fig. 9

Anyway, I myself appreciate very much Furuta's theory from the viewpoint of scientist, and I know many other scientists also appreciate his work. Finally, let me close my talk by saying that *you are now sitting on the ruin of the Kyushu dynasty, the glorious center of ancient Japan.*

Thank you very much.

ACKNOWLEDGEMENTS

The author would like to thank Professor P. Rikvold at Florida State University and Professor K. Kawasaki at Kyushu University for their help in preparing this manuscript. An lecture note by Professor Toru Takemoto (Ref.2) concerning the Furuta's theory on ancient Japan was very helpful in preparing this manuscript.

REFERENCES

1. Takehiko Furuta has published a number of papers and books. The following first three books are the most fundamental and important ones for his theory. The fourth one is the most recently published book which reviews his whole work until 1990.
    T. Furuta, "*Yamataikoku wa Nakatta (There was no Yamatai State)*",
    (Asahi Shinbunsha, Tokyo, 1971; also in the Kadokawa-Bunko Series)
    T. Furuta, "*Ushinawareta Kyushu Ouchou (The Lost Kyushu Dynasty)*",
    (Asahi Shinbunsha; also in the Kadokawa-Bunko Series)
    T. Furuta, "*Ushinawareta Shinwa (The Stolen Myths)*",
    (Asahi Shinbunsha, Tokyo, 1975; also in the Kadokawa-Bunko Series)
    T. Furuta, "*Kodai Shinshi (New History of Ancient Japan)*",
    (Shinsensha, Tokyo, 1991).
2. T. Takemoto, a lecture note entitled "*The Kyushu Dynasty: Furuta's New Theory on Ancient Japan*", unpublished.

ABOUT THE SPEAKER... Professor M. Kamimura is a theoretical nuclear physicist and is internationally well-known for his seminal contributions to the theories of few-body systems, nuclear cluster structure and muon catalyzed fusion. Besides physics, he has a deep interest in the ancient history of northern Kyushu which goes beyond the level of an amateur.

Author Index

A

Abe, K., 475
Ackerson, B. J., 352
Akama, H., 497
Akutsu, N., 507
Akutsu, Y., 507
Alba-Simionesco, C., 3, 75
Angell, C. A., 3
Arai, M., 423
Arima, M., 461
Arita, K., 441
Arzimanoglou, A., 3

B

Balucani, U., 126
Baumgärtner, A., 567
Bedeaux, D., 378
Belloni, L., 211
Bernasconi, A., 207
Bertrand, D., 79
Biljakovic, K., 449
Bocquet, J. L., 533
Böhmer, R., 3
Brady, J. F., 391
Brand, H. R., 253
Butler, S., 183

C

Cametti, C., 322
Chen, S. H., 301
Chen, X. K., 40
Chu, B., 203
Cichocki, B., 384
Codastefano, P., 322
Cohen, E. G. D., 359
Cummins, H. Z., 40

D

Daido, H., 573
Dalbiez, J. P., 211
D'Arrigo, G., 453
Delcourt, O., 71
Descamps, M., 71
Doi, M., 286, 292, 382
Drifford, M., 211
Du, W. M., 40
Düring, E. R., 271
Duffy, S. F., 126

E

Edwards, S. F., 261
Eguchi, T., 441, 537
Elser, V., 541
Ema, K., 473
Endoh, Y., 423
Enomoto, Y., 286, 523
Ewen, B., 193

F

Fan, J., 3
Farago, B., 67, 193
Felderhof, B. U., 370, 384
Fetters, L. J., 193
Frick, B., 67
Fukao, K., 222
Fukazawa, T., 226
Furusaka, M., 423

G

Gorman, B. M., 549
Götze, W., 95, 105
Grest, G. S., 271

H

Habasaki, J., 187
Hagiwara, M., 445
Hamano, Kenzi, 459
Hamano, Katsumi, 471
Hammann, J., 417

Han, C. C., 216
Hara, H., 521
Hara, K., 224, 226, 465
Harada, Y., 320
Harrowell, P., 183
Hasegawa, H., 216
Hashimoto, T., 216, 283
Hayakawa, H., 509, 569
He, Y., 236
Hirano, T., 89
Hiwatari, Y., 115, 155, 187, 189
Holzer, M., 541
Horai, K., 537
Horiuchi, T., 467
Huang, J. S., 193

I

Ichinose, S., 495
Iio, K., 421
Ikeda, H., 433
Imai, M., 230
Ise, T., 327
Ishida, A., 519
Ishida, Y., 63
Ishii, Y., 555
Itagaki, N., 475
Itoh, M., 519
Itoh, S., 189
Iwai, T., 569

J

Jinnai, H., 216

K

Kaburagi, M., 517
Kadono, K., 185
Kaji, K., 91, 230
Kamimura, M., 583
Kanaya, H., 224, 226, 465
Kanaya, T., 91, 230
Kaneko, K., 571
Katayama, T., 327
Kawaguchi, M., 350
Kawaguchi, T., 91
Kawakatsu, T., 267, 283, 331, 503, 553, 565

Kawamura, J., 87
Kawasaki, K., 267, 283, 294, 331, 503, 551, 553, 565, 577
Kawasaki, T., 427
Kawase, Y., 473
Kikuchi, R., 517
Kinugawa, K., 185
Kishi, K., 386
Kitazawa, A., 421
Kjems, J. K., 207
Kobayashi, K., 545
Kobayashi, H., 575
Koga, S., 499
Koga, T., 294
Kojima, S., 83
Komura, Shigehiro, 318, 575
Komura, Shigeyuki, 567
Konishi, T., 571
Koper, G. J. M., 378
Koyama, J., 521
Kremer, K., 271
Kubota, K., 459
Kuwabara, M., 290, 296
Kuwahara, N., 459

L

Lasjaunias, J. C., 449
Lee, S. B., 529
Lefloch, F., 417
Li, G., 40
Limoge, Y., 533
Liss, K. D., 479
Lohfink, M., 30
Lu, Q., 3
Lundgren, L. E., 407

M

Maeda, H., 457
Magerl, A., 479
Mallamace, F., 301
Martin, G., 579
Maruyama, K., 481
Mashita, T., 457
Matsuda, H., 236
Matsumoto, M., 292
Matsushige, K., 224, 226, 465, 467
Matsuura, M., 445

Matsuura, S., 477
Mazenko, G. F., 134
van Megen, W., 335
Meisyo, K., 63
Mezei, F., 53
Miike, H., 469
Minato, T., 495
Mishima, A., 515
Miura, T., 232
Miyagawa, H., 155, 189
Miyaji, H., 222
Miyamoto, Y., 222
Miyashita, S., 427
Miyazima, S., 477, 481
Mizukami, T., 230
Mogi, M., 475
Momii, T., 234
Monceau, P., 449
Mortensen, K., 318
Mountain, R. D., 165
Muguruma, M., 226
Munakata, T., 140
Muranaka, T., 155

N

Nagai, T., 551, 577
Nagata, K., 421
Nakagawa, M., 545
Nakahara, A., 553
Nakanishi, H., 563
Nakayama, T., 279
Nambu, S., 501
Nilgens, H., 401
Nishimori, H., 525
Nishimuta, S., 226
Nomura, S., 234
Novotny, M. A., 549

O

Ochiai, M., 265
Ocio, M., 417
Odagaki, T., 115
Ogawa, A., 565
Ogura, H., 265
Ohta, Shonosuke, 559
Ohta, Shigetoshi, 577
Ohta, T., 286
Ohtsuki, T., 505

Okabe, H., 224, 226, 465
Okada, I., 187
Okumura, K., 481
Okuzono, T., 551
Ono, Y., 290, 296
Onuki, A., 525
Oota, K., 481
Ott, H. R., 207
Ozaki, Y., 511
Ozao, R., 265

P

Pakula, T., 401
Paparelli, A., 453
Paul, W., 145
Pavlovitch, A., 579
Phung, T., 391
Pleiner, H., 253
Posselt, D., 207
Pusey, P. N., 335

R

Richter, D., 67, 193
Rikvold, P. A., 549
Roe, R. J., 288
Rouch, J., 301, 322
Roux, J. N., 173
Ryo, Y., 350

S

Sagala, D. A., 501
Saito, Y., 547
Sakai, A., 40, 461
Sakata, H., 471
Sakurai, S., 234
Sanchez, E., 3
Saruyama, Y., 228
Sasa, S., 509
Sasai, M., 513
Sasaki, S., 457
Sato, Akira, 475
Sato, A., 501
Sato, H., 517
Sato, Hirokazu, 519
Schatzel, K., 352
de Schepper, I. M., 359

Scher, H., 485
Schneider, J. R., 479
Schwahn, D., 318
See, H., 382
Seki, S., 547
Senapati, H., 3
Sengers, J. V., 459
Sethna, J. P., 130, 541
Seto, H., 318
Shigenari, T., 475
Shiiyama, K., 537
Shiwa, Y., 263
Shore, J. D., 541
Shumway, S., 130
Sillescu, H., 30
Sjögren, L., 95, 105
Sleator, T., 207
Soen, T., 63
Song, Y. J., 529
Souletie, J., 79
Spalla, O., 211
Suga, H., 20
Sung, W., 257
Suzuki, A., 517
Suzuki, J., 423
Suzuki, M., 475

T

Tabuchi, M., 320
Taguchi, Y., 557
Taie, K., 234
Takahashi, M., 236
Takamuku, T., 89
Takeda, T., 575
Takenaka, M., 283
Takeuchi, H., 288
Taki, S., 224
Tanaka, F., 261
Tanaka, Hiroshi, 185
Tanaka, Hajime, 232, 238, 240
Taniguchi, T., 503
Tao, N. J., 40
Tartaglia, P., 301, 322
Tatsumisago, M., 3
Tetsuka, A., 286
Thirumalai, D., 165
Tokuyama, M., 523
Tomokiyo, Y., 441
Torcini, A., 126
Tran-Cong, Q., 63

U

Uehara, K., 155
Ueno, S., 575
Umesaki, N., 561
Underwood, S. M., 335
Uwaha, M., 547

V

Vallauri, R., 126
Vincent, E., 417

W

Wada, K., 517
Wakita, H., 89
Walton, D., 122
Wang, Z.-L., 203
Willart, J. F., 71
Witten, T. A., 245

Y

Yagi, T., 461
Yakubo, K., 279
Yamada, T., 85
Yamagami, M., 89
Yamaguchi, M., 327
Yamaguchi, T., 89
Yamamoto, H., 469
Yamamoto, T., 507
Yamashita, S., 459
Yano, O., 63
Yao, H., 473
Yasuda, N., 463
Yoshida, A., 386
Yoshida, H., 236
Yoshida, K., 85
Yoshida, T., 386
Yoshida, Y., 467
Yoshino, S., 354

Z

Zhang, J., 471
Zimmermann, W., 267, 565
Zorn, R., 67
Zulehner, W., 479